CLASSICAL KINETIC THEORY OF WEAKLY TURBULENT NONLINEAR PLASMA PROCESSES

Kinetic theory of weakly turbulent nonlinear processes in plasma helped form the foundation of modern plasma physics. This book provides a systematic overview of the kinetic theory of weak plasma turbulence from a modern perspective. It covers the fundamentals of weak turbulence theory, including the foundational concepts and the mathematical and technical details. Applications to some key space plasma problems are also covered, including the origin of nonthermal charged particle population and radio burst phenomena from the sun. Treating both collective and discrete particle effects, it provides a valuable reference for researchers looking to familiarize themselves with plasma weak turbulence theory.

PETER H. YOON is a senior research scientist at the Institute of Physical Science and Technology, University of Maryland, international scholar professor at Kyung Hee University, Korea and senior research leader at Korea Astronomy and Space Science Institute. He is a leading expert in theoretical space plasma physics and a fellow of the American Physical Society.

CLASSICAL KINETIC THEORY OF WEAKLY TURBULENT NONLINEAR PLASMA PROCESSES

PETER H. YOON
University of Maryland

CAMBRIDGE
UNIVERSITY PRESS

CAMBRIDGE
UNIVERSITY PRESS

University Printing House, Cambridge CB2 8BS, United Kingdom

One Liberty Plaza, 20th Floor, New York, NY 10006, USA

477 Williamstown Road, Port Melbourne, VIC 3207, Australia

314–321, 3rd Floor, Plot 3, Splendor Forum, Jasola District Centre,
New Delhi – 110025, India

79 Anson Road, #06-04/06, Singapore 079906

Cambridge University Press is part of the University of Cambridge.

It furthers the University's mission by disseminating knowledge in the pursuit of education, learning, and research at the highest international levels of excellence.

www.cambridge.org
Information on this title: www.cambridge.org/9781107172005
DOI: 10.1017/9781316771259

First published 2019

Printed in the United Kingdom by TJ International Ltd, Padstow Cornwall

A catalogue record for this publication is available from the British Library.

Library of Congress Cataloging-in-Publication Data
Names: Yoon, Peter H., 1958– author.
Title: Classical kinetic theory of weakly turbulent nonlinear plasma processes / Peter H. Yoon (University of Maryland, College Park).
Description: Cambridge ; New York, NY : Cambridge University Press, 2019. | Includes bibliographical references and index.
Identifiers: LCCN 2019010668 | ISBN 9781107172005 (hardback : alk. paper)
Subjects: LCSH: Plasma turbulence. | Kinetic theory of matter.
Classification: LCC QC718 .Y66 2019 | DDC 530.4/46–dc23
LC record available at https://lccn.loc.gov/2019010668

ISBN 978-1-107-17200-5 Hardback

Contents

Preface

Plasma is an ionized gas in which dynamical processes are governed predominantly by a collective behavior rather than that of individual particle interactions. For this reason, the plasma is often regarded as free of inter-particle collisions. However, the discrete nature of the plasma particles or, equivalently, the single-particle behavior of plasma can be important, and often the description of plasma is incomplete if the discrete particle effects are not properly taken into account. This monograph is concerned with the kinetic theory of weakly turbulent plasma processes in which collective and discrete particle effects are systematically treated. The methodology is classical since dilute high-temperature plasmas in laboratory, near-Earth space, and most astrophysical environments do not require quantum mechanical approaches. The book is exclusively concerned with incoherent nonlinear plasma phenomena. Such processes are known as microscopic or kinetic plasma turbulence, as opposed to macroscopic or fluid turbulence.

The purpose of this book is to present essential ideas for treating the weak plasma turbulence on the basis of kinetic theory. In order to simplify the discussion, the plasma of interest is considered to be free of influence from externally applied electric or magnetic fields. The plasma is also considered as uniform on average, without spatial inhomogeneity associated with it. Despite these simplifying assumptions, basic theoretical tools and concepts developed in the present monograph can, in principle, be applied to situations in which the plasma is immersed in external fields, especially, the ambient magnetic field. Realistic laboratory or space/astrophysical plasmas are magnetized. Consequently, applications of the theory developed in this book are limited to high-frequency phenomena for which the approximation of field-free plasma may be valid.

The weak plasma turbulence theory is a perturbative nonlinear theory, and the essential formalism was developed by the pioneers of modern plasma physics beginning in the late 1950s and 1960s, and continued on through the early 1980s.

It is the purpose of the present author to systematically overview various aspects of this venerable theory, but also certain new developments are included in the discussion.

This monograph comprises four parts. In Part I, the simplest kinetic theory is discussed, where it is assumed that the particles interact primarily through electrostatic force and without collisions. Chapter 1 formulates the perturbative kinetic theory based upon the relatively simple Vlasov–Poisson equation. Chapter 1 establishes conceptual foundations for a statistical description of plasmas, and also includes a detailed discussion of various linear and nonlinear response functions. Chapter 2 takes the formal results developed in Chapter 1 in a further direction, and derives the wave kinetic equation that describes weakly turbulent plasma processes. Chapter 3 derives the corresponding particle kinetic equations for electrons and ions interacting with the waves.

In Part II, the discrete particle effects are discussed within the framework of electrostatic approximation. The conceptual approach is based upon the Klimontovich equation, hence Part II is entitled "Klimontovich Weak Turbulence Theory." Chapter 4 revisits the perturbative nonlinear theory already discussed in earlier chapters, but additional effects that arise from particle discreteness are incorporated. The result is a weak turbulence theory in which various processes are expressed in the balanced form between the so-called spontaneous and induced processes. Chapter 5 considers applications. The first example is on the spontaneous emission of electrostatic fluctuations in thermal equilibrium plasma. The second application relates to plasmas slightly out of equilibrium, where collisional relaxation processes bring such plasmas back to thermal equilibrium. In Chapter 6, further application is made to a situation where collisional processes are not sufficient to relax the system to equilibrium, but rather the relaxation is achieved via excitation of instability. The sample problem considered involves an electron beam interacting with the background plasma, leading to an excitation of the so-called Langmuir turbulence. Numerical solutions of the basic equations show that the long-time evolution of electron beam-plasma instability involves the formation of a nonthermal population in the electron velocity distribution function. Chapter 6 closes with the discussion of the time asymptotic state of the plasma turbulence. Relevance to the near-Earth space environment is also discussed.

Part III returns to the collision-free problem, but it generalizes Part I to fully electromagnetic situation. Chapter 7 formulates the general framework of Vlasov–Maxwell weak turbulence theory, and also discusses the electromagnetic linear and nonlinear response functions. Chapter 8 derives the specific equations of electromagnetic Vlasov weak turbulence theory for plasma normal modes. In Parts III and IV where theories concerning the electromagnetic weak turbulence formalism are discussed, various intermediate steps are deliberately included so that the readers may be able to retrace the derivations of various mathematical formulae.

Part IV generalizes Part III by adding the particle discreteness effects, and it also generalizes Part II by including electromagnetic effects in the Klimontovich weak turbulence formalism, thus bringing all the necessary ingredients together, and completing the generalized formalism of weak turbulence theory. Chapter 9 presents the general formalism, followed by Chapter 10, where we make specific applications, first to the issue of spontaneous emission of magnetic field fluctuations in thermal plasmas, and the radiation emission in thermal plasma. We also discuss the problem of relativistic collisional kinetic equation. Finally, we apply the electromagnetic "Klimontovich" weak turbulence formalism to the problem of plasma emission, which is a physical process that is intimately related to the solar radio bursts phenomena. The backdrop on this problem is also discussed. Various topics and supplemental discussions are included in the appendices.

Part I

Vlasov Weak Turbulence Theory: Electrostatic Approximation

1

Nonlinear Electrostatic Equations for Collisionless Plasmas

This monograph presents a perturbative nonlinear kinetic theory of plasma turbulence, known as the *weak turbulence theory*. At the outset, it should be pointed out that this book does not include the effects of ambient magnetic field. Plasmas in real situations are usually magnetized, so that applications of the method discussed in this book will be somewhat limited, but the purpose is to lay out the fundamental methodology and conceptual foundations so that more general applications for magnetized plasmas may be developed on the basis of this book. This book also limits the discussions to spatially homogeneous plasma.

Plasma kinetic theory has a long history, and many early papers can be found in the literature that discuss the perturbative nonlinear kinetic theory of plasma turbulence – see, for example, papers by Vedenov and Velikhov (1962); Kovrizhnykh and Tsytovich (1964, 1965); Kovrizhnykh (1965); Gorbunov and Silin (1965); Gorbunov et al. (1965); Tsytovich (1967); Rogister and Oberman (1968, 1969), to name just several. These are merely sample papers, among those that personally influenced the author of this book.

If one is interested in the general background on plasma kinetic theory, there are some excellent early monographs, among which may be, for instance, those by Montgomery and Tidman (1964); Kadomtsev (1965); Klimontovich (1967, 1982); Pitaevskii and Lifshitz (1981); Sagdeev and Galeev (1969); Tsytovich (1970, 1977a,b); Davidson (1972); Ichimaru (1973); Krall and Trivelpiece (1973); Akhiezer et al. (1975); Hasegawa (1975); Kaplan and Tsytovich (1973); Sitenko (1967, 1982); Melrose (1980a, 1986); Nicholson (1983); Alexandrov et al. (1984); Chen (1987), etc. This list is incomplete, but they represent some standard works that treat the foundations of plasma kinetic theory and/or weak plasma turbulence theory. More recent books are also available. See, for example, those by Musher et al. (1995); Sitenko and Malnev (1995); Treumann and Baumjohann (1997); Tsytovich (1995); Kono and Škorić (2010); Diamond et al. (2010), etc., which deal with the subject of plasma kinetic theory and nonlinear phenomena.

So, as the readers may appreciate, there is an abundance of resources on the topic of plasma kinetic theory, and one may ask why another book? The rationale for this book is as follows: Discussions of nonlinear plasma theories, particularly those concerning the weak turbulence theory found in many of the above-cited works, are sometimes not so easy to follow, especially for young researchers. Moreover, many of the monographs cover wide-ranging topics with generally brief descriptions for each subject area without going much into in-depth discussions. It is the purpose of this book to focus only on the kinetic theory of weak plasma turbulence, but to present the detailed fundamental discussions and derivations as clearly as possible, without sacrificing the intermediate mathematical steps. Talking of the latter, many authors omit too many intermediate steps, which can be a source of much frustrations for young scientists. This book does not spare the readers the mathematical details. This strategy means that some materials in the book can be a bit lengthy, and casual readers may get lost in the maths. However, if one approaches the material with enough patience, he or she will be rewarded with the intimate knowledge on how the weak turbulence theory actually works, what are the essential assumptions behind the theory, and so forth. Owing to the space devoted to mathematical details, some standard topics often included in the textbooks and monographs on nonlinear plasma theory are left out. For instance, parametric instabilities, solitary wave theory, coherent nonlinear structure formation in plasma, etc., are not covered in this book.

This book is intended for advanced undergraduate, graduate students, or young researchers who are already familiar with the introductory level of plasma kinetic theory, but wishing to familiarize themselves with a more in-depth understanding on nonlinear theory of weak plasma turbulence. In spite of this, this book expounds on foundational principles at the conceptual level as much as possible without assuming too much prior knowledge on the part of the readers.

1.1 Preamble: Fundamental Concepts

We are interested in physical phenomena that are described as turbulent, which loosely means physical quantities that are fluctuating in space and time. In order to characterize such fluctuations, we employ statistical methods and concepts. That is, we deal with averages in time, space, or over hypothetical collection of different possible states called the *ensemble*. One is particularly interested in how fluctuating quantities measured in two or more different times or in two or more different spatial locations are correlated. We begin by considering many-body correlations associated with fluctuating physical quantities, and the spectral transformation of such quantities in space and time.

The statistical correlation is an important concept that characterizes the nature of turbulence. Suppose that one measures a particular physical quantity, say velocity or electromagnetic field, in a turbulent medium at a given time. Suppose also that one measures the same quantity at another time separated by an interval. If one repeats such series of measurements over and over again, then if the physical quantities are uncorrelated, that is, if there is no cause and effect relationship between the two measurements, then on average, the product of two measurements made at two different time intervals may be zero, since by the very nature of turbulence, velocity or field may have random directions. On the other hand, if the first measurement affects the second measurement because there exists an underlying cause-and-effect relationship, then the average of the products may be finite. A systematic way to characterize how the statistical average of the products of physical quantities, or equivalently, their correlation function, behaves in space and time can thus be useful for understanding and characterizing the turbulence. Consequently, in this book we will be concerned with the description of how the statistical average of the correlation of fluctuating (i.e., turbulent) quantity, $\langle \delta a^2 \rangle$, dynamically evolves. Here, δa represents any dynamical quantity, and the symbol $\langle \cdots \rangle$ denotes the statistical average.

The convention adopted in this book for the definition of spatial Fourier transformation and its inverse is

$$f_{\mathbf{k}} = (2\pi)^{-3} \int d\mathbf{r}\, f(\mathbf{r})\, e^{-i\mathbf{k}\cdot\mathbf{r}}, \qquad f(\mathbf{r}) = \int d\mathbf{k}\, f_{\mathbf{k}}\, e^{i\mathbf{k}\cdot\mathbf{r}}. \tag{1.1}$$

Here $f(\mathbf{r})$ is any physical quantity, which is a function of spatial coordinate $\mathbf{r}$, and which is bounded in space. The Fourier transformation of a product of two functions is represented by the convolution

$$(2\pi)^{-3} \int d\mathbf{r}\, f(\mathbf{r})\, g(\mathbf{r})\, e^{-i\mathbf{k}\cdot\mathbf{r}} = \int d\mathbf{k}'\, f_{\mathbf{k}'}\, g_{\mathbf{k}-\mathbf{k}'} = \int d\mathbf{k}'\, f_{\mathbf{k}-\mathbf{k}'}\, g_{\mathbf{k}'}. \tag{1.2}$$

The proof of this "convolution theorem" is straightforward. All one has to do is to insert for $f(\mathbf{r})$ and $g(\mathbf{r})$, their respective Fourier transformations, and make use of the well-known delta function identity

$$\int d\mathbf{r}\, e^{i\mathbf{k}\cdot\mathbf{r}} = \delta(\mathbf{k}). \tag{1.3}$$

Fourier transformation of a function $f(\mathbf{r},t)$ in both space and time can be defined by

$$f_{\mathbf{k},\omega} = (2\pi)^{-4} \int d\mathbf{r} \int dt\, f(\mathbf{r},t)\, e^{-i\mathbf{k}\cdot\mathbf{r}+i\omega t},$$

$$f(\mathbf{r},t) = \int d\mathbf{k} \int d\omega\, f_{\mathbf{k},\omega}\, e^{i\mathbf{k}\cdot\mathbf{r}-i\omega t}. \tag{1.4}$$

Convolution theorem for the spatio-temporal Fourier transformation is

$$\begin{aligned}(2\pi)^{-4}\int d\mathbf{r}\int dt\; f(\mathbf{r},t)\,g(\mathbf{r},t)\,e^{-i\mathbf{k}\cdot\mathbf{r}+i\omega t}\\ =\int d\mathbf{k}'\int d\omega'\; f_{\mathbf{k}',\omega'}\,g_{\mathbf{k}-\mathbf{k}',\omega-\omega'}\\ =\int d\mathbf{k}'\int d\omega'\; f_{\mathbf{k}-\mathbf{k}',\omega-\omega'}\,g_{\mathbf{k}',\omega'}.\end{aligned} \tag{1.5}$$

When the angular frequency ω satisfies the dispersion relation $\omega=\omega_{\mathbf{k}}+i\gamma_{\mathbf{k}}$, that is, when (generally complex) ω is a function of $\mathbf{k}$, then the Fourier representation of function $f(\mathbf{r},t)$ can be re-expressed by virtue of the fact that we may write the spectral amplitude as

$$f_{\mathbf{k},\omega}=f_{\mathbf{k}}\delta(\omega-\omega_{\mathbf{k}}-i\gamma_{\mathbf{k}}), \tag{1.6}$$

or

$$f(\mathbf{r},t)=\int d\mathbf{k}\; f_{\mathbf{k}}\;\exp(i\mathbf{k}\cdot\mathbf{r}-i\omega_{\mathbf{k}}t+\gamma_{\mathbf{k}}t). \tag{1.7}$$

If $f(\mathbf{r},t)$ is real then obviously $f^*(\mathbf{r},t)=f(\mathbf{r},t)$, where the asterisk $*$ represents the complex conjugate. From this it follows that

$$\int d\mathbf{k}\; f_{\mathbf{k}}^*\;\exp(-i\mathbf{k}\cdot\mathbf{r}+i\omega_{\mathbf{k}}t)=\int d\mathbf{k}\; f_{\mathbf{k}}\;\exp(i\mathbf{k}\cdot\mathbf{r}-i\omega_{\mathbf{k}}t), \tag{1.8}$$

which leads to the following symmetry relations:

$$f_{\mathbf{k}}^*=f_{-\mathbf{k}},\qquad \omega_{-\mathbf{k}}=-\omega_{\mathbf{k}},\qquad \gamma_{-\mathbf{k}}=\gamma_{\mathbf{k}}. \tag{1.9}$$

Let $\delta f(\mathbf{r},t)$ represent a fluctuating quantity whose ensemble average is zero:

$$\langle\delta f(\mathbf{r},t)\rangle=0. \tag{1.10}$$

In our notation, any quantity preceded by δ indicates that this quantity is fluctuating in space and time, that is, turbulent. By "ensemble average" we may mean an average over phase, space, or time. Or it could mean an average over all possible configurations. Turbulence is called "homogeneous" if the spatial dependence of the two-body correlation is only upon the relative distance,

$$\langle\delta f(\mathbf{r},t)\,\delta f(\mathbf{r}',t)\rangle=\langle\delta f^2\rangle_{\mathbf{r}-\mathbf{r}',t,t'}=\langle\delta f^2\rangle_{\mathbf{r}'-\mathbf{r},t,t'}, \tag{1.11}$$

and "stationary" if the temporal two-body correlation is a function of relative time difference,

$$\langle\delta f(\mathbf{r},t)\,\delta f(\mathbf{r},t')\rangle=\langle\delta f^2\rangle_{\mathbf{r},\mathbf{r}',t-t'}=\langle\delta f^2\rangle_{\mathbf{r},\mathbf{r}',t'-t}. \tag{1.12}$$

Thus, for homogeneous and stationary turbulence the two-body correlation function is given by

$$\langle \delta f(\mathbf{r},t)\, \delta f(\mathbf{r}',t') \rangle = \langle \delta f^2 \rangle_{\mathbf{r}-\mathbf{r}',t-t'}. \tag{1.13}$$

It should be noted that not all fluctuating quantities in nature satisfy the zero ensemble average property (1.10). Physical processes whose fluctuations satisfy (1.10) are called "incoherent" phenomena, while "coherent" processes may be associated with a nonvanishing ensemble average. For incoherent processes different phases are uncorrelated such that when averaged over them, the result vanishes; hence, such processes are characterized by the zero ensemble average property specified by (1.10).

In a similar way, the three-body correlation function for homogeneous and stationary turbulence is a function of distances between any two points, say $(\mathbf{r},t)$ and $(\mathbf{r}',t')$, among three points $(\mathbf{r},t)$, $(\mathbf{r}',t')$, $(\mathbf{r}'',t'')$, in coordinate-time space:

$$\langle \delta f(\mathbf{r},t)\, \delta f(\mathbf{r}',t')\, \delta f(\mathbf{r}'',t'') \rangle = \langle \delta f^3 \rangle_{\mathbf{r}-\mathbf{r}',\mathbf{r}-\mathbf{r}'';t-t',t-t''}. \tag{1.14}$$

The four-body correlation function for homogeneous and stationary turbulence can be defined likewise:

$$\begin{aligned}
&\langle \delta f(\mathbf{r},t)\, \delta f(\mathbf{r}',t')\, \delta f(\mathbf{r}'',t'')\, \delta f(\mathbf{r}''',t''') \rangle \\
&\quad = \langle \delta f^2 \rangle_{\mathbf{r}-\mathbf{r}';t-t'}\, \langle \delta f^2 \rangle_{\mathbf{r}''-\mathbf{r}''';t''-t'''} + \langle \delta f^2 \rangle_{\mathbf{r}-\mathbf{r}'';t-t''}\, \langle \delta f^2 \rangle_{\mathbf{r}'-\mathbf{r}''';t'-t'''} \\
&\qquad + \langle \delta f^2 \rangle_{\mathbf{r}-\mathbf{r}''';t-t'''}\, \langle \delta f^2 \rangle_{\mathbf{r}'-\mathbf{r}'';t'-t''} \\
&\qquad + \langle \delta f^4 \rangle_{\mathbf{r}-\mathbf{r}',\mathbf{r}'-\mathbf{r}'',\mathbf{r}''-\mathbf{r}''';t-t',t'-t''',t''-t'''}.
\end{aligned} \tag{1.15}$$

Let us represent the two-body correlation function in spectral form:

$$\begin{aligned}
\langle \delta f(\mathbf{r},t)\, \delta f(\mathbf{r}',t') \rangle &= \int d\mathbf{k} \int d\omega\, \langle \delta f^2 \rangle_{\mathbf{k},\omega}\, e^{i\mathbf{k}\cdot(\mathbf{r}-\mathbf{r}')-i\omega(t-t')} \\
&= \int d\mathbf{k} \int d\omega \int d\mathbf{k}' \int d\omega'\, \langle \delta f_{\mathbf{k},\omega}\, \delta f_{\mathbf{k}',\omega'} \rangle\, e^{i\mathbf{k}\cdot\mathbf{r}+i\mathbf{k}'\cdot\mathbf{r}'-i\omega t-i\omega' t'},
\end{aligned} \tag{1.16}$$

where in the second line we have made use of the spectral representations for individual functions $\delta f(\mathbf{r},t)$ and $\delta f(\mathbf{r}',t')$. From this, it is seen that the equality can be obtained if the following condition is satisfied:

$$\langle \delta f_{\mathbf{k},\omega}\, \delta f_{\mathbf{k}',\omega'} \rangle = \delta(\mathbf{k}+\mathbf{k}')\, \delta(\omega+\omega')\, \langle \delta f^2 \rangle_{\mathbf{k},\omega}. \tag{1.17}$$

If we write the spectral component $\delta f_{\mathbf{k},\omega}$ with an explicit phase factor,

$$\delta f_{\mathbf{k},\omega} = \hat{f}_{\mathbf{k},\omega} e^{i\phi_{\mathbf{k},\omega}}, \tag{1.18}$$

where $\phi_{\mathbf{k},\omega}$ represents the phase, then we have

$$\langle \delta f_{\mathbf{k},\omega}\, \delta f_{\mathbf{k}',\omega'} \rangle = \langle \hat{f}_{\mathbf{k},\omega}\, \hat{f}_{\mathbf{k}',\omega'} e^{i\phi_{\mathbf{k},\omega}+i\phi_{\mathbf{k}',\omega'}} \rangle. \tag{1.19}$$

For homogeneous and stationary turbulence the phase is assumed to be random (or uncorrelated). As such, the ensemble average over random phases becomes nonzero only if

$$\phi_{\mathbf{k},\omega} + \phi_{\mathbf{k}',\omega'} = 0, \tag{1.20}$$

which can be satisfied under the assumption that, if for $\mathbf{k} = -\mathbf{k}'$ and $\omega = -\omega'$, the following is also satisfied:

$$\phi_{-\mathbf{k},-\omega} = -\phi_{\mathbf{k},\omega}. \tag{1.21}$$

This is but the rephrasing of condition (1.17). The assumption of homogeneous and stationary turbulence is thus equivalent to the "random phase approximation." In short, the property

$$\langle \delta f^2 \rangle_{\mathbf{k},\omega} = \langle \delta f_{\mathbf{k},\omega}\, \delta f_{-\mathbf{k},-\omega} \rangle \tag{1.22}$$

is a useful spectral characteristic for homogeneous and stationary turbulence, or equivalently, fluctuations with random phases.

Next, consider the three-body correlation, which we may write as

$$\begin{aligned} \langle \delta f(\mathbf{r},t)\, \delta f(\mathbf{r}',t')\, \delta f(\mathbf{r}'',t'') \rangle &= \int d\mathbf{k} \int d\omega \int d\mathbf{k}' \int d\omega'\, \langle \delta f^3 \rangle_{\mathbf{k},\omega;\mathbf{k}',\omega'} \\ &\quad \times e^{i\mathbf{k}\cdot(\mathbf{r}-\mathbf{r}')+i\mathbf{k}'\cdot(\mathbf{r}'-\mathbf{r}'')-i\omega(t-t')-i\omega'(t'-t'')} \\ &= \int d\mathbf{k} \int d\omega \int d\mathbf{k}' \int d\omega' \int d\mathbf{k}'' \int d\omega'' \\ &\quad \times \langle \delta f_{\mathbf{k},\omega}\, \delta f_{\mathbf{k}',\omega'}\, \delta f_{\mathbf{k}'',\omega''} \rangle \\ &\quad \times e^{i\mathbf{k}\cdot\mathbf{r}+i\mathbf{k}'\cdot\mathbf{r}'+i\mathbf{k}''\cdot\mathbf{r}''-i\omega t-i\omega' t'-i\omega'' t''}. \end{aligned} \tag{1.23}$$

From this we obtain the identity

$$\langle \delta f_{\mathbf{k},\omega}\, \delta f_{\mathbf{k}',\omega'}\, \delta f_{\mathbf{k}'',\omega''} \rangle = \delta(\mathbf{k}+\mathbf{k}'+\mathbf{k}'')\, \delta(\omega+\omega'+\omega'')\, \langle \delta f^3 \rangle_{\mathbf{k},\omega;\mathbf{k}+\mathbf{k}',\omega+\omega'}. \tag{1.24}$$

A similar analysis can be carried out for the four-body correlation. The derivation is tedious but straightforward, and is thus omitted.

We summarize the general properties of the many-body correlations, or many-body cumulants for homogeneous and stationary turbulence:

$$\begin{aligned} \langle \delta f_{\mathbf{k},\omega}\, \delta f_{\mathbf{k}',\omega'} \rangle &= \delta(\mathbf{k}+\mathbf{k}')\, \delta(\omega+\omega') \langle \delta f^2 \rangle_{\mathbf{k},\omega}, \\ \langle \delta f_{\mathbf{k},\omega}\, \delta f_{\mathbf{k}',\omega'}\, \delta f_{\mathbf{k}'',\omega''} \rangle &= \delta(\mathbf{k}+\mathbf{k}'+\mathbf{k}'') \delta(\omega+\omega'+\omega'') \langle \delta f^3 \rangle_{\mathbf{k},\omega;\mathbf{k}+\mathbf{k}',\omega+\omega'}, \end{aligned}$$

$$\begin{aligned}\langle \delta f_{\mathbf{k},\omega}\, \delta f_{\mathbf{k}',\omega'}\, \delta f_{\mathbf{k}'',\omega''}\, \delta f_{\mathbf{k}'''\omega'''} \rangle &= \delta(\mathbf{k}+\mathbf{k}'+\mathbf{k}''+\mathbf{k}''')\, \delta(\omega+\omega'+\omega''+\omega''') \\ &\times [\delta(\mathbf{k}+\mathbf{k}')\, \delta(\omega+\omega')\, \langle \delta f^2 \rangle_{\mathbf{k},\omega}\, \langle \delta f^2 \rangle_{\mathbf{k}'',\omega''} \\ &+ \delta(\mathbf{k}+\mathbf{k}'')\, \delta(\omega+\omega'')\, \langle \delta f^2 \rangle_{\mathbf{k},\omega}\, \langle \delta f^2 \rangle_{\mathbf{k}',\omega'} \\ &+ \delta(\mathbf{k}'+\mathbf{k}'')\, \delta(\omega'+\omega'')\, \langle \delta f^2 \rangle_{\mathbf{k},\omega}\, \langle \delta f^2 \rangle_{\mathbf{k}',\omega'} \\ &+ \langle \delta f^4 \rangle_{\mathbf{k},\omega;\, \mathbf{k}+\mathbf{k}',\omega+\omega';\, \mathbf{k}+\mathbf{k}'+\mathbf{k}'',\omega+\omega'+\omega''}]. \end{aligned} \tag{1.25}$$

An important consequence of this result is that an ensemble average of two fluctuating quantities δf and δg, where they are related to each other, can be expressed in terms of their spectral counterparts as follows:

$$\langle \delta f(\mathbf{r},t)\, \delta g(\mathbf{r},t) \rangle = \int d\mathbf{k} \int d\omega\, \langle \delta f_{\mathbf{k},\omega}\, \delta g_{-\mathbf{k},-\omega} \rangle. \tag{1.26}$$

1.2 Electrostatic Vlasov Equation

A simple and intuitive definition of *plasma* is that it is an *ionized gas*. Individual electrons and ions that make up the plasma interact through collective electromagnetic force. Collective behavior of a plasma is described by a statistical means. In this book we are concerned with a fully ionized plasma. For partially ionized plasma, atomic processes such as the recombination and collisions between charged particles and neutrals cannot be ignored, which complicate the matter. Vlasov equation (Vlasov, 1938) describes the statistical property of a plasma governed by collective processes. The system under consideration is a spatially uniform plasma made of single-species ions (protons) and electrons, and there is no net electric or magnetic field. We also assume zero average charge or current in the system. If we make the simplifying approximation that the plasma particles interact primarily through electrostatic field, then the dynamics can be described by the Vlasov–Poisson system of equations

$$\begin{aligned}\left(\frac{\partial}{\partial t} + \mathbf{v}\cdot\nabla + \frac{e_a}{m_a}\mathbf{E}\cdot\frac{\partial}{\partial \mathbf{v}}\right) f_a = 0, \\ \nabla\cdot\mathbf{E} = 4\pi \sum_a e_a \int d\mathbf{v}\, f_a, \end{aligned} \tag{1.27}$$

where e_a and m_a are charge and mass of species a $(= e, i)$ for electrons and ions ($e_a = e$ for protons and $e_a = -e$ for electrons). The one-particle distribution function $f_a(\mathbf{r},\mathbf{v},t)$ is the probability density of finding a collection of plasma particles of species a, at a particular state in phase space $(\mathbf{r},\mathbf{v})$ at a given time t. Consequently, if we integrate $f_a(\mathbf{r},\mathbf{v},t)$ over $\mathbf{v}$, or equivalently, if we collect all possible configuration in velocity space, then the result becomes the density of charged particle species labeled a,

$$\rho_a(\mathbf{r},t) = \int d\mathbf{v}\, f_a(\mathbf{r},\mathbf{v},t). \tag{1.28}$$

Multiplying the charge e_a and summing over all charged particle species leads to the total charge density

$$\rho(\mathbf{r},t) = \sum_a e_a \rho_a(\mathbf{r},t). \tag{1.29}$$

Since $f_a(\mathbf{r},\mathbf{v},t)$ is the probability density, it is normalized to the ambient charged particle number density n_a,

$$\frac{1}{\mathscr{V}} \int d\mathbf{r} \int d\mathbf{v}\, f_a(\mathbf{r},\mathbf{v},t) = n_a, \tag{1.30}$$

where $\mathscr{V}$ is the volume of the system. That is, if we collect all possible configurations in velocity space at a given time, and integrate over the entire volume under consideration and divide by $\mathscr{V}$, that is, take the spatial average, then the result should be the total number of particles per volume, $n_a = N_a/\mathscr{V}$, or equivalently, the ambient density. Since in the absence of source or sink, plasma particles cannot be created or annihilated (that is, no recombination into neutrals or reionization), the one-particle distribution function must be conserved. Hence,

$$\frac{df_a}{dt} = \left(\frac{\partial}{\partial t} + \dot{\mathbf{r}}\cdot\nabla + \dot{\mathbf{v}}\cdot\frac{\partial}{\partial\mathbf{v}}\right) f_a = 0. \tag{1.31}$$

By virtue of the equation of motion,

$$\dot{\mathbf{r}} = \mathbf{v} \text{ and } \dot{\mathbf{v}} = \frac{e_a}{m_a}\mathbf{E}, \tag{1.32}$$

we obtain the Vlasov equation in (1.27). Because of the charge neutrality condition, the ambient density is the same for both ions and electrons,

$$n_e = n_i = n. \tag{1.33}$$

Let us separate the physical quantities into average and fluctuating parts. The average particle distribution function is independent of the spatial coordinate $\mathbf{r}$ since we assume uniform plasma, and there is no average electric field, so that we may write

$$\begin{aligned} f_a(\mathbf{r},\mathbf{v},t) &= n_a F_a(\mathbf{v},t) + \delta f_a(\mathbf{r},\mathbf{v},t), \\ \mathbf{E}(\mathbf{r},t) &= \delta\mathbf{E}(\mathbf{r},t), \end{aligned} \tag{1.34}$$

where δ represents fluctuating quantities whose phases are supposed to be random. When averaged over their phases, these quantities vanish. In (1.34) we have introduced a normalized one-particle distribution function $F_a(\mathbf{v},t)$ $[\int d\mathbf{v}\, F_a(\mathbf{v},t) = 1]$. Inserting (1.34) back into the coupled Vlasov–Poisson equation, we obtain

$$\left(\frac{\partial}{\partial t}+\frac{e_a}{m_a}\,\delta\mathbf{E}\cdot\frac{\partial}{\partial\mathbf{v}}\right)n_aF_a+\left(\frac{\partial}{\partial t}+\mathbf{v}\cdot\nabla+\frac{e_a}{m_a}\,\delta\mathbf{E}\cdot\frac{\partial}{\partial\mathbf{v}}\right)\delta f_a=0,$$
$$\nabla\cdot\delta\mathbf{E}=4\pi\sum_a e_a\int d\mathbf{v}\,\delta f_a. \tag{1.35}$$

Upon averaging (1.35) over random phases of the fluctuations, we obtain the formal particle kinetic equation:

$$\frac{\partial n_aF_a}{\partial t}=-\frac{e_a}{m_a}\frac{\partial}{\partial\mathbf{v}}\cdot\langle\delta f_a\,\delta\mathbf{E}\rangle. \tag{1.36}$$

Let us subtract the formal particle kinetic equation from the original equation in order to obtain the equation for perturbed distribution function:

$$\left(\frac{\partial}{\partial t}+\mathbf{v}\cdot\nabla\right)\delta f_a=-\frac{e_a}{m_a}\,\delta\mathbf{E}\cdot\frac{\partial n_aF_a}{\partial\mathbf{v}}$$
$$-\frac{e_a}{m_a}\frac{\partial}{\partial\mathbf{v}}\cdot\left(\delta f_a\,\delta\mathbf{E}-\langle\delta f_a\,\delta\mathbf{E}\rangle\right). \tag{1.37}$$

Note that (1.37) is nonlinear since it contains terms of order $\mathcal{O}(\delta^2)$.

We assume that the fluctuations can be decomposed in the sense of Fourier–Laplace transformation over the fast-time scales of fluctuations while the spectral amplitudes may vary slowly in time:

$$\delta f_a(\mathbf{r},\mathbf{v},t)=\int d\mathbf{k}\int_L d\omega\,\delta f^a_{\mathbf{k},\omega}(\mathbf{v},t)\,e^{i\mathbf{k}\cdot\mathbf{r}-i\omega t},$$
$$\delta f^a_{\mathbf{k},\omega}(\mathbf{v},t)=\frac{1}{(2\pi)^4}\int d\mathbf{r}\int_0^\infty dt\,\delta f_a(\mathbf{r},\mathbf{v},t)\,e^{-i\mathbf{k}\cdot\mathbf{r}+i\omega t},$$
$$\delta\phi(\mathbf{r},t)=\int d\mathbf{k}\int_L d\omega\,\delta\phi_{\mathbf{k},\omega}(t)\,e^{i\mathbf{k}\cdot\mathbf{r}-i\omega t},$$
$$\delta\phi_{\mathbf{k},\omega}(t)=\frac{1}{(2\pi)^4}\int d\mathbf{r}\int_0^\infty dt\,\delta\phi(\mathbf{r},t)\,e^{-i\mathbf{k}\cdot\mathbf{r}+i\omega t}, \tag{1.38}$$

where we have assumed

$$\delta\mathbf{E}(\mathbf{r},t)=-\nabla\delta\phi(\mathbf{r},t),\qquad\text{or}\qquad\delta\mathbf{E}_{\mathbf{k},\omega}=-i\mathbf{k}\,\delta\phi_{\mathbf{k},\omega}, \tag{1.39}$$

since we work under the electrostatic approximation. In (1.38), the integration $\int_L d\omega$ is taken along the path L that stretches from $\omega=-\infty+i\sigma$ to $\omega=\infty+i\sigma$ ($\sigma>0$ and $\sigma\to 0$). The infinitesimal positive imaginary part σ signifies that we are only interested in causal solutions. The causality requirement is related to the Laplace transformation being defined only over $0<t<\infty$ rather than the entire integral domain, $-\infty<t<\infty$. The reason for the positive infinitesimal imaginary part, $\sigma>0$, in the ω integration along the path L, is to ensure the temporal convergence for $t \to \infty$, hence the causality. By the choice of positive integral $\int_0^\infty dt$, one is effectively breaking the time reversal symmetry, and this forces physical processes to proceed in forward time, $t > 0$. That is, the Laplace transformation in place of the symmetric temporal Fourier transformation is equivalent to imposing the causal relationship to the otherwise time-reversible Vlasov equation. In Appendix A we review the treatment of time-irreversible small amplitude plasma perturbation, as discussed originally by Landau (1946). We also discuss the notion of Landau damping in Appendix A.

In the Fourier–Laplace transformation defined in (1.38), we have made an assumption that the spectral amplitudes $\delta f^a_{\mathbf{k},\omega}(\mathbf{v},t)$ and $\delta\phi_{\mathbf{k},\omega}(t)$ have slow and adiabatic time dependence. These quantities are assumed to vary slowly in time, while the temporal dependence dictated by $\exp(-i\omega t)$ is assumed to be fast. That is, the time dependence of amplitudes is much weaker than that associated with the wave time scale, δf, $\delta\phi \sim \mathcal{O}(t_{\rm slow})$, where $\mathcal{O}(t_{\rm slow}) \gg \mathcal{O}(\omega^{-1})$. These amplitudes are calculated as if they are independent of time on the fast wave time scale $(t \sim \omega^{-1})$.

Employing the transformation (1.38), the equation for fluctuating field, formal particle kinetic equation, and the equation for perturbed distribution function are expressed, respectively, as follows:

$$
\begin{aligned}
\delta\phi_{\mathbf{k},\omega}(t) &= \sum_a \frac{4\pi e_a}{k^2}\int d\mathbf{v}\,\delta f^a_{\mathbf{k},\omega}(\mathbf{v},t),\\
\frac{\partial n_a F_a(\mathbf{v},t)}{\partial t} &= \frac{i e_a}{m_a}\int d\mathbf{k}\int d\omega\int d\mathbf{k}'\int d\omega'\\
&\quad\times\left(\mathbf{k}'\cdot\frac{\partial}{\partial\mathbf{v}}\right)\langle\delta\phi_{\mathbf{k}',\omega'}(t)\delta f^a_{\mathbf{k},\omega}(\mathbf{v},t)\rangle\, e^{i(\mathbf{k}+\mathbf{k}')\cdot\mathbf{r}-i(\omega+\omega')t},\\
\left(\omega-\mathbf{k}\cdot\mathbf{v}+i\frac{\partial}{\partial t}\right)\delta f^a_{\mathbf{k},\omega}(\mathbf{v},t) &= -\frac{e_a}{m_a}\,\delta\phi_{\mathbf{k},\omega}(t)\,\mathbf{k}\cdot\frac{\partial n_a F_a(\mathbf{v},t)}{\partial\mathbf{v}}\\
&\quad-\frac{e_a}{m_a}\int d\mathbf{k}'\int d\omega'\,\mathbf{k}'\cdot\frac{\partial}{\partial\mathbf{v}}\\
&\quad\times\Big[\delta\phi_{\mathbf{k}',\omega'}(t)\,\delta f^a_{\mathbf{k}-\mathbf{k}',\omega-\omega'}(\mathbf{v},t)\\
&\quad-\langle\delta\phi_{\mathbf{k}',\omega'}(t)\,\delta f^a_{\mathbf{k}-\mathbf{k}',\omega-\omega'}(\mathbf{v},t)\rangle\Big].
\end{aligned}
\tag{1.40}
$$

The present treatment of slow and adiabatic time dependence of spectral amplitudes is not rigorous. Mathematically consistent treatment involves the multiple time-scale analysis as employed by Davidson (1972). The present discussion resorts to the shortcut approach, following the method employed in the standard literature, for example, Sitenko (1982), Akhiezer et al. (1975), etc.

1.3 Fast-Time Scale Solution

Formal solutions for the fast-time scale quantities necessitate the inversion of differential operator, $\omega - \mathbf{k}\cdot\mathbf{v} + i\partial/\partial t$. To simplify the matter, we temporarily ignore the adiabatic time derivative by absorbing the time derivative as part of the "new" definition for ω,

$$\omega + i\,\partial/\partial t \to \omega, \tag{1.41}$$

within the definition for $\omega - \mathbf{k}\cdot\mathbf{v} + i0 + i\partial/\partial t$. We will reinstitute the explicit slow-time derivative when we discuss the wave kinetic equation later. Meanwhile, let us define

$$\mathbf{g}_{\mathbf{k},\omega} \equiv -\frac{e_a}{m_a}\frac{1}{\omega - \mathbf{k}\cdot\mathbf{v} + i0}\frac{\partial}{\partial \mathbf{v}}. \tag{1.42}$$

Note that $+i0$ is meant to indicate that the angular frequency ω is always to be interpreted as having an infinitesimally small but positive imaginary part, which is intimately related to the causality condition associated with the Laplace transformation, and discussed in Appendix A. The notion of ω having positive imaginary part arises from the asymptotic treatment associated with the analytic continuation, and it is further discussed in detail in Appendix B. Then the equation for perturbed particle distribution function in (1.40) is re-expressed as follows:

$$\begin{aligned}\delta f^a_{\mathbf{k},\omega} &= \mathbf{k}\cdot\mathbf{g}_{\mathbf{k},\omega}\, n_a F_a\, \delta\phi_{\mathbf{k},\omega} \\ &\quad + \int d\mathbf{k}' \int d\omega'\, \mathbf{k}'\cdot\mathbf{g}_{\mathbf{k},\omega}\left[\delta\phi_{\mathbf{k}',\omega'}\,\delta f^a_{\mathbf{k}-\mathbf{k}',\omega-\omega'} - \langle\delta\phi_{\mathbf{k}',\omega'}\,\delta f^a_{\mathbf{k}-\mathbf{k}',\omega-\omega'}\rangle\right].\end{aligned} \tag{1.43}$$

To solve the equation for $\delta f^a_{\mathbf{k},\omega}$, we expand $\delta f^a_{\mathbf{k},\omega}$ in a formal perturbation series with successive terms proportional to increasing powers of $\delta\phi_{\mathbf{k},\omega}$:

$$\delta f^a_{\mathbf{k},\omega} = \delta f^{a(1)}_{\mathbf{k},\omega} + \delta f^{a(2)}_{\mathbf{k},\omega} + \delta f^{a(3)}_{\mathbf{k},\omega} + \cdots, \tag{1.44}$$

where $\delta f^{a(n)}_{\mathbf{k},\omega}$ scales as $\mathscr{O}(\delta\phi^n_{\mathbf{k},\omega})$. It is not so difficult to see why $\delta f^{a(n)}_{\mathbf{k},\omega}$ should scale as $\mathscr{O}(\delta\phi^n_{\mathbf{k},\omega})$. If we ignore the nonlinear coupling term in (1.43), then it is readily obvious that the perturbed distribution function $\delta f^a_{\mathbf{k},\omega}$ is proportional to the perturbed wave amplitude $\delta\phi_{\mathbf{k},\omega}$. From this it logically follows that the expansion parameter should scale as $\delta\phi_{\mathbf{k},\omega}$. This consideration also leads to the conceptually "necessary" condition for the validity of so-called *weak turbulence* expansion (1.44). If we consider that the characteristic energy associated with the average particle distribution $n_a F_a$ is the thermal energy,

$$\mathscr{E}_{\text{particle}} = \sum_a n_a \int d\mathbf{v}\,\frac{m_a v^2}{2}\,F_a(\mathbf{v}) = \sum_a \frac{n_a m_a v_{Ta}^2}{2} = \sum_a n_a T_a, \tag{1.45}$$

and that the perturbed distribution function, which is defined by $\delta f_a = f_a - n_a F_a$, scales as the wave amplitude, $\delta f_a \propto \delta\phi$, then in order for perturbative expansion (1.44) to converge, the characteristic energy associated with the perturbed distribution function, which is proportional to the wave energy density,

$$\sum_a \int d\mathbf{v}\, \frac{m_a v^2}{2}\, \delta f_a(\mathbf{v}) \propto \mathscr{E}_{\text{wave}} = \frac{\langle \delta E^2 \rangle}{8\pi}, \tag{1.46}$$

where $\delta\mathbf{E} = -\nabla\delta\phi$, must be sufficiently lower than the thermal energy. In short, the following condition must be satisfied for the weak turbulence expansion to be valid:

$$\mathscr{E}_{\text{particle}} \gg \mathscr{E}_{\text{wave}}. \tag{1.47}$$

This is the general requirement for the validity of weak turbulence theory.

Inserting the series expansion (1.44) to the original equation for $\delta f^a_{\mathbf{k},\omega}(\mathbf{v})$, we readily obtain, order by order,

$$\begin{aligned}
\delta f^{a(1)}_{\mathbf{k},\omega} &= \mathbf{k}\cdot\mathbf{g}_{\mathbf{k},\omega}\, n_a F_a\, \delta\phi_{\mathbf{k},\omega},\\
\delta f^{a(2)}_{\mathbf{k},\omega} &= \int d\mathbf{k}' \int d\omega'\, \mathbf{k}'\cdot\mathbf{g}_{\mathbf{k},\omega} \left[\delta\phi_{\mathbf{k}',\omega'}\, \delta f^{a(1)}_{\mathbf{k}-\mathbf{k}',\omega-\omega'} - \langle \delta\phi_{\mathbf{k}',\omega'}\, \delta f^{a(1)}_{\mathbf{k}-\mathbf{k}',\omega-\omega'} \rangle \right],\\
\delta f^{a(3)}_{\mathbf{k},\omega} &= \int d\mathbf{k}' \int d\omega'\, \mathbf{k}'\cdot\mathbf{g}_{\mathbf{k},\omega} \left[\delta\phi_{\mathbf{k}',\omega'}\, \delta f^{a(2)}_{\mathbf{k}-\mathbf{k}',\omega-\omega'} - \langle \delta\phi_{\mathbf{k}',\omega'}\, \delta f^{a(2)}_{\mathbf{k}-\mathbf{k}',\omega-\omega'} \rangle \right],\\
&\cdots
\end{aligned} \tag{1.48}$$

etc. Iterative solutions for $\delta f^{a(2)}_{\mathbf{k},\omega}$ and $\delta f^{a(3)}_{\mathbf{k},\omega}$ are given below:

$$\begin{aligned}
\delta f^{a(2)}_{\mathbf{k},\omega} &= \int d\mathbf{k}' \int d\omega'\, (\mathbf{k}'\cdot\mathbf{g}_{\mathbf{k},\omega}) \left[(\mathbf{k}-\mathbf{k}')\cdot\mathbf{g}_{\mathbf{k}-\mathbf{k}',\omega-\omega'} \right]\\
&\quad \times n_a F_a \left(\delta\phi_{\mathbf{k}',\omega'}\, \delta\phi_{\mathbf{k}-\mathbf{k}',\omega-\omega'} - \langle \delta\phi_{\mathbf{k}',\omega'}\, \delta\phi_{\mathbf{k}-\mathbf{k}',\omega-\omega'} \rangle \right),\\
\delta f^{a(3)}_{\mathbf{k},\omega} &= \int d\mathbf{k}' \int d\omega' \int d\mathbf{k}'' \int d\omega''\, (\mathbf{k}'\cdot\mathbf{g}_{\mathbf{k},\omega})\, (\mathbf{k}''\cdot\mathbf{g}_{\mathbf{k}-\mathbf{k}',\omega-\omega'})\\
&\quad \times \left[(\mathbf{k}-\mathbf{k}'-\mathbf{k}'')\cdot\mathbf{g}_{\mathbf{k}-\mathbf{k}'-\mathbf{k}'',\omega-\omega'-\omega''} \right] n_a F_a\\
&\quad \times \left(\delta\phi_{\mathbf{k}',\omega'}\, \delta\phi_{\mathbf{k}'',\omega''}\, \delta\phi_{\mathbf{k}-\mathbf{k}'-\mathbf{k}'',\omega-\omega'-\omega''} \right.\\
&\quad - \delta\phi_{\mathbf{k}',\omega'}\, \langle \delta\phi_{\mathbf{k}'',\omega''}\, \delta\phi_{\mathbf{k}-\mathbf{k}'-\mathbf{k}'',\omega-\omega'-\omega''} \rangle\\
&\quad \left. - \langle \delta\phi_{\mathbf{k}',\omega'}\, \delta\phi_{\mathbf{k}'',\omega''}\, \delta\phi_{\mathbf{k}-\mathbf{k}'-\mathbf{k}'',\omega-\omega'-\omega''} \rangle \right).
\end{aligned} \tag{1.49}$$

Let us introduce the following simplified notations:

$$\begin{aligned}
\sum_{\mathbf{k}_1+\mathbf{k}_2=\mathbf{k}} &= \int d\mathbf{k}_1 \int d\mathbf{k}_2\, \delta(\mathbf{k}_1+\mathbf{k}_2-\mathbf{k}),\\
\sum_{\omega_1+\omega_2=\omega} &= \int d\omega_1 \int d\omega_2\, \delta(\omega_1+\omega_2-\omega),
\end{aligned}$$

$$\sum_{\mathbf{k}_1+\mathbf{k}_2+\mathbf{k}_3=\mathbf{k}} = \int d\mathbf{k}_1 \int d\mathbf{k}_2 \int d\mathbf{k}_3\, \delta(\mathbf{k}_1+\mathbf{k}_2+\mathbf{k}_3-\mathbf{k}),$$

$$\sum_{\omega_1+\omega_2+\omega_3=\omega} = \int d\omega_1 \int d\omega_2 \int d\omega_3\, \delta(\omega_1+\omega_2+\omega_3-\omega), \tag{1.50}$$

etc. Then, the series solution for $\delta f^a_{\mathbf{k},\omega}$ is given by

$$\begin{aligned}
\delta f^a_{\mathbf{k},\omega} &= \mathbf{k}\cdot\mathbf{g}_{\mathbf{k},\omega}\, n_a F_a\, \delta\phi_{\mathbf{k},\omega} \\
&+ \sum_{\mathbf{k}_1+\mathbf{k}_2=\mathbf{k}} \sum_{\omega_1+\omega_2=\omega} (\mathbf{k}_1\cdot\mathbf{g}_{\mathbf{k},\omega})\,(\mathbf{k}_2\cdot\mathbf{g}_{\mathbf{k}_2,\omega_2})\, n_a F_a \\
&\times (\delta\phi_{\mathbf{k}_1,\omega_1}\,\delta\phi_{\mathbf{k}_2,\omega_2} - \langle\delta\phi_{\mathbf{k}_1,\omega_1}\,\delta\phi_{\mathbf{k}_2,\omega_2}\rangle) \\
&+ \sum_{\mathbf{k}_1+\mathbf{k}_2+\mathbf{k}_3=\mathbf{k}} \sum_{\omega_1+\omega_2+\omega_3=\omega} (\mathbf{k}_1\cdot\mathbf{g}_{\mathbf{k},\omega})\,(\mathbf{k}_2\cdot\mathbf{g}_{\mathbf{k}_2+\mathbf{k}_3,\omega_2+\omega_3}) \\
&\times (\mathbf{k}_3\cdot\mathbf{g}_{\mathbf{k}_3,\omega_3})\, n_a F_a\, (\delta\phi_{\mathbf{k}_1,\omega_1}\,\delta\phi_{\mathbf{k}_2,\omega_2}\,\delta\phi_{\mathbf{k}_3,\omega_3} \\
&- \delta\phi_{\mathbf{k}_1,\omega_1}\,\langle\delta\phi_{\mathbf{k}_2,\omega_2}\,\delta\phi_{\mathbf{k}_3,\omega_3}\rangle - \langle\delta\phi_{\mathbf{k}_1,\omega_1}\,\delta\phi_{\mathbf{k}_2,\omega_2}\,\delta\phi_{\mathbf{k}_3,\omega_3}\rangle). \\
&\ \vdots
\end{aligned} \tag{1.51}$$

The right-hand side of (1.51) is not symmetric with respect to the interchange of dummy variables, $(\mathbf{k}_1,\omega_1)$ and $(\mathbf{k}_2,\omega_2)$. The second-order nonlinear term should be symmetrized with respect to variables $(\mathbf{k}_1,\omega_1)$ and $(\mathbf{k}_2,\omega_2)$, while for the third-order term, the symmetrization should be implemented with respect to $(\mathbf{k}_2,\omega_2)$ and $(\mathbf{k}_3,\omega_3)$. The interchange of these dummy variables leaves (1.51) intact. When the above quantities are symmetrized with respect to these variables, then the correct symmetrized expression emerges,

$$\begin{aligned}
\delta f^a_{\mathbf{k},\omega} &= \alpha(\mathbf{k},\omega)\, n_a F_a\, \delta\phi_{\mathbf{k},\omega} \\
&+ \sum_{\mathbf{k}_1+\mathbf{k}_2=\mathbf{k}} \sum_{\omega_1+\omega_2=\omega} \alpha^{(2)}(\mathbf{k}_1,\omega_1|\mathbf{k}_2,\omega_2)\, n_a F_a \\
&\times (\delta\phi_{\mathbf{k}_1,\omega_1}\,\delta\phi_{\mathbf{k}_2,\omega_2} - \langle\delta\phi_{\mathbf{k}_1,\omega_1}\,\delta\phi_{\mathbf{k}_2,\omega_2}\rangle) \\
&+ \sum_{\mathbf{k}_1+\mathbf{k}_2+\mathbf{k}_3=\mathbf{k}} \sum_{\omega_1+\omega_2+\omega_3=\omega} \alpha^{(3)}(\mathbf{k}_1,\omega_1|\mathbf{k}_2,\omega_2|\mathbf{k}_3,\omega_3)\, n_a F_a \\
&\times \Big(\delta\phi_{\mathbf{k}_1,\omega_1}\,\delta\phi_{\mathbf{k}_2,\omega_2}\,\delta\phi_{\mathbf{k}_3,\omega_3} - \delta\phi_{\mathbf{k}_1,\omega_1}\,\langle\delta\phi_{\mathbf{k}_2,\omega_2}\,\delta\phi_{\mathbf{k}_3,\omega_3}\rangle \\
&- \langle\delta\phi_{\mathbf{k}_1,\omega_1}\,\delta\phi_{\mathbf{k}_2,\omega_2}\,\delta\phi_{\mathbf{k}_3,\omega_3}\rangle\Big),
\end{aligned} \tag{1.52}$$

where

$$\begin{aligned}
\alpha(\mathbf{k},\omega) &= \mathbf{k}\cdot\mathbf{g}_{\mathbf{k},\omega}, \\
\alpha^{(2)}(\mathbf{k}_1,\omega_1|\mathbf{k}_2,\omega_2) &= \frac{1}{2}\,[(\mathbf{k}_1\cdot\mathbf{g}_{\mathbf{k}_1+\mathbf{k}_2,\omega_1+\omega_2})\,(\mathbf{k}_2\cdot\mathbf{g}_{\mathbf{k}_2,\omega_2}) \\
&+ (\mathbf{k}_2\cdot\mathbf{g}_{\mathbf{k}_1+\mathbf{k}_2,\omega_1+\omega_2})\,(\mathbf{k}_1\cdot\mathbf{g}_{\mathbf{k}_1,\omega_1})],
\end{aligned}$$

$$\alpha^{(3)}(\mathbf{k}_1,\omega_1|\mathbf{k}_2,\omega_2|\mathbf{k}_3,\omega_3) = \frac{1}{2}\,(\mathbf{k}_1\cdot\mathbf{g}_{\mathbf{k}_1+\mathbf{k}_2+\mathbf{k}_3,\omega_1+\omega_2+\omega_3}) \times [\,(\mathbf{k}_2\cdot\mathbf{g}_{\mathbf{k}_2+\mathbf{k}_3,\omega_2+\omega_3})\,(\mathbf{k}_3\cdot\mathbf{g}_{\mathbf{k}_3,\omega_3}) + (\mathbf{k}_3\cdot\mathbf{g}_{\mathbf{k}_2+\mathbf{k}_3,\omega_2+\omega_3})\,(\mathbf{k}_2\cdot\mathbf{g}_{\mathbf{k}_2,\omega_2})\,], \tag{1.53}$$

1.4 Perturbed Wave Equation

We insert the iterative solution for $\delta f^a_{\mathbf{k},\omega}$ (1.52) to the right-hand side of perturbed Poisson equation in (1.40),

$$\begin{aligned}0 =& \left(1-\sum_a\frac{4\pi e_a n_a}{k^2}\int d\mathbf{v}\,\alpha(\mathbf{k},\omega)\,F_a\right)\delta\phi_{\mathbf{k},\omega}\\ &-\sum_{\mathbf{k}_1+\mathbf{k}_2=\mathbf{k}}\ \sum_{\omega_1+\omega_2=\omega}\ \sum_a\frac{4\pi e_a n_a}{k^2}\int d\mathbf{v}\,\alpha^{(2)}(\mathbf{k}_1,\omega_1|\mathbf{k}_2,\omega_2)\,F_a\\ &\times\left(\delta\phi_{\mathbf{k}_1,\omega_1}\,\delta\phi_{\mathbf{k}_2,\omega_2}-\langle\delta\phi_{\mathbf{k}_1,\omega_1}\,\delta\phi_{\mathbf{k}_2,\omega_2}\rangle\right)\\ &-\sum_{\mathbf{k}_1+\mathbf{k}_2+\mathbf{k}_3=\mathbf{k}}\ \sum_{\omega_1+\omega_2+\omega_3=\omega}\ \sum_a\frac{4\pi e_a n_a}{k^2}\int d\mathbf{v}\,F_a\,\alpha^{(3)}(\mathbf{k}_1,\omega_1|\mathbf{k}_2,\omega_2|\mathbf{k}_3,\omega_3)\\ &\times\left(\delta\phi_{\mathbf{k}_1,\omega_1}\,\delta\phi_{\mathbf{k}_2,\omega_2}\,\delta\phi_{\mathbf{k}_3,\omega_3}-\langle\delta\phi_{\mathbf{k}_1,\omega_1}\,\delta\phi_{\mathbf{k}_2,\omega_2}\,\delta\phi_{\mathbf{k}_3,\omega_3}\rangle\right.\\ &\left.-\,\delta\phi_{\mathbf{k}_1,\omega_1}\,\langle\delta\phi_{\mathbf{k}_2,\omega_2}\,\delta\phi_{\mathbf{k}_3,\omega_3}\rangle\right).\end{aligned}\tag{1.54}$$

Let us define the linear dielectric response function,

$$\begin{aligned}\epsilon(\mathbf{k},\omega) &= 1+\chi(\mathbf{k},\omega) = 1+\sum_a\chi_a(\mathbf{k},\omega)\\ &= 1+\frac{4\pi}{\omega}\,\sigma(\mathbf{k},\omega) = 1+\frac{4\pi}{\omega}\sum_a\sigma_a(\mathbf{k},\omega),\\ \chi_a(\mathbf{k},\omega) &= \frac{4\pi}{\omega}\,\sigma_a(\mathbf{k},\omega) = -\frac{4\pi e_a n_a}{k^2}\int d\mathbf{v}\,\alpha(\mathbf{k},\omega)\,F_a,\end{aligned}\tag{1.55}$$

where $\chi(\mathbf{k},\omega)$ is the linear dielectric susceptibility and $\sigma(\mathbf{k},\omega)$ is the linear dielectric conductivity. The second-order nonlinear response function, or equivalently, the second-order nonlinear susceptibility, is likewise defined

$$\begin{aligned}\chi^{(2)}(\mathbf{k}_1,\omega_1|\mathbf{k}_2,\omega_2) &= \sum_a\chi_a^{(2)}(\mathbf{k}_1,\omega_1|\mathbf{k}_2,\omega_2),\\ \chi_a^{(2)}(\mathbf{k}_1,\omega_1|\mathbf{k}_2,\omega_2) &= -\frac{4\pi i e_a n_a}{k_1\,k_2\,|\mathbf{k}_1+\mathbf{k}_2|}\int d\mathbf{v}\,\alpha^{(2)}(\mathbf{k}_1,\omega_1|\mathbf{k}_2,\omega_2)\,F_a,\end{aligned}\tag{1.56}$$

and the third-order nonlinear response function, or third-order nonlinear susceptibility, may also be defined

$$
\begin{aligned}
\bar{\chi}^{(3)}(\mathbf{k}_1,\omega_1|\mathbf{k}_2,\omega_2|\mathbf{k}_3,\omega_3) &= \sum_a \bar{\chi}_a^{(3)}(\mathbf{k}_1,\omega_1|\mathbf{k}_2,\omega_2|\mathbf{k}_3,\omega_3),\\
\bar{\chi}_a^{(3)}(\mathbf{k}_1,\omega_1|\mathbf{k}_2,\omega_2|\mathbf{k}_3,\omega_3) &= \frac{4\pi e_a n_a}{k_1\,k_2\,k_3\,|\mathbf{k}_1+\mathbf{k}_2+\mathbf{k}_3|}\\
&\quad\times\int d\mathbf{v}\,\alpha^{(3)}(\mathbf{k}_1,\omega_1|\mathbf{k}_2,\omega_2|\mathbf{k}_3,\omega_3)\,F_a,\\
\chi^{(3)}(\mathbf{k}_1,\omega_1|\mathbf{k}_2,\omega_2|\mathbf{k}_3,\omega_3) &= \frac{1}{3}\,[\,\bar{\chi}^{(3)}(\mathbf{k}_1,\omega_1|\mathbf{k}_2,\omega_2|\mathbf{k}_3,\omega_3)\\
&\quad+\bar{\chi}^{(3)}(\mathbf{k}_2,\omega_2|\mathbf{k}_1,\omega_1|\mathbf{k}_3,\omega_3)\\
&\quad+\bar{\chi}^{(3)}(\mathbf{k}_3,\omega_3|\mathbf{k}_2,\omega_2|\mathbf{k}_1,\omega_1)\,].
\end{aligned}
\tag{1.57}
$$

To sum up, the linear and nonlinear susceptibilities can be expressed as

$$
\chi_a(\mathbf{k},\omega) = -\frac{4\pi e_a n_a}{k^2}\int d\mathbf{v}\,\mathbf{k}\cdot\mathbf{g}_{\mathbf{k},\omega}\,F_a, \tag{1.58}
$$

$$
\begin{aligned}
\chi_a^{(2)}(\mathbf{k}_1,\omega_1|\mathbf{k}_2,\omega_2) &= -\frac{1}{2}\,\frac{4\pi i e_a n_a}{k_1\,k_2\,|\mathbf{k}_1+\mathbf{k}_2|}\int d\mathbf{v}\,\mathbf{g}_{\mathbf{k}_1+\mathbf{k}_2,\omega_1+\omega_2}\\
&\quad\cdot[\,\mathbf{k}_1\,(\mathbf{k}_2\cdot\mathbf{g}_{\mathbf{k}_2,\omega_2})+\mathbf{k}_2\,(\mathbf{k}_1\cdot\mathbf{g}_{\mathbf{k}_1,\omega_1})\,]\,F_a,
\end{aligned}
\tag{1.59}
$$

$$
\begin{aligned}
\bar{\chi}_a^{(3)}(\mathbf{k}_1,\omega_1|\mathbf{k}_2,\omega_2|\mathbf{k}_3,\omega_3) &= \frac{1}{2}\frac{4\pi e_a n_a}{k_1\,k_2\,k_3\,|\mathbf{k}_1+\mathbf{k}_2+\mathbf{k}_3|}\\
&\quad\times\int d\mathbf{v}\,(\mathbf{g}_{\mathbf{k}_1+\mathbf{k}_2+\mathbf{k}_3,\omega_1+\omega_2+\omega_3}\cdot\mathbf{k}_1)\mathbf{g}_{\mathbf{k}_2+\mathbf{k}_3,\omega_2+\omega_3}\\
&\quad\cdot\left[\mathbf{k}_2\,(\mathbf{k}_3\cdot\mathbf{g}_{\mathbf{k}_3,\omega_3})+\mathbf{k}_3\,(\mathbf{k}_2\cdot\mathbf{g}_{\mathbf{k}_2,\omega_2})\right]\,F_a.
\end{aligned}
\tag{1.60}
$$

After explicitly writing out the various objects, making use of their respective definitions, we may also rewrite the susceptibilities in concrete forms as follows:

$$
\chi_a(\mathbf{k},\omega) = \frac{\omega_{pa}^2}{k^2}\int d\mathbf{v}\,\frac{\mathbf{k}\cdot\partial F_a/\partial\mathbf{v}}{\omega-\mathbf{k}\cdot\mathbf{v}+i0}, \tag{1.61}
$$

$$
\begin{aligned}
\chi_a^{(2)}(\mathbf{k}_1,\omega_1|\mathbf{k}_2,\omega_2) &= \frac{-i}{2}\frac{e_a}{m_a}\frac{\omega_{pa}^2}{k_1\,k_2\,|\mathbf{k}_1+\mathbf{k}_2|}\\
&\quad\times\int d\mathbf{v}\,\frac{1}{\omega_1+\omega_2-(\mathbf{k}_1+\mathbf{k}_2)\cdot\mathbf{v}+i0}\\
&\quad\times\left[\mathbf{k}_1\cdot\frac{\partial}{\partial\mathbf{v}}\left(\frac{\mathbf{k}_2\cdot\partial F_a/\partial\mathbf{v}}{\omega_2-\mathbf{k}_2\cdot\mathbf{v}+i0}\right)\right.\\
&\quad\left.+\mathbf{k}_2\cdot\frac{\partial}{\partial\mathbf{v}}\left(\frac{\mathbf{k}_1\cdot\partial F_a/\partial\mathbf{v}}{\omega_1-\mathbf{k}_1\cdot\mathbf{v}+i0}\right)\right],
\end{aligned}
\tag{1.62}
$$

$$\begin{aligned}
\bar{\chi}_a^{(3)}(\mathbf{k}_1,\omega_1|\mathbf{k}_2,\omega_2|\mathbf{k}_3,\omega_3) &= \frac{(-i)^2}{2}\frac{e_a^2}{m_a^2}\frac{\omega_{pa}^2}{k_1\,k_2\,k_3\,|\mathbf{k}_1+\mathbf{k}_2+\mathbf{k}_3|} \\
&\times \int d\mathbf{v}\,\frac{1}{\omega_1+\omega_2+\omega_3-(\mathbf{k}_1+\mathbf{k}_2+\mathbf{k}_3)\cdot\mathbf{v}+i0} \\
&\times \mathbf{k}_1\cdot\frac{\partial}{\partial\mathbf{v}}\left\{\frac{1}{\omega_2+\omega_3-(\mathbf{k}_2+\mathbf{k}_3)\cdot\mathbf{v}+i0}\right. \\
&\times \left[\mathbf{k}_2\cdot\frac{\partial}{\partial\mathbf{v}}\left(\frac{\mathbf{k}_3\cdot\partial F_a/\partial\mathbf{v}}{\omega_3-\mathbf{k}_3\cdot\mathbf{v}+i0}\right)\right. \\
&\left.\left.+\mathbf{k}_3\cdot\frac{\partial}{\partial\mathbf{v}}\left(\frac{\mathbf{k}_2\cdot\partial F_a/\partial\mathbf{v}}{\omega_2-\mathbf{k}_2\cdot\mathbf{v}+i0}\right)\right]\right\},
\end{aligned} \tag{1.63}$$

where

$$\omega_{pa} = \left(\frac{4\pi n_a e_a^2}{m_a}\right)^{1/2} \tag{1.64}$$

is the plasma frequency for species a. In (1.64), we again iterate that the positive infinitesimal imaginary part associated with the angular frequency within the resonant denominators is the result of imposing the causality condition in the Laplace transformation (1.38) – see also Appendix B. Various linear and nonlinear susceptibilities in (1.55)–(1.63) are known by their respective names:

- $\epsilon(\mathbf{k},\omega)$; linear dielectric constant
- $\chi(\mathbf{k},\omega)$; linear dielectric susceptibility
- $\sigma(\mathbf{k},\omega)$; linear dielectric conductivity
- $\chi^{(2)}(\mathbf{k}_1,\omega_1|\mathbf{k}_2,\omega_2)$; second-order nonlinear susceptibility
- $\bar{\chi}^{(3)}(\mathbf{k}_1,\omega_1|\mathbf{k}_2,\omega_2|\mathbf{k}_3,\omega_3)$; partial third-order nonlinear susceptibility
- $\chi^{(3)}(\mathbf{k}_1,\omega_1|\mathbf{k}_2,\omega_2|\mathbf{k}_3,\omega_3)$; fully symmetrized third-order nonlinear susceptibility.

These definitions and designation of conventions (and even the notations) follow Sitenko (1982). In terms of these susceptibilities, (1.54) can be written compactly as

$$\begin{aligned}
0 &= k\,\epsilon(\mathbf{k},\omega)\,\delta\phi_{\mathbf{k},\omega} - i\sum_{\mathbf{k}_1+\mathbf{k}_2=\mathbf{k}}\sum_{\omega_1+\omega_2=\omega} k_1\,k_2\,\chi^{(2)}(\mathbf{k}_1,\omega_1|\mathbf{k}_2,\omega_2) \\
&\times \left(\delta\phi_{\mathbf{k}_1,\omega_1}\,\delta\phi_{\mathbf{k}_2,\omega_2} - \langle\delta\phi_{\mathbf{k}_1,\omega_1}\,\delta\phi_{\mathbf{k}_2,\omega_2}\rangle\right) \\
&- \sum_{\mathbf{k}_1+\mathbf{k}_2+\mathbf{k}_3=\mathbf{k}}\sum_{\omega_1+\omega_2+\omega_3=\omega} k_1\,k_2\,k_3\,\bar{\chi}^{(3)}(\mathbf{k}_1,\omega_1|\mathbf{k}_2,\omega_2|\mathbf{k}_3,\omega_3) \\
&\times \left(\delta\phi_{\mathbf{k}_1,\omega_1}\,\delta\phi_{\mathbf{k}_2,\omega_2}\,\delta\phi_{\mathbf{k}_3,\omega_3} - \langle\delta\phi_{\mathbf{k}_1,\omega_1}\,\delta\phi_{\mathbf{k}_2,\omega_2}\,\delta\phi_{\mathbf{k}_3,\omega_3}\rangle\right. \\
&\left.- \delta\phi_{\mathbf{k}_1,\omega_1}\,\langle\delta\phi_{\mathbf{k}_2,\omega_2}\,\delta\phi_{\mathbf{k}_3,\omega_3}\rangle\right).
\end{aligned} \tag{1.65}$$

Recall that in the statistical theory of turbulence we are interested in the correlation function, $\langle\delta\phi^2\rangle$. We may construct the equation for correlation from (1.65).

Let us therefore multiply $\delta\phi_{\mathbf{k}',\omega'}$ to (1.65), and take the ensemble average of the resulting equation:

$$\begin{aligned}
0 = &\, k\,\epsilon(\mathbf{k},\omega)\,\langle\delta\phi_{\mathbf{k},\omega}\,\delta\phi_{\mathbf{k}',\omega'}\rangle \\
&-i\sum_{\mathbf{k}_1+\mathbf{k}_2=\mathbf{k}}\;\sum_{\omega_1+\omega_2=\omega} k_1\,k_2\,\chi^{(2)}(\mathbf{k}_1,\omega_1|\mathbf{k}_2,\omega_2)\,\langle\delta\phi_{\mathbf{k}_1,\omega_1}\,\delta\phi_{\mathbf{k}_2,\omega_2}\,\delta\phi_{\mathbf{k}',\omega'}\rangle \\
&-\sum_{\mathbf{k}_1+\mathbf{k}_2+\mathbf{k}_3=\mathbf{k}}\;\sum_{\omega_1+\omega_2+\omega_3=\omega} k_1\,k_2\,k_3\,\bar{\chi}^{(3)}(\mathbf{k}_1,\omega_1|\mathbf{k}_2,\omega_2|\mathbf{k}_3,\omega_3) \\
&\times\big(\langle\delta\phi_{\mathbf{k}_1,\omega_1}\,\delta\phi_{\mathbf{k}_2,\omega_2}\,\delta\phi_{\mathbf{k}_3,\omega_3}\,\delta\phi_{\mathbf{k}',\omega'}\rangle \\
&-\langle\delta\phi_{\mathbf{k}_1,\omega_1}\,\delta\phi_{\mathbf{k}',\omega'}\rangle\,\langle\delta\phi_{\mathbf{k}_2,\omega_2}\,\delta\phi_{\mathbf{k}_3,\omega_3}\rangle\big)\,. \qquad (1.66)
\end{aligned}$$

This equation relates the two-body electric field amplitude correlation $\langle\delta\phi^2\rangle$ to three- and four-body correlations $\langle\delta\phi^3\rangle$ and $\langle\delta\phi^4\rangle$. As a consequence, (1.66) is not closed, but instead forms an infinite chain, or hierarchy, of correlations. To break the hierarchy one must introduce certain assumptions. This is the "closure problem." To break the hierarchy, we write the four-body cumulant $\langle\delta\phi^4\rangle$ as the sum of products of two body cumulants while ignoring the irreducible four-body correlation. This is equivalent to ignoring the last term on the right-hand side of (1.15). Let us thus consider the third-order nonlinear term associated with $\bar{\chi}^{(3)}$. Making use of the shorthand notations,

$$\begin{aligned}
&K = (\mathbf{k},\omega), \quad 1 \equiv K_1 = (\mathbf{k}_1,\omega_1), \\
&2 \equiv K_2 = (\mathbf{k}_2,\omega_2), \quad 3 \equiv K_3 = (\mathbf{k}_3,\omega_3), \qquad (1.67)
\end{aligned}$$

we may write

$$\begin{aligned}
&\langle\delta\phi_1\,\delta\phi_2\,\delta\phi_3\,\delta\phi_{K'}\rangle - \langle\delta\phi_1\,\delta\phi_{K'}\rangle\,\langle\delta\phi_2\,\delta\phi_3\rangle \\
&= \delta(1+2+3+K')\,\big[\delta(1+2)\,\langle\delta\phi^2\rangle_1\,\langle\delta\phi^2\rangle_3 \\
&\quad + \delta(1+3)\,\langle\delta\phi^2\rangle_1\,\langle\delta\phi^2\rangle_2\big], \qquad (1.68)
\end{aligned}$$

where we have ignored the irreducible four-body correlation. This leads to

$$\begin{aligned}
&\sum_{1+2+3=K} k_1\,k_2\,k_3\,\bar{\chi}^{(3)}(1|2|3)\,(\langle\delta\phi_1\,\delta\phi_2\,\delta\phi_3\,\delta\phi_{K'}\rangle - \langle\delta\phi_1\,\delta\phi_{K'}\rangle\,\langle\delta\phi_2\,\delta\phi_3\rangle) \\
&= \sum_{1+2+3=K} k_1\,k_2\,k_3\,\bar{\chi}^{(3)}(1|2|3)\,\delta(1+2+3+K') \\
&\quad \times \big(\delta(1+2)\,\langle\delta\phi^2\rangle_1\,\langle\delta\phi^2\rangle_3 + \delta(1+3)\,\langle\delta\phi^2\rangle_1\,\langle\delta\phi^2\rangle_2\big) \\
&= 2\delta(K+K')\sum_1 k\,k_1^2\,\bar{\chi}^{(3)}(1|-1|K)\,\langle\delta\phi^2\rangle_1\,\langle\delta\phi^2\rangle_K. \qquad (1.69)
\end{aligned}$$

Thus, (1.66) simplifies

$$
\begin{aligned}
0 &= k\,\epsilon(\mathbf{k},\omega)\,\langle\delta\phi^2\rangle_{\mathbf{k},\omega}\,\delta(\mathbf{k}+\mathbf{k}')\,\delta(\omega+\omega') \\
&\quad - i\sum_{\mathbf{k}_1+\mathbf{k}_2=\mathbf{k}}\;\sum_{\omega_1+\omega_2=\omega} k_1\,k_2\,\chi^{(2)}(\mathbf{k}_1,\omega_1|\mathbf{k}_2,\omega_2)\,\langle\delta\phi_{\mathbf{k}_1,\omega_1}\,\delta\phi_{\mathbf{k}_2,\omega_2}\,\delta\phi_{\mathbf{k}',\omega'}\rangle \\
&\quad -2\,\delta(\mathbf{k}+\mathbf{k}')\,\delta(\omega+\omega')\sum_{\mathbf{k}',\omega'} k\,k'^2 \\
&\quad \times\bar{\chi}^{(3)}(\mathbf{k}',\omega'|-\mathbf{k}',-\omega'|\mathbf{k},\omega)\,\langle\delta\phi^2\rangle_{\mathbf{k}',\omega'}\,\langle\delta\phi^2\rangle_{\mathbf{k},\omega}.
\end{aligned} \tag{1.70}
$$

In deriving this result, we have made use of the properties of homogeneous and stationary turbulence (1.25) to write

$$
\begin{aligned}
\langle\delta\phi_{\mathbf{k},\omega}\,\delta\phi_{\mathbf{k}',\omega'}\rangle &= \delta(\mathbf{k}+\mathbf{k}')\,\delta(\omega+\omega')\,\langle\delta\phi^2\rangle_{\mathbf{k},\omega},\;\langle\delta\phi_{\mathbf{k},\omega}\,\delta\phi_{\mathbf{k}',\omega'}\,\delta\phi_{\mathbf{k}'',\omega''}\,\delta\phi_{\mathbf{k}''',\omega'''}\rangle \\
&= \delta(\mathbf{k}+\mathbf{k}'+\mathbf{k}''+\mathbf{k}''')\,\delta(\omega+\omega'+\omega''+\omega''') \\
&\quad\times[\,\delta(\mathbf{k}+\mathbf{k}')\,\delta(\omega+\omega')\,\langle\delta\phi^2\rangle_{\mathbf{k},\omega}\,\langle\delta\phi^2\rangle_{\mathbf{k}'',\omega''} \\
&\quad+\delta(\mathbf{k}+\mathbf{k}'')\,\delta(\omega+\omega'')\,\langle\delta\phi^2\rangle_{\mathbf{k},\omega}\,\langle\delta\phi^2\rangle_{\mathbf{k}',\omega'} \\
&\quad+\delta(\mathbf{k}'+\mathbf{k}'')\,\delta(\omega'+\omega'')\,\langle\delta\phi^2\rangle_{\mathbf{k},\omega}\,\langle\delta\phi^2\rangle_{\mathbf{k}',\omega'} \\
&\quad+\langle\delta\phi^4\rangle_{\mathbf{k}\omega;\,\mathbf{k}+\mathbf{k}',\omega+\omega';\,\mathbf{k}+\mathbf{k}'+\mathbf{k}'',\omega+\omega'+\omega''}\,],
\end{aligned} \tag{1.71}
$$

and after having done so, we ignored the irreducible four-body correlation $\langle\delta\phi^4\rangle_{\mathbf{k}\omega;\,\mathbf{k}+\mathbf{k}',\omega+\omega';\,\mathbf{k}+\mathbf{k}'+\mathbf{k}'',\omega+\omega'+\omega''}$ in order to truncate the hierarchy of correlations.

The resultant wave equation (1.70) still contains the three-body correlation $\langle\delta\phi_{\mathbf{k}_1,\omega_1}\,\delta\phi_{\mathbf{k}_2,\omega_2}\,\delta\phi_{-\mathbf{k},-\omega}\rangle$, hence, not completely closed yet. To compute this three-body correlation, we return to (1.65) and consider up to second-order nonlinearity,

$$
\begin{aligned}
0 &= k\,\epsilon(\mathbf{k},\omega)\,\delta\phi_{\mathbf{k},\omega} - i\sum_{\mathbf{k}_1+\mathbf{k}_2=\mathbf{k}}\;\sum_{\omega_1+\omega_2=\omega} k_1\,k_2\,\chi^{(2)}(\mathbf{k}_1,\omega_1|\mathbf{k}_2,\omega_2) \\
&\quad\times\left(\delta\phi_{\mathbf{k}_1,\omega_1}\,\delta\phi_{\mathbf{k}_2,\omega_2} - \langle\delta\phi_{\mathbf{k}_1,\omega_1}\,\delta\phi_{\mathbf{k}_2,\omega_2}\rangle\right).
\end{aligned} \tag{1.72}
$$

The third-order nonlinear term makes contributions to higher-order corrections only, thus can be ignored at the outset. We impose the iterative solution of (1.72),

$$
\delta\phi_{\mathbf{k},\omega} = \delta\phi^{(0)}_{\mathbf{k},\omega} + \delta\phi^{(1)}_{\mathbf{k},\omega} + \cdots \tag{1.73}
$$

If we ignore the nonlinear correction in (1.72), and truncate the solution by $\delta\phi_{\mathbf{k},\omega} = \delta\phi^{(0)}_{\mathbf{k},\omega}$, then we have

$$
0 = \epsilon(\mathbf{k},\omega)\,\delta\phi^{(0)}_{\mathbf{k},\omega}. \tag{1.74}
$$

The solution $\delta\phi^{(0)}_{\mathbf{k},\omega}$ represents a sinusoidal (or plane wave) solution, which is uncorrelated with other spectral components. For such a plane-wave solution, ensemble averages of all odd moments of $\delta\phi^{(0)}_{\mathbf{k},\omega}$ are zero:

$$\langle \delta\phi^{(0)}_{\mathbf{k},\omega} \rangle = 0, \quad \langle \delta\phi^{(0)}_{\mathbf{k},\omega}\, \delta\phi^{(0)}_{\mathbf{k}',\omega'}\, \delta\phi^{(0)}_{\mathbf{k}'',\omega''} \rangle = 0, \ldots \tag{1.75}$$

This is because the plane-wave solution has a phase dependence of the form $\delta\phi^{(0)}_{\mathbf{k},\omega} \sim \hat{\phi}_{\mathbf{k},\omega} e^{i\varphi_{\mathbf{k},\omega}}$, so that odd moments are associated with odd products of sinusoidal functions each with phase $\varphi_{\mathbf{k},\omega}$. When integrated over $\varphi_{\mathbf{k},\omega}$, odd moments thus disappear. When the iterative solution (1.73) is inserted to (1.72), the next order solution emerges:

$$\delta\phi^{(1)}_{\mathbf{k},\omega} = \frac{i}{k\,\epsilon(\mathbf{k},\omega)} \sum_{\mathbf{k}_1+\mathbf{k}_2=\mathbf{k}} \sum_{\omega_1+\omega_2=\omega} k_1\, k_2\, \chi^{(2)}(\mathbf{k}_1,\omega_1|\mathbf{k}_2,\omega_2)$$
$$\times \left(\delta\phi^{(0)}_{\mathbf{k}_1,\omega_1}\, \delta\phi^{(0)}_{\mathbf{k}_2,\omega_2} - \langle \delta\phi^{(0)}_{\mathbf{k}_1,\omega_1}\, \delta\phi^{(0)}_{\mathbf{k}_2,\omega_2} \rangle \right). \tag{1.76}$$

The three-body correlation of interest can be approximately expressed as follows:

$$\begin{aligned} \langle \delta\phi_{\mathbf{k}_1,\omega_1}\, \delta\phi_{\mathbf{k}_2,\omega_2}\, \delta\phi_{\mathbf{k}',\omega'} \rangle &= \langle \delta\phi^{(1)}_{\mathbf{k}_1,\omega_1}\, \delta\phi^{(0)}_{\mathbf{k}_2,\omega_2}\, \delta\phi^{(0)}_{\mathbf{k}',\omega'} \rangle \\ &\quad + \langle \delta\phi^{(0)}_{\mathbf{k}_1,\omega_1}\, \delta\phi^{(1)}_{\mathbf{k}_2,\omega_2}\, \delta\phi^{(0)}_{\mathbf{k}',\omega'} \rangle + \langle \delta\phi^{(0)}_{\mathbf{k}_1,\omega_1}\, \delta\phi^{(0)}_{\mathbf{k}_2,\omega_2}\, \delta\phi^{(1)}_{\mathbf{k}',\omega'} \rangle \\ &\quad + \cdots. \end{aligned} \tag{1.77}$$

Upon making use of solution (1.76) and substituting for $\delta\phi^{(1)}_{\mathbf{k}_1,\omega_1}$, $\delta\phi^{(1)}_{\mathbf{k}_2,\omega_2}$, and $\delta\phi^{(1)}_{\mathbf{k}',\omega'}$, it can be shown that the result is (the intermediate steps are somewhat tedious but straightforward)

$$\begin{aligned} &\langle \delta\phi_{\mathbf{k}_1,\omega_1}\, \delta\phi_{\mathbf{k}_2,\omega_2}\, \delta\phi_{\mathbf{k}',\omega'} \rangle \\ &= \frac{2i\, k\, |\mathbf{k}-\mathbf{k}_1|\, \chi^{(2)}(-\mathbf{k}+\mathbf{k}_1,\, -\omega+\omega_1|\mathbf{k},\omega)}{k_1\, \epsilon(\mathbf{k}_1,\omega_1)} \langle \delta\phi^2 \rangle_{\mathbf{k}-\mathbf{k}_1,\omega-\omega_1} \langle \delta\phi^2 \rangle_{\mathbf{k},\omega} \\ &\quad + \frac{2i\, k\, k_1\, \chi^{(2)}(-\mathbf{k}_1,\, -\omega_1|\mathbf{k},\omega)}{|\mathbf{k}-\mathbf{k}_1|\, \epsilon(\mathbf{k}-\mathbf{k}_1,\omega-\omega_1)} \langle \delta\phi^2 \rangle_{\mathbf{k}_1,\omega_1} \langle \delta\phi^2 \rangle_{\mathbf{k},\omega} \\ &\quad - \frac{2i\, k_1\, |\mathbf{k}-\mathbf{k}_1|\, \chi^{(2)*}(\mathbf{k}_1,\omega_1|\mathbf{k}-\mathbf{k}_1,\omega-\omega_1)}{k\, \epsilon^*(\mathbf{k},\omega)} \langle \delta\phi^2 \rangle_{\mathbf{k}_1,\omega_1} \langle \delta\phi^2 \rangle_{\mathbf{k}-\mathbf{k}_1,\omega-\omega_1}, \end{aligned} \tag{1.78}$$

where we have deleted the superscripts "(0)" after everything is said and done, and have made use of the symmetry properties,

$$\begin{aligned} \epsilon(-\mathbf{k},\, -\omega) &= \epsilon^*(\mathbf{k},\omega), \\ \chi^{(2)}(-\mathbf{k}_1,\, -\omega_1|-\mathbf{k}_2,\, -\omega_2) &= -\chi^{(2)*}(\mathbf{k}_1,\omega_1|\mathbf{k}_2,\omega_2), \\ \chi^{(2)}(\mathbf{k}_1,\omega_1|\mathbf{k}_2,\omega_2) &= \chi^{(2)}(\mathbf{k}_2,\omega_2|\mathbf{k}_1,\omega_1). \end{aligned} \tag{1.79}$$

We have not yet discussed these symmetry properties associated with various susceptibilities, although the complex conjugate properties can be easily deduced from definitions (1.61)–(1.64). In the subsequent section we will discuss the symmetries as well as other properties associated with the susceptibilities in detail.

With (1.78) the wave equation (1.70) is now written as

$$\begin{aligned}
0 = \epsilon(\mathbf{k},\omega)\,\langle k^2\,\delta\phi^2\rangle_{\mathbf{k},\omega} &+ 2\sum_{\mathbf{k}',\omega'}\Bigg[\chi^{(2)}(\mathbf{k}',\omega'|\mathbf{k}-\mathbf{k}',\omega-\omega')\\
&\times\left(\frac{\chi^{(2)}(-\mathbf{k}+\mathbf{k}',\,-\omega+\omega'|\mathbf{k},\omega)}{\epsilon(\mathbf{k}',\omega')}\,\langle\,|\mathbf{k}-\mathbf{k}'|^2\,\delta\phi^2\rangle_{\mathbf{k}-\mathbf{k}',\omega-\omega'}\right.\\
&\left.+\frac{\chi^{(2)}(-\mathbf{k}',\,-\omega'|\mathbf{k},\omega)}{\epsilon(\mathbf{k}-\mathbf{k}',\omega-\omega')}\,\langle k'^2\,\delta\phi^2\rangle_{\mathbf{k}',\omega'}\right)\langle k^2\,\delta\phi^2\rangle_{\mathbf{k},\omega}\\
&-\frac{|\chi^{(2)}(\mathbf{k}',\omega'|\mathbf{k}-\mathbf{k}',\omega-\omega')|^2}{\epsilon^*(\mathbf{k},\omega)}\,\langle k'^2\,\delta\phi^2\rangle_{\mathbf{k}',\omega'}\,\langle\,|\mathbf{k}-\mathbf{k}'|^2\,\delta\phi^2\rangle_{\mathbf{k}-\mathbf{k}',\omega-\omega'}\Bigg]\\
&-2\sum_{\mathbf{k}',\omega'}\bar{\chi}^{(3)}(\mathbf{k}',\omega'|-\mathbf{k}',\,-\omega'|\mathbf{k},\omega)\,\langle k'^2\,\delta\phi^2\rangle_{\mathbf{k}',\omega'}\,\langle k^2\,\delta\phi^2\rangle_{\mathbf{k},\omega}.
\end{aligned} \tag{1.80}$$

This equation is known as the nonlinear spectral balance equation since the linear term, which appears as the first term on the right-hand side of the equality, is balanced by the rest that represents nonlinear response. We may rewrite the (1.80) by noting that the electrostatic potential correlation can be rewritten as the electric field correlation

$$\langle k^2\delta\phi^2\rangle_{\mathbf{k},\omega} = \langle\delta E^2\rangle_{\mathbf{k},\omega}, \tag{1.81}$$

where $\langle\delta E^2\rangle$ is related to the spectral electric field energy density, $\mathscr{E}_{\text{wave}} = \langle\delta E^2\rangle/(8\pi)$. As a consequence, (1.80) is equivalently written as

$$\begin{aligned}
0 = \epsilon(\mathbf{k},\omega)\,\langle\delta E^2\rangle_{\mathbf{k},\omega} &+ 2\sum_{\mathbf{k}',\omega'}\Bigg[\chi^{(2)}(\mathbf{k}',\omega'|\mathbf{k}-\mathbf{k}',\omega-\omega')\\
&\times\left(\frac{\chi^{(2)}(-\mathbf{k}+\mathbf{k}',\,-\omega+\omega'|\mathbf{k},\omega)}{\epsilon(\mathbf{k}',\omega')}\,\langle\delta E^2\rangle_{\mathbf{k}-\mathbf{k}',\omega-\omega'}\right.\\
&\left.+\frac{\chi^{(2)}(-\mathbf{k}',\,-\omega'|\mathbf{k},\omega)}{\epsilon(\mathbf{k}-\mathbf{k}',\omega-\omega')}\,\langle\delta E^2\rangle_{\mathbf{k}',\omega'}\right)\langle\delta E^2\rangle_{\mathbf{k},\omega}\\
&-\frac{|\chi^{(2)}(\mathbf{k}',\omega'|\mathbf{k}-\mathbf{k}',\omega-\omega')|^2}{\epsilon^*(\mathbf{k},\omega)}\,\langle\delta E^2\rangle_{\mathbf{k}',\omega'}\,\langle\delta E^2\rangle_{\mathbf{k}-\mathbf{k}',\omega-\omega'}\Bigg]\\
&-2\sum_{\mathbf{k}',\omega'}\bar{\chi}^{(3)}(\mathbf{k}',\omega'|-\mathbf{k}',\,-\omega'|\mathbf{k},\omega)\,\langle\delta E^2\rangle_{\mathbf{k}',\omega'}\,\langle\delta E^2\rangle_{\mathbf{k},\omega}.
\end{aligned} \tag{1.82}$$

To determine the adiabatic time evolution of the spectral wave energy density, we now reinstitute the slow time dependence implicit in the wave-particle resonance denominator:

$$(\omega - \mathbf{k}\cdot\mathbf{v} + i0 + i\partial/\partial t)^{-1}.$$

Recall that in deriving the final spectral balance equation (1.82), we had "absorbed" the derivative $i\partial/\partial t$ in the "new definition" of ω – see (1.41). This resulted in turning the differential equation (1.40) into an algebraic equation. After the desired equation (1.82) has now been obtained, we reintroduce the factor $i\partial/\partial t$ in the arguments of the response functions. As a consequence, various dielectric susceptibility response functions become operators in the slow time t of the amplitude evolution. However, $i\partial/\partial t$ was present in the original equation (1.40) only on the left-hand side, while the angular frequency ω appeared on both sides. Consequently, when we reintroduce the slow-time derivative $i\partial/\partial t$ to the angular frequency, we do so by treating this object as a small correction, and we introduce it only to the leading term, which is the linear response function. To reiterate, this whole procedure is heuristic, and as we noted already, the proper way to treat this type of problem is via multiple time scale analysis (Davidson, 1972). In this book, we take the present shortcut method nonetheless. When the slow-time derivative is thus reintroduced, the linear term is modified as

$$\begin{aligned}\epsilon(\mathbf{k},\omega)\,\langle\delta E^2\rangle_{\mathbf{k},\omega} &\to \epsilon\left(\mathbf{k},\omega+i\frac{\partial}{\partial t}\right)\,\langle\delta E^2\rangle_{\mathbf{k},\omega}\\ &\to \left(\epsilon(\mathbf{k},\omega)+\frac{i}{2}\frac{\partial\epsilon(\mathbf{k},\omega)}{\partial\omega}\frac{\partial}{\partial t}\right)\,\langle\delta E^2\rangle_{\mathbf{k},\omega},\end{aligned} \tag{1.83}$$

where 1/2 in front of the time derivative in the last expression stems from the fact that

$$\frac{\partial\,\delta E_{-\mathbf{k},-\omega}(t)}{\partial t}=\frac{\partial\,\delta E_{\mathbf{k},\omega}(t)}{\partial t}. \tag{1.84}$$

Recall that the slow-time derivative originally affects only $\delta f^a_{\mathbf{k},\omega}$, which automatically implies that $i\partial/\partial t$ is meant to operate on $\delta E_{\mathbf{k},\omega}$ or $\delta E_{-\mathbf{k},-\omega}$ separately, but not on their products. Consequently, when the slow-time derivative is reintroduced, the proper procedure is as follows:

$$\begin{aligned}\epsilon(\mathbf{k},\omega)\,\langle\delta E^2\rangle_{\mathbf{k},\omega} &= \left\langle\delta E_{-\mathbf{k},-\omega}\,\epsilon(\mathbf{k},\omega)\,\delta E_{\mathbf{k},\omega}\right\rangle\\ &\to \left\langle\delta E_{-\mathbf{k},-\omega}\,\epsilon\left(\mathbf{k},\omega+i\frac{\partial}{\partial t}\right)\delta E_{\mathbf{k},\omega}\right\rangle\\ &= \left\langle\delta E_{-\mathbf{k},-\omega}\left(\epsilon(\mathbf{k},\omega)+i\,\frac{\partial\epsilon(\mathbf{k},\omega)}{\partial\omega}\frac{\partial}{\partial t}\right)\delta E_{\mathbf{k},\omega}\right\rangle\\ &= \epsilon(\mathbf{k},\omega)\left\langle\delta E^2\right\rangle_{\mathbf{k},\omega}\\ &\quad+\frac{i}{2}\frac{\partial\epsilon(\mathbf{k},\omega)}{\partial\omega}\left\langle\delta E_{-\mathbf{k},-\omega}\frac{\partial\delta E_{\mathbf{k},\omega}}{\partial t}+\delta E_{\mathbf{k},\omega}\frac{\partial\delta E_{-\mathbf{k},-\omega}}{\partial t}\right\rangle.\end{aligned}$$

After making use of (1.84), then we readily arrive at (1.83).

This procedure results in the formal wave kinetic equation

$$\begin{aligned}
0 &= \epsilon(\mathbf{k},\omega)\,\langle\delta E^2\rangle_{\mathbf{k},\omega} + \frac{i}{2}\frac{\partial\epsilon(\mathbf{k},\omega)}{\partial\omega}\frac{\partial\,\langle\delta E^2\rangle_{\mathbf{k},\omega}}{\partial t} \\
&+2\sum_{\mathbf{k}',\omega'}\Bigg[\chi^{(2)}(\mathbf{k}',\omega'|\mathbf{k}-\mathbf{k}',\omega-\omega') \\
&\times\left(\frac{\chi^{(2)}(-\mathbf{k}+\mathbf{k}',\,-\omega+\omega'|\mathbf{k},\omega)}{\epsilon(\mathbf{k}',\omega')}\,\langle\delta E^2\rangle_{\mathbf{k}-\mathbf{k}',\omega-\omega'}\right. \\
&+\left.\frac{\chi^{(2)}(-\mathbf{k}',\,-\omega'|\mathbf{k},\omega)}{\epsilon(\mathbf{k}-\mathbf{k}',\omega-\omega')}\,\langle\delta E^2\rangle_{\mathbf{k}',\omega'}\right)\langle\delta E^2\rangle_{\mathbf{k},\omega} \\
&-\frac{|\chi^{(2)}(\mathbf{k}',\omega'|\mathbf{k}-\mathbf{k}',\omega-\omega')|^2}{\epsilon^*(\mathbf{k},\omega)}\,\langle\delta E^2\rangle_{\mathbf{k}',\omega'}\,\langle\delta E^2\rangle_{\mathbf{k}-\mathbf{k}',\omega-\omega'}\Bigg] \\
&-2\sum_{\mathbf{k}',\omega'}\bar{\chi}^{(3)}(\mathbf{k}',\omega'|-\mathbf{k}',\,-\omega'|\mathbf{k},\omega)\,\langle\delta E^2\rangle_{\mathbf{k}',\omega'}\,\langle\delta E^2\rangle_{\mathbf{k},\omega}.
\end{aligned} \tag{1.85}$$

This equation is further manipulated by separating the linear dielectric function $\epsilon(\mathbf{k},\omega)$ into real and imaginary parts,

$$\epsilon(\mathbf{k},\omega) = \mathrm{Re}\,\epsilon(\mathbf{k},\omega) + i\,\mathrm{Im}\,\epsilon(\mathbf{k},\omega), \tag{1.86}$$

where it is assumed that

$$|\mathrm{Im}\,\epsilon(\mathbf{k},\omega)| \ll |\mathrm{Re}\,\epsilon(\mathbf{k},\omega)|. \tag{1.87}$$

This assumption is equivalent to the weak growth/damping approximation. The imaginary part of the derivative, $\partial\,\mathrm{Im}\,\epsilon(\mathbf{k},\omega)/\partial\omega$, which couples with the slow-time derivative $\partial/\partial t$, is also ignored. Before we present the final result, let us invoke another useful symmetry property associated with the second-order nonlinear susceptibility,

$$\chi^{(2)}(\mathbf{k}_1,\omega_1|\mathbf{k}_2,\omega_2) = \chi^{(2)}(\mathbf{k}_1+\mathbf{k}_2,\omega_1+\omega_2|-\mathbf{k}_2,\,-\omega_2), \tag{1.88}$$

which will be derived in the next section. Then (1.85) can be expressed as

$$\begin{aligned}
0 &= \frac{i}{2}\frac{\partial\,\mathrm{Re}\,\epsilon(\mathbf{k},\omega)}{\partial\omega}\frac{\partial}{\partial t}\,\langle\delta E^2\rangle_{\mathbf{k},\omega} + \mathrm{Re}\,\epsilon(\mathbf{k},\omega)\,\langle\delta E^2\rangle_{\mathbf{k},\omega} + i\,\mathrm{Im}\,\epsilon(\mathbf{k},\omega)\,\langle\delta E^2\rangle_{\mathbf{k},\omega} \\
&+2\int d\mathbf{k}'\int d\omega'\,\Bigg[\{\chi^{(2)}(\mathbf{k}',\omega'|\mathbf{k}-\mathbf{k}',\omega-\omega')\}^2 \\
&\times\left(\frac{\langle\delta E^2\rangle_{\mathbf{k}-\mathbf{k}',\omega-\omega'}}{\epsilon(\mathbf{k}',\omega')} + \frac{\langle\delta E^2\rangle_{\mathbf{k}',\omega'}}{\epsilon(\mathbf{k}-\mathbf{k}',\omega-\omega')}\right)\langle\delta E^2\rangle_{\mathbf{k},\omega} \\
&-\frac{|\chi^{(2)}(\mathbf{k}',\omega'|\mathbf{k}-\mathbf{k}',\omega-\omega')|^2}{\epsilon^*(\mathbf{k},\omega)}\,\langle\delta E^2\rangle_{\mathbf{k}',\omega'}\,\langle\delta E^2\rangle_{\mathbf{k}-\mathbf{k}',\omega-\omega'}\Bigg] \\
&-2\int d\mathbf{k}'\int d\omega'\,\bar{\chi}^{(3)}(\mathbf{k}',\omega'|-\mathbf{k}',\,-\omega'|\mathbf{k},\omega)\,\langle\delta E^2\rangle_{\mathbf{k}',\omega'}\,\langle\delta E^2\rangle_{\mathbf{k},\omega}.
\end{aligned} \tag{1.89}$$

The real part of this equation,

$$0 = \Bigg\{ \mathrm{Re}\,\epsilon(\mathbf{k},\omega) + 2\,\mathrm{Re} \int d\mathbf{k}' \int d\omega' \Bigg[\{ \chi^{(2)}(\mathbf{k}',\omega'|\mathbf{k}-\mathbf{k}',\omega-\omega') \}^2$$

$$\times \left(\frac{\langle \delta E^2 \rangle_{\mathbf{k}-\mathbf{k}',\omega-\omega'}}{\epsilon(\mathbf{k}',\omega')} + \frac{\langle \delta E^2 \rangle_{\mathbf{k}',\omega'}}{\epsilon(\mathbf{k}-\mathbf{k}',\omega-\omega')} \right)$$

$$-\bar{\chi}^{(3)}(\mathbf{k}',\omega'|-\mathbf{k}',-\omega'|\mathbf{k},\omega)\,\langle \delta E^2 \rangle_{\mathbf{k}',\omega'} \Bigg] \Bigg\} \langle \delta E^2 \rangle_{\mathbf{k},\omega} \tag{1.90}$$

$$-2\,\mathrm{Re} \int d\mathbf{k}' \int d\omega' \, \frac{|\chi^{(2)}(\mathbf{k}',\omega'|\mathbf{k}-\mathbf{k}',\omega-\omega')|^2}{\epsilon^*(\mathbf{k},\omega)} \langle \delta E^2 \rangle_{\mathbf{k}',\omega'} \langle \delta E^2 \rangle_{\mathbf{k}-\mathbf{k}',\omega-\omega'},$$

determines the wave dispersion relation, $\omega = \omega_{\mathbf{k}}$, while the imaginary part,

$$0 = \frac{1}{2}\frac{\partial\,\mathrm{Re}\,\epsilon(\mathbf{k},\omega)}{\partial\omega}\frac{\partial}{\partial t}\langle \delta E^2 \rangle_{\mathbf{k},\omega} + \mathrm{Im}\,\epsilon(\mathbf{k},\omega)\,\langle \delta E^2 \rangle_{\mathbf{k},\omega}$$

$$+2\,\mathrm{Im} \int d\mathbf{k}' \int d\omega' \Bigg[\{ \chi^{(2)}(\mathbf{k}',\omega'|\mathbf{k}-\mathbf{k}',\omega-\omega') \}^2$$

$$\times \left(\frac{\langle \delta E^2 \rangle_{\mathbf{k}-\mathbf{k}',\omega-\omega'}}{\epsilon(\mathbf{k}',\omega')} + \frac{\langle \delta E^2 \rangle_{\mathbf{k}',\omega'}}{\epsilon(\mathbf{k}-\mathbf{k}',\omega-\omega')} \right) \langle \delta E^2 \rangle_{\mathbf{k},\omega} \tag{1.91}$$

$$-\frac{|\chi^{(2)}(\mathbf{k}',\omega'|\mathbf{k}-\mathbf{k}',\omega-\omega')|^2}{\epsilon^*(\mathbf{k},\omega)} \langle \delta E^2 \rangle_{\mathbf{k}',\omega'} \langle \delta E^2 \rangle_{\mathbf{k}-\mathbf{k}',\omega-\omega'} \Bigg]$$

$$-2\,\mathrm{Im} \int d\mathbf{k}' \int d\omega' \, \bar{\chi}^{(3)}(\mathbf{k}',\omega'|-\mathbf{k}',-\omega'|\mathbf{k},\omega)\,\langle \delta E^2 \rangle_{\mathbf{k}',\omega'} \langle \delta E^2 \rangle_{\mathbf{k},\omega},$$

corresponds to the wave kinetic equation. In general, (1.90) is not quite a "dispersion equation" yet, since in the last term different spectral components are inexorably coupled. However, if we ignore the non-diagonal term, $\langle \delta E^2 \rangle_{\mathbf{k}',\omega'}$ $\langle \delta E^2 \rangle_{\mathbf{k}-\mathbf{k}',\omega-\omega'}$, that is, if we approximate (1.90) by

$$0 = \Bigg\{ \mathrm{Re}\,\epsilon(\mathbf{k},\omega) + 2\,\mathrm{Re} \int d\mathbf{k}' \int d\omega' \Bigg[\{ \chi^{(2)}(\mathbf{k}',\omega'|\mathbf{k}-\mathbf{k}',\omega-\omega') \}^2$$

$$\times \left(\frac{\langle \delta E^2 \rangle_{\mathbf{k}-\mathbf{k}',\omega-\omega'}}{\epsilon(\mathbf{k}',\omega')} + \frac{\langle \delta E^2 \rangle_{\mathbf{k}',\omega'}}{\epsilon(\mathbf{k}-\mathbf{k}',\omega-\omega')} \right)$$

$$-\bar{\chi}^{(3)}(\mathbf{k}',\omega'|-\mathbf{k}',-\omega'|\mathbf{k},\omega)\,\langle \delta E^2 \rangle_{\mathbf{k}',\omega'} \Bigg] \Bigg\} \langle \delta E^2 \rangle_{\mathbf{k},\omega}$$

$$\equiv \mathrm{Re}\,\tilde{\varepsilon}(\mathbf{k},\omega)\,\langle \delta E^2 \rangle_{\mathbf{k},\omega}, \tag{1.92}$$

then (1.92) represents nonlinear dispersion equation in the proper sense. If we are concerned with a situation where the waves excited in the plasma can be characterized by linear dispersion relation, but where we are interested in waves interacting with the particles and among themselves via linear and nonlinear wave-particle and wave-wave interactions, then one may simplify the real part of the spectral balance equation, namely, (1.92), by simply ignoring nonlinear mode coupling terms altogether,

$$\mathrm{Re}\,\epsilon(\mathbf{k},\omega)\,\langle\delta E^2\rangle_{\mathbf{k},\omega}=0. \tag{1.93}$$

In Appendix C, we discuss the result of retaining the nonlinear correction terms in (1.92), but in the main body of this book we focus on linear eigenmodes and nonlinear interactions among the linear modes and the particles.

1.5 Formal Wave Kinetic Equation for Eigenmodes

Linear wave equation (1.93) implies that the angular frequency is a function of wave vector, that is, ω and $\mathbf{k}$ satisfy a "dispersion relation."

$$\omega=\omega_{\mathbf{k}}^{\alpha}, \tag{1.94}$$

where α designates the eigenmode. There could be more than a single solution, hence the superscript α. Equation (1.93) implies that the spectral electric field wave energy density can be represented in terms of the wave intensity

$$\langle\delta E^2\rangle_{\mathbf{k},\omega}=\sum_{\alpha=L,S}[\,I_{\mathbf{k}}^{+\alpha}\,\delta(\omega-\omega_{\mathbf{k}}^{\alpha})+I_{\mathbf{k}}^{-\alpha}\,\delta(\omega+\omega_{\mathbf{k}}^{\alpha})\,]. \tag{1.95}$$

In this equation, $I_{\mathbf{k}}^{\pm\alpha}$ represents the intensity of electrostatic waves associated with eigenmode α, propagating in forward/backward ($\pm$) direction. We will discuss the linear wave properties later, but it is well known that linear electrostatic eigenmodes of a uniform, unmagnetized plasma are high-frequency Langmuir wave ($\alpha=L$) and low-frequency ion-sound (or ion acoustic) wave ($\alpha=S$). It is important to distinguish the forward versus backward propagation, as nonlinear interactions of these modes depend on the wave propagation direction.

Inserting (1.95) to the wave kinetic equation (1.91) we have

$$\begin{aligned}0=&\sum_{\alpha}\sum_{\sigma=\pm1}\left(\frac{1}{2}\frac{\partial\,\mathrm{Re}\,\epsilon(\mathbf{k},\sigma\omega_{\mathbf{k}}^{\alpha})}{\partial\sigma\omega_{\mathbf{k}}^{\alpha}}\frac{\partial I_{\mathbf{k}}^{\sigma\alpha}}{\partial t}+\mathrm{Im}\,\epsilon(\mathbf{k},\sigma\omega_{\mathbf{k}}^{\alpha})\,I_{\mathbf{k}}^{\sigma\alpha}\right)\delta(\omega-\sigma\omega_{\mathbf{k}}^{\alpha})\\&+2\,\mathrm{Im}\sum_{\alpha}\sum_{\sigma=\pm1}\int d\mathbf{k}'\left(\sum_{\gamma}\sum_{\sigma''=\pm1}\right.\\&\times\frac{\{\chi^{(2)}(\mathbf{k}',\sigma\omega_{\mathbf{k}}^{\alpha}-\sigma''\omega_{\mathbf{k}-\mathbf{k}'}^{\gamma}|\mathbf{k}-\mathbf{k}',\sigma''\omega_{\mathbf{k}-\mathbf{k}'}^{\gamma})\}^2}{\epsilon(\mathbf{k}',\sigma\omega_{\mathbf{k}}^{\alpha}-\sigma''\omega_{\mathbf{k}-\mathbf{k}'}^{\gamma})}\,I_{\mathbf{k}-\mathbf{k}'}^{\sigma''\gamma}\end{aligned}$$

$$+\sum_{\beta}\sum_{\sigma'=\pm 1} \frac{\{\chi^{(2)}(\mathbf{k}',\sigma'\omega_{\mathbf{k}'}^{\beta}|\mathbf{k}-\mathbf{k}',\sigma\omega_{\mathbf{k}}^{\alpha}-\sigma'\omega_{\mathbf{k}'}^{\beta})\}^2}{\epsilon(\mathbf{k}-\mathbf{k}',\sigma\omega_{\mathbf{k}}^{\alpha}-\sigma'\omega_{\mathbf{k}'}^{\beta})} I_{\mathbf{k}'}^{\sigma'\beta}\Bigg) I_{\mathbf{k}}^{\sigma\alpha}\,\delta(\omega-\sigma\omega_{\mathbf{k}}^{\alpha})$$

$$-\,2\,\mathrm{Im}\sum_{\beta,\gamma}\sum_{\sigma',\sigma''=\pm 1}\int d\mathbf{k}'\,\frac{|\chi^{(2)}(\mathbf{k}',\sigma'\omega_{\mathbf{k}'}^{\beta}|\mathbf{k}-\mathbf{k}',\sigma''\omega_{\mathbf{k}-\mathbf{k}'}^{\gamma})|^2}{\epsilon^*(\mathbf{k},\sigma'\omega_{\mathbf{k}'}^{\beta}+\sigma''\omega_{\mathbf{k}-\mathbf{k}'}^{\gamma})}$$

$$\times\, I_{\mathbf{k}'}^{\sigma'\beta}\, I_{\mathbf{k}-\mathbf{k}'}^{\sigma''\gamma}\,\delta(\omega-\sigma'\omega_{\mathbf{k}'}^{\beta}-\sigma''\omega_{\mathbf{k}-\mathbf{k}'}^{\gamma})$$

$$-\,2\,\mathrm{Im}\sum_{\alpha,\beta}\sum_{\sigma,\sigma'=\pm 1}\int d\mathbf{k}'\,\bar{\chi}^{(3)}(\mathbf{k}',\sigma'\omega_{\mathbf{k}'}^{\beta}|-\mathbf{k}',\,-\sigma'\omega_{\mathbf{k}'}^{\beta}|\mathbf{k},\sigma\omega_{\mathbf{k}}^{\alpha})$$

$$\times\, I_{\mathbf{k}'}^{\sigma'\beta}\, I_{\mathbf{k}}^{\sigma\alpha}\,\delta(\omega-\sigma\omega_{\mathbf{k}}^{\alpha}),\quad (\sigma=\pm 1). \tag{1.96}$$

Let us focus on the quantity $[\epsilon(\mathbf{k},\omega)]^{-1}$. If ω lies in the vicinity of linear eigenmode, $\omega \sim \sigma\omega_{\mathbf{k}}^{\alpha}$, where $\mathrm{Re}\,\epsilon(\mathbf{k},\sigma\omega_{\mathbf{k}}^{\alpha}) = 0$, since we had assumed that $|\mathrm{Im}\,\epsilon(\mathbf{k},\omega)| \ll |\mathrm{Re}\,\epsilon(\mathbf{k},\omega)|$, we may approximately express

$$\begin{aligned}\frac{1}{\epsilon(\mathbf{k},\omega)} &\approx \frac{1}{\mathrm{Re}\,\epsilon(\mathbf{k},\omega)}\\ &\approx \sum_{\alpha}\sum_{\sigma=\pm 1}\frac{1}{(\omega-\sigma\omega_{\mathbf{k}}^{\alpha}+i0)\,[\partial\mathrm{Re}\,\epsilon(\mathbf{k},\omega)/\partial\omega]_{\omega=\sigma\omega_{\mathbf{k}}^{\alpha}}}\\ &= -\sum_{\alpha}\sum_{\sigma=\pm 1}\frac{i\pi\,\delta(\omega-\sigma\omega_{\mathbf{k}}^{\alpha})}{\partial\mathrm{Re}\,\epsilon(\mathbf{k},\sigma\omega_{\mathbf{k}}^{\alpha})/\partial\sigma\omega_{\mathbf{k}}^{\alpha}},\end{aligned} \tag{1.97}$$

where we have taken the series expansion,

$$\mathrm{Re}\,\epsilon(\mathbf{k},\omega) \approx \sum_{\sigma}\sum_{\alpha}(\omega-\sigma\omega_{\mathbf{k}}^{\alpha}+i0)\,\frac{\partial\mathrm{Re}\,\epsilon(\mathbf{k},\omega)}{\partial\omega}\bigg|_{\omega=\sigma\omega_{\mathbf{k}}^{\alpha}}. \tag{1.98}$$

In the second line of (1.97) we have summed over all possible poles. If we include contributions from those ω's that are sufficiently far away from linear eigenmodes in the complex frequency space, then we must add the principal part contribution to the right-hand side of (1.97) as well:

$$\begin{aligned}\frac{1}{\epsilon(\mathbf{k},\omega)} &= \mathscr{P}\,\frac{1}{\epsilon(\mathbf{k},\omega)} - \sum_{\alpha}\sum_{\sigma=\pm 1}\frac{i\pi\,\delta(\omega-\sigma\omega_{\mathbf{k}}^{\alpha})}{\partial\mathrm{Re}\,\epsilon(\mathbf{k},\sigma\omega_{\mathbf{k}}^{\alpha})/\partial\sigma\omega_{\mathbf{k}}^{\alpha}}.\\ \frac{1}{\epsilon^*(\mathbf{k},\omega)} &= \mathscr{P}\,\frac{1}{\epsilon^*(\mathbf{k},\omega)} + \sum_{\alpha}\sum_{\sigma=\pm 1}\frac{i\pi\,\delta(\omega-\sigma\omega_{\mathbf{k}}^{\alpha})}{\partial\mathrm{Re}\,\epsilon(\mathbf{k},\sigma\omega_{\mathbf{k}}^{\alpha})/\partial\sigma\omega_{\mathbf{k}}^{\alpha}}.\end{aligned} \tag{1.99}$$

Principal value $\mathscr{P}$ is meant to exclude those ω's in the vicinity of linear eigenmodes, $\omega = \pm\omega_{\mathbf{k}}^{\alpha}$.

Making use of (1.99), the wave kinetic equation (1.96) is now expressed as follows:

$$\begin{aligned}
0 &= \sum_{\alpha}\sum_{\sigma=\pm 1}\left(\frac{1}{2}\frac{\partial\,\mathrm{Re}\,\epsilon(\mathbf{k},\sigma\omega_{\mathbf{k}}^{\alpha})}{\partial\sigma\omega_{\mathbf{k}}^{\alpha}}\frac{\partial I_{\mathbf{k}}^{\sigma\alpha}}{\partial t}+\mathrm{Im}\,\epsilon(\mathbf{k},\sigma\omega_{\mathbf{k}}^{\alpha})\,I_{\mathbf{k}}^{\sigma\alpha}\right)\\
&+2\,\mathrm{Im}\sum_{\alpha,\beta}\sum_{\sigma,\sigma'=\pm 1}\int d\mathbf{k}'\left(2\{\chi^{(2)}(\mathbf{k}',\sigma'\omega_{\mathbf{k}'}^{\beta}|\mathbf{k}-\mathbf{k}',\sigma\omega_{\mathbf{k}}^{\alpha}-\sigma'\omega_{\mathbf{k}'}^{\beta})\}^2\right.\\
&\times\mathscr{P}\frac{1}{\epsilon(\mathbf{k}-\mathbf{k}',\sigma\omega_{\mathbf{k}}^{\alpha}-\sigma'\omega_{\mathbf{k}'}^{\beta})}\\
&\left.-\bar{\chi}^{(3)}(\mathbf{k}',\sigma'\omega_{\mathbf{k}'}^{\beta}|-\mathbf{k}',-\sigma'\omega_{\mathbf{k}'}^{\beta}|\mathbf{k},\sigma\omega_{\mathbf{k}}^{\alpha})\right)I_{\mathbf{k}'}^{\sigma'\beta}\,I_{\mathbf{k}}^{\sigma\alpha}\\
&-2\,\mathrm{Im}\sum_{\beta,\gamma}\sum_{\sigma',\sigma''=\pm 1}\int d\mathbf{k}'\,|\chi^{(2)}(\mathbf{k}',\sigma'\omega_{\mathbf{k}'}^{\beta}|\mathbf{k}-\mathbf{k}',\sigma''\omega_{\mathbf{k}-\mathbf{k}'}^{\gamma})|^2\\
&\times\mathscr{P}\frac{1}{\epsilon^*(\mathbf{k},\sigma'\omega_{\mathbf{k}'}^{\beta}+\sigma''\omega_{\mathbf{k}-\mathbf{k}'}^{\gamma})}\,I_{\mathbf{k}'}^{\sigma'\beta}\,I_{\mathbf{k}-\mathbf{k}'}^{\sigma''\gamma}\,\delta(\sigma\omega_{\mathbf{k}}^{\alpha}-\sigma'\omega_{\mathbf{k}'}^{\beta}-\sigma''\omega_{\mathbf{k}-\mathbf{k}'}^{\gamma})\\
&+2\pi\sum_{\alpha,\beta,\gamma}\sum_{\sigma,\sigma',\sigma''=\pm 1}\int d\mathbf{k}'\,|\chi^{(2)}(\mathbf{k}',\sigma'\omega_{\mathbf{k}'}^{\beta}|\mathbf{k}-\mathbf{k}',\sigma\omega_{\mathbf{k}}^{\alpha}-\sigma'\omega_{\mathbf{k}'}^{\beta})|^2\\
&\times\left(\frac{I_{\mathbf{k}-\mathbf{k}'}^{\sigma''\gamma}\,I_{\mathbf{k}}^{\sigma\alpha}}{\partial\mathrm{Re}\,\epsilon(\mathbf{k}',\sigma'\omega_{\mathbf{k}'}^{\beta})/\partial\sigma'\omega_{\mathbf{k}'}^{\beta}}+\frac{I_{\mathbf{k}'}^{\sigma'\beta}\,I_{\mathbf{k}}^{\sigma\alpha}}{\partial\mathrm{Re}\,\epsilon(\mathbf{k}-\mathbf{k}',\sigma''\omega_{\mathbf{k}-\mathbf{k}'}^{\gamma})/\partial\sigma''\omega_{\mathbf{k}-\mathbf{k}'}^{\gamma}}\right.\\
&\left.-\frac{I_{\mathbf{k}'}^{\sigma'\beta}\,I_{\mathbf{k}-\mathbf{k}'}^{\sigma''\gamma}}{\partial\mathrm{Re}\,\epsilon(\mathbf{k},\sigma\omega_{\mathbf{k}}^{\alpha})/\partial\sigma\omega_{\mathbf{k}}^{\alpha}}\right)\delta(\sigma\omega_{\mathbf{k}}^{\alpha}-\sigma'\omega_{\mathbf{k}'}^{\beta}-\sigma''\omega_{\mathbf{k}-\mathbf{k}'}^{\gamma}).
\end{aligned}\tag{1.100}$$

In deriving this result, we have invoked the fact that, to the leading order, $\chi^{(2)}$ is purely imaginary so that

$$\{\chi^{(2)}(\mathbf{k}_1,\omega_1|\mathbf{k}_2,\omega_2)\}^2\approx-|\chi^{(2)}(\mathbf{k}_1,\omega_1|\mathbf{k}_2,\omega_2)|^2,\tag{1.101}$$

which we imposed for the last term that contains the three-wave resonance delta function condition $\delta(\sigma\omega_{\mathbf{k}}^{\alpha}-\sigma'\omega_{\mathbf{k}'}^{\beta}-\sigma''\omega_{\mathbf{k}-\mathbf{k}'}^{\gamma})$. This property will be discussed in the subsequent section. In (1.100), the principal part, $\mathscr{P}[1/\epsilon^*(\mathbf{k},\sigma'\omega_{\mathbf{k}'}^{\beta}+\sigma''\omega_{\mathbf{k}-\mathbf{k}'}^{\gamma})]$, and the condition, $\sigma\omega_{\mathbf{k}}^{\alpha}=\sigma'\omega_{\mathbf{k}'}^{\beta}+\sigma''\omega_{\mathbf{k}-\mathbf{k}'}^{\gamma}$, where $\sigma\omega_{\mathbf{k}}^{\alpha}$ satisfies $\epsilon^*(\mathbf{k},\sigma\omega_{\mathbf{k}}^{\alpha})=0$, are mutually exclusive. Thus, by definition, this term is zero. This leaves us with

$$\begin{aligned}
&\sum_{\alpha}\sum_{\sigma=\pm 1} \frac{\partial\,\mathrm{Re}\,\epsilon(\mathbf{k},\sigma\omega_{\mathbf{k}}^{\alpha})}{\partial\sigma\omega_{\mathbf{k}}^{\alpha}} \frac{\partial I_{\mathbf{k}}^{\sigma\alpha}}{\partial t} \\
&\quad = -\sum_{\alpha}\sum_{\sigma=\pm 1} 2\,\mathrm{Im}\,\epsilon(\mathbf{k},\sigma\omega_{\mathbf{k}}^{\alpha})\, I_{\mathbf{k}}^{\sigma\alpha} \\
&\qquad - 4\sum_{\mathbf{k}'}\sum_{\alpha,\beta}\sum_{\sigma,\sigma'=\pm 1} \mathrm{Im}\Bigg(2\,\{\chi^{(2)}(\mathbf{k}',\sigma'\omega_{\mathbf{k}'}^{\beta}|\mathbf{k}-\mathbf{k}',\sigma\omega_{\mathbf{k}}^{\alpha}-\sigma'\omega_{\mathbf{k}'}^{\beta})\}^2 \\
&\qquad \times \mathscr{P}\,\frac{1}{\epsilon(\mathbf{k}-\mathbf{k}',\sigma\omega_{\mathbf{k}}^{\alpha}-\sigma'\omega_{\mathbf{k}'}^{\beta})} \qquad (1.102)\\
&\qquad - \bar{\chi}^{(3)}(\mathbf{k}',\sigma'\omega_{\mathbf{k}'}^{\beta}|-\mathbf{k}',-\sigma'\omega_{\mathbf{k}'}^{\beta}|\mathbf{k},\sigma\omega_{\mathbf{k}}^{\alpha})\Bigg) I_{\mathbf{k}'}^{\sigma'\beta}\, I_{\mathbf{k}}^{\sigma\alpha} \\
&\qquad - 4\pi\sum_{\mathbf{k}'}\sum_{\alpha,\beta,\gamma}\sum_{\sigma,\sigma',\sigma''=\pm 1} |\chi^{(2)}(\mathbf{k}',\sigma'\omega_{\mathbf{k}'}^{\beta}|\mathbf{k}-\mathbf{k}',\sigma\omega_{\mathbf{k}}^{\alpha}-\sigma'\omega_{\mathbf{k}'}^{\beta})|^2 \\
&\qquad \times \Bigg(\frac{I_{\mathbf{k}-\mathbf{k}'}^{\sigma''\gamma}\, I_{\mathbf{k}}^{\sigma\alpha}}{\partial\mathrm{Re}\,\epsilon(\mathbf{k}',\sigma'\omega_{\mathbf{k}'}^{\beta})/\partial\sigma'\omega_{\mathbf{k}'}^{\beta}} + \frac{I_{\mathbf{k}'}^{\sigma'\beta}\, I_{\mathbf{k}}^{\sigma\alpha}}{\partial\mathrm{Re}\,\epsilon(\mathbf{k}-\mathbf{k}',\sigma''\omega_{\mathbf{k}-\mathbf{k}'}^{\gamma})/\partial\sigma''\omega_{\mathbf{k}-\mathbf{k}'}^{\gamma}} \\
&\qquad - \frac{I_{\mathbf{k}'}^{\sigma'\beta}\, I_{\mathbf{k}-\mathbf{k}'}^{\sigma''\gamma}}{\partial\mathrm{Re}\,\epsilon(\mathbf{k},\sigma\omega_{\mathbf{k}}^{\alpha})/\partial\sigma\omega_{\mathbf{k}}^{\alpha}} \Bigg)\, \delta(\sigma\omega_{\mathbf{k}}^{\alpha}-\sigma'\omega_{\mathbf{k}'}^{\beta}-\sigma''\omega_{\mathbf{k}-\mathbf{k}'}^{\gamma}).
\end{aligned}$$

This equation is the formal wave kinetic equation governing the dynamics of linear eigenmodes as they undergo nonlinear interactions among themselves as well as with the plasma particles. At this stage in the development of formalism, however, the result is not practically useful since the various susceptibilities are expressed at a formal level. These quantities are yet to be explicitly calculated in forms that readily lend themselves to further analysis.

1.6 Formal Particle Kinetic Equation

Formal particle kinetic equation (1.40) can be further manipulated by considering the quantity given in (1.103), which follows from (1.52),

$$\begin{aligned}
\langle\delta f_{\mathbf{k},\omega}^{a}\,\delta\phi_{\mathbf{k}',\omega'}\rangle &= \delta(\mathbf{k}+\mathbf{k}')\,\delta(\omega+\omega')\,\alpha(\mathbf{k},\omega)\,\langle\delta\phi^2\rangle_{\mathbf{k},\omega}\, n_a F_a \\
&\quad + \sum_{\mathbf{k}_1+\mathbf{k}_2=\mathbf{k}}\ \sum_{\omega_1+\omega_2=\omega} \alpha^{(2)}(\mathbf{k}_1,\omega_1|\mathbf{k}_2,\omega_2) \\
&\quad \times \langle\delta\phi_{\mathbf{k}_1,\omega_1}\,\delta\phi_{\mathbf{k}_2,\omega_2}\,\delta\phi_{\mathbf{k}',\omega'}\rangle\, n_a F_a \\
&\quad + 2\,\delta(\mathbf{k}+\mathbf{k}')\,\delta(\omega+\omega')\sum_{\mathbf{k}_1}\sum_{\omega_1} \alpha^{(3)}(\mathbf{k}_1,\omega_1|-\mathbf{k}_1,-\omega_1|\mathbf{k},\omega) \\
&\quad \times \langle\delta\phi^2\rangle_{\mathbf{k}_1,\omega_1}\,\langle\delta\phi^2\rangle_{\mathbf{k},\omega}\, n_a F_a. \qquad (1.103)
\end{aligned}$$

Upon making use of (1.78), we have

$$
\begin{aligned}
\langle \delta f^a_{\mathbf{k},\omega}\, \delta\phi_{\mathbf{k}',\omega'} \rangle = \delta(\mathbf{k}+\mathbf{k}')\,\delta(\omega+\omega') \Bigg\{ & \alpha(\mathbf{k},\omega)\,\langle \delta\phi^2 \rangle_{\mathbf{k},\omega} \\
& + \sum_{\mathbf{k}_1\omega_1} \frac{2i}{k\,k_1\,|\mathbf{k}-\mathbf{k}_1|}\,\alpha^{(2)}(\mathbf{k}_1,\omega_1|\mathbf{k}-\mathbf{k}_1,\omega-\omega_1) \\
& \times \Bigg(\frac{\chi^{(2)}(\mathbf{k}_1,\omega_1|\mathbf{k}-\mathbf{k}_1,\omega-\omega_1)}{\epsilon(\mathbf{k}_1,\omega_1)}\,|\mathbf{k}-\mathbf{k}_1|^2\,k^2 \\
& \times \langle \delta\phi^2 \rangle_{\mathbf{k}-\mathbf{k}_1,\omega-\omega_1}\,\langle \delta\phi^2 \rangle_{\mathbf{k},\omega} \\
& + \frac{\chi^{(2)}(\mathbf{k}_1,\omega_1|\mathbf{k}-\mathbf{k}_1,\omega-\omega_1)}{\epsilon(\mathbf{k}-\mathbf{k}_1,\omega-\omega_1)}\,k_1^2\,k^2\,\langle \delta\phi^2 \rangle_{\mathbf{k}_1,\omega_1}\,\langle \delta\phi^2 \rangle_{\mathbf{k},\omega} \\
& - \frac{\chi^{(2)*}(\mathbf{k}_1,\omega_1|\mathbf{k}-\mathbf{k}_1,\omega-\omega_1)}{\epsilon^*(\mathbf{k},\omega)}\,k_1^2\,|\mathbf{k}-\mathbf{k}_1|^2 \\
& \times \langle \delta\phi^2 \rangle_{\mathbf{k}_1,\omega_1}\,\langle \delta\phi^2 \rangle_{\mathbf{k}-\mathbf{k}_1,\omega-\omega_1} \Bigg) \\
& + 2 \sum_{\mathbf{k}_1,\omega_1} \alpha^{(3)}(\mathbf{k}_1,\omega_1|-\mathbf{k}_1,-\omega_1|\mathbf{k},\omega)\,\langle \delta\phi^2 \rangle_{\mathbf{k}_1,\omega_1}\,\langle \delta\phi^2 \rangle_{\mathbf{k},\omega}.
\end{aligned}
\tag{1.104}
$$

Substituting (1.104) to the right-hand side of formal particle kinetic equation in (1.40), we obtain

$$
\begin{aligned}
\frac{\partial F_a}{\partial t} = -\frac{i e_a}{m_a} \int d\mathbf{k} \int d\omega\, \mathbf{k}\cdot\frac{\partial}{\partial \mathbf{v}} \Bigg\{ & \alpha(\mathbf{k},\omega)\,\frac{\langle \delta E^2 \rangle_{\mathbf{k},\omega}}{k^2} \\
& + 2i \int d\mathbf{k}' \int d\omega'\, \alpha^{(2)}(\mathbf{k}',\omega'|\mathbf{k}-\mathbf{k}',\omega-\omega')\,\frac{M(\mathbf{k}',\omega'|\mathbf{k}-\mathbf{k}',\omega-\omega')}{k\,k'\,|\mathbf{k}-\mathbf{k}'|} \\
& + 2 \int d\mathbf{k}' \int d\omega'\, \alpha^{(3)}(\mathbf{k}',\omega'|-\mathbf{k}',-\omega'|\mathbf{k},\omega)\,\frac{\langle \delta E^2 \rangle_{\mathbf{k}',\omega'}\,\langle \delta E^2 \rangle_{\mathbf{k},\omega}}{k^2\,k'^2} \Bigg\}\, F_a,
\end{aligned}
\tag{1.105}
$$

where

$$
\begin{aligned}
M(\mathbf{k}_1,\omega_1|\mathbf{k}_2,\omega_2) = {} & \chi^{(2)}(\mathbf{k}_1,\omega_1|\mathbf{k}_2,\omega_2) \left(\frac{\langle \delta E^2 \rangle_{\mathbf{k}_2,\omega_2}}{\epsilon(\mathbf{k}_1,\omega_1)} + \frac{\langle \delta E^2 \rangle_{\mathbf{k}_1,\omega_1}}{\epsilon(\mathbf{k}_2,\omega_2)} \right) \langle \delta E^2 \rangle_{\mathbf{k}_1+\mathbf{k}_2,\omega_1+\omega_2} \\
& - \frac{\chi^{(2)*}(\mathbf{k}_1,\omega_1|\mathbf{k}_2,\omega_2)}{\epsilon^*(\mathbf{k}_1+\mathbf{k}_2,\omega_1+\omega_2)}\,\langle \delta E^2 \rangle_{\mathbf{k}_1,\omega_1}\,\langle \delta E^2 \rangle_{\mathbf{k}_2,\omega_2}.
\end{aligned}
\tag{1.106}
$$

It is instructive to rewrite this equation explicitly by making use of (1.53). Taking the real part only we obtain

$$
\begin{aligned}
\frac{\partial F_a}{\partial t} &= -\mathrm{Im}\,\frac{e_a^2}{m_a^2}\int d\mathbf{k}\int d\omega\,\frac{\mathbf{k}}{k}\cdot\frac{\partial}{\partial\mathbf{v}}\frac{\langle\delta E^2\rangle_{\mathbf{k},\omega}}{\omega-\mathbf{k}\cdot\mathbf{v}}\frac{\mathbf{k}}{k}\cdot\frac{\partial F_a}{\partial\mathbf{v}}\\
&\quad+\mathrm{Re}\,\frac{e_a^3}{m_a^3}\int d\mathbf{k}\int d\omega\int d\mathbf{k}'\int d\omega'\,\frac{\mathbf{k}}{k}\cdot\frac{\partial}{\partial\mathbf{v}}\frac{M(\mathbf{k}',\omega'|\mathbf{k}-\mathbf{k}',\omega-\omega')}{\omega-\mathbf{k}\cdot\mathbf{v}}\\
&\quad\times\left(\frac{\mathbf{k}'}{k'}\cdot\frac{\partial}{\partial\mathbf{v}}\frac{1}{\omega-\omega'-(\mathbf{k}-\mathbf{k}')\cdot\mathbf{v}}\frac{(\mathbf{k}-\mathbf{k}')}{|\mathbf{k}-\mathbf{k}'|}\cdot\frac{\partial}{\partial\mathbf{v}}\right.\\
&\quad\left.+\frac{(\mathbf{k}-\mathbf{k}')}{|\mathbf{k}-\mathbf{k}'|}\cdot\frac{\partial}{\partial\mathbf{v}}\frac{1}{\omega'-\mathbf{k}'\cdot\mathbf{v}}\frac{\mathbf{k}'}{k'}\cdot\frac{\partial}{\partial\mathbf{v}}\right)F_a\\
&\quad+\mathrm{Im}\,\frac{e_a^4}{m_a^4}\int d\mathbf{k}\int d\omega\int d\mathbf{k}'\int d\omega'\,\frac{\mathbf{k}}{k}\cdot\frac{\partial}{\partial\mathbf{v}}\frac{\langle\delta E^2\rangle_{\mathbf{k}',\omega'}\langle\delta E^2\rangle_{\mathbf{k},\omega}}{\omega-\mathbf{k}\cdot\mathbf{v}}\frac{\mathbf{k}'}{k'}\cdot\frac{\partial}{\partial\mathbf{v}}\\
&\quad\times\frac{1}{\omega-\omega'-(\mathbf{k}-\mathbf{k}')\cdot\mathbf{v}}\left(\frac{\mathbf{k}'}{k'}\cdot\frac{\partial}{\partial\mathbf{v}}\frac{1}{\omega-\mathbf{k}\cdot\mathbf{v}}\frac{\mathbf{k}}{k}\cdot\frac{\partial}{\partial\mathbf{v}}\right.\\
&\quad\left.-\frac{\mathbf{k}}{k}\cdot\frac{\partial}{\partial\mathbf{v}}\frac{1}{\omega'-\mathbf{k}'\cdot\mathbf{v}}\frac{\mathbf{k}'}{k'}\cdot\frac{\partial}{\partial\mathbf{v}}\right)F_a.
\end{aligned}
\tag{1.107}
$$

We will make explicit use of the spectral wave energy density (1.95) in (1.107), thereby eliminating ω and ω' integrals by virtue of the delta functions. This will be done later in Chapter 4. For now, we treat (1.107) as constituting the formal particle kinetic equation, and as an interlude, we next discuss the properties of various dielectric susceptibility response functions.

1.7 Linear and Nonlinear Susceptibilities

Linear and nonlinear susceptibilities have already been introduced in (1.55)–(1.64). It is instructive to rewrite these response functions by means of partial integrations,

$$
\chi_a(\mathbf{k},\omega) = -\omega_{pa}^2\int d\mathbf{v}\,\frac{F_a}{(\omega-\mathbf{k}\cdot\mathbf{v}+i0)^2},
\tag{1.108}
$$

$$
\begin{aligned}
\chi_a^{(2)}(\mathbf{k}_1,\omega_1|\mathbf{k}_2,\omega_2) &= -\int d\mathbf{v}\,F_a\,\frac{1}{(\omega_1-\mathbf{k}_1\cdot\mathbf{v}+i0)(\omega_2-\mathbf{k}_2\cdot\mathbf{v}+i0)}\\
&\quad\times\frac{1}{\omega_1+\omega_2-(\mathbf{k}_1+\mathbf{k}_2)\cdot\mathbf{v}+i0}\\
&\quad\times\left(\frac{k_1^2\,\mathbf{k}_2\cdot(\mathbf{k}_1+\mathbf{k}_2)}{\omega_1-\mathbf{k}_1\cdot\mathbf{v}+i0}+\frac{k_2^2\,\mathbf{k}_1\cdot(\mathbf{k}_1+\mathbf{k}_2)}{\omega_2-\mathbf{k}_2\cdot\mathbf{v}+i0}\right.\\
&\quad\left.+\frac{(\mathbf{k}_1+\mathbf{k}_2)^2(\mathbf{k}_1\cdot\mathbf{k}_2)}{\omega_1+\omega_2-(\mathbf{k}_1+\mathbf{k}_2)\cdot\mathbf{v}+i0}\right),
\end{aligned}
\tag{1.109}
$$

$$
\begin{aligned}
\bar{\chi}_a^{(3)}(\mathbf{k}_1,\omega_1|\mathbf{k}_2,\omega_2|\mathbf{k}_3,\omega_3) = {} & \frac{1}{2}\frac{e_a^2}{m_a^2}\frac{\omega_{pa}^2\,[\mathbf{k}_1\cdot(\mathbf{k}_1+\mathbf{k}_2+\mathbf{k}_3)]}{k_1\,k_2\,k_3\,|\mathbf{k}_1+\mathbf{k}_2+\mathbf{k}_3|} \\
& \times \int d\mathbf{v}\,F_a\,\frac{1}{(\omega_2-\mathbf{k}_2\cdot\mathbf{v}+i0)\,(\omega_3-\mathbf{k}_3\cdot\mathbf{v}+i0)} \\
& \times \frac{1}{[\omega_1+\omega_2+\omega_3-(\mathbf{k}_1+\mathbf{k}_2+\mathbf{k}_3)\cdot\mathbf{v}+i0]^2} \\
& \times \left[\frac{2}{\omega_1+\omega_2+\omega_3-(\mathbf{k}_1+\mathbf{k}_2+\mathbf{k}_3)\cdot\mathbf{v}+i0}\right. \\
& \times \left(\frac{3\,[\mathbf{k}_2\cdot(\mathbf{k}_1+\mathbf{k}_2+\mathbf{k}_3)]\,[\mathbf{k}_3\cdot(\mathbf{k}_1+\mathbf{k}_2+\mathbf{k}_3)]}{\omega_1+\omega_2+\omega_3-(\mathbf{k}_1+\mathbf{k}_2+\mathbf{k}_3)\cdot\mathbf{v}+i0}\right. \\
& + \frac{k_2^2\,[\mathbf{k}_3\cdot(\mathbf{k}_1+\mathbf{k}_2+\mathbf{k}_3)]}{\omega_2-\mathbf{k}_2\cdot\mathbf{v}+i0} \\
& \left.+ \frac{k_3^2\,[\mathbf{k}_2\cdot(\mathbf{k}_1+\mathbf{k}_2+\mathbf{k}_3)]}{\omega_3-\mathbf{k}_3\cdot\mathbf{v}+i0}\right) \\
& + \frac{1}{\omega_2+\omega_3-(\mathbf{k}_2+\mathbf{k}_3)\cdot\mathbf{v}+i0} \\
& \times \left(\frac{2\,(\mathbf{k}_2\cdot\mathbf{k}_3)\,[(\mathbf{k}_2+\mathbf{k}_3)\cdot(\mathbf{k}_1+\mathbf{k}_2+\mathbf{k}_3)]}{\omega_1+\omega_2+\omega_3-(\mathbf{k}_1+\mathbf{k}_2+\mathbf{k}_3)\cdot\mathbf{v}+i0}\right. \\
& + \frac{k_2^2\,[\mathbf{k}_3\cdot(\mathbf{k}_2+\mathbf{k}_3)]}{\omega_2-\mathbf{k}_2\cdot\mathbf{v}+i0} + \frac{k_3^2\,[\mathbf{k}_2\cdot(\mathbf{k}_2+\mathbf{k}_3)]}{\omega_3-\mathbf{k}_3\cdot\mathbf{v}+i0} \\
& \left.\left. + \frac{(\mathbf{k}_2+\mathbf{k}_3)^2(\mathbf{k}_2\cdot\mathbf{k}_3)}{\omega_2+\omega_3-(\mathbf{k}_2+\mathbf{k}_3)\cdot\mathbf{v}+i0}\right)\right]. \qquad (1.110)
\end{aligned}
$$

1.7.1 Symmetry Relations

The first useful symmetry property involves the permutation of arguments. The second- and third-order susceptibilities, $\chi^{(2)}(\mathbf{k}_1,\omega_1|\mathbf{k}_2,\omega_2)$ and $\chi^{(3)}(\mathbf{k}_1,\omega_1|\mathbf{k}_2,\omega_2|\mathbf{k}_3,\omega_3)$, are fully symmetric with respect to permutations of arguments. However, the partial third-order susceptibility $\bar{\chi}^{(3)}(\mathbf{k}_1,\omega_1|\mathbf{k}_2,\omega_2|\mathbf{k}_3,\omega_3)$ is symmetric only with respect to the permutation of the last two sets of arguments:

$$
\begin{aligned}
\chi^{(2)}(\mathbf{k}_2,\omega_2|\mathbf{k}_1,\omega_1) &= \chi^{(2)}(\mathbf{k}_1,\omega_1|\mathbf{k}_2,\omega_2), \\
\bar{\chi}^{(3)}(\mathbf{k}_1,\omega_1|\mathbf{k}_3,\omega_3|\mathbf{k}_2,\omega_2) &= \bar{\chi}^{(3)}(\mathbf{k}_1,\omega_1|\mathbf{k}_2,\omega_2|\mathbf{k}_3,\omega_3), \qquad (1.111) \\
\chi^{(3)}(\mathbf{k}_1,\omega_1|\mathbf{k}_2,\omega_2|\mathbf{k}_3,\omega_3) &= \chi^{(3)}(\mathbf{k}_1,\omega_1|\mathbf{k}_3,\omega_3|\mathbf{k}_2,\omega_2) \\
&= \chi^{(3)}(\mathbf{k}_2,\omega_2|\mathbf{k}_1,\omega_1|\mathbf{k}_3,\omega_3) = \chi^{(3)}(\mathbf{k}_2,\omega_2|\mathbf{k}_3,\omega_3|\mathbf{k}_1,\omega_1) \\
&= \chi^{(3)}(\mathbf{k}_3,\omega_3|\mathbf{k}_2,\omega_2|\mathbf{k}_1,\omega_1) = \chi^{(3)}(\mathbf{k}_3,\omega_3|\mathbf{k}_1,\omega_1|\mathbf{k}_2,\omega_2).
\end{aligned}
$$

The next useful symmetry property involves arguments of opposite signs:

$$
\begin{aligned}
\epsilon(-\mathbf{k}, -\omega) &= \epsilon^*(\mathbf{k}, \omega), \\
\chi^{(2)}(-\mathbf{k}_1, -\omega_1 | -\mathbf{k}_2, -\omega_2) &= -\chi^{(2)*}(\mathbf{k}_1, \omega_1 | \mathbf{k}_2, \omega_2), \\
\bar{\chi}^{(3)}(-\mathbf{k}_1, -\omega_1 | -\mathbf{k}_2, -\omega_2 | -\mathbf{k}_3, -\omega_3) &= \bar{\chi}^{(3)*}(\mathbf{k}_1, \omega_1 | \mathbf{k}_2, \omega_2 | \mathbf{k}_3, \omega_3), \\
\chi^{(3)}(-\mathbf{k}_1, -\omega_1 | -\mathbf{k}_2, -\omega_2 | -\mathbf{k}_3, -\omega_3) &= \chi^{(3)*}(\mathbf{k}_1, \omega_1 | \mathbf{k}_2, \omega_2 | \mathbf{k}_3, \omega_3).
\end{aligned}
\tag{1.112}
$$

These symmetry relations can easily be checked from definitions (1.62)–(1.64) or (1.108)–(1.110). If the small positive imaginary part in the resonant denominator is ignored, then we obtain a useful approximate symmetry relation from (1.109),

$$
\begin{aligned}
\chi^{(2)}(\mathbf{k}_1, \omega_1 | \mathbf{k}_2, \omega_2) &= \chi^{(2)}(\mathbf{k}_2, \omega_2 | \mathbf{k}_1, \omega_1) \\
&= \chi^{(2)}(\mathbf{k}_1 + \mathbf{k}_2, \omega_1 + \omega_2 | -\mathbf{k}_2, -\omega_2),
\end{aligned}
\tag{1.113}
$$

which we have already invoked in (1.88).

1.7.2 Linear Dielectric Susceptibility

In the long-wavelength limit ($k^2 \to 0$), the linear dielectric function $\chi^a(\mathbf{k}, \omega)$ takes on the limiting form,

$$
\chi^a(0, \omega) = -\frac{\omega_{pa}^2}{\omega^2}. \tag{1.114}
$$

To include thermal corrections, we expand the resonant denominator for small argument, $\mathbf{k} \cdot \mathbf{v}/\omega \ll 1$, to obtain

$$
\begin{aligned}
\chi^a(\mathbf{k}, \omega) \approx & -\frac{\omega_{pa}^2}{\omega^2}\left(1 + \frac{3k^2 T_a}{m_a \omega^2}\right) \\
& - i\pi \frac{\omega_{pa}^2}{k^2} \int d\mathbf{v}\, \mathbf{k} \cdot \frac{\partial F_a}{\partial \mathbf{v}}\, \delta(\omega - \mathbf{k} \cdot \mathbf{v}),
\end{aligned}
\tag{1.115}
$$

where we have defined the kinetic temperature,

$$
T_a = m_a \int d\mathbf{v}\, v^2 F_a = \frac{m_a v_{Ta}^2}{2}. \tag{1.116}
$$

Here, $v_{Ta} = \sqrt{2T_a/m_a}$ represents the thermal speed for species a. In deriving (1.115) we assumed that the plasma species a has zero net drift, $\int d\mathbf{v}\, \mathbf{v} F_a \approx 0$. The expression (1.115) assumes the so-called fast-wave condition, $\omega \gg k v_{Ta}$. For the opposite case when $\omega \ll k v_{Ta}$ is satisfied (the slow-wave condition), we may approximate the principal part of the resonant denominator by

$$
\mathscr{P} \frac{1}{\omega - \mathbf{k} \cdot \mathbf{v} + i0} \approx -\frac{1}{\mathbf{k} \cdot \mathbf{v}}. \tag{1.117}
$$

If we assume that the distribution F_a is given by a quasi-Maxwellian form so that we may write

$$\frac{\partial F_a}{\partial \mathbf{v}} \approx -\frac{2(\mathbf{k}\cdot\mathbf{v})\,F_a}{v_{Ta}^2}, \tag{1.118}$$

then this leads to the top-and-bottom cancelation of the factor $\mathbf{k}\cdot\mathbf{v}$ within the velocity integral. The resulting linear dielectric susceptibility is

$$\chi^a(\mathbf{k},\omega) = \frac{2\omega_{pa}^2}{k^2 v_{Ta}^2} - i\pi\frac{\omega_{pa}^2}{k^2}\int d\mathbf{v}\,\mathbf{k}\cdot\frac{\partial F_a}{\partial \mathbf{v}}\,\delta(\omega - \mathbf{k}\cdot\mathbf{v}). \tag{1.119}$$

For isotropic thermal equilibrium distribution, the linear dielectric function is given in terms of the plasma dispersion function $Z(z)$, as

$$\epsilon(\mathbf{k},\omega) = 1 - \sum_a \frac{\omega_{pa}^2}{\omega^2}\,\zeta_{\mathbf{k},\omega}^{a2}\,Z'(\zeta_{\mathbf{k},\omega}^a), \tag{1.120}$$

where

$$\zeta_{\mathbf{k},\omega}^a = \frac{\omega}{k v_{Ta}}, \quad v_{Ta}^2 = \frac{2T_a}{m_a}, \tag{1.121}$$

and

$$\begin{aligned} Z(z) &= \int_{-\infty}^{\infty}\frac{dx}{\sqrt{\pi}}\,\frac{e^{-x^2}}{x-z} = 2i\,e^{-z^2}\int_{-\infty}^{iz} dt\,e^{t^2}, \\ Z'(z) &= -2[1 + z\,Z(z)]. \end{aligned} \tag{1.122}$$

The plasma dispersion function and its properties are well known, for example, see Huba (2009). Series and asymptotic expansions of $Z(z)$ are given by

$$\begin{aligned} Z(z) &= i\sqrt{\pi}\,e^{-z^2} - 2z + \frac{4z^3}{3} - \frac{8z^5}{15} + \frac{16z^7}{105} + \cdots, \quad (z^2 < 1), \\ Z(z) &= i\sqrt{\pi}\,\sigma\,e^{-z^2} - \frac{1}{z} - \frac{1}{2z^3} - \frac{3}{4z^5} - \frac{15}{8z^7} + \cdots, \quad (z^2 > 1), \end{aligned} \tag{1.123}$$

where $\sigma = 0$ if $\mathrm{Im}z > 1/\mathrm{Re}z$, $\sigma = 1$ if $|\mathrm{Im}z| < 1/\mathrm{Re}z$, and $\sigma = 2$ if $-\mathrm{Im}z > 1/\mathrm{Re}z$. The plasma dispersion or Fried–Conte function (Fried and Conte, 1961) and its properties are further discussed in Appendix D.

Making use of the asymptotic expansion we obtain an approximate expression for the linear dielectric response function for thermal equilibrium,

$$\epsilon(\mathbf{k},\omega) = 1 - \sum_a \frac{\omega_{pa}^2}{\omega^2}\left(1 + \frac{3k^2 T_a}{m_a\omega^2}\right) + 2i\pi^{1/2}\sum_a \frac{\omega_{pa}^2\omega}{k^3 v_{Ta}^3}\exp\left(-\frac{\omega^2}{k^2 v_{Ta}^2}\right). \tag{1.124}$$

As it turns out, this expression, derived under the assumption of fast wave condition, is applicable to high-frequency Langmuir waves where both conditions, $\omega \gg kv_{Te}$ and $\omega \gg kv_{Ti}$, are valid. For ion-sound waves, on the other hand, $\omega \ll kv_{Te}$ (slow-wave condition for electrons), while $\omega \gg kv_{Ti}$ (fast wave condition for protons). In this case, we have

$$\epsilon(\mathbf{k},\omega) = 1 + \frac{2\omega_{pe}^2}{k^2 v_{Te}^2} - \frac{\omega_{pi}^2}{\omega^2}\left(1 + \frac{3k^2 T_i}{m_i\omega^2}\right) + 2i\pi^{1/2}\sum_a \frac{\omega_{pa}^2\omega}{k^3 v_{Ta}^3}\exp\left(-\frac{\omega^2}{k^2 v_{Ta}^2}\right). \tag{1.125}$$

These properties will be used later when we review the linear wave theory.

1.7.3 Second-Order Nonlinear Susceptibility

The following limiting forms for the second-order susceptibility can easily be derived:

$$\chi_a^{(2)}(0,\omega_1|0,\omega_2) = 0,$$
$$\chi_a^{(2)}(\mathbf{k}_1,0|\mathbf{k}_2,0) = -\frac{ie_a}{T_a}\frac{\omega_{pa}^2}{k_1\,k_2\,|\mathbf{k}_1+\mathbf{k}_2|}\frac{1}{v_{Ta}^2}. \tag{1.126}$$

If the fast-wave condition is satisfied for all frequencies and wave numbers,

$$\omega_1 \gg k_1 v_{Ta}, \quad \omega_2 \gg k_2 v_{Ta}, \quad \omega_1+\omega_2 \gg |\mathbf{k}_1+\mathbf{k}_2| v_{Ta}, \tag{1.127}$$

then the approximate second-order nonlinear susceptibility can be obtained by assuming that the temperature is zero ($T_a \to 0$), or equivalently by choosing $F_a(\mathbf{v}) = \delta(\mathbf{v})$:

$$\chi_a^{(2)}(\mathbf{k}_1,\omega_1|\mathbf{k}_2,\omega_2) = \frac{-i\;e_a}{2\;m_a}\frac{\omega_{pa}^2}{\omega_1\,\omega_2\,(\omega_1+\omega_2)}\frac{1}{k_1\,k_2\,|\mathbf{k}_1+\mathbf{k}_2|} \times\left(\frac{k_1^2}{\omega_1}\mathbf{k}_2\cdot(\mathbf{k}_1+\mathbf{k}_2) + \frac{k_2^2}{\omega_2}\mathbf{k}_1\cdot(\mathbf{k}_1+\mathbf{k}_2) + \frac{(\mathbf{k}_1+\mathbf{k}_2)^2}{\omega_1+\omega_2}\mathbf{k}_1\cdot\mathbf{k}_2\right). \tag{1.128}$$

If one of the frequencies, say ω_1, does not satisfy the fast-wave condition while ω_2 and $\omega_1 + \omega_2$ are characterized by $\omega_2 \gg k_2 v_{Ta}$ and $\omega_1 + \omega_2 \gg |\mathbf{k}_1 + \mathbf{k}_2| v_{Ta}$, then we may obtain the following expression:

$$\chi_a^{(2)}(\mathbf{k}_1, \omega_1 | \mathbf{k}_2, \omega_2) = \frac{i}{2} \frac{e_a}{m_a} \frac{\omega_{pa}^2}{\omega_2 (\omega_1 + \omega_2)} \frac{1}{k_1 \, k_2 \, |\mathbf{k}_1 + \mathbf{k}_2|} \times \mathbf{k}_2 \cdot (\mathbf{k}_1 + \mathbf{k}_2) \int d\mathbf{v} \, \frac{\mathbf{k}_1 \cdot \partial F_a / \partial \mathbf{v}}{\omega_1 - \mathbf{k}_1 \cdot \mathbf{v} + i0}, \tag{1.129}$$

where we have made use of the identity

$$\int d\mathbf{v} \frac{F_a}{(\omega_1 - \mathbf{k}_1 \cdot \mathbf{v} + i0)^2} = -\frac{1}{k_1^2} \int d\mathbf{v} \frac{\mathbf{k}_1 \cdot \partial F_a / \partial \mathbf{v}}{\omega_1 - \mathbf{k}_1 \cdot \mathbf{v} + i0}, \tag{1.130}$$

and have invoked the fast wave conditions to ignore terms k_2^2/ω and $(\mathbf{k}_1 + \mathbf{k}_2)^2/(\omega_1 + \omega_2)$. The velocity integral is related to the linear response function so that we have

$$\chi_a^{(2)}(\mathbf{k}_1, \omega_1 | \mathbf{k}_2, \omega_2) = \frac{i}{2} \frac{e_a}{m_a} \frac{k_1}{\omega_2 (\omega_1 + \omega_2)} \frac{\mathbf{k}_2 \cdot (\mathbf{k}_1 + \mathbf{k}_2)}{k_2 \, |\mathbf{k}_1 + \mathbf{k}_2|} \chi_a(\mathbf{k}_1, \omega_1). \tag{1.131}$$

If we further assume that $\omega_1 \ll k_1 v_{Ta}$, that is, the slow-wave condition for ω_1, then we obtain

$$\chi_a^{(2)}(\mathbf{k}_1, \omega_1 | \mathbf{k}_2, \omega_2) = \frac{i}{2} \frac{e_a}{m_a} \frac{\omega_{pa}^2}{\omega_2 (\omega_1 + \omega_2)} \frac{\mathbf{k}_2 \cdot (\mathbf{k}_1 + \mathbf{k}_2)}{k_1 \, k_2 \, |\mathbf{k}_1 + \mathbf{k}_2|} \times \left(\frac{2}{v_{Ta}^2} - i\pi \int d\mathbf{v} \, \mathbf{k}_1 \cdot \frac{\partial F_a}{\partial \mathbf{v}} \, \delta(\omega_1 - \mathbf{k}_1 \cdot \mathbf{v}) \right). \tag{1.132}$$

If ω_2 represents an arbitrary wave, then by following the same steps as in the previous case, we may derive the following result:

$$\begin{aligned} \chi_a^{(2)}(\mathbf{k}_1, \omega_1 | \mathbf{k}_2, \omega_2) &= \frac{i}{2} \frac{e_a}{m_a} \frac{k_2}{\omega_1 (\omega_1 + \omega_2)} \frac{\mathbf{k}_1 \cdot (\mathbf{k}_1 + \mathbf{k}_2)}{k_1 \, |\mathbf{k}_1 + \mathbf{k}_2|} \chi_a(\mathbf{k}_2, \omega_2) \\ &\approx \frac{i}{2} \frac{e_a}{m_a} \frac{\omega_{pa}^2}{\omega_1 (\omega_1 + \omega_2)} \frac{\mathbf{k}_1 \cdot (\mathbf{k}_1 + \mathbf{k}_2)}{k_1 \, k_2 \, |\mathbf{k}_1 + \mathbf{k}_2|} \\ &\quad \times \left(\frac{2}{v_{Ta}^2} - i\pi \int d\mathbf{v} \, \mathbf{k}_2 \cdot \frac{\partial F_a}{\partial \mathbf{v}} \, \delta(\omega_2 - \mathbf{k}_2 \cdot \mathbf{v}) \right). \end{aligned} \tag{1.133}$$

Finally, if $\omega_1 + \omega_2$ represents the low frequency while ω_1 and ω_2 satisfy the fast-wave condition (which becomes possible if ω_1 and ω_2 have opposite signs), then we have

$$\chi_a^{(2)}(\mathbf{k}_1,\omega_1|\mathbf{k}_2,\omega_2) = \frac{i}{2}\frac{e_a}{m_a}\frac{|\mathbf{k}_1+\mathbf{k}_2|}{\omega_1\,\omega_2}\frac{\mathbf{k}_1\cdot\mathbf{k}_2}{k_1\,k_2}\,\chi_a(\mathbf{k}_1+\mathbf{k}_2,\omega_1+\omega_2)$$

$$= \frac{i}{2}\frac{e_a}{m_a}\frac{\omega_{pa}^2}{\omega_1\,\omega_2}\frac{\mathbf{k}_1\cdot\mathbf{k}_2}{k_1\,k_2\,|\mathbf{k}_1+\mathbf{k}_2|}\left(\frac{2}{v_{Ta}^2}\right. \tag{1.134}$$

$$\left. - i\pi\int d\mathbf{v}\,(\mathbf{k}_1+\mathbf{k}_2)\cdot\frac{\partial F_a}{\partial\mathbf{v}}\,\delta[\omega_1+\omega_2-(\mathbf{k}_1+\mathbf{k}_2)\cdot\mathbf{v}]\right).$$

If we ignore the resonant contribution to the second-order nonlinear susceptibility, then it becomes evident that the leading order expression satisfies

$$\{\chi^{(2)}(\mathbf{k}_1,\omega_1|\mathbf{k}_2,\omega_2)\}^2 \approx -|\chi^{(2)}(\mathbf{k}_1,\omega_1|\mathbf{k}_2,\omega_2)|^2, \tag{1.135}$$

which we have invoked in (1.101).

1.7.4 Third-Order Nonlinear Susceptibility

To simplify the discussion, herewith we only consider the third-order susceptibility of the form, $\bar{\chi}_a^{(3)}(\mathbf{k}',\omega'|-\mathbf{k}',-\omega'|\mathbf{k},\omega)$:

$$\bar{\chi}_a^{(3)}(\mathbf{k}',\omega'|-\mathbf{k}',-\omega'|\mathbf{k},\omega) = -\frac{1}{2}\frac{e_a^2}{m_a^2}\frac{\omega_{pa}^2}{k^2\,k'^2}\int d\mathbf{v}\,\frac{1}{\omega-\mathbf{k}\cdot\mathbf{v}+i0}$$

$$\times\,\mathbf{k}'\cdot\frac{\partial}{\partial\mathbf{v}}\left\{\frac{1}{\omega-\omega'-(\mathbf{k}-\mathbf{k}')\cdot\mathbf{v}+i0}\right.$$

$$\times\left[\mathbf{k}\cdot\frac{\partial}{\partial\mathbf{v}}\left(\frac{\mathbf{k}'\cdot\partial F_a/\partial\mathbf{v}}{\omega'-\mathbf{k}'\cdot\mathbf{v}+i0}\right)\right.$$

$$\left.\left. -\,\mathbf{k}'\cdot\frac{\partial}{\partial\mathbf{v}}\left(\frac{\mathbf{k}\cdot\partial F_a/\partial\mathbf{v}}{\omega-\mathbf{k}\cdot\mathbf{v}+i0}\right)\right]\right\}, \tag{1.136}$$

or equivalently (after partial integrations)

$$\bar{\chi}_a^{(3)}(\mathbf{k}',\omega'|-\mathbf{k}',-\omega'|\mathbf{k},\omega) = \frac{1}{2}\frac{e_a^2}{m_a^2}\frac{\omega_{pa}^2\,(\mathbf{k}\cdot\mathbf{k}')}{k^2\,k'^2}$$

$$\times\int d\mathbf{v}\,\frac{F_a}{(\omega-\mathbf{k}\cdot\mathbf{v}+i0)^3(\omega'-\mathbf{k}'\cdot\mathbf{v}+i0)}$$

$$\times\left[\frac{2k^2}{\omega-\mathbf{k}\cdot\mathbf{v}+i0}\left(\frac{4\,(\mathbf{k}\cdot\mathbf{k}')}{\omega-\mathbf{k}\cdot\mathbf{v}+i0}\right.\right.$$

$$\left. +\,\frac{k'^2}{\omega'-\mathbf{k}'\cdot\mathbf{v}+i0}\right) + \frac{1}{\omega-\omega'-(\mathbf{k}-\mathbf{k}')\cdot\mathbf{v}+i0}$$

$$\times \left(\frac{2\,(\mathbf{k}\times\mathbf{k}')^2 + 3\,k^2\,[\mathbf{k}'\cdot(\mathbf{k}-\mathbf{k}')]}{\omega - \mathbf{k}\cdot\mathbf{v} + i0} + \frac{k'^2\,[\mathbf{k}\cdot(\mathbf{k}-\mathbf{k}')]}{\omega' - \mathbf{k}'\cdot\mathbf{v} + i0} + \frac{|\mathbf{k}-\mathbf{k}'|^2(\mathbf{k}\cdot\mathbf{k}')}{\omega - \omega' - (\mathbf{k}-\mathbf{k}')\cdot\mathbf{v} + i0} \right) \Bigg]. \tag{1.137}$$

It is straightforward to obtain the following limiting expressions:

$$\bar{\chi}_a^{(3)}(\mathbf{k}',0|-\mathbf{k}',0|\mathbf{k},0) = -\frac{e_a^2}{T_a^2}\frac{\omega_{pa}^2}{k^2\,k'^2}\frac{\mathbf{k}\cdot\mathbf{k}'}{k^2}\frac{1}{v_{Ta}^2},$$
$$\bar{\chi}_a^{(3)}(0,\omega'|0,-\omega'|0,\omega) = 0. \tag{1.138}$$

The first expression is obtained from (1.136), while the second result follows from (1.137).

When all three frequencies, ω, ω', and $\omega-\omega'$, satisfy the fast-wave conditions $\omega \gg k v_{Ta}$, $\omega' \gg k' v_{Ta}$, $\omega-\omega' \gg |\mathbf{k}-\mathbf{k}'| v_{Ta}$, then we have

$$\bar{\chi}_a^{(3)}(\mathbf{k}',\omega'|-\mathbf{k}',-\omega'|\mathbf{k},\omega) = \frac{1}{2}\frac{e_a^2}{m_a^2}\frac{\omega_{pa}^2}{\omega^3\,\omega'}\frac{\mathbf{k}\cdot\mathbf{k}'}{k^2\,k'^2}\left(\frac{2\,k^2\,k'^2}{\omega\,\omega'} + \frac{k'^2\,[\mathbf{k}\cdot(\mathbf{k}-\mathbf{k}')]}{\omega'\,(\omega-\omega')} + \frac{8\,k^2\,(\mathbf{k}\cdot\mathbf{k}')}{\omega^2} + \frac{2\,(\mathbf{k}\times\mathbf{k}')^2 + 3k^2\,[\mathbf{k}'\cdot(\mathbf{k}-\mathbf{k}')]}{\omega\,(\omega-\omega')} + \frac{(\mathbf{k}-\mathbf{k}')^2(\mathbf{k}\cdot\mathbf{k}')}{(\omega-\omega')^2}\right). \tag{1.139}$$

If ω' represents an arbitrary wave frequency, but the fast-wave condition is applicable for other two frequencies, $\omega \gg k v_{Ta}$ and $\omega-\omega' \gg |\mathbf{k}-\mathbf{k}'|\,v_{Ta}$, then we obtain

$$\begin{aligned}\bar{\chi}_a^{(3)}&(\mathbf{k}',\omega'|-\mathbf{k}',-\omega'|\mathbf{k},\omega)\\ &= -\frac{1}{2}\frac{e_a^2}{m_a^2}\frac{\omega_{pa}^2}{\omega^3}\frac{\mathbf{k}\cdot\mathbf{k}'}{k^2\,k'^2}\left(\frac{2\,k^2}{\omega} + \frac{\mathbf{k}\cdot(\mathbf{k}-\mathbf{k}')}{\omega-\omega'}\right)\int d\mathbf{v}\,\frac{\mathbf{k}'\cdot\partial F_a/\partial\mathbf{v}}{\omega' - \mathbf{k}'\cdot\mathbf{v} + i0}\\ &= -\frac{1}{2\omega^3}\frac{e_a^2}{m_a^2}\frac{\mathbf{k}\cdot\mathbf{k}'}{k^2}\left(\frac{2\,k^2}{\omega} + \frac{\mathbf{k}\cdot(\mathbf{k}-\mathbf{k}')}{\omega-\omega'}\right)\chi_a(\mathbf{k}',\omega').\end{aligned} \tag{1.140}$$

Note that this result can be alternatively expressed as

$$\begin{aligned}\bar{\chi}_a^{(3)}&(\mathbf{k}',\omega'|-\mathbf{k}',-\omega'|\mathbf{k},\omega)\\ &= \frac{i}{\omega^3}\frac{e_a}{m_a}\frac{(\mathbf{k}\cdot\mathbf{k}')\,|\mathbf{k}-\mathbf{k}'|}{k\,k'\,[\mathbf{k}\cdot(\mathbf{k}-\mathbf{k}')]}\{\,2k^2(\omega-\omega') + [\mathbf{k}\cdot(\mathbf{k}-\mathbf{k}')]\,\omega\,\}\\ &\quad\times \chi_a^{(2)}(\mathbf{k}',\omega'|\mathbf{k}-\mathbf{k}',\omega-\omega').\end{aligned} \tag{1.141}$$

Next, let us consider the case when $\omega - \omega'$ represents an arbitrary frequency, while the other two frequencies satisfy the fast-wave condition. In this case, we have

$$
\begin{aligned}
&\bar{\chi}_a^{(3)}(\mathbf{k}',\omega'|-\mathbf{k}',-\omega'|\mathbf{k},\omega) \\
&= -\frac{1}{2}\frac{e_a^2}{m_a^2}\frac{\omega_{pa}^2}{\omega^3\,\omega'}\frac{(\mathbf{k}\cdot\mathbf{k}')^2}{k^2\,k'^2}\int d\mathbf{v}\,\frac{(\mathbf{k}-\mathbf{k}')\cdot\partial F_a/\partial\mathbf{v}}{\omega-\omega'-(\mathbf{k}-\mathbf{k}')\cdot\mathbf{v}+i0} \\
&= -\frac{1}{2}\frac{e_a^2}{m_a^2}\frac{|\mathbf{k}-\mathbf{k}'|^2}{\omega^3\,\omega'}\frac{(\mathbf{k}\cdot\mathbf{k}')^2}{k^2\,k'^2}\,\chi_a(\mathbf{k}-\mathbf{k}',\omega-\omega'),
\end{aligned}
\tag{1.142}
$$

which can be alternatively expressed as

$$
\bar{\chi}_a^{(3)}(\mathbf{k}',\omega'|-\mathbf{k}',-\omega'|\mathbf{k},\omega) = \frac{i e_a}{m_a}\frac{\mathbf{k}\cdot\mathbf{k}'}{k\,k'}\frac{|\mathbf{k}-\mathbf{k}'|}{\omega^2}\,\chi_a^{(2)}(\mathbf{k}',\omega'|\mathbf{k}-\mathbf{k}',\omega-\omega'). \tag{1.143}
$$

1.8 Linear Waves and Weak Instabilities

In this section we review the textbook theory of small amplitude electrostatic (linear) waves and weak instabilities operative in unmagnetized plasmas, which includes the electron beam-plasma (or bump-on-tail) instability. Linear dispersion relation is determined from the solvability condition of (1.93),

$$
\mathrm{Re}\,\epsilon(\mathbf{k},\omega) = 1 + \mathrm{Re}\sum_a \frac{\omega_{pa}^2}{k^2}\int d\mathbf{v}\,\frac{\mathbf{k}\cdot\partial F_a/\partial\mathbf{v}}{\omega-\mathbf{k}\cdot\mathbf{v}+i0} = 0. \tag{1.144}
$$

Let us assume that isotropic thermal Maxwellian forms represent the bulk electron and ion distributions but a tenuous electron beam may also exist. For Langmuir waves satisfying the fast wave condition for both electrons and ions, $|\omega_{\mathbf{k}}^L|/k\,v_{Te} \gg 1$ and $|\omega_{\mathbf{k}}^L|/k\,v_{Ti} \gg 1$, where v_{Ta} is the thermal speed defined via (1.116), upon making use of (1.115) $\mathrm{Re}\,\epsilon(\mathbf{k},\pm\omega_{\mathbf{k}}^L)$ is approximately given by

$$
\mathrm{Re}\,\epsilon(\mathbf{k},\pm\omega_{\mathbf{k}}^L) = 1 - \frac{\omega_{pe}^2}{\omega_{\mathbf{k}}^{L2}}\left(1+\frac{3k^2T_e}{m_e\omega_{\mathbf{k}}^{L2}}\right), \tag{1.145}
$$

where $\omega = \omega_{\mathbf{k}}^L$ denotes the Langmuir wave dispersion relation. By setting $\mathrm{Re}\,\epsilon(\mathbf{k},\pm\omega_{\mathbf{k}}^L)$ equal to zero one readily obtains the following:

$$
\omega_{\mathbf{k}}^L = \omega_{pe}\left(1+\frac{3}{2}k^2\lambda_{De}^2\right), \qquad \omega_{-\mathbf{k}}^L = -\omega_{\mathbf{k}}^L, \tag{1.146}
$$

where T_e is the electron bulk temperature and

$$
\lambda_{De}^2 = \frac{T_e}{4\pi n e^2} = \frac{v_{Te}^2}{2\omega_{pe}^2}, \tag{1.147}
$$

is the square of the electron Debye length. In (1.146) the symmetry property (1.9) is invoked in order to have the relation $\omega^L_{-\mathbf{k}} = -\omega^L_{\mathbf{k}}$.

For ion-sound mode characterized by $|\omega^S_{\mathbf{k}}|/k\,v_{Te} \ll 1$ and $|\omega^S_{\mathbf{k}}|/k\,v_{Ti} \geq 1$, upon combining (1.115) and (1.119) we have

$$\mathrm{Re}\,\epsilon(\mathbf{k},\,\pm\,\omega^S_{\mathbf{k}}) = 1 + \frac{1}{k^2\lambda^2_{De}} - \frac{\omega^2_{pi}}{\omega^{S2}_{\mathbf{k}}}\left(1 + \frac{3k^2T_i}{m_i\omega^{S2}_{\mathbf{k}}}\right), \tag{1.148}$$

where T_i is the ion (proton) temperature and $\omega^S_{\mathbf{k}}$ denotes the ion-sound wave dispersion relation. Setting $\mathrm{Re}\,\epsilon(\mathbf{k},\,\pm\,\omega^S_{\mathbf{k}})$ equal to zero leads to

$$\omega^S_{\mathbf{k}} = \frac{kc_S\,(1+3T_i/T_e)^{1/2}}{(1+k^2\lambda^2_{De})^{1/2}}, \qquad \omega^S_{-\mathbf{k}} = -\omega^S_{\mathbf{k}}, \tag{1.149}$$

where

$$c_S = \sqrt{\frac{T_e}{m_i}} \tag{1.150}$$

is the ion sound (or ion acoustic) speed.

It is useful to evaluate the derivatives of real parts of the dielectric constants,

$$\begin{aligned}
\epsilon'(\mathbf{k},\,\pm\,\omega^L_{\mathbf{k}}) &\equiv \frac{\partial\,\mathrm{Re}\,\epsilon(\mathbf{k},\,\pm\,\omega^L_{\mathbf{k}})}{\partial(\pm\omega^L_{\mathbf{k}})} = \frac{2}{(\pm\omega^L_{\mathbf{k}})},\\
\epsilon'(\mathbf{k},\,\pm\,\omega^S_{\mathbf{k}}) &\equiv \frac{\partial\,\mathrm{Re}\,\epsilon(\mathbf{k},\,\pm\,\omega^S_{\mathbf{k}})}{\partial(\pm\omega^S_{\mathbf{k}})} = \frac{2}{(\pm\omega^L_{\mathbf{k}})}\frac{1}{\mu_{\mathbf{k}}},\\
\mu_{\mathbf{k}} &= k^3\lambda^3_{De}\left(\frac{m_e}{m_i}\right)^{1/2}\left(1+\frac{3T_i}{T_e}\right)^{1/2}.
\end{aligned} \tag{1.151}$$

Figure 1.1 displays the three basic plasma eigenmodes (normal modes) of unmagnetized plasma. These are the transverse modes $\omega^T_{\mathbf{k}} = \sqrt{\omega^2_{pe} + c^2k^2}$, which we did not discuss under the present electrostatic approximation (we will deal with the transverse mode in later chapters), and Langmuir and ion acoustic (or ion sound) modes, which we have already covered.

The instability (or damping) of plasma eigenmodes can be discussed on the basis of the imaginary part of dielectric response function

$$\mathrm{Im}\,\epsilon(\mathbf{k},\omega) = -\sum_a \frac{\pi\,\omega^2_{pa}}{k^2}\int d\mathbf{v}\,\mathbf{k}\cdot\frac{\partial F_a}{\partial \mathbf{v}}\,\delta(\omega - \mathbf{k}\cdot\mathbf{v}). \tag{1.152}$$

Consider the linear term in the formal wave kinetic equation (1.102),

$$\frac{\partial I^{\sigma\alpha}_{\mathbf{k}}}{\partial t} = -2\,\frac{\mathrm{Im}\epsilon(\mathbf{k},\sigma\,\omega^\alpha_{\mathbf{k}})}{\partial\mathrm{Re}\epsilon(\mathbf{k},\sigma\omega^\alpha_{\mathbf{k}})/\partial\sigma\omega^\alpha_{\mathbf{k}}}\,I^{\sigma\alpha}_{\mathbf{k}} \equiv 2\gamma^\alpha_{\mathbf{k}}\,I^{\sigma\alpha}_{\mathbf{k}}. \tag{1.153}$$

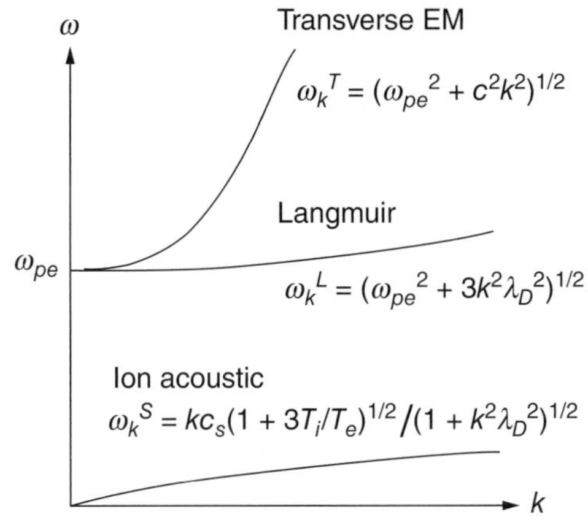

Figure 1.1 Three basic eigenmodes of unmagnetized plasma; transverse mode $\omega_{\mathbf{k}}^T$, which we could not discuss under the present electrostatic treatment, and Langmuir and ion acoustic (or ion sound) modes, $\omega_{\mathbf{k}}^L$ and $\omega_{\mathbf{k}}^S$, respectively.

In this equation we have introduced the linear "growth rate" or equivalently "Landau damping rate," – see Appendix A;

$$\gamma_{\mathbf{k}}^{\alpha} = -\frac{\mathrm{Im}\epsilon(\mathbf{k}, \sigma\omega_{\mathbf{k}}^{\alpha})}{\partial \mathrm{Re}\epsilon(\mathbf{k}, \sigma\omega_{\mathbf{k}}^{\alpha})/\partial\sigma\omega_{\mathbf{k}}^{\alpha}}. \tag{1.154}$$

Depending on whether the sign of $\gamma_{\mathbf{k}}^{\alpha}$ is positive or negative, the quantity $\gamma_{\mathbf{k}}^{\alpha}$ denotes either the growth rate for instabilities or the damping rate associated with waves.

The gentle electron beam-plasma or bump-on-tail instability will play a crucial role as a test bed for weak turbulence theory to be developed in the present monograph. The weak (or gentle) electron beam-plasma instability has been studied since the beginning of modern plasma physics, and there is a substantive body of literature on the topic, so that it is practically impossible to cite them all, but some selective references are those by Vedenov and Velikhov (1962); Drummond and Pines (1962); Frieman and Rutherford (1964); Bernstein and Engelmann (1966); Dawson and Shanny (1968); Morse and Nielson (1969); Vahala and Montgomery (1970); Roberson et al. (1971); Joyce et al. (1971) among earlier works, and Appert et al. (1976); Ivanov et al. (1976); Grognard (1982); Dum (1990); Muschietti and Dum (1991); Tsunoda et al. (1987); Nishikawa and Cairns (1991); Dum and Nishikawa (1994), to cite some representative papers up to the mid-1990s. Since the decade of 1990s, researches on beam-plasma interaction have moved on to more application-oriented topics, so that problems on pure or fundamental aspects of the beam-plasma instability, especially in relations to linear or quasilinear aspects, seem to have been exhausted. However, as we shall see later in this book, certain fundamental aspects associated with the electron beam-plasma interaction in the weak turbulence regime are still being investigated. These relate to the electron acceleration by Langmuir turbulence and the radiation generation.

Let us focus on the Langmuir wave, $\alpha = L$, in (1.154), and restrict ourselves to the forward propagation, $\sigma = +1$. Here, the directionality of forward versus backward is with respect to the beam propagation direction. Making use of (1.151) and (1.152), the growth rate (1.154) for forward-propagating Langmuir wave is given by

$$\gamma_{\mathbf{k}}^L = \frac{\pi \omega_{\mathbf{k}}^L}{2} \frac{\omega_{pe}^2}{k^2} \int d\mathbf{v}\, \mathbf{k} \cdot \frac{\partial F_e}{\partial \mathbf{v}}\, \delta(\omega_{\mathbf{k}}^L - \mathbf{k} \cdot \mathbf{v}), \tag{1.155}$$

where we have omitted the plus sign and have retained the electron terms only in $\text{Im}\,\epsilon(\mathbf{k}, \omega)$. For isotropic Maxwellian, $F_e = (\pi v_{Te})^{-3} \exp(-v^2/v_{Te}^2)$, it is obvious that $\gamma_{\mathbf{k}}^L < 0$. That is, Langmuir waves for thermal plasma is subject to Landau damping. If, on the other hand, we consider that the electron distribution is composed of a core thermal population plus an energetic but tenuous beam, take for instance, the drifting Gaussian beam distribution,

$$F_e(\mathbf{v}) = \left(1 - \frac{n_b}{n_0}\right) \frac{e^{-v^2/v_{Te}^2}}{\pi^{3/2}\, v_{Te}^3} + \frac{n_b}{n_0} \frac{e^{-(\mathbf{v}-\mathbf{V}_b)^2/v_{Tb}^2}}{\pi^{3/2}\, v_{Tb}^3}, \tag{1.156}$$

where $v_{Tb} = \sqrt{2T_b/m_e}$ is the thermal spread (or beam temperature), and where we assume

$$n_b \ll n_0, \tag{1.157}$$

then by assuming that the background population largely determines the real frequency, we may determine the damping (or growth) rate:

$$\begin{aligned} \gamma_{\mathbf{k}}^L = -\pi^{1/2} \omega_{\mathbf{k}}^L \frac{\omega_{pe}^2}{k^2} \Bigg[& \frac{v_\parallel}{v_{Te}^3} \left(1 - \frac{n_b}{n_0}\right) e^{-v_\parallel^2/v_{Te}^2} \\ & + \frac{(v_\parallel - V_b)}{v_{Tb}^3} \frac{n_b}{n_0} e^{-(v_\parallel - V_b)^2/v_{Tb}^2} \Bigg]_{v_\parallel = \omega_{\mathbf{k}}^L/k}. \end{aligned} \tag{1.158}$$

In (1.158) we have decomposed the velocity vector into components perpendicular and parallel to the beam vector, and without loss of generality, we have taken the **k** vector to be directed along the beam.

In the absence of electron beam (1.158) reduces to

$$\begin{aligned} \gamma_{\mathbf{k}}^L &= -\pi^{1/2} \frac{\omega_{pe}^2 (\omega_{\mathbf{k}}^L)^2}{k^3 v_{Te}^3} \exp\left(-\frac{(\omega_{\mathbf{k}}^L)^2}{k^2 v_{Te}^2}\right) \\ &= -\frac{\pi^{1/2} \omega_{pe}^4}{k^3 v_{Te}^3} \left(1 + \frac{3k^2 v_{Te}^2}{2\omega_{pe}^2}\right) \exp\left(-\frac{3}{2} - \frac{v_{Te}^2}{k^2 \omega_{pe}^2}\right). \end{aligned} \tag{1.159}$$

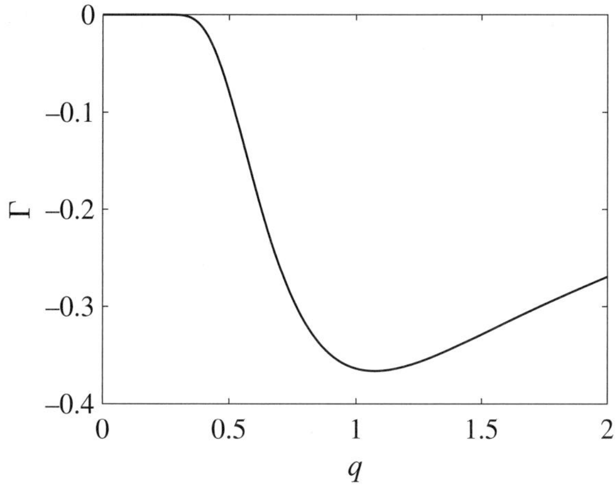

Figure 1.2 Normalized Landau damping rate Γ versus normalized wave number q.

This is the Landau damping rate for thermal plasma. Plotted in Figure 1.2 is the normalized Landau damping rate $\Gamma = \gamma_{\mathbf{k}}^L/\omega_{pe}$ versus normalized wave number $q = kv_{Te}/\omega_{pe}$. Notice that the damping rate exponentially decreases in magnitude for $k \to 0$.

Suppose that the background electrons are cold, $v_{Te} \ll v_{Tb}$. In such a case, the damping rate associated with the background component can be ignored, and the beam electrons determine the net growth/damping rate only,

$$\gamma_{\mathbf{k}}^L \approx \pi^{1/2} \frac{n_b}{n_0} \frac{\omega_{pe}^3}{k^3 v_{Tb}^3} (kV_b - \omega_{pe}) \exp\left(-\frac{(kV_b - \omega_{pe})^2}{k^2 v_{Tb}^2}\right). \tag{1.160}$$

This shows that $\gamma_{\mathbf{k}}^L > 0$ over the unstable range of k corresponding to

$$k > \frac{\omega_{pe}}{V_b}. \tag{1.161}$$

The instability of Langmuir wave driven by a gentle (or weak) electron beam is called the bump-on-tail instability, and (1.160) represents the approximate growth rate for the said instability.

In the low-frequency regime similar processes involving ion-acoustic wave with its associated damping phenomena as well as instabilities may be operative. It is a straightforward exercise to obtain the Landau damping rate for ion acoustic wave,

$$\begin{aligned}\gamma_{\mathbf{k}}^S = &-\omega_{pi} \frac{\pi^{1/2}}{2^{3/2}} \frac{k\lambda_D\,(1+3T_i/T_e)}{(1+k^2\lambda_D^2)^2} \left[\sqrt{\frac{m_e}{m_i}} \exp\left(-\frac{m_e}{2m_i} \frac{1+3T_i/T_e}{1+k^2\lambda_D^2}\right)\right. \\ &\left. + \left(\frac{T_e}{T_i}\right)^{3/2} \exp\left(-\frac{T_e}{2T_i} \frac{1+3T_i/T_e}{1+k^2\lambda_D^2}\right)\right]. \end{aligned} \tag{1.162}$$

If we approximate that $m_e/m_i \approx 0$, and introduce the dimensionless variables,

$$\Gamma = \frac{2^{3/2}}{\pi^{1/2}} \frac{\gamma_k}{\omega_{pi}}, \qquad \kappa = k\lambda_D, \qquad \tau = \frac{T_i}{T_e}, \tag{1.163}$$

then we have

$$\Gamma = -\frac{\kappa\,(1+3\tau)}{(1+\kappa^2)^2} \frac{1}{\tau^{3/2}} \exp\left(-\frac{1}{2\tau}\frac{1+3\tau}{1+\kappa^2}\right). \tag{1.164}$$

Damping generally increases for increasing κ, so let us focus on small κ behavior. If we consider Γ/κ for $\kappa \ll 1$, we have

$$\frac{\Gamma}{\kappa} \approx -\frac{1+3\tau}{\tau^{3/2}} \exp\left(-\frac{1+3\tau}{2\tau}\right). \tag{1.165}$$

Below in Figure 1.3 is the plot of $-\Gamma/\kappa$ versus $\tau = T_i/T_e$ in horizontal logarithmic scale. As one can see, if $T_i \ll T_e$ then the damping rate becomes exponentially small, and ion sound wave may propagate in a plasma undamped. For $T_i \ll T_e$, the ion-sound speed $c_S^2 = T_e/m_i$ is much higher than ion thermal speed $v_{Ti}^2 = 2T_i/m_i$, but lower than electron thermal speed $v_{Te}^2 = 2T_e/m_e$, so that

$$\sqrt{\frac{2T_i}{m_i}} < \frac{\omega_{\mathbf{k}}^S}{k} < \sqrt{\frac{2T_e}{m_e}}. \tag{1.166}$$

In this case the ion-sound mode does not suffer Landau damping by ions as the wave phase speed is sufficiently higher than ion thermal speed. On the other hand,

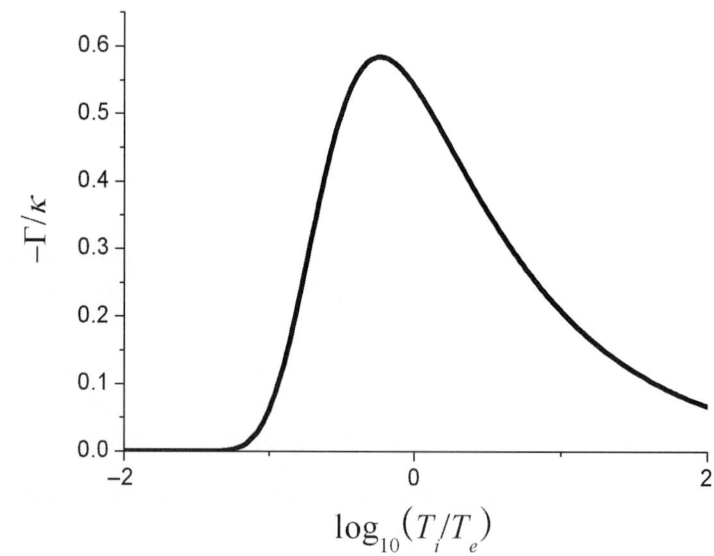

Figure 1.3 Normalized ion acoustic wave damping rate $-\Gamma/\kappa$ versus T_i/T_e.

the wave also does not suffer from electron Landau damping since the electron distribution is practically constant over the velocity range corresponding to the ion acoustic speed. This shows that the excitation and persistence of ion sound wave requires hot electrons. Note that the ion sound damping rate is maximum for $T_i \sim T_e$.

Ion acoustic mode can become unstable when there is a net mild drift between the ions and electrons. Let us consider the stationary ions and drifting Gaussian electrons (without loss of generality, we assume the electron drift direction to be along z axis), where the electron drift speed is significantly lower than electron thermal speed,

$$
\begin{aligned}
F_i(\mathbf{v}) &= \frac{1}{\pi^{3/2}\, v_{Ti}^3} \exp\left(-\frac{v^2}{v_{Ti}^2}\right), \\
F_e(\mathbf{v}) &= \frac{1}{\pi^{3/2}\, v_{Te}^3} \exp\left(-\frac{(\mathbf{v} - V_e\hat{\mathbf{z}})^2}{v_{Te}^2}\right).
\end{aligned} \tag{1.167}
$$

Then, assuming $\mathbf{k} = k\hat{\mathbf{z}}$, and for $T_e/T_i \gg 1$ and $m_e/m_i \ll 1$, we obtain the growth rate for ion acoustic instability,

$$
\gamma_k^S = \frac{\pi^{1/2}}{2^{3/2}} \left(\frac{m_e}{m_i}\right)^{1/2} \left(\frac{V_e}{c_S} - \frac{1}{\sqrt{1 + k^2\lambda_{De}^2}}\right) \frac{k^2 c_S^2}{(1 + k^2\lambda_{De}^2)^2}. \tag{1.168}
$$

If electron drift speed V_e is higher than the ion acoustic speed c_S, the ion sound mode becomes unstable.

2

Electrostatic Vlasov Weak Turbulence Theory

Wave Kinetic Equation

In this chapter and in Chapter 3, we will take the formal wave and particle kinetic equations, (1.102) and (1.107), respectively, and make specific applications to the problem of electrostatic Langmuir and ion-sound modes interacting with the particles via linear wave-particle resonance, interacting among themselves via three-wave resonance, as well as two Langmuir or ion-sound waves resonantly interacting with the particles via nonlinear wave-particle resonance.

In (1.102) let (α, β, γ) denote either L (Langmuir) or S (ion-sound). The delta function condition $\delta(\sigma\omega_{\mathbf{k}}^{\alpha} - \sigma'\omega_{\mathbf{k}'}^{\beta} - \sigma''\omega_{\mathbf{k}-\mathbf{k}'}^{\gamma})$ describes resonant three-wave decay/coalescence processes among L and S modes. A careful consideration leads to the conclusion that the resonance condition, $\sigma\omega_{\mathbf{k}}^{\alpha} - \sigma'\omega_{\mathbf{k}'}^{\beta} - \sigma''\omega_{\mathbf{k}-\mathbf{k}'}^{\gamma} = 0$, cannot be satisfied if all three waves are of the same kind. Langmuir waves possess frequency close to ω_{pe}. If all three waves are Langmuir waves, then one of them must possess wave frequency either close to $2\omega_{pe}$ or to zero, neither of which is possible. For ion-sound waves, the situation is a bit subtle, and we will in fact discuss the three-wave coupling of ion acoustic waves, but a careful analysis later reveals that a similar conclusion can be drawn for ion-sound waves. That is, three-wave decay interaction involving only ion-sound waves can be ignored. Likewise, it is not possible to satisfy the three-wave resonance condition by considering two ion-sound waves and a Langmuir wave, since ion-sound waves are low-frequency waves. Thus, the only possible three-wave process is when two Langmuir waves and an ion-sound wave are involved. It is instructive to write out (1.102) explicitly by taking all permissible interactions into account:

$$\begin{aligned}
&\epsilon'(\mathbf{k},\sigma\omega_{\mathbf{k}}^{L})\,\frac{\partial I_{\mathbf{k}}^{\sigma L}}{\partial t} + \epsilon'(\mathbf{k},\sigma\omega_{\mathbf{k}}^{S})\,\frac{\partial I_{\mathbf{k}}^{\sigma S}}{\partial t} \\
&\quad = -2\,\mathrm{Im}\,\epsilon(\mathbf{k},\sigma\omega_{\mathbf{k}}^{L})\,I_{\mathbf{k}}^{\sigma L} - 2\,\mathrm{Im}\,\epsilon(\mathbf{k},\sigma\omega_{\mathbf{k}}^{S})\,I_{\mathbf{k}}^{\sigma S} \\
&\qquad - 4\,\mathrm{Im}\sum_{\sigma'=\pm1}\int d\mathbf{k}'\,\Big\{\Big[A_{L,L}^{\sigma,\sigma'}(\mathbf{k},\mathbf{k}')\,I_{\mathbf{k}'}^{\sigma' L} + A_{L,S}^{\sigma,\sigma'}(\mathbf{k},\mathbf{k}')\,I_{\mathbf{k}'}^{\sigma' S}\Big]\,I_{\mathbf{k}}^{\sigma L}
\end{aligned}$$

$$
\begin{aligned}
&+ \left[A^{\sigma,\sigma'}_{S,L}(\mathbf{k},\mathbf{k}')\, I^{\sigma' L}_{\mathbf{k}'} + A^{\sigma,\sigma'}_{S,S}(\mathbf{k},\mathbf{k}')\, I^{\sigma' S}_{\mathbf{k}'}\right] I^{\sigma S}_{\mathbf{k}}\Bigg\} \\
&+ 4\pi \sum_{\sigma',\sigma''=\pm 1} \int d\mathbf{k}' \Big[|\chi^{(2)}(\mathbf{k}',\sigma'\omega^L_{\mathbf{k}'}|\mathbf{k}-\mathbf{k}',\sigma\omega^L_{\mathbf{k}} - \sigma'\omega^L_{\mathbf{k}'})|^2 \\
&\times \left(\frac{I^{\sigma' L}_{\mathbf{k}'} I^{\sigma'' S}_{\mathbf{k}-\mathbf{k}'}}{\epsilon'(\mathbf{k},\sigma\omega^L_{\mathbf{k}})} - \frac{I^{\sigma'' S}_{\mathbf{k}-\mathbf{k}'} I^{\sigma L}_{\mathbf{k}}}{\epsilon'(\mathbf{k}',\sigma'\omega^L_{\mathbf{k}'})} - \frac{I^{\sigma' L}_{\mathbf{k}'} I^{\sigma L}_{\mathbf{k}}}{\epsilon'(\mathbf{k}-\mathbf{k}',\sigma''\omega^S_{\mathbf{k}-\mathbf{k}'})} \right) \\
&\times \delta(\sigma\omega^L_{\mathbf{k}} - \sigma'\omega^L_{\mathbf{k}'} - \sigma''\omega^S_{\mathbf{k}-\mathbf{k}'}) \\
&+ |\chi^{(2)}(\mathbf{k}',\sigma'\omega^S_{\mathbf{k}'}|\mathbf{k}-\mathbf{k}',\sigma\omega^L_{\mathbf{k}} - \sigma'\omega^S_{\mathbf{k}'})|^2 \left(\frac{I^{\sigma' S}_{\mathbf{k}'} I^{\sigma'' L}_{\mathbf{k}-\mathbf{k}'}}{\epsilon'(\mathbf{k},\sigma\omega^L_{\mathbf{k}})} \right. \\
&\left. - \frac{I^{\sigma'' L}_{\mathbf{k}-\mathbf{k}'} I^{\sigma L}_{\mathbf{k}}}{\epsilon'(\mathbf{k}',\sigma'\omega^S_{\mathbf{k}'})} - \frac{I^{\sigma' S}_{\mathbf{k}'} I^{\sigma L}_{\mathbf{k}}}{\epsilon'(\mathbf{k}-\mathbf{k}',\sigma''\omega^L_{\mathbf{k}-\mathbf{k}'})} \right) \delta(\sigma\omega^L_{\mathbf{k}} - \sigma'\omega^S_{\mathbf{k}'} - \sigma''\omega^L_{\mathbf{k}-\mathbf{k}'}) \\
&+ |\chi^{(2)}(\mathbf{k}',\sigma'\omega^L_{\mathbf{k}'}|\mathbf{k}-\mathbf{k}',\sigma\omega^S_{\mathbf{k}} - \sigma'\omega^L_{\mathbf{k}'})|^2 \left(\frac{I^{\sigma' L}_{\mathbf{k}'} I^{\sigma'' L}_{\mathbf{k}-\mathbf{k}'}}{\epsilon'(\mathbf{k},\sigma\omega^S_{\mathbf{k}})} \right. \\
&\left. - \frac{I^{\sigma'' L}_{\mathbf{k}-\mathbf{k}'} I^{\sigma S}_{\mathbf{k}}}{\epsilon'(\mathbf{k}',\sigma'\omega^L_{\mathbf{k}'})} - \frac{I^{\sigma' L}_{\mathbf{k}'} I^{\sigma S}_{\mathbf{k}}}{\epsilon'(\mathbf{k}-\mathbf{k}',\sigma''\omega^L_{\mathbf{k}-\mathbf{k}'})} \right) \delta(\sigma\omega^S_{\mathbf{k}} - \sigma'\omega^L_{\mathbf{k}'} - \sigma''\omega^L_{\mathbf{k}-\mathbf{k}'}) \Big],
\end{aligned} \tag{2.1}
$$

where $\sigma = \pm 1$ signify forward/backward propagation, and

$$
\begin{aligned}
A^{\sigma,\sigma'}_{\alpha,\beta}(\mathbf{k},\mathbf{k}') = \mathscr{P}\, &\frac{2\{\chi^{(2)}(\mathbf{k}',\sigma'\omega^\beta_{\mathbf{k}'}|\mathbf{k}-\mathbf{k}',\sigma\omega^\alpha_{\mathbf{k}} - \sigma'\omega^\beta_{\mathbf{k}'})\}^2}{\epsilon(\mathbf{k}-\mathbf{k}',\sigma\omega^\alpha_{\mathbf{k}} - \sigma'\omega^\beta_{\mathbf{k}'})} \\
&-\bar{\chi}^{(3)}(\mathbf{k}',\sigma'\omega^\beta_{\mathbf{k}'}|-\mathbf{k}',-\sigma'\omega^\beta_{\mathbf{k}'}|\mathbf{k},\sigma\omega^\alpha_{\mathbf{k}}).
\end{aligned} \tag{2.2}
$$

We may balance the terms relevant to Langmuir waves and sound waves separately in (2.1). Making use of the fact that $\mathbf{k}'$ and $\mathbf{k}-\mathbf{k}'$ are interchangeable, we obtain

$$
\begin{aligned}
\frac{\partial I^{\sigma L}_{\mathbf{k}}}{\partial t} = &-\frac{2\,\mathrm{Im}\,\epsilon(\mathbf{k},\sigma\omega^L_{\mathbf{k}})}{\epsilon'(\mathbf{k},\sigma\omega^L_{\mathbf{k}})} I^{\sigma L}_{\mathbf{k}} - \frac{4}{\epsilon'(\mathbf{k},\sigma\omega^L_{\mathbf{k}})}\,\mathrm{Im} \sum_{\sigma'=\pm 1} \int d\mathbf{k}' \Big[A^{\sigma,\sigma'}_{L,L}(\mathbf{k},\mathbf{k}')\, I^{\sigma' L}_{\mathbf{k}'} \\
&+ A^{\sigma,\sigma'}_{L,S}(\mathbf{k},\mathbf{k}')\, I^{\sigma' S}_{\mathbf{k}'} \Big] I^{\sigma L}_{\mathbf{k}} + \frac{8\pi}{\epsilon'(\mathbf{k},\sigma\omega^L_{\mathbf{k}})} \sum_{\sigma',\sigma''=\pm 1} \int d\mathbf{k}' \\
&\times |\chi^{(2)}(\mathbf{k}',\sigma'\omega^L_{\mathbf{k}'}|\mathbf{k}-\mathbf{k}',\sigma\omega^L_{\mathbf{k}} - \sigma'\omega^L_{\mathbf{k}'})|^2 \left(\frac{I^{\sigma' L}_{\mathbf{k}'} I^{\sigma'' S}_{\mathbf{k}-\mathbf{k}'}}{\epsilon'(\mathbf{k},\sigma\omega^L_{\mathbf{k}})} - \frac{I^{\sigma'' S}_{\mathbf{k}-\mathbf{k}'} I^{\sigma L}_{\mathbf{k}}}{\epsilon'(\mathbf{k}',\sigma'\omega^L_{\mathbf{k}'})} \right. \\
&\left. - \frac{I^{\sigma' L}_{\mathbf{k}'} I^{\sigma L}_{\mathbf{k}}}{\epsilon'(\mathbf{k}-\mathbf{k}',\sigma''\omega^S_{\mathbf{k}-\mathbf{k}'})} \right) \delta(\sigma\omega^L_{\mathbf{k}} - \sigma'\omega^L_{\mathbf{k}'} - \sigma''\omega^S_{\mathbf{k}-\mathbf{k}'}),
\end{aligned} \tag{2.3}
$$

$$
\begin{aligned}
\frac{\partial I_{\mathbf{k}}^{\sigma S}}{\partial t} &= -\frac{2\,\mathrm{Im}\,\epsilon(\mathbf{k},\sigma\omega_{\mathbf{k}}^{S})}{\epsilon'(\mathbf{k},\sigma\omega_{\mathbf{k}}^{S})} I_{\mathbf{k}}^{\sigma S} \\
&\quad - \frac{4}{\epsilon'(\mathbf{k},\sigma\omega_{\mathbf{k}}^{S})}\,\mathrm{Im} \sum_{\sigma'=\pm 1} \int d\mathbf{k}' \left[A_{S,L}^{\sigma,\sigma'}(\mathbf{k},\mathbf{k}')\, I_{\mathbf{k}'}^{\sigma' L} + A_{S,S}^{\sigma,\sigma'}(\mathbf{k},\mathbf{k}')\, I_{\mathbf{k}'}^{\sigma' S} \right] I_{\mathbf{k}}^{\sigma S} \\
&\quad + \frac{4\pi}{\epsilon'(\mathbf{k},\sigma\omega_{\mathbf{k}}^{S})} \sum_{\sigma',\sigma''=\pm 1} \int d\mathbf{k}'\, |\chi^{(2)}(\mathbf{k}',\sigma'\omega_{\mathbf{k}'}^{L}|\mathbf{k}-\mathbf{k}',\sigma\omega_{\mathbf{k}}^{S}-\sigma'\omega_{\mathbf{k}'}^{L})|^2 \\
&\quad \times \left(\frac{I_{\mathbf{k}'}^{\sigma' L}\, I_{\mathbf{k}-\mathbf{k}'}^{\sigma'' L}}{\epsilon'(\mathbf{k},\sigma\omega_{\mathbf{k}}^{S})} - \frac{I_{\mathbf{k}-\mathbf{k}'}^{\sigma'' L}\, I_{\mathbf{k}}^{\sigma S}}{\epsilon'(\mathbf{k}',\sigma'\omega_{\mathbf{k}'}^{L})} - \frac{I_{\mathbf{k}'}^{\sigma' L}\, I_{\mathbf{k}}^{\sigma S}}{\epsilon'(\mathbf{k}-\mathbf{k}',\sigma''\omega_{\mathbf{k}-\mathbf{k}'}^{L})} \right) \\
&\quad \times \delta(\sigma\omega_{\mathbf{k}}^{S} - \sigma'\omega_{\mathbf{k}'}^{L} - \sigma''\omega_{\mathbf{k}-\mathbf{k}'}^{L}).
\end{aligned} \tag{2.4}
$$

We make use of (1.146), (1.149), and (1.151), namely,

$$
\begin{aligned}
&\frac{1}{\epsilon'(\mathbf{k},\sigma\omega_{\mathbf{k}}^{L})} \approx \frac{(\sigma\omega_{\mathbf{k}}^{L})}{2}, \qquad \frac{1}{\epsilon'(\mathbf{k},\sigma\omega_{\mathbf{k}}^{S})} \approx \mu_{\mathbf{k}} \frac{(\sigma\omega_{\mathbf{k}}^{L})}{2}, \\
&\omega_{\mathbf{k}}^{L} = \omega_{pe}\,(1 + 3k^2\lambda_{De}^2/2), \qquad \omega_{-\mathbf{k}}^{L} = \omega_{\mathbf{k}}^{L}, \\
&\omega_{\mathbf{k}}^{S} = kc_S(1 + 3T_i/T_e)^{1/2}/(1 + k^2\lambda_{De}^2)^{1/2}, \qquad \omega_{-\mathbf{k}}^{S} = \omega_{\mathbf{k}}^{S}, \\
&\mu_{\mathbf{k}} = k^3\lambda_{De}^3\,(m_e/m_i)^{1/2}(1 + 3T_i/T_e)^{1/2}.
\end{aligned} \tag{2.5}
$$

In what follows, we consider various interaction terms on the right-hand sides of (2.3) and (2.4).

2.1 Induced Emission

We begin with the first terms representing linear wave-particle interactions, which depict induced emissions of plasma eigenmodes. The notion of induced emission stems from the following concept: If the particles travel in a wave field and their velocities are in resonance with the wave phase velocities, then the particles may gain or lose energy as they interact with the waves. Waves in turn may also lose or gain energy, in reversal to the particle behavior. This represents a linear wave-particle resonant interaction, but in this picture, a preexisting wave field is assumed. Consequently, the change in the wave intensity by induced emission process is proportional to the wave intensity, with the multiplicative coefficient that dictates the wave-particle resonance affecting the dynamical change also appearing in the formula. The induced emissions are described by the first terms on the right-hand sides of (2.3) and (2.4). The changes in wave intensities due to the influence of linear wave-particle resonance are given in a straightforward manner. For Langmuir waves, we have

$$\left.\frac{\partial I_{\mathbf{k}}^{\sigma L}}{\partial t}\right|_{\text{ind.em.}} = -\frac{2\,\text{Im}\,\epsilon(\mathbf{k},\sigma\omega_{\mathbf{k}}^{L})}{\epsilon'(\mathbf{k},\sigma\omega_{\mathbf{k}}^{L})}\,I_{\mathbf{k}}^{\sigma L}$$
$$= \pi\,(\sigma\omega_{\mathbf{k}}^{L})\,\frac{\omega_{pe}^{2}}{k^{2}}\int d\mathbf{v}\,\delta(\sigma\omega_{\mathbf{k}}^{L}-\mathbf{k}\cdot\mathbf{v})\,\mathbf{k}\cdot\frac{\partial F_{e}}{\partial\mathbf{v}}\,I_{\mathbf{k}}^{\sigma L}, \tag{2.6}$$

while for ion-sound waves, the same is given by

$$\left.\frac{\partial I_{\mathbf{k}}^{\sigma S}}{\partial t}\right|_{\text{ind.em.}} = -\frac{2\,\text{Im}\,\epsilon(\mathbf{k},\sigma\omega_{\mathbf{k}}^{S})}{\epsilon'(\mathbf{k},\sigma\omega_{\mathbf{k}}^{S})}\,I_{\mathbf{k}}^{\sigma S}$$
$$= \pi\,\mu_{\mathbf{k}}\,(\sigma\omega_{\mathbf{k}}^{L})\,\frac{\omega_{pe}^{2}}{k^{2}}\int d\mathbf{v}\,\delta(\sigma\omega_{\mathbf{k}}^{S}-\mathbf{k}\cdot\mathbf{v})$$
$$\times\,\mathbf{k}\cdot\frac{\partial}{\partial\mathbf{v}}\left(F_{e}+\frac{m_{e}}{m_{i}}\,F_{i}\right)I_{\mathbf{k}}^{\sigma S}. \tag{2.7}$$

For L mode only electrons participate in the linear wave-particle resonance, while for S mode both charged particle species, namely, electrons and ions (protons), may resonantly interact with the waves. If we rewrite (2.6) and (2.7) as

$$\left.\frac{\partial I_{\mathbf{k}}^{\sigma\alpha}}{\partial t}\right|_{\text{ind.em.}} = 2\gamma_{\mathbf{k}}^{\sigma\alpha}\,I_{\mathbf{k}}^{\sigma\alpha}, \qquad (\alpha = L, S),$$
$$\gamma_{\mathbf{k}}^{\sigma L} = \frac{\pi}{2}\,(\sigma\omega_{\mathbf{k}}^{L})\,\frac{\omega_{pe}^{2}}{k^{2}}\int d\mathbf{v}\,\delta(\sigma\omega_{\mathbf{k}}^{L}-\mathbf{k}\cdot\mathbf{v})\,\mathbf{k}\cdot\frac{\partial F_{e}}{\partial\mathbf{v}},$$
$$\gamma_{\mathbf{k}}^{\sigma S} = \pi\,\mu_{\mathbf{k}}\,(\sigma\omega_{\mathbf{k}}^{L})\,\frac{\omega_{pe}^{2}}{k^{2}}\int d\mathbf{v}\,\delta(\sigma\omega_{\mathbf{k}}^{S}-\mathbf{k}\cdot\mathbf{v})$$
$$\times\,\mathbf{k}\cdot\frac{\partial}{\partial\mathbf{v}}\left(F_{e}+\frac{m_{e}}{m_{i}}\,F_{i}\right), \tag{2.8}$$

then one may realize that the basic form of evolution equation is none other than that given by (1.153) and that $\gamma_{\mathbf{k}}^{\sigma L}$ and $\gamma_{\mathbf{k}}^{\sigma S}$ are none other than the linear growth/damping rate introduced in (1.154). The delta-function resonance conditions $\omega_{\mathbf{k}}^{\alpha} - \mathbf{k}\cdot\mathbf{v} = 0$ and $\omega_{\mathbf{k}}^{\alpha} + \mathbf{k}\cdot\mathbf{v} = 0$ signify that (2.6) and (2.7) depict linear wave-particle resonant interactions, that is, these depict the particles moving in the wave field with their velocities in resonance with the wave phase velocities. In this process, whether waves gain or lose energy depends on whether more particles move faster or slower than the wave phase velocities. That is, the growth or damping rate depends on the slope of the particle distribution function in velocity space, hence the derivative $\partial F_a/\partial\mathbf{v}$. This form of wave kinetic equation that only includes linear wave-particle resonance is called the "quasilinear" wave kinetic equation. Consequently, the induced emission and quasilinear wave processes are equivalent.

2.2 Spontaneous and Induced Decay/Coalescence

Spontaneous and induced decay/coalescence processes are those terms dictated by the three-wave resonant interaction delta functions.

$$\begin{aligned}\frac{\partial I_{\mathbf{k}}^{\sigma L}}{\partial t}\bigg|_{\text{decay}} &= \frac{8\pi}{\epsilon'(\mathbf{k},\sigma\omega_{\mathbf{k}}^L)}\sum_{\sigma',\sigma''=\pm1}\int d\mathbf{k}'\,|\chi^{(2)}(\mathbf{k}',\sigma'\omega_{\mathbf{k}'}^L|\mathbf{k}-\mathbf{k}',\sigma\omega_{\mathbf{k}}^L-\sigma'\omega_{\mathbf{k}'}^L)|^2\\ &\quad\times\left(\frac{I_{\mathbf{k}'}^{\sigma' L}\,I_{\mathbf{k}-\mathbf{k}'}^{\sigma'' S}}{\epsilon'(\mathbf{k},\sigma\omega_{\mathbf{k}}^L)}-\frac{I_{\mathbf{k}-\mathbf{k}'}^{\sigma'' S}\,I_{\mathbf{k}}^{\sigma L}}{\epsilon'(\mathbf{k}',\sigma'\omega_{\mathbf{k}'}^L)}-\frac{I_{\mathbf{k}'}^{\sigma' L}\,I_{\mathbf{k}}^{\sigma L}}{\epsilon'(\mathbf{k}-\mathbf{k}',\sigma''\omega_{\mathbf{k}-\mathbf{k}'}^S)}\right)\\ &\quad\times\delta(\sigma\omega_{\mathbf{k}}^L-\sigma'\omega_{\mathbf{k}'}^L-\sigma''\omega_{\mathbf{k}-\mathbf{k}'}^S),\end{aligned} \tag{2.9}$$

$$\begin{aligned}\frac{\partial I_{\mathbf{k}}^{\sigma S}}{\partial t}\bigg|_{\text{decay}} &= \frac{4\pi}{\epsilon'(\mathbf{k},\sigma\omega_{\mathbf{k}}^S)}\sum_{\sigma',\sigma''=\pm1}\int d\mathbf{k}'\,|\chi^{(2)}(\mathbf{k}',\sigma'\omega_{\mathbf{k}'}^L|\mathbf{k}-\mathbf{k}',\sigma\omega_{\mathbf{k}}^S-\sigma'\omega_{\mathbf{k}'}^L)|^2\\ &\quad\times\left(\frac{I_{\mathbf{k}'}^{\sigma' L}\,I_{\mathbf{k}-\mathbf{k}'}^{\sigma'' L}}{\epsilon'(\mathbf{k},\sigma\omega_{\mathbf{k}}^S)}-\frac{I_{\mathbf{k}-\mathbf{k}'}^{\sigma'' L}\,I_{\mathbf{k}}^{\sigma S}}{\epsilon'(\mathbf{k}',\sigma'\omega_{\mathbf{k}'}^L)}-\frac{I_{\mathbf{k}'}^{\sigma' L}\,I_{\mathbf{k}}^{\sigma S}}{\epsilon'(\mathbf{k}-\mathbf{k}',\sigma''\omega_{\mathbf{k}-\mathbf{k}'}^L)}\right)\\ &\quad\times\delta(\sigma\omega_{\mathbf{k}}^S-\sigma'\omega_{\mathbf{k}'}^L-\sigma''\omega_{\mathbf{k}-\mathbf{k}'}^L),\end{aligned} \tag{2.10}$$

To compute the various three-wave decay coefficients $|\chi^{(2)}|^2$, we note that

$$\begin{aligned}&\omega_{\mathbf{k}}^L \gg k\,v_{Te}, \quad \omega_{\mathbf{k}'}^L \gg k'\,v_{Te}, \quad \omega_{\mathbf{k}-\mathbf{k}'}^L \gg |\mathbf{k}-\mathbf{k}'|\,v_{Te},\\ &\omega_{\mathbf{k}}^L \gg k\,v_{Ti}, \quad \omega_{\mathbf{k}'}^L \gg k'\,v_{Ti}, \quad \omega_{\mathbf{k}-\mathbf{k}'}^L \gg |\mathbf{k}-\mathbf{k}'|\,v_{Ti},\\ &\omega_{\mathbf{k}}^S \ll k\,v_{Te}, \quad \omega_{\mathbf{k}'}^S \ll k'\,v_{Te}, \quad \omega_{\mathbf{k}-\mathbf{k}'}^S \ll |\mathbf{k}-\mathbf{k}'|\,v_{Te},\\ &\omega_{\mathbf{k}}^S > k\,v_{Ti}, \quad \omega_{\mathbf{k}'}^S > k'\,v_{Ti}, \quad \omega_{\mathbf{k}-\mathbf{k}'}^S > |\mathbf{k}-\mathbf{k}'|\,v_{Ti},\end{aligned} \tag{2.11}$$

which we have already discussed in Section 1.7.3. Note that within the context of three-wave resonance,

$$|\omega_{\mathbf{k}}^L-\omega_{\mathbf{k}'}^L| = |\omega_{\mathbf{k}-\mathbf{k}'}^S| \ll |\mathbf{k}-\mathbf{k}'|\,v_{Te}. \tag{2.12}$$

For ions, we have

$$|\omega_{\mathbf{k}}^L-\omega_{\mathbf{k}'}^L| = |\omega_{\mathbf{k}-\mathbf{k}'}^S| > |\mathbf{k}-\mathbf{k}'|\,v_{Ti}. \tag{2.13}$$

On the basis of these considerations, the following approximate expressions for electron and ion partial second-order nonlinear susceptibilities can be obtained:

$$\begin{aligned}\chi_e^{(2)}(\mathbf{k}',\sigma'\omega_{\mathbf{k}'}^L|\mathbf{k}-\mathbf{k}',\sigma\omega_{\mathbf{k}}^L-\sigma'\omega_{\mathbf{k}'}^L) &= \chi_e^{(2)}(\mathbf{k}',\sigma'\omega_{\mathbf{k}'}^L|\mathbf{k}-\mathbf{k}',\sigma''\omega_{\mathbf{k}-\mathbf{k}'}^S)\\ &\approx -\frac{i}{2}\frac{e}{m_e}\frac{|\mathbf{k}-\mathbf{k}'|}{(\sigma\omega_{\mathbf{k}}^L)(\sigma'\omega_{\mathbf{k}'}^L)}\frac{\mathbf{k}\cdot\mathbf{k}'}{k\,k'}\\ &\quad\times\chi_e(\mathbf{k}-\mathbf{k}',\sigma\omega_{\mathbf{k}}^L-\sigma'\omega_{\mathbf{k}'}^L)\\ &\approx -\frac{i\,e}{2T_e}\frac{\omega_{pe}^2}{(\sigma\omega_{\mathbf{k}}^L)(\sigma'\omega_{\mathbf{k}'}^L)}\frac{\mathbf{k}\cdot\mathbf{k}'}{k\,k'\,|\mathbf{k}-\mathbf{k}'|},\end{aligned}$$

$$\begin{aligned}
\chi_i^{(2)}(\mathbf{k}',\sigma'\omega_{\mathbf{k}'}^L|\mathbf{k}-\mathbf{k}',\sigma\omega_{\mathbf{k}}^L-\sigma'\omega_{\mathbf{k}'}^L) &= \chi_i^{(2)}(\mathbf{k}',\sigma'\omega_{\mathbf{k}'}^L|\mathbf{k}-\mathbf{k}',\sigma''\omega_{\mathbf{k}-\mathbf{k}'}^S)\\
&\approx -\frac{ie}{2m_i}\frac{\omega_{pi}^2}{(\sigma\omega_{\mathbf{k}}^L)(\sigma'\omega_{\mathbf{k}'}^L)(\sigma''\omega_{\mathbf{k}-\mathbf{k}'}^S)}\frac{1}{k\,k'\,|\mathbf{k}-\mathbf{k}'|}\\
&\quad\times\left(\frac{k'^2}{\sigma'\omega_{\mathbf{k}'}^L}\mathbf{k}\cdot(\mathbf{k}-\mathbf{k}')+\frac{|\mathbf{k}-\mathbf{k}'|^2}{\sigma''\omega_{\mathbf{k}-\mathbf{k}'}^S}\mathbf{k}\cdot\mathbf{k}'\right.\\
&\quad\left.+\frac{k^2}{\sigma\omega_{\mathbf{k}}^L}\mathbf{k}'\cdot(\mathbf{k}-\mathbf{k}')\right)\\
&\approx -\frac{ie}{2m_i}\frac{\omega_{pi}^2}{(\sigma\omega_{\mathbf{k}}^L)(\sigma'\omega_{\mathbf{k}'}^L)(\omega_{\mathbf{k}-\mathbf{k}'}^S)^2}\frac{(\mathbf{k}\cdot\mathbf{k}')\,|\mathbf{k}-\mathbf{k}'|}{k\,k'}.
\end{aligned} \tag{2.14}$$

From this, it is seen that

$$\begin{aligned}
&|\chi_e^{(2)}(\mathbf{k}',\sigma'\omega_{\mathbf{k}'}^L|\mathbf{k}-\mathbf{k}',\sigma\omega_{\mathbf{k}}^L-\sigma'\omega_{\mathbf{k}'}^L)|^2\\
&\gg |\chi_i^{(2)}(\mathbf{k}',\sigma'\omega_{\mathbf{k}'}^L|\mathbf{k}-\mathbf{k}',\sigma\omega_{\mathbf{k}}^L-\sigma'\omega_{\mathbf{k}'}^L)|^2,
\end{aligned} \tag{2.15}$$

which shows that the ion response can be ignored.

For the case where the decay is dictated by $\sigma\omega_{\mathbf{k}}^S-\sigma'\omega_{\mathbf{k}'}^L-\sigma''\omega_{\mathbf{k}-\mathbf{k}'}^L=0$, we obtain

$$\begin{aligned}
|\chi^{(2)}(\mathbf{k}',\sigma'\omega_{\mathbf{k}'}^L|\mathbf{k}-\mathbf{k}',\sigma\omega_{\mathbf{k}}^S-\sigma'\omega_{\mathbf{k}'}^L)|^2 &= \frac{e^2}{4T_e^2}\frac{[\mathbf{k}'\cdot(\mathbf{k}-\mathbf{k}')]^2}{k'^2\,|\mathbf{k}-\mathbf{k}'|^2}k^2\lambda_{De}^4\,|\chi_e(\mathbf{k},\sigma\omega_{\mathbf{k}}^S)|^2\\
&\approx \frac{e^2}{4T_e^2}\frac{[\mathbf{k}'\cdot(\mathbf{k}-\mathbf{k}')]^2}{k^2\,k'^2\,|\mathbf{k}-\mathbf{k}'|^2}.
\end{aligned} \tag{2.16}$$

Again, in this result, the ion response has been ignored. Thus, for both Langmuir and ion-sound waves, the three-wave decay processes are dominated by electron nonlinear response.

The changes in the wave energy densities owing to the three-wave decay processes are therefore given by

$$\begin{aligned}
\left.\frac{\partial I_{\mathbf{k}}^{\sigma L}}{\partial t}\right|_{\text{decay}} &= 2\sum_{\sigma',\sigma''=\pm1}\sigma\omega_{\mathbf{k}}^L\int d\mathbf{k}'\,V_{\mathbf{k},\mathbf{k}'}^L\,\delta(\sigma\omega_{\mathbf{k}}^L-\sigma'\omega_{\mathbf{k}'}^L-\sigma''\omega_{\mathbf{k}-\mathbf{k}'}^S)\\
&\quad\times\left(\sigma\omega_{\mathbf{k}}^L\,I_{\mathbf{k}'}^{\sigma'L}\,I_{\mathbf{k}-\mathbf{k}'}^{\sigma''S}-\sigma'\omega_{\mathbf{k}'}^L\,I_{\mathbf{k}-\mathbf{k}'}^{\sigma''S}\,I_{\mathbf{k}}^{\sigma L}-\sigma''\mu_{\mathbf{k}-\mathbf{k}'}\,\omega_{\mathbf{k}-\mathbf{k}'}^L\,I_{\mathbf{k}'}^{\sigma'L}\,I_{\mathbf{k}}^{\sigma L}\right),\\
\left.\frac{\partial I_{\mathbf{k}}^{\sigma S}}{\partial t}\right|_{\text{decay}} &= \sum_{\sigma',\sigma''=\pm1}\sigma\mu_{\mathbf{k}}\,\omega_{\mathbf{k}}^L\int d\mathbf{k}'\,V_{\mathbf{k},\mathbf{k}'}^S\,\delta(\sigma\omega_{\mathbf{k}}^S-\sigma'\omega_{\mathbf{k}'}^L-\sigma''\omega_{\mathbf{k}-\mathbf{k}'}^L)\\
&\quad\times\left(\sigma\mu_{\mathbf{k}}\,\omega_{\mathbf{k}}^L\,I_{\mathbf{k}'}^{\sigma'L}\,I_{\mathbf{k}-\mathbf{k}'}^{\sigma''L}-\sigma'\omega_{\mathbf{k}'}^L\,I_{\mathbf{k}-\mathbf{k}'}^{\sigma''L}\,I_{\mathbf{k}}^{\sigma S}-\sigma''\omega_{\mathbf{k}-\mathbf{k}'}^L\,I_{\mathbf{k}'}^{\sigma'L}\,I_{\mathbf{k}}^{\sigma S}\right),
\end{aligned} \tag{2.17}$$

where

$$
\begin{aligned}
V^L_{\mathbf{k},\mathbf{k}'} &= \frac{\pi}{4}\frac{e^2}{T_e^2}\frac{(\mathbf{k}\cdot\mathbf{k}')^2}{k^2\,k'^2}\,|\mathbf{k}-\mathbf{k}'|^2\lambda_{De}^4\,|\chi_e(\mathbf{k}-\mathbf{k}',\sigma''\omega^S_{\mathbf{k}-\mathbf{k}'})|^2 \\
&\approx \frac{\pi}{4}\frac{e^2}{T_e^2}\frac{(\mathbf{k}\cdot\mathbf{k}')^2}{k^2\,k'^2\,|\mathbf{k}-\mathbf{k}'|^2}, \\
V^S_{\mathbf{k},\mathbf{k}'} &= \frac{\pi}{4}\frac{e^2}{T_e^2}\frac{[\mathbf{k}'\cdot(\mathbf{k}-\mathbf{k}')]^2}{k^2\,k'^2}\,k^2\lambda_{De}^4\,|\chi_e(\mathbf{k},\sigma\omega^S_{\mathbf{k}})|^2 \\
&\approx \frac{\pi}{4}\frac{e^2}{T_e^2}\frac{[\mathbf{k}'\cdot(\mathbf{k}-\mathbf{k}')]^2}{k^2\,k'^2\,|\mathbf{k}-\mathbf{k}'|^2}.
\end{aligned} \tag{2.18}
$$

These two equations describe a Langmuir wave decaying into another Langmuir wave and an ion-sound wave, or a Langmuir wave and an ion-sound wave coalescing (merging) in order to generate another Langmuir wave. The wave frequency and wave number conservation conditions (energy and momentum of the wave mode) are dictated by the delta function. The first terms within the parenthesis on the right-hand sides of (2.17), where the wave intensities have integral variables inexorably coupled via $I_{\mathbf{k}'}$ and $I_{\mathbf{k}-\mathbf{k}'}$, depict the "spontaneous" decay processes, while the second and third terms with the intensity $I_{\mathbf{k}}$, which is independent of the integral variable $\mathbf{k}'$, represent the "induced" decay processes. The spontaneous decay process is conceptually understood as a given wave mode with wave vector (or wave momentum) $\mathbf{k}$ spontaneously decaying into two other waves, one with $\mathbf{k}'$ and the other with $\mathbf{k}-\mathbf{k}'$, while conserving total wave energy and momentum. The induced decay takes place when a wave with $\mathbf{k}$ travels in the background of two other waves with $\mathbf{k}'$ and $\mathbf{k}-\mathbf{k}'$ interacting with each other. The original wave with $\mathbf{k}$ may resonantly interact with the other two waves, during the course of which the original wave with $\mathbf{k}$ may gain or lose energy.

It is interesting to observe that the following conservation relation exists within the context of three-wave processes:

$$
\sum_{\sigma=\pm1}\frac{\partial}{\partial t}\bigg|_{\text{decay}}\int d\mathbf{k}\left(\frac{I^L_{\mathbf{k}}}{\omega^L_{\mathbf{k}}}+\frac{2I^S_{\mathbf{k}}}{\mu_{\mathbf{k}}\,\omega^L_{\mathbf{k}}}\right)=0. \tag{2.19}
$$

This can be directly proven by adding the two quantities $\partial(I^L_{\mathbf{k}}/\omega^L_{\mathbf{k}})/\partial t$ and $\partial(2I^S_{\mathbf{k}}/\mu_{\mathbf{k}}\,\omega^L_{\mathbf{k}})/\partial t$ and integrating over $\mathbf{k}$. The quantities

$$
N^L_{\mathbf{k}}=\frac{I^L_{\mathbf{k}}}{\hbar\,\omega^L_{\mathbf{k}}}\quad\text{and}\quad N^S_{\mathbf{k}}=\frac{I^S_{\mathbf{k}}}{\hbar\,\mu_{\mathbf{k}}\omega^L_{\mathbf{k}}} \tag{2.20}
$$

can be considered as the "plasmon number densities" for eigenmodes, where $\hbar = 1.0545718\times10^{-34}\,\text{m}^2\,\text{kg/s}$ is the Planck constant. This is a quantum mechanical concept. In the literature, sometimes the quantities $N^L_{\mathbf{k}}$ and $N^S_{\mathbf{k}}$ are used as the fundamental quantities instead of $I^L_{\mathbf{k}}$ and $I^S_{\mathbf{k}}$ (Tsytovich, 1970, 1977a,b; Melrose, 1980a, 1986). Equation (2.19) shows that the decay process does not change the total plasmon number densities involved in the three-wave process. The decay process involves two Langmuir waves and an ion-sound wave for a given **k** value. As a consequence, the total number of conserved plasmon number densities equals the total number density of Langmuir modes and twice the total number density of ion-sound modes.

Let us consider the contribution to ion-sound wave kinetic equation that comes from three-wave decay processes involving only ion-sound waves,

$$\begin{aligned}
\frac{\partial I^{\sigma S}_{\mathbf{k}}}{\partial t}\bigg|^{SS}_{\text{decay}} &= \frac{4\pi}{\epsilon'(\mathbf{k},\sigma\omega^S_{\mathbf{k}})} \sum_{\sigma',\sigma''=\pm1} \int d\mathbf{k}'\, |\chi^{(2)}(\mathbf{k}',\sigma'\omega^S_{\mathbf{k}'}|\mathbf{k}-\mathbf{k}',\sigma\omega^S_{\mathbf{k}}-\sigma'\omega^S_{\mathbf{k}'})|^2 \\
&\times \left(\frac{I^{\sigma' S}_{\mathbf{k}'} I^{\sigma'' S}_{\mathbf{k}-\mathbf{k}'}}{\epsilon'(\mathbf{k},\sigma\omega^S_{\mathbf{k}})} - \frac{I^{\sigma'' S}_{\mathbf{k}-\mathbf{k}'} I^{\sigma S}_{\mathbf{k}}}{\epsilon'(\mathbf{k}',\sigma'\omega^S_{\mathbf{k}'})} - \frac{I^{\sigma' S}_{\mathbf{k}'} I^{\sigma S}_{\mathbf{k}}}{\epsilon'(\mathbf{k}-\mathbf{k}',\sigma''\omega^S_{\mathbf{k}-\mathbf{k}'})} \right) \\
&\times \delta(\sigma\omega^S_{\mathbf{k}} - \sigma'\omega^S_{\mathbf{k}'} - \sigma''\omega^S_{\mathbf{k}-\mathbf{k}'}).
\end{aligned} \tag{2.21}$$

Making use of the wave properties

$$\begin{aligned}
&\omega^S_{\mathbf{k}} \ll k v_{Te}, \qquad \omega^S_{\mathbf{k}'} \ll k' v_{Te}, \qquad |\omega^S_{\mathbf{k}} - \omega^S_{\mathbf{k}'}| \ll |\mathbf{k}-\mathbf{k}'| v_{Te}, \\
&\omega^S_{\mathbf{k}} > k v_{Ti}, \qquad \omega^S_{\mathbf{k}'} > k' v_{Ti}, \qquad |\omega^S_{\mathbf{k}} - \omega^S_{\mathbf{k}'}| > |\mathbf{k}-\mathbf{k}'| v_{Ti},
\end{aligned} \tag{2.22}$$

we may proceed to evaluate the second-order susceptibility

$$\begin{aligned}
\chi^{(2)}_e(\mathbf{k}',\sigma'\omega^S_{\mathbf{k}'}|\mathbf{k}-\mathbf{k}',\sigma\omega^S_{\mathbf{k}}-\sigma'\omega^S_{\mathbf{k}'}) &\approx \frac{iem_e}{2T_e^2}\frac{\omega^2_{pe}}{kk'|\mathbf{k}-\mathbf{k}'|}, \\
\chi^{(2)}_i(\mathbf{k}',\sigma'\omega^S_{\mathbf{k}'}|\mathbf{k}-\mathbf{k}',\sigma\omega^S_{\mathbf{k}}-\sigma'\omega^S_{\mathbf{k}'}) &\approx -\frac{iem_e}{2T_e^2}\frac{\omega^2_{pe}}{kk'|\mathbf{k}-\mathbf{k}'|} \left(\sigma'\sigma''\frac{\mathbf{k}'\cdot(\mathbf{k}-\mathbf{k}')}{k'\,|\mathbf{k}-\mathbf{k}'|} \right. \\
&\left. +\sigma\sigma''\frac{\mathbf{k}\cdot(\mathbf{k}-\mathbf{k}')}{k\,|\mathbf{k}-\mathbf{k}'|} + \sigma\sigma'\frac{\mathbf{k}\cdot\mathbf{k}'}{kk'} \right).
\end{aligned} \tag{2.23}$$

Making use of these we obtain

$$\begin{aligned}
\frac{\partial}{\partial t}\bigg|^{SS}_{\text{decay}} \frac{I^{\sigma S}_{\mathbf{k}}}{\mu_{\mathbf{k}}} &= \frac{\pi}{4}\frac{e^2 m_e^2}{T_e^4}\sigma\omega^L_{\mathbf{k}} \sum_{\sigma',\sigma''=\pm1} \int d\mathbf{k}'\, \frac{\mu_{\mathbf{k}}\,\mu_{\mathbf{k}'}\,\mu_{\mathbf{k}-\mathbf{k}'}\,\omega^4_{pe}}{k^2\,k'^2\,|\mathbf{k}-\mathbf{k}'|^2} \\
&\times \left(1 - \sigma'\sigma''\frac{\mathbf{k}'\cdot(\mathbf{k}-\mathbf{k}')}{k'\,|\mathbf{k}-\mathbf{k}'|} - \sigma\sigma''\frac{\mathbf{k}\cdot(\mathbf{k}-\mathbf{k}')}{k\,|\mathbf{k}-\mathbf{k}'|} - \sigma\sigma'\frac{\mathbf{k}\cdot\mathbf{k}'}{kk'} \right)^2 \\
&\times \left(\sigma\omega^L_{\mathbf{k}}\frac{I^{\sigma' S}_{\mathbf{k}'}}{\mu_{\mathbf{k}'}}\frac{I^{\sigma'' S}_{\mathbf{k}-\mathbf{k}'}}{\mu_{\mathbf{k}-\mathbf{k}'}} - \sigma'\omega^L_{\mathbf{k}'}\frac{I^{\sigma'' S}_{\mathbf{k}-\mathbf{k}'}}{\mu_{\mathbf{k}-\mathbf{k}'}}\frac{I^{\sigma S}_{\mathbf{k}}}{\mu_{\mathbf{k}}} - \sigma''\omega^L_{\mathbf{k}-\mathbf{k}'}\frac{I^{\sigma' S}_{\mathbf{k}'}}{\mu_{\mathbf{k}'}}\frac{I^{\sigma S}_{\mathbf{k}}}{\mu_{\mathbf{k}}} \right) \\
&\times \delta(\sigma\omega^S_{\mathbf{k}} - \sigma'\omega^S_{\mathbf{k}'} - \sigma''\omega^S_{\mathbf{k}-\mathbf{k}'}).
\end{aligned} \tag{2.24}$$

The decay process involving only S waves is described by (2.24). In reality, however, contribution to the net decay process from this mechanism is absent. To see this, consider the three-wave resonance condition, which can be expressed explicitly as

$$\omega_{\mathbf{k}}^{S} - \omega_{\mathbf{k}'}^{S} - \omega_{\mathbf{k}-\mathbf{k}'}^{S} = c_S\left(1 + \frac{3T_i}{T_e}\right)^{1/2}\left(\frac{k}{(1+k^2\lambda_{De}^2)^{1/2}} - \frac{k'}{(1+k'^2\lambda_{De}^2)^{1/2}} - \frac{k'}{(1+k'^2\lambda_{De}^2)^{1/2}}\right) = 0. \quad (2.25)$$

This resonance condition is satisfied for $k = k'$, but for such a situation the coupling coefficient in (2.24) identically vanishes by virtue of the factor $\mu_{\mathbf{k}-\mathbf{k}'}$. Consequently, decay processes involving only S modes can be ignored.

2.3 Induced Scattering

We next discuss nonlinear scattering processes, which are described by

$$\left.\frac{\partial I_{\mathbf{k}}^{\sigma L}}{\partial t}\right|_{\text{ind.sc.}} = -\sum_{\mathbf{k}'}\sum_{\sigma'=\pm1}\left[a_{L,L}(\mathbf{k},\mathbf{k}')\,I_{\mathbf{k}'}^{\sigma' L} + a_{L,S}(\mathbf{k},\mathbf{k}')\,I_{\mathbf{k}'}^{\sigma' S}\right] I_{\mathbf{k}}^{\sigma L},$$

$$\left.\frac{\partial I_{\mathbf{k}}^{\sigma S}}{\partial t}\right|_{\text{ind.sc.}} = -\sum_{\mathbf{k}'}\sum_{\sigma'=\pm1}\left[a_{S,L}(\mathbf{k},\mathbf{k}')\,I_{\mathbf{k}'}^{\sigma' L} + a_{S,S}(\mathbf{k},\mathbf{k}')\,I_{\mathbf{k}'}^{\sigma' S}\right] I_{\mathbf{k}}^{\sigma S},$$

$$a_{\alpha,\beta}(\mathbf{k},\mathbf{k}') = \frac{4}{\epsilon'(\mathbf{k},\sigma\omega_{\mathbf{k}}^{\alpha})}\,\mathrm{Im}\left(\mathscr{P}\,\frac{2\,\{\chi^{(2)}(\mathbf{k}',\sigma'\omega_{\mathbf{k}'}^{\beta}|\mathbf{k}-\mathbf{k}',\sigma\omega_{\mathbf{k}}^{\alpha}-\sigma'\omega_{\mathbf{k}'}^{\beta})\}^2}{\epsilon(\mathbf{k}-\mathbf{k}',\sigma\omega_{\mathbf{k}}^{\alpha}-\sigma'\omega_{\mathbf{k}'}^{\beta})} - \bar{\chi}^{(3)}(\mathbf{k}',\sigma'\omega_{\mathbf{k}'}^{\beta}|-\mathbf{k}',-\sigma'\omega_{\mathbf{k}'}^{\beta}|\mathbf{k},\sigma\omega_{\mathbf{k}}^{\alpha})\right). \quad (2.26)$$

2.3.1 Induced Scattering Involving Two Langmuir Waves

The first process of interest involves a Langmuir wave scattering off particles into another Langmuir wave. During the course of such a scattering, the original Langmuir wave may gain or lose energy. This is similar to the induced emission where the particles moving in the background wave field resonantly interacting with the waves give off energy or gain energy. Only this time, the particles resonantly interact with two waves. The scattering coefficient is dictated by $a_{L,L}(\mathbf{k},\mathbf{k}')$. The situation of interest is when the frequency difference $|\omega_{\mathbf{k}}^{L} - \omega_{\mathbf{k}'}^{L}|$ satisfies the slow wave condition. Note that $|\omega_{\mathbf{k}}^{L} - \omega_{\mathbf{k}'}^{L}| \neq |\omega_{\mathbf{k}-\mathbf{k}'}^{S}|$, since the nonlinear scattering process does not involve three-wave decay/coalescence resonance condition. The scattering process is similar to the decay interaction also, except that instead of three waves resonantly interacting, it is the two waves resonantly interacting with a virtual third wave, which is the particles. For electrons, we have

$$\omega_{\mathbf{k}}^L \gg k\, v_{Te}, \quad \omega_{\mathbf{k}'}^L \gg k'\, v_{Te}, \quad \omega_{\mathbf{k}}^L - \omega_{\mathbf{k}'}^L \ll |\mathbf{k} - \mathbf{k}'|\, v_{Te}. \tag{2.27}$$

For ions, we expect that

$$\omega_{\mathbf{k}}^L \gg k\, v_{Ti}, \quad \omega_{\mathbf{k}'}^L \gg k'\, v_{Ti}, \quad |\omega_{\mathbf{k}}^L - \omega_{\mathbf{k}'}^L| \gg |\mathbf{k} - \mathbf{k}'|\, v_{Ti}. \tag{2.28}$$

Nonlinear wave-particle interaction is important for the range of resonant velocity and wave numbers that is inaccessible from the standpoint of linear wave-particle interaction. For instance, linear wave-particle resonance between the electrons and Langmuir waves dictates that the resonant velocity is $v_{\rm res} \sim \pm\omega_{pe}/k$. Nonlinear wave-particle interaction that involves two Langmuir waves leads to $v_{\rm res} \sim \pm(3\omega_{pe}/2)(k^2 - k'^2)\lambda_{De}^2/|\mathbf{k} - \mathbf{k}'|$, which is much lower than the resonant velocity for linear interaction. Thus, in evaluating $a_{LL}(\mathbf{k},\mathbf{k}')$, we are mainly interested in the case where the difference between $\sigma\omega_{\mathbf{k}}^L$ and $\sigma'\omega_{\mathbf{k}'}^L$ is small. If σ and σ' have opposite signs, or even for the same signs of σ and σ', if $\mathbf{k}$ and $\mathbf{k}'$ have opposite signs, then the two quantities can be added up to produce $\sigma\omega_{\mathbf{k}}^L - \sigma'\omega_{\mathbf{k}'}^L \sim 2\omega_{pe}$. Such cases are not of interest, since it implies that all three arguments, $\sigma\omega_{\mathbf{k}}^L$, $\sigma'\omega_{\mathbf{k}'}^L$, and $\sigma\omega_{\mathbf{k}}^L - \sigma'\omega_{\mathbf{k}'}^L$, satisfy the fast-wave condition. When such a situation occurs, then the second- and third-order susceptibilities become exceedingly small, and $a_{L,L}(\mathbf{k},\mathbf{k}')$ will become insignificant. In what follows, although we will consider all signs of σ and σ' in the formal sense, and do not distinguish *a priori* the possibility of two Langmuir waves adding up to twice the plasma frequency, or different signs of $\mathbf{k}$ and $\mathbf{k}'$, which are not of interest, in the end one must discard the resonant velocities such as $v_{\rm res} \approx 2\omega_{pe}/|k \pm k'|$. What is presented subsequently should be interpreted with this caveat.

Imaginary parts of the third-order nonlinear susceptibilities relevant for $a_{LL}(\mathbf{k},\mathbf{k}')$ are given by

$$\begin{aligned}
&\operatorname{Im}\bar{\chi}_e^{(3)}(\mathbf{k}',\sigma'\omega_{\mathbf{k}'}^L| -\mathbf{k}', -\sigma'\omega_{\mathbf{k}'}^L|\mathbf{k},\sigma\omega_{\mathbf{k}}^L) \\
&\quad\approx -\frac{1}{2}\frac{e^2}{m_e^2}\frac{(\mathbf{k}\cdot\mathbf{k}')^2}{k^2\,k'^2}\frac{|\mathbf{k}-\mathbf{k}'|^2}{(\sigma\omega_{\mathbf{k}}^L)^3\,(\sigma'\omega_{\mathbf{k}'}^L)}\operatorname{Im}\chi_e(\mathbf{k}-\mathbf{k}',\sigma\omega_{\mathbf{k}}^L - \sigma'\omega_{\mathbf{k}'}^L), \\
&\operatorname{Im}\bar{\chi}_i^{(3)}(\mathbf{k}',\sigma'\omega_{\mathbf{k}'}^L| -\mathbf{k}', -\sigma'\omega_{\mathbf{k}'}^L|\mathbf{k},\sigma\omega_{\mathbf{k}}^L) \\
&\quad\approx -\frac{1}{2}\frac{e^2}{m_i^2}\frac{(\mathbf{k}\cdot\mathbf{k}')^2}{k^2\,k'^2}\frac{|\mathbf{k}-\mathbf{k}'|^2}{(\sigma\omega_{\mathbf{k}}^L)^3\,(\sigma'\omega_{\mathbf{k}'}^L)}\operatorname{Im}\chi_i(\mathbf{k}-\mathbf{k}',\sigma\omega_{\mathbf{k}}^L - \sigma'\omega_{\mathbf{k}'}^L).
\end{aligned} \tag{2.29}$$

Again, the electron nonlinear response is the dominant term. To reiterate, this approximation assumes that $\sigma\omega_{\mathbf{k}}^L - \sigma'\omega_{\mathbf{k}'}^L \ll |\mathbf{k}-\mathbf{k}'|^2 v_{Tj}^2$ for both $j = e,i$.

Next, we consider the second-order nonlinear susceptibility under the same implicit assumption, $\sigma\omega_{\mathbf{k}}^L - \sigma'\omega_{\mathbf{k}'}^L \ll |\mathbf{k}-\mathbf{k}'|^2 v_{Tj}^2$. Since the electron response is dominant we have

$$\begin{aligned}
&\{\chi^{(2)}(\mathbf{k}',\sigma'\omega_{\mathbf{k}'}^L|\mathbf{k}-\mathbf{k}',\sigma\omega_{\mathbf{k}}^L - \sigma'\omega_{\mathbf{k}'}^L)\}^2 \\
&\quad\approx -\frac{1}{4}\frac{e^2}{m_e^2}\frac{(\mathbf{k}\cdot\mathbf{k}')^2}{k^2\,k'^2}\frac{|\mathbf{k}-\mathbf{k}'|^2}{\omega_{\mathbf{k}}^{L2}\,\omega_{\mathbf{k}'}^{L2}}\{\chi_e(\mathbf{k}-\mathbf{k}',\sigma\omega_{\mathbf{k}}^L - \sigma'\omega_{\mathbf{k}'}^L)\}^2. \qquad (2.30)
\end{aligned}$$

The linear dielectric constant $\epsilon(\mathbf{k}-\mathbf{k}',\sigma\omega_{\mathbf{k}}^L - \sigma'\omega_{\mathbf{k}'}^L)$ that appears within the denominator of the principal value $\mathscr{P}[1/\epsilon(\mathbf{k}-\mathbf{k}',\sigma\omega_{\mathbf{k}}^L - \sigma'\omega_{\mathbf{k}'}^L)]$ can be approximated by

$$\begin{aligned}
\epsilon(\mathbf{k}-\mathbf{k}',\sigma\omega_{\mathbf{k}}^L - \sigma'\omega_{\mathbf{k}'}^L) &\approx \chi_e(\mathbf{k}-\mathbf{k}',\sigma\omega_{\mathbf{k}}^L - \sigma'\omega_{\mathbf{k}'}^L) \\
&\quad + i\,\mathrm{Im}\,\chi_i(\mathbf{k}-\mathbf{k}',\sigma\omega_{\mathbf{k}}^L - \sigma'\omega_{\mathbf{k}'}^L), \qquad (2.31)
\end{aligned}$$

where

$$\begin{aligned}
\chi_e(\mathbf{k}-\mathbf{k}',\sigma\omega_{\mathbf{k}}^L - \sigma'\omega_{\mathbf{k}'}^L) &= \frac{2\omega_{pe}^2}{|\mathbf{k}-\mathbf{k}'|^2 v_{Te}^2} \\
&\quad + i\,\mathrm{Im}\,\chi_e(\mathbf{k}-\mathbf{k}',\sigma\omega_{\mathbf{k}}^L - \sigma'\omega_{\mathbf{k}'}^L). \qquad (2.32)
\end{aligned}$$

As a result, we have all the ingredients to evaluate $a_{L,L}(\mathbf{k},\mathbf{k}')$. Putting everything together we arrive at

$$\begin{aligned}
\left.\frac{\partial I_{\mathbf{k}}^{\sigma L}}{\partial t}\right|_{\text{ind.sc}}^{LL} &= -\sum_{\mathbf{k}'}\sum_{\sigma'=\pm 1} a_{L,L}(\mathbf{k},\mathbf{k}')\,I_{\mathbf{k}'}^{\sigma' L}\,I_{\mathbf{k}}^{\sigma L} \\
&= -\frac{\pi}{\omega_{pe}^2}\frac{e^2}{m_e^2}\sum_{\sigma'=\pm 1}\int d\mathbf{k}'\int d\mathbf{v}\,\frac{(\mathbf{k}\cdot\mathbf{k}')^2}{k^2\,k'^2} \qquad (2.33) \\
&\quad\times\delta[\sigma\omega_{\mathbf{k}}^L - \sigma'\omega_{\mathbf{k}'}^L - (\mathbf{k}-\mathbf{k}')\cdot\mathbf{v}]\,(\mathbf{k}-\mathbf{k}')\cdot\frac{\partial}{\partial\mathbf{v}} \\
&\quad\times\left((\sigma\omega_{\mathbf{k}}^L - \sigma'\omega_{\mathbf{k}'}^L)\,F_e - \frac{m_e}{m_i}\,(\sigma\omega_{\mathbf{k}}^L)\,F_i\right) I_{\mathbf{k}'}^{\sigma' L}\,I_{\mathbf{k}}^{\sigma L}.
\end{aligned}$$

This expression was derived under the specific assumption that the difference in frequencies, $\sigma\omega_{\mathbf{k}}^L - \sigma'\omega_{\mathbf{k}'}^L$, is small when compared with $|\mathbf{k}-\mathbf{k}'|\,v_{Te}^2$. Consequently, the leading term is the induced scattering of Langmuir waves mediated by ions, or scattering off thermal ions,

$$\begin{aligned}
\left.\frac{\partial I_{\mathbf{k}}^{\sigma L}}{\partial t}\right|_{\text{ind.sc}}^{LL} &= \frac{\pi}{\omega_{pe}^2}\frac{e^2}{m_e^2}\sum_{\sigma'=\pm 1}\int d\mathbf{k}'\int d\mathbf{v}\,\frac{(\mathbf{k}\cdot\mathbf{k}')^2}{k^2\,k'^2} \\
&\quad\times\delta[\sigma\omega_{\mathbf{k}}^L - \sigma'\omega_{\mathbf{k}'}^L - (\mathbf{k}-\mathbf{k}')\cdot\mathbf{v}] \\
&\quad\times\frac{m_e}{m_i}\,(\sigma\omega_{\mathbf{k}}^L)\,(\mathbf{k}-\mathbf{k}')\cdot\frac{\partial F_i}{\partial\mathbf{v}}\,I_{\mathbf{k}'}^{\sigma' L}\,I_{\mathbf{k}}^{\sigma L}. \qquad (2.34)
\end{aligned}$$

The induced scattering involving thermal ions can be considered as the nonlinear Landau damping, but unlike the linear Landau damping, it is the ions that mediate the scattering of two Langmuir waves rather than the electrons.

2.3.2 Induced Scattering Involving Langmuir and Ion-Sound Waves

Induced scattering processes that involve Langmuir and ion-sound waves are dictated by coefficients $a_{L,S}(\mathbf{k},\mathbf{k}')$ and $a_{S,L}(\mathbf{k},\mathbf{k}')$. These coefficients include inverse linear dielectric constants with arguments involving shifted frequencies and wave vectors,

$$\frac{1}{\epsilon(\mathbf{k}-\mathbf{k}',\sigma\omega_{\mathbf{k}}^L-\sigma'\omega_{\mathbf{k}'}^S)}, \quad \text{and} \quad \frac{1}{\epsilon(\mathbf{k}-\mathbf{k}',\sigma\omega_{\mathbf{k}}^S-\sigma'\omega_{\mathbf{k}'}^L)}. \tag{2.35}$$

As we have seen in the discussion of the previous subsection that led to (2.34), inverse dielectric constants are associated with resonance delta function conditions $\delta[\sigma\omega_{\mathbf{k}}^L-\sigma'\omega_{\mathbf{k}'}^S-(\mathbf{k}-\mathbf{k}')\cdot\mathbf{v}]$ and $\delta[\sigma\omega_{\mathbf{k}}^S-\sigma'\omega_{\mathbf{k}'}^L-(\mathbf{k}-\mathbf{k}')\cdot\mathbf{v}]$, respectively. However, as we have noted, nonlinear wave-particle resonant interaction and induced scattering processes are supposed to become important in the range of velocity space for which the linear wave-particle resonance $\delta(\sigma\omega_{\mathbf{k}}^L-\mathbf{k}\cdot\mathbf{v})$ or $\delta(\sigma\omega_{\mathbf{k}}^S-\mathbf{k}\cdot\mathbf{v})$ becomes ineffective. In the case of induced scattering involving Langmuir and ion-sound waves because of the fact that $|\omega_{\mathbf{k}}^L| \gg |\omega_{\mathbf{k}}^S|$, nonlinear wave-particle resonance conditions practically overlap with that of linear wave-particle resonance,

$$\begin{aligned}
&\sigma\omega_{\mathbf{k}}^L-\sigma'\omega_{\mathbf{k}'}^S-(\mathbf{k}-\mathbf{k}')\cdot\mathbf{v}\sim\sigma\omega_{\mathbf{k}}^L-(\mathbf{k}-\mathbf{k}')\cdot\mathbf{v},\\
&\sigma\omega_{\mathbf{k}}^S-\sigma'\omega_{\mathbf{k}'}^L-(\mathbf{k}-\mathbf{k}')\cdot\mathbf{v}\sim\sigma'\omega_{\mathbf{k}'}^L-(\mathbf{k}-\mathbf{k}')\cdot\mathbf{v}.
\end{aligned} \tag{2.36}$$

For this reason, the induced scattering involving a high-frequency Langmuir wave and a low-frequency ion-sound wave can at best be treated as a small correction to the essentially linear wave-particle interaction process, and thus can be ignored.

Besides this physical ground for ignoring the coupled L and S induced scattering, such a case has some inherent mathematical problems too. The inverse dielectric constants (2.35) can become extremely large because of the fact that the arguments with shifted frequencies and wave vectors practically satisfy the linear wave-particle resonance conditions, as discussed in (2.36). This indicates the failure of perturbative solution to nonlinear plasma equation. Recall that in the formal derivation of the nonlinear spectral balance equation one had to deal with the three-body cumulant, $\langle\delta\phi_{\mathbf{k}_1,\omega_1}\,\delta\phi_{\mathbf{k}_2,\omega_2}\,\delta\phi_{-\mathbf{k},-\omega}\rangle$. To obtain this quantity, we solved the nonlinear Poisson equation (1.72) iteratively. In doing so the nonlinear correction term to the perturbed electrostatic potential was obtained via (1.76), which contained the inverse dielectric constant on the right-hand side. However, a fundamental assumption there is that the quantity $\epsilon(\mathbf{k},\omega)$ is not small, so that its inverse is well defined and mathematically meaningful. For the case of induced scattering involving two Langmuir waves,

$$\mathscr{P}\,\frac{1}{\epsilon(\mathbf{k}-\mathbf{k}',\sigma\omega_{\mathbf{k}}^{L}-\sigma'\omega_{\mathbf{k}'}^{L})}\ll 1, \tag{2.37}$$

so that the fundamental assumption in the iterative solution is not violated. However, in the present case of combined L and S modes, the denominator, $\epsilon(\mathbf{k}-\mathbf{k}',\sigma\omega_{\mathbf{k}}^{L}-\sigma'\omega_{\mathbf{k}'}^{S})$, can become very small. As a consequence, the iterative solution fails. The principal part must avoid the zeros, but since in this case, the denominator contains possible zeros, the leading contribution must be coming from the residue (i.e., the imaginary part, which is already accounted for in the decay process), and since the precise method of computing the principal part is not available, we must ignore the principal part altogether. This is another reason for ignoring induced scatterings involving L and S modes.

2.3.3 Induced Scattering Involving Two Ion-Sound Waves

For the case of ion-sound waves scattering off particles, the scattering coefficient is governed by $a_{S,S}(\mathbf{k},\mathbf{k}')$. For this case, we have

$$\omega_{\mathbf{k}}^{S}\ll k\,v_{Te},\quad \omega_{\mathbf{k}'}^{S}\ll k'\,v_{Te},\quad |\omega_{\mathbf{k}}^{S}-\omega_{\mathbf{k}'}^{S}|\ll|\mathbf{k}-\mathbf{k}'|\,v_{Te}, \tag{2.38}$$

for electrons, and

$$\omega_{\mathbf{k}}^{S}> k\,v_{Ti},\quad \omega_{\mathbf{k}'}^{S}> k'\,v_{Ti},\quad |\omega_{\mathbf{k}}^{S}-\omega_{\mathbf{k}'}^{S}|>|\mathbf{k}-\mathbf{k}'|\,v_{Ti}, \tag{2.39}$$

for ions. Note that the ion-sound wave can exist in a plasma where $T_e\gg T_i$. As a result, $\omega_{\mathbf{k}}^{S}-\omega_{\mathbf{k}'}^{S}\sim(k-k')\,c_S$, and $c_S/v_{Ti}\sim(T_e/T_i)^{1/2}\gg 1$. This leads to

$$\begin{aligned}\epsilon(\mathbf{k}-\mathbf{k}',\sigma\omega_{\mathbf{k}}^{S}-\sigma'\omega_{\mathbf{k}'}^{S})\approx 1&+\frac{2(\mathbf{k}\cdot\mathbf{k}'-\sigma\sigma'k\,k')}{|\mathbf{k}-\mathbf{k}'|^2(\sigma k-\sigma'k')^2\lambda_{De}^2}\\ &+i\left(\frac{\pi}{2}\frac{m_e}{m_i}\right)^{1/2}\frac{\sigma k-\sigma'k'}{|\mathbf{k}-\mathbf{k}'|^3\,\lambda_{De}^2}\left[\exp\left(-\frac{m_e}{2m_i}\frac{(\sigma k-\sigma'k')^2}{|\mathbf{k}-\mathbf{k}'|^2}\right)\right.\\ &\left.+\left(\frac{m_i}{m_e}\right)^{1/2}\left(\frac{T_e}{T_i}\right)^{3/2}\exp\left(-\frac{T_e}{2T_i}\frac{(\sigma k-\sigma'k')^2}{|\mathbf{k}-\mathbf{k}'|^2}\right)\right].\end{aligned} \tag{2.40}$$

As for the second-order nonlinear susceptibility, unlike the situation involving high-frequency Langmuir waves, both electron and ion nonlinear responses are important,

$$\chi_e^{(2)}(\mathbf{k}',\sigma'\omega_{\mathbf{k}'}^S|\mathbf{k}-\mathbf{k}',\sigma\omega_{\mathbf{k}}^S-\sigma'\omega_{\mathbf{k}'}^S) \approx \frac{ie}{T_e}\frac{\omega_{pe}^2}{k\,k'\,|\mathbf{k}-\mathbf{k}'|}\frac{1}{v_{Te}^2},$$

$$\chi_i^{(2)}(\mathbf{k}',\sigma'\omega_{\mathbf{k}'}^S|\mathbf{k}-\mathbf{k}',\sigma\omega_{\mathbf{k}}^S-\sigma'\omega_{\mathbf{k}'}^S) \approx -\frac{i}{2}\frac{e}{m_i}\frac{\omega_{pi}^2}{\sigma\omega_{\mathbf{k}}^S\sigma'\omega_{\mathbf{k}'}^S(\sigma\omega_{\mathbf{k}}^S-\sigma'\omega_{\mathbf{k}'}^S)}$$
$$\times\frac{1}{k\,k\,|\mathbf{k}-\mathbf{k}'|}\left(\frac{k^2}{\sigma\omega_{\mathbf{k}}^S}\mathbf{k}'\cdot(\mathbf{k}-\mathbf{k}')\right.$$
$$\left.+\frac{k'^2}{\sigma'\omega_{\mathbf{k}'}^S}\mathbf{k}\cdot(\mathbf{k}-\mathbf{k}')+\frac{|\mathbf{k}-\mathbf{k}'|^2}{\sigma\omega_{\mathbf{k}}^S-\sigma'\omega_{\mathbf{k}'}^S}\mathbf{k}\cdot\mathbf{k}'\right), \tag{2.41}$$

from which, we obtain

$$\{\chi^{(2)}(\mathbf{k}',\sigma'\omega_{\mathbf{k}'}^S|\mathbf{k}-\mathbf{k}',\sigma\omega_{\mathbf{k}}^S-\sigma'\omega_{\mathbf{k}'}^S)\}^2$$
$$=-\frac{e^2}{T_e^2}\frac{m_e^2}{4T_e^2}\frac{\omega_{pe}^4\,(\mathbf{k}\cdot\mathbf{k}')^2}{k^4\,k'^4\,|\mathbf{k}-\mathbf{k}'|^2}\left(1+\frac{|\mathbf{k}-\mathbf{k}'|^2}{(\sigma k-\sigma'k')^2}\right)^2. \tag{2.42}$$

The third-order nonlinear response in the present case is dominated by the ions,

$$\begin{aligned}
&\mathrm{Im}\,\bar{\chi}^{(3)}(\mathbf{k}',\sigma'\omega_{\mathbf{k}'}^S|-\mathbf{k}',\,-\sigma'\omega_{\mathbf{k}'}^S|\mathbf{k},\sigma\omega_{\mathbf{k}}^S)\\
&\quad\approx \mathrm{Im}\,\bar{\chi}_i^{(3)}(\mathbf{k}',\sigma'\omega_{\mathbf{k}'}^S|-\mathbf{k}',\,-\sigma'\omega_{\mathbf{k}'}^S|\mathbf{k},\sigma\omega_{\mathbf{k}}^S)\\
&\quad\approx -\frac{\sigma\sigma'}{2}\frac{e^2}{T_e^2}\frac{|\mathbf{k}-\mathbf{k}'|^2(\mathbf{k}\cdot\mathbf{k}')^2}{k^5\,k'^3}\,\mathrm{Im}\,\chi_i(\mathbf{k}-\mathbf{k}',\sigma\omega_{\mathbf{k}}^S-\sigma'\omega_{\mathbf{k}'}^S).
\end{aligned} \tag{2.43}$$

Putting all the results together, we now have

$$\left.\frac{\partial I_{\mathbf{k}}^{\sigma S}}{\partial t}\right|_{\text{ind.sc.}} = \sum_{\mathbf{k}'}\sum_{\sigma'=\pm1} a_{S,S}(\mathbf{k},\mathbf{k}')\,I_{\mathbf{k}'}^{\sigma' S}\,I_{\mathbf{k}}^{\sigma S},$$
$$a_{S,S}(\mathbf{k},\mathbf{k}') = \mu_{\mathbf{k}}(\sigma\omega_{\mathbf{k}}^L)\,\frac{\pi}{\omega_{pe}^2}\,\frac{e^2}{m_e m_i}\,\frac{(\mathbf{k}\cdot\mathbf{k}')^2}{k^4\,k'^4\,\lambda_{De}^4}\int d\mathbf{v}\;W_{\mathbf{k},\mathbf{k}'}$$
$$\times(\mathbf{k}-\mathbf{k}')\cdot\frac{\partial F_i}{\partial\mathbf{v}}\,\delta[\sigma\omega_{\mathbf{k}}^S-\sigma'\omega_{\mathbf{k}'}^S-(\mathbf{k}-\mathbf{k}')\cdot\mathbf{v}], \tag{2.44}$$

where

$$W_{\mathbf{k},\mathbf{k}'} = \left(1+\frac{|\mathbf{k}-\mathbf{k}'|^2}{(\sigma k-\sigma'k')^2}\right)^2\frac{1}{|\mathbf{k}-\mathbf{k}'|^4\,\lambda_{De}^4\,|\epsilon(\mathbf{k}-\mathbf{k}',\sigma\omega_{\mathbf{k}}^S-\sigma'\omega_{\mathbf{k}'}^S)|^2}+\sigma\sigma'\frac{k'}{k},$$
$$|\epsilon(\mathbf{k}-\mathbf{k}',\sigma\omega_{\mathbf{k}}^S-\sigma'\omega_{\mathbf{k}'}^S)|^2 \approx \left(1+\frac{2(\mathbf{k}\cdot\mathbf{k}'-\sigma\sigma'k\,k')}{|\mathbf{k}-\mathbf{k}'|^2(k-\sigma\sigma'k')^2\lambda_{De}^2}\right)^2$$
$$+\frac{\pi}{2}\frac{m_e}{m_i}\frac{(k-\sigma\sigma'k')^2}{|\mathbf{k}-\mathbf{k}'|^6\lambda_{De}^4}\left[\exp\left(-\frac{m_e}{2m_i}\frac{(k-\sigma\sigma'k')^2}{|\mathbf{k}-\mathbf{k}'|^2}\right)\right.$$
$$\left.+\left(\frac{m_i}{m_e}\right)^{1/2}\left(\frac{T_e}{T_i}\right)^{3/2}\exp\left(-\frac{T_e}{2T_i}\frac{(k-\sigma\sigma'k')^2}{|\mathbf{k}-\mathbf{k}'|^2}\right)\right]^2. \tag{2.45}$$

3

Electrostatic Vlasov Weak Turbulence Theory

Particle Kinetic Equation

In this chapter we pick up where we left off in (1.107) and complete the discussion of particle kinetic equation. The simplest form of particle kinetic equation is the quasilinear velocity space diffusion equation, which can be discussed by ignoring nonlinear effects.

3.1 Quasilinear Particle Kinetic Equation

If we ignore nonlinear wave-wave and wave-particle interaction terms in (1.107), we obtain

$$\frac{\partial F_a}{\partial t} = \frac{\partial}{\partial v_i}\left(D_{ij}^a \frac{\partial F_a}{\partial v_j}\right),$$
$$D_{ij}^a = \pi \frac{e_a^2}{m_a^2} \sum_\alpha \sum_{\sigma=\pm 1} \int d\mathbf{k}\, \frac{k_i k_j}{k^2} I_{\mathbf{k}}^{\sigma\alpha}\, \delta(\sigma\omega_{\mathbf{k}}^\alpha - \mathbf{k}\cdot\mathbf{v}), \tag{3.1}$$

where we have made use of the relationship $\langle\delta E^2\rangle_{\mathbf{k}\omega} = \sum_\alpha [\, I_{\mathbf{k}}^{+\alpha}\ \delta(\omega - \omega_{\mathbf{k}}^\alpha) + I_{\mathbf{k}}^{-\alpha}\ \delta(\omega + \omega_{\mathbf{k}}^\alpha)\,]$ and have kept only the imaginary contribution in the resonant denominator,

$$\frac{1}{\omega_{\mathbf{k}}^\alpha - \mathbf{k}\cdot\mathbf{v} + i0} = \mathscr{P}\frac{1}{\omega_{\mathbf{k}}^\alpha - \mathbf{k}\cdot\mathbf{v}} - i\,\pi\,\delta(\omega_{\mathbf{k}}^\alpha - \mathbf{k}\cdot\mathbf{v}). \tag{3.2}$$

Energetic electrons near the tail of velocity distribution interact with Langmuir waves ($\omega_{\mathbf{k}}^L \approx \mathbf{k}\cdot\mathbf{v}$), while the low-energy bulk electrons can, in principle, absorb the ion-sound waves ($\omega_{\mathbf{k}}^S \approx \mathbf{k}\cdot\mathbf{v}$). However, the range of resonant velocities for ion-sound waves is extremely narrow. As such, we can ignore the interaction of electrons with ion-sound waves. Consequently, the electron quasilinear diffusion equation can be discussed by considering only L mode:

$$\frac{\partial F_e}{\partial t} = \frac{\pi e^2}{m_e^2} \frac{\partial}{\partial v_i} \left(\int d\mathbf{k} \, \frac{k_i k_j}{k^2} \left[I_{\mathbf{k}}^{+L} \delta(\omega_{\mathbf{k}}^L - \mathbf{k}\cdot\mathbf{v}) + I_{\mathbf{k}}^{-L} \delta(\omega_{\mathbf{k}}^L + \mathbf{k}\cdot\mathbf{v}) \right] \frac{\partial F_e}{\partial v_j} \right). \tag{3.3}$$

In contrast, the Langmuir wave frequency is too high for ions to respond directly. Thus, the ion kinetic equation is affected only by the presence of ion-sound waves:

$$\frac{\partial F_i}{\partial t} = \frac{\pi e^2}{m_i^2} \frac{\partial}{\partial v_i} \left(\int d\mathbf{k} \, \frac{k_i k_j}{k^2} \left[I_{\mathbf{k}}^{+S} \delta(\omega_{\mathbf{k}}^S - \mathbf{k}\cdot\mathbf{v}) + I_{\mathbf{k}}^{-S} \delta(\omega_{\mathbf{k}}^S + \mathbf{k}\cdot\mathbf{v}) \right] \frac{\partial F_i}{\partial v_j} \right). \tag{3.4}$$

3.2 Null Effects of Decay Processes on the Particles

We next consider the corrections to particle kinetic equations that arise from various nonlinear processes. Let us first consider the influence of three-wave decay/coalescence processes. This effect comes from the second term on the right-hand side of (1.107). In particular, the decay process only involves the residue contributions from inverse linear dielectric functions. Principal part contribution primarily gives rise to nonlinear wave-particle interactions. Therefore, in the present section, we only consider the residue terms within the second term on the right-hand side of (1.107):

$$\begin{aligned}
\left.\frac{\partial F_a}{\partial t}\right|_{\text{decay}} &= \text{Re}\, \frac{e_a^3}{m_a^3} \frac{\partial}{\partial v_i} \int d\mathbf{k} \int d\omega \int d\mathbf{k}' \int d\omega' \, \frac{k_i k_j' (\mathbf{k}-\mathbf{k}')_k}{k k' |\mathbf{k}-\mathbf{k}'|} \frac{1}{\omega - \mathbf{k}\cdot\mathbf{v}} \\
&\quad \times \Bigg[\chi^{(2)}(\mathbf{k}',\omega'|\mathbf{k}-\mathbf{k}',\omega-\omega') \left(\frac{\langle \delta E^2 \rangle_{\mathbf{k}-\mathbf{k}'\omega-\omega'} \langle \delta E^2 \rangle_{\mathbf{k}\omega}}{\epsilon(\mathbf{k}',\omega')} \right. \\
&\quad \left. + \frac{\langle \delta E^2 \rangle_{\mathbf{k}'\omega'} \langle \delta E^2 \rangle_{\mathbf{k}\omega}}{\epsilon(\mathbf{k}-\mathbf{k}',\omega-\omega')} \right) \\
&\quad - \chi^{(2)*}(\mathbf{k}',\omega'|\mathbf{k}-\mathbf{k}',\omega-\omega') \frac{\langle \delta E^2 \rangle_{\mathbf{k}'\omega'} \langle \delta E^2 \rangle_{\mathbf{k}-\mathbf{k}'\omega-\omega'}}{\epsilon^*(\mathbf{k},\omega)} \Bigg] \\
&\quad \times \left(\frac{\partial}{\partial v_j} \frac{1}{\omega - \omega' - (\mathbf{k}-\mathbf{k}')\cdot\mathbf{v}} \frac{\partial}{\partial v_k} + \frac{\partial}{\partial v_k} \frac{1}{\omega' - \mathbf{k}'\cdot\mathbf{v}} \frac{\partial}{\partial v_j} \right) F_a.
\end{aligned} \tag{3.5}$$

Making use of (1.99), explicitly writing the spectral electric field fluctuations in terms of eigenmode intensities (1.95), and expressing the second-order susceptibility as

$$\begin{aligned}
\chi^{(2)}(\mathbf{k}',\sigma'\omega_{\mathbf{k}'}^\beta|\mathbf{k}-\mathbf{k}',\sigma''\omega_{\mathbf{k}-\mathbf{k}'}^\gamma) &\approx i\,\text{Im}\,\chi^{(2)}(\mathbf{k}',\sigma'\omega_{\mathbf{k}'}^\beta|\mathbf{k}-\mathbf{k}',\sigma''\omega_{\mathbf{k}-\mathbf{k}'}^\gamma), \\
\chi^{(2)*}(\mathbf{k}',\sigma'\omega_{\mathbf{k}'}^\beta|\mathbf{k}-\mathbf{k}',\sigma''\omega_{\mathbf{k}-\mathbf{k}'}^\gamma) &\approx -i\,\text{Im}\,\chi^{(2)}(\mathbf{k}',\sigma'\omega_{\mathbf{k}'}^\beta|\mathbf{k}-\mathbf{k}',\sigma''\omega_{\mathbf{k}-\mathbf{k}'}^\gamma),
\end{aligned} \tag{3.6}$$

we have

$$
\begin{aligned}
\left.\frac{\partial F_a}{\partial t}\right|_{\rm decay} &= -{\rm Re}\,\frac{\pi e_a^3}{m_a^3}\frac{\partial}{\partial v_i}\sum_{\sigma,\sigma',\sigma''=\pm 1}\sum_{\alpha,\beta,\gamma}\int d{\bf k}\int d{\bf k}'\,\frac{k_i\,k_j'\,({\bf k}-{\bf k}')_k}{k\,k'\,|{\bf k}-{\bf k}'|} \\
&\times\frac{{\rm Im}\,\chi^{(2)}({\bf k}',\sigma'\omega_{{\bf k}'}^\beta|{\bf k}-{\bf k}',\sigma''\omega_{{\bf k}-{\bf k}'}^\gamma)}{\sigma\omega_{\bf k}^\alpha-{\bf k}\cdot{\bf v}} \\
&\times\left(\frac{I_{{\bf k}'}^{\sigma'\beta}\,I_{{\bf k}-{\bf k}'}^{\sigma''\gamma}}{\epsilon'({\bf k},\sigma\omega_{\bf k}^\alpha)}-\frac{I_{{\bf k}'}^{\sigma'\beta}\,I_{\bf k}^{\sigma\alpha}}{\epsilon'({\bf k}-{\bf k}',\sigma''\omega_{{\bf k}-{\bf k}'}^\gamma)}-\frac{I_{{\bf k}-{\bf k}'}^{\sigma''\gamma}\,I_{\bf k}^{\sigma\alpha}}{\epsilon'({\bf k}',\sigma'\omega_{{\bf k}'}^\beta)}\right) \\
&\times\delta(\sigma\omega_{\bf k}^\alpha-\sigma'\omega_{{\bf k}'}^\beta-\sigma''\omega_{{\bf k}-{\bf k}'}^\gamma) \\
&\times\left(\frac{\partial}{\partial v_j}\frac{1}{\sigma''\omega_{{\bf k}-{\bf k}'}^\gamma-({\bf k}-{\bf k}')\cdot{\bf v}}\frac{\partial}{\partial v_k}+\frac{\partial}{\partial v_k}\frac{1}{\sigma'\omega_{{\bf k}'}^\beta-{\bf k}'\cdot{\bf v}}\frac{\partial}{\partial v_j}\right)F_a.
\end{aligned}
\tag{3.7}
$$

Recalling that the only permissible three-wave interactions are those that correspond to $(\alpha,\beta,\gamma)=(L,L,S)$, (L,S,L), and (S,L,L), and making use of the fact that (L,S,L) can be re-expressed as (L,L,S), thus leading to an overall factor 2, we may explicitly write (3.7) as

$$
\begin{aligned}
\left.\frac{\partial F_a}{\partial t}\right|_{\rm decay} &= {\rm Re}\,\frac{\pi\omega_{pe}^2}{4}\frac{e}{T_e}\frac{e_a^3}{m_a^3}\frac{\partial}{\partial v_i}\sum_{\sigma,\sigma',\sigma''=\pm 1}\int d{\bf k}\int d{\bf k}'\,\frac{k_i\,k_j'\,({\bf k}-{\bf k}')_k}{k^2\,k'^2\,|{\bf k}-{\bf k}'|^2} \\
&\times\left[\frac{2({\bf k}\cdot{\bf k}')}{(\sigma\omega_{\bf k}^L)(\sigma'\omega_{{\bf k}'}^L)}\,\mathscr{P}\,\frac{1}{\sigma\omega_{\bf k}^L-{\bf k}\cdot{\bf v}}\left(\sigma\omega_{\bf k}^L\,I_{{\bf k}'}^{\sigma'L}\,I_{{\bf k}-{\bf k}'}^{\sigma''S}\right.\right. \\
&-\left.\sigma'\omega_{{\bf k}'}^L\,I_{{\bf k}-{\bf k}'}^{\sigma''S}\,I_{\bf k}^{\sigma L}-\sigma''\mu_{{\bf k}-{\bf k}'}\,\omega_{{\bf k}-{\bf k}'}^L\,I_{{\bf k}'}^{\sigma'L}\,I_{\bf k}^{\sigma L}\right) \\
&\times\delta(\sigma\omega_{\bf k}^L-\sigma'\omega_{{\bf k}'}^L-\sigma''\omega_{{\bf k}-{\bf k}'}^S) \\
&\times\left(\frac{\partial}{\partial v_j}\,\mathscr{P}\,\frac{1}{\sigma\omega_{\bf k}^L-\sigma'\omega_{{\bf k}'}^L-({\bf k}-{\bf k}')\cdot{\bf v}}\frac{\partial}{\partial v_k}\right. \\
&-i\pi\,\frac{\partial}{\partial v_j}\,\delta[\sigma\omega_{\bf k}^L-\sigma'\omega_{{\bf k}'}^L-({\bf k}-{\bf k}')\cdot{\bf v}]\,\frac{\partial}{\partial v_k} \\
&+\left.\frac{\partial}{\partial v_k}\,\mathscr{P}\,\frac{1}{\sigma'\omega_{{\bf k}'}^L-{\bf k}'\cdot{\bf v}}\frac{\partial}{\partial v_j}\right) \\
&-\frac{{\bf k}'\cdot({\bf k}-{\bf k}')}{(\sigma'\omega_{{\bf k}'}^L)\,(\sigma''\omega_{{\bf k}-{\bf k}'}^L)}\,\mathscr{P}\,\frac{1}{\sigma\omega_{\bf k}^S-{\bf k}\cdot{\bf v}}\left(\sigma\mu_{\bf k}\,\omega_{\bf k}^L\,I_{{\bf k}'}^{\sigma'L}\,I_{{\bf k}-{\bf k}'}^{\sigma''L}\right. \\
&-\left.\sigma''\omega_{{\bf k}-{\bf k}'}^L\,I_{{\bf k}'}^{\sigma'L}\,I_{\bf k}^{\sigma S}-\sigma'\omega_{{\bf k}'}^L\,I_{{\bf k}-{\bf k}'}^{\sigma''L}\,I_{\bf k}^{\sigma S}\right)\delta(\sigma\omega_{\bf k}^S-\sigma'\omega_{{\bf k}'}^L-\sigma''\omega_{{\bf k}-{\bf k}'}^L) \\
&\times\left(\frac{\partial}{\partial v_j}\,\mathscr{P}\,\frac{1}{\sigma''\omega_{{\bf k}-{\bf k}'}^L-({\bf k}-{\bf k}')\cdot{\bf v}}\frac{\partial}{\partial v_k}\right. \\
&+\left.\left.\frac{\partial}{\partial v_k}\,\mathscr{P}\,\frac{1}{\sigma'\omega_{{\bf k}'}^L-{\bf k}'\cdot{\bf v}}\frac{\partial}{\partial v_j}\right)\right]F_a,
\end{aligned}
\tag{3.8}
$$

where we have made use of approximate forms for the second-order susceptibilities. We are concerned with a situation where linear wave-particle resonant interaction is ineffective, i.e., $\omega_{\mathbf{k}}^L - \mathbf{k}\cdot\mathbf{v} \neq 0$ and $\omega_{\mathbf{k}}^S - \mathbf{k}\cdot\mathbf{v} \neq 0$, while nonlinear wave-particle resonant interactions are effective. As a result, we have retained only the principal parts of the factors containing the wave-particle resonant denominators in (3.8). If we ignore the imaginary contribution, then other terms associated with the principal part contributions are supposed to be small, and thus negligible. From this, it can be seen that the three-wave resonant interaction has little or no contribution to the particle evolution,

$$\left.\frac{\partial F_a}{\partial t}\right|_{\text{decay}} \approx 0. \tag{3.9}$$

That the three-wave decay processes have no effect on the particles is not difficult to understand on physical grounds. Since no energy or momentum is exchanged between the particles and waves, it is only expected that the decay process should not influence the particle evolution.

3.3 Effects of Nonlinear Induced Scattering on the Particles

We next consider the effects that the induced scattering has on the particle dynamics, which come from the principal parts associated with the second term and the last term on the right-hand side of generalized particle kinetic equation (1.107). We write the contribution from the induced scattering in three separate terms for the sake of convenience:

$$\left.\frac{\partial F_a}{\partial t}\right|_{\text{ind.sc.}} = \left.\frac{\partial F_a}{\partial t}\right|_{\text{1st}} + \left.\frac{\partial F_a}{\partial t}\right|_{\text{2nd}} + \left.\frac{\partial F_a}{\partial t}\right|_{\text{3rd}}, \tag{3.10}$$

where

$$\begin{aligned}
\left.\frac{\partial F_a}{\partial t}\right|_{\text{1st}} &= \text{Re}\,\frac{2e_a^3}{m_a^3}\frac{\partial}{\partial v_i}\int d\mathbf{k}\int d\mathbf{k}'\,\frac{k_i\,k_j'\,(\mathbf{k}-\mathbf{k}')_k}{k\,k'\,|\mathbf{k}-\mathbf{k}'|} \\
&\times \sum_{\sigma,\sigma'=\pm1}\sum_{\alpha,\beta}\mathscr{P}\,\frac{\chi^{(2)}(\mathbf{k}',\sigma'\omega_{\mathbf{k}'}^{\beta}|\mathbf{k}-\mathbf{k}',\sigma\omega_{\mathbf{k}}^{\alpha}-\sigma'\omega_{\mathbf{k}'}^{\beta})}{(\sigma\omega_{\mathbf{k}}^{\alpha}-\mathbf{k}\cdot\mathbf{v})\,\epsilon(\mathbf{k}-\mathbf{k}',\sigma\omega_{\mathbf{k}}^{\alpha}-\sigma'\omega_{\mathbf{k}'}^{\beta})} \\
&\times I_{\mathbf{k}'}^{\sigma'\beta}\,I_{\mathbf{k}}^{\sigma\alpha}\left(\frac{\partial}{\partial v_j}\frac{1}{\sigma\omega_{\mathbf{k}}^{\alpha}-\sigma'\omega_{\mathbf{k}'}^{\beta}-(\mathbf{k}-\mathbf{k}')\cdot\mathbf{v}}\frac{\partial}{\partial v_k}\right. \\
&\left. + \frac{\partial}{\partial v_k}\frac{1}{\sigma'\omega_{\mathbf{k}'}^{\beta}-\mathbf{k}'\cdot\mathbf{v}}\frac{\partial}{\partial v_j}\right)F_a,
\end{aligned} \tag{3.11}$$

$$\left.\frac{\partial F_a}{\partial t}\right|_{\rm 2nd} = -\text{Re}\,\frac{e_a^3}{m_a^3}\frac{\partial}{\partial v_i}\int d\mathbf{k}\int d\mathbf{k}'\,\frac{k_i\,k_j'\,(\mathbf{k}-\mathbf{k}')_k}{k\,k'\,|\mathbf{k}-\mathbf{k}'|}$$
$$\times\sum_{\sigma',\sigma''=\pm1}\sum_{\beta,\gamma}\mathscr{P}\,\frac{\chi^{(2)*}(\mathbf{k}',\sigma'\omega_{\mathbf{k}'}^\beta|\mathbf{k}-\mathbf{k}',\sigma''\omega_{\mathbf{k}-\mathbf{k}'}^\gamma)}{(\sigma'\omega_{\mathbf{k}'}^\beta-\sigma''\omega_{\mathbf{k}-\mathbf{k}'}^\gamma-\mathbf{k}\cdot\mathbf{v})\,\epsilon^*(\mathbf{k},\sigma'\omega_{\mathbf{k}'}^\beta-\sigma''\omega_{\mathbf{k}-\mathbf{k}'}^\gamma)}$$
$$\times I_{\mathbf{k}'}^{\sigma'\beta}\,I_{\mathbf{k}-\mathbf{k}'}^{\sigma''\gamma}\left(\frac{\partial}{\partial v_j}\frac{1}{\sigma''\omega_{\mathbf{k}-\mathbf{k}'}^\gamma-(\mathbf{k}-\mathbf{k}')\cdot\mathbf{v}}\frac{\partial}{\partial v_k}\right.$$
$$\left.+\frac{\partial}{\partial v_k}\frac{1}{\sigma'\omega_{\mathbf{k}'}^\beta-\mathbf{k}'\cdot\mathbf{v}}\frac{\partial}{\partial v_j}\right)F_a, \tag{3.12}$$

and

$$\left.\frac{\partial F_a}{\partial t}\right|_{\rm 3rd} = \text{Im}\,\frac{e_a^4}{m_a^4}\frac{\partial}{\partial v_i}\sum_{\sigma,\sigma'=\pm1}\sum_{\alpha,\beta}\int d\mathbf{k}\int d\mathbf{k}'\,\frac{k_i\,k_j'\,k_k'\,k_l}{k^2k'^2}\,\frac{I_{\mathbf{k}'}^{\sigma'\beta}\,I_{\mathbf{k}}^{\sigma\alpha}}{\sigma\omega_{\mathbf{k}}^\alpha-\mathbf{k}\cdot\mathbf{v}}\frac{\partial}{\partial v_j}$$
$$\times\frac{1}{\sigma\omega_{\mathbf{k}}^\alpha-\sigma'\omega_{\mathbf{k}'}^\beta-(\mathbf{k}-\mathbf{k}')\cdot\mathbf{v}}$$
$$\times\left(\frac{\partial}{\partial v_k}\frac{1}{\sigma\omega_{\mathbf{k}}^\alpha-\mathbf{k}\cdot\mathbf{v}}\frac{\partial}{\partial v_l}-\frac{\partial}{\partial v_l}\frac{1}{\sigma'\omega_{\mathbf{k}'}^\beta-\mathbf{k}'\cdot\mathbf{v}}\frac{\partial}{\partial v_k}\right)F_a. \tag{3.13}$$

We evaluate each term. First, $(\partial F_a/\partial t)_{\rm 1st}$. Taking only the linear eigenmodes, L and S, into consideration, we note that the second-order susceptibility has an overall i factor. Consequently, the real-part contributions come from the residue terms associated with various wave-particle resonance denominators. However, when the range of wave-particle interaction overlaps that of linear wave-particle response, then we ignore such contributions. Then it becomes evident that the only surviving terms are those dictated by nonlinear wave-particle interactions involving either two Langmuir waves, $\sigma\omega_{\mathbf{k}}^L-\sigma'\omega_{\mathbf{k}'}^L-(\mathbf{k}-\mathbf{k}')\cdot\mathbf{v}=0$, or two ion-sound waves, $\sigma\omega_{\mathbf{k}}^S-\sigma'\omega_{\mathbf{k}'}^S-(\mathbf{k}-\mathbf{k}')\cdot\mathbf{v}=0$. Noting that the range of velocity space for which linear wave-particle resonance is effective is characterized by the conditions $\omega_{\mathbf{k}}^L-\mathbf{k}\cdot\mathbf{v}\approx 0$ and/or $\omega_{\mathbf{k}}^S-\mathbf{k}\cdot\mathbf{v}\approx 0$, we associate the resonant denominators such as $1/(\sigma\omega_{\mathbf{k}}^L-\mathbf{k}\cdot\mathbf{v})$ and $1/(\sigma\omega_{\mathbf{k}}^S-\mathbf{k}\cdot\mathbf{v})$ with the principal part integrations. Of the various nonlinear wave-particle resonance denominators, the interactions characterized by $1/[\sigma\omega_{\mathbf{k}}^L-\sigma'\omega_{\mathbf{k}'}^S-(\mathbf{k}-\mathbf{k}')\cdot\mathbf{v}]$ or $1/[\sigma\omega_{\mathbf{k}}^S-\sigma'\omega_{\mathbf{k}'}^L-(\mathbf{k}-\mathbf{k}')\cdot\mathbf{v}]$ effectively reduce to the linear wave-particle resonance, $1/[\sigma\omega_{\mathbf{k}}^L-(\mathbf{k}-\mathbf{k}')\cdot\mathbf{v}]$ or $1/[-\sigma'\omega_{\mathbf{k}'}^L-(\mathbf{k}-\mathbf{k}')\cdot\mathbf{v}]$, since $\omega_{\mathbf{k}}^L\gg\omega_{\mathbf{k}}^S$. Consequently, these are not interesting because the resonant velocity regions dictated by these conditions overlap with that dictated by the linear wave-particle resonance conditions.

Making use of approximate properties of various linear and nonlinear response functions, it is possible to simplify $(\partial F_a/\partial t)_{1st}$ as follows:

$$\begin{aligned}\left.\frac{\partial F_a}{\partial t}\right|_{1st} &= -\frac{\pi e\, e_a^3}{m_e\, m_a^3}\frac{\partial}{\partial v_i}\int d\mathbf{k}\int d\mathbf{k}'\,\frac{\mathbf{k}\cdot\mathbf{k}'}{k^2\,k'^2}\,k_i\,k'_j\,(\mathbf{k}-\mathbf{k}')_k\\ &\times\sum_{\sigma,\sigma'=\pm1}\left[\frac{I_{\mathbf{k}'}^{\sigma' L}\,I_{\mathbf{k}}^{\sigma L}}{\sigma\omega_{\mathbf{k}}^L\sigma'\omega_{\mathbf{k}'}^L}\mathscr{P}\,\frac{1}{\sigma\omega_{\mathbf{k}}^L-\mathbf{k}\cdot\mathbf{v}}\frac{\partial}{\partial v_j}\delta[\sigma\omega_{\mathbf{k}}^L-\sigma'\omega_{\mathbf{k}'}^L-(\mathbf{k}-\mathbf{k}')\cdot\mathbf{v}]\right.\\ &-\frac{2m_e}{T_e}\frac{I_{\mathbf{k}'}^{\sigma' S}\,I_{\mathbf{k}}^{\sigma S}}{k\,k'}\,\mathscr{P}\,\frac{1}{\sigma\omega_{\mathbf{k}}^S-\mathbf{k}\cdot\mathbf{v}}\\ &\times\left(\frac{k\,k'-\sigma\sigma'(k^2+k'^2-\mathbf{k}\cdot\mathbf{k}')}{|\mathbf{k}-\mathbf{k}'|^2\,(k-\sigma\sigma' k')^2\,\lambda_{De}^2}\,\frac{\mathrm{Re}\,\epsilon(\mathbf{k}-\mathbf{k}',\sigma\omega_{\mathbf{k}}^S-\sigma'\omega_{\mathbf{k}'}^S)}{|\epsilon(\mathbf{k}-\mathbf{k}',\sigma\omega_{\mathbf{k}}^S-\sigma'\omega_{\mathbf{k}'}^S)|^2}\right)\\ &\times\left.\frac{\partial}{\partial v_j}\,\delta[\sigma\omega_{\mathbf{k}}^S-\sigma'\omega_{\mathbf{k}'}^S-(\mathbf{k}-\mathbf{k}')\cdot\mathbf{v}]\right]\frac{\partial F_a}{\partial v_k}.\end{aligned} \tag{3.14}$$

For electrons we may ignore the contribution from ion sound turbulence:

$$\begin{aligned}\left.\frac{\partial F_e}{\partial t}\right|_{1st} &= \frac{\pi e^4}{m_e^4}\frac{\partial}{\partial v_i}\int d\mathbf{k}\int d\mathbf{k}'\,\frac{\mathbf{k}\cdot\mathbf{k}'}{k^2\,k'^2}\,k_i\,k'_j\,(\mathbf{k}-\mathbf{k}')_k\\ &\times\sum_{\sigma,\sigma'=\pm1}\frac{I_{\mathbf{k}'}^{\sigma' L}\,I_{\mathbf{k}}^{\sigma L}}{\sigma\omega_{\mathbf{k}}^L\sigma'\omega_{\mathbf{k}'}^L}\,\mathscr{P}\,\frac{1}{\sigma\omega_{\mathbf{k}}^L-\mathbf{k}\cdot\mathbf{v}}\frac{\partial}{\partial v_j}\\ &\times\,\delta[\sigma\omega_{\mathbf{k}}^L-\sigma'\omega_{\mathbf{k}'}^L-(\mathbf{k}-\mathbf{k}')\cdot\mathbf{v}]\,\frac{\partial F_e}{\partial v_k}.\end{aligned} \tag{3.15}$$

Interchanging integral variables $\mathbf{k}$ and $\mathbf{k}'$, adding the result to the right-hand side, dividing the result by two, and making use of the delta function condition we may rewrite (3.15) as

$$\begin{aligned}\left.\frac{\partial F_e}{\partial t}\right|_{1st} &= -\frac{\pi e^4}{2m_e^4\,\omega_{pe}^2}\frac{\partial}{\partial v_i}\int d\mathbf{k}\int d\mathbf{k}'\,\frac{(\mathbf{k}\cdot\mathbf{k}')^2}{k^2\,k'^2}\sum_{\sigma,\sigma'=\pm1}(\sigma\sigma')\,I_{\mathbf{k}}^{\sigma L}\,I_{\mathbf{k}'}^{\sigma' L}\\ &\times\mathscr{P}\,\frac{1}{(\sigma\omega_{\mathbf{k}}^L-\mathbf{k}\cdot\mathbf{v})^2}\,(\mathbf{k}-\mathbf{k}')_i\,(\mathbf{k}-\mathbf{k}')_j\\ &\times\delta[\sigma\omega_{\mathbf{k}}^L-\sigma'\omega_{\mathbf{k}'}^L-(\mathbf{k}-\mathbf{k}')\cdot\mathbf{v}]\,\frac{\partial F_e}{\partial v_j}.\end{aligned} \tag{3.16}$$

For ions, the contribution from Langmuir modes can be ignored:

$$\begin{aligned}\left.\frac{\partial F_i}{\partial t}\right|_{1st} &= -\frac{2\pi\, e^4}{c_S^2\, m_i^4}\frac{\partial}{\partial v_i}\int d\mathbf{k}\int d\mathbf{k}'\,\frac{\mathbf{k}\cdot\mathbf{k}'}{k^3\,k'^3}\,k_i\,k'_j\,(\mathbf{k}-\mathbf{k}')_k\sum_{\sigma,\sigma'=\pm1}\\ &\times(\sigma\sigma')\,I_{\mathbf{k}'}^{\sigma' S}\,I_{\mathbf{k}}^{\sigma S}\,\mathscr{P}\,\frac{1}{\sigma\omega_{\mathbf{k}}^S-\mathbf{k}\cdot\mathbf{v}}\,A_{\mathbf{k},\mathbf{k}'}^{\sigma,\sigma'}\,\frac{\partial}{\partial v_j}\end{aligned}$$

$$\times\,\delta[\sigma\omega_{\mathbf{k}}^S-\sigma'\omega_{\mathbf{k}'}^S-(\mathbf{k}-\mathbf{k}')\cdot\mathbf{v}]\,\frac{\partial F_i}{\partial v_k},\tag{3.17}$$

$$A_{\mathbf{k},\mathbf{k}'}^{\sigma,\sigma'}=\frac{k^2+k'^2-\mathbf{k}\cdot\mathbf{k}'-\sigma\sigma'kk'}{|\mathbf{k}-\mathbf{k}'|^2\,(k-\sigma\sigma'k')^2\,\lambda_{De}^2}\,\frac{\mathrm{Re}\,\epsilon(\mathbf{k}-\mathbf{k}',\sigma\omega_{\mathbf{k}}^S-\sigma'\omega_{\mathbf{k}'}^S)}{|\epsilon(\mathbf{k}-\mathbf{k}',\sigma\omega_{\mathbf{k}}^S-\sigma'\omega_{\mathbf{k}'}^S)|^2}.$$

Noting that $A_{\mathbf{k}',\mathbf{k}}^{\sigma,\sigma'}=A_{\mathbf{k},\mathbf{k}'}^{\sigma',\sigma}$, and following the similar steps as in the derivation for (3.16), we obtain the ion equation:

$$\begin{aligned}\left.\frac{\partial F_i}{\partial t}\right|_{1\mathrm{st}}&=\frac{\pi}{c_S^2}\frac{e^4}{m_i^4}\frac{\partial}{\partial v_i}\int d\mathbf{k}\int d\mathbf{k}'\,\frac{(\mathbf{k}\cdot\mathbf{k}')^2}{k^3\,k'^3}\sum_{\sigma,\sigma'=\pm1}(\sigma\sigma')\,I_{\mathbf{k}'}^{\sigma'S}\,I_{\mathbf{k}}^{\sigma S}\\&\quad\times\mathscr{P}\,\frac{1}{(\sigma\omega_{\mathbf{k}}^S-\mathbf{k}\cdot\mathbf{v})^2}\,(\mathbf{k}-\mathbf{k}')_i(\mathbf{k}-\mathbf{k}')_j\,A_{\mathbf{k},\mathbf{k}'}^{\sigma,\sigma'}\\&\quad\times\,\delta[\sigma\omega_{\mathbf{k}}^S-\sigma'\omega_{\mathbf{k}'}^S-(\mathbf{k}-\mathbf{k}')\cdot\mathbf{v}]\,\frac{\partial F_i}{\partial v_j}.\end{aligned}\tag{3.18}$$

We next evaluate the second term $(\partial F_a/\partial t)_{2\mathrm{nd}}$ mentioned in (3.12). By noting that the second-order susceptibility has an overall factor i, we can discard those terms that contain linear wave-particle resonance conditions, or those that effectively reduce to linear wave-particle resonance. For $(\partial F_a/\partial t)_{2\mathrm{nd}}$, a careful examination reveals that the nonlinear wave-particle interaction of the form $\omega\pm\omega'-(\mathbf{k}-\mathbf{k}')\cdot\mathbf{v}=0$ does not exist. Only interactions of the type $\omega\pm\omega'-\mathbf{k}\cdot\mathbf{v}=0$ or $\omega-(\mathbf{k}-\mathbf{k}')\cdot\mathbf{v}=0$ exist, which practically overlap with linear wave-particle resonance conditions. Consequently, we conclude that

$$\left.\frac{\partial F_a}{\partial t}\right|_{2\mathrm{nd}}=0,\tag{3.19}$$

to the lowest order in the description.

Finally, consider the third term, $(\partial F_a/\partial t)_{3\mathrm{rd}}$, mentioned in (3.13). First, the electrons. Retaining only the nonlinear wave-particle interaction terms, and ignoring ion-sound waves, we have

$$\begin{aligned}\left.\frac{\partial F_e}{\partial t}\right|_{3\mathrm{rd}}&=-\frac{\pi e^4}{m_e^4}\int d\mathbf{k}\int d\mathbf{k}'\,\frac{1}{k^2k'^2}\sum_{\sigma,\sigma'=\pm1}\mathbf{k}\cdot\frac{\partial}{\partial\mathbf{v}}\left[\frac{I_{\mathbf{k}'}^{\sigma'L}\,I_{\mathbf{k}}^{\sigma L}}{\sigma\omega_{\mathbf{k}}^L-\mathbf{k}\cdot\mathbf{v}}\mathbf{k}'\cdot\frac{\partial}{\partial\mathbf{v}}\right.\\&\quad\times\,\delta[\sigma\omega_{\mathbf{k}}^L-\sigma'\omega_{\mathbf{k}'}^L-(\mathbf{k}-\mathbf{k}')\cdot\mathbf{v}]\\&\quad\times\left.\left(\mathbf{k}'\cdot\frac{\partial}{\partial\mathbf{v}}\frac{1}{\sigma\omega_{\mathbf{k}}^L-\mathbf{k}\cdot\mathbf{v}}\mathbf{k}\cdot\frac{\partial}{\partial\mathbf{v}}-\mathbf{k}\cdot\frac{\partial}{\partial\mathbf{v}}\frac{1}{\sigma'\omega_{\mathbf{k}'}^L-\mathbf{k}'\cdot\mathbf{v}}\mathbf{k}'\cdot\frac{\partial}{\partial\mathbf{v}}\right)\right]F_e.\end{aligned}\tag{3.20}$$

Making use of the resonance conditions we may manipulate (3.20) and obtain

$$\left.\frac{\partial F_e}{\partial t}\right|_{3\mathrm{rd}} = \frac{\pi e^4}{2m_e^4}\int d\mathbf{k}\int d\mathbf{k}'\,\frac{(\mathbf{k}\cdot\mathbf{k}')^2}{k^2k'^2}\sum_{\sigma,\sigma'=\pm1} I_{\mathbf{k}}^{\sigma L}\,I_{\mathbf{k}'}^{\sigma' L}$$
$$\times\frac{1}{(\sigma\omega_{\mathbf{k}}^L-\mathbf{k}\cdot\mathbf{v})^2}\,(\mathbf{k}-\mathbf{k}')\cdot\frac{\partial}{\partial\mathbf{v}}\frac{1}{(\sigma\omega_{\mathbf{k}}^L-\mathbf{k}\cdot\mathbf{v})^2}$$
$$\times\,\delta[\sigma\omega_{\mathbf{k}}^L-\sigma'\omega_{\mathbf{k}'}^L-(\mathbf{k}-\mathbf{k}')\cdot\mathbf{v}]\,(\mathbf{k}-\mathbf{k}')\cdot\frac{\partial F_e}{\partial\mathbf{v}}. \tag{3.21}$$

Noting that

$$\frac{1}{(\sigma\omega_{\mathbf{k}}^L-\mathbf{k}\cdot\mathbf{v})^2}\,(\mathbf{k}-\mathbf{k}')\cdot\frac{\partial G}{\partial\mathbf{v}} = (\mathbf{k}-\mathbf{k}')\cdot\frac{\partial}{\partial\mathbf{v}}\left(\frac{G}{(\sigma\omega_{\mathbf{k}}^L-\mathbf{k}\cdot\mathbf{v})^2}\right)$$
$$-\frac{k^2-k'^2}{(\sigma\omega_{\mathbf{k}}^L-\mathbf{k}\cdot\mathbf{v})^3}\,G, \tag{3.22}$$

and upon symmetrizing with respect to $\mathbf{k}$ and $\mathbf{k}'$, we have

$$\left.\frac{\partial F_e}{\partial t}\right|_{3\mathrm{rd}} = \frac{\pi e^4}{2m_e^4}\frac{\partial}{\partial v_i}\int d\mathbf{k}\int d\mathbf{k}'\,\frac{(\mathbf{k}\cdot\mathbf{k}')^2}{k^2k'^2}\sum_{\sigma,\sigma'=\pm1} I_{\mathbf{k}}^{\sigma L}\,I_{\mathbf{k}'}^{\sigma' L}$$
$$\times\,\mathscr{P}\frac{1}{(\sigma\omega_{\mathbf{k}}^L-\mathbf{k}\cdot\mathbf{v})^4}\,(\mathbf{k}-\mathbf{k}')_i\,(\mathbf{k}-\mathbf{k}')_j$$
$$\times\,\delta[\sigma\omega_{\mathbf{k}}^L-\sigma'\omega_{\mathbf{k}'}^L-(\mathbf{k}-\mathbf{k}')\cdot\mathbf{v}]\,\frac{\partial F_e}{\partial v_j}. \tag{3.23}$$

Next, consider the ions. For energetic ions, we may ignore linear wave-particle resonances and contributions from Langmuir waves. The structure of the ion equation is the same as that of the electrons, except that Langmuir mode intensities are replaced by ion-sound mode intensities. Thus, without going through the detailed manipulations, the result can be written down readily:

$$\left.\frac{\partial F_i}{\partial t}\right|_{3\mathrm{rd}} = \frac{\pi e^4}{2m_i^4}\frac{\partial}{\partial v_i}\int d\mathbf{k}\int d\mathbf{k}'\,\frac{(\mathbf{k}\cdot\mathbf{k}')^2}{k^2k'^2}\sum_{\sigma,\sigma'=\pm1} I_{\mathbf{k}}^{\sigma S}\,I_{\mathbf{k}'}^{\sigma' S}$$
$$\times\,\mathscr{P}\frac{1}{(\sigma\omega_{\mathbf{k}}^S-\mathbf{k}\cdot\mathbf{v})^4}\,(\mathbf{k}-\mathbf{k}')_i\,(\mathbf{k}-\mathbf{k}')_j$$
$$\times\,\delta[\sigma\omega_{\mathbf{k}}^S-\sigma'\omega_{\mathbf{k}'}^S-(\mathbf{k}-\mathbf{k}')\cdot\mathbf{v}]\,\frac{\partial F_i}{\partial v_j}. \tag{3.24}$$

3.4 Summary of Electrostatic Vlasov Weak Turbulence Theory

In this section, we summarize the basic set of equations relevant for describing weakly turbulent nonlinear processes dictated by electrostatic interactions in unmagnetized plasmas. At this level of description, the plasma is considered as purely collisionless. The effects of discreteness associated with the plasma particles will be discussed in Part II, but the discussion thereof is built upon the findings in Part I. The following are the basic physical quantities and equations that summarize Part I.

- Linear dispersion relations for Langmuir (L) and ion-sound or ion-acoustic (S) modes

$$\omega_{\mathbf{k}}^L = \omega_{pe}\left(1+\frac{3}{2}k^2\lambda_{De}^2\right) = \omega_{pe}\left(1+\frac{3}{4}\frac{k^2v_{Te}^2}{\omega_{pe}^2}\right),$$
$$\omega_{\mathbf{k}}^S = \frac{kc_S\sqrt{1+3T_i/T_e}}{\sqrt{1+k^2\lambda_{De}^2}} = kv_{Te}\sqrt{\frac{m_e}{2m_i}}\frac{\sqrt{1+3T_i/T_e}}{\sqrt{1+k^2v_{Te}^2/2\omega_{pe}^2}}. \tag{3.25}$$

- Fundamental plasma quantities

$$\omega_{pe} = \sqrt{\frac{4\pi ne^2}{m_e}}, \quad \text{plasma frequency;}$$
$$\lambda_{De} = \sqrt{\frac{T_e}{4\pi ne^2}}, \quad \text{Debye length;}$$
$$v_{Te} = \sqrt{\frac{2T_e}{m_e}}, \quad \text{electron thermal speed;}$$
$$v_{Ti} = \sqrt{\frac{2T_i}{m_i}}, \quad \text{ion (proton) thermal speed;}$$
$$c_S = \sqrt{\frac{T_e}{m_i}}, \quad \text{ion sound (ion acoustic) speed.} \tag{3.26}$$

- Wave kinetic equations

$$\frac{\partial I_{\mathbf{k}}^{\sigma L}}{\partial t} = \underbrace{\pi(\sigma\omega_{\mathbf{k}}^L)\frac{\omega_{pe}^2}{k^2}\int d\mathbf{v}\,\delta(\sigma\omega_{\mathbf{k}}^L - \mathbf{k}\cdot\mathbf{v})\,\mathbf{k}\cdot\frac{\partial F_e}{\partial\mathbf{v}}\,I_{\mathbf{k}}^{\sigma L}}_{\text{induced emission (linear growth/Landau damping)}}$$
$$+\,\sigma\omega_{\mathbf{k}}^L\sum_{\sigma',\sigma''=\pm1}\int d\mathbf{k}'\,V_{\mathbf{k},\mathbf{k}'}^{LS}\Bigg[\underbrace{\sigma\omega_{\mathbf{k}}^L\,I_{\mathbf{k}'}^{\sigma' L}\,\frac{I_{\mathbf{k}-\mathbf{k}'}^{\sigma'' S}}{\mu_{\mathbf{k}-\mathbf{k}'}}}_{\text{spontaneous decay}}$$

$$
- \underbrace{\left(\sigma'\omega^L_{\mathbf{k}'}\,\frac{I^{\sigma''S}_{\mathbf{k}-\mathbf{k}'}}{\mu_{\mathbf{k}-\mathbf{k}'}} + \sigma''\omega^L_{\mathbf{k}-\mathbf{k}'}\,I^{\sigma'L}_{\mathbf{k}'}\right) I^{\sigma L}_{\mathbf{k}}}_{\text{induced decay}} \Bigg] \tag{3.27}
$$

$$
+ \underbrace{\frac{m_e}{m_i}\,(\sigma\omega^L_{\mathbf{k}}) \sum_{\sigma'=\pm1} \int d\mathbf{k}' \int d\mathbf{v}\; U^{LL}_{\mathbf{k},\mathbf{k}'}\,(\mathbf{k}-\mathbf{k}')\cdot\frac{\partial F_i}{\partial \mathbf{v}}\,I^{\sigma'L}_{\mathbf{k}'}\,I^{\sigma L}_{\mathbf{k}}}_{\text{induced scattering (of Langmuir wave off thermal ions)}},
$$

$$
\frac{\partial}{\partial t}\frac{I^{\sigma S}_{\mathbf{k}}}{\mu_{\mathbf{k}}} = \underbrace{\pi\,(\sigma\mu_{\mathbf{k}}\omega^L_{\mathbf{k}})\,\frac{\omega^2_{pe}}{k^2}\int d\mathbf{v}\;\delta(\sigma\omega^S_{\mathbf{k}} - \mathbf{k}\cdot\mathbf{v})\left(\mathbf{k}\cdot\frac{\partial}{\partial \mathbf{v}}\right)\left(F_e + \frac{m_e}{m_i}\,F_i\right)\frac{I^{\sigma S}_{\mathbf{k}}}{\mu_{\mathbf{k}}}}_{\text{induced emission for ion-sound mode}}
$$

$$
+ \sum_{\sigma',\sigma''=\pm1} \sigma\omega^L_{\mathbf{k}} \int d\mathbf{k}'\; V^{SL}_{\mathbf{k},\mathbf{k}'} \Bigg[\underbrace{\sigma\omega^L_{\mathbf{k}}\,I^{\sigma'L}_{\mathbf{k}'}\,I^{\sigma''L}_{\mathbf{k}-\mathbf{k}'}}_{\text{spontaneous decay}}
$$

$$
- \underbrace{(\sigma'\omega^L_{\mathbf{k}'}\,I^{\sigma''L}_{\mathbf{k}-\mathbf{k}'} + \sigma''\omega^L_{\mathbf{k}-\mathbf{k}'}\,I^{\sigma'L}_{\mathbf{k}'})\,\frac{I^{\sigma S}_{\mathbf{k}}}{\mu_{\mathbf{k}}}}_{\text{induced decay}} \Bigg] \tag{3.28}
$$

$$
+ \underbrace{(\sigma\omega^L_{\mathbf{k}}) \sum_{\sigma'=\pm1} \int d\mathbf{k}' \int d\mathbf{v}\; U^{SS}_{\mathbf{k},\mathbf{k}'}\,(\mathbf{k}-\mathbf{k}')\cdot\frac{\partial F_i}{\partial \mathbf{v}}\,I^{\sigma'S}_{\mathbf{k}'}\,I^{\sigma S}_{\mathbf{k}}}_{\text{induced scattering (of ion-sound wave)}},
$$

Various processes are indicated by their respective names, and various coupling coefficients are defined by

$$
\mu_{\mathbf{k}} = k^3\lambda^3_{De}\sqrt{\frac{m_e}{m_i}}\sqrt{1+\frac{3T_i}{T_e}} = \frac{k^3 v^3_{Te}}{\omega^3_{pe}}\sqrt{\frac{m_e}{8m_i}}\sqrt{1+\frac{3T_i}{T_e}},
$$

$$
V^{LS}_{\mathbf{k},\mathbf{k}'} = \frac{\pi}{2}\frac{e^2}{T^2_e}\frac{\mu_{\mathbf{k}-\mathbf{k}'}\,(\mathbf{k}\cdot\mathbf{k}')^2}{k^2\,k'^2|\mathbf{k}-\mathbf{k}'|^2}\,\delta(\sigma\omega^L_{\mathbf{k}} - \sigma'\omega^L_{\mathbf{k}'} - \sigma''\omega^S_{\mathbf{k}-\mathbf{k}'}),
$$

$$
V^{SL}_{\mathbf{k},\mathbf{k}'} = \frac{\pi}{4}\frac{e^2}{T^2_e}\frac{\mu_{\mathbf{k}}\,[\mathbf{k}'\cdot(\mathbf{k}-\mathbf{k}')]^2}{k^2\,k'^2\,|\mathbf{k}-\mathbf{k}'|^2}\,\delta(\sigma\omega^S_{\mathbf{k}} - \sigma'\omega^L_{\mathbf{k}'} - \sigma''\omega^L_{\mathbf{k}-\mathbf{k}'}),
$$

$$
U^{LL}_{\mathbf{k},\mathbf{k}'} = \frac{\pi}{\omega^2_{pe}}\frac{e^2}{m^2_e}\frac{(\mathbf{k}\cdot\mathbf{k}')^2}{k^2\,k'^2}\,\delta[\sigma\omega^L_{\mathbf{k}} - \sigma'\omega^L_{\mathbf{k}'} - (\mathbf{k}-\mathbf{k}')\cdot\mathbf{v}],
$$

$$
U^{SS}_{\mathbf{k},\mathbf{k}'} = \frac{\pi}{\omega^2_{pe}}\frac{e^2}{m_e m_i}\frac{(\mathbf{k}\cdot\mathbf{k}')^2}{k^4\,k'^4\,\lambda^4_{De}}\,W_{\mathbf{k},\mathbf{k}'}\;\delta[\sigma\omega^S_{\mathbf{k}} - \sigma'\omega^S_{\mathbf{k}'} - (\mathbf{k}-\mathbf{k}')\cdot\mathbf{v}],
$$

$$W_{\mathbf{k},\mathbf{k}'} = \left(1 + \frac{|\mathbf{k}-\mathbf{k}'|^2}{(\sigma k - \sigma' k')^2}\right)^2 \times \frac{1}{|\mathbf{k}-\mathbf{k}'|^4 \lambda_{De}^4 \, |\epsilon(\mathbf{k}-\mathbf{k}', \sigma\omega_{\mathbf{k}}^S - \sigma'\omega_{\mathbf{k}'}^S)|^2} + \sigma\sigma' \frac{k'}{k},$$

$$|\epsilon(\mathbf{k}-\mathbf{k}', \sigma\omega_{\mathbf{k}}^S - \sigma'\omega_{\mathbf{k}'}^S)|^2 = \left(1 + \frac{2(\mathbf{k}\cdot\mathbf{k}' - \sigma\sigma' k\, k')}{|\mathbf{k}-\mathbf{k}'|^2 (k - \sigma\sigma' k')^2 \lambda_{De}^2}\right)^2 + \frac{\pi}{2}\frac{m_e}{m_i}\frac{(k-\sigma\sigma' k')^2}{|\mathbf{k}-\mathbf{k}'|^6 \lambda_{De}^4}\left[\exp\left(-\frac{m_e}{2m_i}\frac{(k-\sigma\sigma' k')^2}{|\mathbf{k}-\mathbf{k}'|^2}\right) + \left(\frac{m_i}{m_e}\right)^{1/2}\left(\frac{T_e}{T_i}\right)^{3/2}\exp\left(-\frac{T_e}{2T_i}\frac{(k-\sigma\sigma' k')^2}{|\mathbf{k}-\mathbf{k}'|^2}\right)\right]^2. \tag{3.29}$$

- Electron particle kinetic equation

$$\frac{\partial F_e}{\partial t} = \frac{\partial}{\partial v_i}\left(D_{ij}^{\text{el}} \frac{\partial F_e}{\partial v_j}\right),$$

$$D_{ij}^{\text{el}} = \frac{\pi e^2}{m_e^2}\int d\mathbf{k}\, \frac{k_i k_j}{k^2} \sum_{\sigma=\pm1} \delta(\sigma\omega_{\mathbf{k}}^L - \mathbf{k}\cdot\mathbf{v})\, I_{\mathbf{k}}^{\sigma L} + \frac{\pi e^4}{2m_e^4}\int d\mathbf{k}\int d\mathbf{k}'\, \frac{(\mathbf{k}\cdot\mathbf{k}')^2}{k^2 k'^2} \sum_{\sigma,\sigma'=\pm1} \frac{I_{\mathbf{k}'}^{\sigma' L} I_{\mathbf{k}}^{\sigma L}}{(\sigma\omega_{\mathbf{k}}^L - \mathbf{k}\cdot\mathbf{v})^2} (\mathbf{k}-\mathbf{k}')_i (\mathbf{k}-\mathbf{k}')_j \times \left(\frac{1}{(\sigma\omega_{\mathbf{k}}^L - \mathbf{k}\cdot\mathbf{v})^2} - \frac{\sigma\sigma'}{\omega_{pe}^2}\right) \delta[\sigma\omega_{\mathbf{k}}^L - \sigma'\omega_{\mathbf{k}'}^L - (\mathbf{k}-\mathbf{k}')\cdot\mathbf{v}]. \tag{3.30}$$

- Ion particle kinetic equation

$$\frac{\partial F_i}{\partial t} = \frac{\partial}{\partial v_i}\left(D_{ij}^{\text{ion}} \frac{\partial F_i}{\partial v_j}\right),$$

$$D_{ij}^{\text{ion}} = \frac{\pi e^2}{m_i^2}\int d\mathbf{k}\, \frac{k_i k_j}{k^2} \sum_{\sigma=\pm1} \delta(\sigma\omega_{\mathbf{k}}^S - \mathbf{k}\cdot\mathbf{v})\, I_{\mathbf{k}}^{\sigma S} + \frac{\pi e^4}{2m_i^4}\int d\mathbf{k}\int d\mathbf{k}'\, \frac{(\mathbf{k}\cdot\mathbf{k}')^2}{k^2 k'^2} \sum_{\sigma,\sigma'=\pm1} \frac{I_{\mathbf{k}'}^{\sigma' S} I_{\mathbf{k}}^{\sigma S}}{(\sigma\omega_{\mathbf{k}}^S - \mathbf{k}\cdot\mathbf{v})^2} \times \left(\frac{1}{(\sigma\omega_{\mathbf{k}}^S - \mathbf{k}\cdot\mathbf{v})^2} + \frac{2(\sigma\sigma')\, A_{\mathbf{k},\mathbf{k}'}^{\sigma,\sigma'}}{k\, k'\, c_S^2}\right) \times (\mathbf{k}-\mathbf{k}')_i (\mathbf{k}-\mathbf{k}')_j\, \delta[\sigma\omega_{\mathbf{k}}^S - \sigma'\omega_{\mathbf{k}'}^S - (\mathbf{k}-\mathbf{k}')\cdot\mathbf{v}], \tag{3.31}$$

$$A_{\mathbf{k},\mathbf{k}'}^{\sigma,\sigma'} = \frac{k^2 + k'^2 - \mathbf{k}\cdot\mathbf{k}' - \sigma\sigma' k\, k'}{|\mathbf{k}-\mathbf{k}'|^2 (k - \sigma\sigma' k')^2 \lambda_{De}^2} \frac{1}{\epsilon(\mathbf{k}-\mathbf{k}', \sigma\omega_{\mathbf{k}}^S - \sigma'\omega_{\mathbf{k}'}^S)}.$$

In (3.30) and (3.31), we have presented formal electron and ion particle kinetic equations in which the weak turbulence ordering is carried out up to the third order in perturbative expansion (Tsytovich, 1977b; Yoon et al., 2003b). This was done for the sake of being consistent with the same procedure in the wave kinetic equation. However, after the formal theory is being developed and carried out, it becomes clear that the nonlinear correction simply modifies the velocity space diffusion coefficient in the velocity range for which linear wave-particle interaction is not forthcoming. The nonlinear diffusion coefficient is, however, quite insignificant, since the overall magnitude is second order in wave intensity. For a wide range of applications, the simple quasilinear diffusion equation is sufficient. Nevertheless, the present chapter formally discusses the wave and particle kinetic equations with consistent weak turbulence ordering scheme fully implemented, which is done for the sake of consistency.

At the closing of this chapter, we note that the final set of wave kinetic equations contain various terms with designated names indicating different wave-particle and wave-wave interaction processes. Some processes are designated with the term "spontaneous" while others are denoted by "induced." These terms originate from quantum mechanical concepts, and we have also discussed the meaning of these processes on the basis of intuitive concepts. However, the important point is that under the present Vlasov weak turbulence theory, not all processes appear in balanced form where every induced process has a spontaneous counterpart. In order to provide a complete description where spontaneous and induced processes appear in balanced forms, one must turn to the Klimontovich formalism, which is the subject of Part II of this book.

Before we move on to Part II, we discuss a number of caveats inherent in the present weak turbulence formalism. One is that throughout the main part of this monograph we work under the inherent assumption of weak wave growth or damping, which is exemplified by the treatment of resonant denominator, $1/(\omega - \mathbf{k} \cdot \mathbf{v} + i0)$. Such a treatment is applicable when the plasma is unstable to weakly growing instability, such as the bump-on-tail instability, or when one is dealing with a weakly damped mode. However, when the free energy source of the instability is sufficiently high, then it is no longer valid to treat the problem by assuming that ω is essentially real and that the velocity space resonant denominator can be treated by adding an infinitesimally small but positive imaginary part to real ω. For a more general situation, the wave dispersion relation $\omega = \omega_{\mathbf{k}}$ may no longer be determined by considering only the real part of linear dielectric function, $\mathrm{Re}\,\epsilon(\mathbf{k}, \omega_{\mathbf{k}}) = 0$, but instead, for such cases, one must treat ω as complex, and the dispersion relation $\omega = \omega_{\mathbf{k}} + i\gamma_{\mathbf{k}}$ must be solved from the complete plasma dielectric function, $\mathrm{Re}\,\epsilon(\mathbf{k}, \omega_{\mathbf{k}} + i\gamma_{\mathbf{k}}) + i\mathrm{Im}\,\epsilon(\mathbf{k}, \omega_{\mathbf{k}} + i\gamma_{\mathbf{k}}) = 0$. The plasma instability that necessitates this kind of treatment is known as the "reactive instability." The linear theory of waves and instabilities of this type of problem is well established – see, e.g., excellent monographs on linear plasma instability by Mikhailovskii (1974), Cap (1976), Stix (1992), Swanson (1989), Gary (1993), Melrose (1986), or Baumjohann and Treumann (1997) and even quasilinear description of such types of situations are also well developed, e.g., Sagdeev and Galeev (1969) and Davidson (1972). However, the question is whether one may also formulate the perturbative nonlinear theory, i.e., whether the weak turbulence theory may be extended to the reactively unstable problem. This question is not completely addressed in the literature, but Appendix E describes a possible direction – see also the works by Ishihara et al. (1980, 1981) and Yoon (2010).

Another limitation of the fundamental formalism adopted in this book is the assumption that the iterative solution of nonlinear equation is valid and that the truncation of series expansion can effectively be done at the third order. However, it is natural to ask whether the infinite series $\delta f_a = \delta f_a^{(1)} + \delta f_a^{(2)} + \delta f_a^{(3)} + \cdots$ converges or not, and, if so, under what conditions? Or whether one may perform at least a partial summation of infinite series? We have shown that the "necessary" condition for the series convergence is that the wave energy be sufficiently lower than the particle energy, $\mathscr{E}_{\rm wave} \ll \mathscr{E}_{\rm particle}$ – see (1.47). However, this is not the "sufficient" condition, and, under some conditions, the series may appear to diverge. Nevertheless, the *prima facie* divergence may be avoided if one can sum up the most divergent terms in the infinite series. Appendix F overviews this problem, where the iterative solution up to fifth order is explicitly computed. There, it is shown on the basis of the pattern that emerges that the partial summation of leading order terms points to the need for the so-called renormalized resonance denominator. The partial summation idea of the type discussed in Appendix F leads to what is known as the "renormalized" kinetic theory. There are a number of different versions of renormalized kinetic theory, starting from the works by Dupree (1966), Weinstock (1969), etc. – see also works by Kono and Ichikawa (1973); Horton and Choi (1979); Krommes (2002); Itoh and Itoh (2009); Diamond et al. (2010). The main attention of this book is on the standard weak turbulence theory, so that such methods are not systematically reviewed, but Appendix G surveys one such approach. The basic outline of the method overviewed in Appendix G is that of Rudakov and Tsytovich (1971).

Part II

Klimontovich Weak Turbulence Theory: Electrostatic Approximation

4

Electrostatic Klimontovich Weak Turbulence Theory

In Part II of this book we discuss the effects of particle discreteness. In order to include such effects we resort to the statistical mechanical framework pioneered by Klimontovich (1967, 1982). Vlasov equation and Klimontovich equation are mathematically identical, but in the Vlasov approach, the particle discreteness is smoothed out, whereas the Klimontovich formalism retains the individual nature of the plasma particles. As such, the two approaches differ essentially on how the initial state is described. While in the Vlasov approach the initial state is prescribed by a smooth distribution function, in the Klimontovich approach, the initial state is a summation of infinitely sharp delta functions.

4.1 Plasma Equations Based upon Klimontovich Formalism

Let us begin the discussion by considering a collection of N ions (protons) and N electrons, labeled respectively by $a = i$ and e. If the charged particles interact according to the classical nonrelativistic Newtonian dynamics, then their motions, or particle orbits $\mathbf{r}_i^a(t)$ and $\mathbf{v}_i^a(t)$, obey

$$\begin{aligned} \frac{d\mathbf{r}_i^a(t)}{dt} &= \mathbf{v}_i^a(t), \\ m_a \frac{d\mathbf{v}_i^a(t)}{dt} &= e_a \, \mathbf{E}[\mathbf{r}_i^a(t), t] + \frac{e_a}{c} \, \mathbf{v}_i^a(t) \times \mathbf{B}[\mathbf{r}_i^a(t), t], \end{aligned} \tag{4.1}$$

where $i = 1, \ldots, N$. The self-consistent fields $\mathbf{E}(\mathbf{r}, t)$ and $\mathbf{B}(\mathbf{r}, t)$ must be determined from Maxwell's equation:

$$\begin{aligned} \frac{\partial}{\partial \mathbf{r}} \times \mathbf{E}(\mathbf{r}, t) + \frac{1}{c} \frac{\partial}{\partial t} \mathbf{B}(\mathbf{r}, t) &= 0, \\ \frac{\partial}{\partial \mathbf{r}} \cdot \mathbf{B}(\mathbf{r}, t) &= 0, \end{aligned}$$

$$
\begin{aligned}
\frac{\partial}{\partial \mathbf{r}} \cdot \mathbf{E}(\mathbf{r},t) &= 4\pi\,\rho = 4\pi \sum_a e_a \sum_{i=1}^{N} \delta[\mathbf{r} - \mathbf{r}_i^a(t)], \\
\frac{\partial}{\partial \mathbf{r}} \times \mathbf{B}(\mathbf{r},t) - \frac{1}{c}\frac{\partial}{\partial t}\mathbf{E}(\mathbf{r},t) &= \frac{4\pi}{c}\mathbf{j} \\
&= \frac{4\pi}{c} \sum_a e_a \sum_{i=1}^{N} \mathbf{v}_i^a(t)\, \delta[\mathbf{r} - \mathbf{r}_i^a(t)]. \qquad (4.2)
\end{aligned}
$$

The net charge and current densities in (4.2) are computed by collecting contributions from all charged particles. We may rewrite ρ and $\mathbf{j}$ by

$$
\begin{aligned}
\rho &= \sum_a \rho_a, \qquad \mathbf{j} = \sum_a \mathbf{j}_a, \\
\rho_a &= \sum_{i=1}^{N} \delta[\mathbf{r} - \mathbf{r}_i^a(t)] = \int d\mathbf{v} \sum_{i=1}^{N} \delta[\mathbf{r} - \mathbf{r}_i^a(t)]\, \delta[\mathbf{v} - \mathbf{v}_i^a(t)], \qquad (4.3) \\
\mathbf{j}_a &= \sum_{i=1}^{N} \mathbf{v}_i^a(t)\, \delta[\mathbf{r} - \mathbf{r}_i^a(t)] = \int d\mathbf{v}\, \mathbf{v} \sum_{i=1}^{N} \delta[\mathbf{r} - \mathbf{r}_i^a(t)]\, \delta[\mathbf{v} - \mathbf{v}_i^a(t)].
\end{aligned}
$$

From this, the notion of N body phase space density, namely, the "Klimontovich function," emerges:

$$
N_a(\mathbf{r},\mathbf{v},t) = \sum_{i=1}^{N} \delta[\mathbf{r} - \mathbf{r}_i^a(t)]\, \delta[\mathbf{v} - \mathbf{v}_i^a(t)]. \qquad (4.4)
$$

Physically, the function $N_a(\mathbf{r},\mathbf{v},t)$ represents the density in six-dimensional phase space $(\mathbf{r},\mathbf{v})$ at time t. This is because when integrated over $\mathbf{v}$ and $\mathbf{r}$, and averaged over the volume, we obtain the total number of particles. It also represents the phase space probability density of finding a collection of N particles. As is obvious from the definition $N_a(\mathbf{r},\mathbf{v},t)$ is composed of infinitely sharp delta functions.

In terms of $N_a(\mathbf{r},\mathbf{v},t)$, Maxwell's equation is now written as

$$
\begin{aligned}
\frac{\partial}{\partial \mathbf{r}} \times \mathbf{E}(\mathbf{r},t) + \frac{1}{c}\frac{\partial}{\partial t}\mathbf{B}(\mathbf{r},t) &= 0, \\
\frac{\partial}{\partial \mathbf{r}} \cdot \mathbf{B}(\mathbf{r},t) &= 0, \\
\frac{\partial}{\partial \mathbf{r}} \cdot \mathbf{E}(\mathbf{r},t) &= 4\pi \sum_a e_a \int d\mathbf{v}\, N_a(\mathbf{r},\mathbf{v},t), \qquad (4.5) \\
\frac{\partial}{\partial \mathbf{r}} \times \mathbf{B}(\mathbf{r},t) - \frac{1}{c}\frac{\partial}{\partial t}\mathbf{E}(\mathbf{r},t) &= \frac{4\pi}{c} \sum_a e_a \int d\mathbf{v}\, \mathbf{v}\, N_a(\mathbf{r},\mathbf{v},t).
\end{aligned}
$$

The Klimontovich function $N_a(\mathbf{r},\mathbf{v},t)$ is a six-dimensional phase space density. Since particles cannot be created or destroyed, $N_a(\mathbf{r},\mathbf{v},t)$ must be conserved. Consequently, the equation that $N_a(\mathbf{r},\mathbf{v},t)$ obeys is given by

$$\frac{dN}{dt} = 0 = \left(\frac{\partial}{\partial t} + \frac{d\mathbf{r}}{dt}\cdot\frac{\partial}{\partial \mathbf{r}} + \frac{d\mathbf{v}}{dt}\cdot\frac{\partial}{\partial \mathbf{v}}\right) N(\mathbf{r},\mathbf{v},t), \tag{4.6}$$

which can be rewritten as

$$\sum_{i=1}^{N}\left(\frac{\partial}{\partial t} + \frac{d\mathbf{r}_i^a(t)}{dt}\cdot\frac{\partial}{\partial \mathbf{r}} + \frac{d\mathbf{v}_i^a(t)}{dt}\cdot\frac{\partial}{\partial \mathbf{v}}\right)\delta[\mathbf{r}_i^a(t)-\mathbf{r}]\,\delta[\mathbf{v}_i^a(t)-\mathbf{v}] = 0. \tag{4.7}$$

By virtue of the exact microscopic equation of motion (4.1), the dynamical equation (4.6) or (4.7) can be expressed by

$$\left[\frac{\partial}{\partial t} + \mathbf{v}\cdot\frac{\partial}{\partial \mathbf{r}} + \frac{e_a}{m_a}\left(\mathbf{E}(\mathbf{r},t) + \frac{\mathbf{v}}{c}\times\mathbf{B}(\mathbf{r},t)\right)\cdot\frac{\partial}{\partial \mathbf{v}}\right] N_a(\mathbf{r},\mathbf{v},t) = 0. \tag{4.8}$$

Note that this is formally identical to the Vlasov equation, but the difference is that $N_a(\mathbf{r},\mathbf{v},t)$ is not a smooth distribution, but rather it is given by a collection of point-like delta functions (4.4). Both Vlasov and Klimontovich equations advance the phase space distribution, $f_a(\mathbf{r},\mathbf{v},t)$ in the case of Vlasov equation and $N_a(\mathbf{r},\mathbf{v},t)$ for Klimontovich equation, in time, while preserving the total phase space volume. Consequently, both equations represent exact phase space mapping of the initial functions, $f_a(\mathbf{r},\mathbf{v},0)$ and $N_a(\mathbf{r},\mathbf{v},0)$, along their trajectories, $f_a[\mathbf{r}(t),\mathbf{v}(t),t]$ and $N_a[\mathbf{r}(t),\mathbf{v}(t),t]$, but the difference is that for Vlasov system $f_a(\mathbf{r},\mathbf{v},0)$ is a smooth function, while for Klimontovich system, $N_a(\mathbf{r},\mathbf{v},0)$ is made of the summation of infinitely singular delta functions. The singularities associated with the delta functions are what lead to the discreteness effects, as we will see.

The microscopic Klimontovich function (4.4) and the dynamical equation it satisfies, namely, (4.7), contain all the necessary dynamical information to describe the collection of N charged particle, and it is equivalent to solving the coupled Newtonian equations of motion for N particles. However, the Klimontovich function, as with the Newtonian equations of motion, contains too much information. In reality, we cannot deal with all the phase space information, but we can only deal with some gross average quantities. Therefore, we resort to the statistical method. We write the physical quantities into averages and perturbations – see (1.34):

$$\begin{aligned} N_a(\mathbf{r},\mathbf{v},t) &= \langle N_a(\mathbf{r},\mathbf{v},t)\rangle + \delta N_a(\mathbf{r},\mathbf{v},t), \\ \mathbf{E}(\mathbf{r},t) &= \langle \mathbf{E}(\mathbf{r},t)\rangle + \delta\,\mathbf{E}(\mathbf{r},t), \\ \mathbf{B}(\mathbf{r},t) &= \langle \mathbf{B}(\mathbf{r},t)\rangle + \delta\,\mathbf{B}(\mathbf{r},t), \end{aligned} \tag{4.9}$$

where $< \cdots >$ represents the (ensemble) average, and $\delta \cdots$ represents the remainder (i.e., fluctuations). In this book, we are interested in a situation where average quantities are spatially uniform, and where there are no average fields. Thus, we write

$$\langle N_a(\mathbf{r},\mathbf{v},t)\rangle = f_a(\mathbf{v},t), \quad \langle \mathbf{E}(\mathbf{r},t)\rangle = 0, \quad \langle \mathbf{B}(\mathbf{r},t)\rangle = 0. \tag{4.10}$$

Physical interpretation of $\langle N_a(\mathbf{r},\mathbf{v},t)\rangle = f_a(\mathbf{v},t)$ is that the quantity

$$\langle N_a(\mathbf{r},\mathbf{v},t)\rangle = \left\langle \sum_{i=1}^{N} \delta[\mathbf{r}-\mathbf{r}_i^a(t)]\,\delta[\mathbf{v}-\mathbf{v}_i^a(t)]\right\rangle = f_a(\mathbf{v},t) \tag{4.11}$$

represents the average probability distribution of all the particles at a phase space location $(\mathbf{r},\mathbf{v})$ at time t. The quantity $f_a(\mathbf{v},t)$ is the smoothed Klimontovich function, and is the same quantity that one encounters in the Vlasov theory. The idea is that when the ensemble average is taken over the Klimontovich function, which is made of infinitely sharp delta functions, the singularities average out to generate a smooth one-particle distribution function. Since the system is assumed to be spatially uniform, the smoothed probability distribution $f_a(\mathbf{v},t)$ is everywhere equal in $\mathbf{r}$. Making use of the short-hand notation,

$$x = (\mathbf{r},\mathbf{v}), \tag{4.12}$$

we may also consider the ensemble averages of the products of Klimontovich functions. Take, for instance,

$$\begin{aligned}\langle N_a(x,t)\,N_b(x',t)\rangle &= \sum_{i=1}^{N}\sum_{j=1}^{N}\langle \delta(x-x_i^a)\,\delta(x'-x_j^b)\rangle \\ &= \sum_{i=1}^{N}\langle \delta(x-x_i^a)\,\delta(x'-x_i^b)\rangle + \sum_{\substack{i=1\\(i\neq j)}}^{N}\sum_{j=1}^{N}\langle \delta(x-x_i^a)\,\delta(x'-x_j^b)\rangle.\end{aligned} \tag{4.13}$$

In the first term, if $a \neq b$, then we are dealing with a product of two delta functions that belong to two different particle species (say, one electron and one ion). For such a combination, we generally assume that there is no cross correlation. Consequently, we demand that such quantities are zero unless $a = b$:

$$\begin{aligned}\langle N_a(x,t)\,N_b(x',t)\rangle &= \delta_{ab}\,\delta(x-x')\left\langle \sum_{i=1}^{N}\delta(x'-x_i^a)\right\rangle \\ &+ \sum_{\substack{i=1\\(i\neq j)}}^{N}\sum_{j=1}^{N}\langle \delta(x-x_i^a)\,\delta(x'-x_j^a)\rangle.\end{aligned} \tag{4.14}$$

Note that $\left\langle \sum_{i=1}^{N} \delta(x' - x_i^a) \right\rangle = f_a(x,t)$, that is, (4.11). We define the two-body phase space distribution function

$$f_2^{ab}(x,x',t) = \left\langle \sum_{i=1}^{N} \sum_{\substack{j=1 \\ (i \neq j)}}^{N} \delta(x - x_i^a)\, \delta(x' - x_j^a) \right\rangle. \tag{4.15}$$

Then we have

$$\left\langle N_a(x,t)\, N_b(x',t) \right\rangle = \delta_{ab}\, \delta(x - x')\, f_a(x,t) + f_2^{ab}(x,x',t). \tag{4.16}$$

Likewise, we may conceive of the following three-body correlation:

$$\begin{aligned}
\left\langle N_a(x,t)\, N_b(x',t)\, N_c(x'',t) \right\rangle &= \sum_i \sum_j \sum_k \left\langle \delta(x - x_i^a)\, \delta(x' - x_j^b)\, \delta(x'' - x_k^c) \right\rangle \\
&= \sum_i \left\langle \delta(x - x_i^a)\, \delta(x' - x_i^b)\, \delta(x'' - x_i^c) \right\rangle \\
&\quad + \sum_i \sum_{\substack{j \\ (j \neq i)}} \left\langle \delta(x - x_i^a)\, \delta(x' - x_j^b)\, \delta(x'' - x_i^c) \right\rangle \\
&\quad + \sum_i \sum_{\substack{k \\ (k \neq i)}} \left\langle \delta(x - x_i^a)\, \delta(x' - x_i^b)\, \delta(x'' - x_k^c) \right\rangle \\
&\quad + \sum_j \sum_{\substack{k \\ (k \neq j)}} \left\langle \delta(x - x_j^a)\, \delta(x' - x_j^b)\, \delta(x'' - x_k^c) \right\rangle \\
&\quad + \sum_i \sum_j \sum_{\substack{k \\ (i \neq j \neq k)}} \left\langle \delta(x - x_i^a)\, \delta(x' - x_j^b)\, \delta(x'' - x_k^c) \right\rangle \\
&= \delta_{ab}\, \delta_{bc}\, \delta(x - x')\, \delta(x'' - x) \left\langle \sum_i \delta(x' - x_i^a) \right\rangle \\
&\quad + \delta_{ac}\, \delta(x - x'') \left\langle \sum_i \sum_{\substack{j \\ (j \neq i)}} \delta(x - x_i^a)\, \delta(x' - x_j^b) \right\rangle \\
&\quad + \delta_{ab}\, \delta(x - x') \left\langle \sum_i \sum_{\substack{k \\ (k \neq i)}} \delta(x' - x_i^b)\, \delta(x'' - x_k^c) \right\rangle
\end{aligned}$$

$$+\,\delta_{bc}\,\delta(x'-x'')\left\langle\sum_j\sum_{\substack{k\\(k\neq j)}}\delta(x-x_j^a)\,\delta(x''-x_k^c)\right\rangle$$
$$+\left\langle\sum_i\sum_j\sum_{\substack{k\\(i\neq j\neq k)}}\delta(x-x_i^a)\,\delta(x'-x_j^b)\,\delta(x''-x_k^c)\right\rangle. \tag{4.17}$$

Defining the three-body phase space distribution function,

$$f_3^{abc}(x,x',x'',t)=\left\langle\sum_i\sum_j\sum_{\substack{k\\(i\neq j\neq k)}}\delta(x-x_i^a)\,\delta(x'-x_j^b)\,\delta(x''-x_k^c)\right\rangle, \tag{4.18}$$

we have

$$\begin{aligned}\big\langle N_a(x,t)\,N_b(x',t)\,N_c(x'',t)\big\rangle &= \delta_{ab}\,\delta_{bc}\,\delta(x-x')\,\delta(x''-x)\,f_a(x',t)\\ &\quad+\delta_{ac}\,\delta(x-x'')\,f_2^{ab}(x,x',t)\\ &\quad+\delta_{ab}\,\delta(x-x')\,f_2^{bc}(x',x'',t)\\ &\quad+\delta_{bc}\,\delta(x'-x'')\,f_2^{ac}(x,x'',t)+f_3^{abc}(x,x',x'',t).\end{aligned} \tag{4.19}$$

The two- and three-body distributions, $f_2^{ab}(x,x',t)$ and $f_3^{abc}(x,x',x'',t)$, respectively, can be further expressed in an alternative way, in terms of the two- and three-body correlations, $g_2^{ab}(x,x',t)$ and $g_3^{abc}(x,x',x'',t)$, defined respectively as follows:

$$\begin{aligned}f_2^{ab}(x,x',t) &= f_a(x,t)\,f_b(x',t)+g_2^{ab}(x,x',t),\\ f_3^{abc}(x,x',x'',t) &= f_a(x,t)\,f_b(x',t)\,f_c(x'',t)+f_a(x,t)\,g_2^{bc}(x',x'',t)\\ &\quad+f_b(x',t)\,g_2^{ac}(x,x'',t)+f_c(x'',t)\,g_2^{ab}(x,x,',t)\\ &\quad+g_3^{abc}(x,x',x'',t).\end{aligned} \tag{4.20}$$

In terms of these expressions, (4.11), (4.16), and (4.19) can be alternatively written as follows:

$$\begin{aligned}\big\langle N_a(x,t)\big\rangle &= f_a(x,t),\\ \big\langle N_a(x,t)\,N_b(x',t)\big\rangle &= f_a(x,t)\,f_b(x',t)+\delta_{ab}\,\delta(x-x')\,f_a(x,t)\\ &\quad+g_2^{ab}(x,x',t),\big\langle N_a(x,t)\,N_b(x',t)\,N_c(x'',t)\big\rangle\\ &= f_a(x,t)\,f_b(x',t)\,f_c(x'',t)+f_a(x,t)\,g_2^{bc}(x',x'',t)\\ &\quad+f_b(x',t)\,g_2^{ac}(x,x'',t)+f_c(x'',t)\,g_2^{ab}(x,x',t)+g_3^{abc}(x,x',x'',t).\\ &\quad+\delta_{ac}\,\delta(x-x'')\big[f_a(x,t)\,f_b(x',t)+g_2^{ab}(x,x',t)\big]\end{aligned} \tag{4.21}$$

$$+ \delta_{ab}\, \delta(x - x') \left[f_b(x',t)\, f_c(x'',t) + g_2^{bc}(x',x'',t) \right]$$
$$+ \delta_{bc}\, \delta(x' - x'') \left[f_a(x,t)\, f_c(x'',t) + g_2^{ac}(x,x'',t) \right]$$
$$+ \delta_{ab}\, \delta_{bc}\, \delta(x - x')\, \delta(x'' - x)\, f_a(x',t) + \cdots$$

From (4.9) and (4.10), we have

$$\delta N_a(x,t) = N_a(x,t) - \langle N_a(x,t) \rangle = N_a(x,t) - f_a(x,t). \tag{4.22}$$

Making use of the short-hand notation

$$\langle A(x,t)\, B(x',t) \rangle = \langle A\, B \rangle_{x,x',t}, \tag{4.23}$$

let us consider the ensemble averages of correlations of fluctuations. Some straightforward manipulations lead to the following:

$$\begin{aligned}
\langle \delta N_a(x,t) \rangle &= 0, \\
\langle \delta N_a(x,t)\, \delta N_b(x',t) \rangle &= \delta_{ab}\, \delta(x - x')\, f_a(x,t) + g_2^{ab}(x,x',t), \\
\langle \delta N_a(x,t)\, \delta N_b(x',t)\, \delta N_c(x'',t) \rangle &= \delta_{ab}\, \delta_{bc}\, \delta(x - x')\, \delta(x'' - x)\, f_a(x',t) \\
&\quad + \delta_{ab}\, \delta(x - x')\, g_2^{bc}(x',x'',t) \\
&\quad + \delta_{bc}\, \delta(x' - x'') g_2^{ac}(x,x'',t) \\
&\quad + \delta_{ac}\, \delta(x - x'')\, g_2^{ab}(x,x',t) + g_3^{abc}(x,x',x'',t).
\end{aligned} \tag{4.24}$$

The Vlasov approximation is equivalent to ignoring all discrete particle effects. In the Klimontovich formalism, discrete particle effects arise from singular functions associated with delta function terms such as $\delta(x - x')\, \delta_{ab}\, f_a(x,t)$, etc. Even though the ensemble average led to the smoothed Vlasovian one-particle distribution out of infinitely sharp Klimontovich function, the delta function terms mentioned earlier represent what's left over from the smoothing procedure. These delta function terms are intimately related to the discrete particle effects, which we will eventually see. This proves that the singular nature of the infinitely sharp delta functions that make up the Klimontovich function is what eventually leads to the discrete particle effects. If we ignore all such terms,

$$\begin{aligned}
\langle N_a(x,t) \rangle &= f_a(x,t), \\
\langle N_a(x,t)\, N_b(x',t) \rangle &= f_a(x,t)\, f_b(x',t) + g_2^{ab}(x,x',t), \\
\langle N_a(x,t)\, N_b(x',t)\, N_c(x'',t) \rangle &= f_a(x,t)\, f_b(x',t)\, f_c(x'',t) \\
&\quad + f_a(x,t)\, g_2^{bc}(x',x'',t) + f_b(x',t)\, g_2^{ac}(x,x'',t) \\
&\quad + f_c(x'',t)\, g_2^{bc}(x',x'',t) + g_3^{abc}(x,x',x'',t),
\end{aligned} \tag{4.25}$$

etc., and

$$\begin{aligned}\big\langle \delta N_a(x,t) \big\rangle &= 0,\\ \langle \delta N_a\, \delta N_b \rangle_{x,x';t} &= g_2^{ab}(x,x',t),\\ \langle \delta N_a\, \delta N_b\, \delta N_c \rangle_{x,x',x'';t} &= g_3^{abc}(x,x',x'',t),\end{aligned} \tag{4.26}$$

etc., then we obtain the Vlasov turbulence theory. Retaining the discrete particle terms will result in spontaneous emission/scattering and collisions, which will be discussed later. We can clearly see that the Vlasov theory is an approximation since there is no *a priori* justification for ignoring terms such as $\delta(x - x')\,\delta_{ab}\, f_a(x)$, etc., although, later, we will see that all the new terms that arise from retaining the discrete particle effects will have an overall proportionality to a quantity that represents the inverse of total number of plasma particles in a sphere with the radius equal to the Debye length. If the number of particles within the "Debye" sphere is large, then the discrete particle terms will be less important. From this, we will see that the Vlasov approximation is valid if the number of particles in the Debye sphere is infinite. At the moment, however, this is not evident.

Returning to the Klimontovich formalism, Let us now introduce another quantity, that is, the Klimontovich function for free or noninteracting particles:

$$N_a^0(x,t) = \sum_{i=1}^{N} \delta[\mathbf{r} - \mathbf{r}_i^{a0}(t)]\; \delta[\mathbf{v} - \mathbf{v}_i^{a0}(t)], \tag{4.27}$$

where the free-streaming particle orbits $\mathbf{r}_i^{a0}(t)$ and $\mathbf{v}_i^{a0}(t)$ satisfy

$$\frac{d\mathbf{r}_i^{a0}(t)}{dt} = \mathbf{v}_i^{a0}(t), \qquad \frac{d\mathbf{v}_i^{a0}(t)}{dt} = 0. \tag{4.28}$$

Then the Klimontovich function for noninteracting particles obeys a trivial equation,

$$\left(\frac{\partial}{\partial t} + \mathbf{v}\cdot\frac{\partial}{\partial \mathbf{r}}\right) N_a^0(x,t) = 0. \tag{4.29}$$

Upon taking the ensemble average, we have

$$\left(\frac{\partial}{\partial t} + \mathbf{v}\cdot\frac{\partial}{\partial \mathbf{r}}\right) \big\langle N_a^0(x,t) \big\rangle = 0. \tag{4.30}$$

The quantity $N_a(x,t)$ is the exact phase space distribution that describes charged particles interacting with each other via collective electromagnetic force. In contrast, the object $N_a^0(x,t)$ is the phase space distribution that describes noninteracting free particles. The difference between the two, $N_a(x,t) - N_a^0(x,t)$, represents, conceptually, the phase space information in which the collective effects are emphasized while the pure single particle effects are subtracted out. Consequently, it is useful to subtract (4.29) from (4.8), in order to obtain the dynamical equation for the quantity $N_a(x,t) - N_a^0(x,t)$:

$$\left(\frac{\partial}{\partial t} + \mathbf{v} \cdot \frac{\partial}{\partial \mathbf{r}}\right) [N_a(x,t) - N_a^0(x,t)] + \left[\frac{e_a}{m_a}\left(\delta\mathbf{E} + \frac{\mathbf{v}}{c} \times \delta\mathbf{B}\right) \cdot \frac{\partial}{\partial \mathbf{v}}\right] N_a(x,t) = 0. \tag{4.31}$$

Let us define the fluctuations of unperturbed Klimontovich function for free particles:

$$\delta N_a^0(x,t) = N_a^0(x,t) - \langle N_a^0(x,t)\rangle. \tag{4.32}$$

Then (4.31) can be rewritten as

$$\left(\frac{\partial}{\partial t} + \mathbf{v} \cdot \frac{\partial}{\partial \mathbf{r}}\right) [f_a(x,t) + \delta N_a(x,t) - \delta N_a^0(x,t)] + \left[\frac{e_a}{m_a}\left(\delta\mathbf{E} + \frac{\mathbf{v}}{c} \times \delta\mathbf{B}\right) \cdot \frac{\partial}{\partial \mathbf{v}}\right] [f_a(x,t) + \delta N_a(x,t)] = 0, \tag{4.33}$$

where we have made use of (4.30). Maxwell's equation in (4.6) coupled with (4.9) and (4.10) is written as

$$\begin{aligned}
\nabla \cdot \delta\mathbf{E} &= 4\pi \sum_a e_a \int d\mathbf{v} \left[f_a(x,t) + \delta N_a(x,t)\right], \\
\nabla \times \delta\mathbf{E} + \frac{1}{c}\frac{\partial}{\partial t}\delta\mathbf{B} &= 0, \\
\nabla \times \delta\mathbf{B} - \frac{1}{c}\frac{\partial}{\partial t}\delta\mathbf{E} &= \frac{4\pi}{c} \sum_a e_a \int d\mathbf{v}\, \mathbf{v} \left[f_a(x,t) + \delta N_a(x,t)\right], \\
\nabla \cdot \delta\mathbf{B} &= 0.
\end{aligned} \tag{4.34}$$

Taking the ensemble average of (4.33), we obtain the equation for average particle distribution:

$$\left(\frac{\partial}{\partial t} + \mathbf{v} \cdot \nabla\right) f_a(x,t) = -\frac{e_a}{m_a}\left\langle\left(\delta\mathbf{E} + \frac{\mathbf{v}}{c} \times \delta\mathbf{B}\right) \cdot \frac{\partial}{\partial \mathbf{v}} \delta N_a(x,t)\right\rangle. \tag{4.35}$$

Averaging Maxwell's equation leads to the average charge and current neutrality conditions

$$\begin{aligned}
0 &= \sum_a e_a \int d\mathbf{v}\, f_a(x,t), \\
0 &= \sum_a e_a \int d\mathbf{v}\, \mathbf{v}\, f_a(x,t).
\end{aligned} \tag{4.36}$$

Subtracting (4.35) from (4.33), we obtain the equation for perturbed Klimontovich function. Together with the wave equation for perturbed field, we have

$$\begin{aligned}
\left(\frac{\partial}{\partial t}+\mathbf{v}\cdot\nabla\right)\left[\delta N_a(x,t)-\delta N_a^0(x,t)\right] &= -\frac{e_a}{m_a}\left(\delta\mathbf{E}+\frac{\mathbf{v}}{c}\times\delta\mathbf{B}\right)\cdot\frac{\partial f_a(x,t)}{\partial\mathbf{v}} \\
&\quad -\left(\delta\mathbf{a}\cdot\frac{\partial}{\partial\mathbf{v}}\,\delta N_a(x,t)-\left\langle\delta\mathbf{a}\cdot\frac{\partial}{\partial\mathbf{v}}\,\delta N_a(x,t)\right\rangle\right), \\
\delta\mathbf{a} &= \frac{e_a}{m_a}\left(\delta\mathbf{E}+\frac{\mathbf{v}}{c}\times\delta\mathbf{B}\right), \\
\nabla\times\delta\mathbf{B}-\frac{1}{c}\frac{\partial}{\partial t}\,\delta\mathbf{E} &= \frac{4\pi}{c}\sum_a e_a\int d\mathbf{v}\,\mathbf{v}\,\delta N_a(x,t), \\
\nabla\cdot\delta\mathbf{E} &= 4\pi\sum_a e_a\int d\mathbf{v}\,\delta N_a(x,t), \\
\nabla\times\delta\mathbf{E}+\frac{1}{c}\frac{\partial}{\partial t}\,\delta\mathbf{B} &= 0, \qquad \nabla\cdot\delta\mathbf{B}=0.
\end{aligned} \tag{4.37}$$

This is a closed set of equations involving δN_a, $\delta\mathbf{E}$, and $\delta\mathbf{B}$, except that an additional quantity δN_a^0 is also present. One must be able to compute this quantity, which is discussed next.

It is useful to discuss the two-time correlation function of the free particle fluctuations. Subtracting (4.30) from (4.29) and making use of definition (4.32), we have

$$\left(\frac{\partial}{\partial t}+\mathbf{v}\cdot\nabla\right)\delta N_a^0(x,t)=0. \tag{4.38}$$

Multiplying $\delta N_b^0(x',t')$ and taking the ensemble average, we obtain

$$\left(\frac{\partial}{\partial t}+\mathbf{v}\cdot\nabla\right)\left\langle\delta N_a^0\,\delta N_b^0\right\rangle_{x,t;x',t'}=0. \tag{4.39}$$

The following equation is equally valid:

$$\left(\frac{\partial}{\partial t'}+\mathbf{v}'\cdot\nabla'\right)\left\langle\delta N_a^0\,\delta N_b^0\right\rangle_{x,t;x',t'}=0. \tag{4.40}$$

The initial condition of either (4.39) or (4.40) must be

$$\left\langle\delta N_a^0\,\delta N_b^0\right\rangle_{x,t;x',t'=t}=\delta_{ab}\,\delta(x-x')\,f_a(x,t). \tag{4.41}$$

This is because according to (4.24), $\langle\delta N_a\,\delta N_b\rangle_{x,x';t}=g_2^{ab}(x,x',t)+\delta_{ab}\,\delta(x-x')\,f_a(x,t)$. Since δN_a^0 and δN_b^0 are fluctuations for noninteracting particles, their correlation must not possess the correlation function $g_2^{ab}(x,x',t)$. Hence, (4.41).

Another more straightforward way to see this is to directly compute $\langle \delta N_a^0 \, \delta N_b^0 \rangle_{x,t;x',t'}$:

$$\begin{aligned}\langle \delta N_a^0(x,t)\, \delta N_b^0(x',t')\rangle &= \langle N_a^0(x,t)\, N_b^0(x',t')\rangle - \langle N_a^0(x,t)\rangle \langle N_b^0(x',t')\rangle \\ &= \Bigg\langle \sum_{i=1}^N \delta[\mathbf{r} - \mathbf{r}_i^{a0}(t)]\, \delta[\mathbf{v} - \mathbf{v}_i^{a0}(t)] \\ &\quad \times \sum_{j=1}^N \delta[\mathbf{r}' - \mathbf{v}_j^{b0}(t')]\, \delta[\mathbf{v}' - \mathbf{v}_j^{b0}(t')] \Bigg\rangle \\ &\quad - \langle N_a^0(x,t)\rangle \langle N_b^0(x',t')\rangle . \end{aligned} \tag{4.42}$$

Since $\mathbf{v}_i^{a0} = \text{const}$, we have $\mathbf{r}_i^{a0}(t) = \mathbf{v}_i^{a0}\, t + \mathbf{r}_i^{a0}$, with $\mathbf{r}_i^{a0} = \text{const}$. Consequently, (4.42) reduces to

$$\langle \delta N_a^0(x,t)\, \delta N_b^0(x',t')\rangle_{x,t;x',t'} = \delta_{ab}\, \delta[\mathbf{r} - \mathbf{r}' - \mathbf{v}\,(t-t')]\, \delta(\mathbf{v} - \mathbf{v}')\, f_a(x,t), \tag{4.43}$$

which is the desired result (4.41). Here, an assumption is made that the ensemble average of the Klimontovich function for noninteracting, or free, particles is equal to that of the complete Klimontovich function,

$$\langle N_a^0(x,t)\rangle = \langle N_a(x,t)\rangle = f_a(x,t). \tag{4.44}$$

The Fourier transformation of $\langle \delta N_a^0 \, \delta N_b^0 \rangle_{x,t;x',t'}$ with respect to the fast time- and spatial-scales, $t - t'$ and $\mathbf{r} - \mathbf{r}'$, is of interest. In the Laplace transformation (1.38) we restricted ourselves to positive t integral because we are interested in the causal problem. This prescription is equivalent to adding an infinitesimal positive imaginary part to ω. However, in the present case, the spectral representation is to be defined with respect to the time difference, $t - t'$, which may be defined over an entire range $-\infty < t - t' < \infty$. This contrasts to individual time variables, which are defined over $0 < t < \infty$ and $0 < t' < \infty$ (that is, causal). Consequently, in the present case, following Klimontovich (1982), we define

$$\begin{aligned}\langle \delta N_a^0(\mathbf{v})\, \delta N_b^0(\mathbf{v}')\rangle_{\mathbf{k},\omega}^{+} &= \frac{1}{(2\pi)^4} \int d(\mathbf{r} - \mathbf{r}') \int_0^\infty d(t - t') \\ &\quad \times \langle \delta N_a^0 \, \delta N_b^0 \rangle_{x,t;x',t'}\, e^{-i\mathbf{k}\cdot(\mathbf{r}-\mathbf{r}') + i\omega(t-t') - 0^+(t-t')} \\ &= \delta_{ab}\, \delta(\mathbf{v} - \mathbf{v}')\, \frac{1}{(2\pi)^4} \int_0^\infty d\tau\, e^{-i\mathbf{k}\cdot\mathbf{v}\,\tau + i\omega\tau - 0^+\tau}\, f_a(x,t). \end{aligned} \tag{4.45}$$

Then we construct the total spectral representation,

$$\langle \delta N_a^0(\mathbf{v})\, \delta N_b^0(\mathbf{v}')\rangle_{\mathbf{k},\omega} = \langle \delta N_a^0(\mathbf{v})\, \delta N_b^0(\mathbf{v}')\rangle_{\mathbf{k},\omega}^{+} + \Big[\langle \delta N_a^0(\mathbf{v})\, \delta N_b^0(\mathbf{v}')\rangle_{\mathbf{k},\omega}^{+}\Big]^* . \tag{4.46}$$

This quantity reduces to the customary Fourier transformation for $0^+ = 0$ exactly. From this we now have a proper definition, which leads to

$$\langle \delta N_a^0(\mathbf{v})\,\delta N_b^0(\mathbf{v}')\rangle_{\mathbf{k},\omega} = (2\pi)^{-3}\,\delta_{ab}\,\delta(\mathbf{v}-\mathbf{v}')\,\delta(\omega-\mathbf{k}\cdot\mathbf{v})\,f_a(x,t). \tag{4.47}$$

The set of equations, (4.35)–(4.37), with the source fluctuation (4.43) and its spectral representation (4.47), can now be reformulated in terms of Fourier–Laplace transformation (1.38). With the redefinition of angular frequency that includes the slow-time derivative, (1.41), the set of equations that constitute the basis of subsequent analysis readily emerges:

$$\begin{aligned}
\frac{\partial f_a(\mathbf{v},t)}{\partial t} &= -\frac{e_a}{m_a}\frac{\partial}{\partial v_i}\int d\mathbf{k}\int d\omega\left[\left(1-\frac{\mathbf{k}\cdot\mathbf{v}}{\omega}\right)\delta_{ij}+\frac{k_i v_j}{\omega}\right]\\
&\quad\times\left\langle\delta E^j_{-\mathbf{k},-\omega}\,\delta N^a_{\mathbf{k},\omega}(\mathbf{v})\right\rangle,\\
\left[\delta_{ij}-\frac{c^2k^2}{\omega^2}\left(\delta_{ij}-\frac{k_ik_j}{k^2}\right)\right]\delta E^j_{\mathbf{k},\omega} &= -\frac{4\pi i}{\omega}\sum_a e_a\int d\mathbf{v}\;v_i\,\delta N^a_{\mathbf{k},\omega}(\mathbf{v}),\\
&\quad\times i(\omega-\mathbf{k}\cdot\mathbf{v})\left[\delta N^a_{\mathbf{k},\omega}(\mathbf{v})-\delta N^{a0}_{\mathbf{k},\omega}(\mathbf{v})\right]\\
&= \frac{e_a}{m_a}\,\delta E^i_{\mathbf{k},\omega}\left[\left(1-\frac{\mathbf{k}\cdot\mathbf{v}}{\omega}\right)\delta_{ij}+\frac{v_ik_j}{\omega}\right]\frac{\partial f_a(\mathbf{v},t)}{\partial v_j}\\
&\quad+\frac{e_a}{m_a\Omega_a}\frac{\partial}{\partial v_j}\int d\mathbf{k}'\int d\omega'\\
&\quad\times\left[\left(1-\frac{\mathbf{k}'\cdot\mathbf{v}}{\omega'}\right)\delta_{ij}+\frac{v_ik'_j}{\omega'}\right]\\
&\quad\times\left[\delta E^i_{\mathbf{k}',\omega'}\delta N^a_{\mathbf{k}-\mathbf{k}',\omega-\omega'}(\mathbf{v})-\left\langle\delta E^i_{\mathbf{k}',\omega'}\delta N^a_{\mathbf{k}-\mathbf{k}',\omega-\omega'}(\mathbf{v})\right\rangle\right].
\end{aligned} \tag{4.48}$$

In the rest of Part II, we are interested in the electrostatic approximation. Consequently, the system of equations to be solved can be simplified by taking the electrostatic approximation, $\delta\mathbf{B}=0$, $\mathbf{k}\cdot\mathbf{v}/\omega\ll 1$, and $\delta\mathbf{E}=i\mathbf{k}\,\delta\phi$. This simplifies (4.48):

$$\begin{aligned}
\frac{\partial f_a(\mathbf{v},t)}{\partial t} &= \frac{-ie_a}{m_a}\int d\mathbf{k}\int d\omega\,\mathbf{k}\cdot\frac{\partial}{\partial\mathbf{v}}\left\langle\delta\phi_{-\mathbf{k},-\omega'}\,\delta N^a_{\mathbf{k},\omega}(\mathbf{v})\right\rangle,\\
(\omega-\mathbf{k}\cdot\mathbf{v})\left[\delta N^a_{\mathbf{k},\omega}(\mathbf{v})-\delta N^{a0}_{\mathbf{k},\omega}(\mathbf{v})\right] &= -\frac{e_a}{m_a}\,\delta\phi_{\mathbf{k},\omega}\,\mathbf{k}\cdot\frac{\partial f_a(\mathbf{v},t)}{\partial\mathbf{v}}\\
&\quad-\frac{e_a}{m_a}\int d\mathbf{k}'\int d\omega'\,\mathbf{k}'\cdot\frac{\partial}{\partial\mathbf{v}}\left[\delta\phi_{\mathbf{k}',\omega'}\,\delta N^a_{\mathbf{k}-\mathbf{k}',\omega-\omega'}(\mathbf{v})\right.\\
&\quad\left.-\left\langle\delta\phi_{\mathbf{k}',\omega'}\,\delta N^a_{\mathbf{k}-\mathbf{k}',\omega-\omega'}(\mathbf{v},t)\right\rangle\right],\\
\delta\phi_{\mathbf{k},\omega}(t) &= \sum_a\frac{4\pi e_a}{k^2}\int d\mathbf{v}\,\delta N^a_{\mathbf{k},\omega}(\mathbf{v}).
\end{aligned} \tag{4.49}$$

These equations directly generalize (1.40), which formed the basis for Vlasov weak turbulence theory. If we ignore $\delta N^{a0}_{\mathbf{k},\omega}(\mathbf{v})$ and replace $\delta N^{a}_{\mathbf{k},\omega}(\mathbf{v})$ by $\delta f^{a}_{\mathbf{k},\omega}(\mathbf{v})$, then we recover (1.40). Note that the resonance factor $\omega - \mathbf{k}\cdot\mathbf{v} + i0$ in (4.49) contains implicit slow-time derivative, $\omega - \mathbf{k}\cdot\mathbf{v} + i0 + i\,\partial/\partial t$.

4.2 Formal Klimontovich Weak Turbulence Theory

We now reformulate the plasma weak turbulence theory on the basis of (4.49). We follow the same steps as taken in the Vlasov weak turbulence analysis of Chapter 1. We remind the readers of the shorthand notation (1.42),

$$\mathbf{g}^a_{\mathbf{k},\omega} = -\frac{e_a}{m_a}\frac{1}{\omega - \mathbf{k}\cdot\mathbf{v} + i0}\frac{\partial}{\partial \mathbf{v}},$$

with which the equation for perturbed phase space density distribution can be expressed compactly as

$$\begin{aligned}\delta N^a_{\mathbf{k},\omega} &= \delta N^{a0}_{\mathbf{k},\omega} + \mathbf{k}\cdot\mathbf{g}^a_{\mathbf{k},\omega}\, f_a\, \delta\phi_{\mathbf{k},\omega} \\ &+ \int d\mathbf{k}' \int d\omega'\, \mathbf{k}'\cdot\mathbf{g}^a_{\mathbf{k},\omega}\left[\delta\phi_{\mathbf{k}',\omega'}\,\delta N^a_{\mathbf{k}-\mathbf{k}',\omega-\omega'} - \left\langle\delta\phi_{\mathbf{k}',\omega'}\,\delta N^a_{\mathbf{k}-\mathbf{k}',\omega-\omega'}\right\rangle\right].\end{aligned} \tag{4.50}$$

For the sake of notational simplicity, let us introduce a convention,

$$q \equiv (\mathbf{k},\omega). \tag{4.51}$$

We also drop "δ" in front of the field perturbation, $\delta\phi$, and perturbed Klimontovich function, δN. Following the same steps as in arriving at (1.49), the iterative solution can be obtained:

$$\begin{aligned}N^a_q &= N^{a0}_q + \sum_{q_1+q_2=q} (\mathbf{k}_1\cdot\mathbf{g}^a_q)\left[\phi_{q_1}\,N^{a0}_{q_2} - \left\langle\phi_{q_1}\,N^{a0}_{q_2}\right\rangle\right] \\ &+ \sum_{q_1+q_2+q_3=q} (\mathbf{k}_1\cdot\mathbf{g}^a_q)\,(\mathbf{k}_2\cdot\mathbf{g}^a_{q-q_1}) \\ &\times \left[\phi_{q_1}\,\phi_{q_2}\,N^{a0}_{q_3} - \phi_{q_1}\left\langle\phi_{q_2}\,N^{a0}_{q_3}\right\rangle - \left\langle\phi_{q_1}\,\phi_{q_2}\,N^{a0}_{q_3}\right\rangle\right] \\ &+ (\mathbf{k}\cdot\mathbf{g}^a_q)\,f_a\,\phi_q + \sum_{q_1+q_2=q}(\mathbf{k}_1\cdot\mathbf{g}^a_q)\,(\mathbf{k}_2\cdot\mathbf{g}^a_{q_2})\,f_a\left[\phi_{q_1}\,\phi_{q_2} - \left\langle\phi_{q_1}\,\phi_{q_2}\right\rangle\right] \\ &+ \sum_{q_1+q_2+q_3=q} (\mathbf{k}_1\cdot\mathbf{g}^a_q)\,(\mathbf{k}_2\cdot\mathbf{g}^a_{q-q_1})\,(\mathbf{k}_3\cdot\mathbf{g}^a_{q_3})\,f_a \\ &\times \left[\phi_{q_1}\,\phi_{q_2}\,\phi_{q_3} - \phi_{q_1}\left\langle\phi_{q_2}\,\phi_{q_3}\right\rangle - \left\langle\phi_{q_1}\,\phi_{q_2}\,\phi_{q_3}\right\rangle\right].\end{aligned} \tag{4.52}$$

In the present scheme, we take the view that the effects arising from the particle discreteness are supposed to provide the next-order correction to what is essentially Vlasovian nonlinear theory. This is equivalent to the assumption that the plasma parameter $g \propto 1/(n\lambda_{De}^3)$ is small. This allows us to ignore nonlinear responses that arise from N_q^{a0}, and leads to the simplification,

$$\begin{aligned}
N_q^a &= N_q^{a0} + (\mathbf{k}\cdot\mathbf{g}_q^a)\, f_a\, \phi_q \\
&\quad + \sum_{q_1+q_2=q} (\mathbf{k}_1\cdot\mathbf{g}_q^a)\,(\mathbf{k}_2\cdot\mathbf{g}_{q_2}^a)\, f_a \left[\phi_{q_1}\phi_{q_2} - \langle\phi_{q_1}\phi_{q_2}\rangle\right] \\
&\quad + \sum_{q_1+q_2+q_3=q} (\mathbf{k}_1\cdot\mathbf{g}_q^a)\,(\mathbf{k}_2\cdot\mathbf{g}_{q-q_1}^a)\,(\mathbf{k}_3\cdot\mathbf{g}_{q_3}^a)\, f_a \\
&\quad \times \left[\phi_{q_1}\phi_{q_2}\phi_{q_3} - \phi_{q_1}\langle\phi_{q_2}\phi_{q_3}\rangle - \langle\phi_{q_1}\phi_{q_2}\phi_{q_3}\rangle\right].
\end{aligned} \tag{4.53}$$

Fully symmetric version of this iterative solution is

$$\begin{aligned}
N_q^a &= N_q^{a0} + \alpha_a(q)\, f_a\, \phi_q + \sum_{q_1+q_2=q} \alpha_a^{(2)}(q_1|q_2)\, f_a \left[\phi_{q_1}\phi_{q_2} - \langle\phi_{q_1}\phi_{q_2}\rangle\right] \\
&\quad + \sum_{q1+q_2+q_3=q} \alpha_a^{(3)}(q_1|q_2|q_3)\, f_a \\
&\quad \times \left[\phi_{q_1}\phi_{q_2}\phi_{q_3} - \phi_{q_1}\langle\phi_{q_2}\phi_{q_3}\rangle - \langle\phi_{q_1}\phi_{q_2}\phi_{q_3}\rangle\right],
\end{aligned} \tag{4.54}$$

where $\alpha_a(q)$, $\alpha_a^{(2)}(q_1|q_2)$, and $\alpha_a^{(3)}(q_1|q_2|q_3)$ are quantities that we have already encountered in (1.53).

Combining the solution (4.54) and perturbed Poisson's equation in (4.49), we obtain

$$\begin{aligned}
&k^2\,\epsilon(q)\,\phi_q - \sum_{q'} i\,k\,k'\,|\mathbf{k}-\mathbf{k}'|\,\chi^{(2)}(q'|q-q')\left[\phi_{q'}\phi_{q-q'} - \langle\phi_{q'}\phi_{q-q'}\rangle\right] \\
&\quad - \sum_{q'}\sum_{q''} k\,k'\,k''\,|\mathbf{k}-\mathbf{k}'-\mathbf{k}''|\,\bar{\chi}^{(3)}(q'|q''|q-q'-q'')\left[\phi_{q'}\phi_{q''}\phi_{q-q'-q''}\right. \\
&\quad \left. - \phi_{q'}\langle\phi_{q''}\phi_{q-q'-q''}\rangle - \langle\phi_{q'}\phi_{q''}\phi_{q-q'-q''}\rangle\right] \\
&= \sum_a 4\pi e_a \int d\mathbf{v}\, N_q^{a0}(\mathbf{v}),
\end{aligned} \tag{4.55}$$

where linear and nonlinear response functions are defined exactly as in (1.55)–(1.63). Let us multiply ϕ_{-q} to (4.55) and take the ensemble average. Then, we have

$$\begin{aligned}
&k^2\,\epsilon(q)\,\langle\phi^2\rangle_q - i\sum_{q'} k\,k'\,|\mathbf{k}-\mathbf{k}'|\,\chi^{(2)}(q'|q-q')\,\langle\phi_{q'}\phi_{q-q'}\phi_{-q}\rangle \\
&\quad - 2\sum_{q'} k^2\,k'^2\,\bar{\chi}^{(3)}(q'|-q'|q)\,\langle\phi^2\rangle_{q'}\,\langle\phi^2\rangle_q \\
&= \sum_a 4\pi e_a \int d\mathbf{v}\,\langle\phi_{-q}\, N_q^{a0}(\mathbf{v})\rangle.
\end{aligned} \tag{4.56}$$

Next, we multiply $N_{-q}^{a0}(\mathbf{v})$ to (4.55) and take the average. We subsequently replace q by $-q$:

$$k^2\,\epsilon(-q)\,\langle\phi_{-q}\,N_q^{a0}(\mathbf{v})\rangle = -i\sum_{q'} k\,k',|\mathbf{k}-\mathbf{k}'|\,\chi^{(2)*}(q'|q-q')\,\langle\phi_{-q'}\,\phi_{-q+q'}\,N_q^{a0}(\mathbf{v})\rangle$$
$$+2\sum_{q'} k^2\,k'^2\,\bar{\chi}^{(3)*}(q'|-q'|q)\,\langle\phi^2\rangle_{q'}\,\langle\phi_{-q}\,N_q^{a0}(\mathbf{v})\rangle$$
$$+\sum_b 4\pi e_b\int d\mathbf{v}'\,\langle N_{-q}^{b0}(\mathbf{v}')\,N_q^{a0}(\mathbf{v})\rangle. \tag{4.57}$$

Making use of (4.47), we have

$$\langle\phi_{-q}\,N_q^{a0}(\mathbf{v})\rangle = \frac{4\pi e_a}{(2\pi)^3\,k^2\,\epsilon^*(q)}\,\delta(\omega-\mathbf{k}\cdot\mathbf{v})\,f_a(\mathbf{v})$$
$$+\frac{2}{\epsilon^*(q)}\sum_{q'} k'^2\,\bar{\chi}^{(3)*}(q'|-q'|q)\,\langle\phi^2\rangle_{q'}\,\langle\phi_{-q}\,N_q^{a0}(\mathbf{v})\rangle \tag{4.58}$$
$$-\frac{i}{k\,\epsilon^*(q)}\sum_{q'} k'\,|\mathbf{k}-\mathbf{k}'|\,\chi^{(2)*}(q'|q-q')\,\langle\phi_{-q'}\,\phi_{-q+q'}\,N_q^{a0}(\mathbf{v})\rangle.$$

Since the quantity $\langle\phi_{-q}\,N_q^{a0}(\mathbf{v})\rangle$ appears on both sides of (4.58) we iteratively solve for this quantity, making use of the lowest-order solution first:

$$\langle\phi_{-q}\,N_q^{a0}(\mathbf{v})\rangle \approx \frac{4\pi e_a}{(2\pi)^3\,k^2\,\epsilon^*(q)}\,\delta(\omega-\mathbf{k}\cdot\mathbf{v})\,f_a(\mathbf{v}). \tag{4.59}$$

This procedure leads to

$$\langle\phi_{-q}\,N_q^{a0}(\mathbf{v})\rangle = \frac{4\pi e_a}{(2\pi)^3\,k^2\,\epsilon^*(q)} \tag{4.60}$$
$$\times\left(1+\sum_{q'}\frac{2\,\bar{\chi}^{(3)*}(q'|-q'|q)}{\epsilon^*(q)}\,\langle k'^2\phi^2\rangle_{q'}\right)\delta(\omega-\mathbf{k}\cdot\mathbf{v})\,f_a(\mathbf{v})$$
$$-\frac{i}{k\,\epsilon^*(q)}\sum_{q'} k'\,|\mathbf{k}-\mathbf{k}'|\,\chi^{(2)*}(q'|q-q')\,\langle\phi_{-q'}\,\phi_{-q+q'}\,N_q^{a0}(\mathbf{v})\rangle.$$

Inserting (4.60) to the right-hand side of (4.56), we arrive at

$$\epsilon(q)\,\langle E^2\rangle_q - i\int dq'\,\chi^{(2)}(q'|q-q')\,k\,k'\,|\mathbf{k}-\mathbf{k}'|\,\langle\phi_{q'}\,\phi_{q-q'}\,\phi_{-q}\rangle$$
$$-\,2\int dq'\,\bar{\chi}^{(3)}(q'|-q'|q)\,\langle E^2\rangle_{q'}\,\langle E^2\rangle_q = \sum_a\int d\mathbf{v}\,\frac{(4\pi e_a)^2}{(2\pi)^3\,k^2\,\epsilon^*(q)}$$

$$\times\left(1+\int dq'\,\frac{2\,\bar{\chi}^{(3)*}(q'|-q'|q)}{\epsilon^*(q)}\,\langle E^2\rangle_{q'}\right)\delta(\omega-\mathbf{k}\cdot\mathbf{v})\,f_a(\mathbf{v}) \tag{4.61}$$

$$-\,i\sum_a\int dq'\int d\mathbf{v}\,\frac{(4\pi e_a)\,\chi^{(2)*}(q'|q-q')}{k\,\epsilon^*(q)}\,k'\,|\mathbf{k}-\mathbf{k}'|\,\big\langle\phi_{-q'}\,\phi_{-q+q'}\,N_q^{a0}(\mathbf{v})\big\rangle,$$

where we have resorted back to the electric field representation by making use of $\langle E^2\rangle_q=\langle k^2\phi^2\rangle_q$. In (4.61), we note that the first term on the right-hand side of the equality has an overall factor $\delta(\omega-\mathbf{k}\cdot\mathbf{v})$. This is the linear wave-particle resonance delta function condition. This means that the nonlinear correction associated with this factor can be ignored, as it represents a small correction.

To evaluate the third-order cumulants, $\langle\phi_{q'}\,\phi_{q-q'}\,\phi_{-q}\rangle$ and $\langle\phi_{-q'}\,\phi_{-q+q'}N_q^{a0}(\mathbf{v})\rangle$, we make use of (4.55) without the third-order nonlinearity:

$$\phi_{q_1}=\frac{1}{k_1^2\,\epsilon(q_1)}\sum_{q''}i\,k_1\,k''\,|\mathbf{k}_1-\mathbf{k}''|\,\chi^{(2)}(q''|q_1-q'')\left[\phi_{q''}\,\phi_{q_1-q''}-\langle\phi_{q''}\,\phi_{q_1-q''}\rangle\right]$$

$$+\frac{1}{k_1^2\,\epsilon(q_1)}\sum_a 4\pi e_a\int d\mathbf{v}\,N_{q_1}^{a0}(\mathbf{v}). \tag{4.62}$$

Following the same steps as in arriving at (1.78), the quantity $\langle\phi_{q'}\,\phi_{q-q'}\,\phi_{-q}\rangle$ can be constructed by successively making use of the expression (4.62) iteratively for each of $\phi_{q'}$, $\phi_{q-q'}$ and ϕ_{-q}:

$$\begin{aligned}\langle\phi_{q'}\,\phi_{q-q'}\,\phi_{-q}\rangle=2i\,k\,k'\,|\mathbf{k}-\mathbf{k}'|\Bigg(&\frac{\chi^{(2)}(q|-q+q')}{k'^2\,\epsilon(q')}\,\langle\phi^2\rangle_{q-q'}\,\langle\phi^2\rangle_q\\&+\frac{\chi^{(2)}(q|-q')}{|\mathbf{k}-\mathbf{k}'|^2\,\epsilon(q-q')}\,\langle\phi^2\rangle_{q'}\,\langle\phi^2\rangle_q\\&-\frac{\chi^{(2)*}(q'|q-q')}{k^2\,\epsilon^*(q)}\,\langle\phi^2\rangle_{q'}\,\langle\phi^2\rangle_{q-q'}\Bigg)\\&+\sum_a 4\pi e_a\int d\mathbf{v}\Bigg(\frac{\langle\phi_{q-q'}\,\phi_{-q}\,N_{q'}^{a0}(\mathbf{v})\rangle}{k'^2\,\epsilon(q')}\\&+\frac{\langle\phi_{q'}\,\phi_{-q}\,N_{q-q'}^{a0}(\mathbf{v})\rangle}{|\mathbf{k}-\mathbf{k}'|^2\,\epsilon(q-q')}+\frac{\langle\phi_{q'}\,\phi_{q-q'}\,N_{-q}^{a0}(\mathbf{v})\rangle}{k^2\,\epsilon^*(q)}\Bigg),\end{aligned} \tag{4.63}$$

where we made use of the property $\chi^{(2)}(-q_1|-q_2)=-\chi^{(2)*}(q_1|q_2)$ and decomposed the four-body cumulants as products of two-body cumulants, ignoring irreducible components.

Inserting (4.63) to (4.61), we have

$$
\begin{aligned}
&\epsilon(q)\,\langle E^2\rangle_q + 2\int dq'\left[\chi^{(2)}(q'|q-q')\left(\frac{\chi^{(2)}(q|-q+q')}{\epsilon(q')}\,\langle E^2\rangle_{q-q'}\right.\right.\\
&\quad\left.\left. + \frac{\chi^{(2)}(q|-q')}{\epsilon(q-q')}\,\langle E^2\rangle_{q'}\right) - \bar{\chi}^{(3)}(q'|-q'|q)\,\langle E^2\rangle_{q'}\right]\langle E^2\rangle_q\\
&\quad - 2\int dq'\,\frac{|\chi^{(2)}(q'|q-q')|^2}{\epsilon^*(q)}\,\langle E^2\rangle_{q'}\,\langle E^2\rangle_{q-q'}\\
&= \sum_a \frac{(4\pi e_a)^2}{(2\pi)^3\,k^2\,\epsilon^*(q)}\int d\mathbf{v}\,\delta(\omega-\mathbf{k}\cdot\mathbf{v})\,f_a(\mathbf{v})\\
&\quad + i\sum_a 4\pi e_a\int dq'\int d\mathbf{v}\,k\,k'\,|\mathbf{k}-\mathbf{k}'|\left[\chi^{(2)}(q'|q-q')\left(\frac{\langle\phi_{q-q'}\,\phi_{-q}\,N^{a0}_{q'}(\mathbf{v})\rangle}{k'^2\,\epsilon(q')}\right.\right.\\
&\quad\left. + \frac{\langle\phi_{q'}\,\phi_{-q}\,N^{a0}_{q-q'}(\mathbf{v})\rangle}{|\mathbf{k}-\mathbf{k}'|^2\,\epsilon(q-q')} + \frac{\langle\phi_{q'}\,\phi_{q-q'}\,N^{a0}_{-q}(\mathbf{v})\rangle}{k^2\,\epsilon^*(q)}\right)\\
&\quad\left. - \chi^{(2)*}(q'|q-q')\,\frac{\langle\phi_{-q'}\,\phi_{-q+q'}\,N^{a0}_{q}(\mathbf{v})\rangle}{k^2\,\epsilon^*(q)}\right].
\end{aligned} \tag{4.64}
$$

This shows that we need to evaluate the remaining third-order cumulants, $\langle\,\phi_{q-q'}\,\phi_{-q}\,N^{a0}_{q'}(\mathbf{v})\rangle$, $\langle\phi_{q'}\,\phi_{-q}\,N^{a0}_{q-q'}(\mathbf{v})\rangle$, $\langle\phi_{q'}\,\phi_{q-q'}\,N^{a0}_{-q}(\mathbf{v})\rangle$, and $\langle\phi_{-q'}\,\phi_{-q+q'}\,N^{a0}_{q}(\mathbf{v})\rangle$. These quantities are special cases of a generic form $\langle\,\phi_{q_1}\,\phi_{-q_1+q_2}\,N^{a0}_{-q_2}(\mathbf{v})\rangle$. Making use of (4.62) in order to evaluate ϕ_{q_1} and $\phi_{-q_1+q_2}$ successively, we have

$$
\begin{aligned}
\langle\phi_{q_1}\,\phi_{-q_1+q_2}\,N^{a0}_{-q_2}(\mathbf{v})\rangle &= \frac{1}{k_1^2\,\epsilon(q_1)}\sum_{q'} i\,k_1\,k'\,|\mathbf{k}_1-\mathbf{k}'|\,\chi^{(2)}(q'|q_1-q')\\
&\quad\times\left[\langle\phi_{q'}\,\phi_{q_1-q'}\,\phi_{-q_1+q_2}\,N^{a0}_{-q_2}(\mathbf{v})\rangle\right.\\
&\quad\left. - \langle\phi_{q'}\,\phi_{q_1-q'}\rangle\,\langle\phi_{-q_1+q_2}\,N^{a0}_{-q_2}(\mathbf{v})\rangle\right]\\
&\quad + \sum_{q'}\frac{i\,|\mathbf{k}_1-\mathbf{k}_2|\,k'\,|\mathbf{k}_1-\mathbf{k}_2+\mathbf{k}'|\chi^{(2)}(q'|-q_1+q_2-q')}{|\mathbf{k}_1-\mathbf{k}_2|^2\,\epsilon(-q_1+q_2)}\\
&\quad\times\left[\langle\phi_{q'}\,\phi_{-q_1+q_2-q'}\,\phi_{q_1}\,N^{a0}_{-q_2}(\mathbf{v})\rangle\right.\\
&\quad\left. - \langle\phi_{q'}\,\phi_{-q_1+q_2-q'}\rangle\,\langle\phi_{q_1}\,N^{a0}_{-q_2}(\mathbf{v})\rangle\right]\\
&\quad + \frac{1}{k_1^2\,\epsilon(q_1)}\sum_b 4\pi e_b\int d\mathbf{v}'\,\langle N^{b0}_{q_1}(\mathbf{v}')\,\phi_{-q_1+q_2}\,N^{a0}_{-q_2}(\mathbf{v})\rangle\\
&\quad + \frac{1}{|\mathbf{k}_1-\mathbf{k}_2|^2\,\epsilon(-q_1+q_2)}\sum_b 4\pi e_b\int d\mathbf{v}'\\
&\quad\times\langle N^{b0}_{-q_1+q_2}(\mathbf{v}')\,\phi_{q_1}\,N^{a0}_{-q_2}(\mathbf{v})\rangle.
\end{aligned} \tag{4.65}
$$

Of the various terms in (4.65), those that contain three-body cumulants of the form $\langle \phi\, N\, N \rangle$ are all of purely particle effects, with no wave intensity appearing in the expressions. These terms can be ignored since the discrete particle effects are assumed to be a correction to what is essentially a Vlasovian weak turbulence formalism. As a result, we now have a simplification:

$$
\begin{aligned}
\langle \phi_{q_1}\, \phi_{-q_1+q_2}\, N^{a0}_{-q_2}(\mathbf{v}) \rangle &= \frac{2i\, k_2\, \chi^{(2)}(q_2|q_1-q_2)}{k_1\, |\mathbf{k}_1-\mathbf{k}_2|\, \epsilon(q_1)} \langle E^2 \rangle_{q_1-q_2}\, \langle \phi_{q_2}\, N^{a0}_{-q_2}(\mathbf{v}) \rangle \\
&\quad + \frac{2i\, k_2\, \chi^{(2)}(-q_1|q_2)}{k_1\, |\mathbf{k}_1-\mathbf{k}_2|\, \epsilon(-q_1+q_2)} \langle E^2 \rangle_{q_1}\, \langle \phi_{q_2}\, N^{a0}_{-q_2}(\mathbf{v}) \rangle \\
&= \frac{8\pi\, e_a i}{(2\pi)^3\, k_1\, k_2\, |\mathbf{k}_1-\mathbf{k}_2|\, \epsilon(q_2)} \left(\frac{\chi^{(2)}(q_2|q_1-q_2)}{\epsilon(q_1)} \langle E^2 \rangle_{q_1-q_2} \right. \\
&\quad \left. + \frac{\chi^{(2)}(-q_1|q_2)}{\epsilon(-q_1+q_2)} \langle E^2 \rangle_{q_1} \right) \delta(\omega_2 - \mathbf{k}_2 \cdot \mathbf{v})\, f_a(\mathbf{v}). \qquad (4.66)
\end{aligned}
$$

The remaining task is to identify $(q_1, q_2) = (q-q', -q')$, $(q_1, q_2) = (q', -q+q')$, $(q_1, q_2) = (q', q)$, and $(q_1, q_2) = (-q', -q)$ successively. Actual evaluations of the quantity $\langle \phi_{q_1}\, \phi_{-q_1+q_2}\, N^{a0}_{-q_2}(\mathbf{v}) \rangle$ with the corresponding identifications of (q_1, q_2) as specified in (4.66) are omitted, and the intermediate steps can easily be filled by the readers. Inserting the final results to (4.64), we have the desired nonlinear spectral balance equation including the effects of spontaneous fluctuations:

$$
\begin{aligned}
&\epsilon(q)\, \langle E^2 \rangle_q - 2 \int dq' \, \frac{|\, \chi^{(2)}(q'|q-q')\,|^2}{\epsilon^*(q)} \langle E^2 \rangle_{q'}\, \langle E^2 \rangle_{q-q'} \\
&\quad + 2 \int dq' \left[\{ \chi^{(2)}(q'|q-q') \}^2 \left(\frac{\langle E^2 \rangle_{q-q'}}{\epsilon(q')} + \frac{\langle E^2 \rangle_{q'}}{\epsilon(q-q')} \right) \right. \\
&\quad \left. - \bar{\chi}^{(3)}(q'|-q'|q)\, \langle E^2 \rangle_{q'} \right] \langle E^2 \rangle_q \\
&= \sum_a \frac{(4\pi e_a)^2}{(2\pi)^3\, k^2\, \epsilon^*(q)} \int d\mathbf{v}\, \delta(\omega - \mathbf{k}\cdot\mathbf{v})\, f_a(\mathbf{v}) \\
&\quad - \sum_a \int dq' \, \frac{2\,(4\pi e_a)^2}{(2\pi)^3\, k'^2\, |\epsilon(q')|^2} \left(\frac{\{ \chi^{(2)}(q'|q-q') \}^2}{\epsilon(q-q')} \langle E^2 \rangle_q \right. \qquad (4.67) \\
&\quad \left. - \frac{|\, \chi^{(2)}(q'|q-q')\,|^2}{\epsilon^*(q)} \langle E^2 \rangle_{q-q'} \right) \int d\mathbf{v}\, \delta(\omega' - \mathbf{k}'\cdot\mathbf{v})\, f_a(\mathbf{v}) \\
&\quad - \sum_a \int dq' \, \frac{2\,(4\pi e_a)^2}{(2\pi)^3\, |\mathbf{k}-\mathbf{k}'|^2\, |\epsilon(q-q')|^2} \left(\frac{\{ \chi^{(2)}(q'|q-q') \}^2}{\epsilon(q')} \langle E^2 \rangle_q \right. \\
&\quad \left. - \frac{|\, \chi^{(2)}(q'|q-q')\,|^2}{\epsilon^*(q)} \langle E^2 \rangle_{q'} \right) \int d\mathbf{v}\, \delta[\omega - \omega' - (\mathbf{k}-\mathbf{k}')\cdot\mathbf{v}]\, f_a(\mathbf{v}).
\end{aligned}
$$

In (4.67) we have ignored the nonlinear correction terms on the right-hand side of the equality, which are associated with the linear wave-particle resonance factor $\int d\mathbf{v}\,\delta(\omega - \mathbf{k}\cdot\mathbf{v})\,f_a(\mathbf{v})$, since they are but small corrections.

At this point, we reintroduce the slow-time derivative, as in (1.83). This leads to the formal nonlinear spectral balance equation that generalizes (1.89):

$$
\begin{aligned}
&\frac{i}{2}\frac{\partial\,\mathrm{Re}\,\epsilon(\mathbf{k},\omega)}{\partial\omega}\frac{\partial\langle\delta E^2\rangle_{\mathbf{k},\omega}}{\partial t} + \mathrm{Re}\,\epsilon(\mathbf{k},\omega)\,\langle\delta E^2\rangle_{\mathbf{k},\omega} + i\,\mathrm{Im}\,\epsilon(\mathbf{k},\omega)\,\langle\delta E^2\rangle_{\mathbf{k},\omega} \\
&\quad + 2\int d\mathbf{k}'\int d\omega'\left[\{\chi^{(2)}(\mathbf{k}',\omega'|\mathbf{k}-\mathbf{k}',\omega-\omega')\}^2\left(\frac{\langle\delta E^2\rangle_{\mathbf{k}-\mathbf{k}',\omega-\omega'}}{\epsilon(\mathbf{k}',\omega')}\right.\right. \\
&\quad \left.\left. + \frac{\langle\delta E^2\rangle_{\mathbf{k}',\omega'}}{\epsilon(\mathbf{k}-\mathbf{k}',\omega-\omega')}\right) - \bar{\chi}^{(3)}(\mathbf{k}',\omega'|-\mathbf{k}',\,-\omega'|\mathbf{k},\omega)\,\langle\delta E^2\rangle_{\mathbf{k}',\omega'}\right]\langle\delta E^2\rangle_{\mathbf{k},\omega} \\
&\quad - 2\int d\mathbf{k}'\int d\omega'\,\frac{|\,\chi^{(2)}(\mathbf{k}',\omega'|\mathbf{k}-\mathbf{k}',\omega-\omega')\,|^2}{\epsilon^*(\mathbf{k},\omega)}\,\langle\delta E^2\rangle_{\mathbf{k}',\omega'}\,\langle\delta E^2\rangle_{\mathbf{k}-\mathbf{k}',\omega-\omega'} \\
&= \frac{2}{\pi}\frac{1}{k^2\,\epsilon^*(\mathbf{k},\omega)}\sum_a e_a^2\int d\mathbf{v}\;\delta(\omega-\mathbf{k}\cdot\mathbf{v})\;f_a(\mathbf{v}) \\
&\quad - \frac{4}{\pi}\int d\mathbf{k}'\int d\omega'\,\frac{1}{k'^2\,|\epsilon(\mathbf{k}',\omega')|^2}\left(\frac{\{\chi^{(2)}(\mathbf{k}',\omega'|\mathbf{k}-\mathbf{k}',\omega-\omega')\}^2}{\epsilon(\mathbf{k}-\mathbf{k}',\omega-\omega')}\,\langle\delta E^2\rangle_{\mathbf{k},\omega}\right. \\
&\quad \left. - \frac{|\,\chi^{(2)}(\mathbf{k}',\omega'|\mathbf{k}-\mathbf{k}',\omega-\omega')\,|^2}{\epsilon^*(\mathbf{k},\omega)}\,\langle\delta E^2\rangle_{\mathbf{k}-\mathbf{k}',\omega-\omega'}\right) \\
&\quad \times\sum_a e_a^2\int d\mathbf{v}\;\delta(\omega'-\mathbf{k}'\cdot\mathbf{v})\;f_a(\mathbf{v}) \\
&\quad - \frac{4}{\pi}\int d\mathbf{k}'\int d\omega'\,\frac{1}{|\mathbf{k}-\mathbf{k}'|^2\,|\epsilon(\mathbf{k}-\mathbf{k}',\omega-\omega')|^2} \\
&\quad \times\left(\frac{\{\chi^{(2)}(\mathbf{k}',\omega'|\mathbf{k}-\mathbf{k}',\omega-\omega')\}^2}{\epsilon(\mathbf{k}',\omega')}\,\langle\delta E^2\rangle_{\mathbf{k},\omega}\right. \\
&\quad \left. - \frac{|\,\chi^{(2)}(\mathbf{k}',\omega'|\mathbf{k}-\mathbf{k}',\omega-\omega')\,|^2}{\epsilon^*(\mathbf{k},\omega)}\,\langle\delta E^2\rangle_{\mathbf{k}',\omega'}\right) \\
&\quad \times\sum_a e_a^2\int d\mathbf{v}\;\delta[\omega-\omega'-(\mathbf{k}-\mathbf{k}')\cdot\mathbf{v}]\;f_a(\mathbf{v}),
\end{aligned}
\tag{4.68}
$$

where we have resorted back to the longhand notation. If we ignore everything to the right of equality then we recover (1.89).

We next discuss the particle kinetic equation. From (4.49), the particle kinetic equation is given by

$$
\frac{\partial f_a}{\partial t} = -\frac{i e_a}{m_a}\int d\mathbf{k}\int d\omega\,\mathbf{k}\cdot\frac{\partial}{\partial\mathbf{v}}\langle\phi_{-q}\,N_q^a\rangle. \tag{4.69}
$$

To this, we insert the expression for N_q^a as given in (4.54). The quantity of interest is

$$
\begin{aligned}
\langle\phi_{-q}\,N_q^a\rangle &= \langle\phi_{-q}\,N_q^{a0}\rangle + \alpha_a(q)\,f_a\,\langle\phi^2\rangle_q + \sum_{q'}\alpha_a^{(2)}(q'|q-q')\,f_a\,\langle\phi_{-q}\,\phi_{q'}\,\phi_{q-q'}\rangle \\
&\quad + 2\sum_{q'}\alpha_a^{(3)}(q'|-q'|q)\,f_a\,\langle\phi^2\rangle_q\,\langle\phi^2\rangle_{q'}.
\end{aligned}
\tag{4.70}
$$

Making use of (4.60), (4.63), and (4.66), we obtain

$$
\begin{aligned}
\langle\phi_{-q}\,N_q^a\rangle &= \alpha_a(q)\,\langle\phi^2\rangle_q\,f_a + \frac{4\pi e_a}{(2\pi)^3\,k^2\,\epsilon^*(q)} \\
&\quad\times\left(1+\frac{2}{\epsilon^*(q)}\sum_{q'}\bar{\chi}^{(3)*}(q'|-q'|q)\,\langle k'^2\phi^2\rangle_{q'}\right)\delta(\omega-\mathbf{k}\cdot\mathbf{v})\,f_a \\
&\quad + 2i\sum_{q'}\alpha_a^{(2)}(q'|q-q')\,k\,k'\,|\mathbf{k}-\mathbf{k}'|\left(\frac{\chi^{(2)}(q|-q+q')}{k'^2\,\epsilon(q')}\,\langle\phi^2\rangle_{q-q'}\,\langle\phi^2\rangle_q\right. \\
&\quad + \left.\frac{\chi^{(2)}(q|-q')}{|\mathbf{k}-\mathbf{k}'|^2\,\epsilon(q-q')}\langle\phi^2\rangle_{q'}\langle\phi^2\rangle_q - \frac{\chi^{(2)*}(q'|q-q')}{k^2\,\epsilon^*(q)}\,\langle\phi^2\rangle_{q'}\langle\phi^2\rangle_{q-q'}\right)f_a \\
&\quad + \sum_{q'}\alpha_a^{(2)}(q'|q-q')\sum_b 4\pi e_b\int d\mathbf{v}\left(\frac{\langle\,\phi_{q-q'}\,\phi_{-q}\,N_{q'}^{b0}(\mathbf{v})\rangle}{k'^2\,\epsilon(q')}\right. \\
&\quad + \left.\frac{\langle\phi_{q'}\,\phi_{-q}\,N_{q-q'}^{b0}(\mathbf{v})\rangle}{|\mathbf{k}-\mathbf{k}'|^2\,\epsilon(q-q')} + \frac{\langle\phi_{q'}\,\phi_{q-q'}\,N_{-q}^{b0}(\mathbf{v})\rangle}{k^2\,\epsilon^*(q)}\right)f_a \\
&\quad + 2\sum_{q'}\alpha_a^{(3)}(q'|-q'|q)\,f_a\,\langle\phi^2\rangle_q\,\langle\phi^2\rangle_{q'}.
\end{aligned}
\tag{4.71}
$$

Among the terms in (4.71), those terms that contain the factors of the form $\langle\phi\,\phi\,N\rangle$ will be ignored, since these factors are proportional to f_a, leading to an overall contribution of the form f_a^2. These are higher-order terms associated with the single-particle effects, which we ignore. Thus, we simplify (4.71) as

$$
\begin{aligned}
\langle\phi_{-q}\,N_q^a\rangle &= \frac{1}{k^2}\,\alpha_a(q)\,\langle E^2\rangle_q\,f_a + \frac{4\pi e_a}{(2\pi)^3\,k^2\,\epsilon^*(q)} \\
&\quad\times\left(1+\frac{2}{\epsilon^*(q)}\sum_{q'}\bar{\chi}^{(3)*}(q'|-q'|q)\,\langle E^2\rangle_{q'}\right)\delta(\omega-\mathbf{k}\cdot\mathbf{v})\,f_a \\
&\quad + 2i\sum_{q'}\frac{\alpha_a^{(2)}(q'|q-q')}{k\,k'\,|\mathbf{k}-\mathbf{k}'|}\left(\frac{\chi^{(2)}(q'|q-q')}{\epsilon(q')}\,\langle E^2\rangle_{q-q'}\,\langle E^2\rangle_q\right.
\end{aligned}
$$

$$+\frac{\chi^{(2)}(q'|q-q')}{\epsilon(q-q')}\langle E^2\rangle_{q'}\langle E^2\rangle_q-\frac{\chi^{(2)*}(q'|q-q')}{\epsilon^*(q)}\langle E^2\rangle_{q'}\langle E^2\rangle_{q-q'}\bigg)f_a$$
$$+2\sum_{q'}\frac{\alpha_a^{(3)}(q'|-q'|q)}{k^2k'^2}\langle E^2\rangle_q\langle E^2\rangle_{q'}f_a. \tag{4.72}$$

Making explicit definitions for $\alpha_a^{(2)}(q'|q-q')$ and $\alpha_a^{(3)}(q'|-q'|q)$ as defined in (1.53), we have

$$\begin{aligned}
\langle\phi_{-q}N_q^a\rangle &= -\frac{e_a}{m_a}\frac{\langle E^2\rangle_q}{\omega-\mathbf{k}\cdot\mathbf{v}}\frac{\mathbf{k}}{k^2}\cdot\frac{\partial f_a}{\partial\mathbf{v}}\\
&+\frac{4\pi e_a}{(2\pi)^3k^2\epsilon^*(q)}\left(1+\frac{2}{\epsilon^*(q)}\sum_{q'}\bar\chi^{(3)*}(q'|-q'|q)\langle E^2\rangle_{q'}\right)\delta(\omega-\mathbf{k}\cdot\mathbf{v})f_a\\
&+\frac{ie_a^2}{m_a^2}\sum_{q'}\frac{1}{k\,k'\,|\mathbf{k}-\mathbf{k}'|}\left(\frac{\chi^{(2)}(q'|q-q')}{\epsilon(q')}\langle E^2\rangle_{q-q'}\langle E^2\rangle_q\right.\\
&\left.+\frac{\chi^{(2)}(q'|q-q')}{\epsilon(q-q')}\langle E^2\rangle_{q'}\langle E^2\rangle_q-\frac{\chi^{(2)*}(q'|q-q')}{\epsilon^*(q)}\langle E^2\rangle_{q'}\langle E^2\rangle_{q-q'}\right)\\
&\times\frac{1}{\omega-\mathbf{k}\cdot\mathbf{v}}\left(\mathbf{k}'\cdot\frac{\partial}{\partial\mathbf{v}}\frac{1}{\omega-\omega'-(\mathbf{k}-\mathbf{k}')\cdot\mathbf{v}}(\mathbf{k}-\mathbf{k}')\cdot\frac{\partial}{\partial\mathbf{v}}\right.\\
&\left.+(\mathbf{k}-\mathbf{k}')\cdot\frac{\partial}{\partial\mathbf{v}}\frac{1}{\omega'-\mathbf{k}'\cdot\mathbf{v}}\mathbf{k}'\cdot\frac{\partial}{\partial\mathbf{v}}\right)f_a\\
&+\frac{e_a^3}{m_a^3}\sum_{q'}\frac{1}{k^2k'^2}\frac{\langle E^2\rangle_q\langle E^2\rangle_{q'}}{\omega-\mathbf{k}\cdot\mathbf{v}}\mathbf{k}'\cdot\frac{\partial}{\partial\mathbf{v}}\frac{1}{\omega-\omega'-(\mathbf{k}-\mathbf{k}')\cdot\mathbf{v}}\\
&\times\left(\mathbf{k}'\cdot\frac{\partial}{\partial\mathbf{v}}\frac{1}{\omega-\mathbf{k}\cdot\mathbf{v}}\mathbf{k}\cdot\frac{\partial}{\partial\mathbf{v}}-\mathbf{k}\cdot\frac{\partial}{\partial\mathbf{v}}\frac{1}{\omega'-\mathbf{k}'\cdot\mathbf{v}}\mathbf{k}'\cdot\frac{\partial}{\partial\mathbf{v}}\right)f_a.
\end{aligned}\tag{4.73}$$

Inserting this result to (4.69), we obtain

$$\begin{aligned}
\frac{\partial f_a}{\partial t} &= \frac{ie_a^2}{m_a^2}\frac{\partial}{\partial v_i}\int d\mathbf{k}\int d\omega\,\frac{k_ik_j}{k^2}\frac{\langle\delta E^2\rangle_{\mathbf{k},\omega}}{\omega-\mathbf{k}\cdot\mathbf{v}}\frac{\partial f_a}{\partial v_j}\\
&-\frac{i\,(4\pi e_a^2)}{(2\pi)^3m_a}\frac{\partial}{\partial v_i}\int d\mathbf{k}\,d\omega\,\frac{k_i}{k^2\epsilon^*(\mathbf{k},\omega)}\left(1+\frac{2}{\epsilon^*(\mathbf{k},\omega)}\right.\\
&\times\left.\int d\mathbf{k}'\,d\omega'\,\bar\chi^{(3)*}(\mathbf{k}',\omega'|-\mathbf{k}',-\omega'|\mathbf{k},\omega)\langle\delta E^2\rangle_{\mathbf{k}',\omega'}\right)\delta(\omega-\mathbf{k}\cdot\mathbf{v})f_a\\
&+\frac{e_a^3}{m_a^3}\frac{\partial}{\partial v_i}\int d\mathbf{k}\,d\omega\int d\mathbf{k}'\,d\omega'\,\frac{k_ik_j'(\mathbf{k}-\mathbf{k}')_k}{k\,k'\,|\mathbf{k}-\mathbf{k}'|}\Big[\chi^{(2)}(\mathbf{k}',\omega'|\mathbf{k}-\mathbf{k}',\omega-\omega')
\end{aligned}$$

$$\times \left(\frac{\langle \delta E^2 \rangle_{\mathbf{k}-\mathbf{k}',\omega-\omega'}}{\epsilon(\mathbf{k}',\omega')} + \frac{\langle \delta E^2 \rangle_{\mathbf{k}',\omega'}}{\epsilon(\mathbf{k}-\mathbf{k}',\omega-\omega')} \right) \langle \delta E^2 \rangle_{\mathbf{k},\omega} \tag{4.74}$$

$$- \chi^{(2)*}(\mathbf{k}',\omega'|\mathbf{k}-\mathbf{k}',\omega-\omega') \frac{\langle \delta E^2 \rangle_{\mathbf{k}',\omega'} \langle \delta E^2 \rangle_{\mathbf{k}-\mathbf{k}',\omega-\omega'}}{\epsilon^*(\mathbf{k},\omega)} \Bigg]$$

$$\times \frac{1}{\omega - \mathbf{k}\cdot\mathbf{v}} \left(\frac{\partial}{\partial v_j} \frac{1}{\omega-\omega'-(\mathbf{k}-\mathbf{k}')\cdot\mathbf{v}} \frac{\partial}{\partial v_k} + \frac{\partial}{\partial v_k} \frac{1}{\omega'-\mathbf{k}'\cdot\mathbf{v}} \frac{\partial}{\partial v_j} \right) f_a$$

$$- \frac{i e_a^4}{m_a^4} \frac{\partial}{\partial v_i} \int d\mathbf{k} \int d\omega \int d\mathbf{k}' \int d\omega' \frac{k_i k_j' k_k' k_l}{k^2 k'^2} \frac{\langle \delta E^2 \rangle_{\mathbf{k},\omega} \langle \delta E^2 \rangle_{\mathbf{k}',\omega'}}{\omega - \mathbf{k}\cdot\mathbf{v}} \frac{\partial}{\partial v_j}$$

$$\times \frac{1}{\omega-\omega'-(\mathbf{k}-\mathbf{k}')\cdot\mathbf{v}} \left(\frac{\partial}{\partial v_k} \frac{1}{\omega-\mathbf{k}\cdot\mathbf{v}} \frac{\partial}{\partial v_l} - \frac{\partial}{\partial v_l} \frac{1}{\omega'-\mathbf{k}'\cdot\mathbf{v}} \frac{\partial}{\partial v_k} \right) f_a.$$

Taking only the real part of the right-hand side, we obtain the following:

$$\frac{\partial f_a}{\partial t} = \frac{\pi e_a^2}{m_a^2} \int d\mathbf{k} \int d\omega \left(\frac{\mathbf{k}}{k} \cdot \frac{\partial}{\partial \mathbf{v}} \right) \delta(\omega - \mathbf{k}\cdot\mathbf{v})$$

$$\times \left[\boxed{\mathrm{Im} \frac{m_a \, \epsilon(\mathbf{k},\omega)}{2\pi^3 k \, |\epsilon(\mathbf{k},\omega)|^2} f_a} + \langle \delta E^2 \rangle_{\mathbf{k},\omega} \left(\frac{\mathbf{k}}{k} \cdot \frac{\partial f_a}{\partial \mathbf{v}} \right) \right]$$

$$\boxed{\begin{aligned} &+ \frac{e_a^2}{m_a^2} \int d\mathbf{k} \int d\omega \int d\mathbf{k}' \int d\omega' \left(\frac{\mathbf{k}}{k} \cdot \frac{\partial}{\partial \mathbf{v}} \right) \mathrm{Im} \frac{1}{\pi^2 k \, [\epsilon^*(\mathbf{k},\omega)]^2} \\ &\times \bar{\chi}^{(3)*}(\mathbf{k}',\omega'|-\mathbf{k}',-\omega'|\mathbf{k},\omega) \langle \delta E^2 \rangle_{\mathbf{k}',\omega'} \, \delta(\omega - \mathbf{k}\cdot\mathbf{v}) \, f_a \end{aligned}}$$

$$+ \mathrm{Re} \frac{e_a^3}{m_a^3} \frac{\partial}{\partial v_i} \int d\mathbf{k} \int d\omega \int d\mathbf{k}' \int d\omega' \frac{k_i k_j' (\mathbf{k}-\mathbf{k}')_k}{k k' |\mathbf{k}-\mathbf{k}'|} \frac{M(\mathbf{k},\omega|\mathbf{k}',\omega')}{\omega - \mathbf{k}\cdot\mathbf{v}}$$

$$\times \left(\frac{\partial}{\partial v_j} \frac{1}{\omega-\omega'-(\mathbf{k}-\mathbf{k}')\cdot\mathbf{v}} \frac{\partial}{\partial v_k} + \frac{\partial}{\partial v_k} \frac{1}{\omega'-\mathbf{k}'\cdot\mathbf{v}} \frac{\partial}{\partial v_j} \right) f_a$$

$$+ \mathrm{Im} \frac{e_a^4}{m_a^4} \frac{\partial}{\partial v_i} \int d\mathbf{k} \int d\omega \int d\mathbf{k}' \int d\omega' \frac{k_i k_j' k_k' k_l}{k^2 k'^2} \frac{\langle \delta E^2 \rangle_{\mathbf{k},\omega} \langle \delta E^2 \rangle_{\mathbf{k}',\omega'}}{\omega - \mathbf{k}\cdot\mathbf{v}} \frac{\partial}{\partial v_j}$$

$$\times \frac{1}{\omega-\omega'-(\mathbf{k}-\mathbf{k}')\cdot\mathbf{v}} \left(\frac{\partial}{\partial v_k} \frac{1}{\omega-\mathbf{k}\cdot\mathbf{v}} \frac{\partial}{\partial v_l} - \frac{\partial}{\partial v_l} \frac{1}{\omega'-\mathbf{k}'\cdot\mathbf{v}} \frac{\partial}{\partial v_k} \right) f_a, \tag{4.75}$$

where we have indicated the new terms that arise from adding the discrete particle effects by enclosing them within boxes, and $M(\mathbf{k},\omega|\mathbf{k}',\omega')$ is defined in (1.106).

4.3 Weak Turbulence Theory with Discrete Particle Effects

Let us now make use of the formal results (4.68) and (4.75) and derive specific forms of equations of weak turbulence theory in which linear eigenmodes interact among themselves and with the particles, and where discrete particle effects are incorporated. Let us first dispense with the particle kinetic equation (4.75), since it is relatively straightforward.

4.3.1 Particle Kinetic Equation with Discrete Particle Effects

Consider the electron equation first. Recall that Chapter 3 was devoted to the analysis of particle kinetic equation based upon (1.107). In the end, it was shown that the modified kinetic equation for electrons is given by (3.30). In the present Klimontovich formalism the formal particle kinetic equation is generalized to (4.75). It suffices to consider the new, boxed terms in order to address the additional effects that arise from the particle discreteness. Thus, the new terms in the electron kinetic equation are first for the linear wave-particle resonance,

$$\left.\frac{\partial f_e}{\partial t}\right|_{\text{new}}^{\text{lin.}} = \frac{e^2}{4\pi m_e}\int d\mathbf{k}\left(\frac{\mathbf{k}}{k^2}\cdot\frac{\partial}{\partial\mathbf{v}}\right)\sum_{\sigma=\pm 1}\sigma\omega_{\mathbf{k}}^L\,\delta(\sigma\omega_{\mathbf{k}}^L-\mathbf{k}\cdot\mathbf{v})\,f_e, \tag{4.76}$$

while for the nonlinear term,

$$\begin{aligned}\left.\frac{\partial f_e}{\partial t}\right|_{\text{new}}^{\text{nonlin.}} &= \frac{e^2}{m_e^2}\int d\mathbf{k}\int d\omega\int d\mathbf{k}'\int d\omega'\left(\frac{\mathbf{k}}{k}\cdot\frac{\partial}{\partial\mathbf{v}}\right)\text{Im}\,\frac{1}{\pi^2 k[\epsilon^*(\mathbf{k},\omega)]^2}\\ &\quad\times\bar{\chi}^{(3)*}(\mathbf{k}',\omega'|-\mathbf{k}',-\omega'|\mathbf{k},\omega)\,\langle\delta E^2\rangle_{\mathbf{k}',\omega'}\,\delta(\omega-\mathbf{k}\cdot\mathbf{v})\,f_e\\ &= \frac{e^2}{m_e^2}\sum_{\sigma'=\pm 1}\int d\mathbf{k}\int d\omega\int d\mathbf{k}'\left(\frac{\mathbf{k}}{k}\cdot\frac{\partial}{\partial\mathbf{v}}\right)\text{Im}\,\frac{1}{\pi^2 k[\epsilon^*(\mathbf{k},\omega)]^2}\\ &\quad\times\bar{\chi}^{(3)*}(\mathbf{k}',\sigma'\omega_{\mathbf{k}'}^L|-\mathbf{k}',-\sigma'\omega_{\mathbf{k}'}^L|\mathbf{k},\omega)\,I_{\mathbf{k}'}^{\sigma' L}\,\delta(\omega-\mathbf{k}\cdot\mathbf{v})\,f_e.\end{aligned} \tag{4.77}$$

If we make use of

$$\begin{aligned}&\bar{\chi}^{(3)*}(\mathbf{k}',\sigma'\omega_{\mathbf{k}'}^L|-\mathbf{k}',-\sigma'\omega_{\mathbf{k}'}^L|\mathbf{k},\omega)\\ &\quad= -\frac{1}{2}\frac{e^2}{m_e^2}\frac{|\mathbf{k}-\mathbf{k}'|^2}{\omega^3(\sigma'\omega_{\mathbf{k}'}^L)}\frac{(\mathbf{k}\cdot\mathbf{k}')^2}{k^2k'^2}\,\chi_e(\mathbf{k}-\mathbf{k}',\omega-\sigma'\omega_{\mathbf{k}'}^L),\end{aligned} \tag{4.78}$$

and assume that ω is in the vicinity of Langmuir wave dispersion relation, then we may approximate

$$\chi_e(\mathbf{k}-\mathbf{k}',\omega-\sigma'\omega_{\mathbf{k}'}^L)\approx\frac{2\omega_{pe}^2}{|\mathbf{k}-\mathbf{k}'|^2 v_{Te}^2}. \tag{4.79}$$

This leads to

$$\bar{\chi}^{(3)*}(\mathbf{k}',\sigma'\omega_{\mathbf{k}'}^L|-\mathbf{k}',-\sigma'\omega_{\mathbf{k}'}^L|\mathbf{k},\omega) = -\frac{1}{2}\frac{e^2}{m_e}\frac{\omega_{pe}^2}{T_e}\frac{1}{\omega^3(\sigma'\omega_{\mathbf{k}'}^L)}\frac{(\mathbf{k}\cdot\mathbf{k}')^2}{k^2k'^2}, \tag{4.80}$$

resulting in

$$\begin{aligned}\frac{\partial f_e}{\partial t}\bigg|_{\text{new}}^{\text{nonlin.}} &= -\frac{e^4}{m_e^3}\frac{\omega_{pe}^2}{2\pi^2 T_e}\sum_{\sigma'=\pm1}\int d\mathbf{k}\int d\omega\int d\mathbf{k}'\left(\frac{\mathbf{k}}{k}\cdot\frac{\partial}{\partial\mathbf{v}}\right)\text{Im}\,\frac{1}{k[\epsilon^*(\mathbf{k},\omega)]^2}\\ &\quad\times\frac{1}{\omega^3(\sigma'\omega_{\mathbf{k}'}^L)}\frac{(\mathbf{k}\cdot\mathbf{k}')^2}{k^2k'^2}I_{\mathbf{k}'}^{\sigma'L}\,\delta(\omega-\mathbf{k}\cdot\mathbf{v})f_e.\end{aligned} \tag{4.81}$$

Writing

$$\frac{1}{[\epsilon^*(\mathbf{k},\omega)]^2} = \mathscr{P}\frac{1}{\epsilon^*(\mathbf{k},\omega)}\sum_{\sigma=\pm1}\frac{i\pi\,\delta(\omega-\sigma\omega_{\mathbf{k}}^L)}{\epsilon'(\mathbf{k},\sigma\omega_{\mathbf{k}}^L)}, \tag{4.82}$$

we obtain the nonlinear correction term to the electron kinetic equation that stems from the discrete particle effects,

$$\begin{aligned}\frac{\partial f_e}{\partial t}\bigg|_{\text{new}}^{\text{nonlin.}} &= -\frac{e^4}{m_e^3}\frac{\omega_{pe}^2}{4\pi T_e}\sum_{\sigma,\sigma'=\pm1}\int d\mathbf{k}\int d\mathbf{k}'\left(\frac{\mathbf{k}}{k}\cdot\frac{\partial}{\partial\mathbf{v}}\right)\mathscr{P}\frac{1}{\epsilon^*(\mathbf{k},\sigma\omega_{\mathbf{k}}^L)}\\ &\quad\times\frac{1}{(\sigma\omega_{\mathbf{k}}^L)^2(\sigma'\omega_{\mathbf{k}'}^L)}\frac{(\mathbf{k}\cdot\mathbf{k}')^2}{k^3k'^2}I_{\mathbf{k}'}^{\sigma'L}\,\delta(\sigma\omega_{\mathbf{k}}^L-\mathbf{k}\cdot\mathbf{v})F_e.\end{aligned} \tag{4.83}$$

Turning to the ion equation, the new terms are

$$\begin{aligned}\frac{\partial f_i}{\partial t}\bigg|_{\text{new}}^{\text{lin.}} &= \frac{e^2}{4\pi m_i}\int d\mathbf{k}\left(\frac{\mathbf{k}}{k^2}\cdot\frac{\partial}{\partial\mathbf{v}}\right)\sum_{\sigma=\pm1}\sigma\mu_{\mathbf{k}}\omega_{\mathbf{k}}^L\,\delta(\sigma\omega_{\mathbf{k}}^S-\mathbf{k}\cdot\mathbf{v})f_i,\\ \frac{\partial f_i}{\partial t}\bigg|_{\text{new}}^{\text{nonlin.}} &= \frac{e^2}{m_i^2}\frac{1}{2\pi}\sum_{\sigma,\sigma'=\pm1}\int d\mathbf{k}\int d\mathbf{k}'\left(\frac{\mathbf{k}}{k}\cdot\frac{\partial}{\partial\mathbf{v}}\right)\mathscr{P}\frac{1}{\epsilon^*(\mathbf{k},\sigma\omega_{\mathbf{k}}^S)}\frac{\sigma\mu_{\mathbf{k}}\omega_{\mathbf{k}}^L}{k}\\ &\quad\times\bar{\chi}^{(3)*}(\mathbf{k}',\sigma'\omega_{\mathbf{k}'}^S|-\mathbf{k}',-\sigma'\omega_{\mathbf{k}'}^S|\mathbf{k},\sigma\omega_{\mathbf{k}}^S)\,I_{\mathbf{k}'}^{\sigma'S}\,\delta(\sigma\omega_{\mathbf{k}}^S-\mathbf{k}\cdot\mathbf{v})f_i.\end{aligned} \tag{4.84}$$

Since $\omega_{\mathbf{k}}^S > kv_{Ti}$, $\omega_{\mathbf{k}}^S \ll kv_{Te}$, $\omega_{\mathbf{k}'}^S > k'v_{Ti}$, $\omega_{\mathbf{k}'}^S \ll k'v_{Te}$, and $\omega_{\mathbf{k}}^S - \omega_{\mathbf{k}'}^S \ll |\mathbf{k}-\mathbf{k}'|v_{Tj}$, $j=i,e$, we may approximate

$$\begin{aligned}\bar{\chi}^{(3)*}(\mathbf{k}',\sigma'\omega_{\mathbf{k}'}^S|-\mathbf{k}',-\sigma'\omega_{\mathbf{k}'}^S|\mathbf{k},\sigma\omega_{\mathbf{k}}^S) &\approx \bar{\chi}_i^{(3)*}(\mathbf{k}',\sigma'\omega_{\mathbf{k}'}^S|-\mathbf{k}',-\sigma'\omega_{\mathbf{k}'}^S|\mathbf{k},\sigma\omega_{\mathbf{k}}^S)\\ &= -\frac{1}{2}\frac{e^2}{m_i}\frac{\omega_{pi}^2}{T_i}\frac{1}{(\sigma\omega_{\mathbf{k}}^S)^3\sigma'\omega_{\mathbf{k}'}^S}\frac{(\mathbf{k}\cdot\mathbf{k}')^2}{k^2k'^2}.\end{aligned} \tag{4.85}$$

Making use of the this result, we may simplify the nonlinear term in (4.84).

In summary, modified particle kinetic equations, which include the particle discreteness effects, are given by

$$
\begin{aligned}
\frac{\partial f_a}{\partial t} &= \frac{\partial}{\partial \mathbf{v}} \cdot \left(\mathbf{A}_a f_a + \mathsf{D}_a \cdot \frac{\partial f_a}{\partial \mathbf{v}} \right), \qquad (a = e, i), \\
\mathbf{A}_e &= \frac{e^2}{4\pi m_e} \int d\mathbf{k} \, \frac{\mathbf{k}}{k^2} \sum_{\sigma=\pm 1} \sigma \omega_{\mathbf{k}}^L \, \delta(\sigma \omega_{\mathbf{k}}^L - \mathbf{k}\cdot\mathbf{v}) \\
&\quad \times \left(1 - \frac{e^2}{m_e^2} \frac{\omega_{pe}^2}{T_e} \mathscr{P} \frac{1}{\epsilon^*(\mathbf{k}, \sigma\omega_{\mathbf{k}}^L)} \int d\mathbf{k}' \, \frac{(\mathbf{k}\cdot\mathbf{k}')^2}{k^2 k'^2} \sum_{\sigma'=\pm 1} \frac{I_{\mathbf{k}'}^{\sigma' L}}{(\sigma\omega_{\mathbf{k}}^L)^3 (\sigma' \omega_{\mathbf{k}'}^L)} \right), \\
\mathbf{A}_i &= \frac{e^2}{4\pi m_i} \int d\mathbf{k} \, \frac{\mathbf{k}}{k^2} \sum_{\sigma=\pm 1} \sigma \mu_{\mathbf{k}} \omega_{\mathbf{k}}^L \, \delta(\sigma \omega_{\mathbf{k}}^S - \mathbf{k}\cdot\mathbf{v}) \\
&\quad \times \left(1 - \frac{e^2}{m_i^2} \frac{\omega_{pi}^2}{T_i} \mathscr{P} \frac{1}{\epsilon^*(\mathbf{k}, \sigma\omega_{\mathbf{k}}^S)} \int d\mathbf{k}' \, \frac{(\mathbf{k}\cdot\mathbf{k}')^2}{k^2 k'^2} \sum_{\sigma'=\pm 1} \frac{I_{\mathbf{k}'}^{\sigma' S}}{(\sigma\omega_{\mathbf{k}}^S)^3 (\sigma' \omega_{\mathbf{k}'}^S)} \right), \\
\mathsf{D}_e &= \frac{\pi e^2}{m_e^2} \int d\mathbf{k} \, \frac{\mathbf{k}\mathbf{k}}{k^2} \sum_{\sigma=\pm 1} \delta(\sigma \omega_{\mathbf{k}}^L - \mathbf{k}\cdot\mathbf{v}) \, I_{\mathbf{k}}^{\sigma L} \\
&\quad + \frac{\pi e^4}{2 m_e^4} \int d\mathbf{k} \int d\mathbf{k}' \, \frac{(\mathbf{k}\cdot\mathbf{k}')^2}{k^2 k'^2} \sum_{\sigma,\sigma'=\pm 1} \frac{(\mathbf{k}-\mathbf{k}')\,(\mathbf{k}-\mathbf{k}')\, I_{\mathbf{k}'}^{\sigma' L} I_{\mathbf{k}}^{\sigma L}}{(\sigma\omega_{\mathbf{k}}^L - \mathbf{k}\cdot\mathbf{v})^2} \\
&\quad \times \left(\frac{1}{(\sigma\omega_{\mathbf{k}}^L - \mathbf{k}\cdot\mathbf{v})^2} - \frac{\sigma\sigma'}{\omega_{pe}^2} \right) \delta[\sigma\omega_{\mathbf{k}}^L - \sigma'\omega_{\mathbf{k}'}^L - (\mathbf{k}-\mathbf{k}')\cdot\mathbf{v}], \\
\mathsf{D}_i &= \frac{\pi e^2}{m_i^2} \int d\mathbf{k} \, \frac{\mathbf{k}\mathbf{k}}{k^2} \sum_{\sigma=\pm 1} \delta(\sigma \omega_{\mathbf{k}}^S - \mathbf{k}\cdot\mathbf{v}) \, I_{\mathbf{k}}^{\sigma S} \\
&\quad + \frac{\pi e^4}{2 m_i^4} \int d\mathbf{k} \int d\mathbf{k}' \, \frac{(\mathbf{k}\cdot\mathbf{k}')^2}{k^2 \, k'^2} \sum_{\sigma,\sigma'=\pm 1} \frac{(\mathbf{k}-\mathbf{k}')\,(\mathbf{k}-\mathbf{k}')\, I_{\mathbf{k}'}^{\sigma' S} \, I_{\mathbf{k}}^{\sigma S}}{(\sigma\omega_{\mathbf{k}}^S - \mathbf{k}\cdot\mathbf{v})^2} \\
&\quad \times \left(\frac{1}{(\sigma\omega_{\mathbf{k}}^S - \mathbf{k}\cdot\mathbf{v})^2} + \frac{2(\sigma\sigma')\, A_{\mathbf{k},\mathbf{k}'}^{\sigma,\sigma'}}{k\,k'\,c_S^2} \right) \delta[\sigma\omega_{\mathbf{k}}^S - \sigma'\omega_{\mathbf{k}'}^S - (\mathbf{k}-\mathbf{k}')\cdot\mathbf{v}], \\
A_{\mathbf{k},\mathbf{k}'}^{\sigma,\sigma'} &= \frac{k^2 + k'^2 - \mathbf{k}\cdot\mathbf{k}' - \sigma\sigma' k\,k'}{|\mathbf{k}-\mathbf{k}'|^2 (k - \sigma\sigma' k')^2 \lambda_{De}^2} \frac{1}{\epsilon(\mathbf{k}-\mathbf{k}', \sigma\omega_{\mathbf{k}}^S - \sigma'\omega_{\mathbf{k}'}^S)}.
\end{aligned}
\tag{4.86}
$$

This form of particle kinetic equation is given in a balanced form where the velocity space diffusion term, dictated by D, appears with the velocity space friction, or drag, term, governed by $\mathbf{A}$. In steady state these two terms balance each other out. Such a form is reminiscent of the Fokker–Planck equation. The velocity friction term is the result of single (or discrete) particle effects. As it will be shown eventually, this term has an overall inverse proportionality to the number of plasma particles within a Debye sphere. If such a number is infinite, then this term disappears, and we recover the particle kinetic equation of the Vlasov weak turbulence theory.

4.3.2 Wave Kinetic Equation with Discrete Particle Effects

As we have already discussed, taking the real part of nonlinear spectral balance equation (4.68) leads to the dispersion relation, while taking the imaginary part leads to the wave kinetic equation for specific eigenmodes. Let us approximate the dispersion relation by restricting the solution to linear eigenmodes, $\mathrm{Re}\,\epsilon(\mathbf{k},\omega)=0$ (see [1.93]), so that the dispersion relation for each mode is given by $\omega=\omega_{\mathbf{k}}^{\alpha}$ ($\alpha=L,S$). Then we express the wave intensity by $\langle\delta E^2\rangle_{\mathbf{k},\omega}=\sum_{\sigma=\pm1}\sum_{\alpha}I_{\mathbf{k}}^{\sigma\alpha}\,\delta(\omega-\sigma\omega_{\mathbf{k}}^{\alpha})$, as in (1.95). Inserting this to the imaginary part of (4.68), we have

$$
\begin{aligned}
\frac{\partial I_{\mathbf{k}}^{\sigma\alpha}}{\partial t} &= -\frac{2\,\mathrm{Im}\,\epsilon(\mathbf{k},\sigma\omega_{\mathbf{k}}^{\alpha})}{\epsilon'(\mathbf{k},\sigma\omega_{\mathbf{k}}^{\alpha})}\,I_{\mathbf{k}}^{\sigma\alpha} \\
&\quad+\boxed{\sum_a\frac{4e_a^2}{k^2\,[\epsilon'(\mathbf{k},\sigma\omega_{\mathbf{k}}^{\alpha})]^2}\int d\mathbf{v}\,\delta(\sigma\omega_{\mathbf{k}}^{\alpha}-\mathbf{k}\cdot\mathbf{v})\,f_a(\mathbf{v})} \\
&\quad-\frac{4}{\epsilon'(\mathbf{k},\sigma\omega_{\mathbf{k}}^{\alpha})}\sum_{\sigma'=\pm1}\sum_{\beta=L,S}\int d\mathbf{k}' \\
&\quad\times\mathrm{Im}\left(\mathscr{P}\,\frac{2\,\{\chi^{(2)}(\mathbf{k}',\sigma'\omega_{\mathbf{k}'}^{\beta}|\mathbf{k}-\mathbf{k}',\sigma\omega_{\mathbf{k}}^{\alpha}-\sigma'\omega_{\mathbf{k}'}^{\beta})\}^2}{\epsilon(\mathbf{k}-\mathbf{k}',\sigma\omega_{\mathbf{k}}^{\alpha}-\sigma'\omega_{\mathbf{k}'}^{\beta})}\right. \\
&\quad\left.-\bar{\chi}^{(3)}(\mathbf{k}',\sigma'\omega_{\mathbf{k}'}^{\beta}|-\mathbf{k}',-\sigma'\omega_{\mathbf{k}'}^{\beta}|\mathbf{k},\sigma\omega_{\mathbf{k}}^{\alpha})\right)I_{\mathbf{k}'}^{\sigma'\beta}\,I_{\mathbf{k}}^{\sigma\alpha} \\
&\quad\boxed{\begin{aligned}&-\sum_a\frac{16e_a^2}{\epsilon'(\mathbf{k},\sigma\omega_{\mathbf{k}}^{\alpha})}\sum_{\sigma'=\pm1}\sum_{\beta=L,S}\int d\mathbf{k}'\,\frac{|\chi^{(2)}(\mathbf{k}',\sigma'\omega_{\mathbf{k}'}^{\beta}|\mathbf{k}-\mathbf{k}',\sigma\omega_{\mathbf{k}}^{\alpha}-\sigma'\omega_{\mathbf{k}'}^{\beta})\,|^2}{|\mathbf{k}-\mathbf{k}'|^2\,|\epsilon(\mathbf{k}-\mathbf{k}',\sigma\omega_{\mathbf{k}}^{\alpha}-\sigma'\omega_{\mathbf{k}'}^{\beta})|^2}\\&\times\left(\frac{I_{\mathbf{k}}^{\sigma\alpha}}{\epsilon(\mathbf{k}',\sigma'\omega_{\mathbf{k}'}^{\beta})}-\frac{I_{\mathbf{k}'}^{\sigma'\beta}}{\epsilon'(\mathbf{k},\sigma\omega_{\mathbf{k}}^{\alpha})}\right)\int d\mathbf{v}\,\delta[\sigma\omega_{\mathbf{k}}^{\alpha}-\sigma'\omega_{\mathbf{k}'}^{\beta}-(\mathbf{k}-\mathbf{k}')\cdot\mathbf{v}]\,f_a(\mathbf{v})\end{aligned}} \\
&\quad-\frac{4\pi}{\epsilon'(\mathbf{k},\sigma\omega_{\mathbf{k}}^{\alpha})}\sum_{\sigma',\sigma''=\pm1}\sum_{\beta,\gamma=L,S}\int d\mathbf{k}'\,|\,\chi^{(2)}(\mathbf{k}',\sigma'\omega_{\mathbf{k}'}^{\beta}|\mathbf{k}-\mathbf{k}',\sigma''\omega_{\mathbf{k}-\mathbf{k}'}^{\gamma})\,|^2 \\
&\quad\times\left(\frac{I_{\mathbf{k}-\mathbf{k}'}^{\sigma''\gamma}\,I_{\mathbf{k}}^{\sigma\alpha}}{\epsilon'(\mathbf{k}',\sigma'\omega_{\mathbf{k}'}^{\beta})}+\frac{I_{\mathbf{k}'}^{\sigma'\beta}\,I_{\mathbf{k}}^{\sigma\alpha}}{\epsilon'(\mathbf{k}-\mathbf{k}',\sigma''\omega_{\mathbf{k}-\mathbf{k}'}^{\gamma})}-\frac{I_{\mathbf{k}'}^{\sigma'\beta}\,I_{\mathbf{k}-\mathbf{k}'}^{\sigma''\gamma}}{\epsilon'(\mathbf{k},\sigma\omega_{\mathbf{k}}^{\alpha})}\right) \\
&\quad\times\delta(\sigma\omega_{\mathbf{k}}^{\alpha}-\sigma'\omega_{\mathbf{k}'}^{\beta}-\sigma''\omega_{\mathbf{k}-\mathbf{k}'}^{\gamma}),
\end{aligned}
\tag{4.87}
$$

which is a direct generalization of (1.102). The first term on the right-hand side, the *induced emission*, is present in (1.102) but the new term is the second term, which represents the *spontaneous emission* process (encased in a box). The third term, which represents the *induced scattering*, is also present in (1.102), but the fourth term, also encased in a box, which is new, represents the *spontaneous scattering* process. The final term, which depicts the decay processes, is already present in (1.102).

Induced emission terms were already discussed in Section 2.1. We write the velocity space distribution function with the ambient density explicitly taken out of the definition $f_a(\mathbf{v}) = n_a F_a(\mathbf{v}) = n F_a(\mathbf{v})$ (see [1.34]). Induced emissions of L and S modes are described, respectively, with (2.6) and (2.7). Spontaneous emission terms contributing to L and S modes are given by

$$\begin{aligned}
\left.\frac{\partial I_{\mathbf{k}}^{\sigma L}}{\partial t}\right|_{\text{spont.emiss.}} &= \sum_a \frac{4e_a^2}{k^2\,[\epsilon'(\mathbf{k},\sigma\omega_{\mathbf{k}}^L)]^2} \int d\mathbf{v}\; \delta(\sigma\omega_{\mathbf{k}}^L - \mathbf{k}\cdot\mathbf{v})\; f_a(\mathbf{v}) \\
&= \frac{ne^2\omega_{pe}^2}{k^2} \int d\mathbf{v}\; \delta(\sigma\omega_{\mathbf{k}}^L - \mathbf{k}\cdot\mathbf{v})\; F_e, \\
\left.\frac{\partial I_{\mathbf{k}}^{\sigma S}}{\partial t}\right|_{\text{spont.emiss.}} &= \sum_a \frac{4e_a^2}{k^2\,[\epsilon'(\mathbf{k},\sigma\omega_{\mathbf{k}}^S)]^2} \int d\mathbf{v}\; \delta(\sigma\omega_{\mathbf{k}}^S - \mathbf{k}\cdot\mathbf{v})\; f_a(\mathbf{v}) \\
&= \frac{\mu_{\mathbf{k}}^2 ne^2\omega_{pe}^2}{k^2} \int d\mathbf{v}\; \delta(\sigma\omega_{\mathbf{k}}^S - \mathbf{k}\cdot\mathbf{v})\; (F_e + F_i).
\end{aligned} \tag{4.88}$$

The spontaneous emission takes place when randomly moving charged particles emit plasma waves. Such an emission process does not depend on the pre-existing level of waves, which contrasts to the induced emission. Recall that the induced emission is a process that involves charged particles moving in the wave field with their velocities in resonance with the wave phase velocities, and exchanging energy. For such a case, the particles may give off energy to the waves or gain energy from the waves depending on whether more particles move slightly faster or slower than the wave phase velocities; hence, the process is dependent on the derivative of the distribution function. In contrast, the spontaneous emission only requires that the charged particles have finite random velocities, or that the particles have a finite temperature. This means that the spontaneous emission is proportional to the particle distribution function rather than its derivative. And this process does not require pre-existing level of waves, so that the spontaneous emission formula is not proportional to the wave intensity. The spontaneously emitted waves will have phase velocities that match the particle velocities; hence, the process is governed by the linear wave-particle resonance condition. The spontaneous emission is also known as the Cerenkov emission.

Decay processes have already been discussed in Section 2.2, and induced scattering processes have been dealt with in Section 2.3. Spontaneous scattering terms are given by

$$\begin{aligned}\left.\frac{\partial I_{\mathbf{k}}^{\sigma\alpha}}{\partial t}\right|_{\text{spont.scatt.}} &= -\sum_a \frac{16 e_a^2}{\epsilon'(\mathbf{k},\sigma\omega_{\mathbf{k}}^\alpha)} \sum_{\sigma'=\pm1}\sum_{\beta=L,S}\int d\mathbf{k}' \\ &\times \frac{|\,\chi^{(2)}(\mathbf{k}',\sigma'\omega_{\mathbf{k}'}^\beta|\mathbf{k}-\mathbf{k}',\sigma\omega_{\mathbf{k}}^\alpha-\sigma'\omega_{\mathbf{k}'}^\beta)\,|^2}{|\mathbf{k}-\mathbf{k}'|^2\,|\epsilon(\mathbf{k}-\mathbf{k}',\sigma\omega_{\mathbf{k}}^\alpha-\sigma'\omega_{\mathbf{k}'}^\beta)|^2} \\ &\times \left(\frac{I_{\mathbf{k}}^{\sigma\alpha}}{\epsilon(\mathbf{k}',\sigma'\omega_{\mathbf{k}'}^\beta)} - \frac{I_{\mathbf{k}'}^{\sigma'\beta}}{\epsilon'(\mathbf{k},\sigma\omega_{\mathbf{k}}^\alpha)}\right) \\ &\times \int d\mathbf{v}\;\delta[\sigma\omega_{\mathbf{k}}^\alpha-\sigma'\omega_{\mathbf{k}'}^\beta-(\mathbf{k}-\mathbf{k}')\cdot\mathbf{v}]\,f_a(\mathbf{v}).\end{aligned} \tag{4.89}$$

The spontaneous scattering is similar to the spontaneous emission in the sense that the random particle motion leads to the emission of plasma waves, but, in this case, instead of giving off a single wave with the wave phase velocity matching the particle velocity, the scattering process involves two waves with the difference in the wave phase velocities equaling the particle random velocity. Conceptually, the process of spontaneous scattering can be viewed as being triggered by an incoming wave carrying wave momentum $\mathbf{k}$ (or $\mathbf{k}'$), which impinges on thermal distribution of particles. The particles spontaneously give off another wave with $\mathbf{k}'$ (or $\mathbf{k}$), but the particle velocity equals the net difference in the incoming wave phase velocity and the spontaneously emitted wave phase velocity. Consequently, this process is proportional to the incoming wave intensity and the particle distribution function, and the interaction is dictated by nonlinear wave-particle resonance condition.

For Langmuir waves, we retain the interaction involving only another Langmuir mode ($\beta = L$). This is consistent with the discussion of induced scattering. Following the same steps as before we obtain

$$\begin{aligned}\left.\frac{\partial I_{\mathbf{k}}^{\sigma L}}{\partial t}\right|_{\text{spont.scatt.}} &= -(\sigma\omega_{\mathbf{k}}^L)\,\frac{n e^4}{T_e^2}\,\lambda_{De}^4 \sum_{\sigma'=\pm1}\int d\mathbf{k}'\int d\mathbf{v}\,\frac{(\mathbf{k}\cdot\mathbf{k}')^2}{k^2\,k'^2} \\ &\times\left(\sigma'\omega_{\mathbf{k}'}^L\,I_{\mathbf{k}}^{\sigma L}-\sigma\omega_{\mathbf{k}}^L\,I_{\mathbf{k}'}^{\sigma' L}\right) \\ &\times\,\delta[\sigma\omega_{\mathbf{k}}^L-\sigma'\omega_{\mathbf{k}'}^L-(\mathbf{k}-\mathbf{k}')\cdot\mathbf{v}]\,(F_e+F_i).\end{aligned} \tag{4.90}$$

For ion-sound modes, we also follow the earlier discussion and retain the interaction with another ion-sound mode only. Again, such a treatment retains the consistency with the induced scattering process. Following the same steps as taken before we obtain

$$\begin{aligned}\left.\frac{\partial I_{\mathbf{k}}^{\sigma S}}{\partial t}\right|_{\text{spont.scatt.}} &= -\mu_{\mathbf{k}}\,(\sigma\omega_{\mathbf{k}}^{L})\,\frac{ne^4}{T_e^2}\frac{1}{\lambda_{De}^4}\sum_{\sigma'=\pm1}\int d\mathbf{k}' \\ &\times \frac{(\mathbf{k}\cdot\mathbf{k}')^2}{k^4\,k'^4\,|\mathbf{k}-\mathbf{k}'|^4}\left(1+\frac{|\mathbf{k}-\mathbf{k}'|^2}{(\sigma k-\sigma' k')^2}\right)^2 \\ &\times \frac{\mu_{\mathbf{k}'}\,\sigma'\omega_{\mathbf{k}'}^{L}\,I_{\mathbf{k}}^{\sigma S}-\mu_{\mathbf{k}}\,\sigma\omega_{\mathbf{k}}^{L}\,I_{\mathbf{k}'}^{\sigma' S}}{|\epsilon(\mathbf{k}-\mathbf{k}',\sigma\omega_{\mathbf{k}}^{S}-\sigma'\omega_{\mathbf{k}'}^{S})|^2} \\ &\times \int d\mathbf{v}\,\delta[\sigma\omega_{\mathbf{k}}^{S}-\sigma'\omega_{\mathbf{k}'}^{S}-(\mathbf{k}-\mathbf{k}')\cdot\mathbf{v}]\,(F_e+F_i).\end{aligned} \tag{4.91}$$

4.4 Summary of Klimontovich Weak Turbulence Theory

- *Langmuir and Ion-Sound Wave Kinetic Equations*

In the summary of equations we express the ion-sound wave intensity with the factor $\mu_{\mathbf{k}}$ – see (3.29) for definition – explicitly absorbed into the definition:

$$I_{\mathbf{k}}^{\sigma S} \to \frac{I_{\mathbf{k}}^{\sigma S}}{\mu_{\mathbf{k}}}. \tag{4.92}$$

For Langmuir (L) mode, the wave kinetic equation is given by

$$\begin{aligned}\frac{\partial I_{\mathbf{k}}^{\sigma L}}{\partial t} &= \frac{\pi\omega_{pe}^2}{k^2}\int d\mathbf{v}\,\delta(\sigma\omega_{\mathbf{k}}^{L}-\mathbf{k}\cdot\mathbf{v})\Bigg(\underbrace{\frac{ne^2}{\pi}F_e}_{\text{spontaneous}}+\underbrace{\sigma\omega_{\mathbf{k}}^{L}\,I_{\mathbf{k}}^{\sigma L}\,\mathbf{k}\cdot\frac{\partial F_e}{\partial\mathbf{v}}}_{\text{induced emissions}}\Bigg) \quad \text{spontaneous \& induced emissions} \\ &+2\sum_{\sigma',\sigma''=\pm1}\sigma\omega_{\mathbf{k}}^{L}\int d\mathbf{k}'\,V_{\mathbf{k},\mathbf{k}'}^{LS}\Big(\underbrace{\sigma\omega_{\mathbf{k}}^{L}\,I_{\mathbf{k}'}^{\sigma' L}\,I_{\mathbf{k}-\mathbf{k}'}^{\sigma'' S}}_{\text{spontaneous decay}} \\ &\underbrace{-\sigma'\omega_{\mathbf{k}'}^{L}\,I_{\mathbf{k}-\mathbf{k}'}^{\sigma'' S}\,I_{\mathbf{k}}^{\sigma L}-\sigma''\omega_{\mathbf{k}-\mathbf{k}'}^{L}\,I_{\mathbf{k}'}^{\sigma' L}\,I_{\mathbf{k}}^{\sigma L}}_{\text{induced decay}}\Big) \\ &+\sigma\omega_{\mathbf{k}}^{L}\sum_{\sigma'=\pm1}\int d\mathbf{k}'\int d\mathbf{v}\,U_{\mathbf{k},\mathbf{k}'}^{LL}\Bigg[\underbrace{\frac{m_e}{m_i}\,I_{\mathbf{k}'}^{\sigma' L}\,I_{\mathbf{k}}^{\sigma L}\,(\mathbf{k}-\mathbf{k}')\cdot\frac{\partial F_i}{\partial\mathbf{v}}}_{\text{induced scattering (off thermal ions)}} \\ &+\underbrace{\frac{ne^2}{\pi\omega_{pe}^2}\left(\sigma\omega_{\mathbf{k}}^{L}\,I_{\mathbf{k}'}^{\sigma' L}-\sigma'\omega_{\mathbf{k}'}^{L}\,I_{\mathbf{k}}^{\sigma L}\right)(F_e+F_i)}_{\text{spontaneous scattering}}\Bigg].\end{aligned} \tag{4.93}$$

For ion-sound (S) mode, the same is given by

$$
\begin{aligned}
\frac{\partial I_{\mathbf{k}}^{\sigma S}}{\partial t} &= \frac{\pi \mu_{\mathbf{k}} \omega_{pe}^2}{k^2} \int d\mathbf{v}\, \delta(\sigma\omega_{\mathbf{k}}^S - \mathbf{k}\cdot\mathbf{v}) \Bigg[\underbrace{\frac{ne^2}{\pi}(F_e + F_i)}_{\text{spontaneous emission}} \\
&\quad \underbrace{+ \sigma\omega_{\mathbf{k}}^L I_{\mathbf{k}}^{\sigma S} \left(\mathbf{k}\cdot\frac{\partial}{\partial\mathbf{v}}\right)\left(F_e + \frac{m_e}{m_i} F_i\right)}_{\text{induced emission}} \Bigg] \\
&\quad + \sum_{\sigma',\sigma''=\pm1} \sigma\omega_{\mathbf{k}}^L \int d\mathbf{k}'\, V_{\mathbf{k},\mathbf{k}'}^{SL} \Big(\underbrace{\sigma\omega_{\mathbf{k}}^L I_{\mathbf{k}'}^{\sigma' L} I_{\mathbf{k}-\mathbf{k}'}^{\sigma'' L}}_{\text{spontaneous decay}} \\
&\quad \underbrace{- \sigma'\omega_{\mathbf{k}'}^L I_{\mathbf{k}-\mathbf{k}'}^{\sigma'' L} I_{\mathbf{k}}^{\sigma S} - \sigma''\omega_{\mathbf{k}-\mathbf{k}'}^L I_{\mathbf{k}'}^{\sigma' L} I_{\mathbf{k}}^{\sigma S}}_{\text{induced decay}} \Big) \\
&\quad + \mu_{\mathbf{k}} \mu_{\mathbf{k}'} \sigma\omega_{\mathbf{k}}^L \sum_{\sigma'=\pm1} \int d\mathbf{k}' \int d\mathbf{v} \Bigg[\underbrace{U_{\mathbf{k},\mathbf{k}'}^{SS} I_{\mathbf{k}'}^{\sigma' S} I_{\mathbf{k}}^{\sigma S} (\mathbf{k}-\mathbf{k}')\cdot\frac{\partial F_i}{\partial\mathbf{v}}}_{\text{induced scattering}} \\
&\quad \underbrace{+ \frac{ne^2}{\pi\omega_{pe}^2}\frac{m_i}{m_e} Y_{\mathbf{k},\mathbf{k}'}^{SS} \left(\sigma\omega_{\mathbf{k}}^L I_{\mathbf{k}'}^{\sigma' S} - \sigma'\omega_{\mathbf{k}'}^L I_{\mathbf{k}}^{\sigma S}\right)(F_e + F_i)}_{\text{spontaneous scattering}} \Bigg],
\end{aligned}
\tag{4.94}
$$

where the various coefficients are defined in (3.29), except that an additional object is defined by

$$
\begin{aligned}
Y_{\mathbf{k},\mathbf{k}'}^{SS} &= \frac{\pi e^2}{m_e m_i \omega_{pe}^2} \frac{(\mathbf{k}\cdot\mathbf{k}')^2}{k^4 k'^4 \lambda_{De}^4} \left(W_{\mathbf{k},\mathbf{k}'} - \sigma\sigma' \frac{k'}{k} \right) \\
&\quad \times \delta[\sigma\omega_{\mathbf{k}}^S - \sigma'\omega_{\mathbf{k}'}^S - (\mathbf{k}-\mathbf{k}')\cdot\mathbf{v}].
\end{aligned}
\tag{4.95}
$$

The wave equations (4.93) and (4.94) now feature various spontaneous and induced processes appearing in balanced forms.

- *Electron and Ion Particle Kinetic Equations*

For the summary of particle kinetic equation see (4.86).

5

Spontaneous Emission and Collisional Kinetic Equation

One of the most important effects that arise from the particle discreteness is the spontaneous emission of electromagnetic *fluctuations* in thermal plasmas. We have already discussed the process of spontaneous emission of *plasma waves* within the context of wave kinetic equations. However, spontaneous emission of *fluctuations*, which we will discuss in depth, is a more general concept. Fluctuations are not necessarily the same as waves. Waves are defined by their *dispersion relations*, which relate the wave frequency ω with the wave vector $\mathbf{k}$. Fluctuations, on the other hand, do not necessarily satisfy the dispersion relation. It is still possible to define the frequency and wave vector associated with the fluctuations, but they do not have to be related to each other, but, instead, they can remain independent of each other. Those ω's and $\mathbf{k}$'s that are mathematically related, i.e., the dispersion relation, can be termed the "eigenmodes" or, equivalently, "waves." In contrast, when ω's and $\mathbf{k}$'s are independent of each other, they represent simply "fluctuations." These are obviously, "non-eigenmodes." When the plasma contains free energy, which excites eigenmodes, then the non-eigenmodes become ignorable. However, for thermal equilibrium plasma where there is no free energy source, both types of modes become important.

Another important effect that arises from the particle discreteness relates to the collisional kinetic equation. In plasmas, collisions are generally rare, but when the plasma is slightly out of equilibrium, yet the system does not have sufficient free energy to excite instabilities, then the only recourse to bring the system back to equilibrium is through collisions. The collisional relaxation process cannot be discussed if one relies solely on the Vlasovian description of plasma. In contrast, with the discrete particle effects inherent in the Klimontovich formalism, the collisional relaxation process becomes a natural part of the theory.

An interesting point to note is that both Vlasov and Klimontovich equations are time reversible at the microscopic level, before averages are taken over them.

Nevertheless, the Klimontovich formalism, which treats the particle discreteness properly, leads to the collisional kinetic equation, while the Vlasovian approach does not. The collisional processes represent dissipation and time irreversibility, so a conceptual question that may arise is how did the time irreversibility creep in from what is essentially a time-reversible equation. But the answer to this hypothetical question is well known in the non-equilibrium statistical mechanics. That is, at the microscopic level, classical Newtonian equation of motion, or quantum mechanical Schrödinger equation, is time reversible, but at the macroscopic level, physical theories are time irreversible. It is well known that the irreversibility of macroscopic system of equations arises out of statistical averages and subsequent loss of information. In this regard, it is no surprise that the collisional kinetic equation, which results from the statistically averaged Klimontovich theory, i.e., the weak turbulence theory, represents the dissipation. As a matter of fact, the particle and wave kinetic equations derived for the plasma eigenmodes in both Vlasov and Klimontovich weak turbulence theory are also time irreversible, so that the statistical averaging procedure already had produced macroscopically time irreversible theory in both approaches. It is just that, in the Vlasovian theory, the dissipation is not of the collisional variety, but rather it is of the collective dissipation type. For example, the quasilinear diffusion process in velocity space by resonant wave-particle interaction is an irreversible collective dissipation process.

5.1 Spontaneous Emission of Electrostatic Fluctuations

For plasmas in thermal equilibrium, the Maxwell–Boltzmann–Gaussian model defines the velocity distribution function,

$$f_a(v) = \frac{n m_a^{3/2}}{(2\pi T)^{3/2}} \exp\left(-\frac{m_a v^2}{2T}\right). \tag{5.1}$$

This means that we do not have to be concerned with the particle kinetic equation, since the distribution remains constant in time. As for the waves, the spontaneous emission of fluctuations takes place with all possible ω's and $\mathbf{k}$'s, including those ω's that satisfy the dispersion relation $\omega = \omega_{\mathbf{k}}$, but also those so-called non-eigenmodes that do not satisfy the condition $\epsilon(\mathbf{k},\omega) = 0$. These non-eigenmodes characterized by $\omega \neq \omega_{\mathbf{k}}$ are naturally emitted by particles with thermal distribution. When there exists an instability, however, the eigenmodes grow exponentially from the noise level and completely dominate the dynamics. In such cases, the excitation of plasma eigenmodes far outweighs the non-eigenmodes in intensity, so that we may ignore the non-eigenmodes altogether. However, for thermal equilibrium plasmas, the naturally occurring fluctuations must include all possible ω and $\mathbf{k}$ values.

To discuss the spontaneous emission of fluctuations, we return to the formal nonlinear spectral balance equation (4.68). Since for thermal plasmas the electric field spectral energy density does not evolve in time, we may set

$$\frac{\partial}{\partial t}\left\langle \delta E^2 \right\rangle_{\mathbf{k},\omega} = 0. \tag{5.2}$$

We expect the level of $\left\langle \delta E^2 \right\rangle_{\mathbf{k},\omega}$ to be sufficiently low when compared with the particle thermal energy so that we may ignore all nonlinear terms. Consequently, the non-vanishing terms in (4.68) can be collected and written down as

$$\begin{aligned}
\langle \delta E^2 \rangle_{\mathbf{k},\omega} &= \frac{2}{\pi} \sum_a \mathscr{P} \frac{n e_a^2}{k^2 \, |\epsilon(\mathbf{k},\omega)|^2} \int d\mathbf{v} \, \delta(\omega - \mathbf{k}\cdot\mathbf{v}) \, F_a(\mathbf{v}), \\
\epsilon(\mathbf{k},\omega) &= 1 + \sum_a \frac{\omega_{pa}^2}{k^2} \int d\mathbf{v} \, \frac{1}{\omega - \mathbf{k}\cdot\mathbf{v} + i0} \, \mathbf{k}\cdot\frac{\partial F_a(\mathbf{v})}{\partial \mathbf{v}}.
\end{aligned} \tag{5.3}$$

For isothermal plasmas, whose velocity distribution function is given by (5.1), we have

$$\delta(\omega - \mathbf{k}\cdot\mathbf{v}) \, F_a(\mathbf{v}) = -\delta(\omega - \mathbf{k}\cdot\mathbf{v}) \, \mathbf{k}\cdot\frac{\partial F_a(\mathbf{v})}{\partial \mathbf{v}} \frac{T}{m_a \, \omega}. \tag{5.4}$$

This leads to

$$\begin{aligned}
\langle \delta E^2 \rangle_{\mathbf{k},\omega} &= \frac{T}{2\pi^3 \, \omega} \mathscr{P} \frac{\mathrm{Im}\, \epsilon(\mathbf{k},\omega)}{|\epsilon(\mathbf{k},\omega)|^2}, \\
\epsilon(\mathbf{k},\omega) &= 1 + \sum_a \frac{2\omega_{pa}^2}{k^2 v_{Ta}^2} \left[1 + \frac{\omega}{k v_{Ta}} Z\left(\frac{\omega}{k v_{Ta}}\right) \right], \\
v_{Ta} &= \sqrt{\frac{2T}{m_a}}.
\end{aligned} \tag{5.5}$$

Figure 5.1 plots the high-frequency portion of normalized spontaneous thermal fluctuation spectrum that shows enhanced emission near Langmuir wave dispersion relation, $\omega = \omega_{\mathbf{k}}^L$, but the spectrum also shows a broad region of non-vanishing level of thermal fluctuations in the non-eigenmode spectral range ($\omega \neq \omega_{\mathbf{k}}^L$). Figure 5.1 also indicates the enhanced emission in the low-frequency part surrounding the ion acoustic mode, but because of the plotting scales it is not so easy to visually identify the structure, so, in the next figure, we display the blown-up version emphasizing the low-frequency portion of the spectrum.

Thus Figure 5.2 displays the spontaneous emission of electrostatic fluctuations in the low-frequency portion of the angular frequency that covers the range of ion-acoustic wave dispersion relation. In this case, the clear enhancement along the ion-acoustic dispersion curve is not apparent, but instead the fluctuation spectrum

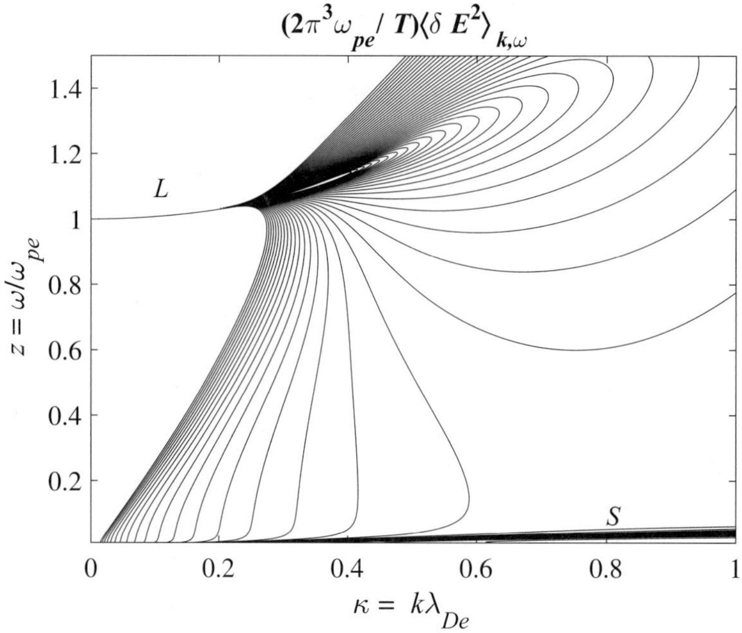

Figure 5.1 Spontaneously emitted electrostatic fluctuations including the singular eigenmode range satisfying the Langmuir wave dispersion relation, $\omega = \omega_{\mathbf{k}}^L$, as well as non-eigenmode spectral range ($\omega \neq \omega_{\mathbf{k}}^L$). In the low-frequency domain, ion acoustic mode range S is also specified.

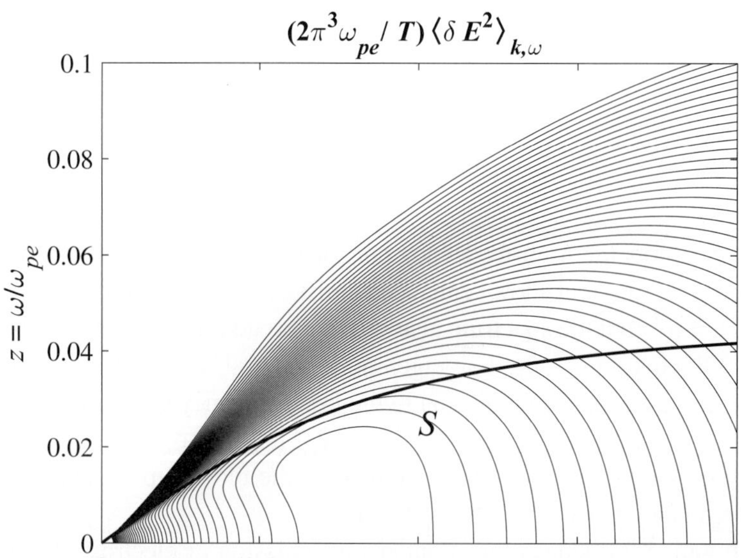

Figure 5.2 Extension of the spontaneously emitted electrostatic fluctuations in the low-frequency portion that shows the enhancement in the vicinity of ion-sound wave dispersion relation, $\omega = \omega_{\mathbf{k}}^S$.

is generally enhanced in the broad vicinity of $\omega = \omega^S_{\mathbf{k}}$. This is because the ion-sound wave for thermal plasmas ($T_e = T_i = T$) is heavily damped, which means that a substantial imaginary part of $\epsilon(\mathbf{k},\omega)$ is associated with the low-frequency range in the vicinity of ion-sound wave. Consequently, the denominator in (5.5) does not remain extremely small even for the eigenmode $\omega = \omega^S_{\mathbf{k}}$. This explains why the spectrum is not highly enhanced for $\omega = \omega^S_{\mathbf{k}}$, which contrast to the spectral behavior near $\omega = \omega^L_{\mathbf{k}}$, especially for low k region. It is well known that Langmuir wave damping rate becomes exponentially small for $k \to 0$ (see Figure 1.2). As a result, the spontaneous emission spectrum strictly diverges for Langmuir waves in the limit of $k \to 0$. However, this does not happen for ion sound mode.

Lastly, we show several cuts along angular frequency for fixed values of wave numbers. The result is shown in Figure 5.3. Note that as $\kappa = k\lambda_{De}$ approaches small value, the peak intensity becomes higher and higher, until for $k = 0$, the peak value becomes infinite. Again, this behavior is intimately related to the disappearance of Landau damping rate for Langmuir waves in the infinite wavelength regime. The fact that the spontaneous emission formula diverges for Langmuir waves in the infinite wavelength regime may indicate that the theory fails as one approaches $k = 0$. However, if we are only interested in the plasma eigenmodes with weak damping rates, then we know how to deal with such a situation. We simply expand the dielectric constant around the pole, $\omega - \omega_{\mathbf{k}}$, in Taylor series, and implement

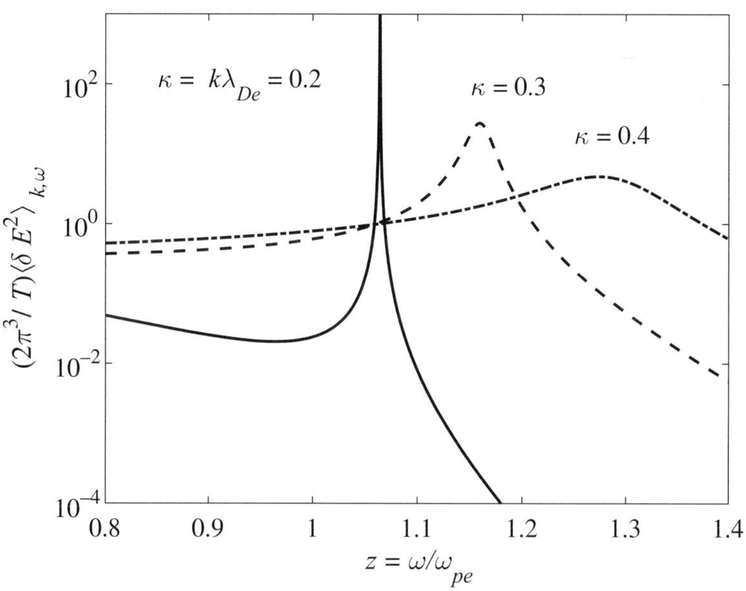

Figure 5.3 Plot of spontaneous emission spectrum along angular frequency for fixed values of wave numbers.

the residue theorem in order to take care of the singularity, which is what we have done in the derivation of particle and wave kinetic equations. Nevertheless, when one applies the spontaneous emission formula to the "fluctuations," then we have a potential problem when the denominator, which contains $\epsilon(\mathbf{k},\omega)$, becomes exceedingly small. This implies that the spontaneous emission theory requires some sort of "renormalization." In regard to this issue, we note that Sitenko (1967), Akhiezer et al. (1975), and Sitenko (1982) discuss the so-called nonlinear theory of spontaneous emission for fluctuations. In the present book, however, we omit such a discussion.

It is interesting to discuss the fluctuation as a function of k only:

$$(2\pi)^3\langle\delta E^2\rangle_{\mathbf{k}} = \int \frac{d\omega}{2\pi}\,(2\pi)^4\langle\delta E^2\rangle_{\mathbf{k},\omega} = 4T\int \frac{d\omega}{\omega}\,\mathscr{P}\,\frac{\mathrm{Im}\,\epsilon(\mathbf{k},\omega)}{|\epsilon(\mathbf{k},\omega)|^2}. \tag{5.6}$$

If we note that for any complex number $z = x + iy$, we may write

$$\frac{\mathrm{Im}\,z}{|z|^2} = \frac{y}{x^2+y^2} = \frac{i}{2}\left(1 - \frac{1}{x-iy} - 1 + \frac{1}{x+iy}\right), \tag{5.7}$$

then we have

$$\frac{\mathrm{Im}\,\epsilon(\mathbf{k},\omega)}{|\epsilon(\mathbf{k},\omega)|^2} = \frac{i}{2}\left(1 - \frac{1}{\epsilon^*(\mathbf{k},\omega)} - 1 + \frac{1}{\epsilon(\mathbf{k},\omega)}\right). \tag{5.8}$$

Let us define

$$f^+(\omega) = 1 - \frac{1}{\epsilon(\mathbf{k},\omega)}, \qquad f^-(\omega) = 1 - \frac{1}{\epsilon^*(\mathbf{k},\omega)}. \tag{5.9}$$

Then we have

$$(2\pi)^3\langle\delta E^2\rangle_{\mathbf{k}} = 2iT\int \frac{d\omega}{\omega}\left[f^-(\omega) - f^+(\omega)\right]. \tag{5.10}$$

We may further write

$$\frac{1}{\omega} = \frac{1}{2}\left(\frac{1}{\omega - i0} + \frac{1}{\omega + i0}\right). \tag{5.11}$$

Consequently, we have

$$(2\pi)^3\langle\delta E^2\rangle_{\mathbf{k}} = iT\int d\omega\left(\frac{1}{\omega - i0} + \frac{1}{\omega + i0}\right)\left[f^-(\omega) - f^+(\omega)\right]. \tag{5.12}$$

The function f^+ is analytic in the upper half ω plane, while f^- is analytic in the lower half complex ω plane. It is also to be noted that

$$\lim_{\omega\to\infty} f^{\pm}(\omega) = 0. \tag{5.13}$$

Making use of the Kramers–Krönig relations,

$$\begin{aligned} -\frac{1}{2\pi i}\int d\omega' \, \frac{f^{\pm}(\omega')}{\omega-\omega'\pm i0} &= \pm f^{\pm}(\omega), \\ -\frac{1}{2\pi i}\int d\omega' \, \frac{f^{\mp}(\omega')}{\omega-\omega'\pm i0} &= 0, \end{aligned} \tag{5.14}$$

we find that

$$(2\pi)^3\langle\delta E^2\rangle_{\mathbf{k}} = 2\pi T\left[f^-(0)+f^+(0)\right] = 4\pi T\left(1-\frac{\mathrm{Re}\,\epsilon(\mathbf{k},0)}{|\epsilon(\mathbf{k},0)|^2}\right). \tag{5.15}$$

For thermal equilibrium plasma we have

$$\epsilon(\mathbf{k},0) = 1+\sum_a \frac{1}{k^2\lambda_{Da}^2} \approx 1+\frac{2}{k^2\lambda_{De}^2}, \tag{5.16}$$

where $\lambda_{Da}^2 = T_a/(4\pi n e_a^2)$. As a consequence, we obtain

$$\langle\delta E^2\rangle_{\mathbf{k}} \approx \frac{T}{2\pi^2}\frac{1}{2+k^2\lambda_{De}^2}. \tag{5.17}$$

So, as one can see, when the spontaneous emission formula, which contains potential singularity associated with the plasma eigenmodes, is integrated over ω, one obtains a well-behaved finite valued function.

For more discussions on the spontaneous emission theory including the nonlinear correction, see Sitenko, 1967, 1982; Akhiezer et al., 1975, for example. Applications of the spontaneous emission theory for space plasmas were pioneered by Meyer-Vernet (1979). Her method, known as the quasi-thermal noise spectroscopy, takes into account the antenna response to the spontaneously emitted electrostatic fluctuations naturally occurring in space plasma, and it is being used for passively diagnosing the underlying space plasma properties (Meyer-Vernet et al., 2017). The basic theory of spontaneous emission of fluctuations in plasmas has been quite extensively discussed in the literature, which include the extension to magnetized plasmas (Meyer-Vernet et al., 1993; Sentman, 1982), generalization for suprathermal plasma distribution (Mace et al., 1998; Le Chat et al., 2009), full extension to electromagnetic formalism in magnetized plasma (Kim et al., 2017; López and Yoon, 2017), etc., although these generalizations and extensions are beyond the scope of the present book.

5.2 Collisional Kinetic Equation

We now consider the situation where the system is slightly out of thermal equilibrium, but not to the extent that any instabilities can be excited in the plasma. For such a case, while the electrostatic field fluctuations can be described by the spontaneous emission theory,

$$\langle \delta E^2 \rangle_{\mathbf{k},\omega} = \frac{2}{\pi} \frac{1}{k^2 \, |\epsilon(\mathbf{k},\omega)|^2} \sum_a e_a^2 \int d\mathbf{v} \, \delta(\omega - \mathbf{k}\cdot\mathbf{v}) \, f_a(\mathbf{v}), \tag{5.18}$$

the dynamical evolution of particle distribution function toward a genuine thermal equilibrium must be discussed on the basis of particle kinetic equation (4.75), but without the nonlinear correction term,

$$\begin{aligned} \frac{\partial f_a}{\partial t} &= \frac{\pi e_a^2}{m_a^2} \int d\mathbf{k} \int d\omega \left(\frac{\mathbf{k}}{k} \cdot \frac{\partial}{\partial \mathbf{v}} \right) \delta(\omega - \mathbf{k}\cdot\mathbf{v}) \\ &\quad \times \left[\operatorname{Im} \frac{m_a \, \epsilon(\mathbf{k},\omega)}{2\pi^3 k \, |\epsilon(\mathbf{k},\omega)|^2} \, f_a + \left\langle \delta E^2 \right\rangle_{\mathbf{k},\omega} \left(\frac{\mathbf{k}}{k} \cdot \frac{\partial f_a}{\partial \mathbf{v}} \right) \right]. \end{aligned} \tag{5.19}$$

When (5.18) is inserted to (5.19), we have

$$\begin{aligned} \frac{\partial f_a}{\partial t} &= \frac{\pi e_a^2}{m_a^2} \int d\mathbf{k} \left(\frac{\mathbf{k}}{k} \cdot \frac{\partial}{\partial \mathbf{v}} \right) \left[\frac{m_a \operatorname{Im} \epsilon(\mathbf{k},\mathbf{k}\cdot\mathbf{v})}{2\pi^3 k \, |\epsilon(\mathbf{k},\mathbf{k}\cdot\mathbf{v})|^2} \, f_a(\mathbf{v}) \right. \\ &\quad \left. + \frac{2}{\pi} \frac{1}{k^2 \, |\epsilon(\mathbf{k},\mathbf{k}\cdot\mathbf{v})|^2} \sum_b e_b^2 \int d\mathbf{v}' \, \delta(\mathbf{k}\cdot\mathbf{v} - \mathbf{k}\cdot\mathbf{v}') \, f_b(\mathbf{v}') \left(\frac{\mathbf{k}}{k} \cdot \frac{\partial f_a(\mathbf{v})}{\partial \mathbf{v}} \right) \right], \end{aligned} \tag{5.20}$$

where we have carried out the ω integration.

Making use of

$$\operatorname{Im} \epsilon(\mathbf{k},\omega) = -\pi \sum_a \frac{4\pi e_a^2}{m_a k^2} \int d\mathbf{v} \, \mathbf{k} \cdot \frac{\partial f_a}{\partial \mathbf{v}} \, \delta(\omega - \mathbf{k}\cdot\mathbf{v}), \tag{5.21}$$

we readily obtain the well-known Balescu–Lenard–Guernsey collisional kinetic equation (Balescu, 1960; Guernsey, 1960; Lenard, 1960),

$$\begin{aligned} \frac{\partial f_a}{\partial t} &= \frac{2 e_a^2}{m_a^2} \sum_b e_b^2 \int d\mathbf{k} \int d\mathbf{v}' \left(\frac{\mathbf{k}}{k} \cdot \frac{\partial}{\partial \mathbf{v}} \right) \frac{\delta(\mathbf{k}\cdot\mathbf{v} - \mathbf{k}\cdot\mathbf{v}')}{k^2 \, |\epsilon(\mathbf{k},\mathbf{k}\cdot\mathbf{v})|^2} \\ &\quad \times \left[\left(\frac{\mathbf{k}}{k} \cdot \frac{\partial f_a(\mathbf{v})}{\partial \mathbf{v}} \right) f_b(\mathbf{v}') - \frac{m_a}{m_b} \left(\frac{\mathbf{k}}{k} \cdot \frac{\partial f_b(\mathbf{v}')}{\partial \mathbf{v}'} \right) f_a(\mathbf{v}) \right]. \end{aligned} \tag{5.22}$$

When we take the ambient density factor n out of the velocity distribution function, $f_a = n F_a$, then we have

$$\frac{\partial F_a(\mathbf{v})}{\partial t} = \sum_b \frac{2ne_a^2 e_b^2}{m_a^2} \frac{\partial}{\partial v_i} \int d\mathbf{k} \int d\mathbf{v}' \, \frac{k_i \, k_j}{k^4} \frac{\delta(\mathbf{k}\cdot\mathbf{v} - \mathbf{k}\cdot\mathbf{v}')}{|\epsilon(\mathbf{k}, \mathbf{k}\cdot\mathbf{v})|^2} \times \left(\frac{\partial}{\partial v_j} - \frac{m_a}{m_b} \frac{\partial}{\partial v_j'} \right) F_a(\mathbf{v}) \, F_b(\mathbf{v}'). \tag{5.23}$$

Equation (5.22) or equivalently (5.23) contains the quantity $\epsilon(\mathbf{k}, \mathbf{k}\cdot\mathbf{v})$, which, in turn, contains the particle distribution function, hence, nonlinearly coupled to the distribution function. A simplifying approximation is to replace $\epsilon(\mathbf{k}, \mathbf{k}\cdot\mathbf{v})$ by 1. The resulting equation is

$$\frac{\partial F_a(\mathbf{v})}{\partial t} = \sum_b \frac{2\hat{n}\, e_a^2 e_b^2}{m_a^2} \frac{\partial}{\partial v_i} \int d\mathbf{k} \int d\mathbf{v}' \, \frac{k_i \, k_j}{k^4} \, \delta(\mathbf{k}\cdot\mathbf{v} - \mathbf{k}\cdot\mathbf{v}') \times \left(\frac{\partial}{\partial v_j} - \frac{m_a}{m_b} \frac{\partial}{\partial v_j'} \right) F_a(\mathbf{v}) \, F_b(\mathbf{v}'). \tag{5.24}$$

The following identity is useful:

$$\int d\mathbf{k} \, \delta(\mathbf{k}\cdot\mathbf{u}) \, \frac{\mathbf{k}\,\mathbf{k}}{k^4} = \mathsf{U}\,\pi \int_0^\infty \frac{dk}{k}, \qquad U_{ij} = \frac{1}{u} \left(\delta_{ij} - \frac{u_i \, u_j}{u^2} \right), \tag{5.25}$$

where $\mathbf{u} = \mathbf{v} - \mathbf{v}'$. To prove this, let $\mathbf{u} = u\,\hat{\mathbf{z}}$ and $\mathbf{k} = k_x\,\hat{\mathbf{x}} + k_y\,\hat{\mathbf{y}} + k_z\,\hat{\mathbf{z}}$, without loss of generality. Then

$$\int_{-\infty}^{\infty} dk_z \, \delta(\mathbf{k}\cdot\mathbf{u}) \, \frac{\mathbf{k}\,\mathbf{k}}{k^4} = \int_{-\infty}^{\infty} dk_z \, \delta(k_z u) \, \frac{\mathbf{k}\,\mathbf{k}}{k^4} = \frac{1}{u} \frac{(k_x\,\hat{\mathbf{x}} + k_y\,\hat{\mathbf{y}})(k_x\,\hat{\mathbf{x}} + k_y\,\hat{\mathbf{y}})}{(k_x^2 + k_y^2)^2}, \tag{5.26}$$

which leads to

$$\int d\mathbf{k} \, \delta(\mathbf{k}\cdot\mathbf{u}) \, \frac{\mathbf{k}\,\mathbf{k}}{k^4} = \frac{1}{u} \int dk_x \int dk_y \, \frac{k_x^2\,\hat{\mathbf{x}}\hat{\mathbf{x}} + k_y^2\,\hat{\mathbf{y}}\hat{\mathbf{y}}}{(k_x^2 + k_y^2)^2}, \tag{5.27}$$

where we have made use of the fact that the cross term $k_y\,k_x$ does not contribute to the integral. When (5.27) is expressed in polar coordinate integrals, then the rest easily leads to the proof

$$\int d\mathbf{k}\,\delta(\mathbf{k}\cdot\mathbf{u})\,\frac{\mathbf{k}\,\mathbf{k}}{k^4} = \frac{1}{u}\int_0^\infty \frac{dk}{k}\int_0^{2\pi} d\phi\,\left(\cos^2\phi\,\hat{\mathbf{x}}\,\hat{\mathbf{x}} + \sin^2\phi\,\hat{\mathbf{y}}\,\hat{\mathbf{y}}\right)$$
$$= \frac{\pi}{u}\int_0^\infty \frac{dk}{k}\,(\hat{\mathbf{x}}\,\hat{\mathbf{x}} + \hat{\mathbf{y}}\,\hat{\mathbf{y}}) = \frac{\pi}{u}\int_0^\infty \frac{dk}{k}\left(\overleftrightarrow{I} - \frac{\mathbf{u}\,\mathbf{u}}{u^2}\right). \tag{5.28}$$

Making use of identity (5.25), we have

$$\frac{\partial F_a(\mathbf{v})}{\partial t} = \sum_b \frac{2\pi\hat{n}\,e_a^2 e_b^2}{m_a^2}\int_0^\infty \frac{dk}{k}\frac{\partial}{\partial v_i}\int d\mathbf{v}'\,\frac{1}{|\mathbf{v}-\mathbf{v}'|}$$
$$\times\left(\delta_{ij} - \frac{(\mathbf{v}-\mathbf{v}')_i\,(\mathbf{v}-\mathbf{v}')_j}{|\mathbf{v}-\mathbf{v}'|^2}\right)\left(\frac{\partial}{\partial v_j} - \frac{m_a}{m_b}\frac{\partial}{\partial v_j'}\right) F_a(\mathbf{v})\,F_b(\mathbf{v}'). \tag{5.29}$$

Since the k integral diverges at both limits, we must introduce cutoffs,

$$\int \frac{dk}{k} = \int_{k_{\rm min}}^{k_{\rm max}} \frac{dk}{k} = \ln\frac{k_{\rm max}}{k_{\rm min}} = \ln\Lambda. \tag{5.30}$$

The resulting equation is called the Landau collisional equation (Landau, 1937):

$$\frac{\partial F_a(\mathbf{v})}{\partial t} = \sum_b \frac{2\pi\hat{n}\,e_a^2 e_b^2\ln\Lambda}{m_a^2}\frac{\partial}{\partial v_i}\int d\mathbf{v}'\,\frac{1}{|\mathbf{v}-\mathbf{v}'|}\left(\delta_{ij} - \frac{(\mathbf{v}-\mathbf{v}')_i\,(\mathbf{v}-\mathbf{v}')_j}{|\mathbf{v}-\mathbf{v}'|^2}\right)$$
$$\times\left(\frac{\partial}{\partial v_j} - \frac{m_a}{m_b}\frac{\partial}{\partial v_j'}\right) F_a(\mathbf{v})\,F_b(\mathbf{v}'). \tag{5.31}$$

To determine the limits, $k_{\rm min}$ and $k_{\rm max}$, we make note of the fact that in Figure 5.1, while the spontaneously emitted fluctuations attain high intensity for ω in the vicinity of $\omega_{\mathbf{k}}^L$, the area (or measure) in (ω, k) space over which this occurs is quite limited. The broad swath of regions where the spontaneous emission is finite actually takes place in the non-eigenmode region, excluding the narrow domain satisfying $\omega = \omega_{\mathbf{k}}^L$. The approximation of replacing $\epsilon(\mathbf{k}, \mathbf{k}\cdot\mathbf{v})$ by 1 is equivalent to ignoring the eigenmode contribution to the spontaneously emitted fluctuations. Mathematically, this approximation is essentially equivalent to restricting the integral range of k to $k\lambda_{De} > 1$. That is, we may choose

$$k_{\rm min} = \frac{1}{\lambda_{De}}. \tag{5.32}$$

For large k, we make note of the fact that we cannot resolve physical processes whose spatial scale is shorter than the atomic scales. Considering that the average particle kinetic energy is the thermal energy, $\mathscr{E} \sim k_B T_a$ or simply $\mathscr{E} \sim T_a$, we estimate that the smallest physical scale can be determined by balancing the kinetic energy with electrostatic potential energy,

$$\frac{|e_a e_b|}{r_{\rm min}} \sim T_a. \tag{5.33}$$

It is the convention of this book to adopt the energy units in eV, thus absorbing the Boltzmann constant $k_B = 1.38064852 \times 10^{-23}$ m^2 kg s^{-2} K^{-1} in the definition for temperature. From this, we have $r_{\rm min} \sim |e_a e_b|/T_a$, which, for $a = b = e$, is the classical radius of electron, or

$$k_{\rm max} = \frac{T_a}{|e_a e_b|}. \tag{5.34}$$

From this we obtain the Coulomb logarithm,

$$\ln \Lambda = \ln \left(\frac{\lambda_{De} T_a}{|e_a e_b|} \right). \tag{5.35}$$

For $a = b = e$ and for equilibrium, we have

$$\ln \Lambda = \ln \left(4\pi n \lambda_{De}^3 \right). \tag{5.36}$$

The quantity

$$g \equiv \frac{1}{n\lambda_{De}^3} \tag{5.37}$$

is known as the "plasma parameter," and it is inversely proportional to the total number of plasma particles in the so-called Debye sphere, that is, a sphere with radius equal to the Debye length, and whose volume is thus, $(4\pi/3)\lambda_{De}^3$. We already encountered this quantity when we mentioned the difference between the Klimontovich versus Vlasov paradigm. Recall that when the number of particles per Debye sphere is infinite, or, equivalently, when $g = 0$, we have the strict Vlasov limit. Obviously, the larger the number of particles in the Debye sphere, the slower the collisional relaxation toward establishing a new thermal equilibrium will proceed, starting from an initial system, which is slightly out of equilibrium. Since the right-hand side of the collisional integral (5.31) is proportional to g, it is evident that the plasma parameter determines the relative importance of inter-particle collisions versus collective processes.

5.3 Properties of Collisional Kinetic Equation

The collisional integral possesses certain invariants. Starting from (5.22), we show that thermal equilibrium Maxwellian–Boltzmann (MB) distribution function is an exact solution for steady state. It is sufficient to consider the following quantity within the velocity integral on the right-hand side of (5.22) for MB distribution:

$$\int d\mathbf{v}'\,\delta(\mathbf{k}\cdot\mathbf{v}-\mathbf{k}\cdot\mathbf{v}')\,\mathbf{k}\cdot\left(\frac{\partial}{\partial\mathbf{v}'}-\frac{m_b}{m_a}\frac{\partial}{\partial\mathbf{v}}\right)f_a(\mathbf{v})\,f_b(\mathbf{v}')$$
$$=-m_b\int d\mathbf{v}'\,\delta(\mathbf{k}\cdot\mathbf{v}-\mathbf{k}\cdot\mathbf{v}')\left(\frac{\mathbf{k}\cdot\mathbf{v}'}{T_b}-\frac{\mathbf{k}\cdot\mathbf{v}}{T_a}\right)f_a(\mathbf{v})\,f_b(\mathbf{v}'), \quad (5.38)$$

which is zero, if $T_a = T_b = T$. This shows that thermal equilibrium distribution satisfies the steady-state kinetic equation. Alternatively, this shows that the collisional kinetic equation should lead any particle distribution slightly out of equilibrium, to evolve toward thermal equilibrium distribution.

Consider now an integral

$$J(\mathbf{r},t)=\sum_a\int d\mathbf{v}\,\Phi(\mathbf{v})\,C_a(\mathbf{r},\mathbf{v},t), \quad (5.39)$$

where $\Phi(\mathbf{v})$ is an arbitrary function and $C_a(\mathbf{r},\mathbf{v},t)$ is the collision integral,

$$C_a(\mathbf{r},\mathbf{v},t)=\sum_b\frac{2\,e_a^2\,e_b^2}{m_a^2}\frac{\partial}{\partial\mathbf{v}}\cdot\int d\mathbf{k}\int d\mathbf{v}'\,\frac{\mathbf{k}\,\mathbf{k}}{k^4}\frac{\delta(\mathbf{k}\cdot\mathbf{v}-\mathbf{k}\cdot\mathbf{v}')}{|\epsilon(\mathbf{k},\mathbf{k}\cdot\mathbf{v})|^2}$$
$$\cdot\left(\frac{\partial}{\partial\mathbf{v}}-\frac{m_a}{m_b}\frac{\partial}{\partial\mathbf{v}'}\right)f_a(\mathbf{v})\,f_b(\mathbf{v}'). \quad (5.40)$$

It can be shown that $J(\mathbf{r},t)=0$ for

$$\Phi(\mathbf{v})=1,\quad \Phi(\mathbf{v})=\mathbf{p}=m_a\mathbf{v},\quad\text{and}\quad \Phi(\mathbf{v})=\frac{m_a v^2}{2}. \quad (5.41)$$

Obviously these three quantities are related to conservations of total number density, momentum, and kinetic energy. To prove this, we proceed as follows:

$$J(\mathbf{r},t)=2\sum_a\sum_b\frac{e_a^2\,e_b^2}{m_a^2}\int d\mathbf{k}\int d\mathbf{v}\int d\mathbf{v}'\,\Phi(\mathbf{v})\frac{\partial}{\partial\mathbf{v}}\cdot\frac{\mathbf{k}\,\mathbf{k}}{k^4}\frac{\delta(\mathbf{k}\cdot\mathbf{v}-\mathbf{k}\cdot\mathbf{v}')}{|\epsilon(\mathbf{k},\mathbf{k}\cdot\mathbf{v})|^2}$$
$$\cdot\left(\frac{\partial}{\partial\mathbf{v}}-\frac{m_a}{m_b}\frac{\partial}{\partial\mathbf{v}'}\right)f_a(\mathbf{v})\,f_b(\mathbf{v}')$$
$$=-\sum_a\sum_b\frac{e_a^2\,e_b^2}{m_a^2}\int d\mathbf{k}\int d\mathbf{v}\int d\mathbf{v}'\left(\frac{\partial\Phi(\mathbf{v})}{\partial\mathbf{v}}-\frac{m_a}{m_b}\frac{\partial\Phi(\mathbf{v}')}{\partial\mathbf{v}'}\right)$$
$$\cdot\frac{\mathbf{k}\,\mathbf{k}}{k^4}\frac{\delta(\mathbf{k}\cdot\mathbf{v}-\mathbf{k}\cdot\mathbf{v}')}{|\epsilon(\mathbf{k},\mathbf{k}\cdot\mathbf{v})|^2}\cdot\left(\frac{\partial}{\partial\mathbf{v}}-\frac{m_a}{m_b}\frac{\partial}{\partial\mathbf{v}'}\right)f_a(\mathbf{v})\,f_b(\mathbf{v}'), \quad (5.42)$$

where we have taken successive partial integrations. It readily follows that $\Phi(\mathbf{v})=1$ and $\Phi(\mathbf{v}')=m_a\mathbf{v}$; $\Phi(\mathbf{v}')=m_b\mathbf{v}'$ renders the right-hand side to vanish. For $\Phi(\mathbf{v})=m_a v^2/2$, $\Phi(\mathbf{v}')=m_b v'^2/2$, the right-hand side becomes

$$\text{rhs} = -\sum_a \sum_b \frac{e_a^2 e_b^2}{m_a} \int d\mathbf{k} \int d\mathbf{v} \int d\mathbf{v}' \, (\mathbf{k}\cdot\mathbf{v} - \mathbf{k}\cdot\mathbf{v}')$$

$$\times \frac{\delta(\mathbf{k}\cdot\mathbf{v} - \mathbf{k}\cdot\mathbf{v}')}{k^4 \, |\epsilon(\mathbf{k}, \mathbf{k}\cdot\mathbf{v})|^2} \mathbf{k}\cdot\left(\frac{\partial}{\partial\mathbf{v}} - \frac{m_a}{m_b}\frac{\partial}{\partial\mathbf{v}'}\right) f_a(\mathbf{v})\, f_b(\mathbf{v}'), \tag{5.43}$$

which again vanishes by virtue of the delta function $\delta(\mathbf{k}\cdot\mathbf{v} - \mathbf{k}\cdot\mathbf{v}')$.

If we choose

$$\Phi(\mathbf{v}) = -\ln f_a(\mathbf{v}, t), \tag{5.44}$$

then we may prove that the total entropy of the plasma, where the entropy is defined by

$$S(t) = -\sum_a \int d\mathbf{v} \; f_a(\mathbf{v}, t) \, \ln f_a(\mathbf{v}, t), \tag{5.45}$$

always increases and tends toward a constant value in the time-asymptotic limit. To see this, we write

$$\begin{aligned} \mathbf{A}(\mathbf{v}, \mathbf{v}', t) &= -\frac{\partial\Phi(\mathbf{v})}{\partial\mathbf{v}} + \frac{m_a}{m_b}\frac{\partial\Phi(\mathbf{v}')}{\partial\mathbf{v}'} \\ &= \frac{\partial}{\partial\mathbf{v}} \ln f_a(\mathbf{v}, t) - \frac{m_a}{m_b}\frac{\partial}{\partial\mathbf{v}'} \ln f_a(\mathbf{v}', t) \\ &= \frac{1}{f_a(\mathbf{v}, t)\, f_a(\mathbf{v}', t)} \left(\frac{\partial}{\partial\mathbf{v}} - \frac{m_a}{m_b}\frac{\partial}{\partial\mathbf{v}'}\right) f_a(\mathbf{v}, t)\, f_a(\mathbf{v}', t). \end{aligned} \tag{5.46}$$

Then we have

$$\begin{aligned} J(t) &= \sum_a \sum_b \frac{e_a^2 e_b^2}{m_a^2} \int d\mathbf{k} \int d\mathbf{v} \int d\mathbf{v}' \, \mathbf{A}(\mathbf{v}, \mathbf{v}', t) \cdot \frac{\mathbf{k}\mathbf{k}}{k^4} \frac{\delta(\mathbf{k}\cdot\mathbf{v} - \mathbf{k}\cdot\mathbf{v}')}{|\epsilon(\mathbf{k}, \mathbf{k}\cdot\mathbf{v})|^2} \\ &\quad \cdot \left(\frac{\partial}{\partial\mathbf{v}} - \frac{m_a}{m_b}\frac{\partial}{\partial\mathbf{v}'}\right) f_a(\mathbf{v})\, f_b(\mathbf{v}') \\ &= \sum_a \sum_b \frac{e_a^2 e_b^2}{m_a^2} \int d\mathbf{k} \int d\mathbf{v} \int d\mathbf{v}' \, \frac{[\mathbf{k}\cdot\mathbf{A}(\mathbf{v}, \mathbf{v}', t)]^2}{k^4 \, |\epsilon(\mathbf{k}, \mathbf{k}\cdot\mathbf{v})|^2} \\ &\quad \times \delta(\mathbf{k}\cdot\mathbf{v} - \mathbf{k}\cdot\mathbf{v}')\, f_a(\mathbf{v})\, f_b(\mathbf{v}') > 0, \end{aligned} \tag{5.47}$$

since the integral is positive definite.

Upon taking the time derivative of $S(t)$, we have

$$\begin{aligned} \frac{dS}{dt} &= -\sum_a \int d\mathbf{v} \, (1 + \ln f_a) \, \frac{\partial f_a}{\partial t} \\ &= -\sum_a \sum_b \int d\mathbf{v} \, (1 + \ln f_a) \, \frac{2e_a^2 e_b^2}{m_a^2} \int d\mathbf{k} \int d\mathbf{v}' \, \frac{\partial}{\partial\mathbf{v}} \end{aligned}$$

$$\times \cdot \frac{\mathbf{k}\mathbf{k}}{k^4} \frac{\delta(\mathbf{k}\cdot\mathbf{v}-\mathbf{k}\cdot\mathbf{v}')}{|\epsilon(\mathbf{k},\mathbf{k}\cdot\mathbf{v})|^2} \cdot \left(\frac{\partial}{\partial \mathbf{v}} - \frac{m_a}{m_b}\frac{\partial}{\partial \mathbf{v}'}\right) f_a(\mathbf{v})\, f_b(\mathbf{v}')$$

$$= \sum_a \sum_b \frac{e_a^2 e_b^2}{m_a^2} \int d\mathbf{k} \int d\mathbf{v} \int d\mathbf{v}' \, \frac{1}{f_a(\mathbf{v})\, f_b(\mathbf{v}')} \left(\frac{\partial}{\partial \mathbf{v}} - \frac{m_a}{m_b}\frac{\partial}{\partial \mathbf{v}'}\right)$$

$$\times \ f_a(\mathbf{v})\, f_b(\mathbf{v}') \cdot \frac{\mathbf{k}\mathbf{k}}{k^4} \frac{\delta(\mathbf{k}\cdot\mathbf{v}-\mathbf{k}\cdot\mathbf{v}')}{|\epsilon(\mathbf{k},\mathbf{k}\cdot\mathbf{v})|^2} \cdot \left(\frac{\partial}{\partial \mathbf{v}} - \frac{m_a}{m_b}\frac{\partial}{\partial \mathbf{v}'}\right) f_a(\mathbf{v})\, f_b(\mathbf{v}')$$

$$= \sum_a \sum_b \frac{e_a^2 e_b^2}{m_a^2} \int d\mathbf{k} \int d\mathbf{v} \int d\mathbf{v}' \, \frac{[\mathbf{k}\cdot\mathbf{A}(\mathbf{v},\mathbf{v}')]^2}{k^4\, |\epsilon(\mathbf{k},\mathbf{k}\cdot\mathbf{v})|^2}$$

$$\times \ \delta(\mathbf{k}\cdot\mathbf{v}-\mathbf{k}\cdot\mathbf{v}')\, f_a(\mathbf{v})\, f_b(\mathbf{v}') = J(t) > 0. \tag{5.48}$$

This shows that

$$\frac{dS}{dt} > 0, \tag{5.49}$$

meaning that the entropy always increases. This is the plasma equivalent of Boltzmann's H-theorem.

5.4 Collisional Kinetic Equation with Maxwellian Background

The collision integrals, either Balescu–Lenard form (5.22) or Landau form (5.31), are nonlinear in their mathematical structures, since the right-hand side involves the product $f_a(\mathbf{v}) f_b(\mathbf{v}')$ such that one must solve the nonlinear integro-differential equation (Pezzi et al., 2016). However, the problem can be substantially simplified if we treat $f_b(\mathbf{v}')$ as a Maxwellian distribution. That is, the problem can be approximately treated as collisions between an arbitrary species (a) and a Maxwellian species (b), which can be taken to be at rest. Let us express the collisional operator as

$$\begin{aligned}
C_{ab}(f_a, f_b) &= \frac{\partial}{\partial v_i}\left(A_i(\mathbf{v}) f_a(\mathbf{v}) + D_{ij}(\mathbf{v})\, \frac{\partial f_a(\mathbf{v})}{\partial v_j}\right),\\
A_i(\mathbf{v}) &= \ln\Lambda \left(\frac{4\pi\, e_a e_b}{m_a}\right)^2 \frac{m_a}{m_b}\frac{\partial \varphi_b(\mathbf{v})}{\partial v_i},\\
D_{ij}(\mathbf{v}) &= -\ln\Lambda \left(\frac{4\pi\, e_a e_b}{m_a}\right)^2 \frac{\partial^2 \psi_b(\mathbf{v})}{\partial v_i \partial v_j},\\
\varphi_b(\mathbf{v}) &= -\frac{1}{4\pi}\int d\mathbf{v}'\, \frac{f_b(\mathbf{v}')}{\mathbf{v}-\mathbf{v}'},\\
\psi_b(\mathbf{v}) &= -\frac{1}{8\pi}\int d\mathbf{v}'\, (\mathbf{v}-\mathbf{v}')\, f_b(\mathbf{v}'),
\end{aligned} \tag{5.50}$$

where f_b given by $f_b(\mathbf{v}) = f_{b0}(v) = n_b(\pi^{1/2}v_{Tb})^{-3}\, e^{-v^2/v_{Tb}^2}$, $v_{Tb}^2 = \sqrt{2T_b/m_b}$. The quantities $\varphi_b(\mathbf{v})$ and $\psi_b(\mathbf{v})$ are known as the Rosenbluth potentials (Rosenbluth et al., 1957). Since f_b is isotropic, the Rosenbluth potentials are also isotropic,

$$\varphi_b(v) = -\frac{1}{4\pi}\int d\mathbf{v}'\,\frac{1}{u}\,f_b(v'),$$

$$\psi_b(v) = -\frac{1}{8\pi}\int d\mathbf{v}'\,u\,f_b(v'), \tag{5.51}$$

where $u = |\mathbf{v}-\mathbf{v}'|$.

Since φ_b and ψ_b are isotropic, we have

$$\frac{\partial\varphi_b}{\partial v_k} = \frac{v_k}{v}\,\varphi_b',$$

$$\frac{\partial^2\psi_b}{\partial v_k\partial v_l} = \frac{\partial}{\partial v_k}\left(\frac{v_l}{v}\,\psi_b'\right) = W_{kl}\,\psi_b' + \frac{v_k v_l}{v^2}\,\psi_b'',$$

$$W_{kl} = \frac{v^2\,\delta_{kl} - v_k v_l}{v^3}. \tag{5.52}$$

This leads to

$$C_{ab}(f_a, f_b) = \ln\Lambda\left(\frac{4\pi\,e_a e_b}{m_a}\right)^2 \frac{\partial}{\partial v_k}\left[\frac{m_a}{m_b}\frac{v_k}{v}\,\varphi_b'\,f_a - \left(W_{kl}\,\psi_b' + \frac{v_k v_l}{v^2}\,\psi_b''\right)\frac{\partial f_a}{\partial v_l}\right]$$
$$= \ln\Lambda\left(\frac{4\pi\,e_a e_b}{m_a}\right)^2\left\{\frac{m_a}{m_b}\frac{\partial}{\partial\mathbf{v}}\cdot\left(\frac{\mathbf{v}}{v}\,\varphi_b'\,f_a\right) - \frac{\partial}{\partial\mathbf{v}}\cdot\left(\frac{1}{v}\frac{\partial f_a}{\partial\mathbf{v}}\,\psi_b'\right)\right.$$
$$\left.+\frac{\partial}{\partial\mathbf{v}}\cdot\left[\frac{\mathbf{v}}{v^3}\left(\mathbf{v}\cdot\frac{\partial f_a}{\partial\mathbf{v}}\right)\psi_b'\right] - \frac{\partial}{\partial\mathbf{v}}\cdot\left[\frac{\mathbf{v}}{v^2}\left(\mathbf{v}\cdot\frac{\partial f_a}{\partial\mathbf{v}}\,\psi_b''\right)\right]\right\}. \tag{5.53}$$

We may rewrite (5.53) in terms of the velocity space spherical coordinate system:

$$\frac{\partial}{\partial\mathbf{v}} = \hat{v}\,\frac{\partial}{\partial v} + \hat{\theta}\,\frac{1}{v}\frac{\partial}{\partial\theta} + \hat{\varphi}\,\frac{1}{v\sin\theta}\frac{\partial}{\partial\varphi},$$

$$\frac{\partial}{\partial\mathbf{v}}\cdot\left[\frac{\mathbf{v}}{v^3}\left(\mathbf{v}\cdot\frac{\partial f_a}{\partial\mathbf{v}}\right)\psi_b'\right] = \frac{1}{v^2}\frac{\partial}{\partial v}\left(v\,\frac{\partial f_a}{\partial v}\,\psi_b'\right),$$

$$\frac{\partial}{\partial\mathbf{v}}\cdot\left[\frac{\mathbf{v}}{v^2}\left(\mathbf{v}\cdot\frac{\partial f_a}{\partial\mathbf{v}}\right)\psi_b''\right] = \frac{1}{v^2}\frac{\partial}{\partial v}\left(v^2\,\frac{\partial f_a}{\partial v}\,\psi_b''\right). \tag{5.54}$$

This leads to

$$C_{ab}(f_a, f_b) = -\frac{2}{v^3}\,\ln\Lambda\left(\frac{4\pi\,e_a e_b}{m_a}\right)^2\psi_b'\,\mathscr{L}(f_a)$$
$$+\frac{1}{v^2}\,\ln\Lambda\left(\frac{4\pi\,e_a e_b}{m_a}\right)^2\frac{\partial}{\partial v}\left(v^2\,\frac{m_a}{m_b}\,\varphi_b'\,f_a - v^2\,\psi_b''\,\frac{\partial f_a}{\partial v}\right),$$

$$\mathscr{L}(f_a) = \frac{1}{2}\left[\frac{1}{\sin\theta}\frac{\partial}{\partial\theta}\left(\sin\theta\,\frac{\partial}{\partial\theta}\right) + \frac{1}{\sin^2\theta}\frac{\partial^2}{\partial\varphi^2}\right]f_a. \tag{5.55}$$

Let us define three basic collision frequencies,

$$\nu_F^{ab}(v) = \ln\Lambda \left(\frac{4\pi e_a e_b}{m_a}\right)^2 \left(1+\frac{m_a}{m_b}\right)\frac{\varphi_b'(v)}{v},$$
$$\nu_D^{ab}(v) = -\frac{2}{v^3}\ln\Lambda\left(\frac{4\pi e_a e_b}{m_a}\right)^2 \psi_b'(v),$$
$$\nu_E^{ab}(v) = -\frac{2}{v^2}\ln\Lambda\left(\frac{4\pi e_a e_b}{m_a}\right)^2 \psi_b''(v), \tag{5.56}$$

where $\nu_F^{ab}(v)$ represents the collisional velocity friction frequency, $\nu_D^{ab}(v)$ designates the collisional frequency associated with diffusion in velocity pitch angle and azimuthal angle space, and $\nu_E^{ab}(v)$ corresponds to the collisional frequency associated with velocity space diffusion, or, equivalently, diffusion in energy space. In terms of these frequencies, we have

$$C_{ab}(f_a, f_b) = \nu_D^{ab}(v)\,\mathscr{L}(f_a) + \frac{1}{v^2}\frac{\partial}{\partial v}v^3\left(\frac{m_a}{m_a+m_b}\nu_F^{ab}(v)\,f_a + \frac{v\nu_E^{ab}(v)}{2}\frac{\partial f_a}{\partial v}\right). \tag{5.57}$$

The next task is to calculate the Rosenbluth potentials:

$$\begin{aligned}\varphi_b(v) &= -\frac{1}{4\pi}\int d\mathbf{v}'\,\frac{f_{b0}(v')}{|\mathbf{v}-\mathbf{v}'|}\\ &= -\frac{n_b}{2\pi^{3/2}v_{Tb}^3}\int_{-1}^{1} d(\cos\theta)\int_0^\infty dv'\,\frac{v'^2\,e^{-v'^2/v_{Tb}^2}}{\sqrt{v^2+v'^2-2vv'\cos\theta}}\\ &= -\frac{n_b}{4\pi v}\frac{2}{\sqrt{\pi}}\int_0^{v/v_{Tb}} dx\,e^{-x^2} = -\frac{n_b}{4\pi v}\,\mathrm{erf}(x_b),\\ x_b &= \frac{v}{v_{Tb}},\end{aligned} \tag{5.58}$$

$$\begin{aligned}\psi_b(v) &= -\frac{1}{8\pi}\int d\mathbf{v}'\ |\mathbf{v}-\mathbf{v}'|\,f_{b0}(v')\\ &= -\frac{n_b}{4\pi^{3/2}v_{Tb}^3}\int_{-1}^{1} d(\cos\theta)\int_0^\infty dv'\,v'^2\sqrt{v^2+v'^2-2vv'\cos\theta}\,e^{-v'^2/v_{Tb}^2}\\ &= -\frac{n_b v_{Tb}}{4\pi^{3/2}}\left[\left(x_b+\frac{1}{2x_b}\right)\int_0^{x_b} dx\,e^{-x^2}+\frac{1}{2}e^{-x_b^2}\right]\\ &= -\frac{n_b v_{Tb}}{8\pi}\left[\left(x_b+\frac{1}{2x_b}\right)\mathrm{erf}(x_b)+\frac{1}{\sqrt{\pi}}e^{-x_b^2}\right],\end{aligned} \tag{5.59}$$

where $\mathrm{erf}(x)$ is the error function.

Quantities of interest are the derivatives of the Rosenbluth potentials:

$$\begin{aligned}
\varphi_b'(v) &= \frac{n_b m_b}{4\pi T_b}\, G(x_b), \\
\psi_b'(v) &= -\frac{n_b}{8\pi}\,[\mathrm{erf}(x_b) - G(x_b)], \\
\psi_b''(v) &= -\frac{n_b}{4\pi v_{Tb}}\frac{G(x_b)}{x_b}, \\
G(x_b) &= \frac{\mathrm{erf}(x_b) - (2/\sqrt{\pi})\, x_b\, e^{-x_b^2}}{2x_b^2}.
\end{aligned} \tag{5.60}$$

The three collision frequencies defined in (5.56) are thus expressed as

$$\begin{aligned}
\nu_F^{ab}(v) &= \hat{\nu}_{ab}\,\frac{2T_a}{T_b}\left(1+\frac{m_b}{m_a}\right)\frac{G(x_b)}{x_a}, \\
\nu_D^{ab}(v) &= \hat{\nu}_{ab}\,\frac{\mathrm{erf}(x_b) - G(x_b)}{x_a^3}, \\
\nu_E^{ab}(v) &= 2\hat{\nu}_{ab}\,\frac{G(x_b)}{x_a^3}, \\
\hat{\nu}_{ab} &= \frac{4\pi n_b\, e_a^2\, e_b^2\, \ln\Lambda}{m_a^2\, v_{Ta}^3}, \qquad x_b = \frac{v}{v_{Tb}}.
\end{aligned} \tag{5.61}$$

The reduced collisional kinetic equation expressed in terms of the Rosenbluth potentials are linear, which greatly facilitates analytical and/or numerical investigations (Tigik et al., 2016). For systematic discussions of collisional relaxation processes in space or laboratory plasmas, the readers are referred to excellent monographs by Helander and Sigmar (2002) and Zank (2014).

6
Langmuir Turbulence and Electron Kappa Distribution

Having formulated the electrostatic weak turbulence theory that includes the effects of discrete nature of plasma particles, which extends the Vlasov weak turbulence theory that assumes collision-free plasma, we now consider an application that involves a plasma instability. Chapter 5 already pertained to applications, but it was for plasmas where collective modes are of no particular significance. In this chapter, in contrast, we consider a situation where the plasma carries significant free energy source so as to excite collective oscillations. Specifically, we consider the problem of Langmuir wave instability excited by gentle (or weak) bump-on-tail instability, and the Langmuir/ion-acoustic turbulence generated during the beam-plasma interaction process. The long-time, quasi-asymptotic state of the Langmuir turbulence state also involves the acceleration/heating of electrons such that the end state corresponds to the celebrated kappa distribution (Olbert, 1968; Vasyliunas, 1968; Tsallis, 2009; Livadiotis, 2017).

Since the 1960s, when the near-Earth space environment was explored with artificial satellites, it became quite evident that the measured electron distribution functions in the solar wind deviate substantially from the Maxwellian thermal distribution. Specifically, while the core part of the electron distribution can be modeled with the Maxwellian distribution function, the high-velocity tail population features a quasi-inverse power law distribution (Vasyliunas, 1968; Feldman et al., 1975; Gosling et al., 1981; Armstrong et al., 1983; Wang et al., 2012a). Olbert (1968) and Vasyliunas (1968) famously introduced the kappa distribution function,

$$f_e \propto \left(1 + \frac{v^2}{\kappa\Theta^2}\right)^{-\kappa-1}, \tag{6.1}$$

to empirically fit the measurement. Note that the kappa distribution approaches the Maxwellian model for $\kappa \to \infty$,

$$\lim_{\kappa\to\infty}\left(1 + \frac{v^2}{\kappa\Theta^2}\right)^{-\kappa-1} = \exp\left(-\frac{v^2}{\Theta^2}\right). \tag{6.2}$$

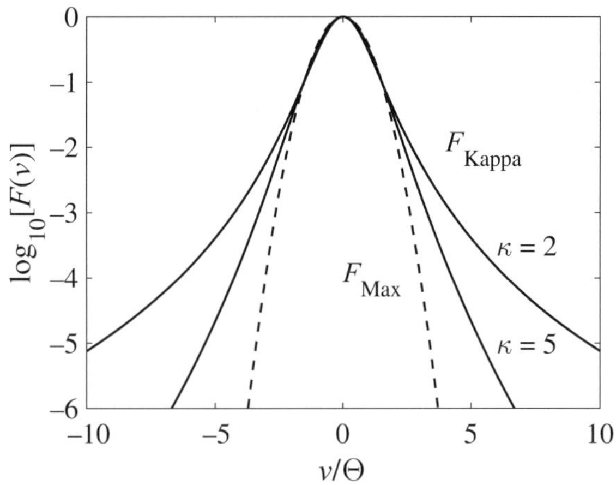

Figure 6.1 Maxwellian versus kappa distributions.

To see this, we display in Figure 6.1 the Maxwellian distribution $F_{\text{Max}} = \exp(-v^2/\Theta^2)$ versus kappa distribution $F_{\text{Kappa}} = (1 + v^2/\kappa\Theta^2)^{-\kappa-1}$ for two different values of κ, which equals to 2 and 5, respectively. As one can see, the case of $\kappa = 5$ is more close to the Maxwellian distribution than that of $\kappa = 2$. It is also obvious that the kappa distribution effectively models the low-energy population characterized by quasi-Maxwellian form, while for high-energy regime, the model describes a supra-thermal tail population. This is why kappa model is widely adopted for space plasmas, where charged particle distributions are often measured to have quasi-inverse power law tail populations.

The empirical fitting of *in situ* spacecraft measurements by kappa distribution, however, did not enjoy any first-principle justifications when it was first introduced. However, in the 1980s, Tsallis proposed the non-extensive thermo-statistical theory, for which the microscopic state of the system that maximizes the non-extensive (or Tsallis) entropy is none other than the kappa distribution (Tsallis, 1988, 2009). This has led to a renewed interest in the observed kappa-like space plasma distribution (Leubner, 2002; Livadiotis and McComas, 2009). Tsallis' original theory is quite general, and not specific to plasmas, but when applied to plasmas, the implication was quite obvious. In a parallel effort, Treumann (1999a,b) attempted to formulate an alternative statistical mechanics based upon Lorentzian distribution function, which is related to the kappa distribution.

Independent of these developments, Yoon et al. (2005) investigated the Langmuir turbulence problem from the standpoint of weak turbulence theory. Numerical solutions of the equations of weak turbulence led to the discovery that the end state of Langmuir turbulence also coincides with the generation of non-thermal tail population of the electron velocity distribution function, which superficially resembles the kappa distribution. This has led the present author to seek the genuine time asymptotic solution of the weakly turbulent state. It was found that the kappa distribution is a unique solution that characterizes the steady-state Langmuir turbulence (Yoon, 2014). The fact that the end state of Langmuir turbulence corresponds to the kappa distribution function is a major theoretical breakthrough in understanding the physical origin of measured electron distribution in the solar wind, and this problem represents a prime example of the success of weak turbulence theory.

Conceptually, the quasi-equilibrium state of Langmuir turbulence represents a collection of electrons interacting among themselves through long-range collective electrostatic fluctuations. Tsallis entropy also describes a system interacting with long-range force such that the entropy of the system exhibits non-extensive thermostatistical characteristics. In this regard, the quasi-equilibrium Langmuir turbulence state and the non-extensive statistical state might be equivalent. The profound interrelationship that may exist between the two approaches has yet to be worked out in a rigorous mathematical sense, but this is an area that is being investigated at a fundamental level – see the recent monograph by Livadiotis (2017), where various aspects related to the kappa distribution and non-extensive statistical mechanics is thoroughly surveyed.

This chapter discusses the application of weak turbulence theory to the beam-plasma instability, dynamical development of the Langmuir turbulence, the steady-state Langmuir turbulence, and the associated electron kappa distribution function.

6.1 Normalized Equations

The electron beam-plasma instability is a classic problem in any standard plasma physics literature. The linear stage of the instability is well understood, and is widely covered in many textbooks, and quasilinear phase of the instability is also well known (Drummond and Pines, 1962; Vedenov and Velikhov, 1962; Frieman and Rutherford, 1964; Bernstein and Engelmann, 1966; Dawson and Shanny, 1968; Morse and Nielson, 1969; Vahala and Montgomery, 1970; Joyce et al., 1971; Roberson et al., 1971; Appert et al., 1976; Ivanov et al., 1976; Grognard, 1982; Tsunoda et al., 1987; Dum, 1990; Muschietti and Dum, 1991; Nishikawa and Cairns, 1991; Dum and Nishikawa, 1994). Many advanced textbooks discuss the saturation of gentle beam-plasma instability via velocity space plateau formation. The nonlinear stage of the instability that takes place after the quasilinear plateau formation, in contrast, is not widely discussed in the elementary textbooks. The weak turbulence theory may be employed in order to discuss various stages of the beam-plasma instability development including the quasilinear saturation by velocity space plateau formation, but also the same theory naturally and quite rigorously applies to processes that take place much beyond the quasilinear saturation stage. The best way to illustrate the nonlinear development is via numerical solution.

We begin the discussion with the particle kinetic equation (4.86), but for the present purpose, let us retain the linear wave-particle interaction terms and focus only on the electrons:

$$\frac{\partial F_e}{\partial t} = \frac{\partial}{\partial v_i}\left(A_i F_e + D_{ij}\frac{\partial F_e}{\partial v_j}\right),$$
$$A_i = \frac{e^2}{4\pi m_e}\int d\mathbf{k}\,\frac{k_i}{k^2}\sum_{\sigma=\pm 1}\sigma\omega_{\mathbf{k}}^L\,\delta(\sigma\omega_{\mathbf{k}}^L - \mathbf{k}\cdot\mathbf{v}),$$
$$D_{ij} = \frac{\pi e^2}{m_e^2}\int d\mathbf{k}\,\frac{k_i k_j}{k^2}\sum_{\sigma=\pm 1}\delta(\sigma\omega_{\mathbf{k}}^L - \mathbf{k}\cdot\mathbf{v})\,I_{\mathbf{k}}^{\sigma L}. \tag{6.3}$$

We employ the wave kinetic equation summarized by (4.93) and (4.94), but in doing so, we neglect the scattering terms involving ion-sound waves, as such processes are extremely slow ones. Thus, the wave kinetic equations for Langmuir (L) and ion-sound (S) modes adopted in the present discussion are

$$\begin{aligned}\frac{\partial I_{\mathbf{k}}^{\sigma L}}{\partial t} &= \frac{\pi\omega_{pe}^2}{k^2}\int d\mathbf{v}\,\delta(\sigma\omega_{\mathbf{k}}^L - \mathbf{k}\cdot\mathbf{v})\left(\frac{ne^2}{\pi}F_e + \sigma\omega_{\mathbf{k}}^L I_{\mathbf{k}}^{\sigma L}\,\mathbf{k}\cdot\frac{\partial F_e}{\partial \mathbf{v}}\right)\\
&\quad + 2\sum_{\sigma',\sigma''=\pm 1}\sigma\omega_{\mathbf{k}}^L\int d\mathbf{k}'\,V_{\mathbf{k},\mathbf{k}'}^{LS}\Big(\sigma\omega_{\mathbf{k}}^L I_{\mathbf{k}'}^{\sigma' L} I_{\mathbf{k}-\mathbf{k}'}^{\sigma'' S}\\
&\quad - \sigma'\omega_{\mathbf{k}'}^L I_{\mathbf{k}-\mathbf{k}'}^{\sigma'' S} I_{\mathbf{k}}^{\sigma L} - \sigma''\omega_{\mathbf{k}-\mathbf{k}'}^L I_{\mathbf{k}'}^{\sigma' L} I_{\mathbf{k}}^{\sigma L}\Big)\\
&\quad + \sigma\omega_{\mathbf{k}}^L\sum_{\sigma'=\pm 1}\int d\mathbf{k}'\int d\mathbf{v}\,U_{\mathbf{k},\mathbf{k}'}^{LL}\left[\frac{m_e}{m_i} I_{\mathbf{k}'}^{\sigma' L} I_{\mathbf{k}}^{\sigma L}\,(\mathbf{k}-\mathbf{k}')\cdot\frac{\partial F_i}{\partial \mathbf{v}}\right.\\
&\quad \left. + \frac{ne^2}{\pi\omega_{pe}^2}\left(\sigma\omega_{\mathbf{k}}^L I_{\mathbf{k}'}^{\sigma' L} - \sigma'\omega_{\mathbf{k}'}^L I_{\mathbf{k}}^{\sigma L}\right)(F_e + F_i)\right],\\
\frac{\partial I_{\mathbf{k}}^{\sigma S}}{\partial t} &= \frac{\pi\mu_{\mathbf{k}}\omega_{pe}^2}{k^2}\int d\mathbf{v}\,\delta(\sigma\omega_{\mathbf{k}}^S - \mathbf{k}\cdot\mathbf{v})\left[\frac{ne^2}{\pi}(F_e + F_i)\right.\\
&\quad \left. + \sigma\omega_{\mathbf{k}}^L I_{\mathbf{k}}^{\sigma S}\left(\mathbf{k}\cdot\frac{\partial}{\partial \mathbf{v}}\right)\left(F_e + \frac{m_e}{m_i}F_i\right)\right]\\
&\quad + \sum_{\sigma',\sigma''=\pm 1}\sigma\omega_{\mathbf{k}}^L\int d\mathbf{k}'\,V_{\mathbf{k},\mathbf{k}'}^{SL}\Big(\sigma\omega_{\mathbf{k}}^L I_{\mathbf{k}'}^{\sigma' L} I_{\mathbf{k}-\mathbf{k}'}^{\sigma'' L}\\
&\quad - \sigma'\omega_{\mathbf{k}'}^L I_{\mathbf{k}-\mathbf{k}'}^{\sigma'' L} I_{\mathbf{k}}^{\sigma S} - \sigma''\omega_{\mathbf{k}-\mathbf{k}'}^L I_{\mathbf{k}'}^{\sigma' L} I_{\mathbf{k}}^{\sigma S}\Big),\end{aligned}$$

(6.4)

(6.5)

where

$$\omega_{\mathbf{k}}^L = \omega_{pe}\left(1+\frac{3}{2}k^2\lambda_{De}^2\right) = \omega_{pe}\left(1+\frac{3}{4}\frac{k^2v_{Te}^2}{\omega_{pe}^2}\right),$$

$$\omega_{\mathbf{k}}^S = \frac{kc_S\sqrt{1+3T_i/T_e}}{\sqrt{1+k^2\lambda_{De}^2}} = kv_{Te}\sqrt{\frac{m_e}{2m_i}}\frac{\sqrt{1+3T_i/T_e}}{\sqrt{1+k^2v_{Te}^2/2\omega_{pe}^2}}.$$

$$\mu_{\mathbf{k}} = k^3\lambda_{De}^3\sqrt{\frac{m_e}{m_i}}\sqrt{1+\frac{3T_i}{T_e}} = \frac{k^3v_{Te}^3}{\omega_{pe}^3}\sqrt{\frac{m_e}{8m_i}}\sqrt{1+\frac{3T_i}{T_e}},$$

$$V_{\mathbf{k},\mathbf{k}'}^{LS} = \frac{\pi}{2}\frac{e^2}{T_e^2}\frac{\mu_{\mathbf{k}-\mathbf{k}'}(\mathbf{k}\cdot\mathbf{k}')^2}{k^2k'^2|\mathbf{k}-\mathbf{k}'|^2}\delta(\sigma\omega_{\mathbf{k}}^L-\sigma'\omega_{\mathbf{k}'}^L-\sigma''\omega_{\mathbf{k}-\mathbf{k}'}^S),$$

$$V_{\mathbf{k},\mathbf{k}'}^{SL} = \frac{\pi}{4}\frac{e^2}{T_e^2}\frac{\mu_{\mathbf{k}}[\mathbf{k}'\cdot(\mathbf{k}-\mathbf{k}')]^2}{k^2k'^2|\mathbf{k}-\mathbf{k}'|^2}\delta(\sigma\omega_{\mathbf{k}}^S-\sigma'\omega_{\mathbf{k}'}^L-\sigma''\omega_{\mathbf{k}-\mathbf{k}'}^L),$$

$$U_{\mathbf{k},\mathbf{k}'}^{LL} = \frac{\pi}{\omega_{pe}^2}\frac{e^2}{m_e^2}\frac{(\mathbf{k}\cdot\mathbf{k}')^2}{k^2k'^2}\delta[\sigma\omega_{\mathbf{k}}^L-\sigma'\omega_{\mathbf{k}'}^L-(\mathbf{k}-\mathbf{k}')\cdot\mathbf{v}]. \tag{6.6}$$

The initial exponential growth (or inverse Landau damping rate) was already discussed in Section 1.8. After the onset of bump-on-tail instability, the saturation is achieved via velocity space plateau formation. The velocity space diffusion is effective only in the range where the electron distribution has a positive slope. This range corresponds to the positive growth of the Langmuir waves. It is convenient to consider one-dimensional situation. If we approximate $\omega_k \approx \omega_{pe}$, then upon ignoring the velocity friction term for now, the particle velocity space diffusion equation may be approximately written as

$$\frac{\partial F_e(v,t)}{\partial t} = \frac{\pi e^2}{m_e^2}\frac{\partial}{\partial v}\left[\frac{1}{v}I_{\omega_{pe}/v}(t)\frac{\partial}{\partial v}F_e(v,t)\right]. \tag{6.7}$$

In the time asymptotic limit, $\partial/\partial t \to 0$, we must have

$$I_{\omega_{pe}/v}(t)\left.\frac{\partial F_e(v,t)}{\partial v}\right|_{t\to\infty} \to 0, \tag{6.8}$$

over the resonant velocity space. Outside the resonant velocity space, the diffusion coefficient is zero, so that F_e does not change in time. The asymptotic condition (6.8) can be satisfied if

$$\left.I_{\omega_{pe}/v}(t)\right|_{t\to\infty} \to 0, \tag{6.9}$$

or

$$\left.\frac{\partial F_e(v,t)}{\partial v}\right|_{t\to\infty} \to 0. \tag{6.10}$$

Condition (6.9) is generally not satisfied, since at saturation the wave level will be finite. Consequently, the second condition (6.10) must be true. This means that, in the resonant velocity space, the electron distribution must form a plateau. Since the growth rate is proportional to $\partial F_e/\partial v$, but, in the time asymptotic regime, $\partial F_e/\partial v \to 0$ in the resonant velocity range, the asymptotic growth rate must be zero. The wave intensity, which grows from some initially low noise level, must saturate to some finite value:

$$\begin{aligned}
&\gamma_k \to 0, \qquad t \to \infty, \\
&I_k(t) \to \text{finite value}, \qquad t \to \infty.
\end{aligned} \tag{6.11}$$

This overall picture of the quasilinear development of the gentle beam-plasma, or bump-in-tail instability, is a well-known velocity plateau formation by quasilinear process.

Physical processes that take place beyond the quasilinear saturation stage involve wave-wave interaction dictated by nonlinear decay interaction terms in L and S mode equations, and nonlinear wave-particle interaction dictated by induced and spontaneous scattering processes in L mode equation. Toward the end of the mode coupling stage, the electron velocity distribution function develops a quasi-inverse power law energetic tail population. This involves energy redistribution among the electrons, which, in the end, resembles the so-called kappa distribution function. Before we discuss the various stages of bump-on-tail instability in detail, it is useful to discuss the dimensionless form of weak turbulence equations (6.3)–(6.6). Let us introduce the following normalization scheme:

$$\begin{aligned}
&T = \omega_{pe} t, \quad x_{\mathbf{q}}^{\alpha} = \frac{\omega_{\mathbf{k}}^{\alpha}}{\omega_{pe}}, \quad \mathbf{q} = \frac{\mathbf{k}\, v_{Te}}{\omega_{pe}}, \quad \mathbf{u} = \frac{\mathbf{v}}{v_{Te}}, \\
&\Phi_a(\mathbf{u}) = v_{Te}^3\, F_a(\mathbf{v}), \quad \mathscr{E}_{\mathbf{q}}^{\sigma\alpha} = (2\pi)^2 g\, \frac{I_{\mathbf{k}}^{\sigma\alpha}}{m_e\, v_{Te}^2}, \\
&M = \frac{m_i}{m_e}, \quad \tau = \frac{T_i}{T_e}, \quad g = \frac{1}{2^{3/2}\,(4\pi)^2\, n\, \lambda_{De}^3}.
\end{aligned} \tag{6.12}$$

In (6.12), $v_{Te} = \sqrt{2T_e/m_e}$ represents the thermal speed for bulk quasi-Maxwellian electrons. The quantity $3/(4\pi n \lambda_{De}^3)$ represents the inverse of the total number of plasma particles in a spherical volume with the radius equal to the Debye length, the so-called Debye sphere. This quantity is the plasma parameter already introduced in (5.37). As we already discussed in Section 5.2, the plasma parameter is associated with the collisional process. As we will shortly see, the plasma parameter is also related to the discrete particle effects. This will become obvious once we rewrite the basic equations (6.3)–(6.5) in normalized form. All the terms that are associated with the particle discreteness, namely, velocity friction term in the electron kinetic equation and spontaneous emission and spontaneous scattering terms in the wave kinetic equation, will turn out to have an overall proportionality to the factor $3/(4\pi n\lambda_{De}^3)$ or, equivalently, g defined in (6.12). Note that the definition of g in (6.12) is the same as that of (5.37), except for a trivial numerical factor. If $g \to 0$, then all the terms associated with the particle discreteness will disappear, and the system reduces to that of Vlasov weak turbulence theory. Recall that in the same limit of $g \to 0$, collisional processes also disappear, so that both binary collisions and discrete particle effects are controlled by the same g parameter.

Employing the normalization convention (6.12), the relevant equations (6.5)–(6.6) can be shown to reduce to the following. First, the electron particle kinetic equation is given by

$$\frac{\partial \Phi_e(\mathbf{u})}{\partial T} = \frac{\partial}{\partial u_i}\left(A_i^{\rm el}(\mathbf{u})\,\Phi_e(\mathbf{u}) + D_{ij}^{\rm el}(\mathbf{u})\,\frac{\partial \Phi_e(\mathbf{u})}{\partial u_j}\right),$$

$$A_i^{\rm el}(\mathbf{u}) = g\int d\mathbf{q}\,\frac{q_i}{q^2}\sum_{\sigma=\pm1}\sigma x_{\mathbf{q}}^L\,\delta(\sigma x_{\mathbf{q}}^L - \mathbf{q}\cdot\mathbf{u}),$$

$$D_{ij}^{\rm el}(\mathbf{u}) = \int d\mathbf{q}\,\frac{q_i\,q_j}{q^2}\sum_{\sigma=\pm1}\delta(\sigma x_{\mathbf{q}}^L - \mathbf{q}\cdot\mathbf{u})\,\mathscr{E}_{\mathbf{q}}^{\sigma L}. \tag{6.13}$$

Note that the velocity space friction (or drag) term has an overall proportionality to g, which proves our previous point that the finiteness of the g parameter is intimately related to the particle discreteness effect.

Second, the wave kinetic equations for Langmuir and ion-sound waves are given in dimensionless form by

$$\begin{aligned}
\frac{\partial \mathscr{E}_{\mathbf{q}}^{\sigma L}}{\partial T} &= \frac{\pi}{q^2}\int d\mathbf{u}\,\delta(\sigma x_{\mathbf{q}}^L - \mathbf{q}\cdot\mathbf{u})\left(g\,\Phi_e(\mathbf{u}) + \sigma x_{\mathbf{q}}^L\,\mathscr{E}_{\mathbf{q}}^{\sigma L}\,\mathbf{q}\cdot\frac{\partial\Phi_e(\mathbf{u})}{\partial\mathbf{u}}\right)\\
&\quad+2\sum_{\sigma',\sigma''=\pm1}\sigma x_{\mathbf{q}}^L\int d\mathbf{q}'\,\frac{\mu_{\mathbf{q}-\mathbf{q}'}\,(\mathbf{q}\cdot\mathbf{q}')^2}{q^2\,q'^2\,|\mathbf{q}-\mathbf{q}'|^2}\,\delta(\sigma x_{\mathbf{q}}^L - \sigma' x_{\mathbf{q}'}^L - \sigma'' x_{\mathbf{q}-\mathbf{q}'}^S)\\
&\quad\times\left(\sigma x_{\mathbf{q}}^L\,\mathscr{E}_{\mathbf{q}'}^{\sigma' L}\,\mathscr{E}_{\mathbf{q}-\mathbf{q}'}^{\sigma'' S} - \sigma' x_{\mathbf{q}'}^L\,\mathscr{E}_{\mathbf{q}-\mathbf{q}'}^{\sigma'' S}\,\mathscr{E}_{\mathbf{q}}^{\sigma L} - \sigma'' x_{\mathbf{q}-\mathbf{q}'}^L\,\mathscr{E}_{\mathbf{q}'}^{\sigma' L}\,\mathscr{E}_{\mathbf{q}}^{\sigma L}\right)\\
&\quad-\sum_{\sigma'=\pm1}\int d\mathbf{q}'\int d\mathbf{u}\,\frac{(\mathbf{q}\cdot\mathbf{q}')^2}{q^2\,q'^2}\,\delta[\sigma x_{\mathbf{q}}^L - \sigma' x_{\mathbf{q}'}^L - (\mathbf{q}-\mathbf{q}')\cdot\mathbf{u}]\\
&\quad\times\Bigg[g\,\sigma x_{\mathbf{q}}^L\left(\sigma' x_{\mathbf{q}'}^L\,\mathscr{E}_{\mathbf{q}}^{\sigma L} - \sigma x_{\mathbf{q}}^L\,\mathscr{E}_{\mathbf{q}'}^{\sigma' L}\right)[\Phi_e(\mathbf{u}) + \Phi_i(\mathbf{u})] \qquad (6.14)\\
&\quad+\mathscr{E}_{\mathbf{q}'}^{\sigma' L}\,\mathscr{E}_{\mathbf{q}}^{\sigma L}\,(\mathbf{q}-\mathbf{q}')\cdot\frac{\partial}{\partial\mathbf{u}}\left((\sigma x_{\mathbf{q}}^L - \sigma' x_{\mathbf{q}'}^L)\,\Phi_e(\mathbf{u}) - \frac{\sigma x_{\mathbf{q}}^L}{M}\,\Phi_i(\mathbf{u})\right)\Bigg],
\end{aligned}$$

$$\begin{aligned}\frac{\partial \mathscr{E}_{\mathbf{q}}^{\sigma S}}{\partial T} &= \frac{\pi \mu_{\mathbf{q}}}{q^2} \int d\mathbf{u}\, \delta(\sigma x_{\mathbf{q}}^S - \mathbf{q}\cdot\mathbf{u}) \big[g\,[\Phi_e(\mathbf{u}) + \Phi_i(\mathbf{u})] \\ &\quad + \sigma x_{\mathbf{q}}^L\, \mathscr{E}_{\mathbf{q}}^{\sigma S} \left(\mathbf{q}\cdot\frac{\partial}{\partial \mathbf{u}}\right)\left(\Phi_e(\mathbf{u}) + \frac{\Phi_i(\mathbf{u})}{M}\right)\Big] \\ &\quad + \sum_{\sigma',\sigma''=\pm 1} \sigma x_{\mathbf{q}}^L \int d\mathbf{q}'\, \frac{\mu_{\mathbf{q}}\,[\mathbf{q}'\cdot(\mathbf{q}-\mathbf{q}')]^2}{q^2\, q'^2\, |\mathbf{q}-\mathbf{q}'|^2}\, \delta(\sigma x_{\mathbf{q}}^S - \sigma' x_{\mathbf{q}'}^L - \sigma'' x_{\mathbf{q}-\mathbf{q}'}^L) \\ &\quad \times \left(\sigma x_{\mathbf{q}}^L\, \mathscr{E}_{\mathbf{q}'}^{\sigma' L}\, \mathscr{E}_{\mathbf{q}-\mathbf{q}'}^{\sigma'' L} - \sigma' x_{\mathbf{q}'}^L\, \mathscr{E}_{\mathbf{q}-\mathbf{q}'}^{\sigma'' L}\, \mathscr{E}_{\mathbf{q}}^{\sigma S} - \sigma'' x_{\mathbf{q}-\mathbf{q}'}^L\, \mathscr{E}_{\mathbf{q}'}^{\sigma' L}\, \mathscr{E}_{\mathbf{q}}^{\sigma S}\right).\end{aligned} \tag{6.15}$$

The dispersion relations and the auxiliary quantity $\mu_{\mathbf{k}}$ defined in (6.6) are given in dimensionless form by

$$x_{\mathbf{q}}^L = 1 + \frac{3q^2}{4}, \qquad x_{\mathbf{q}}^S = q\sqrt{\frac{1+3\tau}{M(2+q^2)}}, \qquad \mu_{\mathbf{q}} = q^3\sqrt{\frac{1+3\tau}{8M}}. \tag{6.16}$$

Again, note that in both wave kinetic equations in (6.14) and (6.15), spontaneous emission and scattering terms have an overall proportionality $\propto g$, which demonstrates that the particle discreteness determines the relative importance of these terms in relations to their induced counterparts. In the textbooks, it is stated that one of the conditions for ionized gas to qualify as the plasma is that g should be sufficiently small. The limit of $g \to 0$ represents the Vlasov limit, and, in such a case, the number of plasma particles within a Debye sphere is infinite. In general, even for a finite g, its value must be sufficiently low in order for the tools of plasma physics to apply in a rigorous sense. If g is not so low, or, equivalently, the number of charged particles in the Debye sphere is not high enough, then the problem reduces to that of a small collection of charged particle interacting with electromagnetic fields and among themselves in a vacuum.

The dimensionless equations (6.13)–(6.15) have been numerically solved by many researchers (Grognard, 1982; Muschietti and Dum, 1991; Kontar and Pécseli, 2002), but also by the present author and his colleagues in one dimension and for two dimensions (and with cylindrical symmetry, three dimensions) (Ziebell et al., 2001, 2008b). The relatively simple one-dimensional limit may be of interest to the readers. In Appendix H, we further discuss the one-dimensional version of the normalized equations, and present the derivations of concrete forms of equations that readily lend themselves to the numerical finite difference analysis. Here, we mention a salient point regarding the limit of one dimension. In order to reduce (6.13)–(6.15) to one-dimensional form, one may simply consider one-dimensional velocity and wave vector, $\mathbf{u} \to u$, $\mathbf{q} \to q$, but, in doing so, one should bear in mind that the parameter g is a result of normalizing the three-dimensional equation. For a one-dimensional world, the plasma parameter, or g, should no longer be interpreted as the inverse of number of particles in a Debye sphere, but rather it should be the inverse of number of particles in a line segment equal to the Debye length,

$$g_{1D} \propto \frac{1}{n\lambda_{De}}. \tag{6.17}$$

Since the number of particles in a sphere is obviously much higher than the same in a line segment, the effective plasma parameter in one dimension must be significantly higher than the three-dimensional plasma parameter,

$$g_{1D} \gg g_{3D}. \tag{6.18}$$

Roughly, we expect that $g_{3D} \propto g_{1D}^3$, such that if g_{1D}, which is much less than unity, is given by a value, then g_{3D} is smaller still by many orders of magnitude.

6.2 Numerical Analysis of Weak Langmuir Turbulence

Dynamical evolution of weak electron beam and Langmuir/ion-sound turbulence as described by the equations of weak turbulence theory is best discussed by numerical solutions. Figure 6.2 plots the result of numerical analysis based upon (6.13)–(6.15), except that the equations are reduced to one dimension – see Appendix H. The initial electron distribution and the initial Langmuir wave intensity profile are specified by

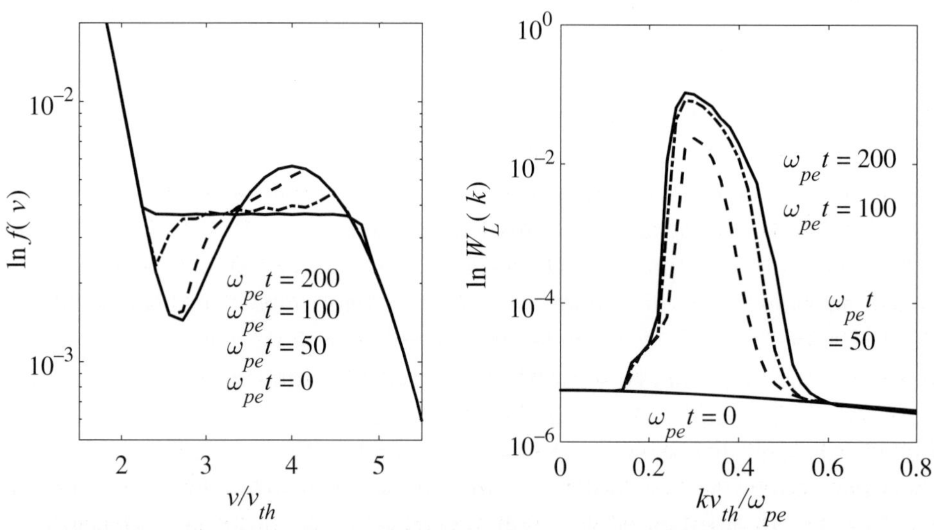

Figure 6.2 Excitation of bump-on-tail Langmuir wave instability followed by subsequent quasilinear saturation during the relatively early time, $0 < t < 200\omega_{pe}^{-1}$.

$$F_e(v) = \left(1 - \frac{n_b}{n_0}\right) \frac{e^{-v^2/v_{Te}^2}}{\pi^{1/2} v_{Te}} + \frac{n_b}{n_0} \frac{e^{-(v-V_b)^2/v_{Tb}^2}}{\pi^{1/2} v_{Tb}},$$

$$I_k^L = \frac{T_e}{4\pi^2} \frac{1}{1 + 3k^2\lambda_{De}^2}, \tag{6.19}$$

where we assumed $n_b/n_0 = 10^{-2}$, $T_b = T_e$, $V_b/v_{Te} = 4$, and the dimensionless plasma parameter of $g = (n\lambda_{De})^{-1} = 10^{-3}$ is adopted. Note that this is a one-dimensional plasma parameter. The three-dimensional value should be roughly $g_{3D} \propto 10^{-9}$. The normalization of Langmuir wave spectral energy density is $gI_{\mathbf{k}}^{\sigma L}/(8m_e v_{Te}^2)$. For $t = 0$, there exists a positive velocity gradient associated with the beam electrons – see (6.18). The gentle, or weak, beam with positive gradient in velocity space, $\partial F_e/\partial v > 0$, over the range $\sim 2.5 < v/v_{Te} < \sim 4$ (see the left-hand panel, for $t = 0$), corresponds to the unstable wave number range $\sim 0.2 < kv_{Te}/\omega_{pe} < \sim 0.55$. Over this range of normalized k, the bump-on-tail instability operates so that the initially low Langmuir wave intensity (shown on the right-hand panel) exponentially increases. Between $t = 0$ and $50\omega_{pe}^{-1}$, it is seen that the wave intensity is greatly amplified ($\omega_{pe}t = 50$), but by the time the system evolves to $\omega_{pe}t = 100$, the instability is practically saturated, and the Langmuir energy density does not show much increase between $\omega_{pe}t = 100$ and 200. During the same time period it is seen that the positive velocity gradient associated with the resonant portion of electron distribution function gradually disappears until when $\omega_{pe}t = 200$, the velocity space plateau formation is almost complete. This is the well-known textbook example of linear instability excitation and subsequent quasilinear saturation.

Physical processes that take place after the quasilinear saturation stage is dictated by the nonlinear wave dynamics. Beyond the velocity space plateau formation stage, the electron velocity distribution function does not undergo much change in the bulk velocity and the flattened beam range. However, near the energetic suprathermal region with v much higher than v_{Te}, quasi-inverse power law tail population is gradually and spontaneously generated. We will discuss such an acceleration later, but for now let us focus on the nonlinear wave dynamics. Nonlinear processes are of two types. One is the resonant three-wave interaction, or decay process, and the other is the nonlinear wave-particle interaction, or scattering.

In Figures 6.3 and 6.4, we show the time development of wave intensities. First, Figure 6.3 shows the dynamic spectrum of Langmuir wave intensity. In the plot, positive k region corresponds to the forward-propagating Langmuir wave (signified by $\sigma = 1$), while the negative k region designates the backward-propagating L mode (corresponding to $\sigma = -1$). We combined the two modes into a single figure, plotting I_k^{+L} over positive k range, and I_k^{-L} in $k < 0$ space. Actual numerical computation was done over $k > 0$ space for both I_k^{+L} and I_k^{-L}. For early time

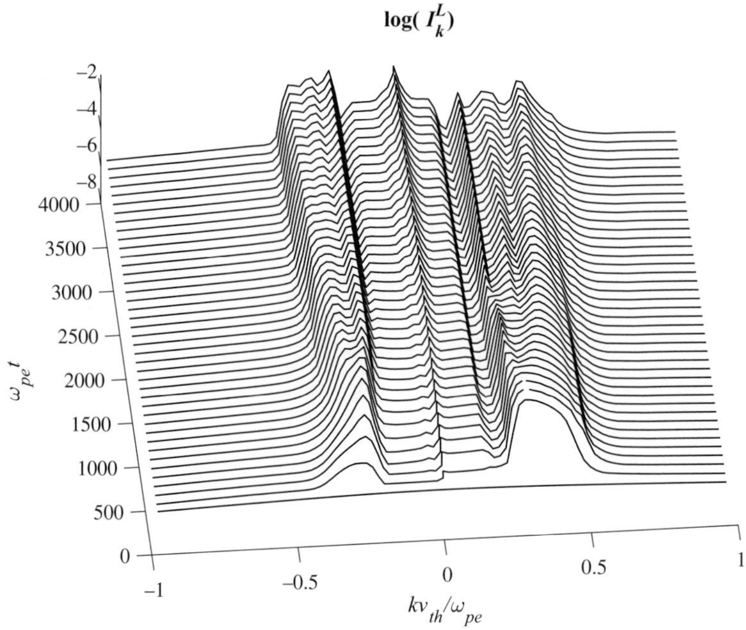

Figure 6.3 Nonlinear progression of Langmuir wave turbulence in the mode coupling stage.

periods between $t = 0$ and $\omega_{pe}t = 200$, or so, the Langmuir wave dynamics is simply dictated by the exponential growth and subsequent quasilinear saturation, which is already shown in Figure 6.2. During this stage we begin to see the growth associated with the backward Langmuir waves ($k < 0$ region). This is the result of combined three-wave decay process (that is, forward or primary L mode decaying into backward L mode and an ion-sound mode S) and scattering of forward L mode off thermal ions. It is also seen that Langmuir waves near $k \sim 0$ slowly but steadily grow in intensity. This is again the result of combined decay and scattering, and such a long wavelength Langmuir mode generation is known as the Langmuir condensation effect. Nonlinear mode coupling processes involve multiple back-and-forth wave-wave interaction processes and scattering off thermal ions, and the entire mode coupling processes continue on well beyond the quasilinear saturation phase.

Figure 6.4 is the corresponding dynamic spectrum associated with the ion-sound mode turbulence. In the early stage, between $t = 0$ and $\omega_{pe}t = 200$ or so, no ion-sound mode is apparent above initial noise level in both positive and negative k regions. However, around the time when the backward Langmuir wave begins to appear, it can be see that, first, the forward-propagating S mode wave becomes slightly enhanced, followed by the backward ($k < 0$) ion-sound mode waves.

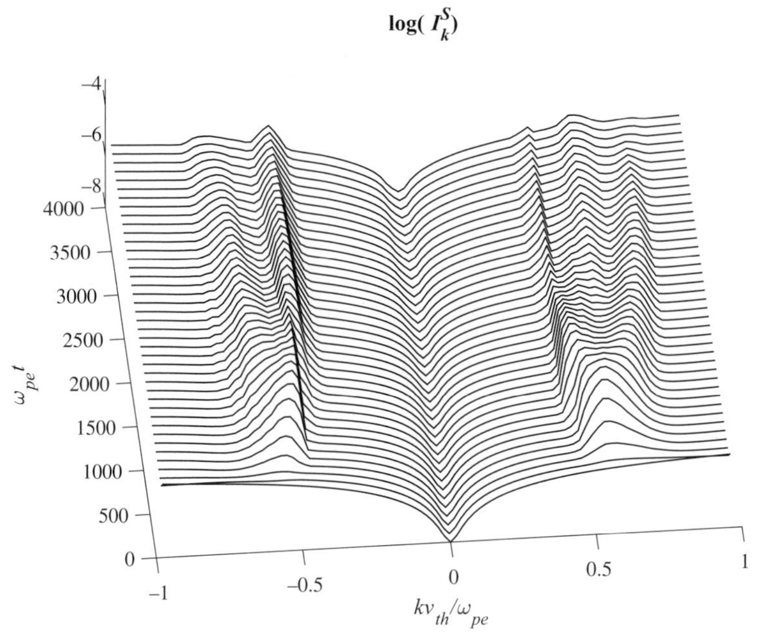

Figure 6.4 Nonlinear progression of ion-sound wave turbulence in the mode coupling stage.

The production of S mode turbulence is entirely owing to the decay process since the scattering process does not involve ion-sound waves. In the solar wind, the ratio T_i/T_e varies (Gurnett et al., 1979; Newbury et al., 1998), but in general, unless T_i/T_e is small, ion-sound waves may heavily damp thus prohibiting the three-wave resonant interaction. However, numerical solutions of the equations of electrostatic weak turbulence theory as well as electrostatic particle-in-cell simulation show that the three-wave interaction may still be operative even when T_i/T_e is as high as an order unity (Yoon et al., 2012a; Rha et al., 2013). It is interesting to note that the ion-sound turbulence is a transient phenomenon. As the system evolves toward the end of the present numerical computation, that is, $\omega_{pe}t = 4000$, it is seen that the S mode wave intensity gradually settles down back toward the initial noise level. This is important for later discussion of asymptotically steady-state solution, or, equivalently, the so-called turbulent quasi-equilibrium (Treumann, 1999a,b). In the asymptotically steady state, $t \to \infty$, we will solve for the steady-state particle distribution function as well as for the steady-state Langmuir wave spectral intensity, which form a coupled set of solutions. The notion of turbulent quasi-equilibrium is the state of plasma in which the electrons and Langmuir fluctuations continuously exchange momentum and energy but, on average, the whole system is in dynamical equilibrium. Since the wave-wave or decay processes exchange

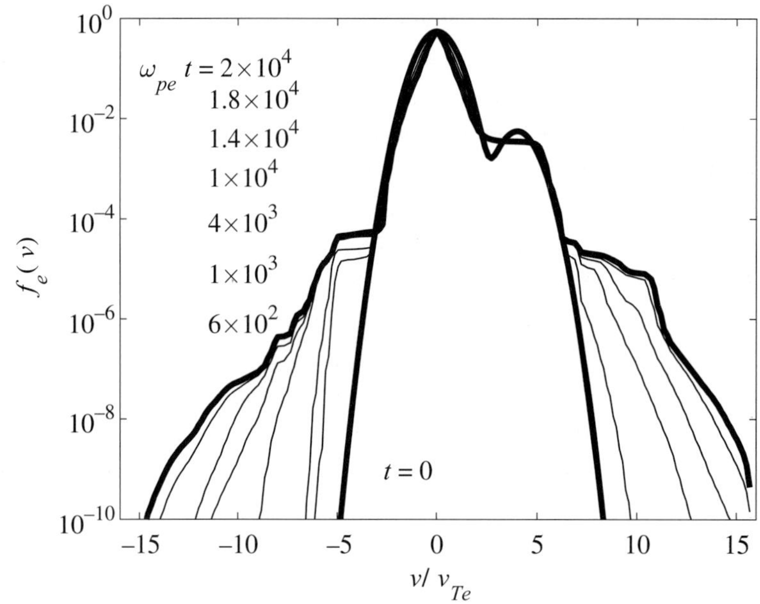

Figure 6.5 Development of energetic tail during the nonlinear mode-coupling stage.

momentum and energy only among the waves, asymptotically they are assumed to be insignificant. The present numerical calculation bears out such a feature.

Returning to the Langmuir mode dynamic spectrum, the Langmuir condensation is intimately related to the acceleration of small amount of electrons to form an energetic tail population. This is because the long wavelength, or small k mode, leads to the resonant electron speed given by $v_{\rm res} \approx \omega_{pe}/k$, such that, for small k, suprathermal electrons can resonate with the long wavelength Langmuir mode to form the high-energy tail distribution. In Figure 6.5, we display the long-time evolution of electron distribution function much beyond the mode coupling stage. Recall that Figures 6.3 and 6.4 show turbulent wave dynamics only up to $\omega_{pe}t = 4000$. In contrast, the electron velocity distribution function shown in Figure 6.5 is calculated up to $\omega_{pe}t = 2 \times 10^4$. Observe the formation of energetic tail population in the suprathermal energy range, which superficially resembles the kappa distribution. Ryu et al. (2007) confirmed the present findings based upon the weak turbulence formalism by means of particle-in-cell (PIC) simulation. That is, they demonstrated by means of PIC code simulation that the generation of kappa-like tail population in the late stages of beam-plasma instability process is consistent with the theoretical calculation. As we noted already, the kappa-like electron distribution function is pervasively observed in the near-Earth space environment. The question that naturally arises is whether the genuine electron kappa distribution function corresponds to the time asymptotic state of weak Langmuir turbulence or not. Or could it be that the numerical solution continually flattens the suprathermal portion of the electron velocity distribution with asymptotically steady-state solution? These questions cannot be addressed by the brute-force numerical solution. Instead, one must attack this issue with some form of analytical method. To address this issue, we next turn to an analytical approach and consider the steady-state solution for both electrons and Langmuir turbulence. As already noted, in the time asymptotic state, the ion-sound turbulence may be ignored since the S mode generation is only part of a transient phenomenon.

The analytical approach to understand the formation of electron kappa distribution function in the asymptotically steady stage of Langmuir turbulence is useful since the numerical demonstration presented herewith is based upon one-dimensional system. The actual solar wind electron distribution is observed to be highly isotropic in three velocity dimensions. It is desirable to carry out the numerical analysis for fully three-dimensional system and demonstrate the formation of genuine three-dimensional electron kappa distribution function, but, thus far, such a demonstration has not been done. Instead, we bypass the issue of numerical demonstration, and directly attack the problem of three-dimensional kappa distribution by means of analytical approach with the assumption of three-dimensional isotropy at the outset. The next section discusses such a solution.

6.3 Langmuir Turbulence and Electron Kappa Distribution

6.3.1 Steady-State Electron Distribution

Consider the particle kinetic equation for electrons (6.3). In what follows, we assume that the wave dispersion relation depends only on the magnitude of $\mathbf{k}$, and that the forward and backward wave intensities are identical and isotropic, $\omega_{\mathbf{k}}^L = \omega_k^L$ and $I_{\mathbf{k}}^{\sigma L} = I_k^L$, which are valid assumptions, provided the electron distribution function is isotropic, $F_e(\mathbf{v}) = F_e(v)$. We assume the steady state, $\partial F_e/\partial t = 0$, by virtue of the velocity friction and diffusion terms balancing each other out,

$$\begin{aligned}
\frac{\partial F_e}{\partial t} &= 0 = \frac{\partial}{\partial v_i} A_i F_e + \frac{\partial}{\partial v_i} D_{ij} \frac{\partial F_e}{\partial v_j}, \\
A_i &= \frac{e^2}{4\pi m_e} \int d\mathbf{k} \, \frac{k_i}{k^2} \sum_{\sigma=\pm 1} \sigma \omega_k^L \, \delta(\sigma \omega_k^L - \mathbf{k}\cdot\mathbf{v}), \\
D_{ij} &= \frac{\pi e^2}{m_e^2} \int d\mathbf{k} \, \frac{k_i k_j}{k^2} \sum_{\sigma=\pm 1} \delta(\sigma \omega_k^L - \mathbf{k}\cdot\mathbf{v}) \, I_k^L .
\end{aligned} \tag{6.20}$$

We may decompose the wave vector into components perpendicular and parallel to the velocity vector:

$$\begin{aligned}\mathbf{k} &= \frac{(\mathbf{v}\times\mathbf{k})\times\mathbf{v}}{v^2} + \frac{(\mathbf{v}\cdot\mathbf{k})\,\mathbf{v}}{v^2} = \mathbf{k}_\perp + \mathbf{k}_\parallel \\ &= k_\perp\,\hat{\mathbf{e}}_\perp + k_\parallel\,\hat{\mathbf{e}}_\parallel = k_\perp\,\hat{\mathbf{e}}_\perp + k_\parallel\,\frac{\mathbf{v}}{|v|}.\end{aligned} \tag{6.21}$$

With this decomposition, we may write the drag (or friction) coefficient as follows:

$$\begin{aligned}\mathbf{A} &= \frac{e^2}{2m_e}\sum_{\sigma=\pm1}\int_0^\infty dk_\perp k_\perp \int_{-\infty}^\infty dk_\parallel\,\frac{k_\perp\hat{\mathbf{e}}_\perp + k_\parallel\hat{\mathbf{e}}_\parallel}{k_\perp^2+k_\parallel^2}\,k_\parallel v\;\delta(\sigma\omega_k^L - k_\parallel v) \\ &\approx \hat{\mathbf{e}}_\parallel\,\frac{e^2}{2m_e}\sum_{\sigma=\pm1}\int_0^\infty dk_\perp\,k_\perp\int_{-\infty}^\infty dk_\parallel\,\frac{k_\parallel^2\,v}{k_\perp^2+k_\parallel^2}\;\delta(\sigma\omega_{pe} - k_\parallel v) \\ &= \hat{\mathbf{e}}_\parallel\,\frac{e^2}{4\pi\,m_e\,v}\sum_{\sigma=\pm1}\int d\mathbf{k}\,\frac{(\mathbf{k}\cdot\mathbf{v})^2}{k^2}\;\delta(\sigma\omega_{pe} - \mathbf{k}\cdot\mathbf{v}) \\ &= \hat{\mathbf{e}}_\parallel\,\frac{e^2\,\omega_{pe}^2}{4\pi\,m_e\,v}\sum_{\sigma=\pm1}\int\frac{d\mathbf{k}}{k^2}\;\delta(\sigma\omega_{pe} - \mathbf{k}\cdot\mathbf{v}),\end{aligned} \tag{6.22}$$

where the perpendicular component of $\mathbf{A}$ vanishes upon summing over both signs of $\sigma = \pm 1$. Upon making use of the notation $\hat{\mathbf{e}}_\parallel = \mathbf{v}/v$, we thus obtain

$$\begin{aligned}\mathbf{A} &= \mathbf{v}\,G(v), \\ G(v) &= \frac{e^2\,\omega_{pe}^2}{4\pi\,m_e\,v^2}\sum_{\sigma=\pm1}\int\frac{d\mathbf{k}}{k^2}\;\delta(\sigma\omega_{pe} - \mathbf{k}\cdot\mathbf{v}).\end{aligned} \tag{6.23}$$

Consider next the diffusion coefficient,

$$\begin{aligned}D_{ij} &= \frac{\pi e^2}{m_e^2\,v^2}\sum_{\sigma=\pm1}\int d\mathbf{k}\,\frac{\hat{\mathbf{e}}_\perp\,\hat{\mathbf{e}}_\perp\,k_\perp^2 v^2 + \hat{\mathbf{e}}_\parallel\,\hat{\mathbf{e}}_\parallel\,k_\parallel^2 v^2}{k^2}\;\delta(\sigma\omega_{pe} - \mathbf{k}\cdot\mathbf{v})\,I_L(k) \\ &= \frac{\pi e^2}{m_e^2\,v^2}\sum_{\sigma=\pm1}\int d\mathbf{k}\,\frac{\hat{\mathbf{e}}_\perp\,\hat{\mathbf{e}}_\perp\,[1-(\mathbf{k}\cdot\mathbf{v})^2] + \hat{\mathbf{e}}_\parallel\,\hat{\mathbf{e}}_\parallel\,(\mathbf{k}\cdot\mathbf{v})^2}{k^2} \\ &\quad\times\;\delta(\sigma\omega_{pe} - \mathbf{k}\cdot\mathbf{v})\,I_L(k) \\ &= \frac{\pi e^2}{m_e^2\,v^2}\sum_{\sigma=\pm1}\int d\mathbf{k}\,\frac{\hat{\mathbf{e}}_\perp\,\hat{\mathbf{e}}_\perp\,(1-\omega_{pe}^2) + \hat{\mathbf{e}}_\parallel\,\hat{\mathbf{e}}_\parallel\,\omega_{pe}^2}{k^2}\;\delta(\sigma\omega_{pe} - \mathbf{k}\cdot\mathbf{v})\,I_L(k),\end{aligned} \tag{6.24}$$

where, again, the mixed terms disappear when summed over both signs of $\sigma = \pm 1$. We may rewrite this result as (Hasegawa et al., 1985)

$$
\begin{aligned}
D_{ij} &= D_\perp \left(\delta_{ij} - \frac{v_i\, v_j}{v^2}\right) + D_\parallel \frac{v_i\, v_j}{v^2}, \\
D_\perp &= \frac{\pi e^2 (1 - \omega_{pe}^2)}{m_e^2\, v^2} \sum_{\sigma=\pm 1} \int \frac{d\mathbf{k}}{k^2}\, \delta(\sigma\omega_{pe} - \mathbf{k}\cdot\mathbf{v})\, I_L(k), \\
D_\parallel &= \frac{\pi e^2\, \omega_{pe}^2}{m_e^2\, v^2} \sum_{\sigma=\pm 1} \int \frac{d\mathbf{k}}{k^2}\, \delta(\sigma\omega_{pe} - \mathbf{k}\cdot\mathbf{v})\, I_L(k).
\end{aligned}
\tag{6.25}
$$

It is instructive to rewrite the electron kinetic equation (6.20) in spherical velocity space coordinate system by making use of the following transformation properties between cylindrical and spherical coordinate representations:

$$
\begin{aligned}
A_v &= A_\perp \sin\theta + A_\parallel \cos\theta, \\
A_\theta &= A_\perp \cos\theta - A_\parallel \sin\theta, \\
D_{vv} &= D_{\perp\perp} \sin^2\theta + 2D_{\perp\parallel} \sin\theta\cos\theta + D_{\parallel\parallel}\cos^2\theta, \\
D_{\theta\theta} &= D_{\perp\perp} \cos^2\theta - 2D_{\perp\parallel} \sin\theta\cos\theta + D_{\parallel\parallel}\sin^2\theta, \\
D_{v\theta} &= D_{\theta v} = (D_{\perp\perp} - D_{\parallel\parallel})\sin\theta\cos\theta - D_{\perp\parallel}(\sin^2\theta - \cos^2\theta).
\end{aligned}
\tag{6.26}
$$

This leads to

$$
\begin{aligned}
\frac{\partial F_e}{\partial t} = 0 = {} & \frac{1}{v^2}\frac{\partial}{\partial v}(v^2 A_v F_e) + \frac{1}{v\sin\theta}\frac{\partial}{\partial\theta}(A_\theta F_e) \\
& + \frac{1}{v^2}\frac{\partial}{\partial v}\left(v^2 D_{vv}\frac{\partial F_e}{\partial v}\right) + \frac{1}{v^2}\frac{\partial}{\partial v}\left(v D_{v\theta}\frac{\partial F_e}{\partial\theta}\right) \\
& + \frac{1}{v\sin\theta}\frac{\partial}{\partial\theta}\left(\sin\theta\, D_{\theta v}\frac{\partial F_e}{\partial v}\right) + \frac{1}{v^2\sin\theta}\frac{\partial}{\partial\theta}\left(\sin\theta\, D_{\theta\theta}\frac{\partial F_e}{\partial\theta}\right),
\end{aligned}
\tag{6.27}
$$

where $v = \sqrt{v_\perp^2 + v_\parallel^2}$ and $\cos\theta = v_\parallel/v$. If we denote

$$
\mu = \cos\theta,
\tag{6.28}
$$

then we may alternatively write

$$
\begin{aligned}
\frac{\partial F_e}{\partial t} = 0 = {} & \frac{1}{v^2}\frac{\partial}{\partial v}(v^2 A_v F_e) - \frac{1}{v}\frac{\partial}{\partial\mu}(A_\mu F_e) \\
& + \frac{1}{v^2}\frac{\partial}{\partial v}\left[v^2\left(D_{vv}\frac{\partial F_e}{\partial v} - \frac{(1-\mu^2)^{1/2}}{v} D_{v\mu}\frac{\partial F_e}{\partial\mu}\right)\right] \\
& - \frac{1}{v}\frac{\partial}{\partial\mu}\left[(1-\mu^2)^{1/2}\left(D_{\mu v}\frac{\partial F_e}{\partial v} - \frac{(1-\mu^2)^{1/2}}{v} D_{\mu\mu}\frac{\partial F_e}{\partial\mu}\right)\right].
\end{aligned}
\tag{6.29}
$$

From (6.23) and (6.24), we have

$$
\begin{aligned}
&\mathbf{A} = \hat{\mathbf{e}}_v\, v\, G = \hat{\mathbf{e}}_v\, A_v, \\
&D_{vv} = D_\parallel, \qquad D_{v\mu} = D_{\mu v} = 0, \qquad D_{\mu\mu} = D_\perp,
\end{aligned}
\tag{6.30}
$$

which reduces (6.29) to

$$\frac{\partial F_e}{\partial t} = 0 = \frac{1}{v^2}\frac{\partial}{\partial v}(v^2 A_v F_e) + \frac{1}{v^2}\frac{\partial}{\partial v}\left(v^2 D_{vv}\frac{\partial F_e}{\partial v}\right) + \frac{1}{v^2}\frac{\partial}{\partial \mu}\left((1-\mu^2)\, D_{\mu\mu}\frac{\partial F_e}{\partial \mu}\right). \tag{6.31}$$

For isotropic F_e, we further reduce (6.31) as

$$\frac{\partial F_e}{\partial t} = \frac{1}{v^2}\frac{\partial}{\partial v}\left[v^2\left(A_v F_e + D_{vv}\frac{\partial F_e}{\partial v}\right)\right]. \tag{6.32}$$

The steady-state solution is thus given by

$$F_e = C\,\exp\left(-\int dv\,\frac{A_v}{D_{vv}}\right) = C\,\exp\left(-\int dv\,\frac{vG}{D_\parallel}\right) \tag{6.33}$$
$$= C\,\exp\left(-\int dv\,\frac{m_e v}{4\pi^2}\frac{\sum_{\sigma=\pm 1}\int \frac{d\mathbf{k}}{k^2}\,\delta(\sigma\omega_{pe} - \mathbf{k}\cdot\mathbf{v})}{\sum_{\sigma=\pm 1}\int \frac{d\mathbf{k}}{k^2}\,\delta(\sigma\omega_{pe} - \mathbf{k}\cdot\mathbf{v})\, I_L(k)}\right).$$

To compute the quantity $vG/D_\parallel$, we define the angle between $\mathbf{k}$ and $\mathbf{v}$ by γ, or $\alpha = \cos\gamma$,

$$d\mathbf{k} = 4\pi k^2 dk\, d\cos\gamma = 4\pi k^2 dk\, d\alpha. \tag{6.34}$$

Then we have

$$\frac{vG}{D_\parallel} = \frac{m_e v}{4\pi^2}\frac{\sum_{\sigma=\pm 1}\int_{-1}^{1} d\alpha \int_0^\infty dk\,\delta(\sigma\omega_{pe} - kv\alpha)}{\sum_{\sigma=\pm 1}\int_{-1}^{1} d\alpha \int_0^\infty dk\,\delta(\sigma\omega_{pe} - kv\alpha)\, I_L(k)}$$
$$= \frac{m_e v}{4\pi^2}\frac{\displaystyle\int_{\omega_{pe}/v}^\infty \frac{dk}{k}}{\displaystyle\int_{\omega_{pe}/v}^\infty \frac{dk}{k}\, I_L(k)}. \tag{6.35}$$

In (6.35) the integral $\int_{\omega_{pe}/v}^\infty dk/k$ formally diverges for $k \to \infty$. In order to avoid such a problem, we invoke the following approach. We formally define

$$\mathscr{H}(v) = \int_{\omega_{pe}/v}^\infty \frac{dk}{k}, \qquad \mathscr{H}(v)\,\mathscr{I}(v) = \int_{\omega_{pe}/v}^\infty \frac{dk}{k}\, I_L(k). \tag{6.36}$$

Then we may formally remove the divergence, so that we have

$$\frac{vG}{D_\parallel} = \frac{m_e v}{4\pi^2}\frac{1}{\mathscr{I}(v)}. \tag{6.37}$$

At this point, we anticipate the final result in that the electron distribution function in the asymptotic state of weak Langmuir turbulence is the kappa distribution. We thus force our steady-state solution to be the kappa-like solution,

$$F_e = \frac{C}{(1 + m_e v^2/2\kappa'\,\theta_e)^{\kappa+1}}. \tag{6.38}$$

The normalization constant C can be obtained by requiring the condition

$$1 = 4\pi \int_0^\infty dv\, v^2 F_e = \frac{4\pi C\,(2\kappa'\,\theta_e)^{3/2}}{m_e^{3/2}} \int_0^\infty dx\, \frac{x^2}{(1+x^2)^{\kappa+1}}. \tag{6.39}$$

Making use of the integral identity

$$\int_0^\infty \frac{x^{\mu-1}\,dx}{(1+x^2)^\kappa} = \frac{\Gamma(\mu/2)\,\Gamma(\kappa-\mu/2)}{2\,\Gamma(\kappa)}, \tag{6.40}$$

we readily obtain the normalized F_e,

$$F_e = \frac{m_e^{3/2}}{(2\pi\,\theta_e)^{3/2}} \frac{\Gamma(\kappa+1)}{\kappa'^{3/2}\,\Gamma(\kappa-1/2)} \frac{1}{(1+m_e v^2/2\kappa'\,\theta_e)^{\kappa+1}}. \tag{6.41}$$

The effective temperature for this kappa-like model can be computed on the basis of definition $T_a = \int d\mathbf{v}(m_e v^2/3)F_e$. The result is

$$T_e = \theta_e \frac{\kappa'}{\kappa - 3/2}. \tag{6.42}$$

The proof is straightforward, which is omitted.

The relevant question is what kind of wave spectrum $I_L(k)$ will be consistent with the kappa-like distribution (6.41). To answer this question, we consider the relationship

$$\frac{1}{(1+m_e v^2/2\kappa'\,\theta_e)^{\kappa+1}} = \exp\left(-\int dv\, \frac{m_e\, v}{4\pi^2} \frac{1}{\mathscr{I}(v)}\right). \tag{6.43}$$

Taking the logarithm of both sides, making use of the integral identity

$$\int dv\, \frac{2v}{v^2/b+1} = b\,\ln\left(1+\frac{v^2}{b}\right), \tag{6.44}$$

and identifying $b = 2\kappa'\,\theta_e/m_e$, we obtain

$$\frac{\kappa+1}{\kappa'}\int dv\, \frac{m_e v/\theta_e}{m_e v^2/(2\kappa'\,\theta_e)+1} = \int dv\, \frac{m_e\, v}{4\pi^2}\frac{1}{\mathscr{I}(v)}, \tag{6.45}$$

or, equivalently,

$$\mathscr{I}(v) = \frac{\theta_e}{4\pi^2}\frac{\kappa'}{\kappa+1}\left(1+\frac{m_e v^2}{2\kappa'\theta_e}\right). \tag{6.46}$$

From (6.36), which can be rewritten upon replacing ω_{pe}/v by k,

$$\mathscr{H}(k) = \int_k^\infty \frac{dk'}{k'}, \qquad \mathscr{H}(k)\,\mathscr{I}(k) = \int_k^\infty \frac{dk'}{k'}\,I_L(k'), \tag{6.47}$$

we obtain

$$\mathscr{I}(k) = \frac{1}{\int_k^\infty dk''/k''}\int_k^\infty \frac{dk'}{k'}\,I_L(k'). \tag{6.48}$$

The quantity $\mathscr{I}(k)$ on the left-hand side is already given by (6.46). As a consequence, we have

$$\frac{\theta_e}{4\pi^2}\frac{\kappa'}{\kappa+1}\int_k^\infty \frac{dk'}{k'}\left(1+\frac{m_e\omega_{pe}^2}{2\kappa' k^2\theta_e}\right) = \int_k^\infty \frac{dk'}{k'}\,I_L(k'). \tag{6.49}$$

Taking the derivative d/dk to both sides,

$$\frac{\theta_e}{4\pi^2}\frac{\kappa'}{\kappa+1}\left[-\frac{1}{k}\left(1+\frac{m_e\omega_{pe}^2}{2\kappa' k^2\theta_e}\right) - \frac{m_e\omega_{pe}^2}{\kappa' k^3\theta_e}\int_k^\infty \frac{dk'}{k'}\right] = -\frac{1}{k}\,I_L(k), \tag{6.50}$$

we readily obtain

$$I_L(k) = \frac{\theta_e}{4\pi^2}\frac{\kappa'}{\kappa+1}\left[1+\frac{m_e\omega_{pe}^2}{2\kappa' k^2\theta_e}\left(1+2\int_k^\infty \frac{dk'}{k'}\right)\right]. \tag{6.51}$$

Thus, we obtain the turbulence spectrum that is consistent with the kappa-like electron distribution function (6.41):

$$\begin{aligned} I_L(k) &= \frac{\theta_e}{4\pi^2}\frac{\kappa'}{\kappa+1}\left(1+\frac{m_e\omega_{pe}^2}{2\kappa' k^2\theta_e}\left[1+2\mathscr{H}(k)\right]\right), \\ \mathscr{I}(k) &= \frac{1}{\mathscr{H}(k)}\int_k^\infty \frac{dk}{k}\,I_L(k) = \frac{\theta_e}{4\pi^2}\frac{\kappa'}{\kappa+1}\left(1+\frac{m_e\omega_{pe}^2}{2\kappa' k^2\theta_e}\right), \\ \mathscr{H}(k) &= \int_k^\infty \frac{dk}{k}. \end{aligned} \tag{6.52}$$

It should be noted that the forced solution (6.41) and the corresponding Langmuir turbulence intensity (6.52) are by no means unique solutions. In fact, at this stage, the coupled solutions (6.41) and (6.52) are but one in a class of infinitely many solutions. Even if we accept that these are the only acceptable coupled solutions, the indices κ and κ' are free parameters at this point. In order to determine the specific values for κ and κ' as well as to prove the uniqueness theorem for the kappa-like distribution function (6.41), we need to turn to the steady-state wave equations. As it turns out, when we consider the balance of spontaneous and induced processes in the steady-state wave kinetic equation, the resulting constraints on possible solutions drastically gets reduced, and the coupled solutions (6.41) and (6.52) become the only acceptable solution, provided κ and κ' are given by specific values. To show this, we next consider the steady-state wave kinetic equation. Before we do that, however, let us revisit the problem of wave dispersion relations when the kappa-like model defines the electron distribution function.

6.3.2 Electrostatic Waves with Kappa Distribution

Consider the electrostatic dispersion relation

$$\epsilon(\mathbf{k},\omega) = 0 = 1 + \sum_a \frac{\omega_{pa}^2}{k^2} \int d\mathbf{v} \, \frac{\mathbf{k}\cdot\partial F_a/\partial\mathbf{v}}{\omega - \mathbf{k}\cdot\mathbf{v} + i0}. \tag{6.53}$$

Decomposing $\mathbf{v}$ into component perpendicular and parallel to $\mathbf{k}$,

$$\mathbf{v} = \mathbf{v}_\perp + \mathbf{v}_\parallel = \frac{(\mathbf{k}\times\mathbf{v})\times\mathbf{k}}{k^2} + \frac{(\mathbf{k}\cdot\mathbf{v})\,\mathbf{k}}{k^2}, \tag{6.54}$$

we have

$$\epsilon(\mathbf{k},\omega) = 0 = 1 + \sum_a \frac{\omega_{pa}^2}{k} \int d\mathbf{v} \, \frac{\partial f_a/\partial v_\parallel}{\omega - k v_\parallel + i0}. \tag{6.55}$$

For model distribution (6.41) we obtain the reduced distribution and its derivative,

$$\begin{aligned} 2\pi \int_0^\infty dv_\perp v_\perp F_a &= \frac{m_e^{1/2}}{(2\pi\theta_e)^{1/2}} \frac{\Gamma(\kappa)}{\kappa'^{1/2}\,\Gamma(\kappa-1/2)} \left(1 + \frac{m_e v_\parallel^2}{2\kappa'\theta_e}\right)^{-\kappa}, \\ 2\pi \int_0^\infty dv_\perp v_\perp \frac{\partial F_a}{\partial v_\parallel} &= -\frac{m_e^{3/2}}{(2\pi)^{1/2}\theta_e^{3/2}} \frac{\kappa\Gamma(\kappa)}{\kappa'^{3/2}\,\Gamma(\kappa-1/2)} v_\parallel \left(1 + \frac{m_e v_\parallel^2}{2\kappa'\theta_e}\right)^{-\kappa-1}. \end{aligned} \tag{6.56}$$

For high-frequency Langmuir waves, one may expand the resonant denominator by

$$\frac{1}{\omega - k v_\parallel} \approx \frac{1}{\omega}\left(1 + \frac{k v_\parallel}{\omega} + \frac{k^2 v_\parallel^2}{\omega^2} + \frac{k^3 v_\parallel^3}{\omega^3} + \cdots\right). \tag{6.57}$$

Under this assumption and making use of (6.56), one may obtain

$$0 = 1 - \frac{\omega_{pe}^2}{\omega^2}\left(1 + \frac{\kappa'}{\kappa - 3/2} \frac{3k^2\theta_e}{m_e\omega^2}\right), \tag{6.58}$$

which leads to the desired Langmuir wave dispersion relation for kappa-like distribution (6.41),

$$\omega_k^L = \omega_{pe}\left(1+\frac{3}{2}\frac{\kappa'}{\kappa-3/2}\frac{k^2\theta_e}{m_e\omega_{pe}^2}\right). \tag{6.59}$$

The ion-sound mode dispersion relation can be obtained by approximating the resonant denominator by

$$\frac{1}{\omega-kv_\parallel}\approx-\frac{1}{kv_\parallel}, \tag{6.60}$$

for electrons, but the ion response can be treated in the same way as we did for the Langmuir wave dispersion relation. For ions, we assume Mawellian distribution. Then it is straightforward to show that (6.55) reduces to

$$0=1-\frac{\omega_{pi}^2}{\omega^2}\left(1+\frac{3k^2T_i}{m_i\omega^2}\right)+\frac{m_e\omega_{pe}^2}{k^2\theta_e}\frac{\kappa-1/2}{\kappa'}, \tag{6.61}$$

from which we obtain

$$\begin{aligned}\omega_k^S &= kc_S\sqrt{\frac{1+(3T_i/\theta_e)[(\kappa-1/2)/\kappa']}{(\kappa-1/2)/\kappa'+k^2\lambda_{De}^2}},\\ c_S^2 &= \frac{\theta_e}{m_i}, \qquad \lambda_{De}=\frac{\theta_e}{4\pi ne^2}.\end{aligned} \tag{6.62}$$

It is useful for later purposes to consider specific forms of dielectric constants corresponding to each wave mode, and the derivatives of dielectric constants for Langmuir and ion-acoustic waves:

$$\begin{aligned}\epsilon(k,\sigma\omega_k^L) &= 1-\frac{\omega_{pe}^2}{(\sigma\omega_k^L)^2}\left(1+\frac{3}{2}\frac{\kappa'}{\kappa-3/2}\frac{m_ek^2\theta_e}{m_e\omega_{pe}^2}\right),\\ \epsilon(k,\sigma\omega_k^S) &= 1-\frac{\omega_{pi}^2}{(\sigma\omega_k^S)^2}\left(1+\frac{3T_i}{\theta_e}\frac{\kappa-1/2}{\kappa'}\right)+\frac{1}{k^2\lambda_{De}^2}\frac{\kappa-1/2}{\kappa'},\\ \frac{1}{\epsilon'(k,\sigma\omega_k^L)} &= \frac{\sigma\omega_k^L}{2}, \qquad \frac{1}{\epsilon'(k,\sigma\omega_k^S)}=\mu_k\frac{\sigma\omega_k^L}{2},\\ \mu_k &= k^3\lambda_{De}^3\left(\frac{m_e}{m_i}\right)^{1/2}\left(\frac{\kappa'}{\kappa-1/2}+\frac{3T_i}{\theta_e}\right)^{1/2}.\end{aligned} \tag{6.63}$$

6.3.3 Steady-State Wave Kinetic Equation

The starting point is the nonlinear spectral balance equation (4.87). The only difference in the customary weak turbulence theory that assumes a Maxwellian background versus the present formalism that assumes a kappa-like state is that the wave dispersive properties are modified via dispersion relations (6.59) and (6.62), as well as the properties (6.63). We find that the induced and spontaneous emission terms are formally identical to the previous results, except for definitions (6.59), (6.62), and (6.63).

To compute the various three-wave decay coefficients, we must evaluate the quantity $|\chi^{(2)}|^2$, which is given by

$$
\begin{aligned}
|\chi_e^{(2)}(\mathbf{k}',\sigma'\omega_{\mathbf{k}'}^L|\mathbf{k}-\mathbf{k}',\sigma\omega_{\mathbf{k}}^L-\sigma'\omega_{\mathbf{k}'}^L)|^2 &\approx \left(\frac{\kappa-1/2}{\kappa'}\right)^2 \frac{e^2}{4\theta_e^2} \frac{(\mathbf{k}\cdot\mathbf{k}')^2}{k^2\,k'^2\,|\mathbf{k}-\mathbf{k}'|^2}, \\
|\chi^{(2)}(\mathbf{k}',\sigma'\omega_{\mathbf{k}'}^L|\mathbf{k}-\mathbf{k}',\sigma\omega_{\mathbf{k}}^S-\sigma'\omega_{\mathbf{k}'}^L)|^2 &\approx \left(\frac{\kappa-1/2}{\kappa'}\right)^2 \frac{e^2}{4\theta_e^2} \frac{[\mathbf{k}'\cdot(\mathbf{k}-\mathbf{k}')]^2}{k^2\,k'^2\,|\mathbf{k}-\mathbf{k}'|^2}.
\end{aligned}
\tag{6.64}
$$

With these modified modified results, the equations that describe the decay processes are identical as before, except that the coupling coefficients are given by

$$
\begin{aligned}
V_{\mathbf{k},\mathbf{k}'}^L &= \left(\frac{\kappa-1/2}{\kappa'}\right)^2 \frac{\pi}{4}\frac{e^2}{\theta_e^2} \frac{(\mathbf{k}\cdot\mathbf{k}')^2}{k^2\,k'^2\,|\mathbf{k}-\mathbf{k}'|^2}, \\
V_{\mathbf{k},\mathbf{k}'}^S &= \left(\frac{\kappa-1/2}{\kappa'}\right)^2 \frac{\pi}{4}\frac{e^2}{\theta_e^2} \frac{[\mathbf{k}'\cdot(\mathbf{k}-\mathbf{k}')]^2}{k^2\,k'^2\,|\mathbf{k}-\mathbf{k}'|^2}.
\end{aligned}
\tag{6.65}
$$

Similarly, equations describing the induced and spontaneous scattering processes are formally identical except that the coupling coefficient is modified to

$$
\begin{aligned}
U_{\mathbf{k},\mathbf{k}'}^{LL} &= \left(\frac{\kappa-1/2}{\kappa'}\right)^2 \frac{\pi}{\omega_{pe}^2}\frac{e^2}{m_e^2} \frac{(\mathbf{k}\cdot\mathbf{k}')^2}{k^2\,k'^2} \\
&\quad\times \delta[\sigma\omega_{\mathbf{k}}^L-\sigma'\omega_{\mathbf{k}'}^L-(\mathbf{k}-\mathbf{k}')\cdot\mathbf{v}].
\end{aligned}
\tag{6.66}
$$

Putting all the results together, we obtain the steady-state L mode weak turbulence kinetic equation for the kappa-like model:

$$
\begin{aligned}
0 &= \frac{\pi\omega_{pe}^2}{k^2}\int d\mathbf{v}\,\delta(\sigma\omega_{\mathbf{k}}^L-\mathbf{k}\cdot\mathbf{v})\left(\frac{ne^2}{\pi}F_e+\sigma\omega_{\mathbf{k}}^L\,I_{\mathbf{k}}^{\sigma L}\,\mathbf{k}\cdot\frac{\partial F_e}{\partial\mathbf{v}}\right) \\
&\quad-\left(\frac{\kappa-1/2}{\kappa'}\right)^2 \frac{\pi}{2}\frac{e^2}{\theta_e^2}\sum_{\sigma',\sigma''=\pm1}\sigma\omega_{\mathbf{k}}^L\int d\mathbf{k}'\,\frac{\mu_{\mathbf{k}-\mathbf{k}'}\,(\mathbf{k}\cdot\mathbf{k}')^2}{k^2\,k'^2\,|\mathbf{k}-\mathbf{k}'|^2} \\
&\quad\times\delta(\sigma\omega_{\mathbf{k}}^L-\sigma'\omega_{\mathbf{k}'}^L-\sigma''\omega_{\mathbf{k}-\mathbf{k}'}^S)\,\sigma''\omega_{\mathbf{k}-\mathbf{k}'}^L\,I_{\mathbf{k}'}^{\sigma'L}\,I_{\mathbf{k}}^{\sigma L} \\
&\quad-\left(\frac{\kappa-1/2}{\kappa'}\right)^2 \frac{\pi e^2}{m_e^2\,\omega_{pe}^2}\sum_{\sigma'=\pm1}\int d\mathbf{k}'\int d\mathbf{v}\,\frac{(\mathbf{k}\cdot\mathbf{k}')^2}{k^2\,k'^2}
\end{aligned}
$$

$$\times\,\delta[\sigma\omega_{\mathbf{k}}^L-\sigma'\omega_{\mathbf{k}'}^L-(\mathbf{k}-\mathbf{k}')\cdot\mathbf{v}]\left[\frac{ne^2}{\pi\omega_{pe}^2}\sigma\omega_{\mathbf{k}}^L\left(\sigma'\omega_{\mathbf{k}'}^L I_{\mathbf{k}}^{\sigma L}-\sigma\omega_{\mathbf{k}}^L I_{\mathbf{k}'}^{\sigma' L}\right)F_i\right.$$
$$\left.-\frac{m_e}{m_i}\sigma\omega_{\mathbf{k}}^L I_{\mathbf{k}'}^{\sigma' L} I_{\mathbf{k}}^{\sigma L}(\mathbf{k}-\mathbf{k}')\cdot\frac{\partial F_i}{\partial\mathbf{v}}\right]. \tag{6.67}$$

Let us assume that the forward- and backward-propagating Langmuir waves have the same isotropic spectral forms, $I_k^{+L}=I_k^{-L}=I_k$, as before, thus, omitting the subscript L from the wave intensity and Langmuir wave dispersion relation for the sake of simplicity. In deriving the result in (6.67), we have ignored the S mode contribution in the decay process. This is because, as we have seen in Figure 6.4, the ion-sound turbulence generation process by decay instability is a transient phenomenon, so that in the asymptotic stage, ion-sound turbulence is at the level of initial noise, hence S mode intensity can be ignored. Since the wave intensity is symmetric in $\mathbf{k}$, it is sufficient to consider only $\sigma=1$ in the wave kinetic equation. Thus, the relevant equation further reduces to

$$0=\frac{\pi\omega_{pe}^2}{k^2}\int d\mathbf{v}\,\delta(\omega_k-\mathbf{k}\cdot\mathbf{v})\left(\frac{ne^2}{\pi}F_e+\omega_k I_h\,\mathbf{k}\cdot\frac{\partial F_e}{\partial\mathbf{v}}\right)$$
$$-\left(\frac{\kappa-1/2}{\kappa'}\right)^2\frac{\pi}{2}\frac{e^2}{\theta_e^2}\sum_{\sigma',\sigma''=\pm1}\omega_k\int d\mathbf{k}'\,\frac{\mu_{\mathbf{k}-\mathbf{k}'}(\mathbf{k}\cdot\mathbf{k}')^2}{k^2k'^2|\mathbf{k}-\mathbf{k}'|^2}$$
$$\times\,\delta(\omega_k-\sigma'\omega_{k'}-\sigma''\omega_{|\mathbf{k}-\mathbf{k}'|}^S)\,\sigma''\omega_{|\mathbf{k}-\mathbf{k}'|}I_{k'}I_k$$
$$-\left(\frac{\kappa-1/2}{\kappa'}\right)^2\frac{\omega_k}{4\pi nT_i}\sum_{+,-}\int d\mathbf{k}'\int d\mathbf{v}\,\frac{(\mathbf{k}\cdot\mathbf{k}')^2}{k^2k'^2}\,\delta[\omega_k\mp\omega_{k'}-(\mathbf{k}-\mathbf{k}')\cdot\mathbf{v}]$$
$$\times\left(\frac{T_i}{4\pi^2}(\pm\omega_{k'}I_k-\omega_kI_{k'})+I_{k'}I_k(\omega_k\mp\omega_{k'})\right)F_i, \tag{6.68}$$

where we have taken advantage of the fact that the ion distribution is assumed to remain stationary and given by the Maxwellian form. The steady-state equation (6.68) must be solved for Langmuir wave intensity, I_k, but, at first sight, the general solution appears to be extremely difficult to obtain. However, the three basic processes, namely, spontaneous versus induced emission, decay, and scattering, can be separately considered. This is physically meaningful, since spontaneous and induced processes are supposed to balance each other out, according to the principle of detailed balance in the steady state. We thus consider the balance of each process.

Balance of Induced and Spontaneous Emission

Making use of

$$\frac{\partial F_e}{\partial\mathbf{v}}=-\frac{m_e(\kappa+1)\mathbf{v}}{\kappa'\theta_e}\frac{1}{1+m_ev^2/2\kappa'\theta_e}F_e, \tag{6.69}$$

we may express the balanced equation of spontaneous and induced emission terms within (6.68),

$$0=\frac{\pi\omega_{pe}^2}{k^2}\frac{ne^2}{\pi}\int d\mathbf{v}\,\delta(\omega_{pe}-kv\mu)$$
$$\times\left(1-\frac{\kappa+1}{\kappa'}\frac{4\pi^2}{\theta_e}\frac{I_L(k)}{1+m_ev^2/2\kappa'\theta_e}\right)F_e. \tag{6.70}$$

Let us integrate this equation in positive k space,

$$\int_0^\infty dk\int d\mathbf{v}\,\delta(\omega_{pe}-kv\mu)\left(1-\frac{\kappa+1}{\kappa'}\frac{4\pi^2}{\theta_e}\frac{I_L(k)}{1+m_ev^2/2\kappa'\theta_e}\right)F_e$$
$$=4\pi\int_0^\infty dv\,v\int_{-1}^1 d\mu\int_0^\infty\frac{dk}{k}\,\delta\left(\mu-\frac{\omega_{pe}}{kv}\right)$$
$$\times\left(1-\frac{\kappa+1}{\kappa'}\frac{4\pi^2}{\theta_e}\frac{I_L(k)}{1+m_ev^2/2\kappa\theta_e}\right)F_e$$
$$=4\pi\int_0^\infty dv\,v\left(\mathscr{H}(v)-\frac{\kappa+1}{\kappa'}\frac{4\pi^2}{\theta_e}\frac{\mathscr{H}(v)\,\mathscr{I}(v)}{1+m_ev^2/2\kappa\theta_e}\right)F_e, \tag{6.71}$$

from which we obtain

$$\mathscr{I}(v)=\frac{\kappa'}{\kappa+1}\frac{\theta_e}{4\pi^2}\left(1+\frac{m_ev^2}{2\kappa\theta_e}\right), \tag{6.72}$$

which is none other than (6.52), upon replacing v by ω_{pe}/k. We thus see that the balance of induced and spontaneous emissions leads to the same solution as before. Let us next consider the balance of nonlinear terms.

Absence of Contributions from Decay Process

Consider the decay term,

$$\left(\frac{\kappa-1/2}{\kappa'}\right)^2\frac{\pi}{2}\frac{e^2}{\theta_e^2}\sum_{\sigma',\sigma''=\pm1}\omega_{\mathbf{k}}\int d\mathbf{k}'\,\frac{\mu_{\mathbf{k}-\mathbf{k}'}\,(\mathbf{k}\cdot\mathbf{k}')^2}{k^2\,k'^2\,|\mathbf{k}-\mathbf{k}'|^2}$$
$$\times\,\delta(\omega_k-\sigma'\omega_{k'}-\sigma''\omega^S_{|\mathbf{k}-\mathbf{k}'|})\,\sigma''\omega_{|\mathbf{k}-\mathbf{k}'|}\,I_{k'}\,I_k$$
$$=\left(\frac{\kappa-1/2}{\kappa'}\right)^2\frac{\pi}{2}\frac{e^2}{\theta_e^2}\omega_k\int d\mathbf{k}'\,\frac{\mu_{\mathbf{k}-\mathbf{k}'}\,(\mathbf{k}\cdot\mathbf{k}')^2}{k^2\,k'^2\,|\mathbf{k}-\mathbf{k}'|^2}\big[\delta(\omega_k-\omega_{k'}-\omega^S_{|\mathbf{k}-\mathbf{k}'|})$$
$$-\,\delta(\omega_k-\omega_{k'}+\omega^S_{|\mathbf{k}-\mathbf{k}'|})+\delta(\omega_k+\omega_{k'}-\omega^S_{|\mathbf{k}-\mathbf{k}'|})$$
$$-\,\delta(\omega_k+\omega_{k'}+\omega^S_{|\mathbf{k}-\mathbf{k}'|})\big]\,\omega_{|\mathbf{k}-\mathbf{k}'|}\,I_{k'}\,I_k. \tag{6.73}$$

Of the four delta function terms in (6.73), those involving $\omega_k+\omega_{k'}$ obviously, cannot be satisfied. Ignoring these terms, we have

$$\left(\frac{\kappa-1/2}{\kappa'}\right)^2 \frac{\pi}{2}\frac{e^2}{\theta_e^2}\,\omega_k \int d\mathbf{k}'\, \frac{\mu_{\mathbf{k}-\mathbf{k}'}\,(\mathbf{k}\cdot\mathbf{k}')^2}{k^2\,k'^2\,|\mathbf{k}-\mathbf{k}'|^2}\,\omega_{|\mathbf{k}-\mathbf{k}'|}\,I_{k'}\,I_k$$
$$\times\left[\delta(\omega_k-\omega_{k'}-\omega^S_{|\mathbf{k}-\mathbf{k}'|})-\delta(\omega_k-\omega_{k'}+\omega^S_{|\mathbf{k}-\mathbf{k}'|})\right]. \tag{6.74}$$

The remaining two delta function terms cancel each other. To see this, let us consider the following:

$$\begin{aligned}
&\delta(\omega_k-\omega_{k'}\pm\omega^S_{|\mathbf{k}-\mathbf{k}'|})\\
&=\delta\Bigg[\omega_{pe}\left(1+\frac{3}{2}\frac{\kappa'}{\kappa-3/2}k^2\lambda_{De}^2\right)-\omega_{pe}\left(1+\frac{3}{2}\frac{\kappa'}{\kappa-3/2}k'^2\lambda_{De}^2\right)\\
&\quad\pm\sqrt{\frac{m_e}{m_i}}\,\omega_{pe}\,|\mathbf{k}-\mathbf{k}'|\,\lambda_{De}\left(\frac{1+(3T_i/\theta_e)[(\kappa-1/2)/\kappa']}{(\kappa-1/2)/\kappa'+|\mathbf{k}-\mathbf{k}'|^2\lambda_{De}^2}\right)^{1/2}\Bigg]\\
&=\frac{1}{\omega_{pe}}\,\delta\left[\frac{3}{2}\frac{\kappa'}{\kappa-3/2}(k^2-k'^2)\lambda_{De}^2\pm\sqrt{\frac{m_e}{m_i}}\frac{|\mathbf{k}-\mathbf{k}'|\,\lambda_{De}}{[(\kappa-1/2)/\kappa']^{1/2}}\right]\\
&\approx\frac{1}{\omega_{pe}}\frac{2}{3}\frac{\kappa-3/2}{\kappa'}\,\delta[(k^2-k'^2)\lambda_{De}^2].
\end{aligned}\tag{6.75}$$

This shows that the two decay terms cancel each other upon ignoring terms of order $\sqrt{m_e/m_i}$, and hence proves that the three-wave term can be ignored.

Balance of Induced and Spontaneous Scattering

We next consider the balance of induced and spontaneous scattering terms (off thermal ions):

$$0=-\left(\frac{\kappa-1/2}{\kappa'}\right)^2\frac{\omega_k}{4\pi nT_i}\sum_{+,-}\int d\mathbf{k}'\int d\mathbf{v}\,\frac{(\mathbf{k}\cdot\mathbf{k}')^2}{k^2\,k'^2}\,\delta[\omega_k\mp\omega_{k'}-(\mathbf{k}-\mathbf{k}')\cdot\mathbf{v}]$$
$$\times\left(\frac{T_i}{4\pi^2}(\pm\omega_{k'}\,I_k-\omega_k\,I_{k'})+I_{k'}\,I_k\,(\omega_k\mp\omega_{k'})\right)F_i. \tag{6.76}$$

Explicitly writing down the summation over two signs and ignoring resonance conditions with the factor $\omega_k+\omega_{k'}$, we obtain

$$0=\int d\mathbf{k}'\int d\mathbf{v}\,\frac{(\mathbf{k}\cdot\mathbf{k}')^2}{k^2\,k'^2}\,\delta[\omega_k-\omega_{k'}-(\mathbf{k}-\mathbf{k}')\cdot\mathbf{v}]$$
$$\times\left(\frac{T_i}{4\pi^2}(\omega_{k'}\,I_k-\omega_k\,I_{k'})+I_{k'}\,I_k\,(\omega_k-\omega_{k'})\right)F_i. \tag{6.77}$$

Let us write

$$\mathbf{k}'=\mathbf{k}+\delta\mathbf{k}. \tag{6.78}$$

The nonlinear wave-particle resonance condition $\omega_k - \omega_{k'} - (\mathbf{k}-\mathbf{k}')\cdot\mathbf{v} = 0$ can be satisfied only if the magnitude of resonant velocity is on the order of ion thermal speed, $v_{\rm res} \sim (\omega_k - \omega_{k'})/|\mathbf{k}-\mathbf{k}'| \sim v_{Ti} \ll v_{Te}$. Making use of the Langmuir wave dispersion relation, it can be shown that the resonant velocity is roughly given by $v_{\rm res} \sim 3\left(k-k'\right)\theta_e/(2m_e\omega_{pe})$, when $\mathbf{k}$ and $\mathbf{k}'$ are in opposite directions. This shows that in order for $v_{\rm res} \sim v_{Ti} \ll v_{Te}$ to be satisfied, $\mathbf{k}$ and $\mathbf{k}'$ must be sufficiently close to each other, or $|\mathbf{k}-\mathbf{k}'| \sim |\delta\mathbf{k}| \ll 1$. Making use of this consideration, we may employ the Taylor series expansion to obtain

$$\begin{aligned}
\omega_{k'}\, I_k - \omega_k\, I_{k'} &= \delta\mathbf{k}\cdot\frac{d\omega(k)}{d\mathbf{k}}\,I(k) - \omega(k)\,\delta\mathbf{k}\cdot\frac{dI(k)}{d\mathbf{k}},\\
I_{k'}\, I_k\,(\omega_k - \omega_{k'}) &= -[I(k)]^2\,\delta\mathbf{k}\cdot\frac{d\omega(k)}{d\mathbf{k}},
\end{aligned} \tag{6.79}$$

from which we have

$$\begin{aligned}
0 = &\int d(\delta\mathbf{k})\int d\mathbf{v}\,\frac{(\mathbf{k}\cdot\mathbf{k}')^2}{k^2\,k'^2}\,\delta[\omega_k - \omega_{k'} - (\mathbf{k}-\mathbf{k}')\cdot\mathbf{v}] \\
&\times(\delta k)\left(\omega(k)\,\frac{dI(k)}{dk} + \frac{4\pi^2}{T_i}\,\frac{d\omega(k)}{dk}\,[I(k)]^2 - \frac{d\omega(k)}{dk}\,I(k)\right)F_i.
\end{aligned} \tag{6.80}$$

Making use of dispersion relation (6.59) and its derivative with respect to k, we obtain the necessary condition for equality:

$$\begin{aligned}
0 &= \omega(k)\,\frac{dI(k)}{dk} + \frac{4\pi^2}{T_i}\,\frac{d\omega(k)}{dk}\,[I(k)]^2 - \frac{d\omega(k)}{dk}\,I(k)\\
&= \frac{3k\theta_e}{m_e\omega_{pe}}\left[\left(1+\frac{\kappa'}{\kappa-3/2}\,x\right)\frac{dI(x)}{dx}\right.\\
&\quad\left. + \frac{4\pi^2}{T_i}\,\frac{\kappa'}{\kappa-3/2}\,[I(x)]^2 - \frac{\kappa'}{\kappa-3/2}\,I(x)\right],
\end{aligned} \tag{6.81}$$

where we have defined a dimensionless variable

$$x \equiv \frac{3k^2\theta_e}{2m_e\omega_{pe}^2}. \tag{6.82}$$

Equation (6.81) is exactly satisfied for

$$I(k) = \frac{T_i}{4\pi^2}\left(1 + \frac{2}{3}\,\frac{\kappa - 3/2}{\kappa'\,k^2\theta_e/m_e\omega_{pe}^2}\right). \tag{6.83}$$

Upon comparing (6.83) with the earlier solution (6.52), it immediately becomes obvious that the two expressions are identical, provided we identify

$$T_i = \theta_e\,\frac{\kappa'}{\kappa+1}, \qquad \frac{4}{3}\left(\kappa - \frac{3}{2}\right) = 1 + 2\mathscr{H}, \tag{6.84}$$

or, equivalently,

$$\kappa = \frac{9}{4} + \frac{3}{2}\mathscr{H} = 2.25 + 1.5\mathscr{H}, \qquad \kappa'\theta_e = (\kappa + 1)\,T_i, \tag{6.85}$$

where we have treated the correction factor $\mathscr{H}$ as constant. This reconciliation between (6.52) and (6.83) would not have been possible had we assumed any other form of electron distribution than the kappa-like model (6.41). Note that the notion of electron distribution function did not enter in the derivation of (6.83). Rather, (6.83) was derived on the basis of balancing the nonlinear scattering terms, which involved ions only. Nevertheless, the end result is mathematically identical to (6.52). Had we assumed some other form of electron velocity distribution function, then one would have obtained a totally different form of wave intensity when compared with (6.52), which would have made the reconciliation impossible. This shows that the kappa distribution is the only solution that is allowed in the asymptotically steady-state Langmuir turbulence system, thus proving the uniqueness theorem for kappa distribution.

Here, we should note that if the system is initially placed at an isotropic thermal equilibrium state (that is, Maxwellian distribution), then the system will never spontaneously evolve toward the turbulent equilibrium state characterized by kappa distribution function. The system must be first perturbed, say, by an instability (in this case, bump-on-tail instability), or subject the system to some other forms of large amplitude disturbance. Then one must "wait and watch," as it were, before the system resettles to a new quasi-equilibrium. In the case of Langmuir turbulence, the new quasi-equilibrium is characterized by the electron kappa velocity distribution function, but it is important to note that an initially Maxwellian electron distribution can never spontaneously make the transition to kappa distribution without some disturbances. Of course, one such disturbance is the gentle electron beam, and the mechanism that allows the transition is the bump-on-tail instability. The discussion presented in this section did not address how the initial system is perturbed, but the numerical analysis of bump-on-tail instability in the previous section already provided some clues as to how the initial system can be perturbed, and how the perturbed system gradually tends toward the turbulent quasi-equilibrium. In the present section we did not address the issue of how long it takes for the system to make the transition to quasi-equilibrium either, but again, the numerical demonstration of the previous section provides some clues as to the time scales involved. Instead, in the present section, we have addressed the issue of how to describe the final state when the dust is settled, and the system has finally managed to make the transition to the new turbulent quasi-equilibrium.

Even though, strictly speaking, we do not quite have any exact way to estimate the turbulent quasi-equilibrium time scale, we may surmise on the basis of numerical solution of the previous section that the nonlinear development of bump-on-tail instability and Langmuir turbulence proceeds over distinct time scales, that is, quasilinear saturation time scale, $t_{\rm QL}$; nonlinear mode coupling time scale, $t_{\rm NL}$; turbulent quasi-equilibrium time scale, $t_{\rm turb}$; and finally collisional relaxation time scale, $t_{\rm coll}$. On the basis of numerical demonstration of the previous section, for the given set of parameters adopted in the numerical example, we estimate the following:

$$\begin{aligned}
&\omega_{pe} t_{\rm QL} \sim \mathcal{O}(10^2),\\
&\omega_{pe} t_{\rm NL} \sim \mathcal{O}(10^3) \text{ to } \mathcal{O}(10^4),\\
&\omega_{pe} t_{\rm turb} \sim \mathcal{O}(10^4) \text{ to } \mathcal{O}(10^5),\\
&\omega_{pe} t_{\rm coll} \gg \mathcal{O}(10^5).
\end{aligned}$$

Another point we should make is that the kappa distribution function is a quasi-equilibrium state, rather than a true thermodynamic equilibrium. In the derivation of time asymptotic solution, we have relied on the particle kinetic equation in which the velocity space diffusion process driven by the wave intensity is balanced by the velocity space drag effect. The kappa distribution exactly satisfies the steady-state condition for the particle kinetic equation. However, over time scales much longer than even the turbulent quasi-equilibrium time scale, eventually binary collisional processes will kick in. Processes that are operative over collisional time scales have not been included in the present discussion – but the recent work by Tigik et al. (2016) addresses some aspects of this problem. In any case, once we subject the kappa distribution function to collisional kinetic equation, then, over much longer collisional time scale, the system will ultimately evolve toward a genuine thermodynamic equilibrium state.

Before we close, let us address the issue of $\mathscr{H}$:

$$\mathscr{H} = \int_k^\infty \frac{dk'}{k'} = \ln k' \Big|_{k'=k}^\infty . \tag{6.86}$$

The quantity $\mathscr{H}$ formally diverges for $k' = \infty$ as well as for $k' = 0$. To remedy the divergent situation, we introduce the lower and upper cutoffs, $k_{\rm min}$ and $k_{\rm max}$:

$$\mathscr{H} = \ln \frac{k_{\rm max}}{k_{\rm min}}. \tag{6.87}$$

If we choose these quantities as

$$k_{\rm min} = \frac{1}{\lambda_{De}}, \qquad k_{\rm max} = \frac{\theta_e}{e^2}, \tag{6.88}$$

in analogy with (5.34) and (5.35), then we obtain the customary Coulomb logarithm

$$\mathscr{H} = \ln \frac{\lambda_{De}\theta_e}{e^2} \equiv \ln \Lambda, \qquad \Lambda = \frac{\theta_e^{3/2}}{(4\pi n)^{1/2} e^3} = \frac{1}{g}, \qquad g = \frac{1}{4\pi n \lambda_{De}^3}. \tag{6.89}$$

Alternatively, we may simply consider $k_{\min}$ and $k_{\max}$ as undetermined and treat $\mathscr{H}$ as a free adjustable parameter.

6.3.4 Electron Kappa Distribution and Steady-State Langmuir Turbulence

Let us summarize our findings concerning the quasi-equilibrium state of weak Langmuir turbulence and electron kappa distribution function. Since according to (6.42),

$$\theta_e = T_e \frac{\kappa - 3/2}{\kappa'},$$

we may express our final solution as

$$\begin{aligned} F_e(v) &= \frac{m_e^{3/2}}{(2\pi T_e)^{3/2}} \frac{\Gamma(\kappa+1)}{(\kappa - 3/2)^{3/2}\Gamma(\kappa - 1/2)} \left(1 + \frac{1}{\kappa - 3/2}\frac{m_e v^2}{2T_e}\right)^{-\kappa-1}, \\ I(k) &= \frac{T_e}{4\pi^2}\frac{\kappa - 3/2}{\kappa + 1}\left(1 + \frac{1 + 2\mathscr{H}}{\kappa - 3/2}\frac{2\pi n e^2}{k^2 T_e}\right), \\ \kappa &= \frac{9}{4} + \frac{3\mathscr{H}}{2} = 2.25 + 1.5\mathscr{H}, \qquad \frac{T_i}{T_e} = \frac{\kappa - 3/2}{\kappa + 1} = \frac{3 + 6\mathscr{H}}{13 + 6\mathscr{H}}. \end{aligned} \tag{6.90}$$

This solution, which was obtained by Yoon (2014), represents the steady-state Langmuir turbulence and electron kappa distribution that together form a quasi-equilibrium physical state in which the electrons and Langmuir waves exchange momenta and energies in a dynamically steady state fashion. The steady-state Langmuir turbulence characterized by kappa electron velocity distribution function may be highly relevant to the solar wind plasma.

The kappa electron distribution (6.90) predicts that for suprathermal energy range, the velocity distribution should behave as an inverse power law,

$$F_e \sim v^{-6.5}, \qquad \text{for } v \gg v_{Te}, \tag{6.91}$$

since $\kappa \approx 9/4 = 2.25$, assuming $\mathscr{H}$ can be ignored. Such a velocity power-law spectral index compares favorably against observations made in the solar wind. Typically, the solar wind electrons can be modeled as being comprised of three components, the Maxwellian "core," a suprathermal "halo," and a highly field-aligned component called the "strahl" that is observed to be streaming away from the Sun. The strahl is typically associated with the high-speed or fast solar wind, while the core-halo two-component feature is more common in the slow-speed or slow solar wind. While customary discussions of the solar wind electrons often do not include the fourth component known as the "superhalo," these highly energetic electrons are observed to possess nearly isotropic angular distribution and an inverse power law velocity spectrum (Lin, 1998). The characteristic energy for the Maxwellian core is tens of eV, while the halo component is defined up to $\sim 10^2$–10^3 eV or so. The highly energetic superhalo is characterized by 10^3–10^5 eV. The superhalo electrons are observed in all solar wind conditions, whether active or quiet, and their velocity power law indices do not show much dependence on solar conditions (Wang et al., 2012a). The strahl component occupies a similar energy range as the halo ($\sim 10^2$–10^3 eV). These electron components are distinguishable from each other on the basis of their morphology in velocity (or energy) space distribution functions. It is generally known from observations that at 1 AU ($= 149.6 \times 10^6$ km, the astronomical unit, which is the distance from the Sun to Earth), the Maxwellian component makes up about $\sim 95\%$, and, in terms of the number density, the halo is roughly estimated to possess $\sim 4\%$ density, and the strahl component takes up, approximately, the remaining 1%. The superhalo component is very low in terms of number density (almost negligible, on the order of 10^{-9}–10^{-6} when compared with the core density), yet owing to their high energy, their presence can be clearly identified in the velocity distribution function.

Instead of fitting the solar wind electron velocity distribution function by various components, it is possible to model the entire solar wind electrons by a single kappa model (Olbert, 1968; Vasyliunas, 1968; Maksimovic et al., 1997, 2005). Measured electron velocity distribution functions for suprathermal energy range (the superhalo component of the solar wind) near Earth orbit behave as $F_e \sim v^{-5.0}$ to $v^{-8.7}$ with average behavior (Wang et al., 2012a)

$$F_e^{\text{obs}} \sim v^{-6.69}, \qquad v \gg v_{Te}. \tag{6.92}$$

Note that (6.92) agrees quite well with (6.91). Moreover, taking into account only the so-called halo electron component, Štverák et al. (2009) analyzed *Helios*, *Cluster*, and *Ulysses* spacecraft data in order to show that the value of κ decreases from ~ 9 near 0.3 AU to ~ 4 near 1 AU, to ~ 2.25 near ~ 5 AU. This shows that as the solar wind evolves radially, the solar wind electrons seem to increasingly evolve toward the turbulent quasi-equilibrium state predicted by Yoon (2014).

As for the steady-state Langmuir turbulence, the solar wind is also replete with pervasive electrostatic fluctuations with peak frequency near plasma frequency. Such an electrostatic fluctuation is known in the literature as quasi-thermal noise fluctuations (Meyer-Vernet, 1979). The quasi-thermal noise represents the spontaneously emitted electromagnetic fluctuations that take place in a plasma, which may be close to but not necessarily at thermal equilibrium state (Sitenko, 1967).

A theoretical tool to analyze quasi-thermal noise is known as the quasi-thermal noise "spectroscopy," which was developed by Meyer-Vernet (1979) in the context of solar wind. The quasi-thermal noise spectroscopy is a versatile research tool that can be employed to indirectly determine not only the total electron density and temperature (Meyer-Vernet et al., 1986; Maksimovic et al., 1995) but also properties of energetic and, often, nonthermal electrons (Zouganelis, 2008). In this regard, the thermal noise spectrum ubiquitously observed in the solar wind may be none other than the steady-state Langmuir turbulence intensity as specified by (6.90).

Part III

Vlasov Weak Turbulence Theory: Electromagnetic Formalism

7

Nonlinear Electromagnetic Equations in Vlasov Plasmas

In Part III, we extend the formalism outlined in Part I to the fully electromagnetic case. One of the most prominent examples demonstrating the usefulness of electromagnetic weak turbulence theory is on the radiation emission with frequency close to the plasma frequency and/or harmonics. Such a process is known as plasma emission, and it is the basic mechanism for type II and type III solar radio bursts. It is the process that is intimately related to the beam-plasma instability. We have seen in Chapter 6 that the bump-on-tail instability leads to the excitation of Langmuir waves, which undergo nonlinear mode coupling process involving ion-sound waves. If one includes electromagnetic effects, then mode coupling also involves the transverse radiation. The emission at the plasma frequency and/or its second harmonic is the consequence of the mode coupling process involving transverse mode. The electromagnetic formalism to be developed in Parts III and IV of this book thus can be applied to the problem of plasma emission. Before we do that, however, let us proceed with the fundamental development of formalism.

7.1 Nonlinear Electrodynamic Equations in Plasmas

The starting point is the Vlasov–Maxwell system of equations for unmagnetized plasmas:

$$\begin{aligned}
&\left[\frac{\partial}{\partial t} + \mathbf{v}\cdot\frac{\partial}{\partial \mathbf{r}} + \frac{e_a}{m_a}\Big(\mathbf{E}(\mathbf{r},t) + \frac{\mathbf{v}}{c}\times\mathbf{B}(\mathbf{r},t)\Big)\cdot\frac{\partial}{\partial \mathbf{v}}\right] f_a(\mathbf{r},\mathbf{v},t) = 0,\\
&\frac{\partial}{\partial \mathbf{r}}\times\mathbf{E}(\mathbf{r},t) + \frac{1}{c}\frac{\partial \mathbf{B}(\mathbf{r},t)}{\partial t} = 0,\\
&\frac{\partial}{\partial \mathbf{r}}\cdot\mathbf{E}(\mathbf{r},t) = 4\pi\sum_a e_a\int d\mathbf{v}\, f_a(\mathbf{r},\mathbf{v},t), \qquad \frac{\partial}{\partial \mathbf{r}}\cdot\mathbf{B}(\mathbf{r},t) = 0,\\
&\frac{\partial}{\partial \mathbf{r}}\times\mathbf{B}(\mathbf{r},t) - \frac{1}{c}\frac{\partial \mathbf{E}(\mathbf{r},t)}{\partial t} = \frac{4\pi}{c}\sum_a e_a\int d\mathbf{v}\,\mathbf{v}\, f_a(\mathbf{r},\mathbf{v},t).
\end{aligned} \tag{7.1}$$

We again separate physical quantities into the customary average and fluctuating parts:

$$
\begin{aligned}
&f_a(\mathbf{r},\mathbf{v},t) = nF_a(\mathbf{v},t) + \delta f_a(\mathbf{r},\mathbf{v},t),\\
&\mathbf{E}(\mathbf{r},t) = \delta\mathbf{E}(\mathbf{r},t), \qquad \mathbf{B}(\mathbf{r},t) = \delta\mathbf{B}(\mathbf{r},t).
\end{aligned}
\tag{7.2}
$$

This leads to the following:

$$
\begin{aligned}
&\frac{\partial nF_a(\mathbf{v},t)}{\partial t} + \frac{e_a}{m_a}\left(\delta\mathbf{E}(\mathbf{r},t) + \frac{\mathbf{v}}{c}\times\delta\mathbf{B}(\mathbf{r},t)\right)\cdot\frac{\partial nF_a(\mathbf{v},t)}{\partial\mathbf{v}}\\
&\quad + \left(\frac{\partial}{\partial t} + \mathbf{v}\cdot\frac{\partial}{\partial\mathbf{r}}\right)\delta f_a(\mathbf{r},\mathbf{v},t)\\
&\quad + \frac{e_a}{m_a}\left(\delta\mathbf{E}(\mathbf{r},t) + \frac{\mathbf{v}}{c}\times\delta\mathbf{B}(\mathbf{r},t)\right)\cdot\frac{\partial}{\partial\mathbf{v}}\delta f_a(\mathbf{r},\mathbf{v},t) = 0,\\
&\frac{\partial}{\partial\mathbf{r}}\times\delta\mathbf{E}(\mathbf{r},t) + \frac{1}{c}\frac{\partial\,\delta\mathbf{B}(\mathbf{r},t)}{\partial t} = 0,\\
&\frac{\partial}{\partial\mathbf{r}}\cdot\delta\mathbf{E}(\mathbf{r},t) = 4\pi\sum_a e_a\int d\mathbf{v}\,\delta f_a(\mathbf{r},\mathbf{v},t),\\
&\frac{\partial}{\partial\mathbf{r}}\cdot\delta\mathbf{B}(\mathbf{r},t) = 0,\\
&\frac{\partial}{\partial\mathbf{r}}\times\delta\mathbf{B}(\mathbf{r},t) - \frac{1}{c}\frac{\partial\,\delta\mathbf{E}(\mathbf{r},t)}{\partial t} = \frac{4\pi}{c}\sum_a e_a\int d\mathbf{v}\,\mathbf{v}\,\delta f_a(\mathbf{r},\mathbf{v},t).
\end{aligned}
\tag{7.3}
$$

Upon averaging (7.3) over the random phases of the fluctuations, we obtain the formal particle kinetic equation:

$$
\frac{\partial nF_a(\mathbf{v},t)}{\partial t} = -\frac{e_a}{m_a}\frac{\partial}{\partial\mathbf{v}}\cdot\left(\langle\delta\mathbf{E}(\mathbf{r},t)\,\delta f_a(\mathbf{r},\mathbf{v},t)\rangle + \frac{\mathbf{v}}{c}\times\langle\delta\mathbf{B}(\mathbf{r},t)\,\delta f_a(\mathbf{r},\mathbf{v},t)\rangle\right).
\tag{7.4}
$$

Subtracting (7.4) from (7.3), we obtain the equation for perturbed distribution function:

$$
\begin{aligned}
\left(\frac{\partial}{\partial t} + \mathbf{v}\cdot\frac{\partial}{\partial\mathbf{r}}\right)\delta f_a(\mathbf{r},\mathbf{v},t) &= -\frac{e_a}{m_a}\left(\delta\mathbf{E}(\mathbf{r},t) + \frac{\mathbf{v}}{c}\times\delta\mathbf{B}(\mathbf{r},t)\right)\cdot\frac{\partial nF_a(\mathbf{v},t)}{\partial\mathbf{v}}\\
&\quad - \frac{e_a}{m_a}\frac{\partial}{\partial\mathbf{v}}\cdot\Big(\delta\mathbf{E}(\mathbf{r},t)\,\delta f_a(\mathbf{r},\mathbf{v},t) + \frac{\mathbf{v}}{c}\times\delta\mathbf{B}(\mathbf{r},t)\,\delta f_a(\mathbf{r},\mathbf{v},t)\\
&\quad - \langle\delta\mathbf{E}(\mathbf{r},t)\,\delta f_a(\mathbf{r},\mathbf{v},t)\rangle - \frac{\mathbf{v}}{c}\times\langle\delta\mathbf{B}(\mathbf{r},t)\,\delta f_a(\mathbf{r},\mathbf{v},t)\rangle\Big).
\end{aligned}
\tag{7.5}
$$

Taking spectral transformation for the fluctuations following the definitions in (1.38), but including now the magnetic field fluctuation, and including the slow and adiabatic time dependence for spectral amplitudes, $\delta f^a_{\mathbf{k},\omega}(\mathbf{v},t)$, $\delta\mathbf{E}_{\mathbf{k},\omega}(t)$, and $\delta\mathbf{B}_{\mathbf{k},\omega}(t)$, we now have the formal particle kinetic equation

$$\frac{\partial n F_a(\mathbf{v},t)}{\partial t} = -\frac{e_a}{m_a}\frac{\partial}{\partial v_i}\int d\mathbf{k}\int d\omega\left[\left(1-\frac{\mathbf{k}\cdot\mathbf{v}}{\omega}\right)\delta_{ij}+\frac{v_j k_i}{\omega}\right]$$
$$\times\left\langle\delta E^j_{-\mathbf{k},-\omega}(t)\,\delta f^a_{\mathbf{k},\omega}(\mathbf{v},t)\right\rangle, \tag{7.6}$$

the wave equation

$$\left(\omega+i\frac{\partial}{\partial t}\right)\delta\mathbf{B}_{\mathbf{k},\omega}(t) = c\mathbf{k}\times\delta\mathbf{E}_{\mathbf{k},\omega}(t),$$
$$\mathbf{k}\cdot\delta\mathbf{E}_{\mathbf{k},\omega}(t) = -4\pi i\sum_a e_a\int d\mathbf{v}\,\delta f^a_{\mathbf{k},\omega}(\mathbf{v},t),$$
$$c\mathbf{k}\times\delta\mathbf{B}_{\mathbf{k},\omega}(t)+\left(\omega+i\frac{\partial}{\partial t}\right)\delta\mathbf{E}_{\mathbf{k},\omega}(t) = -4\pi i\sum_a e_a\int d\mathbf{v}\,\mathbf{v}\,\delta f^a_{\mathbf{k},\omega}(\mathbf{v},t),$$
$$\mathbf{k}\cdot\delta\mathbf{B}_{\mathbf{k},\omega}(t) = 0, \tag{7.7}$$

and the equation for perturbed distribution function

$$\left(\omega-\mathbf{k}\cdot\mathbf{v}+i\frac{\partial}{\partial t}\right)\delta f^a_{\mathbf{k},\omega}(\mathbf{v},t)$$
$$= -i\,\frac{e_a}{m_a}\,\delta E^i_{\mathbf{k},\omega}(t)\left[\left(1-\frac{\mathbf{k}\cdot\mathbf{v}}{\omega}\right)\delta_{ij}+\frac{v_i k_j}{\omega}\right]\frac{\partial n F_a(\mathbf{v},t)}{\partial v_j}$$
$$-i\,\frac{e_a}{m_a}\frac{\partial}{\partial v_j}\int d\mathbf{k}'\int d\omega'\left[\left(1-\frac{\mathbf{k}'\cdot\mathbf{v}}{\omega'}\right)\delta_{ij}+\frac{v_i k'_j}{\omega'}\right] \tag{7.8}$$
$$\times\left[\delta E^i_{\mathbf{k}',\omega'}(t)\,\delta f^a_{\mathbf{k}-\mathbf{k}',\omega-\omega'}(\mathbf{v},t)-\left\langle\delta E^i_{\mathbf{k}',\omega'}(t)\,\delta f^a_{\mathbf{k}-\mathbf{k}',\omega-\omega'}(\mathbf{v},t)\right\rangle\right].$$

On the right-hand side of (7.8), we have ignored the adiabatic time-dependence in converting the magnetic field fluctuation to electric field fluctuation. That is, we have made use of the relation

$$\delta\mathbf{E}_{\mathbf{k},\omega}+\frac{\mathbf{v}}{c}\times\delta\mathbf{B}_{\mathbf{k},\omega} = \delta\mathbf{E}_{\mathbf{k},\omega}+\frac{\mathbf{v}}{\omega}\times(\mathbf{k}\times\delta\mathbf{E}_{\mathbf{k},\omega})$$
$$= \left(1-\frac{\mathbf{k}\cdot\mathbf{v}}{\omega}\right)\delta\mathbf{E}_{\mathbf{k},\omega}+\frac{\mathbf{k}}{\omega}(\mathbf{v}\cdot\delta\mathbf{E}_{\mathbf{k},\omega}). \tag{7.9}$$

Note that, in (7.8), the arguments $(\mathbf{k}',\omega')$ and $(\mathbf{k}-\mathbf{k}',\omega-\omega')$ are not symmetrized yet. Eventually, these arguments should be fully symmetrized with respect to dummy integral variables.

Let us consider the fast time-scale solution by absorbing the slow time derivative as part of the new definition for ω, as in (1.41). Let us also define the shorthand notation as in (1.42), except that we now generalize to electromagnetic formalism:

$$g_{\mathbf{k},\omega} = \frac{1}{\omega-\mathbf{k}\cdot\mathbf{v}+i0},$$
$$L^i_{\mathbf{k},\omega} = -\frac{ie_a}{m_a}\left[\left(1-\frac{\mathbf{k}\cdot\mathbf{v}}{\omega}\right)\delta_{ij}+\frac{v_i k_j}{\omega}\right]\frac{\partial}{\partial v_j}. \tag{7.10}$$

Then, omitting δ in front of perturbed quantities for the sake of simplicity, (7.8) is re-expressed as follows:

$$f^a_{\mathbf{k},\omega} = g_{\mathbf{k},\omega} L^i_{\mathbf{k},\omega} F_a E^i_{\mathbf{k},\omega} + g_{\mathbf{k},\omega} \int d\mathbf{k}' \int d\omega' \, L^i_{\mathbf{k}',\omega'} \times \left(f^a_{\mathbf{k}-\mathbf{k}',\omega-\omega'} E^i_{\mathbf{k}'\omega'} - \langle f^a_{\mathbf{k}-\mathbf{k}',\omega-\omega'} E^i_{\mathbf{k}'\omega'} \rangle \right). \tag{7.11}$$

We again solve (7.11) for $f^a_{\mathbf{k},\omega}$ in an iterative fashion, as in (1.44):

$$f^{a(1)}_{\mathbf{k},\omega} = g_{\mathbf{k},\omega} L^i_{\mathbf{k},\omega} F_a E^i_{\mathbf{k},\omega}, \tag{7.12}$$

$$f^{a(2)}_{\mathbf{k},\omega} = g_{\mathbf{k},\omega} \int d\mathbf{k}' \int d\omega' \, L^i_{\mathbf{k}',\omega'} \left(f^{a(1)}_{\mathbf{k}-\mathbf{k}',\omega-\omega'} E^i_{\mathbf{k}',\omega'} - \left\langle f^{a(1)}_{\mathbf{k}-\mathbf{k}',\omega-\omega'} E^i_{\mathbf{k}',\omega'} \right\rangle \right),$$

$$f^{a(3)}_{\mathbf{k},\omega} = g_{\mathbf{k},\omega} \int d\mathbf{k}' \int d\omega' \, L^i_{\mathbf{k}',\omega'} \left(f^{a(2)}_{\mathbf{k}-\mathbf{k}',\omega-\omega'} E^i_{\mathbf{k}',\omega'} - \left\langle f^{a(2)}_{\mathbf{k}-\mathbf{k}',\omega-\omega'} E^i_{\mathbf{k}',\omega'} \right\rangle \right),$$

etc., or, inserting lower-order solution to successively higher-order equation, we have

$$f^{a(1)}_{\mathbf{k},\omega} = g_{\mathbf{k},\omega} L^i_{\mathbf{k},\omega} F_a E^i_{\mathbf{k},\omega},$$

$$f^{a(2)}_{\mathbf{k},\omega} = g_{\mathbf{k},\omega} \int d\mathbf{k}' \int d\omega' \, L^i_{\mathbf{k}',\omega'} g_{\mathbf{k}-\mathbf{k}',\omega-\omega'} L^j_{\mathbf{k}-\mathbf{k}',\omega-\omega'} n F_a \times \left(E^i_{\mathbf{k}',\omega'} E^j_{\mathbf{k}-\mathbf{k}',\omega-\omega'} - \left\langle E^i_{\mathbf{k}',\omega'} E^j_{\mathbf{k}-\mathbf{k}',\omega-\omega'} \right\rangle \right),$$

$$\begin{aligned} f^{a(3)}_{\mathbf{k},\omega} = {} & g_{\mathbf{k},\omega} \int d\mathbf{k}' \int d\omega' \int d\mathbf{k}'' \int d\omega'' \, L^i_{\mathbf{k}',\omega'} g_{\mathbf{k}-\mathbf{k}',\omega-\omega'} L^j_{\mathbf{k}'',\omega''} \\ & \times g_{\mathbf{k}-\mathbf{k}'-\mathbf{k}'',\omega-\omega'-\omega''} L^l_{\mathbf{k}-\mathbf{k}'-\mathbf{k}'',\omega-\omega'-\omega''} n F_a \left(E^i_{\mathbf{k}',\omega'} E^j_{\mathbf{k}'',\omega''} E^l_{\mathbf{k}-\mathbf{k}'-\mathbf{k}'',\omega-\omega'-\omega''} \right. \\ & \left. - E^i_{\mathbf{k}',\omega'} \left\langle E^j_{\mathbf{k}'',\omega''} E^l_{\mathbf{k}-\mathbf{k}'-\mathbf{k}'',\omega-\omega'-\omega''} \right\rangle - \left\langle E^i_{\mathbf{k}',\omega'} E^j_{\mathbf{k}'',\omega''} E^l_{\mathbf{k}-\mathbf{k}'-\mathbf{k}'',\omega-\omega'-\omega''} \right\rangle \right), \end{aligned} \tag{7.13}$$

etc. The series solution for $\delta f^a_{\mathbf{k},\omega}$ is given by adding each component, $f^a_{\mathbf{k},\omega} = f^{a(1)}_{\mathbf{k},\omega} + f^{a(2)}_{\mathbf{k},\omega} + f^{a(3)}_{\mathbf{k},\omega}$. We reiterate that the iterative solution (7.13) is not yet symmetrized. The symmetrized expression will be given below in (7.16), but before we do that, let us introduce some further simplified notations:

$$K = (\mathbf{k},\omega), \quad K_1 = (\mathbf{k}_1,\omega_1), \quad K_2 = (\mathbf{k}_2,\omega_2), \quad K_3 = (\mathbf{k}_3,\omega_3),$$

$$\sum_{K_1+K_2=K} = \int d\mathbf{k}_1 \int d\mathbf{k}_2 \, \delta(\mathbf{k}_1+\mathbf{k}_2-\mathbf{k}) \int d\omega_1 \int d\omega_2 \, \delta(\omega_1+\omega_2-\omega),$$

$$\sum_{K_1+K_2+K_3=K} = \int d\mathbf{k}_1 \int d\mathbf{k}_2 \int d\mathbf{k}_3 \, \delta(\mathbf{k}_1+\mathbf{k}_2+\mathbf{k}_3-\mathbf{k}) \times \int d\omega_1 \int d\omega_2 \int d\omega_3 \, \delta(\omega_1+\omega_2+\omega_3-\omega),$$

$$
\begin{aligned}
&g(K) = g_{\mathbf{k},\omega}, \quad g(K_1) = g_{\mathbf{k}_1,\omega_1},\\
&g(K_2) = g_{\mathbf{k}_2,\omega_2}, \quad g(K_1+K_2) = g_{\mathbf{k}_1+\mathbf{k}_2,\omega_1+\omega_2},\\
&L_i(K) = L^i_{\mathbf{k},\omega}, \quad L_i(K_1) = L^i_{\mathbf{k}_1,\omega_1},\\
&L_i(K_2) = L^i_{\mathbf{k}_2,\omega_2}, \quad L_i(K_1+K_2) = L^i_{\mathbf{k}_1+\mathbf{k}_2,\omega_1+\omega_2},\\
&E_i(K_1) = E^i_{\mathbf{k}_1,\omega_1}, \quad E_i(K_2) = E^i_{\mathbf{k}_2,\omega_2},\\
&E_i(K_1+K_2) = E^i_{\mathbf{k}_1+\mathbf{k}_2,\omega_1+\omega_2}.
\end{aligned} \tag{7.14}
$$

Sometimes we will use even more abbreviated notations, such as

$$
g(1) = g(K_1), \quad L(1+2) = L(K_1+K_2), \quad \text{etc.} \tag{7.15}
$$

In terms of these notations, the fully symmetrized expression for iterative solution to (7.8) finally emerges,

$$
\begin{aligned}
f_a(K) = {}& g(K)\, L_i(K)\, n F_a\, E_i(K)\\
&+ \frac{1}{2} \sum_{1+2=K} g(1+2)\left[L_i(1)\, g(2)\, L_j(2) + L_j(2)\, g(1)\, L_i(1)\right]\\
&\times nF_a\left[E_i(1)\, E_j(2) - \langle E_i(1)\, E_j(2)\rangle\right]\\
&+ \frac{1}{2} \sum_{1+2+3=K} g(1+2+3) L_i(1)\, g(2+3)\\
&\times \left[L_j(2)\, g(3)\, L_k(3) + L_k(3)\, g(2)\, L_j(2)\right] nF_a\\
&\times \left[E_i(1)\, E_j(2)\, E_k(3) - E_i(1)\langle E_j(2)\, E_k(3)\rangle - \langle E_i(1)\, E_j(2)\, E_k(3)\rangle\right].
\end{aligned} \tag{7.16}
$$

We next insert (7.16) to the right-hand side of wave equation (7.7), which, upon absorbing the slow-time derivative to ω, is expressed as

$$
\left[\delta_{ij} - \frac{c^2k^2}{\omega^2}\left(\delta_{ij} - \frac{k_ik_j}{k^2}\right)\right] \delta E^j_{\mathbf{k},\omega} = -\frac{4\pi i}{\omega} \sum_a e_a \int d\mathbf{v}\, v_i\, \delta f^a_{\mathbf{k},\omega}. \tag{7.17}
$$

The result, expressed in terms of shorthand notations, is the following:

$$
\begin{aligned}
0 = {}& \left[\delta_{ij} - \frac{c^2k^2}{\omega^2}\left(\delta_{ij} - \frac{k_ik_j}{k^2}\right)\right] E_j(K)\\
&+ i \sum_a \frac{4\pi e_a n}{\omega} \int d\mathbf{v}\, v_i\, g(K)\, L_j(K)\, F_a\, E_j(K)\\
&+ \frac{i}{2} \sum_{1+2=K} \sum_a \frac{4\pi e_a n}{\omega} \int d\mathbf{v}\, v_i\, g(1+2)
\end{aligned} \tag{7.18}
$$

$$\times \left[L_j(1)\, g(2)\, L_k(2) + L_k(2)\, g(1)\, L_j(1)\right] F_a \left[E_j(1)\, E_k(2) - \langle E_j(1)\, E_k(2)\rangle\right]$$

$$+ \frac{i}{2} \sum_{1+2+3=K} \sum_a \frac{4\pi e_a n}{\omega} \int d\mathbf{v}\, v_i\, g(1+2+3)\, L_j(1)\, g(2+3)$$

$$\times \left[L_k(2)\, g(3)\, L_l(3) + L_l(3)\, g(2)\, L_k(2)\right] F_a$$

$$\times \left[E_j(1)\, E_k(2)\, E_l(3) - E_j(1)\, \langle E_k(2)\, E_l(3)\rangle - \langle E_j(1)\, E_k(2)\, E_l(3)\rangle\right].$$

This result can be compactly rewritten as follows:

$$\begin{aligned} 0 = \Lambda_{ij}(K)\, E_j(K) &+ \sum_{1+2=K} \chi^{(2)}_{ijk}(1|2) \left[E_j(1)\, E_k(2) - \langle E_j(1)\, E_k(2)\rangle\right] \\ &+ \sum_{1+2+3=K} \chi^{(3)}_{ijkl}(1|2|3) \left[E_j(1)\, E_k(2)\, E_l(3)\right. \\ &\left. - E_j(1)\, \langle E_k(2)\, E_l(3)\rangle - \langle E_j(1)\, E_k(2)\, E_l(3)\rangle\right], \end{aligned} \tag{7.19}$$

where

$$\begin{aligned} \Lambda_{ij}(K) &= \epsilon_{ij}(K) - \frac{c^2k^2}{\omega^2}\left(\delta_{ij} - \frac{k_i k_j}{k^2}\right), \\ \epsilon_{ij}(K) &= \delta_{ij} + \chi_{ij}(K) = \delta_{ij} + \sum_a \chi^a_{ij}(K), \\ \chi^a_{ij}(K) &= \frac{4\pi i e_a n}{\omega} \int d\mathbf{v}\, v_i\, g(K)\, L_j(K)\, F_a, \\ \chi^{(2)}_{ijk}(1|2) &= \sum_a \chi^{a(2)}_{ijk}(1|2), \\ \chi^{a(2)}_{ijk}(1|2) &= \frac{i}{2} \frac{4\pi e_a n}{\omega_1 + \omega_2} \int d\mathbf{v}\, v_i\, g(1+2) \\ &\quad \times \left[L_j(1)\, g(2)\, L_k(2) + L_k(2)\, g(1)\, L_j(1)\right] F_a, \\ \chi^{(3)}_{ijkl}(1|2|3) &= \sum_a \chi^{a(3)}_{ijkl}(1|2|3), \\ \chi^{a(3)}_{ijkl}(1|2|3) &= \frac{i}{2} \frac{4\pi e_a n}{\omega_1 + \omega_2 + \omega_3} \int d\mathbf{v}\, v_i\, g(1+2+3) \\ &\quad \times L_j(1)\, g(2+3) \left[L_k(2)\, g(3)\, L_l(3) + L_l(3)\, g(2)\, L_k(2)\right] F_a. \end{aligned} \tag{7.20}$$

Equation (7.19) directly generalizes (1.65) to fully electromagnetic circumstances. The electromagnetic response functions defined in (7.20) also directly generalize the electrostatic counterparts (1.55)–(1.57). It is instructive to write the various response functions in explicit forms:

$$\chi^a_{ij}(\mathbf{k},\omega) = \frac{\omega_{pa}^2}{\omega} \int d\mathbf{v}\, \frac{v_i}{\omega - \mathbf{k}\cdot\mathbf{v} + i0} \left[\left(1 - \frac{\mathbf{k}\cdot\mathbf{v}}{\omega}\right)\frac{\partial}{\partial v_j} + \frac{v_j}{\omega}\mathbf{k}\cdot\frac{\partial}{\partial \mathbf{v}}\right] F_a,$$

$$\chi^{a(2)}_{ijk}(\mathbf{k}_1,\omega_1|\mathbf{k}_2,\omega_2) = -\frac{i}{2}\frac{e_a}{m_a}\frac{\omega_{pa}^2}{\omega_1+\omega_2}\int d\mathbf{v}\,\frac{v_i}{\omega_1+\omega_2-(\mathbf{k}_1+\mathbf{k}_2)\cdot\mathbf{v}+i0}$$
$$\times\left\{\left[\left(1-\frac{\mathbf{k}_1\cdot\mathbf{v}}{\omega_1}\right)\frac{\partial}{\partial v_j}+\frac{v_j}{\omega_1}\mathbf{k}_1\cdot\frac{\partial}{\partial\mathbf{v}}\right]\frac{1}{\omega_2-\mathbf{k}_2\cdot\mathbf{v}+i0}\right.$$
$$\times\left[\left(1-\frac{\mathbf{k}_2\cdot\mathbf{v}}{\omega_2}\right)\frac{\partial}{\partial v_k}+\frac{v_k}{\omega_2}\mathbf{k}_2\cdot\frac{\partial}{\partial\mathbf{v}}\right]$$
$$+\left[\left(1-\frac{\mathbf{k}_2\cdot\mathbf{v}}{\omega_2}\right)\frac{\partial}{\partial v_k}+\frac{v_k}{\omega_2}\mathbf{k}_2\cdot\frac{\partial}{\partial\mathbf{v}}\right]\frac{1}{\omega_1-\mathbf{k}_1\cdot\mathbf{v}+i0}$$
$$\left.\times\left[\left(1-\frac{\mathbf{k}_1\cdot\mathbf{v}}{\omega_1}\right)\frac{\partial}{\partial v_j}+\frac{v_j}{\omega_1}\mathbf{k}_1\cdot\frac{\partial}{\partial\mathbf{v}}\right]\right\}F_a,$$

$$\chi^{a(3)}_{ijkl}(\mathbf{k}_1,\omega_1|\mathbf{k}_2,\omega_2|\mathbf{k}_3,\omega_3) = \frac{(-i)^2}{2}\frac{e_a^2}{m_a^2}\frac{\omega_{pa}^2}{\omega_1+\omega_2+\omega_3}$$
$$\times\int d\mathbf{v}\,\frac{v_i}{\omega_1+\omega_2+\omega_3-(\mathbf{k}_1+\mathbf{k}_2+\mathbf{k}_3)\cdot\mathbf{v}}$$
$$\times\left[\left(1-\frac{\mathbf{k}_1\cdot\mathbf{v}}{\omega_1}\right)\frac{\partial}{\partial v_j}+\frac{v_j}{\omega_1}\mathbf{k}_1\cdot\frac{\partial}{\partial\mathbf{v}}\right]$$
$$\times\frac{1}{\omega_2+\omega_3-(\mathbf{k}_2+\mathbf{k}_3)\cdot\mathbf{v}+i0}$$
$$\times\left\{\left[\left(1-\frac{\mathbf{k}_2\cdot\mathbf{v}}{\omega_2}\right)\frac{\partial}{\partial v_k}+\frac{v_k}{\omega_2}\mathbf{k}_2\cdot\frac{\partial}{\partial\mathbf{v}}\right]\frac{1}{\omega_3-\mathbf{k}_3\cdot\mathbf{v}+i0}\right.$$
$$\times\left[\left(1-\frac{\mathbf{k}_3\cdot\mathbf{v}}{\omega_3}\right)\frac{\partial}{\partial v_l}+\frac{v_l}{\omega_3}\mathbf{k}_3\cdot\frac{\partial}{\partial\mathbf{v}}\right]$$
$$+\left[\left(1-\frac{\mathbf{k}_3\cdot\mathbf{v}}{\omega_3}\right)\frac{\partial}{\partial v_l}+\frac{v_l}{\omega_3}\mathbf{k}_3\cdot\frac{\partial}{\partial\mathbf{v}}\right]\frac{1}{\omega_2-\mathbf{k}_2\cdot\mathbf{v}+i0}$$
$$\left.\times\left[\left(1-\frac{\mathbf{k}_2\cdot\mathbf{v}}{\omega_2}\right)\frac{\partial}{\partial v_k}+\frac{v_k}{\omega_2}\mathbf{k}_2\cdot\frac{\partial}{\partial\mathbf{v}}\right]\right\}F_a. \qquad (7.21)$$

These are now linear dielectric susceptibility tensor, second-order nonlinear susceptibility tensor, and (the partial) third-order nonlinear susceptibility tensor, rather than constants.

As with (1.66), we take the dot product of (7.19) with $E_i(K')$, and take the ensemble average:

$$0 = \Lambda_{ij}(K)\langle E_i(K')\,E_j(K)\rangle + \sum_{1+2=K}\chi^{(2)}_{ijk}(1|2)\langle E_i(K')\,E_j(1)\,E_k(2)\rangle$$
$$+\sum_{1+2+3=K}\chi^{(3)}_{ijkl}(1|2|3)\left[\langle E_i(K')\,E_j(1)\,E_k(2)\,E_l(3)\rangle\right.$$
$$\left.-\langle E_i(K')\,E_j(1)\rangle\langle E_k(2)\,E_l(3)\rangle\right]$$

$$
\begin{aligned}
&= \Lambda_{ij}(K) \langle E_i\, E_j \rangle_K \;\delta(K+K') + \sum_{1+2=K} \chi^{(2)}_{ijk}(1|2) \langle E_i(K')\, E_j(1)\, E_k(2) \rangle \\
&\quad + \delta(K+K') \sum_{1+2+3=K} \chi^{(3)}_{ijkl}(1|2|3) \left[\delta(K'+2) \langle E_i\, E_k \rangle_{K'} \langle E_j\, E_l \rangle_1 \right. \\
&\quad \left. + \;\delta(K_1+2) \langle E_i\, E_l \rangle_{K'} \langle E_j\, E_k \rangle_1 \right] \\
&= \Lambda_{ij}(K) \langle E_i\, E_j \rangle_K \;\delta(K+K') \\
&\quad + \sum_1 \chi^{(2)}_{ijk}(1|K-1) \langle E_i(K')\, E_j(1)\, E_k(K-1) \rangle \\
&\quad + 2\,\delta(K+K') \sum_1 \chi^{(3)}_{ijkl}(1|-1|K) \langle E_i\, E_l \rangle_K \langle E_j\, E_k \rangle_1 ,
\end{aligned}
\tag{7.22}
$$

where in the last equality, 1, 2, 3, etc. stand for K_1, K_2, K_3, etc.

To determine the third-body correlation, we approach the problem in the same way as in (1.73), and take the steps, which we have taken thereafter, until we arrive at (1.78). In this regard, it is useful to consider the solution to (7.18) without the third-order nonlinear terms, and express it in direct analogy with and generalization of (1.76),

$$
\begin{aligned}
E_i^{(1)}(K') &= -\Lambda^{-1}_{il}(K) \sum_2 \chi^{(2)}_{lmn}(2|K'-2) \\
&\quad \times \left[E_m^{(0)}(2)\, E_n^{(0)}(K'-2) - \langle E_m^{(0)}(2)\, E_n^{(0)}(K'-2) \rangle \right], \\
E_j^{(1)}(1) &= -\Lambda^{-1}_{jl}(1) \sum_2 \chi^{(2)}_{lmn}(2|1-2) \\
&\quad \times \left[E_m^{(0)}(2)\, E_n^{(0)}(1-2) - \langle E_m^{(0)}(2)\, E_n^{(0)}(1-2) \rangle \right], \\
E_k^{(1)}(K-1) &= -\Lambda^{-1}_{kl}(K-1) \sum_2 \chi^{(2)}_{lmn}(2|K-1-2) \\
&\quad \times \left[E_m^{(0)}(2)\, E_n^{(0)}(K-1-2) - \langle E_m^{(0)}(2)\, E_n^{(0)}(K-1-2) \rangle \right],
\end{aligned}
\tag{7.23}
$$

and upon writing

$$
\begin{aligned}
\langle E_i(K')\, E_j(1)\, E_k(K-1) \rangle &\approx \left\langle E_i^{(1)}(K')\, E_j^{(0)}(1)\, E_k^{(0)}(K-1) \right\rangle \\
&\quad + \left\langle E_i^{(0)}(K')\, E_j^{(1)}(1)\, E_k^{(0)}(K-1) \right\rangle \\
&\quad + \left\langle E_i^{(0)}(K')\, E_j^{(0)}(1)\, E_k^{(1)}(K-1) \right\rangle ,
\end{aligned}
\tag{7.24}
$$

we may obtain the third-body cumulant by virtue of (7.23). After everything is said and done, as it were, we then drop the superscript (0). This results in the following generalization of (1.78):

$$
\begin{aligned}
&\big\langle E_i(K')\, E_j(1)\, E_k(K-1)\big\rangle \\
&= -\Lambda_{il}^{-1}(K') \sum_2 \chi_{lmn}^{(2)}(2|K'-2)\, \Big[\big\langle E_m(2)\, E_n(K'-2)\, E_j(1)\, E_k(K-1)\big\rangle \\
&\quad - \big\langle E_m(2)\, E_n(K'-2)\big\rangle \big\langle E_j(1)\, E_k(K-1)\big\rangle\Big] \\
&\quad - \Lambda_{jl}^{-1}(1) \sum_2 \chi_{lmn}^{(2)}(2|1-2)\, \Big[\big\langle E_i(K')\, E_k(K-1)\, E_m(2)\, E_n(1-2)\big\rangle \\
&\quad - \big\langle E_i(K')\, E_k(K-1)\big\rangle \langle E_m(2)\, E_n(1-2)\rangle\Big] \\
&\quad - \Lambda_{kl}^{-1}(K-1) \sum_2 \chi_{lmn}^{(2)}(2|K-1-2) \\
&\quad \times \Big[\big\langle E_i(K')\, E_j(1)\, E_m(2)\, E_n(K-1-2)\big\rangle \\
&\quad - \big\langle E_i(K')\, E_j(1)\big\rangle \langle E_m(2)\, E_n(K-1-2)\rangle\Big] \\
&= -2\,\delta(K+K') \Big[\Lambda_{il}^{-1}(-K)\, \chi_{lmn}^{(2)}(-1|-K+1)\big\langle E_j\, E_m\big\rangle_1 \langle E_k\, E_n\rangle_{K-1} \\
&\quad + \Lambda_{jl}^{-1}(1)\, \chi_{lmn}^{(2)}(K|1-K)\, \langle E_i\, E_n\rangle_K \langle E_k\, E_m\rangle_{K-1} \\
&\quad + \Lambda_{kl}^{-1}(K-1)\, \chi_{lmn}^{(2)}(K|-1)\, \langle E_i\, E_n\rangle_K \big\langle E_j\, E_m\big\rangle_1\Big].
\end{aligned}
\tag{7.25}
$$

Inserting (7.25) to (7.22), we obtain the formal nonlinear spectral balance equation,

$$
\begin{aligned}
0 &= \Lambda_{ij}(K)\big\langle E_i\, E_j\big\rangle_K \\
&\quad - 2\sum_{K'} \chi_{ijk}^{(2)}(K'|K-K') \Big\{\Big[\Lambda_{jl}^{-1}(K')\, \chi_{lmn}^{(2)}(-K+K'|K)\, \langle E_k\, E_m\rangle_{K-K'} \\
&\quad + \Lambda_{kl}^{-1}(K-K')\, \chi_{lmn}^{(2)}(-K'|K)\big\langle E_j\, E_m\big\rangle_{K'}\Big] \langle E_i\, E_n\rangle_K \\
&\quad + \Lambda_{il}^{-1}(-K)\, \chi_{lmn}^{(2)}(-K'|-K+K')\big\langle E_j\, E_m\big\rangle_{K'} \langle E_k\, E_n\rangle_{K-K'}\Big\} \\
&\quad + 2\sum_{K'} \chi_{ijkl}^{(3)}(K'|-K'|K)\, \langle E_i\, E_l\rangle_K \big\langle E_j\, E_k\big\rangle_{K'}.
\end{aligned}
\tag{7.26}
$$

7.1.1 Formal Nonlinear Wave Kinetic Equation

To determine the adiabatic time evolution of wave energy, we reintroduce the slow time dependence to $\Lambda_{ij}(K)$, as in (1.83), except that we now deal with the electromagnetic linear response function

$$\Lambda_{ij}(K) \to \Lambda_{ij}\left(K + i\frac{\partial}{\partial t}\right) \approx \Lambda_{ij}(K) + \frac{i}{2}\frac{\partial \Lambda_{ij}(K)}{\partial \omega}\frac{\partial}{\partial t}. \tag{7.27}$$

If we further assume that $|\text{Im}\,\Lambda_{ij}(K)| \ll |\text{Re}\,\Lambda_{ij}(K)|$, as before, which implies weakly growing or damped mode, then we rewrite (7.26) in long-hand notation by

$$\begin{aligned}
0 = &\frac{i}{2}\frac{\partial\,\text{Re}\,\Lambda_{ij}(\mathbf{k},\omega)}{\partial \omega}\frac{\partial}{\partial t}\langle E_i\,E_j\rangle_{\mathbf{k},\omega} \\
&+ \text{Re}\,\Lambda_{ij}(\mathbf{k},\omega)\,\langle E_i\,E_j\rangle_{\mathbf{k},\omega} + i\,\text{Im}\,\Lambda_{ij}(\mathbf{k},\omega)\,\langle E_i\,E_j\rangle_{\mathbf{k},\omega} \\
&-2\int d\mathbf{k}'\int d\omega' \left\{ \chi^{(2)}_{ijk}(\mathbf{k}',\omega'|\mathbf{k}-\mathbf{k}',\omega-\omega') \right. \\
&\times \left[\Lambda^{-1}_{jl}(\mathbf{k}',\omega')\,\chi^{(2)}_{lmn}(-\mathbf{k}+\mathbf{k}',\,-\omega+\omega'|\mathbf{k},\omega)\,\langle E_k\,E_m\rangle_{\mathbf{k}-\mathbf{k}',\omega-\omega'}\right. \\
&\left. + \Lambda^{-1}_{kl}(\mathbf{k}-\mathbf{k}',\omega-\omega')\,\chi^{(2)}_{lmn}(-\mathbf{k}',\,-\omega'|\mathbf{k},\omega)\,\langle E_j\,E_m\rangle_{\mathbf{k}',\omega'}\right]\langle E_i\,E_n\rangle_{\mathbf{k},\omega} \\
&+ \chi^{(2)}_{ijk}(\mathbf{k}',\omega'|\mathbf{k}-\mathbf{k}',\omega-\omega')\,\Lambda^{-1}_{il}(-\mathbf{k},\,-\omega) \\
&\left.\times\, \chi^{(2)}_{lmn}(-\mathbf{k}',\,-\omega'|-\mathbf{k}+\mathbf{k}',\,-\omega+\omega')\,\langle E_j\,E_m\rangle_{\mathbf{k}',\omega'}\,\langle E_k\,E_n\rangle_{\mathbf{k}-\mathbf{k}',\omega-\omega'}\right\} \\
&+2\int d\mathbf{k}'\int d\omega'\,\chi^{(3)}_{ijkl}(\mathbf{k}',\omega'|-\mathbf{k}',\,-\omega'|\mathbf{k},\omega)\,\langle E_i\,E_l\rangle_{\mathbf{k},\omega}\,\langle E_j\,E_k\rangle_{\mathbf{k}',\omega'}.
\end{aligned} \tag{7.28}$$

Note that the following symmetry relations can be immediately deduced from definitions (7.21):

$$\begin{aligned}
\chi^{a}_{ij}(-\mathbf{k},\,-\omega) &= \chi^{a*}_{ij}(\mathbf{k},\omega), \\
\chi^{a(2)}_{ijk}(-\mathbf{k}_1,\,-\omega_1|-\mathbf{k}_2,\,-\omega_2) &= \chi^{a(2)*}_{ijk}(\mathbf{k}_1,\omega_1|\mathbf{k}_2,\omega_2), \\
\chi^{a(3)}_{ijkl}(-\mathbf{k}_1,\,-\omega_1|-\mathbf{k}_2,\,-\omega_2|-\mathbf{k}_3,\,-\omega_3) &= \chi^{a(3)*}_{ijkl}(\mathbf{k}_1,\omega_1|\mathbf{k}_2,\omega_2|\mathbf{k}_3,\omega_3).
\end{aligned} \tag{7.29}$$

Moreover, the following symmetries hold for nonlinear susceptibilities upon interchange of indices and arguments (more on the symmetries will be discussed subsequently):

$$\begin{aligned}
\chi^{a(2)}_{ijk}(\mathbf{k}_1,\omega_1|\mathbf{k}_2,\omega_2) &= \chi^{a(2)}_{ikj}(\mathbf{k}_2,\omega_2|\mathbf{k}_1,\omega_1), \\
\chi^{a(3)}_{ijkl}(\mathbf{k}_1,\omega_1|\mathbf{k}_2,\omega_2|\mathbf{k}_3,\omega_3) &= \chi^{a(3)}_{ijlk}(\mathbf{k}_1,\omega_1|\mathbf{k}_3,\omega_3|\mathbf{k}_2,\omega_2).
\end{aligned} \tag{7.30}$$

The second-order nonlinear susceptibility enjoys an additional (approximate) symmetry property, which is a direct generalization of (1.113) (we will discuss this property in more detail later also),

$$\chi^{a(2)}_{ijk}(\mathbf{k}_1+\mathbf{k}_2,\omega_1+\omega_2|-\mathbf{k}_2,\,-\omega_2) = -\chi^{a(2)}_{jik}(\mathbf{k}_1,\omega_1|\mathbf{k}_2,\omega_2). \tag{7.31}$$

It also follows from the definition that the linear susceptibility tensor and its inverse are expressed as

$$\Lambda_{ij}(\mathbf{k},\omega) = \frac{k_i\,k_j}{k^2}\,\epsilon_{\|}(\mathbf{k},\omega) + \left(\delta_{ij} - \frac{k_i\,k_j}{k^2}\right)\Lambda_{\perp}(\mathbf{k},\omega),$$

$$\Lambda_{ij}^{-1}(\mathbf{k},\omega) = \frac{k_i\,k_j}{k^2}\frac{1}{\epsilon_{\|}(\mathbf{k},\omega)} + \left(\delta_{ij} - \frac{k_i\,k_j}{k^2}\right)\frac{1}{\Lambda_{\perp}(\mathbf{k},\omega)},$$

$$\Lambda_{\perp}(\mathbf{k},\omega) = \epsilon_{\perp}(\mathbf{k},\omega) - \frac{c^2k^2}{\omega^2}. \tag{7.32}$$

From this it follows that the spectral wave energy density tensor can be decomposed into longitudinal and transverse parts in diagonal form,

$$\langle E_i\,E_j\rangle_{\mathbf{k},\omega} = \frac{k_i\,k_j}{k^2}\,\langle E_{\|}^2\rangle_{\mathbf{k},\omega} + \frac{1}{2}\left(\delta_{ij} - \frac{k_i\,k_j}{k^2}\right)\langle E_{\perp}^2\rangle_{\mathbf{k},\omega}. \tag{7.33}$$

The properties (7.32) and (7.33) can be easily proven if we take, without loss of generality, $\mathbf{k} = k\hat{\mathbf{z}}$, for instance, and proceed to evaluate the tensor quantity $\Lambda_{ij}(\mathbf{k},\omega)$. The factor $\frac{1}{2}$ associated with $\langle E_{\perp}^2\rangle_{\mathbf{k},\omega}$ is related to the fact that the transverse mode has two independent polarizations. That is, the transverse mode in unmagnetized plasmas is degenerate. This is intuitively obvious. Since there are two orthogonal axes perpendicular to the wave vector, the transverse mode electric field can be linearly polarized in either directions, yet they both satisfy the same dispersion relation. The factor $\frac{1}{2}$ ensures that the total electric field energy, which is the determinant, $\langle E_i\,E_j\rangle_{\mathbf{k},\omega}\delta_{ij}$, is properly partitioned between the longitudinal and transverse components, $\langle E^2\rangle_{\mathbf{k},\omega} = \langle E_{\|}^2\rangle_{\mathbf{k},\omega} + \langle E_{\perp}^2\rangle_{\mathbf{k},\omega}$.

These properties lead to

$$\begin{aligned}
0 = {} & \frac{i}{2}\left[\frac{k_i\,k_j}{k^2}\frac{\partial\,\mathrm{Re}\,\epsilon_{\|}(\mathbf{k},\omega)}{\partial\omega} + \left(\delta_{ij} - \frac{k_i\,k_j}{k^2}\right)\frac{\partial\,\mathrm{Re}\,\Lambda_{\perp}(\mathbf{k},\omega)}{\partial\omega}\right] \\
& \times \left[\frac{k_i\,k_j}{k^2}\frac{\partial}{\partial t}\,\langle E_{\|}^2\rangle_{\mathbf{k},\omega} + \frac{1}{2}\left(\delta_{ij} - \frac{k_i\,k_j}{k^2}\right)\frac{\partial}{\partial t}\,\langle E_{\perp}^2\rangle_{\mathbf{k},\omega}\right] \\
& + \left[\frac{k_i\,k_j}{k^2}\,\mathrm{Re}\,\epsilon_{\|}(\mathbf{k},\omega) + \left(\delta_{ij} - \frac{k_i\,k_j}{k^2}\right)\mathrm{Re}\,\Lambda_{\perp}(\mathbf{k},\omega)\right] \\
& \times \left[\frac{k_i\,k_j}{k^2}\,\langle E_{\|}^2\rangle_{\mathbf{k},\omega} + \frac{1}{2}\left(\delta_{ij} - \frac{k_i\,k_j}{k^2}\right)\langle E_{\perp}^2\rangle_{\mathbf{k},\omega}\right] \\
& + i\left[\frac{k_i\,k_j}{k^2}\,\mathrm{Im}\,\epsilon_{\|}(\mathbf{k},\omega) + \left(\delta_{ij} - \frac{k_i\,k_j}{k^2}\right)\mathrm{Im}\,\Lambda_{\perp}(\mathbf{k},\omega)\right] \\
& \times \left[\frac{k_i\,k_j}{k^2}\,\langle E_{\|}^2\rangle_{\mathbf{k},\omega} + \frac{1}{2}\left(\delta_{ij} - \frac{k_i\,k_j}{k^2}\right)\langle E_{\perp}^2\rangle_{\mathbf{k},\omega}\right]
\end{aligned}$$

$$
\begin{aligned}
&- 2\int d\mathbf{k}' \int d\omega' \left(\chi^{(2)}_{ijk}(\mathbf{k}',\omega'|\mathbf{k}-\mathbf{k}',\omega-\omega') \left\{ - \left[\frac{k'_j\, k'_l}{k'^2} \frac{1}{\epsilon_\parallel(\mathbf{k}',\omega')} \right.\right.\right. \\
&+ \left. \left(\delta_{jl} - \frac{k'_j\, k'_l}{k'^2} \right) \frac{1}{\Lambda_\perp(\mathbf{k}',\omega')} \right] \chi^{(2)}_{nlm}(\mathbf{k}',\omega'|\mathbf{k}-\mathbf{k}',\omega-\omega') \\
&\times \left[\frac{(\mathbf{k}-\mathbf{k}')_k\, (\mathbf{k}-\mathbf{k}')_m}{|\mathbf{k}-\mathbf{k}'|^2} \langle E_\parallel^2 \rangle_{\mathbf{k}-\mathbf{k}',\omega-\omega'} \right. \\
&+ \left. \frac{1}{2} \left(\delta_{km} - \frac{(\mathbf{k}-\mathbf{k}')_k\, (\mathbf{k}-\mathbf{k}')_m}{|\mathbf{k}-\mathbf{k}'|^2} \right) \langle E_\perp^2 \rangle_{\mathbf{k}-\mathbf{k}',\omega-\omega'} \right] \\
&- \left[\frac{(\mathbf{k}-\mathbf{k}')_k\, (\mathbf{k}-\mathbf{k}')_l}{|\mathbf{k}-\mathbf{k}'|^2} \frac{1}{\epsilon_\parallel(\mathbf{k}-\mathbf{k}',\omega-\omega')} \right. \\
&+ \left. \left(\delta_{kl} - \frac{(\mathbf{k}-\mathbf{k}')_k\, (\mathbf{k}-\mathbf{k}')_l}{|\mathbf{k}-\mathbf{k}'|^2} \right) \frac{1}{\Lambda_\perp(\mathbf{k}-\mathbf{k}',\omega-\omega')} \right] \\
&\times \chi^{(2)}_{nml}(\mathbf{k}',\omega'|\mathbf{k}-\mathbf{k}',\omega-\omega') \\
&\times \left. \left[\frac{k'_j\, k'_m}{k'^2} \langle E_\parallel^2 \rangle_{\mathbf{k}',\omega'} + \frac{1}{2} \left(\delta_{jm} - \frac{k'_j\, k'_m}{k'^2} \right) \langle E_\perp^2 \rangle_{\mathbf{k}',\omega'} \right] \right\} \\
&\times \left[\frac{k_i\, k_n}{k^2} \langle E_\parallel^2 \rangle_{\mathbf{k},\omega} + \frac{1}{2} \left(\delta_{in} - \frac{k_i\, k_n}{k^2} \right) \langle E_\perp^2 \rangle_{\mathbf{k},\omega} \right] \\
&+ \chi^{(2)}_{ijk}(\mathbf{k}',\omega'|\mathbf{k}-\mathbf{k}',\omega-\omega') \left[\frac{k_i\, k_l}{k^2} \frac{1}{\epsilon^*_\parallel(\mathbf{k},\omega)} + \left(\delta_{il} - \frac{k_i\, k_l}{k^2} \right) \frac{1}{\Lambda^*_\perp(\mathbf{k},\omega)} \right] \\
&\times \chi^{(2)*}_{lmn}(\mathbf{k}',\omega'|\mathbf{k}-\mathbf{k}',\omega-\omega') \left[\frac{k'_j\, k'_m}{k'^2} \langle E_\parallel^2 \rangle_{\mathbf{k}',\omega'} + \frac{1}{2} \left(\delta_{jm} - \frac{k'_j\, k'_m}{k'^2} \right) \langle E_\perp^2 \rangle_{\mathbf{k}',\omega'} \right] \\
&\times \left[\frac{(\mathbf{k}-\mathbf{k}')_k\, (\mathbf{k}-\mathbf{k}')_n}{|\mathbf{k}-\mathbf{k}'|^2} \langle E_\parallel^2 \rangle_{\mathbf{k}-\mathbf{k}',\omega-\omega'} \right. \\
&+ \left.\left. \frac{1}{2} \left(\delta_{kn} - \frac{(\mathbf{k}-\mathbf{k}')_k\, (\mathbf{k}-\mathbf{k}')_n}{|\mathbf{k}-\mathbf{k}'|^2} \right) \langle E_\perp^2 \rangle_{\mathbf{k}-\mathbf{k}',\omega-\omega'} \right] \right) \\
&+ 2\int d\mathbf{k}' \int d\omega' \, \chi^{(3)}_{ijkl}(\mathbf{k}',\omega'|-\mathbf{k}',-\omega'|\mathbf{k},\omega) \\
&\times \left[\frac{k_i\, k_l}{k^2} \langle E_\parallel^2 \rangle_{\mathbf{k},\omega} + \frac{1}{2} \left(\delta_{il} - \frac{k_i\, k_l}{k^2} \right) \langle E_\perp^2 \rangle_{\mathbf{k},\omega} \right] \\
&\times \left[\frac{k'_j\, k'_k}{k'^2} \langle E_\parallel^2 \rangle_{\mathbf{k}',\omega'} + \frac{1}{2} \left(\delta_{jk} - \frac{k'_j\, k'_k}{k'^2} \right) \langle E_\perp^2 \rangle_{\mathbf{k}',\omega'} \right].
\end{aligned}
\tag{7.34}
$$

Carrying out the various matrix multiplications, one may obtain

$$
\begin{aligned}
0 = {} & \frac{i}{2}\frac{\partial \operatorname{Re}\epsilon_{\parallel}(\mathbf{k},\omega)}{\partial\omega}\frac{\partial}{\partial t}\langle E_{\parallel}^2\rangle_{\mathbf{k},\omega} + \frac{i}{2}\frac{\partial \operatorname{Re}\Lambda_{\perp}(\mathbf{k},\omega)}{\partial\omega}\frac{\partial}{\partial t}\langle E_{\perp}^2\rangle_{\mathbf{k},\omega} \\
& + \langle E_{\parallel}^2\rangle_{\mathbf{k},\omega}\operatorname{Re}\epsilon_{\parallel}(\mathbf{k},\omega) + \langle E_{\perp}^2\rangle_{\mathbf{k},\omega}\operatorname{Re}\Lambda_{\perp}(\mathbf{k},\omega) \\
& + i\,\langle E_{\parallel}^2\rangle_{\mathbf{k},\omega}\operatorname{Im}\epsilon_{\parallel}(\mathbf{k},\omega) + i\,\langle E_{\perp}^2\rangle_{\mathbf{k},\omega}\operatorname{Im}\Lambda_{\perp}(\mathbf{k},\omega) + 2\int d\mathbf{k}'\int d\omega' \\
& \times\left\{\left[\frac{k'_j\,k'_l}{k'^2}\,\frac{\chi^{(2)}_{ijk}(\mathbf{k}',\omega'|\mathbf{k}-\mathbf{k}',\omega-\omega')\,\chi^{(2)}_{nlm}(\mathbf{k}',\omega'|\mathbf{k}-\mathbf{k}',\omega-\omega')}{\epsilon_{\parallel}(\mathbf{k}',\omega')}\right.\right. \\
& \left.+\left(\delta_{jl}-\frac{k'_j\,k'_l}{k'^2}\right)\frac{\chi^{(2)}_{ijk}(\mathbf{k}',\omega'|\mathbf{k}-\mathbf{k}',\omega-\omega')\,\chi^{(2)}_{nlm}(\mathbf{k}',\omega'|\mathbf{k}-\mathbf{k}',\omega-\omega')}{\Lambda_{\perp}(\mathbf{k}',\omega')}\right] \\
& \times\left[\frac{k_i\,k_n}{k^2}\,\frac{(\mathbf{k}-\mathbf{k}')_k\,(\mathbf{k}-\mathbf{k}')_m}{|\mathbf{k}-\mathbf{k}'|^2}\,\langle E_{\parallel}^2\rangle_{\mathbf{k}-\mathbf{k}',\omega-\omega'}\,\langle E_{\parallel}^2\rangle_{\mathbf{k},\omega}\right. \\
& + \frac{1}{2}\frac{k_i\,k_n}{k^2}\left(\delta_{km}-\frac{(\mathbf{k}-\mathbf{k}')_k\,(\mathbf{k}-\mathbf{k}')_m}{|\mathbf{k}-\mathbf{k}'|^2}\right)\langle E_{\perp}^2\rangle_{\mathbf{k}-\mathbf{k}',\omega-\omega'}\,\langle E_{\parallel}^2\rangle_{\mathbf{k},\omega} \\
& + \frac{1}{2}\left(\delta_{in}-\frac{k_i\,k_n}{k^2}\right)\frac{(\mathbf{k}-\mathbf{k}')_k\,(\mathbf{k}-\mathbf{k}')_m}{|\mathbf{k}-\mathbf{k}'|^2}\,\langle E_{\parallel}^2\rangle_{\mathbf{k}-\mathbf{k}',\omega-\omega'}\,\langle E_{\perp}^2\rangle_{\mathbf{k},\omega} \\
& + \frac{1}{4}\left(\delta_{in}-\frac{k_i\,k_n}{k^2}\right)\left(\delta_{km}-\frac{(\mathbf{k}-\mathbf{k}')_k\,(\mathbf{k}-\mathbf{k}')_m}{|\mathbf{k}-\mathbf{k}'|^2}\right) \\
& \left.\times\,\langle E_{\perp}^2\rangle_{\mathbf{k}-\mathbf{k}',\omega-\omega'}\,\langle E_{\perp}^2\rangle_{\mathbf{k},\omega}\right] + \left[\frac{(\mathbf{k}-\mathbf{k}')_k\,(\mathbf{k}-\mathbf{k}')_l}{|\mathbf{k}-\mathbf{k}'|^2}\right. \\
& \times\frac{\chi^{(2)}_{ijk}(\mathbf{k}',\omega'|\mathbf{k}-\mathbf{k}',\omega-\omega')\,\chi^{(2)}_{nml}(\mathbf{k}',\omega'|\mathbf{k}-\mathbf{k}',\omega-\omega')}{\epsilon_{\parallel}(\mathbf{k}-\mathbf{k}',\omega-\omega')} \\
& + \left(\delta_{kl}-\frac{(\mathbf{k}-\mathbf{k}')_k\,(\mathbf{k}-\mathbf{k}')_l}{|\mathbf{k}-\mathbf{k}'|^2}\right) \\
& \left.\times\frac{\chi^{(2)}_{ijk}(\mathbf{k}',\omega'|\mathbf{k}-\mathbf{k}',\omega-\omega')\,\chi^{(2)}_{nml}(\mathbf{k}',\omega'|\mathbf{k}-\mathbf{k}',\omega-\omega')}{\Lambda_{\perp}(\mathbf{k}-\mathbf{k}',\omega-\omega')}\right] \\
& \times\left[\frac{k_i\,k_n}{k^2}\,\frac{k'_j\,k'_m}{k'^2}\,\langle E_{\parallel}^2\rangle_{\mathbf{k}',\omega'}\,\langle E_{\parallel}^2\rangle_{\mathbf{k},\omega}\right. \\
& + \frac{1}{2}\frac{k_i\,k_n}{k^2}\left(\delta_{jm}-\frac{k'_j\,k'_m}{k'^2}\right)\langle E_{\perp}^2\rangle_{\mathbf{k}',\omega'}\,\langle E_{\parallel}^2\rangle_{\mathbf{k},\omega} \\
& + \frac{1}{2}\left(\delta_{in}-\frac{k_i\,k_n}{k^2}\right)\frac{k'_j\,k'_m}{k'^2}\,\langle E_{\parallel}^2\rangle_{\mathbf{k}',\omega'}\,\langle E_{\perp}^2\rangle_{\mathbf{k},\omega} \\
& \left.+ \frac{1}{4}\left(\delta_{in}-\frac{k_i\,k_n}{k^2}\right)\left(\delta_{jm}-\frac{k'_j\,k'_m}{k'^2}\right)\langle E_{\perp}^2\rangle_{\mathbf{k}',\omega'}\,\langle E_{\perp}^2\rangle_{\mathbf{k},\omega}\right]
\end{aligned}
$$

$$
\begin{aligned}
&-\left[\frac{k_i\,k_l}{k^2}\frac{1}{\epsilon_\parallel^*(\mathbf{k},\omega)}\,\chi_{ijk}^{(2)}(\mathbf{k}',\omega'|\mathbf{k}-\mathbf{k}',\omega-\omega')\,\chi_{lmn}^{(2)*}(\mathbf{k}',\omega'|\mathbf{k}-\mathbf{k}',\omega-\omega')\right.\\
&+\left(\delta_{il}-\frac{k_i\,k_l}{k^2}\right)\frac{1}{\Lambda_\perp^*(\mathbf{k},\omega)}\\
&\left.\times\,\chi_{ijk}^{(2)}(\mathbf{k}',\omega'|\mathbf{k}-\mathbf{k}',\omega-\omega')\,\chi_{lmn}^{(2)*}(\mathbf{k}',\omega'|\mathbf{k}-\mathbf{k}',\omega-\omega')\right]\\
&\times\left[\frac{k_j'\,k_m'}{k'^2}\frac{(\mathbf{k}-\mathbf{k}')_k\,(\mathbf{k}-\mathbf{k}')_n}{|\mathbf{k}-\mathbf{k}'|^2}\,\langle E_\parallel^2\rangle_{\mathbf{k}-\mathbf{k}',\omega-\omega'}\,\langle E_\parallel^2\rangle_{\mathbf{k}',\omega'}\right.\\
&+\frac{1}{2}\frac{k_j'\,k_m'}{k'^2}\left(\delta_{kn}-\frac{(\mathbf{k}-\mathbf{k}')_k\,(\mathbf{k}-\mathbf{k}')_n}{|\mathbf{k}-\mathbf{k}'|^2}\right)\langle E_\parallel^2\rangle_{\mathbf{k}',\omega'}\langle E_\perp^2\rangle_{\mathbf{k}-\mathbf{k}',\omega-\omega'}\\
&+\frac{1}{2}\left(\delta_{jm}-\frac{k_j'\,k_m'}{k'^2}\right)\frac{(\mathbf{k}-\mathbf{k}')_k\,(\mathbf{k}-\mathbf{k}')_n}{|\mathbf{k}-\mathbf{k}'|^2}\,\langle E_\perp^2\rangle_{\mathbf{k}',\omega'}\,\langle E_\parallel^2\rangle_{\mathbf{k}-\mathbf{k}',\omega-\omega'}\\
&+\frac{1}{4}\left(\delta_{jm}-\frac{k_j'\,k_m'}{k'^2}\right)\left(\delta_{kn}-\frac{(\mathbf{k}-\mathbf{k}')_k\,(\mathbf{k}-\mathbf{k}')_n}{|\mathbf{k}-\mathbf{k}'|^2}\right)\\
&\left.\left.\times\,\langle E_\perp^2\rangle_{\mathbf{k}',\omega'}\,\langle E_\perp^2\rangle_{\mathbf{k}-\mathbf{k}',\omega-\omega'}\right]\right\}\\
&+2\int d\mathbf{k}'\int d\omega'\,\chi_{ijkl}^{(3)}(\mathbf{k}',\omega'|-\mathbf{k}',-\omega'|\mathbf{k},\omega)\\
&\times\left[\frac{k_i\,k_l}{k^2}\frac{k_j'\,k_k'}{k'^2}\,\langle E_\parallel^2\rangle_{\mathbf{k},\omega}\,\langle E_\parallel^2\rangle_{\mathbf{k}',\omega'}\right.\\
&+\frac{1}{2}\frac{k_i\,k_l}{k^2}\left(\delta_{jk}-\frac{k_j'\,k_k'}{k'^2}\right)\langle E_\perp^2\rangle_{\mathbf{k}',\omega'}\,\langle E_\parallel^2\rangle_{\mathbf{k},\omega}\\
&+\frac{1}{2}\left(\delta_{il}-\frac{k_i\,k_l}{k^2}\right)\frac{k_j'\,k_k'}{k'^2}\,\langle E_\parallel^2\rangle_{\mathbf{k}',\omega'}\,\langle E_\perp^2\rangle_{\mathbf{k},\omega}\\
&\left.+\frac{1}{4}\left(\delta_{il}-\frac{k_i\,k_l}{k^2}\right)\left(\delta_{jk}-\frac{k_j'\,k_k'}{k'^2}\right)\langle E_\perp^2\rangle_{\mathbf{k}',\omega'}\,\langle E_\perp^2\rangle_{\mathbf{k},\omega}\right].
\end{aligned}
\tag{7.35}
$$

In deriving this result, we have made use of the relations

$$
\begin{aligned}
&\frac{k_i\,k_j}{k^2}\frac{k_i\,k_j}{k^2}=1,\quad \frac{k_i\,k_j}{k^2}\left(\delta_{ij}-\frac{k_i\,k_j}{k^2}\right)=0,\\
&\left(\delta_{ij}-\frac{k_i\,k_j}{k^2}\right)\left(\delta_{ij}-\frac{k_i\,k_j}{k^2}\right)=\delta_{ii}-1=2.
\end{aligned}
\tag{7.36}
$$

Equation (7.35) represents the formal nonlinear wave kinetic equation, which directly generalizes (1.85).

7.1.2 Formal Particle Kinetic Equation

In Section 1.6, we discussed the formal particle kinetic equation under electrostatic approximation in which the effects of nonlinear mode coupling were included. This was done for the sake of being consistent with the weak turbulence ordering in the wave kinetic equation. It turns out, however, that the nonlinear mode coupling term could be ignored for reasons, which we have already discussed at the end of Chapter 3. On such a basis, in the present electromagnetic formalism, we take the simple view that quasilinear approximation is valid as far as the particle kinetic equation is concerned. Thus, we simply write

$$\frac{\partial F_a}{\partial t} = -\mathrm{Im}\,\frac{e_a^2}{m_a^2}\frac{\partial}{\partial v_i}\int d\mathbf{k}\int d\omega\left[\left(1-\frac{\mathbf{k}\cdot\mathbf{v}}{\omega}\right)\delta_{ij}+\frac{v_j\,k_i}{\omega}\right]\langle E_j\,E_k\rangle_{\mathbf{k},\omega}$$
$$\times\frac{1}{\omega-\mathbf{k}\cdot\mathbf{v}+i0}\left[\left(1-\frac{\mathbf{k}\cdot\mathbf{v}}{\omega}\right)\delta_{kl}+\frac{v_k\,k_l}{\omega}\right]\frac{\partial F_a}{\partial v_l}. \qquad (7.37)$$

Decomposition of $\langle E_i\,E_j\rangle_{\mathbf{k},\omega}$ into longitudinal and transverse components as in (7.33) leads to

$$\frac{\partial F_a}{\partial t} = -\mathrm{Im}\,\frac{e_a^2}{m_a^2}\frac{\partial}{\partial v_i}\int d\mathbf{k}\int d\omega\left[\left(1-\frac{\mathbf{k}\cdot\mathbf{v}}{\omega}\right)\delta_{ij}+\frac{v_j\,k_i}{\omega}\right]$$
$$\times\left[\frac{k_j\,k_k}{k^2}\langle E_\parallel^2\rangle_{\mathbf{k},\omega}+\frac{1}{2}\left(\delta_{jk}-\frac{k_j\,k_k}{k^2}\right)\langle E_\perp^2\rangle_{\mathbf{k},\omega}\right]$$
$$\times\frac{1}{\omega-\mathbf{k}\cdot\mathbf{v}+i0}\left[\left(1-\frac{\mathbf{k}\cdot\mathbf{v}}{\omega}\right)\delta_{kl}+\frac{v_k\,k_l}{\omega}\right]\frac{\partial F_a}{\partial v_l}. \qquad (7.38)$$

Note that

$$\left[\left(1-\frac{\mathbf{k}\cdot\mathbf{v}}{\omega}\right)\delta_{ij}+\frac{v_i\,k_j}{\omega}\right]\frac{k_j\,k_k}{k^2}\left[\left(1-\frac{\mathbf{k}\cdot\mathbf{v}}{\omega}\right)\delta_{kl}+\frac{v_k\,k_l}{\omega}\right]=\frac{k_i\,k_l}{k^2}. \qquad (7.39)$$

Note also the following:

$$\left[\left(1-\frac{\mathbf{k}\cdot\mathbf{v}}{\omega}\right)\delta_{ij}+\frac{v_j\,k_i}{\omega}\right]\left(\delta_{jk}-\frac{k_j\,k_k}{k^2}\right)\left[\left(1-\frac{\mathbf{k}\cdot\mathbf{v}}{\omega}\right)\delta_{kl}+\frac{v_k\,k_l}{\omega}\right]$$
$$=\left[\left(1-\frac{\mathbf{k}\cdot\mathbf{v}}{\omega}\right)\left(\delta_{ik}-\frac{k_i\,k_k}{k^2}\right)+\frac{k_i}{\omega}\frac{[(\mathbf{k}\times\mathbf{v})\times\mathbf{k}]_k}{k^2}\right]$$
$$\times\left[\left(1-\frac{\mathbf{k}\cdot\mathbf{v}}{\omega}\right)\left(\delta_{kl}-\frac{k_k\,k_l}{k^2}\right)+\frac{[(\mathbf{k}\times\mathbf{v})\times\mathbf{k}]_k}{k^2}\frac{k_l}{\omega}\right], \qquad (7.40)$$

where we have made use of (7.36), and have also made use of the following:

$$k_k\,[(\mathbf{k}\times\mathbf{v})\times\mathbf{k}]_k=\mathbf{k}\cdot\mathbf{v}_\perp=0. \qquad (7.41)$$

In (7.41), we have decomposed the velocity vector into components parallel and perpendicular to the wave vector – see (6.54),

$$\mathbf{v} = \mathbf{v}_\parallel + \mathbf{v}_\perp, \qquad \mathbf{v}_\parallel = \frac{\mathbf{k}\,(\mathbf{k}\cdot\mathbf{v})}{k^2}, \qquad \mathbf{v}_\perp = \frac{(\mathbf{k}\times\mathbf{v})\times\mathbf{k}}{k^2}.$$

Making use of (7.39) and (7.40), the particle kinetic equation (7.37) reduces to

$$\begin{aligned}\frac{\partial F_a}{\partial t} &= -\mathrm{Im}\,\frac{e_a^2}{m_a^2}\frac{\partial}{\partial v_i}\int d\mathbf{k}\int d\omega\,\frac{1}{\omega-\mathbf{k}\cdot\mathbf{v}+i0}\left\{\frac{k_i\,k_j}{k^2}\,\langle E_\parallel^2\rangle_{\mathbf{k},\omega}\right.\\ &\quad+\frac{1}{2}\,\langle E_\perp^2\rangle_{\mathbf{k},\omega}\left[\left(1-\frac{\mathbf{k}\cdot\mathbf{v}}{\omega}\right)\left(\delta_{ik}-\frac{k_i\,k_k}{k^2}\right)+\frac{k_i\,[(\mathbf{k}\times\mathbf{v})\times\mathbf{k}]_k}{\omega\,k^2}\right]\\ &\quad\left.\times\left[\left(1-\frac{\mathbf{k}\cdot\mathbf{v}}{\omega}\right)\left(\delta_{kj}-\frac{k_k\,k_j}{k^2}\right)+\frac{[(\mathbf{k}\times\mathbf{v})\times\mathbf{k}]_k\,k_j}{\omega\,k^2}\right]\right\}\frac{\partial F_a}{\partial v_j},\end{aligned} \tag{7.42}$$

or, equivalently,

$$\begin{aligned}\frac{\partial F_a}{\partial t} &= \frac{\partial}{\partial v_i}\,D_{ij}\,\frac{\partial F_a}{\partial v_j},\\ D_{ij} &= -\mathrm{Im}\,\frac{e_a^2}{m_a^2}\int d\mathbf{k}\int d\omega\,\frac{1}{\omega-\mathbf{k}\cdot\mathbf{v}+i0}\\ &\quad\times\left(\frac{k_i\,k_j}{k^2}\,\langle E_\parallel^2\rangle_{\mathbf{k},\omega}+\frac{1}{2}\,a_{ki}\,a_{kj}\,\langle E_\perp^2\rangle_{\mathbf{k},\omega}\right),\\ a_{ij}(\mathbf{k},\omega;\mathbf{v}) &= \left(1-\frac{\mathbf{k}\cdot\mathbf{v}}{\omega}\right)\left(\delta_{ij}-\frac{k_i\,k_j}{k^2}\right)+\frac{[(\mathbf{k}\times\mathbf{v})\times\mathbf{k}]_i}{k^2}\frac{k_j}{\omega}.\end{aligned} \tag{7.43}$$

7.2 Formal Electromagnetic Kinetic Equations for Eigenmodes

In this section, we discuss further manipulations of formal theory developed in the previous section, namely, formal wave kinetic equation (7.35) and formal particle kinetic equation (7.42), and render them applicable for linear eigenmodes. Since the particle equation (7.42) is relatively simple, major portion of the discussions will be devoted to the wave kinetic equation. The difference between electrostatic and electromagnetic theory is the addition of a new eigenmode, namely, transverse polarized mode. Electrostatic linear eigenmodes have already been discussed in Part I, of this book, and they satisfy the formal longitudinal wave dispersion relation,

$$\mathrm{Re}\,\epsilon_\parallel(\mathbf{k},\omega^\alpha) = 0. \tag{7.44}$$

We already discussed that $\alpha = L$ and S correspond to Langmuir and ion-sound modes, and these are the two longitudinal modes satisfying (7.44). The electromagnetic mode satisfies the transverse wave dispersion relation, which is formally expressed by

$$\mathrm{Re}\,\Lambda_\perp(\mathbf{k},\omega^T) = 0. \tag{7.45}$$

As we already discussed in (1.95), the longitudinal mode can be represented by

$$\langle E_\parallel^2\rangle_{\mathbf{k},\omega} = \sum_{\alpha=L,S}\sum_{\sigma=\pm1} I_{\mathbf{k}}^{\sigma\alpha}\,\delta(\omega-\sigma\omega_{\mathbf{k}}^{\alpha}). \tag{7.46}$$

Likewise, the transverse mode can be represented by

$$\langle E_\perp^2\rangle_{\mathbf{k},\omega} = \sum_{\sigma=\pm1} I_{\mathbf{k}}^{\sigma T}\,\delta(\omega-\sigma\omega_{\mathbf{k}}^{T}). \tag{7.47}$$

The dispersion relations (7.44) and (7.45) effectively replace the real part of formal wave kinetic equation (7.35). We thus need only to consider the imaginary part. In explicitly writing down the imaginary part of (7.35), we make use of the shorthand notations – see (1.151),

$$\begin{aligned}
\epsilon_\parallel'(\mathbf{k},\omega) &= \frac{\partial \mathrm{Re}\,\epsilon_\parallel(\mathbf{k},\omega)}{\partial\omega},\\
\Lambda_\perp'(\mathbf{k},\omega) &= \frac{\partial \mathrm{Re}\,\Lambda_\perp(\mathbf{k},\omega)}{\partial\omega}.
\end{aligned}$$

The result is

$$\begin{aligned}
0 &= \epsilon_\parallel'(\mathbf{k},\omega)\sum_{\alpha=L,S}\sum_{\sigma=\pm1}\frac{\partial I_{\mathbf{k}}^{\sigma\alpha}}{\partial t}\,\delta(\omega-\sigma\omega_{\mathbf{k}}^{\alpha}) + \Lambda_\perp'(\mathbf{k},\omega)\sum_{\sigma=\pm1}\frac{\partial I_{\mathbf{k}}^{\sigma T}}{\partial t}\,\delta(\omega-\sigma\omega_{\mathbf{k}}^{T})\\
&+ 2\,\mathrm{Im}\,\epsilon_\parallel(\mathbf{k},\omega)\sum_{\alpha}\sum_{\sigma=\pm1} I_{\mathbf{k}}^{\sigma\alpha} + 2\,\mathrm{Im}\,\Lambda_\perp(\mathbf{k},\omega)\sum_{\sigma=\pm1} I_{\mathbf{k}}^{\sigma T}\,\delta(\omega-\sigma\omega_{\mathbf{k}}^{T})\\
&+ 4\,\mathrm{Im}\int d\mathbf{k}'\int d\omega'\,\Bigg\{\Bigg[\frac{k_j'\,k_l'}{k'^2}\,\frac{\chi_{ijk}^{(2)}(\mathbf{k}',\omega'|\mathbf{k}-\mathbf{k}',\omega-\omega')}{\epsilon_\parallel(\mathbf{k}',\omega')}\\
&\times\chi_{nlm}^{(2)}(\mathbf{k}',\omega'|\mathbf{k}-\mathbf{k}',\omega-\omega') + \left(\delta_{jl}-\frac{k_j'\,k_l'}{k'^2}\right)\frac{\chi_{ijk}^{(2)}(\mathbf{k}',\omega'|\mathbf{k}-\mathbf{k}',\omega-\omega')}{\Lambda_\perp(\mathbf{k}',\omega')}\\
&\times\chi_{nlm}^{(2)}(\mathbf{k}',\omega'|\mathbf{k}-\mathbf{k}',\omega-\omega')\Bigg]\frac{k_i\,k_n}{k^2}\frac{(\mathbf{k}-\mathbf{k}')_k\,(\mathbf{k}-\mathbf{k}')_m}{|\mathbf{k}-\mathbf{k}'|^2}\sum_{\alpha,\gamma=L,S}\sum_{\sigma,\sigma''=\pm1}\\
&\times I_{\mathbf{k}-\mathbf{k}'}^{\sigma''\gamma}\,I_{\mathbf{k}}^{\sigma\alpha}\,\delta(\omega-\omega'-\sigma''\omega_{\mathbf{k}-\mathbf{k}'}^{\gamma})\,\delta(\omega-\sigma\omega_{\mathbf{k}}^{\alpha})\\
&+\Bigg[\frac{k_j'\,k_l'}{k'^2}\,\frac{\chi_{ijk}^{(2)}(\mathbf{k}',\omega'|\mathbf{k}-\mathbf{k}',\omega-\omega')}{\epsilon_\parallel(\mathbf{k}',\omega')}\,\chi_{nlm}^{(2)}(\mathbf{k}',\omega'|\mathbf{k}-\mathbf{k}',\omega-\omega')\\
&+\left(\delta_{jl}-\frac{k_j'\,k_l'}{k'^2}\right)\frac{\chi_{ijk}^{(2)}(\mathbf{k}',\omega'|\mathbf{k}-\mathbf{k}',\omega-\omega')}{\Lambda_\perp(\mathbf{k}',\omega')}\,\chi_{nlm}^{(2)}(\mathbf{k}',\omega'|\mathbf{k}-\mathbf{k}',\omega-\omega')\Bigg]\\
&\times\frac{1}{2}\frac{k_i\,k_n}{k^2}\left(\delta_{km}-\frac{(\mathbf{k}-\mathbf{k}')_k\,(\mathbf{k}-\mathbf{k}')_m}{|\mathbf{k}-\mathbf{k}'|^2}\right)\sum_{\alpha}\sum_{\sigma,\sigma''}\\
&\times I_{\mathbf{k}-\mathbf{k}'}^{\sigma''T}\,I_{\mathbf{k}}^{\sigma\alpha}\,\delta(\omega-\omega'-\sigma''\omega_{\mathbf{k}-\mathbf{k}'}^{T})\,\delta(\omega-\sigma\omega_{\mathbf{k}}^{\alpha})
\end{aligned}$$

$$
\begin{aligned}
&+\Bigg[\frac{k'_j\,k'_l}{k'^2}\,\frac{\chi^{(2)}_{ijk}(\mathbf{k}',\omega'|\mathbf{k}-\mathbf{k}',\omega-\omega')}{\epsilon_{\parallel}(\mathbf{k}',\omega')}\,\chi^{(2)}_{nlm}(\mathbf{k}',\omega'|\mathbf{k}-\mathbf{k}',\omega-\omega')\\
&+\left(\delta_{jl}-\frac{k'_j\,k'_l}{k'^2}\right)\frac{\chi^{(2)}_{ijk}(\mathbf{k}',\omega'|\mathbf{k}-\mathbf{k}',\omega-\omega')}{\Lambda_{\perp}(\mathbf{k}',\omega')}\\
&\times\,\chi^{(2)}_{nlm}(\mathbf{k}',\omega'|\mathbf{k}-\mathbf{k}',\omega-\omega')\Bigg]\frac{1}{2}\left(\delta_{in}-\frac{k_i\,k_n}{k^2}\right)\frac{(\mathbf{k}-\mathbf{k}')_k\,(\mathbf{k}-\mathbf{k}')_m}{|\mathbf{k}-\mathbf{k}'|^2}\\
&\times\sum_{\gamma}\sum_{\sigma,\sigma''}I^{\sigma''\gamma}_{\mathbf{k}-\mathbf{k}'}\,I^{\sigma T}_{\mathbf{k}}\,\delta(\omega-\omega'-\sigma''\omega^{\gamma}_{\mathbf{k}-\mathbf{k}'})\,\delta(\omega-\sigma\omega^{T}_{\mathbf{k}})\\
&+\Bigg[\frac{k'_j\,k'_l}{k'^2}\,\frac{\chi^{(2)}_{ijk}(\mathbf{k}',\omega'|\mathbf{k}-\mathbf{k}',\omega-\omega')}{\epsilon_{\parallel}(\mathbf{k}',\omega')}\,\chi^{(2)}_{nlm}(\mathbf{k}',\omega'|\mathbf{k}-\mathbf{k}',\omega-\omega')\\
&+\left(\delta_{jl}-\frac{k'_j\,k'_l}{k'^2}\right)\frac{\chi^{(2)}_{ijk}(\mathbf{k}',\omega'|\mathbf{k}-\mathbf{k}',\omega-\omega')}{\Lambda_{\perp}(\mathbf{k}',\omega')}\,\chi^{(2)}_{nlm}(\mathbf{k}',\omega'|\mathbf{k}-\mathbf{k}',\omega-\omega')\Bigg]\\
&\times\frac{1}{4}\left(\delta_{in}-\frac{k_i\,k_n}{k^2}\right)\left(\delta_{km}-\frac{(\mathbf{k}-\mathbf{k}')_k\,(\mathbf{k}-\mathbf{k}')_m}{|\mathbf{k}-\mathbf{k}'|^2}\right)\sum_{\sigma,\sigma''}I^{\sigma''T}_{\mathbf{k}-\mathbf{k}'}\,I^{\sigma T}_{\mathbf{k}}\\
&\times\,\delta(\omega-\omega'-\sigma''\omega^{T}_{\mathbf{k}-\mathbf{k}'})\,\delta(\omega-\sigma\omega^{T}_{\mathbf{k}})+\Bigg[\frac{(\mathbf{k}-\mathbf{k}')_k\,(\mathbf{k}-\mathbf{k}')_l}{|\mathbf{k}-\mathbf{k}'|^2}\\
&\times\,\frac{\chi^{(2)}_{ijk}(\mathbf{k}',\omega'|\mathbf{k}-\mathbf{k}',\omega-\omega')}{\epsilon_{\parallel}(\mathbf{k}-\mathbf{k}',\omega-\omega')}\,\chi^{(2)}_{nml}(\mathbf{k}',\omega'|\mathbf{k}-\mathbf{k}',\omega-\omega')\\
&+\left(\delta_{kl}-\frac{(\mathbf{k}-\mathbf{k}')_k\,(\mathbf{k}-\mathbf{k}')_l}{|\mathbf{k}-\mathbf{k}'|^2}\right)\frac{\chi^{(2)}_{ijk}(\mathbf{k}',\omega'|\mathbf{k}-\mathbf{k}',\omega-\omega')}{\Lambda_{\perp}(\mathbf{k}-\mathbf{k}',\omega-\omega')}\\
&\times\,\chi^{(2)}_{nml}(\mathbf{k}',\omega'|\mathbf{k}-\mathbf{k}',\omega-\omega')\Bigg]\frac{k_i\,k_n}{k^2}\,\frac{k'_j\,k'_m}{k'^2}\sum_{\alpha,\beta}\sum_{\sigma,\sigma'}I^{\sigma'\beta}_{\mathbf{k}'}\,I^{\sigma\alpha}_{\mathbf{k}}\\
&\times\,\delta(\omega'-\sigma'\omega^{\beta}_{\mathbf{k}'})\,\delta(\omega-\sigma\omega^{\alpha}_{\mathbf{k}})+\Bigg[\frac{(\mathbf{k}-\mathbf{k}')_k\,(\mathbf{k}-\mathbf{k}')_l}{|\mathbf{k}-\mathbf{k}'|^2}\\
&\times\,\frac{\chi^{(2)}_{ijk}(\mathbf{k}',\omega'|\mathbf{k}-\mathbf{k}',\omega-\omega')}{\epsilon_{\parallel}(\mathbf{k}-\mathbf{k}',\omega-\omega')}\,\chi^{(2)}_{nml}(\mathbf{k}',\omega'|\mathbf{k}-\mathbf{k}',\omega-\omega')\\
&+\left(\delta_{kl}-\frac{(\mathbf{k}-\mathbf{k}')_k\,(\mathbf{k}-\mathbf{k}')_l}{|\mathbf{k}-\mathbf{k}'|^2}\right)\frac{\chi^{(2)}_{ijk}(\mathbf{k}',\omega'|\mathbf{k}-\mathbf{k}',\omega-\omega')}{\Lambda_{\perp}(\mathbf{k}-\mathbf{k}',\omega-\omega')}\\
&\times\,\chi^{(2)}_{nml}(\mathbf{k}',\omega'|\mathbf{k}-\mathbf{k}',\omega-\omega')\Bigg]\frac{1}{2}\,\frac{k_i\,k_n}{k^2}\left(\delta_{jm}-\frac{k'_j\,k'_m}{k'^2}\right)\sum_{\alpha}\sum_{\sigma,\sigma'}I^{\sigma'T}_{\mathbf{k}'}\,I^{\sigma\alpha}_{\mathbf{k}}\\
&\times\,\delta(\omega'-\sigma'\omega^{T}_{\mathbf{k}'})\,\delta(\omega-\sigma\omega^{\alpha}_{\mathbf{k}})+\Bigg[\frac{(\mathbf{k}-\mathbf{k}')_k\,(\mathbf{k}-\mathbf{k}')_l}{|\mathbf{k}-\mathbf{k}'|^2}\\
&\times\,\frac{\chi^{(2)}_{ijk}(\mathbf{k}',\omega'|\mathbf{k}-\mathbf{k}',\omega-\omega')}{\epsilon_{\parallel}(\mathbf{k}-\mathbf{k}',\omega-\omega')}\,\chi^{(2)}_{nml}(\mathbf{k}',\omega'|\mathbf{k}-\mathbf{k}',\omega-\omega')
\end{aligned}
$$

$$+\left(\delta_{kl}-\frac{(\mathbf{k}-\mathbf{k}')_k\,(\mathbf{k}-\mathbf{k}')_l}{|\mathbf{k}-\mathbf{k}'|^2}\right)\frac{\chi^{(2)}_{ijk}(\mathbf{k}',\omega'|\mathbf{k}-\mathbf{k}',\omega-\omega')}{\Lambda_\perp(\mathbf{k}-\mathbf{k}',\omega-\omega')}$$

$$\times\ \chi^{(2)}_{nml}(\mathbf{k}',\omega'|\mathbf{k}-\mathbf{k}',\omega-\omega')\Bigg]\frac{1}{2}\left(\delta_{in}-\frac{k_i\,k_n}{k^2}\right)\frac{k'_j\,k'_m}{k'^2}\sum_\beta\sum_{\sigma,\sigma'}I^{\sigma'\beta}_{\mathbf{k}'}\,I^{\sigma T}_{\mathbf{k}}$$

$$\times\ \delta(\omega'-\sigma'\omega^\beta_{\mathbf{k}'})\,\delta(\omega-\sigma\omega^T_{\mathbf{k}})+\Bigg[\frac{(\mathbf{k}-\mathbf{k}')_k\,(\mathbf{k}-\mathbf{k}')_l}{|\mathbf{k}-\mathbf{k}'|^2}$$

$$\times\ \frac{\chi^{(2)}_{ijk}(\mathbf{k}',\omega'|\mathbf{k}-\mathbf{k}',\omega-\omega')}{\epsilon_\parallel(\mathbf{k}-\mathbf{k}',\omega-\omega')}\,\chi^{(2)}_{nml}(\mathbf{k}',\omega'|\mathbf{k}-\mathbf{k}',\omega-\omega')$$

$$+\left(\delta_{kl}-\frac{(\mathbf{k}-\mathbf{k}')_k\,(\mathbf{k}-\mathbf{k}')_l}{|\mathbf{k}-\mathbf{k}'|^2}\right)\frac{\chi^{(2)}_{ijk}(\mathbf{k}',\omega'|\mathbf{k}-\mathbf{k}',\omega-\omega')}{\Lambda_\perp(\mathbf{k}-\mathbf{k}',\omega-\omega')}$$

$$\times\ \chi^{(2)}_{nml}(\mathbf{k}',\omega'|\mathbf{k}-\mathbf{k}',\omega-\omega')\Bigg]\frac{1}{4}\left(\delta_{in}-\frac{k_i\,k_n}{k^2}\right)\left(\delta_{jm}-\frac{k'_j\,k'_m}{k'^2}\right)$$

$$\times\sum_{\sigma,\sigma'}I^{\sigma'T}_{\mathbf{k}'}\,I^{\sigma T}_{\mathbf{k}}\,\delta(\omega'-\sigma'\omega^T_{\mathbf{k}'})\,\delta(\omega-\sigma\omega^T_{\mathbf{k}})-\Bigg[\frac{k_i\,k_l}{k^2}$$

$$\times\ \frac{\chi^{(2)}_{ijk}(\mathbf{k}',\omega'|\mathbf{k}-\mathbf{k}',\omega-\omega')}{\epsilon^*_\parallel(\mathbf{k},\omega)}\,\chi^{(2)*}_{lmn}(\mathbf{k}',\omega'|\mathbf{k}-\mathbf{k}',\omega-\omega')$$

$$+\left(\delta_{il}-\frac{k_i\,k_l}{k^2}\right)\frac{\chi^{(2)}_{ijk}(\mathbf{k}',\omega'|\mathbf{k}-\mathbf{k}',\omega-\omega')}{\Lambda^*_\perp(\mathbf{k},\omega)}\,\chi^{(2)*}_{lmn}(\mathbf{k}',\omega'|\mathbf{k}-\mathbf{k}',\omega-\omega')\Bigg]$$

$$\times\ \frac{k'_j\,k'_m}{k'^2}\frac{(\mathbf{k}-\mathbf{k}')_k\,(\mathbf{k}-\mathbf{k}')_n}{|\mathbf{k}-\mathbf{k}'|^2}\sum_{\beta,\gamma}\sum_{\sigma',\sigma''}I^{\sigma''\gamma}_{\mathbf{k}-\mathbf{k}'}\,I^{\sigma'\beta}_{\mathbf{k}'}$$

$$\times\ \delta(\omega-\omega'-\sigma''\omega^\gamma_{\mathbf{k}-\mathbf{k}'})\,\delta(\omega'-\sigma'\omega^\beta_{\mathbf{k}'})-\Bigg[\frac{k_i\,k_l}{k^2}$$

$$\times\ \frac{\chi^{(2)}_{ijk}(\mathbf{k}',\omega'|\mathbf{k}-\mathbf{k}',\omega-\omega')}{\epsilon^*_\parallel(\mathbf{k},\omega)}\,\chi^{(2)*}_{lmn}(\mathbf{k}',\omega'|\mathbf{k}-\mathbf{k}',\omega-\omega')$$

$$+\left(\delta_{il}-\frac{k_i\,k_l}{k^2}\right)\frac{\chi^{(2)}_{ijk}(\mathbf{k}',\omega'|\mathbf{k}-\mathbf{k}',\omega-\omega')}{\Lambda^*_\perp(\mathbf{k},\omega)}\,\chi^{(2)*}_{lmn}(\mathbf{k}',\omega'|\mathbf{k}-\mathbf{k}',\omega-\omega')\Bigg]$$

$$\times\ \frac{1}{2}\frac{k'_j\,k'_m}{k'^2}\left(\delta_{kn}-\frac{(\mathbf{k}-\mathbf{k}')_k\,(\mathbf{k}-\mathbf{k}')_n}{|\mathbf{k}-\mathbf{k}'|^2}\right)\sum_\beta\sum_{\sigma',\sigma''}I^{\sigma'\beta}_{\mathbf{k}'}\,I^{\sigma''T}_{\mathbf{k}-\mathbf{k}'}$$

$$\times\ \delta(\omega'-\sigma'\omega^\beta_{\mathbf{k}'})\,\delta(\omega-\omega'-\sigma''\omega^T_{\mathbf{k}-\mathbf{k}'})-\Bigg[\frac{k_i\,k_l}{k^2}\frac{\chi^{(2)}_{ijk}(\mathbf{k}',\omega'|\mathbf{k}-\mathbf{k}',\omega-\omega')}{\epsilon^*_\parallel(\mathbf{k},\omega)}$$

$$\times\ \chi^{(2)*}_{lmn}(\mathbf{k}',\omega'|\mathbf{k}-\mathbf{k}',\omega-\omega')+\left(\delta_{il}-\frac{k_i\,k_l}{k^2}\right)\frac{\chi^{(2)}_{ijk}(\mathbf{k}',\omega'|\mathbf{k}-\mathbf{k}',\omega-\omega')}{\Lambda^*_\perp(\mathbf{k},\omega)}$$

$$\times\ \chi^{(2)*}_{lmn}(\mathbf{k}',\omega'|\mathbf{k}-\mathbf{k}',\omega-\omega')\Bigg]\frac{1}{2}\left(\delta_{jm}-\frac{k'_j\,k'_m}{k'^2}\right)\frac{(\mathbf{k}-\mathbf{k}')_k\,(\mathbf{k}-\mathbf{k}')_n}{|\mathbf{k}-\mathbf{k}'|^2}$$

$$\times\sum_{\gamma}\sum_{\sigma',\sigma''} I^{\sigma' T}_{\mathbf{k}'}\,I^{\sigma''\gamma}_{\mathbf{k}-\mathbf{k}'}\,\delta(\omega'-\sigma'\omega^T_{\mathbf{k}'})\,\delta(\omega-\omega'-\sigma''\omega^\gamma_{\mathbf{k}-\mathbf{k}'})$$

$$-\left[\frac{k_i\,k_l}{k^2}\,\frac{\chi^{(2)}_{ijk}(\mathbf{k}',\omega'|\mathbf{k}-\mathbf{k}',\omega-\omega')\,\chi^{(2)*}_{lmn}(\mathbf{k}',\omega'|\mathbf{k}-\mathbf{k}',\omega-\omega')}{\epsilon^*_\parallel(\mathbf{k},\omega)}\right.$$

$$\left.+\left(\delta_{il}-\frac{k_i\,k_l}{k^2}\right)\frac{\chi^{(2)}_{ijk}(\mathbf{k}',\omega'|\mathbf{k}-\mathbf{k}',\omega-\omega')}{\Lambda^*_\perp(\mathbf{k},\omega)}\,\chi^{(2)*}_{lmn}(\mathbf{k}',\omega'|\mathbf{k}-\mathbf{k}',\omega-\omega')\right]$$

$$\times\frac{1}{4}\left(\delta_{jm}-\frac{k'_j\,k'_m}{k'^2}\right)\left(\delta_{kn}-\frac{(\mathbf{k}-\mathbf{k}')_k\,(\mathbf{k}-\mathbf{k}')_n}{|\mathbf{k}-\mathbf{k}'|^2}\right)$$

$$\left.\times\sum_{\sigma',\sigma''} I^{\sigma' T}_{\mathbf{k}'}\,I^{\sigma'' T}_{\mathbf{k}-\mathbf{k}'}\,\delta(\omega'-\sigma'\omega^T_{\mathbf{k}'})\,\delta(\omega-\omega'-\sigma''\omega^T_{\mathbf{k}-\mathbf{k}'})\right\}$$

$$+\,4\,\mathrm{Im}\int d\mathbf{k}'\int d\omega'\,\chi^{(3)}_{ijkl}(\mathbf{k}',\omega'|-\mathbf{k}',-\omega'|\mathbf{k},\omega)$$

$$\times\left[\frac{k_i\,k_l}{k^2}\frac{k'_j\,k'_k}{k'^2}\sum_{\alpha,\beta}\sum_{\sigma,\sigma'} I^{\sigma\alpha}_{\mathbf{k}}\,I^{\sigma'\beta}_{\mathbf{k}'}\,\delta(\omega-\sigma\omega^\alpha_{\mathbf{k}})\,\delta(\omega'-\sigma'\omega^\beta_{\mathbf{k}'})\right.$$

$$+\frac{1}{2}\frac{k_i\,k_l}{k^2}\left(\delta_{jk}-\frac{k'_j\,k'_k}{k'^2}\right)\sum_{\alpha}\sum_{\sigma,\sigma'} I^{\sigma' T}_{\mathbf{k}'}\,I^{\sigma\alpha}_{\mathbf{k}}\,\delta(\omega'-\sigma'\omega^T_{\mathbf{k}'})\,\delta(\omega-\sigma\omega^\alpha_{\mathbf{k}})$$

$$+\frac{1}{2}\left(\delta_{il}-\frac{k_i\,k_l}{k^2}\right)\frac{k'_j\,k'_k}{k'^2}\sum_{\beta}\sum_{\sigma,\sigma'} I^{\sigma'\beta}_{\mathbf{k}'}\,I^{\sigma T}_{\mathbf{k}}\,\delta(\omega'-\sigma'\omega^\beta_{\mathbf{k}'})\,\delta(\omega-\sigma\omega^T_{\mathbf{k}})$$

$$+\frac{1}{4}\left(\delta_{il}-\frac{k_i\,k_l}{k^2}\right)\left(\delta_{jk}-\frac{k'_j\,k'_k}{k'^2}\right)$$

$$\left.\times\sum_{\sigma,\sigma'} I^{\sigma' T}_{\mathbf{k}'}\,I^{\sigma T}_{\mathbf{k}}\,\delta(\omega'-\sigma'\omega^T_{\mathbf{k}'})\,\delta(\omega-\sigma\omega^T_{\mathbf{k}})\right]. \tag{7.48}$$

We make use of the following – see (1.97) and (1.99):

$$\begin{aligned}
\frac{1}{\epsilon_\parallel(\mathbf{k},\omega)} &= \mathscr{P}\,\frac{1}{\epsilon_\parallel(\mathbf{k},\omega)}-i\pi\sum_\alpha\sum_\sigma\frac{\delta(\omega-\sigma\omega^\alpha_{\mathbf{k}})}{\epsilon'_\parallel(\mathbf{k},\sigma\omega^\alpha_{\mathbf{k}})},\\
\frac{1}{\epsilon^*_\parallel(\mathbf{k},\omega)} &= \mathscr{P}\,\frac{1}{\epsilon_\parallel(\mathbf{k},\omega)}+i\pi\sum_\alpha\sum_\sigma\frac{\delta(\omega-\sigma\omega^\alpha_{\mathbf{k}})}{\epsilon'_\parallel(\mathbf{k},\sigma\omega^\alpha_{\mathbf{k}})},\\
\frac{1}{\Lambda_\perp(\mathbf{k},\omega)} &= \mathscr{P}\,\frac{1}{\Lambda_\perp(\mathbf{k},\omega)}-i\pi\sum_\sigma\frac{\delta(\omega-\sigma\omega^T_{\mathbf{k}})}{\Lambda'_\perp(\mathbf{k},\sigma\omega^T_{\mathbf{k}})},\\
\frac{1}{\Lambda^*_\perp(\mathbf{k},\omega)} &= \mathscr{P}\,\frac{1}{\Lambda_\perp(\mathbf{k},\omega)}+i\pi\sum_\sigma\frac{\delta(\omega-\sigma\omega^T_{\mathbf{k}})}{\Lambda'_\perp(\mathbf{k},\sigma\omega^T_{\mathbf{k}})}.
\end{aligned} \tag{7.49}$$

Then (7.48) can be further manipulated by separating the principal parts and the residues. In writing the final result, it turns out that the final form can be separated into terms that are associated with an overall delta function $\delta(\omega - \sigma\omega_{\mathbf{k}}^{\alpha})$, and terms that have an overall factor $\delta(\omega - \sigma\omega_{\mathbf{k}}^{T})$. (Actual demonstration is left as an exercise for the readers.) This shows that those terms with the factor $\delta(\omega - \sigma\omega_{\mathbf{k}}^{\alpha})$ constitute the formal wave kinetic equation for longitudinal modes, while those terms with the factor $\delta(\omega - \sigma\omega_{\mathbf{k}}^{T})$ collectively constitute the formal wave kinetic equation for transverse mode. Thus, we may separate the equations for longitudinal mode $\omega = \sigma\omega_{\mathbf{k}}^{\alpha}$ and transverse mode $\omega = \sigma\omega_{\mathbf{k}}^{T}$.

7.2.1 Formal Wave Kinetic Equation for Longitudinal Modes

Collecting all the terms associated with $\delta(\omega - \sigma\omega_{\mathbf{k}}^{\alpha})$, we have

$$
\begin{aligned}
0 &= \epsilon_{\parallel}'(\mathbf{k},\sigma\omega_{\mathbf{k}}^{\alpha})\frac{\partial I_{\mathbf{k}}^{\sigma\alpha}}{\partial t} + 2\,\mathrm{Im}\,\epsilon_{\parallel}(\mathbf{k},\sigma\omega_{\mathbf{k}}^{\alpha})\,I_{\mathbf{k}}^{\sigma\alpha} \\
&+ 4\,\mathrm{Im}\int d\mathbf{k}'\int d\omega'\left\{\left[\frac{k_j' k_l'}{k'^2}\,\chi_{ijk}^{(2)}(\mathbf{k}',\omega'|\mathbf{k}-\mathbf{k}',\sigma\omega_{\mathbf{k}}^{\alpha}-\omega')\right.\right. \\
&\times \chi_{nlm}^{(2)}(\mathbf{k}',\omega'|\mathbf{k}-\mathbf{k}',\sigma\omega_{\mathbf{k}}^{\alpha}-\omega')\left(\mathscr{P}\frac{1}{\epsilon_{\parallel}(\mathbf{k}',\omega')}\right. \\
&- i\pi\sum_{\beta}\sum_{\sigma'}\left.\frac{\delta(\omega'-\sigma'\omega_{\mathbf{k}'}^{\beta})}{\epsilon_{\parallel}'(\mathbf{k}',\sigma'\omega_{\mathbf{k}'}^{\beta})}\right) + \left(\delta_{jl} - \frac{k_j' k_l'}{k'^2}\right)\chi_{ijk}^{(2)}(\mathbf{k}',\omega'|\mathbf{k}-\mathbf{k}',\sigma\omega_{\mathbf{k}}^{\alpha}-\omega') \\
&\times \chi_{nlm}^{(2)}(\mathbf{k}',\omega'|\mathbf{k}-\mathbf{k}',\sigma\omega_{\mathbf{k}}^{\alpha}-\omega')\left.\left(\mathscr{P}\frac{1}{\Lambda_{\perp}(\mathbf{k}',\omega')} - i\pi\sum_{\sigma'}\frac{\delta(\omega'-\sigma'\omega_{\mathbf{k}'}^{T})}{\Lambda_{\perp}'(\mathbf{k}',\sigma'\omega_{\mathbf{k}'}^{T})}\right)\right] \\
&\times \frac{k_i\,k_n}{k^2}\frac{(\mathbf{k}-\mathbf{k}')_k\,(\mathbf{k}-\mathbf{k}')_m}{|\mathbf{k}-\mathbf{k}'|^2}\sum_{\gamma}\sum_{\sigma''} I_{\mathbf{k}-\mathbf{k}'}^{\sigma''\gamma}\,I_{\mathbf{k}}^{\sigma\alpha}\,\delta(\sigma\omega_{\mathbf{k}}^{\alpha}-\omega'-\sigma''\omega_{\mathbf{k}-\mathbf{k}'}^{\gamma}) \\
&+ \left[\frac{k_j' k_l'}{k'^2}\,\chi_{ijk}^{(2)}(\mathbf{k}',\omega'|\mathbf{k}-\mathbf{k}',\sigma\omega_{\mathbf{k}}^{\alpha}-\omega')\,\chi_{nlm}^{(2)}(\mathbf{k}',\omega'|\mathbf{k}-\mathbf{k}',\sigma\omega_{\mathbf{k}}^{\alpha}-\omega')\right. \\
&\times \left(\mathscr{P}\frac{1}{\epsilon_{\parallel}(\mathbf{k}',\omega')} - i\pi\sum_{\beta}\sum_{\sigma'}\frac{\delta(\omega'-\sigma'\omega_{\mathbf{k}'}^{\beta})}{\epsilon_{\parallel}'(\mathbf{k}',\sigma'\omega_{\mathbf{k}'}^{\beta})}\right) + \left(\delta_{jl} - \frac{k_j' k_l'}{k'^2}\right) \\
&\times \chi_{ijk}^{(2)}(\mathbf{k}',\omega'|\mathbf{k}-\mathbf{k}',\sigma\omega_{\mathbf{k}}^{\alpha}-\omega')\,\chi_{nlm}^{(2)}(\mathbf{k}',\omega'|\mathbf{k}-\mathbf{k}',\sigma\omega_{\mathbf{k}}^{\alpha}-\omega') \\
&\times \left.\left(\mathscr{P}\frac{1}{\Lambda_{\perp}(\mathbf{k}',\omega')} - i\pi\sum_{\sigma'}\frac{\delta(\omega'-\sigma'\omega_{\mathbf{k}'}^{T})}{\Lambda_{\perp}'(\mathbf{k}',\sigma'\omega_{\mathbf{k}'}^{T})}\right)\right]\frac{1}{2}\frac{k_i\,k_n}{k^2} \\
&\times \left(\delta_{km} - \frac{(\mathbf{k}-\mathbf{k}')_k\,(\mathbf{k}-\mathbf{k}')_m}{|\mathbf{k}-\mathbf{k}'|^2}\right)\sum_{\sigma''} I_{\mathbf{k}-\mathbf{k}'}^{\sigma''T}\,I_{\mathbf{k}}^{\sigma\alpha}\,\delta(\sigma\omega_{\mathbf{k}}^{\alpha}-\omega'-\sigma''\omega_{\mathbf{k}-\mathbf{k}'}^{T})
\end{aligned}
$$

$$
+\left[\frac{(\mathbf{k}-\mathbf{k}')_k\,(\mathbf{k}-\mathbf{k}')_l}{|\mathbf{k}-\mathbf{k}'|^2}\,\chi^{(2)}_{ijk}(\mathbf{k}',\omega'|\mathbf{k}-\mathbf{k}',\sigma\omega^\alpha_{\mathbf{k}}-\omega')\right.
$$

$$
\times\,\chi^{(2)}_{nml}(\mathbf{k}',\omega'|\mathbf{k}-\mathbf{k}',\sigma\omega^\alpha_{\mathbf{k}}-\omega')\left(\mathscr{P}\,\frac{1}{\epsilon_\parallel(\mathbf{k}-\mathbf{k}',\sigma\omega^\alpha_{\mathbf{k}}-\omega')}\right.
$$

$$
\left.-\,i\pi\sum_\gamma\sum_{\sigma''}\frac{\delta(\sigma\omega^\alpha_{\mathbf{k}}-\omega'-\sigma''\omega^\gamma_{\mathbf{k}-\mathbf{k}'})}{\epsilon'_\parallel(\mathbf{k}-\mathbf{k}',\sigma''\omega^\gamma_{\mathbf{k}-\mathbf{k}'})}\right)+\left(\delta_{kl}-\frac{(\mathbf{k}-\mathbf{k}')_k\,(\mathbf{k}-\mathbf{k}')_l}{|\mathbf{k}-\mathbf{k}'|^2}\right)
$$

$$
\times\,\chi^{(2)}_{ijk}(\mathbf{k}',\omega'|\mathbf{k}-\mathbf{k}',\sigma\omega^\alpha_{\mathbf{k}}-\omega')\,\chi^{(2)}_{nml}(\mathbf{k}',\omega'|\mathbf{k}-\mathbf{k}',\sigma\omega^\alpha_{\mathbf{k}}-\omega')
$$

$$
\left.\times\left(\mathscr{P}\,\frac{1}{\Lambda_\perp(\mathbf{k}-\mathbf{k}',\sigma\omega^\alpha_{\mathbf{k}}-\omega')}-i\pi\sum_{\sigma''}\frac{\delta(\sigma\omega^\alpha_{\mathbf{k}}-\omega'-\sigma''\omega^T_{\mathbf{k}-\mathbf{k}'})}{\Lambda'_\perp(\mathbf{k}-\mathbf{k}',\sigma''\omega^T_{\mathbf{k}-\mathbf{k}'})}\right)\right]
$$

$$
\times\,\frac{k_i\,k_n}{k^2}\frac{k'_j\,k'_m}{k'^2}\sum_\beta\sum_{\sigma'}I^{\sigma'\beta}_{\mathbf{k}'}\,I^{\sigma\alpha}_{\mathbf{k}}\,\delta(\omega'-\sigma'\omega^\beta_{\mathbf{k}'})+\left[\frac{(\mathbf{k}-\mathbf{k}')_k\,(\mathbf{k}-\mathbf{k}')_l}{|\mathbf{k}-\mathbf{k}'|^2}\right.
$$

$$
\times\,\chi^{(2)}_{ijk}(\mathbf{k}',\omega'|\mathbf{k}-\mathbf{k}',\sigma\omega^\alpha_{\mathbf{k}}-\omega')\,\chi^{(2)}_{nml}(\mathbf{k}',\omega'|\mathbf{k}-\mathbf{k}',\sigma\omega^\alpha_{\mathbf{k}}-\omega')
$$

$$
\times\left(\mathscr{P}\,\frac{1}{\epsilon_\parallel(\mathbf{k}-\mathbf{k}',\sigma\omega^\alpha_{\mathbf{k}}-\omega')}-i\pi\sum_\gamma\sum_{\sigma''}\frac{\delta(\sigma\omega^\alpha_{\mathbf{k}}-\omega'-\sigma''\omega^\gamma_{\mathbf{k}-\mathbf{k}'})}{\epsilon'_\parallel(\mathbf{k}-\mathbf{k}',\sigma''\omega^\gamma_{\mathbf{k}-\mathbf{k}'})}\right)
$$

$$
+\left(\delta_{kl}-\frac{(\mathbf{k}-\mathbf{k}')_k\,(\mathbf{k}-\mathbf{k}')_l}{|\mathbf{k}-\mathbf{k}'|^2}\right)\chi^{(2)}_{ijk}(\mathbf{k}',\omega'|\mathbf{k}-\mathbf{k}',\sigma\omega^\alpha_{\mathbf{k}}-\omega')
$$

$$
\times\,\chi^{(2)}_{nml}(\mathbf{k}',\omega'|\mathbf{k}-\mathbf{k}',\sigma\omega^\alpha_{\mathbf{k}}-\omega')\left(\mathscr{P}\,\frac{1}{\Lambda_\perp(\mathbf{k}-\mathbf{k}',\sigma\omega^\alpha_{\mathbf{k}}-\omega')}\right.
$$

$$
\left.\left.-\,i\pi\sum_{\sigma''}\frac{\delta(\sigma\omega^\alpha_{\mathbf{k}}-\omega'-\sigma''\omega^T_{\mathbf{k}-\mathbf{k}'})}{\Lambda'_\perp(\mathbf{k}-\mathbf{k}',\sigma''\omega^T_{\mathbf{k}-\mathbf{k}'})}\right)\right]\frac{1}{2}\frac{k_i\,k_n}{k^2}\left(\delta_{jm}-\frac{k'_j\,k'_m}{k'^2}\right)
$$

$$
\times\sum_{\sigma'}I^{\sigma'T}_{\mathbf{k}'}\,I^{\sigma\alpha}_{\mathbf{k}}\,\delta(\omega'-\sigma'\omega^T_{\mathbf{k}'})-i\pi\,\frac{k_i\,k_l}{k^2}\frac{\chi^{(2)}_{ijk}(\mathbf{k}',\omega'|\mathbf{k}-\mathbf{k}',\sigma\omega^\alpha_{\mathbf{k}}-\omega')}{\epsilon'_\parallel(\mathbf{k},\sigma\omega^\alpha_{\mathbf{k}})}
$$

$$
\times\,\chi^{(2)*}_{lmn}(\mathbf{k}',\omega'|\mathbf{k}-\mathbf{k}',\sigma\omega^\alpha_{\mathbf{k}}-\omega')\,\frac{k'_j\,k'_m}{k'^2}\frac{(\mathbf{k}-\mathbf{k}')_k\,(\mathbf{k}-\mathbf{k}')_n}{|\mathbf{k}-\mathbf{k}'|^2}
$$

$$
\times\sum_{\beta,\gamma}\sum_{\sigma',\sigma''}I^{\sigma''\gamma}_{\mathbf{k}-\mathbf{k}'}\,I^{\sigma'\beta}_{\mathbf{k}'}\,\delta(\sigma\omega^\alpha_{\mathbf{k}}-\omega'-\sigma''\omega^\gamma_{\mathbf{k}-\mathbf{k}'})\,\delta(\omega'-\sigma'\omega^\beta_{\mathbf{k}'})
$$

$$
-\,i\pi\,\frac{k_i\,k_l}{k^2}\frac{\chi^{(2)}_{ijk}(\mathbf{k}',\omega'|\mathbf{k}-\mathbf{k}',\sigma\omega^\alpha_{\mathbf{k}}-\omega')}{\epsilon'_\parallel(\mathbf{k},\sigma\omega^\alpha_{\mathbf{k}})}\,\chi^{(2)*}_{lmn}(\mathbf{k}',\omega'|\mathbf{k}-\mathbf{k}',\sigma\omega^\alpha_{\mathbf{k}}-\omega')
$$

$$
\times\,\frac{1}{2}\frac{k'_j\,k'_m}{k'^2}\left(\delta_{kn}-\frac{(\mathbf{k}-\mathbf{k}')_k\,(\mathbf{k}-\mathbf{k}')_n}{|\mathbf{k}-\mathbf{k}'|^2}\right)\sum_\beta\sum_{\sigma',\sigma''}I^{\sigma'\beta}_{\mathbf{k}'}\,I^{\sigma''T}_{\mathbf{k}-\mathbf{k}'}
$$

$$
\times\,\delta(\omega'-\sigma'\omega^\beta_{\mathbf{k}'})\,\delta(\sigma\omega^\alpha_{\mathbf{k}}-\omega'-\sigma''\omega^T_{\mathbf{k}-\mathbf{k}'})-i\pi\,\frac{k_i\,k_l}{k^2}
$$

$$\times \frac{\chi^{(2)}_{ijk}(\mathbf{k}',\omega'|\mathbf{k}-\mathbf{k}',\sigma\omega^\alpha_\mathbf{k}-\omega')}{\epsilon'_\parallel(\mathbf{k},\sigma\omega^\alpha_\mathbf{k})}\,\chi^{(2)*}_{lmn}(\mathbf{k}',\omega'|\mathbf{k}-\mathbf{k}',\sigma\omega^\alpha_\mathbf{k}-\omega')$$
$$\times \frac{1}{2}\left(\delta_{jm}-\frac{k'_j k'_m}{k'^2}\right)\frac{(\mathbf{k}-\mathbf{k}')_k(\mathbf{k}-\mathbf{k}')_n}{|\mathbf{k}-\mathbf{k}'|^2}\sum_\gamma\sum_{\sigma',\sigma''} I^{\sigma' T}_{\mathbf{k}'}\, I^{\sigma''\gamma}_{\mathbf{k}-\mathbf{k}'}$$
$$\times\,\delta(\omega'-\sigma'\omega^T_{\mathbf{k}'})\,\delta(\sigma\omega^\alpha_\mathbf{k}-\omega'-\sigma''\omega^\gamma_{\mathbf{k}-\mathbf{k}'})$$
$$-\,i\pi\,\frac{k_i k_l}{k^2}\frac{\chi^{(2)}_{ijk}(\mathbf{k}',\omega'|\mathbf{k}-\mathbf{k}',\sigma\omega^\alpha_\mathbf{k}-\omega')}{\epsilon'_\parallel(\mathbf{k},\sigma\omega^\alpha_\mathbf{k})}\,\chi^{(2)*}_{lmn}(\mathbf{k}',\omega'|\mathbf{k}-\mathbf{k}',\sigma\omega^\alpha_\mathbf{k}-\omega')$$
$$\times\frac{1}{4}\left(\delta_{jm}-\frac{k'_j k'_m}{k'^2}\right)\left(\delta_{kn}-\frac{(\mathbf{k}-\mathbf{k}')_k(\mathbf{k}-\mathbf{k}')_n}{|\mathbf{k}-\mathbf{k}'|^2}\right) \tag{7.50}$$
$$\times\sum_{\sigma',\sigma''} I^{\sigma' T}_{\mathbf{k}'}\, I^{\sigma'' T}_{\mathbf{k}-\mathbf{k}'}\,\delta(\omega'-\sigma'\omega^T_{\mathbf{k}'})\,\delta(\sigma\omega^\alpha_\mathbf{k}-\omega'-\sigma''\omega^T_{\mathbf{k}-\mathbf{k}'})\Bigg\}$$
$$+\,4\sum_{\sigma'}\int d\mathbf{k}'\left[\frac{k_i k_l}{k^2}\frac{k'_j k'_k}{k'^2}\,\mathrm{Im}\,\chi^{(3)}_{ijkl}(\mathbf{k}',\sigma'\omega^\beta_{\mathbf{k}'}|-\mathbf{k}',\,-\sigma'\omega^\beta_{\mathbf{k}'}|\mathbf{k},\sigma\omega^\alpha_\mathbf{k})\sum_\beta I^{\sigma'\beta}_{\mathbf{k}'}\right.$$
$$\left.+\,\frac{1}{2}\frac{k_i k_l}{k^2}\left(\delta_{jk}-\frac{k'_j k'_k}{k'^2}\right)\mathrm{Im}\,\chi^{(3)}_{ijkl}(\mathbf{k}',\sigma'\omega^T_{\mathbf{k}'}|-\mathbf{k}',\,-\sigma'\omega^T_{\mathbf{k}'}|\mathbf{k},\sigma\omega^\alpha_\mathbf{k})I^{\sigma' T}_{\mathbf{k}'}\right]I^{\sigma\alpha}_\mathbf{k}.$$

Proceeding from (7.50), one may make use of various delta functions in order to carry out the ω' integrations, thus leaving the results with only $\mathbf{k}'$ integrals. The result is that we obtain the following formal wave kinetic equation for longitudinal modes ($\alpha = L, S$). The result is only formal at this point since the various nonlinear coupling coefficients that are expressed in terms of the susceptibility tensors are yet to be evaluated. In what follows, we present the result together with brief descriptive texts encased in boxes:

$$0=\epsilon'_\parallel(\mathbf{k},\sigma\omega^\alpha_\mathbf{k})\frac{\partial I^{\sigma\alpha}_\mathbf{k}}{\partial t}+2\,\mathrm{Im}\,\epsilon_\parallel(\mathbf{k},\sigma\omega^\alpha_\mathbf{k})\,I^{\sigma\alpha}_\mathbf{k} \qquad \boxed{\text{Induced emission}}$$
$$+\,4\pi\,\mathrm{Im}\int d\mathbf{k}'\left\{\sum_{\beta,\gamma}\sum_{\sigma',\sigma''}\frac{k_i k_n}{k^2}\frac{k'_j k'_m}{k'^2}\frac{(\mathbf{k}-\mathbf{k}')_k(\mathbf{k}-\mathbf{k}')_l}{|\mathbf{k}-\mathbf{k}'|^2}\right.$$
$$\times\,\chi^{(2)}_{ijk}(\mathbf{k}',\sigma'\omega^\beta_{\mathbf{k}'}|\mathbf{k}-\mathbf{k}',\sigma''\omega^\gamma_{\mathbf{k}-\mathbf{k}'})\,\chi^{(2)*}_{nml}(\mathbf{k}',\sigma'\omega^\beta_{\mathbf{k}'}|\mathbf{k}-\mathbf{k}',\sigma''\omega^\gamma_{\mathbf{k}-\mathbf{k}'})$$
$$\times\left[\frac{I^{\sigma''\gamma}_{\mathbf{k}-\mathbf{k}'}\,I^{\sigma\alpha}_\mathbf{k}}{\epsilon'_\parallel(\mathbf{k}',\sigma'\omega^\beta_{\mathbf{k}'})}+\frac{I^{\sigma'\beta}_{\mathbf{k}'}\,I^{\sigma\alpha}_\mathbf{k}}{\epsilon'_\parallel(\mathbf{k}-\mathbf{k}',\sigma''\omega^\gamma_{\mathbf{k}-\mathbf{k}'})}\right.$$
$$\left.-\,\frac{I^{\sigma''\gamma}_{\mathbf{k}-\mathbf{k}'}\,I^{\sigma'\beta}_{\mathbf{k}'}}{\epsilon'_\parallel(\mathbf{k},\sigma\omega^\alpha_\mathbf{k})}\right]\delta(\sigma\omega^\alpha_\mathbf{k}-\sigma'\omega^\beta_{\mathbf{k}'}-\sigma''\omega^\gamma_{\mathbf{k}-\mathbf{k}'}) \qquad \boxed{\text{Decay }(\alpha,\beta,\gamma)}$$
$$+\sum_\beta\sum_{\sigma',\sigma''}\frac{k_i k_n}{k^2}\frac{k'_j k'_m}{k'^2}\left(\delta_{kl}-\frac{(\mathbf{k}-\mathbf{k}')_k(\mathbf{k}-\mathbf{k}')_l}{|\mathbf{k}-\mathbf{k}'|^2}\right)$$

$$\times\, \chi^{(2)}_{ijk}(\mathbf{k}',\sigma'\omega^{\beta}_{\mathbf{k}'}|\mathbf{k}-\mathbf{k}',\sigma''\omega^{T}_{\mathbf{k}-\mathbf{k}'})\,\chi^{(2)*}_{nml}(\mathbf{k}',\sigma'\omega^{\beta}_{\mathbf{k}'}|\mathbf{k}-\mathbf{k}',\sigma''\omega^{T}_{\mathbf{k}-\mathbf{k}'})$$

$$\times\left[\frac{I^{\sigma''T}_{\mathbf{k}-\mathbf{k}'}\,I^{\sigma\alpha}_{\mathbf{k}}}{\epsilon'_{\parallel}(\mathbf{k}',\sigma'\omega^{\beta}_{\mathbf{k}'})}+\frac{2\,I^{\sigma'\beta}_{\mathbf{k}'}\,I^{\sigma\alpha}_{\mathbf{k}}}{\Lambda'_{\perp}(\mathbf{k}-\mathbf{k}',\sigma''\omega^{T}_{\mathbf{k}-\mathbf{k}'})}\right.$$

$$\left.-\frac{I^{\sigma'\beta}_{\mathbf{k}'}\,I^{\sigma''T}_{\mathbf{k}-\mathbf{k}'}}{\epsilon'_{\parallel}(\mathbf{k},\sigma\omega^{\alpha}_{\mathbf{k}})}\right]\delta(\sigma\omega^{\alpha}_{\mathbf{k}}-\sigma'\omega^{\beta}_{\mathbf{k}'}-\sigma''\omega^{T}_{\mathbf{k}-\mathbf{k}'})\qquad\boxed{\text{Decay }(\alpha,\beta,T)}$$

$$+\sum_{\sigma',\sigma''}\frac{1}{4}\frac{k_i\,k_n}{k^2}\left(\delta_{jm}-\frac{k'_j\,k'_m}{k'^2}\right)\left(\delta_{kl}-\frac{(\mathbf{k}-\mathbf{k}')_k\,(\mathbf{k}-\mathbf{k}')_l}{|\mathbf{k}-\mathbf{k}'|^2}\right)$$

$$\times\, \chi^{(2)}_{ijk}(\mathbf{k}',\sigma'\omega^{T}_{\mathbf{k}'}|\mathbf{k}-\mathbf{k}',\sigma''\omega^{T}_{\mathbf{k}-\mathbf{k}'})\,\chi^{(2)*}_{nml}(\mathbf{k}',\sigma'\omega^{T}_{\mathbf{k}'}|\mathbf{k}-\mathbf{k}',\sigma''\omega^{T}_{\mathbf{k}-\mathbf{k}'})$$

$$\times\left[\frac{2\,I^{\sigma''T}_{\mathbf{k}-\mathbf{k}'}\,I^{\sigma\alpha}_{\mathbf{k}}}{\Lambda'_{\perp}(\mathbf{k}',\sigma'\omega^{T}_{\mathbf{k}'})}+\frac{2\,I^{\sigma'T}_{\mathbf{k}'}\,I^{\sigma\alpha}_{\mathbf{k}}}{\Lambda'_{\perp}(\mathbf{k}-\mathbf{k}',\sigma''\omega^{T}_{\mathbf{k}-\mathbf{k}'})}\right.$$

$$\left.\left.-\frac{I^{\sigma'T}_{\mathbf{k}'}\,I^{\sigma''T}_{\mathbf{k}-\mathbf{k}'}}{\epsilon'_{\parallel}(\mathbf{k},\sigma\omega^{\alpha}_{\mathbf{k}})}\right]\delta(\sigma\omega^{\alpha}_{\mathbf{k}}-\sigma'\omega^{T}_{\mathbf{k}'}-\sigma''\omega^{T}_{\mathbf{k}-\mathbf{k}'})\right\}\qquad\boxed{\text{Decay }(\alpha,T,T)}$$

$$+4\,\mathrm{Im}\int d\mathbf{k}'\left\{\sum_{\beta}\sum_{\sigma'}\frac{k_i\,k_n}{k^2}\frac{k'_j\,k'_m}{k'^2}\left[\frac{(\mathbf{k}-\mathbf{k}')_k\,(\mathbf{k}-\mathbf{k}')_l}{|\mathbf{k}-\mathbf{k}'|^2}\right.\right.$$

$$\times\, \chi^{(2)}_{ijk}(\mathbf{k}',\sigma'\omega^{\beta}_{\mathbf{k}'}|\mathbf{k}-\mathbf{k}',\sigma\omega^{\alpha}_{\mathbf{k}}-\sigma'\omega^{\beta}_{\mathbf{k}'})\,\chi^{(2)}_{nml}(\mathbf{k}',\sigma'\omega^{\beta}_{\mathbf{k}'}|\mathbf{k}-\mathbf{k}',\sigma\omega^{\alpha}_{\mathbf{k}}-\sigma'\omega^{\beta}_{\mathbf{k}'})$$

$$\times\,\mathscr{P}\,\frac{2}{\epsilon_{\parallel}(\mathbf{k}-\mathbf{k}',\sigma\omega^{\alpha}_{\mathbf{k}}-\sigma'\omega^{\beta}_{\mathbf{k}'})}+\left(\delta_{kl}-\frac{(\mathbf{k}-\mathbf{k}')_k\,(\mathbf{k}-\mathbf{k}')_l}{|\mathbf{k}-\mathbf{k}'|^2}\right)$$

$$\times\, \chi^{(2)}_{ijk}(\mathbf{k}',\sigma'\omega^{\beta}_{\mathbf{k}'}|\mathbf{k}-\mathbf{k}',\sigma\omega^{\alpha}_{\mathbf{k}}-\sigma'\omega^{\beta}_{\mathbf{k}'})\,\chi^{(2)}_{nml}(\mathbf{k}',\sigma'\omega^{\beta}_{\mathbf{k}'}|\mathbf{k}-\mathbf{k}',\sigma\omega^{\alpha}_{\mathbf{k}}-\sigma'\omega^{\beta}_{\mathbf{k}'})$$

$$\left.\times\,\mathscr{P}\,\frac{2}{\Lambda_{\perp}(\mathbf{k}-\mathbf{k}',\sigma\omega^{\alpha}_{\mathbf{k}}-\sigma'\omega^{\beta}_{\mathbf{k}'})}+\chi^{(3)}_{ijmn}(\mathbf{k}',\sigma'\omega^{\beta}_{\mathbf{k}'}|-\mathbf{k}',-\sigma'\omega^{\beta}_{\mathbf{k}'}|\mathbf{k},\sigma\omega^{\alpha}_{\mathbf{k}})\right]$$

$$\times\, I^{\sigma'\beta}_{\mathbf{k}'}\,I^{\sigma\alpha}_{\mathbf{k}}\qquad\boxed{\text{Induced scattering }\alpha,\beta}$$

$$+\frac{1}{2}\sum_{\sigma'}\frac{k_i\,k_n}{k^2}\left(\delta_{jm}-\frac{k'_j\,k'_m}{k'^2}\right)\left[\frac{(\mathbf{k}-\mathbf{k}')_k\,(\mathbf{k}-\mathbf{k}')_l}{|\mathbf{k}-\mathbf{k}'|^2}\right.$$

$$\times\, \chi^{(2)}_{ijk}(\mathbf{k}',\sigma'\omega^{T}_{\mathbf{k}'}|\mathbf{k}-\mathbf{k}',\sigma\omega^{\alpha}_{\mathbf{k}}-\sigma'\omega^{T}_{\mathbf{k}'})\,\chi^{(2)}_{nml}(\mathbf{k}',\sigma'\omega^{T}_{\mathbf{k}'}|\mathbf{k}-\mathbf{k}',\sigma\omega^{\alpha}_{\mathbf{k}}-\sigma'\omega^{T}_{\mathbf{k}'})$$

$$\times\,\mathscr{P}\,\frac{2}{\epsilon_{\parallel}(\mathbf{k}-\mathbf{k}',\sigma\omega^{\alpha}_{\mathbf{k}}-\sigma'\omega^{T}_{\mathbf{k}'})}+\left(\delta_{kl}-\frac{(\mathbf{k}-\mathbf{k}')_k\,(\mathbf{k}-\mathbf{k}')_l}{|\mathbf{k}-\mathbf{k}'|^2}\right)$$

$$\times\, \chi^{(2)}_{ijk}(\mathbf{k}',\sigma'\omega^{T}_{\mathbf{k}'}|\mathbf{k}-\mathbf{k}',\sigma\omega^{\alpha}_{\mathbf{k}}-\sigma'\omega^{T}_{\mathbf{k}'})\,\chi^{(2)}_{nml}(\mathbf{k}',\sigma'\omega^{T}_{\mathbf{k}'}|\mathbf{k}-\mathbf{k}',\sigma\omega^{\alpha}_{\mathbf{k}}-\sigma'\omega^{T}_{\mathbf{k}'})$$

$$\left.\times\,\mathscr{P}\,\frac{2}{\Lambda_{\perp}(\mathbf{k}-\mathbf{k}',\sigma\omega^{\alpha}_{\mathbf{k}}-\sigma'\omega^{T}_{\mathbf{k}'})}+\chi^{(3)}_{ijmn}(\mathbf{k}',\sigma'\omega^{T}_{\mathbf{k}'}|-\mathbf{k}',-\sigma'\omega^{T}_{\mathbf{k}'}|\mathbf{k},\sigma\omega^{\alpha}_{\mathbf{k}})\right]$$

$$\left.\times\, I^{\sigma'T}_{\mathbf{k}'}\,I^{\sigma\alpha}_{\mathbf{k}}\right\}.\qquad\boxed{\text{Induced scattering }\alpha,T}.\tag{7.51}$$

When compared with the electrostatic Vlasov weak turbulence formalism, the present electromagnetic generalization shows that the time evolution of longitudinal modes is modified by the presence of additional nonlinear processes. These are the decay processes involving two longitudinal modes and a transverse mode, as well as the three-wave decay involving two transverse modes and a single longitudinal mode. They also include an additional induced scattering process involving a longitudinal mode and a transverse mode.

In (7.51), we have made use of the fact that the leading term associated with the second-order susceptibility in the context of decay processes is the imaginary part,

$$\chi_{ijk}^{(2)}(\mathbf{k}_1,\omega_1|\mathbf{k}_2,\omega_2) \approx i\,\mathrm{Im}\,\chi_{ijk}^{(2)}(\mathbf{k}_1,\omega_1|\mathbf{k}_2,\omega_2) = -\chi_{ijk}^{(2)*}(\mathbf{k}_1,\omega_1|\mathbf{k}_2,\omega_2). \quad (7.52)$$

7.2.2 Formal Wave Kinetic Equation for Transverse Mode

Collecting all the terms associated with $\delta(\omega-\sigma\omega_{\mathbf{k}}^T)$ in (7.48), one may show that the resulting equation is

$$\begin{aligned}
0 &= \Lambda_\perp'(\mathbf{k},\sigma\omega_{\mathbf{k}}^T)\,\frac{\partial I_{\mathbf{k}}^{\sigma T}}{\partial t} + 2\,\mathrm{Im}\,\Lambda_\perp(\mathbf{k},\sigma\omega_{\mathbf{k}}^T)\,I_{\mathbf{k}}^{\sigma T} \\
&+ 4\,\mathrm{Im}\int d\mathbf{k}'\int d\omega'\left\{\left[\frac{k_j'\,k_l'}{k'^2}\,\chi_{ijk}^{(2)}(\mathbf{k}',\omega'|\mathbf{k}-\mathbf{k}',\sigma\omega_{\mathbf{k}}^T-\omega')\right.\right. \\
&\times \chi_{nlm}^{(2)}(\mathbf{k}',\omega'|\mathbf{k}-\mathbf{k}',\sigma\omega_{\mathbf{k}}^T-\omega')\left(\mathscr{P}\frac{1}{\epsilon_\parallel(\mathbf{k}',\omega')} - i\pi\sum_\beta\sum_{\sigma'}\frac{\delta(\omega'-\sigma'\omega_{\mathbf{k}'}^\beta)}{\epsilon_\parallel'(\mathbf{k}',\sigma'\omega_{\mathbf{k}'}^\beta)}\right) \\
&+ \left(\delta_{jl}-\frac{k_j'\,k_l'}{k'^2}\right)\chi_{ijk}^{(2)}(\mathbf{k}',\omega'|\mathbf{k}-\mathbf{k}',\sigma\omega_{\mathbf{k}}^T-\omega')\,\chi_{nlm}^{(2)}(\mathbf{k}',\omega'|\mathbf{k}-\mathbf{k}',\sigma\omega_{\mathbf{k}}^T-\omega') \\
&\times \left.\left(\mathscr{P}\frac{1}{\Lambda_\perp(\mathbf{k}',\omega')} - i\pi\sum_{\sigma'}\frac{\delta(\omega'-\sigma'\omega_{\mathbf{k}'}^T)}{\Lambda_\perp'(\mathbf{k}',\sigma'\omega_{\mathbf{k}'}^T)}\right)\right]\frac{1}{2}\left(\delta_{in}-\frac{k_i\,k_n}{k^2}\right) \\
&\times \frac{(\mathbf{k}-\mathbf{k}')_k\,(\mathbf{k}-\mathbf{k}')_m}{|\mathbf{k}-\mathbf{k}'|^2}\sum_\gamma\sum_{\sigma''} I_{\mathbf{k}-\mathbf{k}'}^{\sigma''\gamma}\,I_{\mathbf{k}}^{\sigma T}\,\delta(\sigma\omega_{\mathbf{k}}^T-\omega'-\sigma''\omega_{\mathbf{k}-\mathbf{k}'}^\gamma) \\
&+ \left[\frac{k_j'\,k_l'}{k'^2}\,\chi_{ijk}^{(2)}(\mathbf{k}',\omega'|\mathbf{k}-\mathbf{k}',\sigma\omega_{\mathbf{k}}^T-\omega')\,\chi_{nlm}^{(2)}(\mathbf{k}',\omega'|\mathbf{k}-\mathbf{k}',\sigma\omega_{\mathbf{k}}^T-\omega')\right. \\
&\times \left(\mathscr{P}\frac{1}{\epsilon_\parallel(\mathbf{k}',\omega')} - i\pi\sum_\beta\sum_{\sigma'}\frac{\delta(\omega'-\sigma'\omega_{\mathbf{k}'}^\beta)}{\epsilon_\parallel'(\mathbf{k}',\sigma'\omega_{\mathbf{k}'}^\beta)}\right) + \left(\delta_{jl}-\frac{k_j'\,k_l'}{k'^2}\right) \\
&\times \chi_{ijk}^{(2)}(\mathbf{k}',\omega'|\mathbf{k}-\mathbf{k}',\sigma\omega_{\mathbf{k}}^T-\omega')\,\chi_{nlm}^{(2)}(\mathbf{k}',\omega'|\mathbf{k}-\mathbf{k}',\sigma\omega_{\mathbf{k}}^T-\omega') \\
&\times \left.\left(\mathscr{P}\frac{1}{\Lambda_\perp(\mathbf{k}',\omega')} - i\pi\sum_{\sigma'}\frac{\delta(\omega'-\sigma'\omega_{\mathbf{k}'}^T)}{\Lambda_\perp'(\mathbf{k}',\sigma'\omega_{\mathbf{k}'}^T)}\right)\right]
\end{aligned}$$

$$
\begin{aligned}
&\times \frac{1}{4}\left(\delta_{in} - \frac{k_i\, k_n}{k^2}\right)\left(\delta_{km} - \frac{(\mathbf{k}-\mathbf{k}')_k\, (\mathbf{k}-\mathbf{k}')_m}{|\mathbf{k}-\mathbf{k}'|^2}\right) \\
&\times \sum_{\sigma''} I^{\sigma'' T}_{\mathbf{k}-\mathbf{k}'}\, I^{\sigma T}_{\mathbf{k}}\, \delta(\sigma\omega^T_{\mathbf{k}} - \omega' - \sigma''\omega^T_{\mathbf{k}-\mathbf{k}'}) + \left[\frac{(\mathbf{k}-\mathbf{k}')_k\, (\mathbf{k}-\mathbf{k}')_l}{|\mathbf{k}-\mathbf{k}'|^2}\right. \\
&\times \chi^{(2)}_{ijk}(\mathbf{k}',\omega'|\mathbf{k}-\mathbf{k}',\sigma\omega^T_{\mathbf{k}} - \omega')\, \chi^{(2)}_{nml}(\mathbf{k}',\omega'|\mathbf{k}-\mathbf{k}',\sigma\omega^T_{\mathbf{k}} - \omega') \\
&\times \left(\mathscr{P}\,\frac{1}{\epsilon_{\parallel}(\mathbf{k}-\mathbf{k}',\sigma\omega^T_{\mathbf{k}} - \omega')} - i\pi \sum_{\gamma}\sum_{\sigma''} \frac{\delta(\sigma\omega^T_{\mathbf{k}} - \omega' - \sigma''\omega^{\gamma}_{\mathbf{k}-\mathbf{k}'})}{\epsilon'_{\parallel}(\mathbf{k}-\mathbf{k}',\sigma''\omega^{\gamma}_{\mathbf{k}-\mathbf{k}'})}\right) \\
&+ \left(\delta_{kl} - \frac{(\mathbf{k}-\mathbf{k}')_k\, (\mathbf{k}-\mathbf{k}')_l}{|\mathbf{k}-\mathbf{k}'|^2}\right) \chi^{(2)}_{ijk}(\mathbf{k}',\omega'|\mathbf{k}-\mathbf{k}',\sigma\omega^T_{\mathbf{k}} - \omega') \\
&\times \chi^{(2)}_{nml}(\mathbf{k}',\omega'|\mathbf{k}-\mathbf{k}',\sigma\omega^T_{\mathbf{k}} - \omega')\left(\mathscr{P}\,\frac{1}{\Lambda_{\perp}(\mathbf{k}-\mathbf{k}',\sigma\omega^T_{\mathbf{k}} - \omega')}\right. \\
&\left.\left. - i\pi \sum_{\sigma''} \frac{\delta(\sigma\omega^T_{\mathbf{k}} - \omega' - \sigma''\omega^T_{\mathbf{k}-\mathbf{k}'})}{\Lambda'_{\perp}(\mathbf{k}-\mathbf{k}',\sigma''\omega^T_{\mathbf{k}-\mathbf{k}'})}\right)\right]\frac{1}{2}\left(\delta_{in} - \frac{k_i\, k_n}{k^2}\right) \\
&\times \frac{k'_j\, k'_m}{k'^2} \sum_{\beta}\sum_{\sigma'} I^{\sigma'\beta}_{\mathbf{k}'}\, I^{\sigma T}_{\mathbf{k}}\, \delta(\omega' - \sigma'\omega^{\beta}_{\mathbf{k}'}) + \left[\frac{(\mathbf{k}-\mathbf{k}')_k\, (\mathbf{k}-\mathbf{k}')_l}{|\mathbf{k}-\mathbf{k}'|^2}\right. \\
&\times \chi^{(2)}_{ijk}(\mathbf{k}',\omega'|\mathbf{k}-\mathbf{k}',\sigma\omega^T_{\mathbf{k}} - \omega')\, \chi^{(2)}_{nml}(\mathbf{k}',\omega'|\mathbf{k}-\mathbf{k}',\sigma\omega^T_{\mathbf{k}} - \omega') \\
&\times \left(\mathscr{P}\,\frac{1}{\epsilon_{\parallel}(\mathbf{k}-\mathbf{k}',\sigma\omega^T_{\mathbf{k}} - \omega')} - i\pi \sum_{\gamma}\sum_{\sigma''} \frac{\delta(\sigma\omega^T_{\mathbf{k}} - \omega' - \sigma''\omega^{\gamma}_{\mathbf{k}-\mathbf{k}'})}{\epsilon'_{\parallel}(\mathbf{k}-\mathbf{k}',\sigma''\omega^{\gamma}_{\mathbf{k}-\mathbf{k}'})}\right) \\
&+ \left(\delta_{kl} - \frac{(\mathbf{k}-\mathbf{k}')_k\, (\mathbf{k}-\mathbf{k}')_l}{|\mathbf{k}-\mathbf{k}'|^2}\right) \chi^{(2)}_{ijk}(\mathbf{k}',\omega'|\mathbf{k}-\mathbf{k}',\sigma\omega^T_{\mathbf{k}} - \omega') \\
&\times \chi^{(2)}_{nml}(\mathbf{k}',\omega'|\mathbf{k}-\mathbf{k}',\sigma\omega^T_{\mathbf{k}} - \omega')\left(\mathscr{P}\,\frac{1}{\Lambda_{\perp}(\mathbf{k}-\mathbf{k}',\sigma\omega^T_{\mathbf{k}} - \omega')}\right. \\
&\left.\left. - i\pi \sum_{\sigma''} \frac{\delta(\sigma\omega^T_{\mathbf{k}} - \omega' - \sigma''\omega^T_{\mathbf{k}-\mathbf{k}'})}{\Lambda'_{\perp}(\mathbf{k}-\mathbf{k}',\sigma''\omega^T_{\mathbf{k}-\mathbf{k}'})}\right)\right]\frac{1}{4}\left(\delta_{in} - \frac{k_i\, k_n}{k^2}\right) \\
&\times \left(\delta_{jm} - \frac{k'_j\, k'_m}{k'^2}\right) \sum_{\sigma'} I^{\sigma' T}_{\mathbf{k}'}\, I^{\sigma T}_{\mathbf{k}}\, \delta(\omega' - \sigma'\omega^T_{\mathbf{k}'}) \\
&- i\pi \left(\delta_{il} - \frac{k_i\, k_l}{k^2}\right) \frac{\chi^{(2)}_{ijk}(\mathbf{k}',\omega'|\mathbf{k}-\mathbf{k}',\sigma\omega^T_{\mathbf{k}} - \omega')}{\Lambda'_{\perp}(\mathbf{k},\sigma\omega^T_{\mathbf{k}})}\, \chi^{(2)*}_{lmn}(\mathbf{k}',\omega'|\mathbf{k}-\mathbf{k}',\sigma\omega^T_{\mathbf{k}} - \omega') \\
&\times \frac{k'_j\, k'_m}{k'^2}\, \frac{(\mathbf{k}-\mathbf{k}')_k\, (\mathbf{k}-\mathbf{k}')_n}{|\mathbf{k}-\mathbf{k}'|^2} \sum_{\beta,\gamma}\sum_{\sigma',\sigma''} I^{\sigma''\gamma}_{\mathbf{k}-\mathbf{k}'}\, I^{\sigma'\beta}_{\mathbf{k}'} \\
&\times \delta(\sigma\omega^T_{\mathbf{k}} - \omega' - \sigma''\omega^{\gamma}_{\mathbf{k}-\mathbf{k}'})\, \delta(\omega' - \sigma'\omega^{\beta}_{\mathbf{k}'}) \\
&- i\pi \left(\delta_{il} - \frac{k_i\, k_l}{k^2}\right) \frac{\chi^{(2)}_{ijk}(\mathbf{k}',\omega'|\mathbf{k}-\mathbf{k}',\sigma\omega^T_{\mathbf{k}} - \omega')}{\Lambda'_{\perp}(\mathbf{k},\sigma\omega^T_{\mathbf{k}})}\, \chi^{(2)*}_{lmn}(\mathbf{k}',\omega'|\mathbf{k}-\mathbf{k}',\sigma\omega^T_{\mathbf{k}} - \omega')
\end{aligned}
$$

$$
\begin{aligned}
&\times \frac{1}{2}\frac{k'_j\,k'_m}{k'^2}\left(\delta_{kn} - \frac{(\mathbf{k}-\mathbf{k}')_k\,(\mathbf{k}-\mathbf{k}')_n}{|\mathbf{k}-\mathbf{k}'|^2}\right) \\
&\times \sum_{\beta}\sum_{\sigma',\sigma''} I^{\sigma'\beta}_{\mathbf{k}'}\,I^{\sigma''T}_{\mathbf{k}-\mathbf{k}'}\,\delta(\omega'-\sigma'\omega^{\beta}_{\mathbf{k}'})\,\delta(\sigma\omega^{T}_{\mathbf{k}}-\omega'-\sigma''\omega^{T}_{\mathbf{k}-\mathbf{k}'}) \\
&- i\pi\left(\delta_{il} - \frac{k_i\,k_l}{k^2}\right)\frac{\chi^{(2)}_{ijk}(\mathbf{k}',\omega'|\mathbf{k}-\mathbf{k}',\sigma\omega^{T}_{\mathbf{k}}-\omega')}{\Lambda'_{\perp}(\mathbf{k},\sigma\omega^{T}_{\mathbf{k}})}\,\chi^{(2)*}_{lmn}(\mathbf{k}',\omega'|\mathbf{k}-\mathbf{k}',\sigma\omega^{T}_{\mathbf{k}}-\omega') \\
&\times \frac{1}{2}\left(\delta_{jm} - \frac{k'_j\,k'_m}{k'^2}\right)\frac{(\mathbf{k}-\mathbf{k}')_k\,(\mathbf{k}-\mathbf{k}')_n}{|\mathbf{k}-\mathbf{k}'|^2}\sum_{\gamma}\sum_{\sigma',\sigma''} I^{\sigma'T}_{\mathbf{k}'}\,I^{\sigma''\gamma}_{\mathbf{k}-\mathbf{k}'} \\
&\times \delta(\omega'-\sigma'\omega^{T}_{\mathbf{k}'})\,\delta(\sigma\omega^{T}_{\mathbf{k}}-\omega'-\sigma''\omega^{\gamma}_{\mathbf{k}-\mathbf{k}'}) \\
&- i\pi\left(\delta_{il} - \frac{k_i\,k_l}{k^2}\right)\frac{\chi^{(2)}_{ijk}(\mathbf{k}',\omega'|\mathbf{k}-\mathbf{k}',\sigma\omega^{T}_{\mathbf{k}}-\omega')}{\Lambda'_{\perp}(\mathbf{k},\sigma\omega^{T}_{\mathbf{k}})}\,\chi^{(2)*}_{lmn}(\mathbf{k}',\omega'|\mathbf{k}-\mathbf{k}',\sigma\omega^{T}_{\mathbf{k}}-\omega') \\
&\times \frac{1}{4}\left(\delta_{jm} - \frac{k'_j\,k'_m}{k'^2}\right)\left(\delta_{kn} - \frac{(\mathbf{k}-\mathbf{k}')_k\,(\mathbf{k}-\mathbf{k}')_n}{|\mathbf{k}-\mathbf{k}'|^2}\right) \\
&\times \sum_{\sigma',\sigma''} I^{\sigma'T}_{\mathbf{k}'}\,I^{\sigma''T}_{\mathbf{k}-\mathbf{k}'}\,\delta(\omega'-\sigma'\omega^{T}_{\mathbf{k}'})\,\delta(\sigma\omega^{T}_{\mathbf{k}}-\omega'-\sigma''\omega^{T}_{\mathbf{k}-\mathbf{k}'})\Bigg\} \\
&+ 4\sum_{\sigma'}\int d\mathbf{k}'\left[\frac{1}{2}\left(\delta_{il} - \frac{k_i\,k_l}{k^2}\right)\frac{k'_j\,k'_k}{k'^2}\,\mathrm{Im}\,\chi^{(3)}_{ijkl}(\mathbf{k}',\sigma'\omega^{\beta}_{\mathbf{k}'}|-\mathbf{k}',-\sigma'\omega^{\beta}_{\mathbf{k}'}|\mathbf{k},\sigma\omega^{T}_{\mathbf{k}})\right. \\
&\times \sum_{\beta}\sum_{\sigma'} I^{\sigma'\beta}_{\mathbf{k}'}\,I^{\sigma T}_{\mathbf{k}} + \frac{1}{4}\left(\delta_{il} - \frac{k_i\,k_l}{k^2}\right)\left(\delta_{jk} - \frac{k'_j\,k'_k}{k'^2}\right) \\
&\left.\times\, \mathrm{Im}\,\chi^{(3)}_{ijkl}(\mathbf{k}',\sigma'\omega^{T}_{\mathbf{k}'}|-\mathbf{k}',-\sigma'\omega^{T}_{\mathbf{k}'}|\mathbf{k},\sigma\omega^{T}_{\mathbf{k}})\,I^{\sigma'T}_{\mathbf{k}'}\,I^{\sigma T}_{\mathbf{k}}\right].
\end{aligned}
\tag{7.53}
$$

Proceeding from (7.53) by focusing on eigenmodes, and carrying out the ω' integrations by virtue of the delta functions, we obtain the formal wave kinetic equation for transverse wave, which is to be considered in conjunction with (7.51). Brief descriptive texts, which appear in boxes, are also added to each term:

$$
\begin{aligned}
0 = {}& \Lambda'_{\perp}(\mathbf{k},\sigma\omega^{T}_{\mathbf{k}})\,\frac{\partial I^{\sigma T}_{\mathbf{k}}}{\partial t} + 2\,\mathrm{Im}\,\Lambda_{\perp}(\mathbf{k},\sigma\omega^{T}_{\mathbf{k}})\,I^{\sigma T}_{\mathbf{k}} \qquad \boxed{\text{Induced emission}} \\
&+ 4\pi\,\mathrm{Im}\int d\mathbf{k}'\left\{\sum_{\beta,\gamma}\sum_{\sigma',\sigma''}\frac{1}{2}\left(\delta_{in} - \frac{k_i\,k_n}{k^2}\right)\frac{k'_j\,k'_m}{k'^2}\frac{(\mathbf{k}-\mathbf{k}')_k\,(\mathbf{k}-\mathbf{k}')_l}{|\mathbf{k}-\mathbf{k}'|^2}\right. \\
&\times \chi^{(2)}_{ijk}(\mathbf{k}',\sigma'\omega^{\beta}_{\mathbf{k}'}|\mathbf{k}-\mathbf{k}',\sigma''\omega^{\gamma}_{\mathbf{k}-\mathbf{k}'})\,\chi^{(2)*}_{nml}(\mathbf{k}',\sigma'\omega^{\beta}_{\mathbf{k}'}|\mathbf{k}-\mathbf{k}',\sigma''\omega^{\gamma}_{\mathbf{k}-\mathbf{k}'}) \\
&\times \left[\frac{I^{\sigma''\gamma}_{\mathbf{k}-\mathbf{k}'}\,I^{\sigma T}_{\mathbf{k}}}{\epsilon'_{\parallel}(\mathbf{k}',\sigma'\omega^{\beta}_{\mathbf{k}'})} + \frac{I^{\sigma'\beta}_{\mathbf{k}'}\,I^{\sigma T}_{\mathbf{k}}}{\epsilon'_{\parallel}(\mathbf{k}-\mathbf{k}',\sigma''\omega^{\gamma}_{\mathbf{k}-\mathbf{k}'})}\right.
\end{aligned}
$$

$$
\begin{aligned}
&- \frac{2\, I^{\sigma''\gamma}_{\mathbf{k}-\mathbf{k}'}\, I^{\sigma'\beta}_{\mathbf{k}'}}{\Lambda'_\perp(\mathbf{k},\sigma\omega^T_{\mathbf{k}})}\Bigg]\, \delta(\sigma\omega^T_{\mathbf{k}} - \sigma'\omega^\beta_{\mathbf{k}'} - \sigma''\omega^\gamma_{\mathbf{k}-\mathbf{k}'}) \qquad \boxed{\text{Decay } (T,\beta,\gamma)} \\
&+ \sum_\gamma \sum_{\sigma',\sigma''} \frac{1}{2}\left(\delta_{in} - \frac{k_i\, k_n}{k^2}\right)\left(\delta_{jm} - \frac{k'_j\, k'_m}{k'^2}\right) \frac{(\mathbf{k}-\mathbf{k}')_k\, (\mathbf{k}-\mathbf{k}')_l}{|\mathbf{k}-\mathbf{k}'|^2} \\
&\times \chi^{(2)}_{ijk}(\mathbf{k}',\sigma'\omega^T_{\mathbf{k}'}|\mathbf{k}-\mathbf{k}',\sigma''\omega^\gamma_{\mathbf{k}-\mathbf{k}'})\, \chi^{(2)*}_{nml}(\mathbf{k}',\sigma'\omega^T_{\mathbf{k}'}|\mathbf{k}-\mathbf{k}',\sigma''\omega^\gamma_{\mathbf{k}-\mathbf{k}'}) \\
&\times \Bigg[\frac{2\, I^{\sigma''\gamma}_{\mathbf{k}-\mathbf{k}'}\, I^{\sigma T}_{\mathbf{k}}}{\Lambda'_\perp(\mathbf{k}',\sigma'\omega^T_{\mathbf{k}'})} + \frac{I^{\sigma' T}_{\mathbf{k}'}\, I^{\sigma T}_{\mathbf{k}}}{\epsilon'_\parallel(\mathbf{k}-\mathbf{k}',\sigma''\omega^\gamma_{\mathbf{k}-\mathbf{k}'})} \\
&- \frac{2\, I^{\sigma''\gamma}_{\mathbf{k}-\mathbf{k}'}\, I^{\sigma' T}_{\mathbf{k}'}}{\Lambda'_\perp(\mathbf{k},\sigma\omega^T_{\mathbf{k}})}\Bigg]\, \delta(\sigma\omega^T_{\mathbf{k}} - \sigma'\omega^T_{\mathbf{k}'} - \sigma''\omega^\gamma_{\mathbf{k}-\mathbf{k}'})\Bigg\} \qquad \boxed{\text{Decay } (T,T,\gamma)} \\
&+ 4\,\mathrm{Im} \int d\mathbf{k}' \left\{ \sum_\beta \sum_{\sigma'} \frac{1}{2}\left(\delta_{in} - \frac{k_i\, k_n}{k^2}\right) \frac{k'_j\, k'_m}{k'^2} \Bigg[\frac{(\mathbf{k}-\mathbf{k}')_k\, (\mathbf{k}-\mathbf{k}')_l}{|\mathbf{k}-\mathbf{k}'|^2} \right. \\
&\times \chi^{(2)}_{ijk}(\mathbf{k}',\sigma'\omega^\beta_{\mathbf{k}'}|\mathbf{k}-\mathbf{k}',\sigma\omega^T_{\mathbf{k}} - \sigma'\omega^\beta_{\mathbf{k}'})\, \chi^{(2)}_{nml}(\mathbf{k}',\sigma'\omega^\beta_{\mathbf{k}'}|\mathbf{k}-\mathbf{k}',\sigma\omega^T_{\mathbf{k}} - \sigma'\omega^\beta_{\mathbf{k}'}) \\
&\times \mathscr{P}\, \frac{2}{\epsilon_\parallel(\mathbf{k}-\mathbf{k}',\sigma\omega^T_{\mathbf{k}} - \sigma'\omega^\beta_{\mathbf{k}'})} + \left(\delta_{kl} - \frac{(\mathbf{k}-\mathbf{k}')_k\, (\mathbf{k}-\mathbf{k}')_l}{|\mathbf{k}-\mathbf{k}'|^2}\right) \\
&\times \chi^{(2)}_{ijk}(\mathbf{k}',\sigma'\omega^\beta_{\mathbf{k}'}|\mathbf{k}-\mathbf{k}',\sigma\omega^T_{\mathbf{k}} - \sigma'\omega^\beta_{\mathbf{k}'})\, \chi^{(2)}_{nml}(\mathbf{k}',\sigma'\omega^\beta_{\mathbf{k}'}|\mathbf{k}-\mathbf{k}',\sigma\omega^T_{\mathbf{k}} - \sigma'\omega^\beta_{\mathbf{k}'}) \\
&\times \mathscr{P}\, \frac{2}{\Lambda_\perp(\mathbf{k}-\mathbf{k}',\sigma\omega^T_{\mathbf{k}} - \sigma'\omega^\beta_{\mathbf{k}'})} + \chi^{(3)}_{ijmn}(\mathbf{k}',\sigma'\omega^\beta_{\mathbf{k}'}|-\mathbf{k}',-\sigma'\omega^\beta_{\mathbf{k}'}|\mathbf{k},\sigma\omega^T_{\mathbf{k}})\Bigg] \\
&\times I^{\sigma'\beta}_{\mathbf{k}'}\, I^{\sigma T}_{\mathbf{k}} \qquad \boxed{\text{Induced scattering } \beta, T} \\
&+ \sum_{\sigma'} \frac{1}{4}\left(\delta_{in} - \frac{k_i\, k_n}{k^2}\right)\left(\delta_{jm} - \frac{k'_j\, k'_m}{k'^2}\right) \Bigg[\frac{(\mathbf{k}-\mathbf{k}')_k\, (\mathbf{k}-\mathbf{k}')_l}{|\mathbf{k}-\mathbf{k}'|^2} \\
&\times \chi^{(2)}_{ijk}(\mathbf{k}',\sigma'\omega^T_{\mathbf{k}'}|\mathbf{k}-\mathbf{k}',\sigma\omega^T_{\mathbf{k}} - \sigma'\omega^T_{\mathbf{k}'})\, \chi^{(2)}_{nml}(\mathbf{k}',\sigma'\omega^T_{\mathbf{k}'}|\mathbf{k}-\mathbf{k}',\sigma\omega^T_{\mathbf{k}} - \sigma'\omega^T_{\mathbf{k}'}) \\
&\times \mathscr{P}\, \frac{2}{\epsilon_\parallel(\mathbf{k}-\mathbf{k}',\sigma\omega^T_{\mathbf{k}} - \sigma'\omega^T_{\mathbf{k}'})} + \left(\delta_{kl} - \frac{(\mathbf{k}-\mathbf{k}')_k\, (\mathbf{k}-\mathbf{k}')_l}{|\mathbf{k}-\mathbf{k}'|^2}\right) \\
&\times \chi^{(2)}_{ijk}(\mathbf{k}',\sigma'\omega^T_{\mathbf{k}'}|\mathbf{k}-\mathbf{k}',\sigma\omega^T_{\mathbf{k}} - \sigma'\omega^T_{\mathbf{k}'})\, \chi^{(2)}_{nml}(\mathbf{k}',\sigma'\omega^T_{\mathbf{k}'}|\mathbf{k}-\mathbf{k}',\sigma\omega^T_{\mathbf{k}} - \sigma'\omega^T_{\mathbf{k}'}) \\
&\times \mathcal{P}\, \frac{2}{\Lambda_\perp(\mathbf{k}-\mathbf{k}',\sigma\omega^T_{\mathbf{k}} - \sigma'\omega^T_{\mathbf{k}'})} + \chi^{(3)}_{ijmn}(\mathbf{k}',\sigma'\omega^T_{\mathbf{k}'}|-\mathbf{k}',-\sigma'\omega^T_{\mathbf{k}'}|\mathbf{k},\sigma\omega^T_{\mathbf{k}})\Bigg] \\
&\times I^{\sigma' T}_{\mathbf{k}'}\, I^{\sigma T}_{\mathbf{k}}. \qquad \boxed{\text{Induced scattering } T, T} \qquad (7.54)
\end{aligned}
$$

The wave kinetic equation for T mode is an entirely new equation that could not be discussed within the context of electrostatic approximation.

7.3 Nonlinear Susceptibilities

In this section, we revisit the formal definitions of various linear and nonlinear susceptibility tensors, defined in (7.21), and evaluate specific forms in approximate manners, in order to facilitate further analysis.

7.3.1 Linear Susceptibility Tensor

The linear susceptibility tensor is given by

$$\chi_{ij}^{a}(\mathbf{k},\omega) = \frac{\omega_{pa}^2}{\omega}\int d\mathbf{v}\,\frac{v_i}{\omega-\mathbf{k}\cdot\mathbf{v}+i0}\left[\left(1-\frac{\mathbf{k}\cdot\mathbf{v}}{\omega}\right)\frac{\partial}{\partial v_j}+\frac{v_j}{\omega}\mathbf{k}\cdot\frac{\partial}{\partial\mathbf{v}}\right]F_a$$
$$= \frac{\omega_{pa}^2}{\omega}\int d\mathbf{v}\,\frac{v_i}{\omega-\mathbf{k}\cdot\mathbf{v}+i0}\left[\frac{\partial F_a}{\partial\mathbf{v}}+\frac{\mathbf{k}}{\omega}\times\left(\mathbf{v}\times\frac{\partial F_a}{\partial\mathbf{v}}\right)\right]_j. \tag{7.55}$$

It is instructive to rewrite this expression by means of partial integration,

$$\chi_{ij}^{a}(\mathbf{k},\omega) = -\frac{\omega_{pa}^2}{\omega^2}\int d\mathbf{v}\,F_a\left(\delta_{ij}+\frac{k_i v_j+k_j v_i}{\omega-\mathbf{k}\cdot\mathbf{v}}+\frac{k^2 v_i\,v_j}{(\omega-\mathbf{k}\cdot\mathbf{v})^2}\right). \tag{7.56}$$

The linear susceptibility satisfies the following property:

$$\chi_{ij}^{a}(-\mathbf{k},\,-\omega) = \chi_{ij}^{a*}(\mathbf{k},\omega). \tag{7.57}$$

From the first expression (7.55), the imaginary (or anti-Hermitian) part of linear susceptibility tensor is given by

$$\mathrm{Im}\,\chi_{ij}^{a}(\mathbf{k},\omega) = -i\pi\,\frac{\omega_{pa}^2}{\omega^2}\int d\mathbf{v}\;v_i\;v_j\;\delta(\omega-\mathbf{k}\cdot\mathbf{v})\,\mathbf{k}\cdot\frac{\partial F_a}{\partial\mathbf{v}}. \tag{7.58}$$

For fast waves ($\omega \gg \mathbf{k}\cdot\mathbf{v}$), assuming a quasi-Gaussian form of F_a, we obtain

$$\mathrm{Re}\,\chi_{ij}^{a}(\mathbf{k},\omega) = -\frac{\omega_{pa}^2}{\omega^2}\left[\frac{k_i\,k_j}{k^2}\left(1+\frac{3k^2 v_{Ta}^2}{2\omega^2}\right)\right.$$
$$\left.+\left(\delta_{ij}-\frac{k_i\,k_j}{k^2}\right)\left(1+\frac{k^2 v_{Ta}^2}{2\omega^2}\right)\right]. \tag{7.59}$$

In (7.59), we remind the readers that $v_{Ta}^2 = 2T_a/m_a$ is the square of thermal speed, T_a being the temperature defined by $T_a = m_a \int d\mathbf{v}\; v^2\; F_a$.

For slow waves ($\omega \ll \mathbf{k}\cdot\mathbf{v}$), upon making use of the property of quasi-Gaussian distribution,

$$\frac{\partial F_a}{\partial\mathbf{v}} \approx -\frac{2\mathbf{v}}{v_{Ta}^2}\,F_a,$$

we obtain

$$\mathrm{Re}\,\chi_{ij}^{a}(\mathbf{k},\omega) = -\frac{2\omega_{pa}^2}{v_{Ta}^2\omega}\int d\mathbf{v}\,\frac{v_i\,v_j}{\omega-\mathbf{k}\cdot\mathbf{v}}\,F_a. \tag{7.60}$$

We may decompose $\mathbf{v}$ into components parallel and transverse to $\mathbf{k}$, as is customary,

$$\mathbf{v} = \frac{\mathbf{k}\,(\mathbf{k}\cdot\mathbf{v})}{k^2} + \frac{(\mathbf{k}\times\mathbf{v})\times\mathbf{k}}{k^2} = \frac{\mathbf{k}}{k}\,v_\parallel + \mathbf{v}_\perp,$$

where

$$v_\parallel = \frac{\mathbf{k}\cdot\mathbf{v}}{k}, \qquad \mathbf{v}_\perp = \frac{(\mathbf{k}\times\mathbf{v})\times\mathbf{k}}{k^2}. \tag{7.61}$$

As a consequence, we may write

$$\int d\mathbf{v}\,\frac{v_i\,v_j}{\omega-\mathbf{k}\cdot\mathbf{v}}\,F_a = \int d\mathbf{v}\,\frac{1}{\omega-k\,v_\parallel}\left[\frac{k_i\,k_j}{k^2}\,v_\parallel^2 + \left(\delta_{ij}-\frac{k_i\,k_j}{k^2}\right)v_\perp^2\right]F_a, \tag{7.62}$$

where we have made use of the fact that non-diagonal elements vanish since the velocity integrals associated with those terms are odd. For the parallel component, we may expand the denominator,

$$\frac{1}{\omega-k\,v_\parallel} \approx -\frac{1}{k\,v_\parallel}\left(1+\frac{\omega}{k\,v_\parallel}\right). \tag{7.63}$$

The first term cannot contribute to the integral since it is odd. The second term leads to the cancellation of $v_\parallel^2$ from top and bottom. This leads to

$$\mathrm{Re}\,\chi_{ij}^{a}(\mathbf{k},\omega) = \frac{2\omega_{pa}^2}{k^2\,v_{Ta}^2}\frac{k_i\,k_j}{k^2} - \frac{2\omega_{pa}^2}{v_{Ta}^2\omega}\left(\delta_{ij}-\frac{k_i\,k_j}{k^2}\right)\int d\mathbf{v}\,\frac{v_\perp^2}{\omega-k\,v_\parallel}\,F_a. \tag{7.64}$$

For Gaussian distribution, we have

$$\int d\mathbf{v}\,\frac{v_\perp^2}{\omega-k\,v_\parallel}\,F_a = -\frac{v_{Ta}}{2k}\,Z\left(\frac{\omega}{k\,v_{Ta}}\right). \tag{7.65}$$

For slow waves, we may approximate the real part of the Z function by $Z(\omega/k\,v_{Ta}) \approx -2(\omega/k\,v_{Ta})$. Thus, we have

$$\mathrm{Re}\,\chi_{ij}^{a}(\mathbf{k},\omega) = \frac{2\omega_{pa}^2}{k^2\,v_{Ta}^2}\left[\frac{k_i\,k_j}{k^2} - \left(\delta_{ij}-\frac{k_i\,k_j}{k^2}\right)\right]. \tag{7.66}$$

Note that the imaginary part of linear susceptibility tensor (7.58) can be more specifically written as

$$\begin{aligned}
\mathrm{Im}\,\chi^a_{ij}(\mathbf{k},\omega) &= \frac{k_i\,k_j}{k^2}\,\mathrm{Im}\,\chi^a_{\parallel}(\mathbf{k},\omega) + \left(\delta_{ij} - \frac{k_i\,k_j}{k^2}\right)\mathrm{Im}\,\chi^a_{\perp}(\mathbf{k},\omega), \\
\mathrm{Im}\,\chi^a_{\parallel}(\mathbf{k},\omega) &= -\pi\,\frac{\omega^2_{pa}}{\omega^2}\int d\mathbf{v}\,\frac{(\mathbf{k}\cdot\mathbf{v})^2}{k^2}\,\delta(\omega-\mathbf{k}\cdot\mathbf{v})\,\mathbf{k}\cdot\frac{\partial F_a}{\partial\mathbf{v}}, \\
\mathrm{Im}\,\chi^a_{\perp}(\mathbf{k},\omega) &= -\frac{\pi}{2}\,\frac{\omega^2_{pa}}{\omega^2}\int d\mathbf{v}\left(\mathbf{v} - \frac{(\mathbf{k}\cdot\mathbf{v})\,\mathbf{k}}{k^2}\right)^2\delta(\omega-\mathbf{k}\cdot\mathbf{v})\,\mathbf{k}\cdot\frac{\partial F_a}{\partial\mathbf{v}}.
\end{aligned} \tag{7.67}$$

We have already discussed the general diagonal property in (7.32), but for Gaussian thermal distribution, one can show that

$$\begin{aligned}
\chi^a_{ij}(\mathbf{k},\omega) &= \frac{k_i\,k_j}{k^2}\,\chi^a_{\parallel}(\mathbf{k},\omega) + \left(\delta_{ij} - \frac{k_i\,k_j}{k^2}\right)\chi^a_{\perp}(\mathbf{k},\omega), \\
\chi^a_{\parallel}(\mathbf{k},\omega) &= -\frac{\omega^2_{pa}}{\omega^2}\,\xi^2_a\,Z'(\xi_a), \quad \chi^a_{\perp}(\mathbf{k},\omega) = \frac{\omega^2_{pa}}{\omega^2}\,\xi_a\,Z(\xi_a), \\
\xi_a &= \frac{\omega}{k\,v_{Ta}}.
\end{aligned} \tag{7.68}$$

Let us discuss some further alternative expressions. From (7.55), we have

$$\begin{aligned}
\epsilon^a_{ij}(\mathbf{k},\omega) = \delta_{ij} &+ \frac{\omega^2_{pa}}{\omega}\int d\mathbf{v}\,\frac{1}{\omega-\mathbf{k}\cdot\mathbf{v}+i0}\left(\frac{k_i\,(\mathbf{k}\cdot\mathbf{v})}{k^2} + \frac{[(\mathbf{k}\times\mathbf{v})\times\mathbf{k}]_i}{k^2}\right) \\
&\times\left[\frac{\partial F_a}{\partial\mathbf{v}} + \frac{\mathbf{k}}{\omega}\times\left(\mathbf{v}\times\frac{\partial F_a}{\partial\mathbf{v}}\right)\right]_j.
\end{aligned} \tag{7.69}$$

From the property (7.32), the tensor ϵ_{ij} is given by the diagonal form

$$\epsilon_{ij} = \frac{k_i\,k_j}{k^2}\,\epsilon_{\parallel} + \left(\delta_{ij} - \frac{k_i\,k_j}{k^2}\right)\epsilon_{\perp}. \tag{7.70}$$

From this, we see that

$$\epsilon_{\parallel} = \frac{k_i\,k_j}{k^2}\,\epsilon_{ij}, \quad \epsilon_{\perp} = \frac{1}{2}\left(\delta_{ij} - \frac{k_i\,k_j}{k^2}\right)\epsilon_{ij}. \tag{7.71}$$

Now,

$$\epsilon_{\parallel} = \frac{k_i\,k_j}{k^2}\,\epsilon_{ij} = 1 + \frac{\omega^2_{pa}}{\omega\,k^2}\int d\mathbf{v}\,\frac{\mathbf{k}\cdot\mathbf{v}}{\omega-\mathbf{k}\cdot\mathbf{v}+i0}\,\mathbf{k}\cdot\frac{\partial F_a}{\partial\mathbf{v}}.$$

This allows us to obtain the familiar expression for longitudinal dielectric constant

$$\epsilon_{\parallel}(\mathbf{k},\omega) = 1 + \chi_{\parallel}(\mathbf{k},\omega) = 1 + \frac{\omega^2_{pa}}{k^2}\int d\mathbf{v}\,\frac{\mathbf{k}\cdot\partial F_a/\partial\mathbf{v}}{\omega-\mathbf{k}\cdot\mathbf{v}+i0}. \tag{7.72}$$

The transverse dielectric tensor can be obtained in a similar way:

$$\begin{aligned}\epsilon_\perp &= \frac{1}{2}\left(\delta_{ij} - \frac{k_i k_j}{k^2}\right)\epsilon_{ij} \\ &= 1 + \frac{\omega_{pa}^2}{2\omega}\int d\mathbf{v}\,\frac{1}{\omega - \mathbf{k}\cdot\mathbf{v} + i0}\frac{(\mathbf{k}\times\mathbf{v})\times\mathbf{k}}{k^2} \\ &\quad\cdot\left[\frac{\partial F_a}{\partial \mathbf{v}} + \frac{\mathbf{k}}{\omega}\times\left(\mathbf{v}\times\frac{\partial F_a}{\partial \mathbf{v}}\right)\right].\end{aligned} \tag{7.73}$$

We may rewrite this result as

$$\begin{aligned}\epsilon_\perp(\mathbf{k},\omega) &= 1 + \chi_\perp(\mathbf{k},\omega) \\ &= 1 + \frac{\omega_{pa}^2}{2\omega}\int d\mathbf{v}\,\frac{[(\mathbf{k}\times\mathbf{v})\times\mathbf{k}]_i}{k^2} \\ &\quad\times\left[\left(1 - \frac{\mathbf{k}\cdot\mathbf{v}}{\omega}\right)\delta_{ij} + \frac{v_i k_j}{\omega}\right]\frac{\partial F_a/\partial v_j}{\omega - \mathbf{k}\cdot\mathbf{v} + i0}.\end{aligned} \tag{7.74}$$

7.3.2 Second-Order Nonlinear Susceptibility Tensor

The second-order susceptibility is given by (7.21), which is repeated here for the sake of convenience:

$$\begin{aligned}\chi_{ijk}^{a(2)}(\mathbf{k}_1,\omega_1|\mathbf{k}_2,\omega_2) &= -\frac{i}{2}\frac{e_a}{m_a}\frac{\omega_{pa}^2}{\omega_1+\omega_2}\int d\mathbf{v}\,\frac{v_i}{\omega_1+\omega_2-(\mathbf{k}_1+\mathbf{k}_2)\cdot\mathbf{v}+i0} \\ &\quad\times\left\{\left[\left(1-\frac{\mathbf{k}_1\cdot\mathbf{v}}{\omega_1}\right)\frac{\partial}{\partial v_j} + \frac{v_j}{\omega_1}\mathbf{k}_1\cdot\frac{\partial}{\partial \mathbf{v}}\right]\frac{1}{\omega_2-\mathbf{k}_2\cdot\mathbf{v}+i0}\right. \\ &\quad\times\left[\left(1-\frac{\mathbf{k}_2\cdot\mathbf{v}}{\omega_2}\right)\frac{\partial}{\partial v_k} + \frac{v_k}{\omega_2}\mathbf{k}_2\cdot\frac{\partial}{\partial \mathbf{v}}\right] \\ &\quad+\left[\left(1-\frac{\mathbf{k}_2\cdot\mathbf{v}}{\omega_2}\right)\frac{\partial}{\partial v_k} + \frac{v_k}{\omega_2}\mathbf{k}_2\cdot\frac{\partial}{\partial \mathbf{v}}\right] \\ &\quad\left.\times\frac{1}{\omega_1-\mathbf{k}_1\cdot\mathbf{v}+i0}\left[\left(1-\frac{\mathbf{k}_1\cdot\mathbf{v}}{\omega_1}\right)\frac{\partial}{\partial v_j} + \frac{v_j}{\omega_1}\mathbf{k}_1\cdot\frac{\partial}{\partial \mathbf{v}}\right]\right\}F_a.\end{aligned}$$

Let us take partial integrations in order to obtain an alternative expression. Let us adopt the shorthand notation,

$$\omega = \omega_1 + \omega_2, \quad \mathbf{k} = \mathbf{k}_1 + \mathbf{k}_2. \tag{7.75}$$

Then we have

$$
\begin{aligned}
&\chi^{a(2)}_{ijk}(\mathbf{k}_1,\omega_1|\mathbf{k}_2,\omega_2)\\
&\quad = -\frac{i}{2}\frac{e_a}{m_a}\frac{\omega_{pa}^2}{\omega\omega_1\omega_2}\int d\mathbf{v}\,\frac{v_i}{\omega-\mathbf{k}\cdot\mathbf{v}+i0}\\
&\qquad \times\left[\left((\omega_1-\mathbf{k}_1\cdot\mathbf{v})\frac{\partial}{\partial v_j}+v_j\,\mathbf{k}_1\cdot\frac{\partial}{\partial \mathbf{v}}\right)\left(\frac{\partial}{\partial v_k}+\frac{v_k}{\omega_2-\mathbf{k}_2\cdot\mathbf{v}+i0}\mathbf{k}_2\cdot\frac{\partial}{\partial\mathbf{v}}\right)F_a\right.\\
&\qquad \left.+\,(1\leftrightarrow 2, j\leftrightarrow k)\vphantom{\frac{\partial}{\partial}}\right]\\
&\quad = \frac{i}{2}\frac{e_a}{m_a}\frac{\omega_{pa}^2}{\omega\omega_1\omega_2}\int d\mathbf{v}\left[\left(\frac{\partial F_a}{\partial v_k}+\frac{v_k}{\omega_2-\mathbf{k}_2\cdot\mathbf{v}}\mathbf{k}_2\cdot\frac{\partial F_a}{\partial\mathbf{v}}\right)\right.\\
&\qquad \times\left(\frac{(\omega_1-\mathbf{k}_1\cdot\mathbf{v})\,\delta_{ij}+k_{1i}\,v_j}{\omega-\mathbf{k}\cdot\mathbf{v}}+\frac{(\omega_1-\mathbf{k}_1\cdot\mathbf{v})\,k_j\,v_i+\mathbf{k}_1\cdot\mathbf{k}\,v_i\,v_j}{(\omega-\mathbf{k}\cdot\mathbf{v})^2}\right)\\
&\qquad \left.+\,(1\leftrightarrow 2, j\leftrightarrow k)\vphantom{\frac{\partial}{\partial}}\right]\\
&\quad = -\frac{i}{2}\frac{e_a}{m_a}\frac{\omega_{pa}^2}{\omega\omega_1\omega_2}\int d\mathbf{v}\,F_a\left[\left(\frac{\partial}{\partial v_k}+\mathbf{k}_2\cdot\frac{\partial}{\partial\mathbf{v}}\,\frac{v_k}{\omega_2-\mathbf{k}_2\cdot\mathbf{v}}\right)\right.\\
&\qquad \times\left(\delta_{ij}-\frac{\omega_2-\mathbf{k}_2\cdot\mathbf{v}}{\omega-\mathbf{k}\cdot\mathbf{v}}\,\delta_{ij}+\frac{k_{1i}\,v_j+v_i\,k_j}{\omega-\mathbf{k}\cdot\mathbf{v}}\right.\\
&\qquad \left.\left.-\,\frac{\omega_2-\mathbf{k}_2\cdot\mathbf{v}}{(\omega-\mathbf{k}\cdot\mathbf{v})^2}\,v_i\,k_j+\frac{\mathbf{k}_1\cdot\mathbf{k}}{(\omega-\mathbf{k}\cdot\mathbf{v})^2}\,v_i\,v_j\right)+(1\leftrightarrow 2, j\leftrightarrow k)\right]\\
&\quad = -\frac{i}{2}\frac{e_a}{m_a}\frac{\omega_{pa}^2}{\omega\,\omega_1\,\omega_2}\int d\mathbf{v}\,F_a\left(\frac{\delta_{ik}\,k_{1j}}{\omega_1-\mathbf{k}_1\cdot\mathbf{v}}+\frac{\delta_{ij}\,k_{2k}}{\omega_2-\mathbf{k}_2\cdot\mathbf{v}}+\frac{\delta_{jk}\,k_i}{\omega-\mathbf{k}\cdot\mathbf{v}}\right.\\
&\qquad +\frac{\delta_{ik}\,k_1^2\,v_j}{(\omega_1-\mathbf{k}_1\cdot\mathbf{v})^2}+\frac{\delta_{ij}\,k_2^2\,v_k}{(\omega_2-\mathbf{k}_2\cdot\mathbf{v})^2}+\frac{\delta_{jk}\,k^2\,v_i}{(\omega-\mathbf{k}\cdot\mathbf{v})^2}\\
&\qquad +\frac{k_{2i}\,v_k\,k_{1i}+v_i\,k_k\,k_{1j}+k_{2i}\,k_{1k}\,v_j+k_{1i}\,k_k\,v_j}{(\omega_1-\mathbf{k}_1\cdot\mathbf{v})(\omega-\mathbf{k}\cdot\mathbf{v})}\\
&\qquad +\frac{k_{1i}\,v_j\,k_{2k}+v_i\,k_j\,k_{2k}+k_{1i}\,k_{2j}\,v_k+k_{2i}\,k_j\,v_k}{(\omega_2-\mathbf{k}_2\cdot\mathbf{v})(\omega-\mathbf{k}\cdot\mathbf{v})}\\
&\qquad +\frac{k_1^2\,(k_{2i}\,v_j\,v_k+k_k\,v_i\,v_j)}{(\omega_1-\mathbf{k}_1\cdot\mathbf{v})^2(\omega-\mathbf{k}\cdot\mathbf{v})}+\frac{k_2^2\,(k_{1i}\,v_j\,v_k+k_j\,v_i\,v_k)}{(\omega_2-\mathbf{k}_2\cdot\mathbf{v})^2(\omega-\mathbf{k}\cdot\mathbf{v})}\\
&\qquad +\frac{(\mathbf{k}_1\cdot\mathbf{k})\,(k_{2i}\,v_j\,v_k+k_k\,v_i\,v_j)+(\mathbf{k}_2\cdot\mathbf{k})\,(k_{1i}\,v_j\,v_k+k_{1j}\,v_i\,v_k+k_{1k}\,v_i\,v_j)}{(\omega_1-\mathbf{k}_1\cdot\mathbf{v})(\omega-\mathbf{k}\cdot\mathbf{v})^2}\\
&\qquad +\frac{(\mathbf{k}_2\cdot\mathbf{k})\,(k_{1i}\,v_j\,v_k+k_j\,v_i\,v_k)+(\mathbf{k}_1\cdot\mathbf{k})\,(k_{2i}\,v_j\,v_k+k_{2j}\,v_i\,v_k+k_{2k}\,v_i\,v_j)}{(\omega_2-\mathbf{k}_2\cdot\mathbf{v})(\omega-\mathbf{k}\cdot\mathbf{v})^2}\\
&\qquad +\frac{k_1^2\,(\mathbf{k}_2\cdot\mathbf{k})\,v_i\,v_j\,v_k}{(\omega_1-\mathbf{k}_1\cdot\mathbf{v})^2(\omega-\mathbf{k}\cdot\mathbf{v})^2}+\frac{k_2^2\,(\mathbf{k}_1\cdot\mathbf{k})\,v_i\,v_j\,v_k}{(\omega_2-\mathbf{k}_2\cdot\mathbf{v})^2(\omega-\mathbf{k}\cdot\mathbf{v})^2}\\
&\qquad \left.+\frac{2(\mathbf{k}_1\cdot\mathbf{k})(\mathbf{k}_2\cdot\mathbf{k})\,v_i\,v_j\,v_k}{(\omega_1-\mathbf{k}_1\cdot\mathbf{v})(\omega-\mathbf{k}\cdot\mathbf{v})^3}+\frac{2(\mathbf{k}_1\cdot\mathbf{k})(\mathbf{k}_2\cdot\mathbf{k})\,v_i\,v_j\,v_k}{(\omega_2-\mathbf{k}_2\cdot\mathbf{v})(\omega-\mathbf{k}\cdot\mathbf{v})^3}\right),\\
&\omega=\omega_1+\omega_2,\quad \mathbf{k}=\mathbf{k}_1+\mathbf{k}_2.
\end{aligned}
\tag{7.76}
$$

It is possible to further rearrange (7.76) into a slightly more compact form,

$$\begin{aligned}
\chi^{a(2)}_{ijk}(\mathbf{k}_1,\omega_1|\mathbf{k}_2,\omega_2) = &-\frac{i}{2}\frac{e_a}{m_a}\frac{\omega_{pa}^2}{\omega\,\omega_1\,\omega_2}\int d\mathbf{v}\;F_a \\
&\times\left[\frac{\delta_{ik}\,k_{1j}}{\omega_1-\mathbf{k}_1\cdot\mathbf{v}}+\frac{\delta_{ij}\,k_{2k}}{\omega_2-\mathbf{k}_2\cdot\mathbf{v}}+\frac{\delta_{jk}\,k_i}{\omega-\mathbf{k}\cdot\mathbf{v}}\right.\\
&+\frac{\delta_{ik}\,k_1^2\,v_j}{(\omega_1-\mathbf{k}_1\cdot\mathbf{v})^2}+\frac{\delta_{ij}\,k_2^2\,v_k}{(\omega_2-\mathbf{k}_2\cdot\mathbf{v})^2}+\frac{\delta_{jk}\,k^2\,v_i}{(\omega-\mathbf{k}\cdot\mathbf{v})^2}\\
&+\frac{k_{1i}\,v_j\,k_{2k}+k_{2i}\,k_{1j}\,v_k}{(\omega_1-\mathbf{k}_1\cdot\mathbf{v})(\omega_2-\mathbf{k}_2\cdot\mathbf{v})}+\frac{v_i\,k_j\,k_{2k}+k_i\,k_{2j}\,v_k}{(\omega_2-\mathbf{k}_2\cdot\mathbf{v})(\omega-\mathbf{k}\cdot\mathbf{v})}\\
&+\frac{v_i\,k_{1j}\,k_k+k_i\,v_j\,k_{1k}}{(\omega_1-\mathbf{k}_1\cdot\mathbf{v})(\omega-\mathbf{k}\cdot\mathbf{v})}\\
&+\frac{1}{(\omega_1-\mathbf{k}_1\cdot\mathbf{v})(\omega_2-\mathbf{k}_2\cdot\mathbf{v})(\omega-\mathbf{k}\cdot\mathbf{v})}\\
&\times\left(k_1^2\,k_k\,v_i\,v_j\,\frac{\omega_2-\mathbf{k}_2\cdot\mathbf{v}}{\omega_1-\mathbf{k}_1\cdot\mathbf{v}}+k_2^2\,k_j\,v_i\,v_k\,\frac{\omega_1-\mathbf{k}_1\cdot\mathbf{v}}{\omega_2-\mathbf{k}_2\cdot\mathbf{v}}\right.\\
&+k^2\,k_{1k}\,v_i\,v_j\,\frac{\omega_2-\mathbf{k}_2\cdot\mathbf{v}}{\omega-\mathbf{k}\cdot\mathbf{v}}+k^2\,k_{2j}\,v_i\,v_k\,\frac{\omega_1-\mathbf{k}_1\cdot\mathbf{v}}{\omega-\mathbf{k}\cdot\mathbf{v}}\\
&+k_1^2\,k_{2i}\,v_j\,v_k\,\frac{\omega-\mathbf{k}\cdot\mathbf{v}}{\omega_1-\mathbf{k}_1\cdot\mathbf{v}}+k_2^2\,k_{1i}\,v_j\,v_k\,\frac{\omega-\mathbf{k}\cdot\mathbf{v}}{\omega_2-\mathbf{k}_2\cdot\mathbf{v}}\\
&\left.+(\mathbf{k}_1\cdot\mathbf{k}_2)\,k_i\,v_j\,v_k+(\mathbf{k}_1\cdot\mathbf{k})\,k_{2k}\,v_i\,v_j+(\mathbf{k}_2\cdot\mathbf{k})\,k_{1j}\,v_i\,v_k\right)\\
&+\frac{v_i\,v_j\,v_k}{(\omega_1-\mathbf{k}_1\cdot\mathbf{v})(\omega_2-\mathbf{k}_2\cdot\mathbf{v})(\omega-\mathbf{k}\cdot\mathbf{v})}\\
&\left.\times\left(\frac{k_1^2\,(\mathbf{k}_2\cdot\mathbf{k})}{\omega_1-\mathbf{k}_1\cdot\mathbf{v}}+\frac{k_2^2\,(\mathbf{k}_1\cdot\mathbf{k})}{\omega_2-\mathbf{k}_2\cdot\mathbf{v}}+\frac{k^2\,(\mathbf{k}_1\cdot\mathbf{k}_2)}{\omega-\mathbf{k}\cdot\mathbf{v}}\right)\right].
\end{aligned}\tag{7.77}$$

In (7.77), one may replace the first argument by $(\mathbf{k}_1,\omega_1)\rightarrow(\mathbf{k}_1+\mathbf{k}_2,\omega_1+\omega_2)$ and the second argument by $(\mathbf{k}_2,\omega_2)\rightarrow(-\mathbf{k}_2,\,-\omega_2)$. Then one can easily show the following symmetry property,

$$\chi^{a(2)}_{ijk}(\mathbf{k}_1+\mathbf{k}_2,\omega_1+\omega_2|-\mathbf{k}_2,\,-\omega_2)=-\chi^{a(2)}_{jik}(\mathbf{k}_1,\omega_1|\mathbf{k}_2,\omega_2),\tag{7.78}$$

which we have already invoked in (7.31).

Infinite Wavelength Limit

In this limiting case, all the $\mathbf{k}$ vectors may be set equal to zero. From the alternative expression (7.77), we can immediately deduce that

$$\chi^a_{ijk}(0,\omega_1|0,\omega_2)=0.\tag{7.79}$$

This can also be seen directly from definition (7.21). If we take all k vectors to zero, then we are left with

$$\chi^{a(2)}_{ijk}(0,\omega_1|0,\omega_2) = -\frac{i}{2}\frac{e_a}{m_a}\frac{\omega_{pa}^2}{\omega_1\,\omega_2\,(\omega_1+\omega_2)}\int d\mathbf{v}\; v_i\,\frac{\partial^2 F_a}{\partial v_j\,\partial v_k}$$
$$= \delta_{ij}\,\frac{i}{2}\frac{e_a}{m_a}\frac{\omega_{pa}^2}{\omega_1\,\omega_2\,(\omega_1+\omega_2)}\int d\mathbf{v}\,\frac{\partial F_a}{\partial v_k} = 0.$$

Fast Wave Limit

Suppose that

$$\omega_1 \gg k_1 v_{Ta}, \quad \omega_2 \gg k_2 v_{Ta}, \quad \text{and} \quad (\omega_1+\omega_2) \gg |\mathbf{k}_1+\mathbf{k}_2|\,v_{Ta}. \tag{7.80}$$

Then we have from (7.77)

$$\chi^{a}_{ijk}(\mathbf{k}_1,\omega_1|\mathbf{k}_2,\omega_2) = -\frac{i}{2}\frac{e_a}{m_a}\frac{\omega_{pa}^2}{\omega_1\,\omega_2\,(\omega_1+\omega_2)}\left(\frac{\delta_{ik}\,k_{1j}}{\omega_1} + \frac{\delta_{ij}\,k_{2k}}{\omega_2} + \frac{\delta_{jk}\,(\mathbf{k}_1+\mathbf{k}_2)_i}{\omega_1+\omega_2}\right). \tag{7.81}$$

One could also obtain (7.81) directly from the original definition (7.21).

One Slow Wave and Two Fast Waves

If ω_1 remains arbitrary but the combined frequency $\omega = \omega_1 + \omega_2$ as well as ω_2 satisfy the fast-wave condition, then retaining the leading terms only, we obtain

$$\chi^{a}_{ijk}(\mathbf{k}_1,\omega_1|\mathbf{k}_2,\omega_2) = -\frac{i}{2}\frac{e_a}{m_a}\frac{\omega_{pa}^2}{\omega\,\omega_1\,\omega_2}\int d\mathbf{v}\; F_a \times\left(\frac{\delta_{ik}\,k_{1j}}{\omega_1-\mathbf{k}_1\cdot\mathbf{v}} + \frac{\delta_{ij}\,k_{2k}}{\omega_2} + \frac{\delta_{jk}\,k_i}{\omega} + \frac{\delta_{ik}\,k_1^2\,v_j}{(\omega_1-\mathbf{k}_1\cdot\mathbf{v})^2}\right). \tag{7.82}$$

Consider the quantity

$$\int d\mathbf{v}\; F_a\,\frac{k_1\,v_j}{(\omega_1-\mathbf{k}_1\cdot\mathbf{v})^2}.$$

This is zero unless v_j is parallel to $\mathbf{k}_1$. Consequently, we may write

$$\int d\mathbf{v}\; F_a\,\frac{k_1\,v_j}{(\omega_1-\mathbf{k}_1\cdot\mathbf{v})^2} = \frac{k_{1j}}{k_1}\int d\mathbf{v}\; F_a\,\frac{k_1\,v_\parallel}{(\omega_1-k_1 v_\parallel)^2}$$
$$= \frac{k_{1j}}{k_1}\int d\mathbf{v}\; F_a\left(-\frac{1}{\omega_1-k_1 v_\parallel} + \frac{\omega_1}{(\omega_1-k_1 v_\parallel)^2}\right).$$

This leads to

$$\chi^a_{ijk}(\mathbf{k}_1,\omega_1|\mathbf{k}_2,\omega_2) = -\frac{i}{2}\frac{e_a}{m_a}\frac{\omega_{pa}^2}{\omega\,\omega_1\,\omega_2}\int d\mathbf{v}\; F_a \times\left(\frac{\delta_{ij}\,k_{2k}}{\omega_2}+\frac{\delta_{jk}\,k_i}{\omega}+\frac{\delta_{ik}\,k_{1j}\,\omega_1}{(\omega_1-\mathbf{k}_1\cdot\mathbf{v})^2}\right). \tag{7.83}$$

If we make use of

$$\int d\mathbf{v}\; F_a\,\frac{1}{(\omega_1-\mathbf{k}\cdot\mathbf{v})^2} = -\frac{1}{k_1^2}\int d\mathbf{v}\;\frac{\mathbf{k}_1\cdot\partial F_a/\partial\mathbf{v}}{\omega_1-\mathbf{k}_1\cdot\mathbf{v}},$$

and retain the most important term, then we have

$$\chi^a_{ijk}(\mathbf{k}_1,\omega_1|\mathbf{k}_2,\omega_2) = \frac{i}{2}\frac{e_a}{m_a}\frac{\omega_{pa}^2}{\omega_2\,(\omega_1+\omega_2)}\,\delta_{ik}\,\frac{k_{1j}}{k_1^2}\int d\mathbf{v}\;\frac{\mathbf{k}_1\cdot\partial F_a/\partial\mathbf{v}}{\omega_1-\mathbf{k}_1\cdot\mathbf{v}}. \tag{7.84}$$

If ω_2 is a slow wave, then we may simply interchange 1 and 2 as well as j and k in (7.84) in order to arrive at

$$\chi^a_{ijk}(\mathbf{k}_1,\omega_1|\mathbf{k}_2,\omega_2) = \frac{i}{2}\frac{e_a}{m_a}\frac{\omega_{pa}^2}{\omega_1\,(\omega_1+\omega_2)}\,\delta_{ij}\,\frac{k_{2k}}{k_2^2}\int d\mathbf{v}\;\frac{\mathbf{k}_2\cdot\partial F_a/\partial\mathbf{v}}{\omega_2-\mathbf{k}_2\cdot\mathbf{v}}. \tag{7.85}$$

If ω_1 and ω_2 are fast waves but $\omega = \omega_1+\omega_2$ is a slow mode, then starting from the expression

$$\chi^a_{ijk}(\mathbf{k}_1,\omega_1|\mathbf{k}_2,\omega_2) = -\frac{i}{2}\frac{e_a}{m_a}\frac{\omega_{pa}^2}{\omega\,\omega_1\,\omega_2}\int d\mathbf{v}\; F_a\left(\frac{\delta_{jk}\,k_i}{\omega-\mathbf{k}\cdot\mathbf{v}}+\frac{\delta_{jk}\,k^2\,v_i}{(\omega-\mathbf{k}\cdot\mathbf{v})^2}\right)$$

and following the same steps as before, we obtain

$$\chi^a_{ijk}(\mathbf{k}_1,\omega_1|\mathbf{k}_2,\omega_2) = \frac{i}{2}\frac{e_a}{m_a}\frac{\omega_{pa}^2}{\omega_1\,\omega_2}\,\delta_{jk}\,\frac{(\mathbf{k}_1+\mathbf{k}_2)_i}{|\mathbf{k}_1+\mathbf{k}_2|^2}\int d\mathbf{v}\;\frac{(\mathbf{k}_1+\mathbf{k}_2)\cdot\partial F_a/\partial\mathbf{v}}{\omega_1+\omega_2-(\mathbf{k}_1+\mathbf{k}_2)\cdot\mathbf{v}}. \tag{7.86}$$

7.3.3 Third-Order Nonlinear Susceptibility Tensor

Third-order nonlinear susceptibility tensor is defined by (7.21), which is again repeated here for the sake of convenience:

$$\chi^{a(3)}_{ijkl}(\mathbf{k}_1,\omega_1|\mathbf{k}_2,\omega_2|\mathbf{k}_3,\omega_3) = \frac{(-i)^2}{2}\frac{e_a^2}{m_a^2}\frac{\omega_{pa}^2}{\omega_1+\omega_2+\omega_3}\int d\mathbf{v} \times\frac{v_i}{\omega_1+\omega_2+\omega_3-(\mathbf{k}_1+\mathbf{k}_2+\mathbf{k}_3)\cdot\mathbf{v}}$$

$$\times \left[\left(1-\frac{\mathbf{k}_1\cdot\mathbf{v}}{\omega_1}\right)\frac{\partial}{\partial v_j}+\frac{v_j}{\omega_1}\mathbf{k}_1\cdot\frac{\partial}{\partial\mathbf{v}}\right]$$
$$\times \frac{1}{\omega_2+\omega_3-(\mathbf{k}_2+\mathbf{k}_3)\cdot\mathbf{v}+i0}$$
$$\times \left\{\left[\left(1-\frac{\mathbf{k}_2\cdot\mathbf{v}}{\omega_2}\right)\frac{\partial}{\partial v_k}+\frac{v_k}{\omega_2}\mathbf{k}_2\cdot\frac{\partial}{\partial\mathbf{v}}\right]\right.$$
$$\times \frac{1}{\omega_3-\mathbf{k}_3\cdot\mathbf{v}+i0}\left[\left(1-\frac{\mathbf{k}_3\cdot\mathbf{v}}{\omega_3}\right)\frac{\partial}{\partial v_l}+\frac{v_l}{\omega_3}\mathbf{k}_3\cdot\frac{\partial}{\partial\mathbf{v}}\right]$$
$$+\left[\left(1-\frac{\mathbf{k}_3\cdot\mathbf{v}}{\omega_3}\right)\frac{\partial}{\partial v_l}+\frac{v_l}{\omega_3}\mathbf{k}_3\cdot\frac{\partial}{\partial\mathbf{v}}\right]\frac{1}{\omega_2-\mathbf{k}_2\cdot\mathbf{v}+i0}$$
$$\left.\times \left[\left(1-\frac{\mathbf{k}_2\cdot\mathbf{v}}{\omega_2}\right)\frac{\partial}{\partial v_k}+\frac{v_k}{\omega_2}\mathbf{k}_2\cdot\frac{\partial}{\partial\mathbf{v}}\right]\right\}F_a.$$

Since we are only interested in the situation where $(\mathbf{k}_1,\omega_1)=(\mathbf{k}',\omega')$, $(\mathbf{k}_2,\omega_2)=(-\mathbf{k}',-\omega')$, and $(\mathbf{k}_3,\omega_3)=(\mathbf{k},\omega)$, we henceforth consider

$$\chi^{a(3)}_{ijkl}(\mathbf{k}',\omega'|-\mathbf{k}',\omega'|\mathbf{k},\omega)=\frac{(-i)^2}{2}\frac{e_a^2}{m_a^2}\frac{\omega_{pa}^2}{\omega^2\,\omega'^2}\int d\mathbf{v}$$
$$\times\frac{v_i}{\omega-\mathbf{k}\cdot\mathbf{v}}\left((\omega'-\mathbf{k}'\cdot\mathbf{v})\frac{\partial}{\partial v_j}+v_j\,\mathbf{k}'\cdot\frac{\partial}{\partial\mathbf{v}}\right)$$
$$\times\frac{1}{\omega-\omega'-(\mathbf{k}-\mathbf{k}')\cdot\mathbf{v}}$$
$$\times\left[\left((\omega'-\mathbf{k}'\cdot\mathbf{v})\frac{\partial}{\partial v_k}+v_k\,\mathbf{k}'\cdot\frac{\partial}{\partial\mathbf{v}}\right)\right.$$
$$\times\left(\frac{\partial}{\partial v_l}+\frac{v_l}{\omega-\mathbf{k}\cdot\mathbf{v}}\mathbf{k}\cdot\frac{\partial}{\partial\mathbf{v}}\right)$$
$$-\left((\omega-\mathbf{k}\cdot\mathbf{v})\frac{\partial}{\partial v_l}+v_l\,\mathbf{k}\cdot\frac{\partial}{\partial\mathbf{v}}\right)$$
$$\left.\times\left(\frac{\partial}{\partial v_k}+\frac{v_k}{\omega'-\mathbf{k}'\cdot\mathbf{v}}\mathbf{k}'\cdot\frac{\partial}{\partial\mathbf{v}}\right)\right]F_a. \tag{7.87}$$

The third-order susceptibilities of actual relevance, that is, combinations that appear in the wave kinetic equation, are always given in the following forms:

$$T_1\equiv\frac{k_i\,k_l}{k^2}\frac{k'_j\,k'_k}{k'^2}\,\chi^{a(3)}_{ijkl}(\mathbf{k}',\omega'|-\mathbf{k}',\omega'|\mathbf{k},\omega),$$
$$T_2\equiv\frac{k_i\,k_l}{k^2}\left(\delta_{jk}-\frac{k'_j\,k'_k}{k'^2}\right)\chi^{a(3)}_{ijkl}(\mathbf{k}',\omega'|-\mathbf{k}',\omega'|\mathbf{k},\omega),$$
$$T_3\equiv\left(\delta_{il}-\frac{k_i\,k_l}{k^2}\right)\frac{k'_j\,k'_k}{k'^2}\,\chi^{a(3)}_{ijkl}(\mathbf{k}',\omega'|-\mathbf{k}',\omega'|\mathbf{k},\omega). \tag{7.88}$$

Moreover, the actual situation corresponds to the case when both ω and ω' satisfy the fast wave condition, but $\omega - \omega'$ remains arbitrary. Writing out the quantities in (7.88) explicitly, we obtain, first for T_1,

$$\begin{aligned}
T_1 = & \frac{(-i)^2}{2} \frac{e_a^2}{m_a^2} \frac{\omega_{pa}^2}{\omega^2 \omega'^2} \int d\mathbf{v} \frac{k_i\, k_l}{k^2} \frac{k'_j\, k'_k}{k'^2} \frac{v_i}{\omega - \mathbf{k}\cdot\mathbf{v}} \\
& \times \left((\omega' - \mathbf{k}'\cdot\mathbf{v}) \frac{\partial}{\partial v_j} + v_j\, \mathbf{k}' \cdot \frac{\partial}{\partial \mathbf{v}} \right) \frac{1}{\omega - \omega' - (\mathbf{k} - \mathbf{k}')\cdot\mathbf{v}} \\
& \times \left[\left((\omega' - \mathbf{k}'\cdot\mathbf{v}) \frac{\partial}{\partial v_k} + v_k\, \mathbf{k}' \cdot \frac{\partial}{\partial \mathbf{v}} \right) \left(\frac{\partial}{\partial v_l} + \frac{v_l}{\omega - \mathbf{k}\cdot\mathbf{v}} \mathbf{k} \cdot \frac{\partial}{\partial \mathbf{v}} \right) \right. \\
& \left. - \left((\omega - \mathbf{k}\cdot\mathbf{v}) \frac{\partial}{\partial v_l} + v_l\, \mathbf{k} \cdot \frac{\partial}{\partial \mathbf{v}} \right) \left(\frac{\partial}{\partial v_k} + \frac{v_k}{\omega' - \mathbf{k}'\cdot\mathbf{v}} \mathbf{k}' \cdot \frac{\partial}{\partial \mathbf{v}} \right) \right] F_a \\
= & \frac{1}{2} \frac{e_a^2}{m_a^2} \frac{\omega_{pa}^2}{\omega} \frac{1}{k^2 k'^2} \int d\mathbf{v} \left(1 - \frac{\omega}{\omega - \mathbf{k}\cdot\mathbf{v}} \right) \left\{ \mathbf{k}' \cdot \frac{\partial}{\partial \mathbf{v}} \frac{1}{\omega - \omega' - (\mathbf{k} - \mathbf{k}')\cdot\mathbf{v}} \right. \\
& \left. \times \left[\mathbf{k}' \cdot \frac{\partial}{\partial \mathbf{v}} \left(\frac{\mathbf{k}\cdot \partial F_a/\partial \mathbf{v}}{\omega - \mathbf{k}\cdot\mathbf{v}} \right) - \mathbf{k} \cdot \frac{\partial}{\partial \mathbf{v}} \left(\frac{\mathbf{k}'\cdot \partial F_a/\partial \mathbf{v}}{\omega' - \mathbf{k}'\cdot\mathbf{v}} \right) \right] \right\} \\
= & -\frac{1}{2} \frac{e_a^2}{m_a^2} \frac{\omega_{pa}^2}{k^2 k'^2} \int d\mathbf{v} \frac{1}{\omega - \mathbf{k}\cdot\mathbf{v}} \left\{ \mathbf{k}' \cdot \frac{\partial}{\partial \mathbf{v}} \frac{1}{\omega - \omega' - (\mathbf{k} - \mathbf{k}')\cdot\mathbf{v}} \right. \\
& \left. \times \left[\mathbf{k}' \cdot \frac{\partial}{\partial \mathbf{v}} \left(\frac{\mathbf{k}\cdot \partial F_a/\partial \mathbf{v}}{\omega - \mathbf{k}\cdot\mathbf{v}} \right) - \mathbf{k} \cdot \frac{\partial}{\partial \mathbf{v}} \left(\frac{\mathbf{k}'\cdot \partial F_a/\partial \mathbf{v}}{\omega' - \mathbf{k}'\cdot\mathbf{v}} \right) \right] \right\}.
\end{aligned} \tag{7.89}$$

In the last inequality, we have made use of the fact that

$$\int d\mathbf{v}\, \mathbf{k}' \cdot \frac{\partial}{\partial \mathbf{v}} G(\mathbf{v}) = 0,$$

for bounded function $G(\mathbf{v})$. The result (7.89) shows that

$$T_1 = -\bar{\chi}_a^{(3)}(\mathbf{k}', \omega' | -\mathbf{k}', -\omega' | \mathbf{k}, \omega), \tag{7.90}$$

where $\bar{\chi}_a^{(3)}(\mathbf{k}', \omega' | -\mathbf{k}', -\omega' | \mathbf{k}, \omega)$ is given by (1.136), and is the partial third-order nonlinear susceptibility associated with the purely electrostatic case. Consequently, the approximate expression already derived in (1.142) for

$$\omega \gg k v_{Te}, \qquad \omega' \gg k' v_{Te}, \qquad (\omega - \omega') < |\mathbf{k} - \mathbf{k}'|\, v_{Te},$$

is immediately applicable for the electrons:

$$\begin{aligned}
T_1 = & \frac{k_i\, k_l}{k^2} \frac{k'_j\, k'_k}{k'^2} \chi_{ijkl}^{e(3)}(\mathbf{k}', \omega' | -\mathbf{k}', \omega' | \mathbf{k}, \omega) \\
\approx & \frac{e^2}{2 m_e^2} \frac{(\mathbf{k}\cdot\mathbf{k}')^2}{k^2 k'^2} \frac{(\mathbf{k} - \mathbf{k}')^2}{\omega^3 \omega'} \chi_{\parallel}^e(\mathbf{k} - \mathbf{k}', \omega - \omega').
\end{aligned} \tag{7.91}$$

Next, we move on to evaluating two other related quantities. First, the quantity T_2. Upon a series of partial integrations, we have

$$
\begin{aligned}
T_2 = & \frac{k_i\,k_l}{k^2}\left(\delta_{jk} - \frac{k'_j\,k'_k}{k'^2}\right)\frac{(-i)^2}{2}\frac{e_a^2}{m_a^2}\frac{\omega_{pa}^2}{\omega^2\,\omega'^2}\int d\mathbf{v} \\
& \times \frac{v_i}{\omega - \mathbf{k}\cdot\mathbf{v}}\left((\omega' - \mathbf{k}'\cdot\mathbf{v})\frac{\partial}{\partial v_j} + v_j\,\mathbf{k}'\cdot\frac{\partial}{\partial \mathbf{v}}\right)\frac{1}{\omega - \omega' - (\mathbf{k}-\mathbf{k}')\cdot\mathbf{v}} \\
& \times \left[\left((\omega' - \mathbf{k}'\cdot\mathbf{v})\frac{\partial}{\partial v_k} + v_k\,\mathbf{k}'\cdot\frac{\partial}{\partial \mathbf{v}}\right)\left(\frac{\partial}{\partial v_l} + \frac{v_l}{\omega - \mathbf{k}\cdot\mathbf{v}}\mathbf{k}\cdot\frac{\partial}{\partial \mathbf{v}}\right)\right. \\
& \left. - \left((\omega - \mathbf{k}\cdot\mathbf{v})\frac{\partial}{\partial v_l} + v_l\,\mathbf{k}\cdot\frac{\partial}{\partial \mathbf{v}}\right)\left(\frac{\partial}{\partial v_k} + \frac{v_k}{\omega' - \mathbf{k}'\cdot\mathbf{v}}\mathbf{k}'\cdot\frac{\partial}{\partial \mathbf{v}}\right)\right] F_a \\
= & -\frac{1}{2}\frac{e_a^2}{m_a^2}\frac{\omega_{pa}^2}{\omega'^2\,k^2}\int d\mathbf{v}\left(\delta_{jk} - \frac{k'_j\,k'_k}{k'^2}\right)\frac{1}{\omega - \mathbf{k}\cdot\mathbf{v}} \\
& \times \left((\omega' - \mathbf{k}'\cdot\mathbf{v})\frac{\partial}{\partial v_j} + v_j\,\mathbf{k}'\cdot\frac{\partial}{\partial \mathbf{v}}\right)\frac{1}{\omega - \omega' - (\mathbf{k}-\mathbf{k}')\cdot\mathbf{v}} \\
& \times \left[\left((\omega' - \mathbf{k}'\cdot\mathbf{v})\frac{\partial}{\partial v_k} + v_k\,\mathbf{k}'\cdot\frac{\partial}{\partial \mathbf{v}}\right)\frac{1}{\omega - \mathbf{k}\cdot\mathbf{v}}\mathbf{k}\cdot\frac{\partial}{\partial \mathbf{v}}\right. \\
& \left. - \mathbf{k}\cdot\frac{\partial}{\partial \mathbf{v}}\left(\frac{\partial}{\partial v_k} + \frac{v_k}{\omega' - \mathbf{k}'\cdot\mathbf{v}}\mathbf{k}'\cdot\frac{\partial}{\partial \mathbf{v}}\right)\right] F_a.
\end{aligned}
\tag{7.92}
$$

This result can be re-expressed in terms of T_1, which we have already discussed in (7.90) and (7.91), and a new quantity P, as follows:

$$
\begin{aligned}
T_2 = & P - T_1, \\
P = & -\frac{1}{2}\frac{e_a^2}{m_a^2}\frac{\omega_{pa}^2}{\omega'^2\,k^2}\int d\mathbf{v}\,\frac{1}{\omega - \mathbf{k}\cdot\mathbf{v}}\left((\omega' - \mathbf{k}'\cdot\mathbf{v})\frac{\partial}{\partial v_j} + v_j\,\mathbf{k}'\cdot\frac{\partial}{\partial \mathbf{v}}\right) \\
& \times \left\{\frac{1}{\omega - \omega' - (\mathbf{k}-\mathbf{k}')\cdot\mathbf{v}}\left[\left((\omega' - \mathbf{k}'\cdot\mathbf{v})\frac{\partial}{\partial v_j} + v_j\,\mathbf{k}'\cdot\frac{\partial}{\partial \mathbf{v}}\right)\right.\right. \\
& \left.\left. \times \frac{1}{\omega - \mathbf{k}\cdot\mathbf{v}}\mathbf{k}\cdot\frac{\partial}{\partial \mathbf{v}} - \mathbf{k}\cdot\frac{\partial}{\partial \mathbf{v}}\left(\frac{\partial}{\partial v_j} + \frac{v_j}{\omega' - \mathbf{k}'\cdot\mathbf{v}}\mathbf{k}'\cdot\frac{\partial}{\partial \mathbf{v}}\right)\right] F_a\right\}.
\end{aligned}
\tag{7.93}
$$

The third quantity T_2 can similarly be manipulated as follows in terms of T_1 and a new quantity, Q:

$$
\begin{aligned}
T_3 = & Q - T_1 \\
Q = & -\frac{1}{2}\frac{e_a^2}{m_a^2}\frac{\omega_{pa}^2}{\omega^2\,k'^2}\int d\mathbf{v}\,\frac{v_i}{\omega - \mathbf{k}\cdot\mathbf{v}}\mathbf{k}'\cdot\frac{\partial}{\partial \mathbf{v}}\left\{\frac{1}{\omega - \omega' - (\mathbf{k}-\mathbf{k}')\cdot\mathbf{v}}\right. \\
& \times \left[\mathbf{k}'\cdot\frac{\partial}{\partial \mathbf{v}}\left(\frac{\partial}{\partial v_i} + \frac{v_i}{\omega - \mathbf{k}\cdot\mathbf{v}}\mathbf{k}\cdot\frac{\partial}{\partial \mathbf{v}}\right)\right. \\
& \left.\left. - \left((\omega - \mathbf{k}\cdot\mathbf{v})\frac{\partial}{\partial v_i} + v_i\,\mathbf{k}\cdot\frac{\partial}{\partial \mathbf{v}}\right)\frac{1}{\omega' - \mathbf{k}'\cdot\mathbf{v}}\mathbf{k}'\cdot\frac{\partial}{\partial \mathbf{v}}\right] F_a\right\}.
\end{aligned}
\tag{7.94}
$$

To compute the quantities P and Q, we introduce shorthand notations

$$r = \omega - \mathbf{k}\cdot\mathbf{v}, \qquad r' = \omega' - \mathbf{k}'\cdot\mathbf{v},$$
$$\partial_i = \frac{\partial}{\partial v_i}, \qquad R = \omega - \omega' - (\mathbf{k}-\mathbf{k}')\cdot\mathbf{v}. \tag{7.95}$$

Then, we have

$$\begin{aligned} P = &-\frac{1}{2}\frac{e_a^2}{m_a^2}\frac{\omega_{pa}^2}{\omega'^2 k^2}\int d\mathbf{v}\,\frac{1}{r}\,(r'\,\partial_j + v_j\,\mathbf{k}'\cdot\partial) \\ &\times\left\{\frac{1}{R}\left[(r'\,\partial_j + v_j\,\mathbf{k}'\cdot\partial)\left(\frac{1}{r}\,(\mathbf{k}\cdot\partial)\,F_a\right)\right.\right. \\ &\left.\left.-(\mathbf{k}\cdot\partial)\left(\partial_j\,F_a + \frac{v_j}{r'}\,(\mathbf{k}'\cdot\partial)\,F_a\right)\right]\right\}, \end{aligned} \tag{7.96}$$

where we may manipulate the various terms in (7.96) as shown below in (7.97),

$$\begin{aligned} r'\,\partial_j\left(\frac{1}{r}\,(\mathbf{k}\cdot\partial)\,F_a\right) &= \frac{r'}{r}\,\partial_j\,(\mathbf{k}\cdot\partial)\,F_a + k_j\,\frac{r'}{r^2}\,(\mathbf{k}\cdot\partial)\,F_a, \\ v_j\,(\mathbf{k}'\cdot\partial)\left(\frac{1}{r}\,(\mathbf{k}\cdot\partial)\,F_a\right) &= \frac{v_j}{r}\,(\mathbf{k}'\cdot\partial)(\mathbf{k}\cdot\partial)\,F_a + \frac{(\mathbf{k}\cdot\mathbf{k}')\,v_j}{r^2}\,(\mathbf{k}\cdot\partial)\,F_a, \\ (\mathbf{k}\cdot\partial)\left(\frac{v_j}{r'}\,(\mathbf{k}'\cdot\partial)\,F_a\right) &= \frac{v_j}{r'}\,(\mathbf{k}\cdot\partial)(\mathbf{k}'\cdot\partial)\,F_a \\ &\quad + \frac{(\mathbf{k}\cdot\mathbf{k}')\,v_j}{r'^2}\,(\mathbf{k}'\cdot\partial)\,F_a + \frac{k_j}{r'}\,(\mathbf{k}'\cdot\partial)\,F_a. \end{aligned} \tag{7.97}$$

Making use of these results, we obtain

$$\begin{aligned} P = &-\frac{1}{2}\frac{e_a^2}{m_a^2}\frac{\omega_{pa}^2}{\omega'^2 k^2}\int d\mathbf{v}\,\frac{1}{r}\,(r'\,\partial_j + v_j\,\mathbf{k}'\cdot\partial) \\ &\times\left\{\frac{1}{R}\left[r'\,\partial_j\left(\frac{1}{r}\,(\mathbf{k}\cdot\partial)\,F_a\right) + v_j\,(\mathbf{k}'\cdot\partial)\left(\frac{1}{r}\,(\mathbf{k}\cdot\partial)\,F_a\right)\right.\right. \\ &\left.\left.-(\mathbf{k}\cdot\partial)\,\partial_j\,F_a - (\mathbf{k}\cdot\partial)\left(\frac{v_j}{r'}\,(\mathbf{k}'\cdot\partial)\,F_a\right)\right]\right\} \\ = &\,\frac{1}{2}\frac{e_a^2}{m_a^2}\frac{\omega_{pa}^2}{\omega'^2 k^2}\int d\mathbf{v}\left(\partial_j\,\frac{r'}{r} + (\mathbf{k}'\cdot\partial)\,v_j\,\frac{1}{r}\right) \\ &\times\left\{-\frac{1}{r}\,\partial_j\,(\mathbf{k}\cdot\partial)\,F_a - k_j\,\frac{1}{r^2}\,(\mathbf{k}\cdot\partial)\,F_a - v_j\,\frac{1}{r\,r'}\,(\mathbf{k}\cdot\partial)(\mathbf{k}'\cdot\partial)\,F_a\right. \\ &+k_j\,\frac{1}{r\,R}\,(\mathbf{k}\cdot\partial)\,F_a - k_j\,\frac{1}{r'\,R}\,(\mathbf{k}'\cdot\partial)\,F_a \\ &\left.+\,v_j\,\frac{(\mathbf{k}\cdot\mathbf{k}')}{r^2\,R}\,(\mathbf{k}\cdot\partial)\,F_a - v_j\,\frac{(\mathbf{k}\cdot\mathbf{k}')}{r'^2\,R}\,(\mathbf{k}'\cdot\partial)\,F_a\right\}. \end{aligned} \tag{7.98}$$

Making use of

$$\partial_j \frac{r'}{r} + (\mathbf{k}' \cdot \partial)\, v_j \frac{1}{r} = k_j \frac{r'}{r^2} + v_j \frac{(\mathbf{k} \cdot \mathbf{k}')}{r^2}, \tag{7.99}$$

we have

$$\begin{aligned} P = \frac{1}{2}\frac{e_a^2}{m_a^2}\frac{\omega_{pa}^2}{\omega'^2 k^2} \int d\mathbf{v} \Bigg\{ & -\frac{k_j r' + v_j (\mathbf{k}\cdot\mathbf{k}')}{r^3} (\mathbf{k}\cdot\partial)\,\partial_j F_a \\ & - \frac{k^2 r' + (\mathbf{k}\cdot\mathbf{k}')(\mathbf{k}\cdot\mathbf{v})}{r^4} (\mathbf{k}\cdot\partial) F_a - \frac{(\mathbf{k}\cdot\mathbf{v})\, r' + (\mathbf{k}\cdot\mathbf{k}')\, v^2}{r^3 r'} (\mathbf{k}\cdot\partial)(\mathbf{k}'\cdot\partial) F_a \\ & - \frac{k^2}{r^3} (\mathbf{k}\cdot\partial) F_a + \frac{k^2}{r^2 r'} (\mathbf{k}'\cdot\partial) F_a - \frac{(\mathbf{k}\cdot\mathbf{k}')(\mathbf{k}\cdot\mathbf{v})}{r^4} (\mathbf{k}\cdot\partial) F_a \\ & + \frac{(\mathbf{k}\cdot\mathbf{k}')(\mathbf{k}\cdot\mathbf{v})}{r^2 r'^2} (\mathbf{k}'\cdot\partial) F_a + \frac{k^2}{r^2 R} (\mathbf{k}\cdot\partial) F_a - \frac{k^2}{r\, r' R} (\mathbf{k}'\cdot\partial) F_a \\ & + \frac{(\mathbf{k}\cdot\mathbf{k}')(\mathbf{k}\cdot\mathbf{v})}{r^3 R} (\mathbf{k}\cdot\partial) F_a - \frac{(\mathbf{k}\cdot\mathbf{k}')(\mathbf{k}\cdot\mathbf{v})}{r\, r'^2 R} (\mathbf{k}'\cdot\partial) F_a \\ & + \frac{(\mathbf{k}\cdot\mathbf{k}')(\mathbf{k}\cdot\mathbf{v})}{r^3 R} (\mathbf{k}\cdot\partial) F_a - \frac{(\mathbf{k}\cdot\mathbf{k}')(\mathbf{k}\cdot\mathbf{v})}{r^2 r' R} (\mathbf{k}'\cdot\partial) F_a \\ & + \frac{(\mathbf{k}\cdot\mathbf{k}')^2 v^2}{r^4 R} (\mathbf{k}\cdot\partial) F_a - \frac{(\mathbf{k}\cdot\mathbf{k}')^2 v^2}{r^2 r'^2 R} (\mathbf{k}'\cdot\partial) F_a \Bigg\}. \end{aligned} \tag{7.100}$$

If we retain only the leading terms, that is, those terms that contain R in the denominator and without the velocity-dependent terms in the numerator, then we have

$$P = \frac{1}{2}\frac{e_a^2}{m_a^2}\frac{\omega_{pa}^2}{\omega'^2 k^2} \int d\mathbf{v} \left\{ \frac{k^2}{r^2 R} (\mathbf{k}\cdot\partial) F_a - \frac{k^2}{r\, r' R} (\mathbf{k}'\cdot\partial) F_a \right\}. \tag{7.101}$$

Upon approximating $r \approx \omega$ and $r' \approx \omega'$, we further simplify (7.101) as

$$\begin{aligned} P = & \frac{1}{2}\frac{e_a^2}{m_a^2}\frac{\omega_{pa}^2}{\omega\,\omega'^2} \int d\mathbf{v} \frac{1}{\omega - \omega' - (\mathbf{k}-\mathbf{k}')\cdot\mathbf{v}} \\ & \times \left(\frac{1}{\omega}\mathbf{k}\cdot\frac{\partial F_a}{\partial \mathbf{v}} - \frac{1}{\omega'}\mathbf{k}'\cdot\frac{\partial F_a}{\partial \mathbf{v}} \right). \end{aligned} \tag{7.102}$$

For actual applications of this formula to the induced scattering problem, we may further approximate $\omega \sim \omega' \sim \omega_{pe}$. In what follows, we are only interested in the electron response. Then, we finally have

$$\begin{aligned} P &= \frac{1}{2}\frac{e^2}{m_e^2}\frac{1}{\omega_{pe}^2} \int d\mathbf{v} \frac{(\mathbf{k}-\mathbf{k}')\cdot\partial F_e/\partial\mathbf{v}}{\omega - \omega' - (\mathbf{k}-\mathbf{k}')\cdot\mathbf{v}} \\ &= \frac{1}{2}\frac{e^2}{m_e^2}\frac{(\mathbf{k}-\mathbf{k}')^2}{\omega_{pe}^4} \chi_\parallel^e(\mathbf{k}-\mathbf{k}', \omega-\omega'). \end{aligned} \tag{7.103}$$

Inserting (7.103) to (7.93), we arrive at

$$T_2 = \frac{k_i\,k_l}{k^2}\left(\delta_{jk} - \frac{k'_j\,k'_k}{k'^2}\right)\chi^{e(3)}_{ijkl}(\mathbf{k}',\omega'|-\mathbf{k}',\omega'|\mathbf{k},\omega)$$
$$= \frac{e^2}{2m_e^2}\frac{(\mathbf{k}\times\mathbf{k}')^2}{k^2\,k'^2}\frac{(\mathbf{k}-\mathbf{k}')^2}{\omega_{pe}^4}\,\chi^e_\parallel(\mathbf{k}-\mathbf{k}',\omega-\omega'). \tag{7.104}$$

We proceed in a similar manner for Q:

$$Q = -\frac{1}{2}\frac{e_a^2}{m_a^2}\frac{\omega_{pa}^2}{\omega^2\,k'^2}\int d\mathbf{v}\,\frac{v_i}{r}\,(\mathbf{k}'\cdot\partial)\left\{\frac{1}{R}\left[(\mathbf{k}'\cdot\partial)\left(\partial_i\,F_a + \frac{v_i}{r}\,(\mathbf{k}\cdot\partial)\,F_a\right)\right.\right.$$
$$\left.\left. -(r\,\partial_i + v_i\,\mathbf{k}\cdot\partial)\left(\frac{1}{r'}\,(\mathbf{k}'\cdot\partial)\,F_a\right)\right]\right\}. \tag{7.105}$$

Making use of

$$(\mathbf{k}'\cdot\partial)\left(\frac{v_i}{r}\,(\mathbf{k}\cdot\partial)\,F_a\right) = \frac{v_i}{r}\,(\mathbf{k}\cdot\partial)(\mathbf{k}'\cdot\partial)\,F_a$$
$$+\frac{(\mathbf{k}\cdot\mathbf{k}')\,v_i}{r^2}\,(\mathbf{k}\cdot\partial)\,F_a + \frac{k'_i}{r}\,(\mathbf{k}\cdot\partial)\,F_a,$$
$$r\,\partial_i\left(\frac{1}{r'}\,(\mathbf{k}'\cdot\partial)\,F_a\right) = \frac{r}{r'}\,\partial_i\,(\mathbf{k}'\cdot\partial)\,F_a + k'_i\,\frac{r}{r'^2}\,(\mathbf{k}'\cdot\partial)\,F_a, \tag{7.106}$$
$$v_i\,(\mathbf{k}\cdot\partial)\left(\frac{1}{r'}\,(\mathbf{k}'\cdot\partial)\,F_a\right) = \frac{v_i}{r'}\,(\mathbf{k}\cdot\partial)(\mathbf{k}'\cdot\partial)\,F_a + \frac{(\mathbf{k}\cdot\mathbf{k}')\,v_i}{r'^2}\,(\mathbf{k}'\cdot\partial)\,F_a,$$

it is possible to show that

$$Q = -\frac{1}{2}\frac{e_a^2}{m_a^2}\frac{\omega_{pa}^2}{\omega^2\,k'^2}\int d\mathbf{v}\,\frac{v_i}{r}\,(\mathbf{k}'\cdot\partial)$$
$$\times\left\{\frac{1}{R}\left[(\mathbf{k}'\cdot\partial)\,\partial_i\,F_a + (\mathbf{k}'\cdot\partial)\left(\frac{v_i}{r}\,(\mathbf{k}\cdot\partial)\,F_a\right)\right.\right.$$
$$\left.\left. - r\partial_i\left(\frac{1}{r'}\,(\mathbf{k}'\cdot\partial)\,F_a\right) - v_i\,(\mathbf{k}\cdot\partial)\left(\frac{1}{r'}\,(\mathbf{k}'\cdot\partial)\,F_a\right)\right]\right\}$$
$$= \frac{1}{2}\frac{e_a^2}{m_a^2}\frac{\omega_{pa}^2}{\omega^2\,k'^2}\int d\mathbf{v}\left(\frac{k'_i}{r} + \frac{(\mathbf{k}\cdot\mathbf{k}')\,v_i}{r^2}\right)$$
$$\times\left\{-\frac{1}{r'}\,(\mathbf{k}'\cdot\partial)\,\partial_i\,F_a - k'_i\,\frac{1}{r'^2}\,(\mathbf{k}'\cdot\partial)\,F_a - v_i\,\frac{1}{r\,r'}\,(\mathbf{k}\cdot\partial)(\mathbf{k}'\cdot\partial)\,F_a\right.$$
$$+\frac{1}{R}\left[\frac{k'_i}{r}\,(\mathbf{k}\cdot\partial)\,F_a - \frac{k'_i}{r'}\,(\mathbf{k}'\cdot\partial)\,F_a\right.$$
$$\left.\left.+\frac{(\mathbf{k}\cdot\mathbf{k}')\,v_i}{r^2}\,(\mathbf{k}\cdot\partial)\,F_a - \frac{(\mathbf{k}\cdot\mathbf{k}')\,v_i}{r'^2}\,(\mathbf{k}'\cdot\partial)\,F_a\right]\right\}$$
$$\approx \frac{1}{2}\frac{e_a^2}{m_a^2}\frac{\omega_{pa}^2}{\omega^2}\int d\mathbf{v}\,\frac{1}{r\,R}\left(\frac{1}{r}\,(\mathbf{k}\cdot\partial)\,F_a - \frac{1}{r'}\,(\mathbf{k}'\cdot\partial)\,F_a\right). \tag{7.107}$$

For electrons in the context of induced scattering, we have

$$Q = \frac{e^2}{2m_e^2\,\omega_{pe}^2}\int d\mathbf{v}\,\frac{(\mathbf{k}-\mathbf{k}')\cdot\partial F_e/\partial\mathbf{v}}{\omega-\omega'-(\mathbf{k}-\mathbf{k}')\cdot\mathbf{v}}$$
$$= \frac{e^2}{2m_e^2}\frac{(\mathbf{k}-\mathbf{k}')^2}{\omega_{pe}^2}\,\chi_\parallel^e(\mathbf{k}-\mathbf{k}',\omega-\omega') = P, \tag{7.108}$$

which leads to

$$T_3 = \left(\delta_{il}-\frac{k_i\,k_l}{k^2}\right)\frac{k'_j\,k'_k}{k'^2}\,\chi_{ijkl}^{e(3)}(\mathbf{k}',\omega'|-\mathbf{k}',\omega'|\mathbf{k},\omega)$$
$$= \frac{e^2}{2m_e^2}\frac{(\mathbf{k}\times\mathbf{k}')^2}{k^2\,k'^2}\frac{(\mathbf{k}-\mathbf{k}')^2}{\omega_{pe}^4}\,\chi_\parallel^e(\mathbf{k}-\mathbf{k}',\omega-\omega'). \tag{7.109}$$

In summary, the third-order nonlinear susceptibility of interest is given by

$$T_1 \equiv \frac{k_i\,k_l}{k^2}\frac{k'_j\,k'_k}{k'^2}\,\chi_{ijkl}^{a(3)}(\mathbf{k}',\omega'|-\mathbf{k}',\omega'|\mathbf{k},\omega)$$
$$= \frac{e^2}{2m_e^2}\frac{(\mathbf{k}\cdot\mathbf{k}')^2}{k^2\,k'^2}\frac{(\mathbf{k}-\mathbf{k}')^2}{\omega^3\,\omega'}\,\chi_\parallel^e(\mathbf{k}-\mathbf{k}',\omega-\omega'),$$
$$T_2 \equiv \frac{k_i\,k_l}{k^2}\left(\delta_{jk}-\frac{k'_j\,k'_k}{k'^2}\right)\chi_{ijkl}^{a(3)}(\mathbf{k}',\omega'|-\mathbf{k}',\omega'|\mathbf{k},\omega)$$
$$= \frac{e^2}{2m_e^2}\frac{(\mathbf{k}\times\mathbf{k}')^2}{k^2\,k'^2}\frac{(\mathbf{k}-\mathbf{k}')^2}{\omega_{pe}^4}\,\chi_\parallel^e(\mathbf{k}-\mathbf{k}',\omega-\omega'),$$
$$T_3 \equiv \left(\delta_{il}-\frac{k_i\,k_l}{k^2}\right)\frac{k'_j\,k'_k}{k'^2}\,\chi_{ijkl}^{a(3)}(\mathbf{k}',\omega'|-\mathbf{k}',\omega'|\mathbf{k},\omega)$$
$$= \frac{e^2}{2m_e^2}\frac{(\mathbf{k}\times\mathbf{k}')^2}{k^2\,k'^2}\frac{(\mathbf{k}-\mathbf{k}')^2}{\omega_{pe}^4}\,\chi_\parallel^e(\mathbf{k}-\mathbf{k}',\omega-\omega'). \tag{7.110}$$

8

Electromagnetic Vlasov Weak Turbulence Theory

In this chapter, we formulate the detailed weak turbulence theory for unmagnetized plasmas interacting through fully electromagnetic force. Unlike Chapter 7, where longitudinal and transverse mode equations are given in formal representations, the results to be derived in this chapter relate directly to the plasma eigenmodes, and the mathematical expressions of the governing equations are explicit in that they can be solved by direct numerical means, or may be analyzed in detail from a theoretical standpoint. The development of the theory is based upon formal equations (7.51) and (7.54) for wave kinetic equations, as well as the simplified quasilinear particle kinetic equation (7.42) or, equivalently, (7.43). The discussion also makes use of specific forms of linear and nonlinear susceptibilities, which were extensively analyzed in Section 7.3. We begin with the discussion of linear theory of plasma eigenmodes.

8.1 Linear Theory of Electromagnetic Plasma Eigenmodes

8.1.1 Linear Dispersion Relation

Linear dispersion relations for longitudinal and transverse waves determine their respective wave properties,

$$\mathrm{Re}\,\epsilon_{\parallel}(\mathbf{k},\omega_{\mathbf{k}}^{\alpha}) = 0, \quad (\alpha = L, S), \qquad \mathrm{Re}\,\Lambda_{\perp}(\mathbf{k},\omega_{\mathbf{k}}^{T}) = 0. \tag{8.1}$$

For $\alpha = L, S$, we have already discussed the linear dispersive properties in Section 1.8. Specifically, the Langmuir and ion-acoustic mode dispersion relations are given, respectively, by

$$\omega_{\mathbf{k}}^{L} = \omega_{pe}\left(1 + \frac{3}{2}k^2\lambda_{De}^2\right), \qquad \omega_{-\mathbf{k}}^{L} = -\omega_{\mathbf{k}}^{L},$$
$$\omega_{\mathbf{k}}^{S} = \frac{kc_S\,(1+3T_i/T_e)^{1/2}}{(1+k^2\lambda_{De}^2)^{1/2}}, \qquad \omega_{-\mathbf{k}}^{S} = -\omega_{\mathbf{k}}^{S}. \tag{8.2}$$

In (8.2), Debye length and ion sound speed are defined by $\lambda_{De}^2 = T_e/(4\pi\hat{n}e^2) = v_{Te}^2/(2\omega_{pe}^2)$ and $c_S^2 = T_e/m_i$, as already discussed previously – see, e.g., (3.26). We have also derived in (1.151) the following useful properties associated with longitudinal modes:

$$\frac{\partial\,\mathrm{Re}\,\epsilon_\parallel(\mathbf{k},\sigma\omega_\mathbf{k}^L)}{\partial(\sigma\omega_\mathbf{k}^L)} \approx \frac{2}{\sigma\omega_\mathbf{k}^L}, \qquad \frac{\partial\,\mathrm{Re}\,\epsilon_\parallel(\mathbf{k},\sigma\omega_\mathbf{k}^S)}{\partial(\sigma\omega_\mathbf{k}^S)} \approx \frac{2}{\sigma\mu_\mathbf{k}\omega_\mathbf{k}^L}, \tag{8.3}$$

where

$$\mu_\mathbf{k} = k^3\lambda_{De}^3\sqrt{\frac{m_e}{m_i}}\sqrt{1+\frac{3T_i}{T_e}} = \left(\frac{kv_{Te}}{\omega_{pe}}\right)^3\sqrt{\frac{m_e}{8m_i}}\sqrt{1+\frac{3T_i}{T_e}}$$

is defined in (3.29).

In the present electromagnetic theory of plasma waves, we now have an additional transverse wave dispersion relation, which, under the cold-plasma approximation, is given by

$$\mathrm{Re}\,\Lambda_\perp(\mathbf{k},\omega_\mathbf{k}^T) \approx 1 - \frac{\omega_{pe}^2}{\omega^2} - \frac{c^2k^2}{\omega^2}. \tag{8.4}$$

Solving for the transverse mode dispersion relation, we obtain the well-known result

$$\omega_\mathbf{k}^T = \sqrt{\omega_{pe}^2 + c^2k^2}, \qquad \omega_{-\mathbf{k}}^T = -\omega_\mathbf{k}^T. \tag{8.5}$$

It is also useful to compute the following ($\sigma = \pm 1$):

$$\frac{\partial\,\mathrm{Re}\,\Lambda_\perp(\mathbf{k},\sigma\omega_\mathbf{k}^T)}{\partial(\sigma\omega_\mathbf{k}^T)} \approx \frac{2}{\sigma\omega_\mathbf{k}^T}. \tag{8.6}$$

8.1.2 Linear Instabilities for Unmagnetized Plasmas

We next discuss some well-known electrostatic and electromagnetic linear instabilities for unmagnetized plasmas. Weak electrostatic electron beam-plasma, or bump-on-tail instability, was already discussed in Section 1.8. The present section further extends the discussion to include other instabilities. It is convenient to start from linear dielectric constants for longitudinal and transverse modes, $\epsilon_\parallel(\mathbf{k},\omega)$ and $\epsilon_\perp(\mathbf{k},\omega)$. The waves and instabilities correspond to the solutions of complex algebraic (generally transcendental) equations,

$$0 = \epsilon_\parallel(\mathbf{k},\omega) = 1 + \sum_a \frac{\omega_{pa}^2}{k^2}\int d\mathbf{v}\,\frac{\mathbf{k}\cdot\partial F_a/\partial\mathbf{v}}{\omega - \mathbf{k}\cdot\mathbf{v}},$$

$$0 = \Lambda_\perp(\mathbf{k},\omega) = \epsilon_\perp(\mathbf{k},\omega) - \frac{c^2k^2}{\omega^2} = 1 - \frac{c^2k^2}{\omega^2} \tag{8.7}$$
$$+ \sum_a \frac{\omega_{pa}^2}{2\omega^2} \int d\mathbf{v}\, \frac{[(\mathbf{k}\times\mathbf{v})\times\mathbf{k}]_i}{k^2} \left(\frac{\partial F_a}{\partial v_l} + \frac{v_i}{\omega - \mathbf{k}\cdot\mathbf{v}}\, \mathbf{k}\cdot \frac{\partial F_a}{\partial \mathbf{v}} \right).$$

Without loss of generality, we may take $\mathbf{k} = \hat{\mathbf{z}}k$. Then $\mathbf{k}\times\mathbf{v} = \hat{\mathbf{x}}k^2 v_x + \hat{\mathbf{y}}k^2 v_y$. If we assume that F_a depends only on v_z and $v_\perp^2 = v_x^2 + v_y^2$, then we have

$$0 = 1 + \sum_a \frac{\omega_{pa}^2}{k} \int d\mathbf{v}\, \frac{\partial F_a/\partial v_z}{\omega - k\,v_z},$$
$$0 = 1 - \frac{c^2k^2}{\omega^2} - \sum_a \frac{\omega_{pa}^2}{\omega^2} + \sum_a \frac{\omega_{pa}^2}{2\omega^2} \int d\mathbf{v}\, \frac{kv_\perp^2}{\omega - k\,v_z} \frac{\partial F_a}{\partial v_z}, \tag{8.8}$$

where we have made use of $\int d\mathbf{v}\; v_\perp^2\, \partial F_a/\partial v_\perp^2 = -1$.

Electrostatic Instabilities

For electrostatic instabilities, the dependence of F_a on $v_\perp^2$ is irrelevant. We may simply define the reduced distribution,

$$\hat{f}_a(v_z) = 2\pi \int_0^\infty dv_\perp\, v_\perp\, F_a(v_\perp^2, v_z), \tag{8.9}$$

and rewrite the dispersion equation as

$$0 = \epsilon_\parallel(\omega,k) = 1 + \sum_a \frac{\omega_{pa}^2}{k} \int_{-\infty}^{\infty} dv_z\, \frac{1}{\omega - k\,v_z} \frac{\partial \hat{f}_a(v_z)}{\partial v_z}. \tag{8.10}$$

In the bulk part of this book, we take the approach of treating ω as real, and discuss the growth of unstable mode by wave kinetic equation. However, in the customary approach of linear theory of plasma waves, the angular frequency is generally treated as complex, $\omega = \mathrm{Re}(\omega) + i\mathrm{Im}(\omega) = \omega_r + i\gamma$. If the growth rate or imaginary part, $\mathrm{Im}(\omega)$, is much lower than the real frequency, $\mathrm{Re}(\omega)$, then the growth rate γ is computed on the basis of the weak growth rate formula (1.154). Such a situation is known as the "dissipative" instabilities in the literature. The weak turbulence theory of this book is thus applicable to dissipative instabilities. The growth rate (1.154) is equivalent to the induced emission terms in the wave kinetic equation, except for a factor 2. When the condition $\mathrm{Im}(\omega) \ll \mathrm{Re}(\omega)$ is not satisfied, the instability is called "reactive," and such instabilities generally do not lend themselves to the present weak turbulence theoretical treatment. Rather, more precisely speaking, the weak turbulence theory of this book must be generalized to treat the reactive instability, a short discussion of which is given in Appendix E.

An example of reactive instabilities may be the electromagnetic Weibel instability, to be discussed later.

In this section, we will take the approach of treating the growth rate via (1.154). The first term on the right-hand side of wave kinetic equation (3.27), (3.28), (4.93), or (4.94) is equivalent to twice the growth rate (1.154):

$$2\omega_i \equiv -\frac{2\,\mathrm{Im}\,\epsilon_\parallel(\omega,k)}{\partial \mathrm{Re}\,\epsilon_\parallel(\omega,k)/\partial\omega}$$
$$= \frac{2\pi}{\partial \mathrm{Re}\,\epsilon_\parallel(\omega,k)/\partial\omega}\sum_a \frac{\omega_{pa}^2}{k}\int_{-\infty}^{\infty} dv_z\,\delta(\omega - k\,v_z)\,\frac{\partial \hat{f}_a(v_z)}{\partial v_z}. \tag{8.11}$$

An example of weakly growing, or dissipative instability, is the bump-on-tail instability, already discussed Section 1.8. For Langmuir waves, the bump-on-tail instability growth rate (1.155) is given by

$$\omega_i = \frac{\pi\,\omega}{2}\sum_a \frac{\omega_{pa}^2}{k^2}\frac{\partial \hat{f}_a(v_z)}{\partial v_z}\bigg|_{v_z=\omega/k}. \tag{8.12}$$

For a combination of thermal plus a tenuous drifting Gaussian model (1.156),

$$\hat{f}_a(v_z) = \frac{e^{-v_z^2/v_{Te}^2}}{\pi^{1/2}\,v_{Te}} + \frac{\delta e^{-(v_z-U_b)^2/v_{Tb}^2}}{\pi^{1/2}\,v_{Tb}}, \tag{8.13}$$

where $\delta \ll 1$ represents the density ratio between the beam and background electrons, and U_b designates the drift associated with tenuous energetic beam, the weak growth rate for Langmuir wave is given by

$$\frac{\omega_i}{\omega_{pe}} = -\frac{\pi^{1/2}}{2k^2\lambda_{De}^2}\left[\frac{1}{\sqrt{2}\,k\,\lambda_{De}}\exp\left(-\frac{1+3k^2\lambda_{De}^2}{2k^2\lambda_{De}^2}\right)\right.$$
$$\left. -\,\delta\,\frac{T_e}{T_b}\frac{U_b-\omega_{pe}/k}{v_{Tb}}\exp\left(-\frac{(\omega_{pe}-k\,U_b)^2}{k^2\,v_{Tb}^2}\right)\right]. \tag{8.14}$$

This result was already discussed in (1.158), and the first term on the right-hand side depicts Landau damping, while the second term indicates bump-on-tail instability. If the beam speed U_b is higher than the Langmuir wave phase speed, $v_\varphi = \omega_{pe}/k$, then the instability ($\omega_i > 0$) is possible, provided the growth rate exceeds the Landau damping rate. Chapter 6 was devoted to the analysis of bump-on-tail instability and its nonlinear consequence.

Another example of electrostatic dissipative instability is the ion-acoustic instability. For slow ion-sound waves, whose dispersion relation is given by (8.2), or

$$\omega^2 = k^2\,\frac{T_e}{m_i}\left(1+\frac{3T_i}{T_e}\right)\frac{1}{1+k^2\,v_{Te}^2/2\,\omega_{pe}^2}, \tag{8.15}$$

it follows from (1.155) that the growth rate is given by

$$\omega_i = \frac{\pi\omega^4}{2\omega_{pi}^2}\sum_a \frac{\omega_{pa}^2}{k^2}\frac{\partial \hat{f}_a(v_z)}{\partial v_z}\bigg|_{v_z=\omega/k}. \tag{8.16}$$

In writing this equation, we have already assumed that T_i/T_e is small. It is well known that the ion-acoustic mode is heavily damped if T_i/T_e is large – see Figure 1.3. For stationary ions and drifting Gaussian electrons with the drift speed significantly lower than electron thermal speed, we already discussed the ion-acoustic instability, whose growth rate in the limiting case of $T_e/T_i \gg 1$ is given by (1.168), which is reproduced here, for the sake of convenience,

$$\omega_i = \frac{\pi^{1/2}}{2^{3/2}}\left(\frac{m_e}{m_i}\right)^{1/2}\left(\frac{V_e}{c_S} - \frac{1}{\sqrt{1+k^2\lambda_{De}^2}}\right)\frac{k^2c_S^2}{(1+k^2\lambda_{De}^2)^2}. \tag{8.17}$$

It is possible to apply the weak turbulence formalism for studying the nonlinear phase of ion-acoustic instability (Ishihara and Hirose, 1981), as we have done for the bump-in-tail instability, but we do not discuss such a problem in this book.

Electromagnetic Instability

Electromagnetic branch of the dispersion relation in (8.8) does not support any dissipative instabilities. Instead, it enjoys a purely growing unstable solution, that is, the Weibel instability (Weibel, 1959), which is a prime example of reactive instability, since the purely growing (or aperiod) mode satisfies $\omega_i \neq 0$ and $\omega_r = 0$. As we already noted, Weibel instability does not lend itself to the weak turbulence formalism developed in this book. However, we may discuss the Weibel instability for the sake of completeness. Consider the electromagnetic dispersion equation given by the second line of (8.8),

$$0 = 1 - \frac{c^2k^2}{\omega^2} - \sum_a \frac{\omega_{pa}^2}{\omega^2} + \sum_a \frac{\omega_{pa}^2}{2\omega^2}\int d\mathbf{v}\,\frac{kv_\perp^2}{\omega - kv_z}\frac{\partial F_a}{\partial v_z}. \tag{8.18}$$

The Weibel instability is characterized by zero real frequency, and it is characterized by $c^2k^2/|\omega_i|^2 \gg 1$. This is a generalized condition for slow mode, and thus (8.18) can be simplified by ignoring the unity term (the displacement current term),

$$c^2k^2 + \sum_a \omega_{pa}^2 = \sum_a \frac{\omega_{pa}^2 k}{2}\int d\mathbf{v}\,\frac{v_\perp^2}{\omega - k\,v_z}\frac{\partial F_a}{\partial v_z}. \tag{8.19}$$

If we define the perpendicular temperature

$$2\pi\int_0^\infty dv_\perp\, v_\perp^3\, F_a(\mathbf{v}) = \frac{2T_{\perp a}}{m_a}\,\hat{f}_a(v_z), \tag{8.20}$$

then (8.19) can be expressed as

$$c^2k^2 + \sum_a \omega_{pa}^2 = -\sum_a \omega_{pa}^2 \frac{k^2 T_{\perp a}}{m_a} \int_{-\infty}^{\infty} dv_z \frac{\hat{f}_a(v_z)}{(\omega - k v_z)^2}. \tag{8.21}$$

This equation supports a purely growing or damped solution, $\omega = i\omega_i$, with zero real frequency (that is, an aperiod mode). If we assume that $\hat{f}_a(v_z)$ is given by the Maxwellian form,

$$\hat{f}_a(v_z) = \frac{e^{-v_z^2/v_{Ta}^2}}{\pi^{1/2} v_{Ta}} \qquad v_{Ta}^2 = \frac{2T_{\parallel a}}{m_a}, \tag{8.22}$$

then upon inserting the model distribution, we obtain

$$0 = -c^2k^2 - \sum_a \omega_{pa}^2 + \sum_a \omega_{pa}^2 \frac{T_{\perp a}}{T_{\parallel a}} \left[1 + \frac{\omega}{kv_{Ta}} Z\left(\frac{\omega}{kv_{Ta}} \right) \right], \tag{8.23}$$

where $Z(\zeta)$ is the plasma dispersion function (1.122). The purely growing unstable mode is called the Weibel instability (Weibel, 1959). If we assume that

$$\left| \frac{\omega}{kv_z} \right| = \left| \frac{\omega_i}{kv_z} \right| \ll 1, \tag{8.24}$$

then we may approximately write (8.23) as

$$0 \approx -c^2k^2 - \sum_a \omega_{pa}^2 \left(1 - \frac{k^2 T_{\perp a}}{m_a \omega_i^2} \right), \tag{8.25}$$

and solving for ω_i^2, we obtain

$$\omega_i^2 = \frac{k^2 \sum_a \omega_{pa}^2 T_{\perp a}/m_a}{c^2k^2 + \sum_a \omega_{pa}^2}. \tag{8.26}$$

The maximum growth occurs for $k \to \infty$,

$$\omega_i^{\max} = \left(\sum_a \frac{T_{\perp a}^2}{m_a c^2} \omega_{pa}^2 \right)^{1/2}. \tag{8.27}$$

To add finite k effects, we include thermal correction term associated with the plasma dispersion function – see the series expansion (1.123),

$$0 \approx -c^2k^2 - \sum_a \omega_{pa}^2 + \sum_a \omega_{pa}^2 \frac{T_{\perp a}}{T_{\parallel a}} \left(1 - \frac{\sqrt{\pi}}{kv_{Ta}} \gamma_k \right). \tag{8.28}$$

This leads to

$$\omega_i = \gamma \frac{k}{k_0}\left(1 - \frac{k^2}{k_0^2}\right),$$

$$\gamma = \frac{k_0 v_{Ta}}{\sqrt{\pi}} \frac{c^2 k_0^2}{\sum_a \omega_{pa}^2 (T_{\perp a}/T_{\parallel a})},$$

$$k_0^2 = \sum_a \frac{\omega_{pa}^2}{c^2}\left(\frac{T_{\perp a}}{T_{\parallel a}} - 1\right). \tag{8.29}$$

According to this approximate analytical formula, the maximum growth should take place at $k^2/k_0^2 = 1/3$, and the maximum growth rate is given by

$$\omega_i^{\max} = \frac{2\gamma}{3^{3/2}}. \tag{8.30}$$

The instability also has a finite range of wave numbers over which the mode grows, $0 < k < k_0$.

An interesting extension of the aperiodic mode solution is when the plasma is isotropic, $T_{\perp a} = T_{\parallel a} = T_a$. Then the dispersion relation (8.28) leads to the solution

$$\gamma_k = -\frac{c^2 k^3}{\sqrt{\pi}\sum_a(\omega_{pa}^2/v_{Ta})}, \tag{8.31}$$

which is a purely damped mode with zero real frequency.

8.2 Electromagnetic Vlasov Weak Turbulence Theory for Eigenmodes

We next move on to the discussion of electromagnetic weak turbulence theory that involves linear eigenmodes, that is, two longitudinal modes, L and S, and the transverse mode, T. Linear wave properties determine the characteristics of the eigenmodes. We start from the formal wave kinetic equations (7.51) and (7.54), and quasilinear particle kinetic equation (7.42) or (7.43), which we derived in Chapter 7. For Langmuir and ion-sound waves, the nonlinear wave kinetic equation at the formal level is given by (7.51). If we take $\alpha = L$ explicitly, then we obtain the formal wave kinetic equation for L mode. For ion-sound wave, we take $\alpha = S$ explicitly. Similarly, the formal wave kinetic equation for transverse wave is already given by (7.53). Formal wave kinetic equations already obtained can be further manipulated by making use of various approximate but explicit forms of linear and nonlinear susceptibility response tensors discussed in Section 7.3.

8.2.1 Induced Emission

Induced emission terms are given by

$$\begin{aligned}
\left.\frac{\partial I_{\mathbf{k}}^{\sigma L}}{\partial t}\right|_{\text{ind.em.}} &= -\frac{2\,\text{Im}\,\epsilon_{\parallel}(\mathbf{k},\sigma\omega_{\mathbf{k}}^{L})}{\epsilon_{\parallel}'(\mathbf{k},\sigma\omega_{\mathbf{k}}^{L})}\,I_{\mathbf{k}}^{\sigma L},\\
\left.\frac{\partial I_{\mathbf{k}}^{\sigma S}}{\partial t}\right|_{\text{ind.em.}} &= -\frac{2\,\text{Im}\,\epsilon_{\parallel}(\mathbf{k},\sigma\omega_{\mathbf{k}}^{S})}{\epsilon_{\parallel}'(\mathbf{k},\sigma\omega_{\mathbf{k}}^{S})}\,I_{\mathbf{k}}^{\sigma S},\\
\left.\frac{\partial I_{\mathbf{k}}^{\sigma T}}{\partial t}\right|_{\text{ind.em.}} &= -\frac{2\,\text{Im}\,\Lambda_{\perp}(\mathbf{k},\sigma\omega_{\mathbf{k}}^{T})}{\Lambda_{\perp}'(\mathbf{k},\sigma\omega_{\mathbf{k}}^{T})}\,I_{\mathbf{k}}^{\sigma T}.
\end{aligned} \tag{8.32}$$

Making use of

$$\begin{aligned}
\text{Im}\,\epsilon_{\parallel}(\mathbf{k},\sigma\omega_{\mathbf{k}}^{L}) &= -\pi\sum_a\frac{\omega_{pa}^2}{k^2}\int d\mathbf{v}\;\delta(\sigma\omega_{\mathbf{k}}^{L}-\mathbf{k}\cdot\mathbf{v})\;\mathbf{k}\cdot\frac{\partial F_a}{\partial\mathbf{v}},\\
\text{Im}\,\epsilon_{\parallel}(\mathbf{k},\sigma\omega_{\mathbf{k}}^{S}) &= -\pi\sum_a\frac{\omega_{pa}^2}{k^2}\int d\mathbf{v}\;\delta(\sigma\omega_{\mathbf{k}}^{S}-\mathbf{k}\cdot\mathbf{v})\;\mathbf{k}\cdot\frac{\partial F_a}{\partial\mathbf{v}},\\
\text{Im}\,\Lambda_{\perp}(\mathbf{k},\sigma\omega_{\mathbf{k}}^{T}) &= -\pi\sum_a\frac{\omega_{pa}^2}{(\omega_{\mathbf{k}}^{T})^2}\int d\mathbf{v}\;\frac{(\mathbf{k}\times\mathbf{v})^2}{k^2}\;\delta(\sigma\omega_{\mathbf{k}}^{T}-\mathbf{k}\cdot\mathbf{v})\;\mathbf{k}\cdot\frac{\partial F_a}{\partial\mathbf{v}},
\end{aligned} \tag{8.33}$$

and

$$\begin{aligned}
\epsilon_{\parallel}'(\mathbf{k},\sigma\omega_{\mathbf{k}}^{L}) &= \frac{2}{\sigma\omega_{\mathbf{k}}^{L}},\qquad \epsilon_{\parallel}'(\mathbf{k},\sigma\omega_{\mathbf{k}}^{S}) = \frac{2}{\sigma\mu_{\mathbf{k}}\omega_{\mathbf{k}}^{L}},\\
\mu_{\mathbf{k}} &= k^3\lambda_{De}^3\left(\frac{m_e}{m_i}\right)^{1/2}\left(1+\frac{3T_i}{T_e}\right)^{1/2},\\
\Lambda_{\perp}'(\mathbf{k},\sigma\omega_{\mathbf{k}}^{T}) &= \frac{2}{\sigma\omega_{\mathbf{k}}^{T}},
\end{aligned} \tag{8.34}$$

we obtain the desired expressions describing induced emissions of plasma eigenmodes,

$$\begin{aligned}
\left.\frac{\partial I_{\mathbf{k}}^{\sigma L}}{\partial t}\right|_{\text{ind.em.}} &= \pi\,(\sigma\omega_{\mathbf{k}}^{L})\sum_a\frac{\omega_{pa}^2}{k^2}\int d\mathbf{v}\;\delta(\sigma\omega_{\mathbf{k}}^{L}-\mathbf{k}\cdot\mathbf{v})\;\mathbf{k}\cdot\frac{\partial F_a}{\partial\mathbf{v}}\,I_{\mathbf{k}}^{\sigma L},\\
\left.\frac{\partial I_{\mathbf{k}}^{\sigma S}}{\partial t}\right|_{\text{ind.em.}} &= \pi\,(\sigma\mu_{\mathbf{k}}\omega_{\mathbf{k}}^{L})\sum_a\frac{\omega_{pa}^2}{k^2}\int d\mathbf{v}\;\delta(\sigma\omega_{\mathbf{k}}^{S}-\mathbf{k}\cdot\mathbf{v})\;\mathbf{k}\cdot\frac{\partial F_a}{\partial\mathbf{v}}\,I_{\mathbf{k}}^{\sigma S},\\
\left.\frac{\partial I_{\mathbf{k}}^{\sigma T}}{\partial t}\right|_{\text{ind.em.}} &= \pi\sum_a\frac{\omega_{pa}^2}{\sigma\omega_{\mathbf{k}}^{T}}\int d\mathbf{v}\;\frac{(\mathbf{k}\times\mathbf{v})^2}{k^2}\;\delta(\sigma\omega_{\mathbf{k}}^{T}-\mathbf{k}\cdot\mathbf{v})\;\mathbf{k}\cdot\frac{\partial F_a}{\partial\mathbf{v}}\,I_{\mathbf{k}}^{\sigma T}.
\end{aligned} \tag{8.35}$$

In (8.35), we note that the induced emission for transverse electromagnetic wave is practically zero since the resonance condition cannot be satisfied for fast T mode. Nevertheless, we include the expression for the sake of completeness.

8.2.2 Decay/Coalescence

L Mode

Decay processes for L mode involve the following terms:

$$
\begin{aligned}
\left.\frac{\partial I_{\mathbf{k}}^{\sigma L}}{\partial t}\right|_{\text{decay}} = & -\pi\,(\sigma\omega_{\mathbf{k}}^{L})\,\mathrm{Im}\int d\mathbf{k}'\sum_{\sigma',\sigma''}\Big\{Q_1\big(\mu_{\mathbf{k}'}\,\sigma'\omega_{\mathbf{k}'}^{L}\,I_{\mathbf{k}-\mathbf{k}'}^{\sigma''L}\,I_{\mathbf{k}}^{\sigma L} \\
& +\sigma''\omega_{\mathbf{k}-\mathbf{k}'}^{L}\,I_{\mathbf{k}'}^{\sigma'S}\,I_{\mathbf{k}}^{\sigma L}-\sigma\omega_{\mathbf{k}}^{L}\,I_{\mathbf{k}-\mathbf{k}'}^{\sigma''L}\,I_{\mathbf{k}'}^{\sigma'S}\big)\,\delta(\sigma\omega_{\mathbf{k}}^{L}-\sigma'\omega_{\mathbf{k}'}^{S}-\sigma''\omega_{\mathbf{k}-\mathbf{k}'}^{L}) \\
& +Q_2\big(\sigma'\omega_{\mathbf{k}'}^{L}\,I_{\mathbf{k}-\mathbf{k}'}^{\sigma''S}\,I_{\mathbf{k}}^{\sigma L}+\mu_{\mathbf{k}-\mathbf{k}'}\,\sigma''\omega_{\mathbf{k}-\mathbf{k}'}^{L}\,I_{\mathbf{k}'}^{\sigma'L}\,I_{\mathbf{k}}^{\sigma L} \\
& -\sigma\omega_{\mathbf{k}}^{L}\,I_{\mathbf{k}-\mathbf{k}'}^{\sigma''S}\,I_{\mathbf{k}'}^{\sigma'L}\big)\,\delta(\sigma\omega_{\mathbf{k}}^{L}-\sigma'\omega_{\mathbf{k}'}^{L}-\sigma''\omega_{\mathbf{k}-\mathbf{k}'}^{S}) \\
& +Q_3\big(\sigma'\omega_{\mathbf{k}'}^{L}\,I_{\mathbf{k}-\mathbf{k}'}^{\sigma''T}\,I_{\mathbf{k}}^{\sigma L}+\sigma''\omega_{\mathbf{k}-\mathbf{k}'}^{T}\,2\,I_{\mathbf{k}'}^{\sigma'L}\,I_{\mathbf{k}}^{\sigma L} \\
& -\sigma\omega_{\mathbf{k}}^{L}\,I_{\mathbf{k}'}^{\sigma'L}\,I_{\mathbf{k}-\mathbf{k}'}^{\sigma''T}\big)\,\delta(\sigma\omega_{\mathbf{k}}^{L}-\sigma'\omega_{\mathbf{k}'}^{L}-\sigma''\omega_{\mathbf{k}-\mathbf{k}'}^{T}) \\
& +Q_4\big(\mu_{\mathbf{k}'}\,\sigma'\omega_{\mathbf{k}'}^{L}\,I_{\mathbf{k}-\mathbf{k}'}^{\sigma''T}\,I_{\mathbf{k}}^{\sigma L}+\sigma''\omega_{\mathbf{k}-\mathbf{k}'}^{T}\,2\,I_{\mathbf{k}'}^{\sigma'S}\,I_{\mathbf{k}}^{\sigma L} \\
& -\sigma\omega_{\mathbf{k}}^{L}\,I_{\mathbf{k}'}^{\sigma'S}\,I_{\mathbf{k}-\mathbf{k}'}^{\sigma''T}\big)\,\delta(\sigma\omega_{\mathbf{k}}^{L}-\sigma'\omega_{\mathbf{k}'}^{S}-\sigma''\omega_{\mathbf{k}-\mathbf{k}'}^{T}) \\
& +Q_5\big(\sigma'\omega_{\mathbf{k}'}^{T}\,2\,I_{\mathbf{k}-\mathbf{k}'}^{\sigma''T}\,I_{\mathbf{k}}^{\sigma L}+\sigma''\omega_{\mathbf{k}-\mathbf{k}'}^{T}\,2\,I_{\mathbf{k}'}^{\sigma'T}\,I_{\mathbf{k}}^{\sigma L} \\
& -\sigma\omega_{\mathbf{k}}^{L}\,I_{\mathbf{k}'}^{\sigma'T}\,I_{\mathbf{k}-\mathbf{k}'}^{\sigma''T}\big)\,\delta(\sigma\omega_{\mathbf{k}}^{L}-\sigma'\omega_{\mathbf{k}'}^{T}-\sigma''\omega_{\mathbf{k}-\mathbf{k}'}^{T})\Big\},
\end{aligned}
\tag{8.36}
$$

where the coupling coefficients are given by

$$
\begin{aligned}
Q_1 = & \frac{k_i\,k_n}{k^2}\frac{k'_j\,k'_m}{k'^2}\frac{(\mathbf{k}-\mathbf{k}')_k\,(\mathbf{k}-\mathbf{k}')_l}{|\mathbf{k}-\mathbf{k}'|^2} \\
& \times\chi_{ijk}^{(2)}(\mathbf{k}',\sigma'\omega_{\mathbf{k}'}^{S}|\mathbf{k}-\mathbf{k}',\sigma''\omega_{\mathbf{k}-\mathbf{k}'}^{L})\,\chi_{nml}^{(2)*}(\mathbf{k}',\sigma'\omega_{\mathbf{k}'}^{S}|\mathbf{k}-\mathbf{k}',\sigma''\omega_{\mathbf{k}-\mathbf{k}'}^{L}), \\
Q_2 = & \frac{k_i\,k_n}{k^2}\frac{k'_j\,k'_m}{k'^2}\frac{(\mathbf{k}-\mathbf{k}')_k\,(\mathbf{k}-\mathbf{k}')_l}{|\mathbf{k}-\mathbf{k}'|^2} \\
& \times\chi_{ijk}^{(2)}(\mathbf{k}',\sigma'\omega_{\mathbf{k}'}^{L}|\mathbf{k}-\mathbf{k}',\sigma''\omega_{\mathbf{k}-\mathbf{k}'}^{S})\,\chi_{nml}^{(2)*}(\mathbf{k}',\sigma'\omega_{\mathbf{k}'}^{L}|\mathbf{k}-\mathbf{k}',\sigma''\omega_{\mathbf{k}-\mathbf{k}'}^{S}), \\
Q_3 = & \frac{k_i\,k_n}{k^2}\frac{k'_j\,k'_m}{k'^2}\left(\delta_{kl}-\frac{(\mathbf{k}-\mathbf{k}')_k\,(\mathbf{k}-\mathbf{k}')_l}{|\mathbf{k}-\mathbf{k}'|^2}\right) \\
& \times\chi_{ijk}^{(2)}(\mathbf{k}',\sigma'\omega_{\mathbf{k}'}^{L}|\mathbf{k}-\mathbf{k}',\sigma''\omega_{\mathbf{k}-\mathbf{k}'}^{T})\,\chi_{nml}^{(2)*}(\mathbf{k}',\sigma'\omega_{\mathbf{k}'}^{L}|\mathbf{k}-\mathbf{k}',\sigma''\omega_{\mathbf{k}-\mathbf{k}'}^{T}), \\
Q_4 = & \frac{k_i\,k_n}{k^2}\frac{k'_j\,k'_m}{k'^2}\left(\delta_{kl}-\frac{(\mathbf{k}-\mathbf{k}')_k\,(\mathbf{k}-\mathbf{k}')_l}{|\mathbf{k}-\mathbf{k}'|^2}\right) \\
& \times\chi_{ijk}^{(2)}(\mathbf{k}',\sigma'\omega_{\mathbf{k}'}^{S}|\mathbf{k}-\mathbf{k}',\sigma''\omega_{\mathbf{k}-\mathbf{k}'}^{T})\,\chi_{nml}^{(2)*}(\mathbf{k}',\sigma'\omega_{\mathbf{k}'}^{S}|\mathbf{k}-\mathbf{k}',\sigma''\omega_{\mathbf{k}-\mathbf{k}'}^{T}),
\end{aligned}
$$

$$Q_5 = \frac{1}{4}\frac{k_i\,k_n}{k^2}\left(\delta_{jm} - \frac{k'_j\,k'_m}{k'^2}\right)\left(\delta_{kl} - \frac{(\mathbf{k}-\mathbf{k}')_k\,(\mathbf{k}-\mathbf{k}')_l}{|\mathbf{k}-\mathbf{k}'|^2}\right) \times \chi^{(2)}_{ijk}(\mathbf{k}',\sigma'\omega^T_{\mathbf{k}'}|\mathbf{k}-\mathbf{k}',\sigma''\omega^T_{\mathbf{k}-\mathbf{k}'})\,\chi^{(2)*}_{nml}(\mathbf{k}',\sigma'\omega^T_{\mathbf{k}'}|\mathbf{k}-\mathbf{k}',\sigma''\omega^T_{\mathbf{k}-\mathbf{k}'}). \tag{8.37}$$

In nonlinear response functions, only the electron contribution is of importance, since lighter electrons rapidly respond to finite amplitude perturbations, while the heavier ions lag behind. The approximate expressions for various second-order nonlinear susceptibility tensors in the context of decay processes in (8.36) are given by

$$\chi^{(2)}_{ijk}(\mathbf{k}',\sigma'\omega^S_{\mathbf{k}'}|\mathbf{k}-\mathbf{k}',\sigma''\omega^L_{\mathbf{k}-\mathbf{k}'}) = -\frac{i}{2}\frac{e}{m_e}\frac{\delta_{ik}\,k'_j\,\chi^e_\parallel(\mathbf{k}',\sigma'\omega^S_{\mathbf{k}'})}{(\sigma''\omega^L_{\mathbf{k}-\mathbf{k}'})(\sigma\omega^L_{\mathbf{k}})},$$

$$\chi^{(2)*}_{nml}(\mathbf{k}',\sigma'\omega^S_{\mathbf{k}'}|\mathbf{k}-\mathbf{k}',\sigma''\omega^L_{\mathbf{k}-\mathbf{k}'}) = \frac{i}{2}\frac{e}{m_e}\frac{\delta_{nl}\,k'_m\,\chi^e_\parallel(\mathbf{k}',\sigma'\omega^S_{\mathbf{k}'})}{(\sigma''\omega^L_{\mathbf{k}-\mathbf{k}'})(\sigma\omega^L_{\mathbf{k}})},$$

$$\chi^{(2)}_{ijk}(\mathbf{k}',\sigma'\omega^L_{\mathbf{k}'}|\mathbf{k}-\mathbf{k}',\sigma''\omega^S_{\mathbf{k}-\mathbf{k}'}) = -\frac{i}{2}\frac{e}{m_e}\frac{\delta_{ij}\,(\mathbf{k}-\mathbf{k}')_k\,\chi^e_\parallel(\mathbf{k}-\mathbf{k}',\sigma''\omega^S_{\mathbf{k}-\mathbf{k}'})}{(\sigma'\omega^L_{\mathbf{k}'})(\sigma\omega^L_{\mathbf{k}})},$$

$$\chi^{(2)*}_{nml}(\mathbf{k}',\sigma'\omega^L_{\mathbf{k}'}|\mathbf{k}-\mathbf{k}',\sigma''\omega^S_{\mathbf{k}-\mathbf{k}'}) = \frac{i}{2}\frac{e}{m_e}\frac{\delta_{nm}\,(\mathbf{k}-\mathbf{k}')_l\,\chi^e_\parallel(\mathbf{k}-\mathbf{k}',\sigma''\omega^S_{\mathbf{k}-\mathbf{k}'})}{(\sigma'\omega^L_{\mathbf{k}'})(\sigma\omega^L_{\mathbf{k}})},$$

$$\chi^{(2)}_{ijk}(\mathbf{k}',\sigma'\omega^L_{\mathbf{k}'}|\mathbf{k}-\mathbf{k}',\sigma''\omega^T_{\mathbf{k}-\mathbf{k}'}) = \frac{i}{2}\frac{e}{m_e}\frac{\omega^2_{pe}}{(\sigma'\omega^L_{\mathbf{k}'})(\sigma''\omega^T_{\mathbf{k}-\mathbf{k}'})(\sigma\omega^L_{\mathbf{k}})} \times \left(\delta_{ik}\frac{k'_j}{\sigma'\omega^L_{\mathbf{k}'}} + \delta_{ij}\frac{(\mathbf{k}-\mathbf{k}')_k}{\sigma''\omega^T_{\mathbf{k}-\mathbf{k}'}} + \delta_{jk}\frac{k_i}{\sigma\omega^L_{\mathbf{k}}}\right),$$

$$\chi^{(2)*}_{nml}(\mathbf{k}',\sigma'\omega^L_{\mathbf{k}'}|\mathbf{k}-\mathbf{k}',\sigma''\omega^T_{\mathbf{k}-\mathbf{k}'}) = -\frac{i}{2}\frac{e}{m_e}\frac{\omega^2_{pe}}{(\sigma'\omega^L_{\mathbf{k}'})(\sigma''\omega^T_{\mathbf{k}-\mathbf{k}'})(\sigma\omega^L_{\mathbf{k}})} \times \left(\delta_{nl}\frac{k'_m}{\sigma'\omega^L_{\mathbf{k}'}} + \delta_{nm}\frac{(\mathbf{k}-\mathbf{k}')_l}{\sigma''\omega^T_{\mathbf{k}-\mathbf{k}'}} + \delta_{ml}\frac{k_n}{\sigma\omega^L_{\mathbf{k}}}\right),$$

$$\chi^{(2)}_{ijk}(\mathbf{k}',\sigma'\omega^S_{\mathbf{k}'}|\mathbf{k}-\mathbf{k}',\sigma''\omega^T_{\mathbf{k}-\mathbf{k}'}) = -\frac{i}{2}\frac{e}{m_e}\frac{\delta_{ik}\,k'_j\,\chi^e_\parallel(\mathbf{k}',\sigma'\omega^S_{\mathbf{k}'})}{(\sigma''\omega^T_{\mathbf{k}-\mathbf{k}'})(\sigma\omega^L_{\mathbf{k}})},$$

$$\chi^{(2)*}_{nml}(\mathbf{k}',\sigma'\omega^S_{\mathbf{k}'}|\mathbf{k}-\mathbf{k}',\sigma''\omega^T_{\mathbf{k}-\mathbf{k}'}) = \frac{i}{2}\frac{e}{m_e}\frac{\delta_{nl}\,k'_m\,\chi^e_\parallel(\mathbf{k}',\sigma'\omega^S_{\mathbf{k}'})}{(\sigma''\omega^T_{\mathbf{k}-\mathbf{k}'})(\sigma\omega^L_{\mathbf{k}})},$$

$$\chi^{(2)}_{ijk}(\mathbf{k}',\sigma'\omega^T_{\mathbf{k}'}|\mathbf{k}-\mathbf{k}',\sigma''\omega^T_{\mathbf{k}-\mathbf{k}'}) = \frac{i}{2}\frac{e}{m_e}\frac{\omega^2_{pe}}{(\sigma'\omega^T_{\mathbf{k}'})(\sigma''\omega^T_{\mathbf{k}-\mathbf{k}'})(\sigma\omega^L_{\mathbf{k}})} \times \left(\delta_{ik}\frac{k'_j}{\sigma'\omega^T_{\mathbf{k}'}} + \delta_{ij}\frac{(\mathbf{k}-\mathbf{k}')_k}{\sigma''\omega^T_{\mathbf{k}-\mathbf{k}'}} + \delta_{jk}\frac{k_i}{\sigma\omega^L_{\mathbf{k}}}\right),$$

$$\chi^{(2)*}_{nml}(\mathbf{k}',\sigma'\omega^T_{\mathbf{k}'}|\mathbf{k}-\mathbf{k}',\sigma''\omega^T_{\mathbf{k}-\mathbf{k}'}) = -\frac{i}{2}\frac{e}{m_e}\frac{\omega_{pe}^2}{(\sigma'\omega^T_{\mathbf{k}'})(\sigma''\omega^T_{\mathbf{k}-\mathbf{k}'})(\sigma\omega^L_{\mathbf{k}})} \times \left(\delta_{nl}\frac{k'_m}{\sigma'\omega^T_{\mathbf{k}'}} + \delta_{nm}\frac{(\mathbf{k}-\mathbf{k}')_l}{\sigma''\omega^T_{\mathbf{k}-\mathbf{k}'}} + \delta_{ml}\frac{k_n}{\sigma\omega^L_{\mathbf{k}}}\right). \tag{8.38}$$

Substituting (8.38) to (8.37), we obtain

$$\begin{aligned}
Q_1 &= \frac{e^2}{4m_e^2}\frac{k'^2\,[\mathbf{k}\cdot(\mathbf{k}-\mathbf{k}')]^2}{k^2\,|\mathbf{k}-\mathbf{k}'|^2}\frac{[\chi^e_\parallel(\mathbf{k}',\sigma'\omega^S_{\mathbf{k}'})]^2}{(\omega^L_{\mathbf{k}})^2\,(\omega^L_{\mathbf{k}-\mathbf{k}'})^2},\\
Q_2 &= \frac{e^2}{4m_e^2}\frac{|\mathbf{k}-\mathbf{k}'|^2\,(\mathbf{k}\cdot\mathbf{k}')^2}{k^2\,k'^2}\frac{[\chi^e_\parallel(\mathbf{k}-\mathbf{k}',\sigma''\omega^S_{\mathbf{k}-\mathbf{k}'})]^2}{(\omega^L_{\mathbf{k}})^2\,(\omega^L_{\mathbf{k}'})^2},\\
Q_3 &= \frac{e^2}{4m_e^2}\frac{(\mathbf{k}\times\mathbf{k}')^2}{k^2\,k'^2\,|\mathbf{k}-\mathbf{k}'|^2}\frac{\omega_{pe}^4}{(\omega^L_{\mathbf{k}'})^2\,(\omega^T_{\mathbf{k}-\mathbf{k}'})^2\,(\omega^L_{\mathbf{k}})^2}\left(\frac{k^2}{\sigma\omega^L_{\mathbf{k}}}+\frac{k'^2}{\sigma'\omega^L_{\mathbf{k}'}}\right)^2,\\
Q_4 &= \frac{e^2}{4m_e^2}\frac{k'^2\,(\mathbf{k}\times\mathbf{k}')^2}{k^2\,|\mathbf{k}-\mathbf{k}'|^2}\frac{[\chi^e_\parallel(\mathbf{k}',\sigma'\omega^S_{\mathbf{k}'})]^2}{(\omega^T_{\mathbf{k}-\mathbf{k}'})^2\,(\omega^L_{\mathbf{k}})^2},\\
Q_5 &= \frac{e^2}{16m_e^2}\frac{k^2\,\{k'^2\,|\mathbf{k}-\mathbf{k}'|^2+[\mathbf{k}'\cdot(\mathbf{k}-\mathbf{k}')]^2\}}{k'^2\,|\mathbf{k}-\mathbf{k}'|^2}\frac{\omega_{pe}^4}{(\omega^T_{\mathbf{k}'})^2\,(\omega^T_{\mathbf{k}-\mathbf{k}'})^2\,(\omega^L_{\mathbf{k}})^4}.
\end{aligned} \tag{8.39}$$

Inserting the coupling coefficients (8.39) to (8.36), we obtain the expressions for decay processes in the wave kinetic equation,

$$\begin{aligned}
\left.\frac{\partial I^{\sigma L}_{\mathbf{k}}}{\partial t}\right|_{\text{decay}} &= -\pi\,(\sigma\omega^L_{\mathbf{k}})\sum_{\sigma',\sigma''}\int d\mathbf{k}'\left\{\frac{e^2}{4T_e^2}\frac{\mu_{\mathbf{k}'}\,[\mathbf{k}\cdot(\mathbf{k}-\mathbf{k}')]^2}{k^2\,k'^2\,|\mathbf{k}-\mathbf{k}'|^2}\right.\\
&\quad\times\left(\sigma'\omega^L_{\mathbf{k}'}\,I^{\sigma''L}_{\mathbf{k}-\mathbf{k}'}\,I^{\sigma L}_{\mathbf{k}}+\sigma''\omega^L_{\mathbf{k}-\mathbf{k}'}\frac{I^{\sigma'S}_{\mathbf{k}'}}{\mu_{\mathbf{k}'}}\,I^{\sigma L}_{\mathbf{k}}-\sigma\omega^L_{\mathbf{k}}\,I^{\sigma''L}_{\mathbf{k}-\mathbf{k}'}\frac{I^{\sigma'S}_{\mathbf{k}'}}{\mu_{\mathbf{k}'}}\right)\\
&\quad\times\delta(\sigma\omega^L_{\mathbf{k}}-\sigma'\omega^S_{\mathbf{k}'}-\sigma''\omega^L_{\mathbf{k}-\mathbf{k}'})\\
&\quad+\frac{e^2}{4T_e^2}\frac{\mu_{\mathbf{k}-\mathbf{k}'}\,(\mathbf{k}\cdot\mathbf{k}')^2}{k^2\,k'^2|\,\mathbf{k}-\mathbf{k}'|^2}\left(\sigma'\omega^L_{\mathbf{k}'}\frac{I^{\sigma''S}_{\mathbf{k}-\mathbf{k}'}}{\mu_{\mathbf{k}-\mathbf{k}'}}\,I^{\sigma L}_{\mathbf{k}}\right.\\
&\quad\left.+\sigma''\omega^L_{\mathbf{k}-\mathbf{k}'}\,I^{\sigma'L}_{\mathbf{k}'}\,I^{\sigma L}_{\mathbf{k}}-\sigma\omega^L_{\mathbf{k}}\frac{I^{\sigma''S}_{\mathbf{k}-\mathbf{k}'}}{\mu_{\mathbf{k}-\mathbf{k}'}}\,I^{\sigma'L}_{\mathbf{k}'}\right)\delta(\sigma\omega^L_{\mathbf{k}}-\sigma'\omega^L_{\mathbf{k}'}-\sigma''\omega^S_{\mathbf{k}-\mathbf{k}'})\\
&\quad+\frac{e^2}{2T_e^2}\frac{\mu_{\mathbf{k}'}\,(\mathbf{k}\times\mathbf{k}')^2}{k^2\,k'^2\,|\mathbf{k}-\mathbf{k}'|^2}\left(\sigma'\omega^L_{\mathbf{k}'}\frac{I^{\sigma''T}_{\mathbf{k}-\mathbf{k}'}}{2}\,I^{\sigma L}_{\mathbf{k}}\right.\\
&\quad\left.+\sigma''\omega^T_{\mathbf{k}-\mathbf{k}'}\frac{I^{\sigma'S}_{\mathbf{k}'}}{\mu_{\mathbf{k}'}}\,I^{\sigma L}_{\mathbf{k}}-\sigma\omega^L_{\mathbf{k}}\frac{I^{\sigma'S}_{\mathbf{k}'}}{\mu_{\mathbf{k}'}}\frac{I^{\sigma''T}_{\mathbf{k}-\mathbf{k}'}}{2}\right)\delta(\sigma\omega^L_{\mathbf{k}}-\sigma'\omega^S_{\mathbf{k}'}-\sigma''\omega^T_{\mathbf{k}-\mathbf{k}'})
\end{aligned}$$

$$+\frac{e^2}{8\,m_e^2\,\omega_{pe}^2}\frac{(\mathbf{k}\times\mathbf{k}')^2}{k^2\,k'^2\,|\mathbf{k}-\mathbf{k}'|^2}\left(\frac{k^2}{\sigma\omega_{\mathbf{k}}^L}+\frac{k'^2}{\sigma'\omega_{\mathbf{k}'}^L}\right)^2\left(\sigma'\omega_{\mathbf{k}'}^L\,\frac{I_{\mathbf{k}-\mathbf{k}'}^{\sigma''T}}{2}\,I_{\mathbf{k}}^{\sigma L}\right.$$

$$\left.+\sigma''\omega_{\mathbf{k}-\mathbf{k}'}^T\,I_{\mathbf{k}'}^{\sigma' L}\,I_{\mathbf{k}}^{\sigma L}-\sigma\omega_{\mathbf{k}}^L\,I_{\mathbf{k}'}^{\sigma' L}\,\frac{I_{\mathbf{k}-\mathbf{k}'}^{\sigma''T}}{2}\right)\delta(\sigma\omega_{\mathbf{k}}^L-\sigma'\omega_{\mathbf{k}'}^L-\sigma''\omega_{\mathbf{k}-\mathbf{k}'}^T)$$

$$+\frac{e^2}{4m_e^2}\frac{k^2\,\{k'^2\,|\mathbf{k}-\mathbf{k}'|^2+[\mathbf{k}'\cdot(\mathbf{k}-\mathbf{k}')]^2\}}{k'^2\,|\mathbf{k}-\mathbf{k}'|^2}\frac{1}{(\omega_{\mathbf{k}'}^T)^2\,(\omega_{\mathbf{k}-\mathbf{k}'}^T)^2}$$

$$\times\left(\sigma'\omega_{\mathbf{k}'}^T\,\frac{I_{\mathbf{k}-\mathbf{k}'}^{\sigma''T}}{2}\,I_{\mathbf{k}}^{\sigma L}+\sigma''\omega_{\mathbf{k}-\mathbf{k}'}^T\,\frac{I_{\mathbf{k}'}^{\sigma' T}}{2}\,I_{\mathbf{k}}^{\sigma L}-\sigma\omega_{\mathbf{k}}^L\,\frac{I_{\mathbf{k}'}^{\sigma' T}}{2}\,\frac{I_{\mathbf{k}-\mathbf{k}'}^{\sigma''T}}{2}\right)$$

$$\left.\times\,\delta(\sigma\omega_{\mathbf{k}}^L-\sigma'\omega_{\mathbf{k}'}^T-\sigma''\omega_{\mathbf{k}-\mathbf{k}'}^T)\right\}. \tag{8.40}$$

The first two terms on the right-hand side dictated by three-wave resonance conditions, $\delta(\sigma\omega_{\mathbf{k}}^L-\sigma'\omega_{\mathbf{k}'}^S-\sigma''\omega_{\mathbf{k}-\mathbf{k}'}^L)$ and $\delta(\sigma\omega_{\mathbf{k}}^L-\sigma'\omega_{\mathbf{k}'}^L-\sigma''\omega_{\mathbf{k}-\mathbf{k}'}^S)$, are the same processes already present in the electrostatic theory, and they describe a Langmuir wave decaying into another Langmuir wave and an ion-sound wave ($L\to L+S$). The third decay term dictated by $\delta(\sigma\omega_{\mathbf{k}}^L-\sigma'\omega_{\mathbf{k}'}^S-\sigma''\omega_{\mathbf{k}-\mathbf{k}'}^T)$ describes a Langmuir wave decaying into an ion-sound wave and a transverse wave ($L\to S+T$), and this process cannot be described by the electrostatic theory. The term dictated by $\delta(\sigma\omega_{\mathbf{k}}^L-\sigma'\omega_{\mathbf{k}'}^L-\sigma''\omega_{\mathbf{k}-\mathbf{k}'}^T)$ describes the decay of a Langmuir wave into another Langmuir wave and a transverse mode ($L\to L+T$). The final term associated with the resonance delta function $\delta(\sigma\omega_{\mathbf{k}}^L-\sigma'\omega_{\mathbf{k}'}^T-\sigma''\omega_{\mathbf{k}-\mathbf{k}'}^T)$ dictates the decay of an L mode into two T modes, or $L\to T+T$. All the processes pertaining to T modes are new, which could not be discussed with the electrostatic theory.

S Mode

Contribution to the S mode wave kinetic equation from decay processes is given by

$$\left.\frac{\partial I_{\mathbf{k}}^{\sigma S}}{\partial t}\right|_{\text{decay}}=-\pi\,\mu_{\mathbf{k}}\,\sigma\omega_{\mathbf{k}}^L\,\text{Im}\sum_{\sigma',\sigma''}\int d\mathbf{k}'\left\{Q_1\left(\sigma'\omega_{\mathbf{k}'}^L\,I_{\mathbf{k}-\mathbf{k}'}^{\sigma''L}\,I_{\mathbf{k}}^{\sigma S}\right.\right.$$

$$\left.+\sigma''\omega_{\mathbf{k}-\mathbf{k}'}^L\,I_{\mathbf{k}'}^{\sigma' L}\,I_{\mathbf{k}}^{\sigma S}-\mu_{\mathbf{k}}\,\sigma\omega_{\mathbf{k}}^L\,I_{\mathbf{k}-\mathbf{k}'}^{\sigma''L}\,I_{\mathbf{k}'}^{\sigma' L}\right)\delta(\sigma\omega_{\mathbf{k}}^S-\sigma'\omega_{\mathbf{k}'}^L-\sigma''\omega_{\mathbf{k}-\mathbf{k}'}^L)$$

$$+Q_2\left(\sigma'\omega_{\mathbf{k}'}^L\,I_{\mathbf{k}-\mathbf{k}'}^{\sigma''T}\,I_{\mathbf{k}}^{\sigma S}+2\,\sigma''\omega_{\mathbf{k}-\mathbf{k}'}^T\,I_{\mathbf{k}'}^{\sigma' L}\,I_{\mathbf{k}}^{\sigma S}\right.$$

$$\left.-\mu_{\mathbf{k}}\,\sigma\omega_{\mathbf{k}}^L\,I_{\mathbf{k}'}^{\sigma' L}\,I_{\mathbf{k}-\mathbf{k}'}^{\sigma''T}\right)\,\delta(\sigma\omega_{\mathbf{k}}^S-\sigma'\omega_{\mathbf{k}'}^L-\sigma''\omega_{\mathbf{k}-\mathbf{k}'}^T)$$

$$+Q_3\left(2\,\sigma'\omega_{\mathbf{k}'}^T\,I_{\mathbf{k}-\mathbf{k}'}^{\sigma''T}\,I_{\mathbf{k}}^{\sigma S}+2\,\sigma''\omega_{\mathbf{k}-\mathbf{k}'}^T\,I_{\mathbf{k}'}^{\sigma' T}\,I_{\mathbf{k}}^{\sigma S}\right.$$

$$\left.\left.-\mu_{\mathbf{k}}\,\sigma\omega_{\mathbf{k}}^L\,I_{\mathbf{k}'}^{\sigma' T}\,I_{\mathbf{k}-\mathbf{k}'}^{\sigma''T}\right)\,\delta(\sigma\omega_{\mathbf{k}}^S-\sigma'\omega_{\mathbf{k}'}^T-\sigma''\omega_{\mathbf{k}-\mathbf{k}'}^T)\right\}, \tag{8.41}$$

where

$$
\begin{aligned}
Q_1 &= \frac{k_i\,k_n}{k^2}\frac{k'_j\,k'_m}{k'^2}\frac{(\mathbf{k}-\mathbf{k}')_k\,(\mathbf{k}-\mathbf{k}')_l}{|\mathbf{k}-\mathbf{k}'|^2} \\
&\quad\times \chi^{(2)}_{ijk}(\mathbf{k}',\sigma'\omega^L_{\mathbf{k}'}|\mathbf{k}-\mathbf{k}',\sigma''\omega^L_{\mathbf{k}-\mathbf{k}'})\,\chi^{(2)*}_{nml}(\mathbf{k}',\sigma'\omega^L_{\mathbf{k}'}|\mathbf{k}-\mathbf{k}',\sigma''\omega^L_{\mathbf{k}-\mathbf{k}'}), \\
Q_2 &= \frac{k_i\,k_n}{k^2}\frac{k'_j\,k'_m}{k'^2}\left(\delta_{kl}-\frac{(\mathbf{k}-\mathbf{k}')_k\,(\mathbf{k}-\mathbf{k}')_l}{|\mathbf{k}-\mathbf{k}'|^2}\right) \\
&\quad\times \chi^{(2)}_{ijk}(\mathbf{k}',\sigma'\omega^L_{\mathbf{k}'}|\mathbf{k}-\mathbf{k}',\sigma''\omega^T_{\mathbf{k}-\mathbf{k}'})\,\chi^{(2)*}_{nml}(\mathbf{k}',\sigma'\omega^L_{\mathbf{k}'}|\mathbf{k}-\mathbf{k}',\sigma''\omega^T_{\mathbf{k}-\mathbf{k}'}), \\
Q_3 &= \frac{1}{4}\frac{k_i\,k_n}{k^2}\left(\delta_{jm}-\frac{k'_j\,k'_m}{k'^2}\right)\left(\delta_{kl}-\frac{(\mathbf{k}-\mathbf{k}')_k\,(\mathbf{k}-\mathbf{k}')_l}{|\mathbf{k}-\mathbf{k}'|^2}\right) \\
&\quad\times \chi^{(2)}_{ijk}(\mathbf{k}',\sigma'\omega^T_{\mathbf{k}'}|\mathbf{k}-\mathbf{k}',\sigma''\omega^T_{\mathbf{k}-\mathbf{k}'})\,\chi^{(2)*}_{nml}(\mathbf{k}',\sigma'\omega^T_{\mathbf{k}'}|\mathbf{k}-\mathbf{k}',\sigma''\omega^T_{\mathbf{k}-\mathbf{k}'}).
\end{aligned}
\tag{8.42}
$$

The relevant susceptibilities are

$$
\begin{aligned}
\chi^{(2)}_{ijk}(\mathbf{k}',\sigma'\omega^L_{\mathbf{k}'}|\mathbf{k}-\mathbf{k}',\sigma''\omega^L_{\mathbf{k}-\mathbf{k}'}) &= -\frac{i}{2}\frac{e}{m_e}\frac{\delta_{jk}\,k_i\,\chi^e_\parallel(\mathbf{k},\sigma\omega^S_{\mathbf{k}})}{(\sigma'\omega^L_{\mathbf{k}'})(\sigma''\omega^L_{\mathbf{k}-\mathbf{k}'})}, \\
\chi^{(2)*}_{nml}(\mathbf{k}',\sigma'\omega^L_{\mathbf{k}'}|\mathbf{k}-\mathbf{k}',\sigma''\omega^L_{\mathbf{k}-\mathbf{k}'}) &= \frac{i}{2}\frac{e}{m_e}\frac{\delta_{ml}\,k_n\,\chi^e_\parallel(\mathbf{k},\sigma\omega^S_{\mathbf{k}})}{(\sigma'\omega^L_{\mathbf{k}'})(\sigma''\omega^L_{\mathbf{k}-\mathbf{k}'})}, \\
\chi^{(2)}_{ijk}(\mathbf{k}',\sigma'\omega^L_{\mathbf{k}'}|\mathbf{k}-\mathbf{k}',\sigma''\omega^T_{\mathbf{k}-\mathbf{k}'}) &= -\frac{i}{2}\frac{e}{m_e}\frac{\delta_{jk}\,k_i\,\chi^e_\parallel(\mathbf{k},\sigma\omega^S_{\mathbf{k}})}{(\sigma'\omega^L_{\mathbf{k}'})(\sigma''\omega^T_{\mathbf{k}-\mathbf{k}'})}, \\
\chi^{(2)*}_{nml}(\mathbf{k}',\sigma'\omega^L_{\mathbf{k}'}|\mathbf{k}-\mathbf{k}',\sigma''\omega^T_{\mathbf{k}-\mathbf{k}'}) &= \frac{i}{2}\frac{e}{m_e}\frac{\delta_{ml}\,k_n\,\chi^e_\parallel(\mathbf{k},\sigma\omega^S_{\mathbf{k}})}{(\sigma'\omega^L_{\mathbf{k}'})(\sigma''\omega^T_{\mathbf{k}-\mathbf{k}'})}, \\
\chi^{(2)}_{ijk}(\mathbf{k}',\sigma'\omega^T_{\mathbf{k}'}|\mathbf{k}-\mathbf{k}',\sigma''\omega^T_{\mathbf{k}-\mathbf{k}'}) &= -\frac{i}{2}\frac{e}{m_e}\frac{\delta_{jk}\,k_i\,\chi^e_\parallel(\mathbf{k},\sigma\omega^S_{\mathbf{k}})}{(\sigma'\omega^T_{\mathbf{k}'})(\sigma''\omega^T_{\mathbf{k}-\mathbf{k}'})}, \\
\chi^{(2)*}_{nml}(\mathbf{k}',\sigma'\omega^T_{\mathbf{k}'}|\mathbf{k}-\mathbf{k}',\sigma''\omega^T_{\mathbf{k}-\mathbf{k}'}) &= \frac{i}{2}\frac{e}{m_e}\frac{\delta_{ml}\,k_n\,\chi^e_\parallel(\mathbf{k},\sigma\omega^S_{\mathbf{k}})}{(\sigma'\omega^T_{\mathbf{k}'})(\sigma''\omega^T_{\mathbf{k}-\mathbf{k}'})}.
\end{aligned}
\tag{8.43}
$$

Making use of (8.43), the quantities Q_1, Q_2, and Q_3 defined in (8.42) become

$$
\begin{aligned}
Q_1 &= \frac{e^2}{4T_e^2}\frac{[\mathbf{k}'\cdot(\mathbf{k}-\mathbf{k}')]^2}{k^2\,k'^2\,|\mathbf{k}-\mathbf{k}'|^2}, \\
Q_2 &= \frac{e^2}{4T_e^2}\frac{(\mathbf{k}\times\mathbf{k}')^2}{k^2\,k'^2\,|\mathbf{k}-\mathbf{k}'|^2}, \\
Q_3 &= \frac{e^2}{16\,T_e^2}\frac{1}{k^2}\left(1+\frac{[\mathbf{k}'\cdot(\mathbf{k}-\mathbf{k}')]^2}{k'^2\,|\mathbf{k}-\mathbf{k}'|^2}\right),
\end{aligned}
\tag{8.44}
$$

which leads to

$$
\begin{aligned}
\frac{\partial}{\partial t}\bigg|_{\text{decay}} \frac{I_{\mathbf{k}}^{\sigma S}}{\mu_{\mathbf{k}}} = &-\pi\,\sigma\omega_{\mathbf{k}}^{L} \sum_{\sigma',\sigma''} \int d\mathbf{k}' \left\{ \frac{e^2}{4T_e^2} \frac{\mu_{\mathbf{k}}\,[\mathbf{k}'\cdot(\mathbf{k}-\mathbf{k}')]^2}{k^2\,k'^2\,|\mathbf{k}-\mathbf{k}'|^2} \left(\sigma'\omega_{\mathbf{k}'}^{L}\, I_{\mathbf{k}-\mathbf{k}'}^{\sigma''L} \frac{I_{\mathbf{k}}^{\sigma S}}{\mu_{\mathbf{k}}} \right.\right. \\
&\left. + \sigma''\omega_{\mathbf{k}-\mathbf{k}'}^{L}\, I_{\mathbf{k}'}^{\sigma' L} \frac{I_{\mathbf{k}}^{\sigma S}}{\mu_{\mathbf{k}}} - \sigma\omega_{\mathbf{k}}^{L}\, I_{\mathbf{k}-\mathbf{k}'}^{\sigma''L}\, I_{\mathbf{k}'}^{\sigma' L} \right) \delta(\sigma\omega_{\mathbf{k}}^{S} - \sigma'\omega_{\mathbf{k}'}^{L} - \sigma''\omega_{\mathbf{k}-\mathbf{k}'}^{L}) \\
&+ \frac{e^2}{2T_e^2} \frac{\mu_{\mathbf{k}}\,(\mathbf{k}\times\mathbf{k}')^2}{k^2\,k'^2\,|\mathbf{k}-\mathbf{k}'|^2} \left(\sigma'\omega_{\mathbf{k}'}^{L} \frac{I_{\mathbf{k}-\mathbf{k}'}^{\sigma''T}}{2} \frac{I_{\mathbf{k}}^{\sigma S}}{\mu_{\mathbf{k}}} \right. \\
&\left.\left. + \sigma''\omega_{\mathbf{k}-\mathbf{k}'}^{T}\, I_{\mathbf{k}'}^{\sigma' L} \frac{I_{\mathbf{k}}^{\sigma S}}{\mu_{\mathbf{k}}} - \sigma\omega_{\mathbf{k}}^{L}\, I_{\mathbf{k}'}^{\sigma' L} \frac{I_{\mathbf{k}-\mathbf{k}'}^{\sigma''T}}{2} \right) \delta(\sigma\omega_{\mathbf{k}}^{S} - \sigma'\omega_{\mathbf{k}'}^{L} - \sigma''\omega_{\mathbf{k}-\mathbf{k}'}^{T}) \right\}.
\end{aligned}
\tag{8.45}
$$

In deriving this result, we have ignored $T+T \to S$ term since such a process cannot satisfy the wave energy and momentum conservation conditions. Note that the first term on the right-hand side is the same as the electrostatic counterpart, but the second term that involves a T mode is new.

T Mode

Next, we consider decay processes contributing to the wave kinetic equation for T mode. In doing so, let us ignore $T+T \to S$ process at the outset:

$$
\begin{aligned}
\frac{\partial}{\partial t}\bigg|_{\text{decay}} \frac{I_{\mathbf{k}}^{\sigma T}}{2} = &-\pi\sigma\omega_{\mathbf{k}}^{T} \sum_{\sigma',\sigma''} \mathrm{Im} \int d\mathbf{k}' \left\{ Q_1 \left(\sigma'\omega_{\mathbf{k}'}^{L}\, I_{\mathbf{k}-\mathbf{k}'}^{\sigma''L} \frac{I_{\mathbf{k}}^{\sigma T}}{2} \right.\right. \\
&\left. + \sigma''\omega_{\mathbf{k}-\mathbf{k}'}^{L}\, I_{\mathbf{k}'}^{\sigma' L} \frac{I_{\mathbf{k}}^{\sigma T}}{2} - \sigma\omega_{\mathbf{k}}^{T}\, I_{\mathbf{k}-\mathbf{k}'}^{\sigma''L}\, I_{\mathbf{k}'}^{\sigma' L} \right) \delta(\sigma\omega_{\mathbf{k}}^{T} - \sigma'\omega_{\mathbf{k}'}^{L} - \sigma''\omega_{\mathbf{k}-\mathbf{k}'}^{L}) \\
&+ Q_2 \left(\sigma'\mu_{\mathbf{k}}\omega_{\mathbf{k}'}^{L}\, I_{\mathbf{k}-\mathbf{k}'}^{\sigma''L} \frac{I_{\mathbf{k}}^{\sigma T}}{2} + \sigma''\omega_{\mathbf{k}-\mathbf{k}'}^{L}\, I_{\mathbf{k}'}^{\sigma' S} \frac{I_{\mathbf{k}}^{\sigma T}}{2} \right. \\
&\left. - \sigma\omega_{\mathbf{k}}^{T}\, I_{\mathbf{k}-\mathbf{k}'}^{\sigma''L}\, I_{\mathbf{k}'}^{\sigma' S} \right) \delta(\sigma\omega_{\mathbf{k}}^{T} - \sigma'\omega_{\mathbf{k}'}^{S} - \sigma''\omega_{\mathbf{k}-\mathbf{k}'}^{L}) \\
&+ Q_3 \left(\sigma'\omega_{\mathbf{k}'}^{L}\, I_{\mathbf{k}-\mathbf{k}'}^{\sigma''S} \frac{I_{\mathbf{k}}^{\sigma T}}{2} + \sigma''\mu_{\mathbf{k}-\mathbf{k}'}\omega_{\mathbf{k}-\mathbf{k}'}^{L}\, I_{\mathbf{k}'}^{\sigma' L} \frac{I_{\mathbf{k}}^{\sigma T}}{2} \right. \\
&\left. - \sigma\omega_{\mathbf{k}}^{T}\, I_{\mathbf{k}-\mathbf{k}'}^{\sigma''S}\, I_{\mathbf{k}'}^{\sigma' L} \right) \delta(\sigma\omega_{\mathbf{k}}^{T} - \sigma'\omega_{\mathbf{k}'}^{L} - \sigma''\omega_{\mathbf{k}-\mathbf{k}'}^{S}) \\
&+ Q_4 \left(\sigma'\omega_{\mathbf{k}'}^{T}\, I_{\mathbf{k}-\mathbf{k}'}^{\sigma''L} \frac{I_{\mathbf{k}}^{\sigma T}}{2} + \sigma''\omega_{\mathbf{k}-\mathbf{k}'}^{L} \frac{I_{\mathbf{k}'}^{\sigma' T}}{2} \frac{I_{\mathbf{k}}^{\sigma T}}{2} \right. \\
&\left.\left. - \sigma\omega_{\mathbf{k}}^{T}\, I_{\mathbf{k}-\mathbf{k}'}^{\sigma''L} \frac{I_{\mathbf{k}'}^{\sigma' T}}{2} \right) \delta(\sigma\omega_{\mathbf{k}}^{T} - \sigma'\omega_{\mathbf{k}'}^{T} - \sigma''\omega_{\mathbf{k}-\mathbf{k}'}^{L}) \right\},
\end{aligned}
\tag{8.46}
$$

where

$$
\begin{aligned}
Q_1 &= \frac{1}{2}\left(\delta_{in} - \frac{k_i\,k_n}{k^2}\right)\frac{k'_j\,k'_m}{k'^2}\frac{(\mathbf{k}-\mathbf{k}')_k\,(\mathbf{k}-\mathbf{k}')_l}{|\mathbf{k}-\mathbf{k}'|^2} \\
&\quad \times \chi^{(2)}_{ijk}(\mathbf{k}',\sigma'\omega^L_{\mathbf{k}'}|\mathbf{k}-\mathbf{k}',\sigma''\omega^L_{\mathbf{k}-\mathbf{k}'})\,\chi^{(2)*}_{nml}(\mathbf{k}',\sigma'\omega^L_{\mathbf{k}'}|\mathbf{k}-\mathbf{k}',\sigma''\omega^L_{\mathbf{k}-\mathbf{k}'}) \\
&\approx \frac{e^2}{32\,m_e^2\,\omega_{pe}^2}\frac{(\mathbf{k}\times\mathbf{k}')^2}{k^2\,k'^2\,|\mathbf{k}-\mathbf{k}'|^2}\left(\frac{k'^2}{\sigma'\omega^L_{\mathbf{k}'}} - \frac{|\mathbf{k}-\mathbf{k}'|^2}{\sigma''\omega^L_{\mathbf{k}-\mathbf{k}'}}\right)^2, \\
Q_2 &= \frac{1}{2}\left(\delta_{in} - \frac{k_i\,k_n}{k^2}\right)\frac{k'_j\,k'_m}{k'^2}\frac{(\mathbf{k}-\mathbf{k}')_k\,(\mathbf{k}-\mathbf{k}')_l}{|\mathbf{k}-\mathbf{k}'|^2} \\
&\quad \times \chi^{(2)}_{ijk}(\mathbf{k}',\sigma'\omega^S_{\mathbf{k}'}|\mathbf{k}-\mathbf{k}',\sigma''\omega^L_{\mathbf{k}-\mathbf{k}'})\,\chi^{(2)*}_{nml}(\mathbf{k}',\sigma'\omega^S_{\mathbf{k}'}|\mathbf{k}-\mathbf{k}',\sigma''\omega^L_{\mathbf{k}-\mathbf{k}'}) \\
&\approx \frac{e^2}{8T_e^2}\frac{(\mathbf{k}\times\mathbf{k}')^2}{k^2\,k'^2\,|\mathbf{k}-\mathbf{k}'|^2}, \\
Q_3 &= \frac{1}{2}\left(\delta_{in} - \frac{k_i\,k_n}{k^2}\right)\frac{k'_j\,k'_m}{k'^2}\frac{(\mathbf{k}-\mathbf{k}')_k\,(\mathbf{k}-\mathbf{k}')_l}{|\mathbf{k}-\mathbf{k}'|^2} \\
&\quad \times \chi^{(2)}_{ijk}(\mathbf{k}',\sigma'\omega^L_{\mathbf{k}'}|\mathbf{k}-\mathbf{k}',\sigma''\omega^S_{\mathbf{k}-\mathbf{k}'})\,\chi^{(2)*}_{nml}(\mathbf{k}',\sigma'\omega^L_{\mathbf{k}'}|\mathbf{k}-\mathbf{k}',\sigma''\omega^S_{\mathbf{k}-\mathbf{k}'}) \\
&\approx \frac{e^2}{8T_e^2}\frac{(\mathbf{k}\times\mathbf{k}')^2}{k^2\,k'^2\,|\mathbf{k}-\mathbf{k}'|^2}, \\
Q_4 &= \left(\delta_{in} - \frac{k_i\,k_n}{k^2}\right)\left(\delta_{jm} - \frac{k'_j\,k'_m}{k'^2}\right)\frac{(\mathbf{k}-\mathbf{k}')_k\,(\mathbf{k}-\mathbf{k}')_l}{|\mathbf{k}-\mathbf{k}'|^2} \\
&\quad \times \chi^{(2)}_{ijk}(\mathbf{k}',\sigma'\omega^T_{\mathbf{k}'}|\mathbf{k}-\mathbf{k}',\sigma''\omega^L_{\mathbf{k}-\mathbf{k}'})\,\chi^{(2)*}_{nml}(\mathbf{k}',\sigma'\omega^T_{\mathbf{k}'}|\mathbf{k}-\mathbf{k}',\sigma''\omega^L_{\mathbf{k}-\mathbf{k}'}) \\
&\approx \frac{1}{4}\frac{e^2}{m_e^2}\frac{|\mathbf{k}-\mathbf{k}'|^2}{(\sigma\omega^T_{\mathbf{k}})^2(\sigma'\omega^T_{\mathbf{k}'})^2}\left(1 + \frac{(\mathbf{k}\cdot\mathbf{k}')^2}{k'^4}\right).
\end{aligned}
\tag{8.47}
$$

Inserting the Q coefficients to the basic equation, we arrive at

$$
\begin{aligned}
\frac{\partial}{\partial t}\bigg|_{\text{decay}} \frac{I^{\sigma T}_{\mathbf{k}}}{2} &= -\pi\,\sigma\omega^T_{\mathbf{k}} \sum_{\sigma',\sigma''}\int d\mathbf{k}' \left\{ \frac{e^2}{32\,m_e^2\,\omega_{pe}^2}\frac{(\mathbf{k}\times\mathbf{k}')^2}{k^2\,k'^2\,|\mathbf{k}-\mathbf{k}'|^2} \right. \\
&\quad \times \left(\frac{k'^2}{\sigma'\omega^L_{\mathbf{k}'}} - \frac{|\mathbf{k}-\mathbf{k}'|^2}{\sigma''\omega^L_{\mathbf{k}-\mathbf{k}'}}\right)^2 \left(\sigma'\omega^L_{\mathbf{k}'}\,I^{\sigma''L}_{\mathbf{k}-\mathbf{k}'}\,\frac{I^{\sigma T}_{\mathbf{k}}}{2} + \sigma''\omega^L_{\mathbf{k}-\mathbf{k}'}\,I^{\sigma' L}_{\mathbf{k}'}\,\frac{I^{\sigma T}_{\mathbf{k}}}{2}\right. \\
&\quad \left. - \sigma\omega^T_{\mathbf{k}}\,I^{\sigma''L}_{\mathbf{k}-\mathbf{k}'}\,I^{\sigma' L}_{\mathbf{k}'}\right)\delta(\sigma\omega^T_{\mathbf{k}} - \sigma'\omega^L_{\mathbf{k}'} - \sigma''\omega^L_{\mathbf{k}-\mathbf{k}'}) \\
&\quad + \frac{e^2}{8T_e^2}\frac{\mu_{\mathbf{k}'}\,(\mathbf{k}\times\mathbf{k}')^2}{k^2\,k'^2\,|\mathbf{k}-\mathbf{k}'|^2}\left(\sigma'\omega^L_{\mathbf{k}'}\,I^{\sigma''L}_{\mathbf{k}-\mathbf{k}'}\,\frac{I^{\sigma T}_{\mathbf{k}}}{2} + \sigma''\omega^L_{\mathbf{k}-\mathbf{k}'}\,\frac{I^{\sigma' S}_{\mathbf{k}'}}{\mu_{\mathbf{k}'}}\,\frac{I^{\sigma T}_{\mathbf{k}}}{2}\right. \\
&\quad \left. - \sigma\omega^T_{\mathbf{k}}\,I^{\sigma''L}_{\mathbf{k}-\mathbf{k}'}\,\frac{I^{\sigma' S}_{\mathbf{k}'}}{\mu_{\mathbf{k}'}}\right)\delta(\sigma\omega^T_{\mathbf{k}} - \sigma'\omega^S_{\mathbf{k}'} - \sigma''\omega^L_{\mathbf{k}-\mathbf{k}'})
\end{aligned}
$$

$$+\frac{e^2}{8T_e^2}\frac{\mu_{\mathbf{k}-\mathbf{k}'}(\mathbf{k}\times\mathbf{k}')^2}{k^2\,k'^2\,|\mathbf{k}-\mathbf{k}'|^2}\left(\sigma'\omega_{\mathbf{k}'}^L\frac{I_{\mathbf{k}-\mathbf{k}'}^{\sigma''S}}{\mu_{\mathbf{k}-\mathbf{k}'}}\frac{I_{\mathbf{k}}^{\sigma T}}{2}+\sigma''\omega_{\mathbf{k}-\mathbf{k}'}^L I_{\mathbf{k}'}^{\sigma' L}\frac{I_{\mathbf{k}}^{\sigma T}}{2}\right.$$
$$\left.-\sigma\omega_{\mathbf{k}}^T\frac{I_{\mathbf{k}-\mathbf{k}'}^{\sigma''S}}{\mu_{\mathbf{k}-\mathbf{k}'}}I_{\mathbf{k}'}^{\sigma' L}\right)\delta(\sigma\omega_{\mathbf{k}}^T-\sigma'\omega_{\mathbf{k}'}^L-\sigma''\omega_{\mathbf{k}-\mathbf{k}'}^S)$$
$$+\frac{1}{4}\frac{e^2}{m_e^2}\frac{|\mathbf{k}-\mathbf{k}'|^2}{(\sigma\omega_{\mathbf{k}}^T)^2(\sigma'\omega_{\mathbf{k}'}^T)^2}\left(1+\frac{(\mathbf{k}\cdot\mathbf{k}')^2}{k'^4}\right)\left(\sigma'\omega_{\mathbf{k}'}^T I_{\mathbf{k}-\mathbf{k}'}^{\sigma'' L}\frac{I_{\mathbf{k}}^{\sigma T}}{2}\right. \tag{8.48}$$
$$\left.\left.+\sigma''\omega_{\mathbf{k}-\mathbf{k}'}^L\frac{I_{\mathbf{k}'}^{\sigma' T}}{2}\frac{I_{\mathbf{k}}^{\sigma T}}{2}-\sigma\omega_{\mathbf{k}}^T I_{\mathbf{k}-\mathbf{k}'}^{\sigma'' L}\frac{I_{\mathbf{k}'}^{\sigma' T}}{2}\right)\delta(\sigma\omega_{\mathbf{k}}^T-\sigma'\omega_{\mathbf{k}'}^T-\sigma''\omega_{\mathbf{k}-\mathbf{k}'}^L)\right\}.$$

8.2.3 Induced Scattering

L Mode

As for Langmuir waves, we will ignore induced scattering involving S mode. The rationale for such a treatment is already discussed in Section 2.3.2. Consequently, contribution to the L mode wave kinetic equation from induced scattering processes includes the following terms:

$$\frac{\partial I_{\mathbf{k}}^{\sigma L}}{\partial t}\bigg|_{\text{ind.scatt.}}=-\frac{4}{\epsilon_{\parallel}'(\mathbf{k},\sigma\omega_{\mathbf{k}}^L)}\,\text{Im}\int d\mathbf{k}'\left\{\sum_{\sigma'}\frac{k_i\,k_n}{k^2}\frac{k_j'\,k_m'}{k'^2}\right.$$
$$\times\left[\frac{(\mathbf{k}-\mathbf{k}')_k\,(\mathbf{k}-\mathbf{k}')_l}{|\mathbf{k}-\mathbf{k}'|^2}\chi_{ijk}^{(2)}(\mathbf{k}',\sigma'\omega_{\mathbf{k}'}^L|\mathbf{k}-\mathbf{k}',\sigma\omega_{\mathbf{k}}^L-\sigma'\omega_{\mathbf{k}'}^L)\right.$$
$$\times\chi_{nml}^{(2)}(\mathbf{k}',\sigma'\omega_{\mathbf{k}'}^L|\mathbf{k}-\mathbf{k}',\sigma\omega_{\mathbf{k}}^L-\sigma'\omega_{\mathbf{k}'}^L)\,\mathscr{P}\frac{2}{\epsilon_{\parallel}(\mathbf{k}-\mathbf{k}',\sigma\omega_{\mathbf{k}}^L-\sigma'\omega_{\mathbf{k}'}^L)}$$
$$+\left(\delta_{kl}-\frac{(\mathbf{k}-\mathbf{k}')_k\,(\mathbf{k}-\mathbf{k}')_l}{|\mathbf{k}-\mathbf{k}'|^2}\right)\chi_{ijk}^{(2)}(\mathbf{k}',\sigma'\omega_{\mathbf{k}'}^L|\mathbf{k}-\mathbf{k}',\sigma\omega_{\mathbf{k}}^L-\sigma'\omega_{\mathbf{k}'}^L)$$
$$\times\chi_{nml}^{(2)}(\mathbf{k}',\sigma'\omega_{\mathbf{k}'}^L|\mathbf{k}-\mathbf{k}',\sigma\omega_{\mathbf{k}}^L-\sigma'\omega_{\mathbf{k}'}^L)\,\mathscr{P}\frac{2}{\Lambda_{\perp}(\mathbf{k}-\mathbf{k}',\sigma\omega_{\mathbf{k}}^L-\sigma'\omega_{\mathbf{k}'}^L)}$$
$$\left.+\chi_{ijmn}^{(3)}(\mathbf{k}',\sigma'\omega_{\mathbf{k}'}^L|-\mathbf{k}',-\sigma'\omega_{\mathbf{k}'}^L|\mathbf{k},\sigma\omega_{\mathbf{k}}^L)\right]I_{\mathbf{k}'}^{\sigma' L}\,I_{\mathbf{k}}^{\sigma L}$$
$$+\frac{1}{2}\sum_{\sigma'}\frac{k_i\,k_n}{k^2}\left(\delta_{jm}-\frac{k_j'\,k_m'}{k'^2}\right)\left[\frac{(\mathbf{k}-\mathbf{k}')_k\,(\mathbf{k}-\mathbf{k}')_l}{|\mathbf{k}-\mathbf{k}'|^2}\right.$$
$$\times\chi_{ijk}^{(2)}(\mathbf{k}',\sigma'\omega_{\mathbf{k}'}^T|\mathbf{k}-\mathbf{k}',\sigma\omega_{\mathbf{k}}^L-\sigma'\omega_{\mathbf{k}'}^T)$$
$$\times\chi_{nml}^{(2)}(\mathbf{k}',\sigma'\omega_{\mathbf{k}'}^T|\mathbf{k}-\mathbf{k}',\sigma\omega_{\mathbf{k}}^L-\sigma'\omega_{\mathbf{k}'}^T)$$
$$\times\mathscr{P}\frac{2}{\epsilon_{\parallel}(\mathbf{k}-\mathbf{k}',\sigma\omega_{\mathbf{k}}^L-\sigma'\omega_{\mathbf{k}'}^T)}+\left(\delta_{kl}-\frac{(\mathbf{k}-\mathbf{k}')_k\,(\mathbf{k}-\mathbf{k}')_l}{|\mathbf{k}-\mathbf{k}'|^2}\right)$$
$$\times\chi_{ijk}^{(2)}(\mathbf{k}',\sigma'\omega_{\mathbf{k}'}^T|\mathbf{k}-\mathbf{k}',\sigma\omega_{\mathbf{k}}^L-\sigma'\omega_{\mathbf{k}'}^T)$$

$$
\begin{aligned}
&\times \chi^{(2)}_{nml}(\mathbf{k}',\sigma'\omega^T_{\mathbf{k}'}|\mathbf{k}-\mathbf{k}',\sigma\omega^L_{\mathbf{k}}-\sigma'\omega^T_{\mathbf{k}'})\\
&\times \mathscr{P}\frac{2}{\Lambda_\perp(\mathbf{k}-\mathbf{k}',\sigma\omega^L_{\mathbf{k}}-\sigma'\omega^T_{\mathbf{k}'})}\\
&+\chi^{(3)}_{ijmn}(\mathbf{k}',\sigma'\omega^T_{\mathbf{k}'}|-\mathbf{k}',-\sigma'\omega^T_{\mathbf{k}'}|\mathbf{k},\sigma\omega^L_{\mathbf{k}})\Big]\, I^{\sigma' T}_{\mathbf{k}'}\, I^{\sigma L}_{\mathbf{k}}\Big\}.
\end{aligned}
\tag{8.49}
$$

Induced scattering involving transverse wave is relevant only for fundamental T mode. We thus have the following inequalities:

$$
\begin{aligned}
&\omega^L_{\mathbf{k}} \gg k v_{Te}, \quad &&\omega^L_{\mathbf{k}'} \gg k' v_{Te}, \quad &&\omega^L_{\mathbf{k}} - \omega^L_{\mathbf{k}'} \ll |\mathbf{k}-\mathbf{k}'|\, v_{Te},\\
&\omega^L_{\mathbf{k}} \gg k v_{Ti}, &&\omega^L_{\mathbf{k}'} \gg k' v_{Ti}, &&\omega^L_{\mathbf{k}} - \omega^L_{\mathbf{k}'} \gg |\mathbf{k}-\mathbf{k}'|\, v_{Ti},\\
&\omega^L_{\mathbf{k}} \gg k v_{Te}, &&\omega^T_{\mathbf{k}'} \gg k' v_{Te}, &&\omega^L_{\mathbf{k}} - \omega^T_{\mathbf{k}'} \ll |\mathbf{k}-\mathbf{k}'|\, v_{Te},\\
&\omega^L_{\mathbf{k}} \gg k v_{Ti}, &&\omega^T_{\mathbf{k}'} \gg k' v_{Ti}, &&\omega^L_{\mathbf{k}} - \omega^T_{\mathbf{k}'} \gg |\mathbf{k}-\mathbf{k}'|\, v_{Ti},\\
&c^2|\mathbf{k}-\mathbf{k}'|^2 \gg (\sigma\omega^L_{\mathbf{k}} - \sigma'\omega^L_{\mathbf{k}'})^2, && &&c^2|\mathbf{k}-\mathbf{k}'|^2 \gg (\sigma\omega^L_{\mathbf{k}} - \sigma'\omega^T_{\mathbf{k}'})^2.
\end{aligned}
\tag{8.50}
$$

Because of the inequality in the last line of (8.50), the induced scattering term that involves $\Lambda_\perp$ is unimportant. This means that we can simplify (8.49) by

$$
\begin{aligned}
\frac{\partial I^{\sigma L}_{\mathbf{k}}}{\partial t}\bigg|_{\text{ind.scatt.}} &= -\frac{4}{\epsilon'_\parallel(\mathbf{k},\sigma\omega^L_{\mathbf{k}})}\\
&\times \text{Im}\int d\mathbf{k}' \Bigg\{\sum_{\sigma'}\Bigg[\mathcal{I}_1\, \mathscr{P}\frac{2}{\epsilon_\parallel(\mathbf{k}-\mathbf{k}',\sigma\omega^L_{\mathbf{k}}-\sigma'\omega^L_{\mathbf{k}'})}+\mathcal{J}_1\Bigg] I^{\sigma' L}_{\mathbf{k}'}\, I^{\sigma L}_{\mathbf{k}}\\
&+\frac{1}{2}\sum_{\sigma'}\Bigg[\mathcal{I}_2\, \mathscr{P}\frac{2}{\epsilon_\parallel(\mathbf{k}-\mathbf{k}',\sigma\omega^L_{\mathbf{k}}-\sigma'\omega^T_{\mathbf{k}'})}+\mathcal{J}_2\Bigg] I^{\sigma' T}_{\mathbf{k}'}\, I^{\sigma L}_{\mathbf{k}}\Bigg\},
\end{aligned}
\tag{8.51}
$$

where the quantities of relevance are defined by

$$
\begin{aligned}
\mathcal{I}_1 &= \frac{k_i\,k_n}{k^2}\frac{k'_j\,k'_m}{k'^2}\frac{(\mathbf{k}-\mathbf{k}')_k\,(\mathbf{k}-\mathbf{k}')_l}{|\mathbf{k}-\mathbf{k}'|^2}\\
&\times \chi^{(2)}_{ijk}(\mathbf{k}',\sigma'\omega^L_{\mathbf{k}'}|\mathbf{k}-\mathbf{k}',\sigma\omega^L_{\mathbf{k}}-\sigma'\omega^L_{\mathbf{k}'})\\
&\times \chi^{(2)}_{nml}(\mathbf{k}',\sigma'\omega^L_{\mathbf{k}'}|\mathbf{k}-\mathbf{k}',\sigma\omega^L_{\mathbf{k}}-\sigma'\omega^L_{\mathbf{k}'}),\\
\mathcal{I}_2 &= \frac{k_i\,k_n}{k^2}\left(\delta_{jm}-\frac{k'_j\,k'_m}{k'^2}\right)\frac{(\mathbf{k}-\mathbf{k}')_k\,(\mathbf{k}-\mathbf{k}')_l}{|\mathbf{k}-\mathbf{k}'|^2}\\
&\times \chi^{(2)}_{ijk}(\mathbf{k}',\sigma'\omega^T_{\mathbf{k}'}|\mathbf{k}-\mathbf{k}',\sigma\omega^L_{\mathbf{k}}-\sigma'\omega^T_{\mathbf{k}'})\\
&\times \chi^{(2)}_{nml}(\mathbf{k}',\sigma'\omega^T_{\mathbf{k}'}|\mathbf{k}-\mathbf{k}',\sigma\omega^L_{\mathbf{k}}-\sigma'\omega^T_{\mathbf{k}'}),\\
\mathcal{J}_1 &= \frac{k_i\,k_n}{k^2}\frac{k'_j\,k'_m}{k'^2}\,\chi^{(3)}_{ijmn}(\mathbf{k}',\sigma'\omega^L_{\mathbf{k}'}|-\mathbf{k}',-\sigma'\omega^L_{\mathbf{k}'}|\mathbf{k},\sigma\omega^L_{\mathbf{k}}),\\
\mathcal{J}_2 &= \frac{k_i\,k_n}{k^2}\left(\delta_{jm}-\frac{k'_j\,k'_m}{k'^2}\right)\chi^{(3)}_{ijmn}(\mathbf{k}',\sigma'\omega^T_{\mathbf{k}'}|-\mathbf{k}',-\sigma'\omega^T_{\mathbf{k}'}|\mathbf{k},\sigma\omega^L_{\mathbf{k}}).
\end{aligned}
\tag{8.52}
$$

Making use of approximate forms of second-order nonlinear susceptibilities,

$$
\begin{aligned}
&\chi^{(2)}_{ijk}(\mathbf{k}',\sigma'\omega^L_{\mathbf{k}'}|\mathbf{k}-\mathbf{k}',\sigma\omega^L_{\mathbf{k}}-\sigma'\omega^L_{\mathbf{k}'})\\
&\quad\approx -\frac{i}{2}\frac{e}{m_e\,\omega^2_{pe}}\,(\mathbf{k}-\mathbf{k}')_k\,\delta_{ij}\,\chi^e_\parallel(\mathbf{k}-\mathbf{k}',\sigma\omega^L_{\mathbf{k}}-\sigma'\omega^L_{\mathbf{k}'}),\\
&\chi^{(2)}_{nml}(\mathbf{k}',\sigma'\omega^L_{\mathbf{k}'}|\mathbf{k}-\mathbf{k}',\sigma\omega^L_{\mathbf{k}}-\sigma'\omega^L_{\mathbf{k}'})\\
&\quad\approx -\frac{i}{2}\frac{e}{m_e\,\omega^2_{pe}}\,(\mathbf{k}-\mathbf{k}')_l\,\delta_{nm}\,\chi^e_\parallel(\mathbf{k}-\mathbf{k}',\sigma\omega^L_{\mathbf{k}}-\sigma'\omega^L_{\mathbf{k}'}),\\
&\chi^{(2)}_{ijk}(\mathbf{k}',\sigma'\omega^T_{\mathbf{k}'}|\mathbf{k}-\mathbf{k}',\sigma\omega^L_{\mathbf{k}}-\sigma'\omega^T_{\mathbf{k}'})\\
&\quad\approx -\frac{i}{2}\frac{e}{m_e\,\omega^2_{pe}}\,(\mathbf{k}-\mathbf{k}')_k\,\delta_{ij}\,\chi^e_\parallel(\mathbf{k}-\mathbf{k}',\sigma\omega^L_{\mathbf{k}}-\sigma'\omega^T_{\mathbf{k}'}),\\
&\chi^{(2)}_{nml}(\mathbf{k}',\sigma'\omega^T_{\mathbf{k}'}|\mathbf{k}-\mathbf{k}',\sigma\omega^L_{\mathbf{k}}-\sigma'\omega^T_{\mathbf{k}'})\\
&\quad\approx -\frac{i}{2}\frac{e}{m_e\,\omega^2_{pe}}\,(\mathbf{k}-\mathbf{k}')_l\,\delta_{nm}\,\chi^e_\parallel(\mathbf{k}-\mathbf{k}',\sigma\omega^L_{\mathbf{k}}-\sigma'\omega^T_{\mathbf{k}'}),
\end{aligned}
\tag{8.53}
$$

we obtain

$$
\begin{aligned}
\mathcal{I}_1 &= -\frac{e^2}{4m_e^2}\frac{(\mathbf{k}\cdot\mathbf{k}')^2}{k^2\,k'^2}\frac{(\mathbf{k}-\mathbf{k}')^2}{\omega^4_{pe}}\,[\chi^e_\parallel(\mathbf{k}-\mathbf{k}',\sigma\omega^L_{\mathbf{k}}-\sigma'\omega^L_{\mathbf{k}'})]^2,\\
\mathcal{I}_2 &= -\frac{e^2}{4m_e^2}\frac{(\mathbf{k}\times\mathbf{k}')^2}{k^2\,k'^2}\frac{(\mathbf{k}-\mathbf{k}')^2}{\omega^4_{pe}}\,[\chi^e_\parallel(\mathbf{k}-\mathbf{k}',\sigma\omega^L_{\mathbf{k}}-\sigma'\omega^T_{\mathbf{k}'})]^2.
\end{aligned}
\tag{8.54}
$$

Similarly, by making use of approximate third-order nonlinear susceptibilities discussed in Section 7.3.3, we obtain

$$
\begin{aligned}
\mathcal{J}_1 &\approx \frac{e^2}{2m_e^2}\frac{(\mathbf{k}\cdot\mathbf{k}')^2}{k^2\,k'^2}\frac{(\mathbf{k}-\mathbf{k}')^2}{\omega^4_{pe}}\,\chi^e_\parallel(\mathbf{k}-\mathbf{k}',\sigma\omega^L_{\mathbf{k}}-\sigma'\omega^L_{\mathbf{k}'}),\\
\mathcal{J}_2 &\approx \frac{e^2}{2m_e^2}\frac{(\mathbf{k}\times\mathbf{k}')^2}{k^2\,k'^2}\frac{(\mathbf{k}-\mathbf{k}')^2}{\omega^4_{pe}}\,\chi^e_\parallel(\mathbf{k}-\mathbf{k}',\sigma\omega^L_{\mathbf{k}}-\sigma'\omega^T_{\mathbf{k}'}).
\end{aligned}
\tag{8.55}
$$

This leads to

$$
\begin{aligned}
\left.\frac{\partial I^{\sigma L}_{\mathbf{k}}}{\partial t}\right|_{\text{ind.scatt.}} &= \frac{e^2}{m_e^2}\,(\sigma\omega^L_{\mathbf{k}})\,\mathrm{Im}\int d\mathbf{k}'\left\{\frac{(\mathbf{k}\cdot\mathbf{k}')^2}{k^2\,k'^2}\frac{(\mathbf{k}-\mathbf{k}')^2}{\omega^4_{pe}}\right.\\
&\quad\times\sum_{\sigma'}\left[\mathscr{P}\,\frac{[\chi^e_\parallel(\mathbf{k}-\mathbf{k}',\sigma\omega^L_{\mathbf{k}}-\sigma'\omega^L_{\mathbf{k}'})]^2}{\epsilon_\parallel(\mathbf{k}-\mathbf{k}',\sigma\omega^L_{\mathbf{k}}-\sigma'\omega^L_{\mathbf{k}'})}\right.\\
&\quad\left.-\,\chi^e_\parallel(\mathbf{k}-\mathbf{k}',\sigma\omega^L_{\mathbf{k}}-\sigma'\omega^L_{\mathbf{k}'})\right]I^{\sigma' L}_{\mathbf{k}'}\,I^{\sigma L}_{\mathbf{k}}
\end{aligned}
$$

$$+\frac{1}{2}\frac{(\mathbf{k}\times\mathbf{k}')^2}{k^2\,k'^2}\frac{(\mathbf{k}-\mathbf{k}')^2}{\omega_{pe}^4}\sum_{\sigma'}\left[\mathscr{P}\,\frac{[\chi_\parallel^e(\mathbf{k}-\mathbf{k}',\sigma\omega_\mathbf{k}^L-\sigma'\omega_{\mathbf{k}'}^T)]^2}{\epsilon_\parallel(\mathbf{k}-\mathbf{k}',\sigma\omega_\mathbf{k}^L-\sigma'\omega_{\mathbf{k}'}^T)}\right.$$
$$\left.-\chi_\parallel^e(\mathbf{k}-\mathbf{k}',\sigma\omega_\mathbf{k}^L-\sigma'\omega_{\mathbf{k}'}^T)\right]I_{\mathbf{k}'}^{\sigma'T}\,I_\mathbf{k}^{\sigma L}\Bigg\}. \tag{8.56}$$

The remaining terms to evaluate are related to the principal parts of inverse dielectric constants, which are treated as shown in (8.57):

$$\begin{aligned}
\operatorname{Im}\mathscr{P}\,\frac{[\chi_\parallel^e]^2}{\epsilon_\parallel} &\approx \operatorname{Im}\mathscr{P}\,\frac{[\chi_\parallel^e]^2}{\chi_\parallel^e+i\operatorname{Im}\chi_\parallel^i}\approx\operatorname{Im}\mathscr{P}\,\chi_\parallel^e\left(1-i\,\frac{\operatorname{Im}\chi_\parallel^i}{\chi_\parallel^e}\right)\\
&=\operatorname{Im}(\chi_\parallel^e-i\operatorname{Im}\chi_\parallel^i)=\operatorname{Im}\chi_\parallel^e-\operatorname{Im}\chi_\parallel^i \\
&=-\sum_{a=e,i}\frac{\pi\,\omega_{pa}^2}{(\mathbf{k}-\mathbf{k}')^2}\int d\mathbf{v}\,(\mathbf{k}-\mathbf{k}')\\
&\quad\cdot\frac{\partial F_a}{\partial\mathbf{v}}\,\delta[\sigma\omega_\mathbf{k}-\sigma'\omega_{\mathbf{k}'}-(\mathbf{k}-\mathbf{k}')\cdot\mathbf{v}].
\end{aligned}\tag{8.57}$$

Of the two terms associated with the summation over particle species, the ion term is the dominant term since it is related to the scattering off thermal ions – see the discussion in Section 2.3.1. Substituting (8.57) to (8.56), we finally obtain

$$\begin{aligned}
\left.\frac{\partial I_\mathbf{k}^{\sigma L}}{\partial t}\right|_{\text{ind.scatt.}} &= \frac{\pi}{\omega_{pe}^2}\frac{e^2}{m_e m_i}\,\sigma\omega_\mathbf{k}^L\sum_{\sigma'=\pm1}\int d\mathbf{k}'\int d\mathbf{v}\,(\mathbf{k}-\mathbf{k}')\cdot\frac{\partial F_i}{\partial\mathbf{v}}\\
&\times\left(\frac{(\mathbf{k}\cdot\mathbf{k}')^2}{k^2\,k'^2}\,\delta[\sigma\omega_\mathbf{k}^L-\sigma'\omega_{\mathbf{k}'}^L-(\mathbf{k}-\mathbf{k}')\cdot\mathbf{v}]\,I_{\mathbf{k}'}^{\sigma'L}\right.\\
&+\left.\frac{(\mathbf{k}\times\mathbf{k}')^2}{k^2\,k'^2}\,\delta[\sigma\omega_\mathbf{k}^L-\sigma'\omega_{\mathbf{k}'}^T-(\mathbf{k}-\mathbf{k}')\cdot\mathbf{v}]\,\frac{I_{\mathbf{k}'}^{\sigma'T}}{2}\right)I_\mathbf{k}^{\sigma L}.
\end{aligned}\tag{8.58}$$

T Mode

The induced scattering process involving ion-sound modes, which is discussed in Section 2.3.3, will not be considered here. It is a slow process, and, as we have seen in Chapter 6, especially Figure 6.4, the ion-sound mode excitation by decay processes during the beam-plasma instability development is generally a transient process with low intensity. Consequently, the scattering of protons mediated by interaction with ion-sound wave can be ignored, which is a process that is expected to take place, if at all, over a time scale even longer than the ion-sound decay processes. This is because the ion-sound decay involves the Langmuir wave intensity such that the time scale of relevance is governed by the products of the wave intensities, $t_{S\,\text{decay}}\propto\mathscr{O}(I_S I_L)$. In contrast, the scattering process involving two ion-sound modes is expected to be characterized by $t_{S\,\text{scatt.}}\propto\mathscr{O}(I_S^2)\ll\mathscr{O}(I_S I_L)$.

We thus consider the induced scattering process involving transverse waves. For this process, we only consider L–T mode coupling.

$$\begin{aligned}
\left.\frac{\partial I_{\mathbf{k}}^{\sigma T}}{\partial t}\right|_{\text{ind.scatt.}} &= -\frac{4}{\Lambda_{\perp}'(\mathbf{k},\sigma\omega_{\mathbf{k}}^{T})}\,\mathrm{Im}\int d\mathbf{k}'\sum_{\sigma'}\frac{1}{2}\left(\delta_{in}-\frac{k_i\,k_n}{k^2}\right)\frac{k_j'\,k_m'}{k'^2}\\
&\times\left[\frac{(\mathbf{k}-\mathbf{k}')_k\,(\mathbf{k}-\mathbf{k}')_l}{|\mathbf{k}-\mathbf{k}'|^2}\,\chi_{ijk}^{(2)}(\mathbf{k}',\sigma'\omega_{\mathbf{k}'}^{L}|\mathbf{k}-\mathbf{k}',\sigma\omega_{\mathbf{k}}^{T}-\sigma'\omega_{\mathbf{k}'}^{L})\right.\\
&\times\chi_{nml}^{(2)}(\mathbf{k}',\sigma'\omega_{\mathbf{k}'}^{L}|\mathbf{k}-\mathbf{k}',\sigma\omega_{\mathbf{k}}^{T}-\sigma'\omega_{\mathbf{k}'}^{L})\,\mathscr{P}\,\frac{2}{\epsilon_{\parallel}(\mathbf{k}-\mathbf{k}',\sigma\omega_{\mathbf{k}}^{T}-\sigma'\omega_{\mathbf{k}'}^{L})}\\
&+\left(\delta_{kl}-\frac{(\mathbf{k}-\mathbf{k})_k\,(\mathbf{k}-\mathbf{k})_l}{|\mathbf{k}-\mathbf{k}|^2}\right)\chi_{ijk}^{(2)}(\mathbf{k}',\sigma'\omega_{\mathbf{k}'}^{L}|\mathbf{k}-\mathbf{k}',\sigma\omega_{\mathbf{k}}^{T}-\sigma'\omega_{\mathbf{k}'}^{L})\\
&\times\chi_{nml}^{(2)}(\mathbf{k}',\sigma'\omega_{\mathbf{k}'}^{L}|\mathbf{k}-\mathbf{k}',\sigma\omega_{\mathbf{k}}^{T}-\sigma'\omega_{\mathbf{k}'}^{L})\,\mathscr{P}\,\frac{2}{\Lambda_{\perp}(\mathbf{k}-\mathbf{k}',\sigma\omega_{\mathbf{k}}^{T}-\sigma'\omega_{\mathbf{k}'}^{L})}\\
&\left.+\chi_{ijmn}^{(3)}(\mathbf{k}',\sigma'\omega_{\mathbf{k}'}^{L}|-\mathbf{k}',-\sigma'\omega_{\mathbf{k}'}^{L}|\mathbf{k},\sigma\omega_{\mathbf{k}}^{T})\right]I_{\mathbf{k}'}^{\sigma' L}\,I_{\mathbf{k}}^{\sigma T}.
\end{aligned}\tag{8.59}$$

Of these, contribution from the second term that contains the inverse of $\Lambda_\perp$ in the denominator can again be ignored, since the frequency- and wave number–shifted arguments satisfy inequalities (8.50). The remaining terms for induced scattering of transverse waves involving the second-order nonlinear susceptibility are relevant only if the shifted frequency, $\sigma\omega_{\mathbf{k}}^{T}-\sigma'\omega_{\mathbf{k}'}^{L}$ remains small, which implies that this term is important only if the transverse mode has angular frequency close to the plasma frequency.

We make use of approximate expressions for the second-order susceptibilities to simplify the following quantity:

$$\begin{aligned}
&\left(\delta_{in}-\frac{k_i\,k_n}{k^2}\right)\frac{k_j'\,k_m'}{k'^2}\,\frac{(\mathbf{k}-\mathbf{k}')_k\,(\mathbf{k}-\mathbf{k}')_l}{|\mathbf{k}-\mathbf{k}'|^2}\\
&\quad\times\chi_{ijk}^{(2)}(\mathbf{k}',\sigma'\omega_{\mathbf{k}'}^{L}|\mathbf{k}-\mathbf{k}',\sigma\omega_{\mathbf{k}}^{T}-\sigma'\omega_{\mathbf{k}'}^{L})\,\chi_{nml}^{(2)}(\mathbf{k}',\sigma'\omega_{\mathbf{k}'}^{L}|\mathbf{k}-\mathbf{k}',\sigma\omega_{\mathbf{k}}^{T}-\sigma'\omega_{\mathbf{k}'}^{L})\\
&\approx-\frac{e^2}{4m_e^2}\,\frac{(\mathbf{k}-\mathbf{k}')^2}{\omega_{pe}^4}\,\frac{(\mathbf{k}\times\mathbf{k}')^2}{k^2\,k'^2}\,[\chi_{\parallel}^{e}(\mathbf{k}-\mathbf{k}',\sigma\omega_{\mathbf{k}}^{T}-\sigma'\omega_{\mathbf{k}'}^{L})]^2.
\end{aligned}\tag{8.60}$$

Likewise, making use of approximate third-order nonlinear susceptibility, we have

$$\begin{aligned}
&\left(\delta_{in}-\frac{k_i\,k_n}{k^2}\right)\frac{k_j'\,k_m'}{k'^2}\,\chi_{ijmn}^{(3)}(\mathbf{k}',\sigma'\omega_{\mathbf{k}'}^{L}|-\mathbf{k}',-\sigma'\omega_{\mathbf{k}'}^{L}|\mathbf{k},\sigma\omega_{\mathbf{k}}^{T})\\
&\approx\frac{e^2}{2m_e^2}\,\frac{(\mathbf{k}\times\mathbf{k}')^2}{k^2\,k'^2}\,\frac{(\mathbf{k}-\mathbf{k}')^2}{\omega_{pe}^4}\,\chi_{\parallel}^{e}(\mathbf{k}-\mathbf{k}',\sigma\omega_{\mathbf{k}}^{T}-\sigma'\omega_{\mathbf{k}'}^{L}).
\end{aligned}\tag{8.61}$$

From this, we arrive at

$$\frac{\partial}{\partial t}\bigg|_{\text{ind.scatt.}} \frac{I_{\mathbf{k}}^{\sigma T}}{2} = \frac{1}{2}\frac{\pi}{\omega_{pe}^2}\frac{e^2}{m_e m_i} \sum_{\sigma'=\pm 1} \int d\mathbf{k}' \int d\mathbf{v}\, (\mathbf{k}-\mathbf{k}')\cdot\frac{\partial F_i}{\partial \mathbf{v}}$$
$$\times \frac{(\mathbf{k}\times\mathbf{k}')^2}{k^2 k'^2}\,\delta[\sigma\omega_{\mathbf{k}}^T - \sigma'\omega_{\mathbf{k}'}^L - (\mathbf{k}-\mathbf{k}')\cdot\mathbf{v}]\, I_{\mathbf{k}'}^{\sigma' L}\,\frac{I_{\mathbf{k}}^{\sigma T}}{2}. \tag{8.62}$$

8.3 Summary of Equations

Let us summarize the equations of electromagnetic Vlasov weak turbulence theory within the framework of Vlasovian (or collisionless) approximation. In this summary, the final set of equations directly generalize the equations of electrostatic Vlasov weak turbulence theory, but, as in that case, the present electromagnetic Vlasov theory is incomplete in that various terms representing induced processes do not all have their spontaneous counterparts. To complete the theory requires the consideration of Klimontovich formalism, which is done in Part IV, but the discussion thereof builds upon the findings in Part III, which is summarized next.

8.3.1 Particle Kinetic Equation

From (7.43), upon taking the eigenmodes to express the spectral wave electric field wave energy densities,

$$\langle E_{\parallel}^2\rangle_{\mathbf{k},\omega} = \sum_{\alpha=L,S}\sum_{\sigma=\pm 1} I_{\mathbf{k}}^{\sigma\alpha}\,\delta(\omega-\sigma\omega_{\mathbf{k}}^{\alpha}),$$
$$\langle E_{\perp}^2\rangle_{\mathbf{k},\omega} = \sum_{\sigma=\pm 1} I_{\mathbf{k}}^{\sigma T}\,\delta(\omega-\sigma\omega_{\mathbf{k}}^{T}),$$

we have the particle kinetic equation

$$\frac{\partial F_a}{\partial t} = \frac{\partial}{\partial v_i} D_{ij}\frac{\partial F_a}{\partial v_j},$$
$$D_{ij} = -\mathrm{Im}\,\frac{e_a^2}{m_a^2}\int d\mathbf{k}\left(\sum_{\alpha}\sum_{\sigma}\frac{k_i k_j}{k^2}\frac{I_{\mathbf{k}}^{\sigma\alpha}}{\sigma\omega_{\mathbf{k}}^{\alpha}-\mathbf{k}\cdot\mathbf{v}+i0}\right.$$
$$\left.+\frac{1}{2}\sum_{\sigma} a_{k_i} a_{kj}\,\frac{I_{\mathbf{k}}^{\sigma T}}{\sigma\omega_{\mathbf{k}}^{T}-\mathbf{k}\cdot\mathbf{v}+i0}\right),$$
$$a_{ij} = \left(1-\frac{\mathbf{k}\cdot\mathbf{v}}{\omega}\right)\left(\delta_{ij}-\frac{k_i k_j}{k^2}\right)+\frac{[(\mathbf{k}\times\mathbf{v})\times\mathbf{k}]_i}{k^2}\frac{k_j}{\omega}. \tag{8.63}$$

Note that the contribution from transverse mode to the diffusion coefficient tensor is included only for the sake of completeness. In reality, because of the fact that the transverse mode is a fast mode (superluminal mode with the phase speed higher than the speed of light *in vacuo*, or $\omega/k > c$), the resonance condition cannot be satisfied, so that for all practical purposes, the T mode has no contribution to the velocity space diffusion.

8.3.2 L Mode

Wave kinetic equation for Langmuir wave is given in (8.64), together with brief descriptions for each process, displayed as boxed texts, and indicated with underbraces:

$$\begin{aligned}
\frac{\partial I_{\mathbf{k}}^{\sigma L}}{\partial t} &= \pi\,(\sigma\omega_{\mathbf{k}}^L)\,\frac{\omega_{pe}^2}{k^2}\int d\mathbf{v}\,\underbrace{\delta(\sigma\omega_{\mathbf{k}}^L-\mathbf{k}\cdot\mathbf{v})}_{\boxed{\text{ind. emiss. } L}}\,\mathbf{k}\cdot\frac{\partial F_e}{\partial\mathbf{v}}\,I_{\mathbf{k}}^{\sigma L} \\
&-\pi\,(\sigma\omega_{\mathbf{k}}^L)\sum_{\sigma',\sigma''}\int d\mathbf{k}'\,\frac{e^2}{4T_e^2}\,\frac{\mu_{\mathbf{k}'}\,[\mathbf{k}\cdot(\mathbf{k}-\mathbf{k}')]^2}{k^2\,k'^2\,|\mathbf{k}-\mathbf{k}'|^2}\bigg(\sigma'\omega_{\mathbf{k}'}^L\,I_{\mathbf{k}-\mathbf{k}'}^{\sigma''L}\,I_{\mathbf{k}}^{\sigma L} \\
&+\sigma''\omega_{\mathbf{k}-\mathbf{k}'}^L\,\frac{I_{\mathbf{k}'}^{\sigma'S}}{\mu_{\mathbf{k}'}}\,I_{\mathbf{k}}^{\sigma L}-\sigma\omega_{\mathbf{k}}^L\,I_{\mathbf{k}-\mathbf{k}'}^{\sigma''L}\,\frac{I_{\mathbf{k}'}^{\sigma'S}}{\mu_{\mathbf{k}'}}\bigg)\,\underbrace{\delta(\sigma\omega_{\mathbf{k}}^L-\sigma'\omega_{\mathbf{k}'}^S-\sigma''\omega_{\mathbf{k}-\mathbf{k}'}^L)}_{\boxed{\text{decay } L\leftrightarrow L+S}} \\
&-\pi\,(\sigma\omega_{\mathbf{k}}^L)\sum_{\sigma',\sigma''}\int d\mathbf{k}'\,\frac{e^2}{4T_e^2}\,\frac{\mu_{\mathbf{k}-\mathbf{k}'}\,(\mathbf{k}\cdot\mathbf{k}')^2}{k^2\,k'^2|\,\mathbf{k}-\mathbf{k}'|^2}\bigg(\sigma'\omega_{\mathbf{k}'}^L\,\frac{I_{\mathbf{k}-\mathbf{k}'}^{\sigma''S}}{\mu_{\mathbf{k}-\mathbf{k}'}}\,I_{\mathbf{k}}^{\sigma L} \\
&+\sigma''\omega_{\mathbf{k}-\mathbf{k}'}^L\,I_{\mathbf{k}'}^{\sigma'L}\,I_{\mathbf{k}}^{\sigma L}-\sigma\omega_{\mathbf{k}}^L\,\frac{I_{\mathbf{k}-\mathbf{k}'}^{\sigma''S}}{\mu_{\mathbf{k}-\mathbf{k}'}}\,I_{\mathbf{k}'}^{\sigma'L}\bigg)\,\underbrace{\delta(\sigma\omega_{\mathbf{k}}^L-\sigma'\omega_{\mathbf{k}'}^L-\sigma''\omega_{\mathbf{k}-\mathbf{k}'}^S)}_{\boxed{\text{decay } L\leftrightarrow L+S}} \\
&-\pi\,(\sigma\omega_{\mathbf{k}}^L)\sum_{\sigma',\sigma''}\int d\mathbf{k}'\,\frac{e^2}{2T_e^2}\,\frac{\mu_{\mathbf{k}'}\,(\mathbf{k}\times\mathbf{k}')^2}{k^2\,k'^2\,|\mathbf{k}-\mathbf{k}'|^2}\bigg(\sigma'\omega_{\mathbf{k}'}^L\,\frac{I_{\mathbf{k}-\mathbf{k}'}^{\sigma''T}}{2}\,I_{\mathbf{k}}^{\sigma L} \\
&+\sigma''\omega_{\mathbf{k}-\mathbf{k}'}^T\,\frac{I_{\mathbf{k}'}^{\sigma'S}}{\mu_{\mathbf{k}'}}\,I_{\mathbf{k}}^{\sigma L}-\sigma\omega_{\mathbf{k}}^L\,\frac{I_{\mathbf{k}'}^{\sigma'S}}{\mu_{\mathbf{k}'}}\,\frac{I_{\mathbf{k}-\mathbf{k}'}^{\sigma''T}}{2}\bigg)\,\underbrace{\delta(\sigma\omega_{\mathbf{k}}^L-\sigma'\omega_{\mathbf{k}'}^S-\sigma''\omega_{\mathbf{k}-\mathbf{k}'}^T)}_{\boxed{\text{decay } L\leftrightarrow T+S}} \\
&-\pi\,(\sigma\omega_{\mathbf{k}}^L)\sum_{\sigma',\sigma''}\int d\mathbf{k}'\,\frac{e^2}{8\,m_e^2\,\omega_{pe}^2}\,\frac{(\mathbf{k}\times\mathbf{k}')^2}{k^2\,k'^2\,|\mathbf{k}-\mathbf{k}'|^2}\left(\frac{k^2}{\sigma\omega_{\mathbf{k}}^L}+\frac{k'^2}{\sigma'\omega_{\mathbf{k}'}^L}\right)^2 \\
&\times\bigg(\sigma'\omega_{\mathbf{k}'}^L\,\frac{I_{\mathbf{k}-\mathbf{k}'}^{\sigma''T}}{2}\,I_{\mathbf{k}}^{\sigma L}+\sigma''\omega_{\mathbf{k}-\mathbf{k}'}^T\,I_{\mathbf{k}'}^{\sigma'L}\,I_{\mathbf{k}}^{\sigma L}
\end{aligned}$$

$$
- \sigma \omega_{\mathbf{k}}^{L} \, I_{\mathbf{k}'}^{\sigma' L} \, \frac{I_{\mathbf{k}-\mathbf{k}'}^{\sigma'' T}}{2} \Bigg) \underbrace{\delta(\sigma \omega_{\mathbf{k}}^{L} - \sigma' \omega_{\mathbf{k}'}^{L} - \sigma'' \omega_{\mathbf{k}-\mathbf{k}'}^{T})}_{\boxed{\text{decay } L \leftrightarrow T + L}}
$$

$$
- \pi \, (\sigma \omega_{\mathbf{k}}^{L}) \sum_{\sigma', \sigma''} \int d\mathbf{k}' \, \frac{e^2}{4 m_e^2} \frac{k^2 \, \{k'^2 \, |\mathbf{k} - \mathbf{k}'|^2 + [\mathbf{k}' \cdot (\mathbf{k} - \mathbf{k}')]^2\}}{k'^2 \, |\mathbf{k} - \mathbf{k}'|^2 \, (\omega_{\mathbf{k}'}^{T})^2 \, (\omega_{\mathbf{k}-\mathbf{k}'}^{T})^2}
$$

$$
\times \left(\sigma' \omega_{\mathbf{k}'}^{T} \, \frac{I_{\mathbf{k}-\mathbf{k}'}^{\sigma'' T}}{2} \, I_{\mathbf{k}}^{\sigma L} + \sigma'' \omega_{\mathbf{k}-\mathbf{k}'}^{T} \, \frac{I_{\mathbf{k}'}^{\sigma' T}}{2} \, I_{\mathbf{k}}^{\sigma L} - \sigma \omega_{\mathbf{k}}^{L} \, \frac{I_{\mathbf{k}'}^{\sigma' T}}{2} \, \frac{I_{\mathbf{k}-\mathbf{k}'}^{\sigma'' T}}{2} \right)
$$

$$
\times \underbrace{\delta(\sigma \omega_{\mathbf{k}}^{L} - \sigma' \omega_{\mathbf{k}'}^{T} - \sigma'' \omega_{\mathbf{k}-\mathbf{k}'}^{T})}_{\boxed{\text{decay } L \leftrightarrow T + T}}
$$

$$
+ \frac{\pi}{\omega_{pe}^2} \frac{e^2}{m_e m_i} \sigma \omega_{\mathbf{k}}^{L} \sum_{\sigma' = \pm 1} \int d\mathbf{k}' \int d\mathbf{v} \, (\mathbf{k} - \mathbf{k}') \cdot \frac{\partial F_i}{\partial \mathbf{v}}
$$

$$
\times \left(\frac{(\mathbf{k} \cdot \mathbf{k}')^2}{k^2 \, k'^2} \underbrace{\delta[\sigma \omega_{\mathbf{k}}^{L} - \sigma' \omega_{\mathbf{k}'}^{L} - (\mathbf{k} - \mathbf{k}') \cdot \mathbf{v}]}_{\boxed{\text{ind. scatt. } L \leftrightarrow L + i}} I_{\mathbf{k}'}^{\sigma' L} \right.
$$

$$
\left. + \frac{(\mathbf{k} \times \mathbf{k}')^2}{k^2 \, k'^2} \underbrace{\delta[\sigma \omega_{\mathbf{k}}^{L} - \sigma' \omega_{\mathbf{k}'}^{T} - (\mathbf{k} - \mathbf{k}') \cdot \mathbf{v}]}_{\boxed{\text{ind. scatt. } L \leftrightarrow T + i}} \frac{I_{\mathbf{k}'}^{\sigma' T}}{2} \right) I_{\mathbf{k}}^{\sigma L}. \tag{8.64}
$$

The L mode equation mentioned in (8.64) is to be considered in conjunction with wave kinetic equations for S and T modes as well as with the particle kinetic equations.

8.3.3 S Mode

The equation that dictates the time evolution of S mode intensity is given by

$$
\frac{\partial}{\partial t} \frac{I_{\mathbf{k}}^{\sigma S}}{\mu_{\mathbf{k}}} = \pi \, (\sigma \mu_{\mathbf{k}} \omega_{\mathbf{k}}^{L}) \, \frac{\omega_{pe}^2}{k^2} \int d\mathbf{v} \, \underbrace{\delta(\sigma \omega_{\mathbf{k}}^{S} - \mathbf{k} \cdot \mathbf{v})}_{\boxed{\text{ind. emiss. } S}} \, \mathbf{k} \cdot \frac{\partial (F_e + F_i)}{\partial \mathbf{v}} \, \frac{I_{\mathbf{k}}^{\sigma S}}{\mu_{\mathbf{k}}}
$$

$$
- \pi \, \sigma \omega_{\mathbf{k}}^{L} \sum_{\sigma', \sigma''} \int d\mathbf{k}' \, \frac{e^2}{4 T_e^2} \frac{\mu_{\mathbf{k}} \, [\mathbf{k}' \cdot (\mathbf{k} - \mathbf{k}')]^2}{k^2 \, k'^2 \, |\mathbf{k} - \mathbf{k}'|^2} \left(\sigma' \omega_{\mathbf{k}'}^{L} \, I_{\mathbf{k}-\mathbf{k}'}^{\sigma'' L} \, \frac{I_{\mathbf{k}}^{\sigma S}}{\mu_{\mathbf{k}}} \right.
$$

$$
\left. + \sigma'' \omega_{\mathbf{k}-\mathbf{k}'}^{L} \, I_{\mathbf{k}'}^{\sigma' L} \, \frac{I_{\mathbf{k}}^{\sigma S}}{\mu_{\mathbf{k}}} - \sigma \omega_{\mathbf{k}}^{L} \, I_{\mathbf{k}-\mathbf{k}'}^{\sigma'' L} \, I_{\mathbf{k}'}^{\sigma' L} \right) \underbrace{\delta(\sigma \omega_{\mathbf{k}}^{S} - \sigma' \omega_{\mathbf{k}'}^{L} - \sigma'' \omega_{\mathbf{k}-\mathbf{k}'}^{L})}_{\boxed{\text{decay } S \leftrightarrow L + L}}
$$

$$
-\pi\,\sigma\omega_{\mathbf{k}}^{L}\sum_{\sigma',\sigma''}\int d\mathbf{k}'\,\frac{e^2}{2T_e^2}\frac{\mu_{\mathbf{k}}\,(\mathbf{k}\times\mathbf{k}')^2}{k^2\,k'^2\,|\mathbf{k}-\mathbf{k}'|^2}\left(\sigma'\omega_{\mathbf{k}'}^{L}\,\frac{I_{\mathbf{k}-\mathbf{k}'}^{\sigma''T}}{2}\,\frac{I_{\mathbf{k}}^{\sigma S}}{\mu_{\mathbf{k}}}\right. \tag{8.65}
$$

$$
\left.+\,\sigma''\omega_{\mathbf{k}-\mathbf{k}'}^{T}\,I_{\mathbf{k}'}^{\sigma' L}\,\frac{I_{\mathbf{k}}^{\sigma S}}{\mu_{\mathbf{k}}}-\sigma\omega_{\mathbf{k}}^{L}\,I_{\mathbf{k}'}^{\sigma' L}\,\frac{I_{\mathbf{k}-\mathbf{k}'}^{\sigma''T}}{2}\right)\underbrace{\delta(\sigma\omega_{\mathbf{k}}^{S}-\sigma'\omega_{\mathbf{k}'}^{L}-\sigma''\omega_{\mathbf{k}-\mathbf{k}'}^{T})}_{\boxed{\text{decay } S\leftrightarrow L+T}}.
$$

8.3.4 T Mode

The transverse mode wave intensity dynamically evolves according to

$$
\frac{\partial}{\partial t}\frac{I_{\mathbf{k}}^{\sigma T}}{2}=\pi\sum_{a}\frac{\omega_{pa}^2}{\sigma\omega_{\mathbf{k}}^{T}}\int d\mathbf{v}\,\frac{(\mathbf{k}\times\mathbf{v})^2}{k^2}\,\underbrace{\delta(\sigma\omega_{\mathbf{k}}^{T}-\mathbf{k}\cdot\mathbf{v})}_{\boxed{\text{ind. emiss. } T}}\,\mathbf{k}\cdot\frac{\partial F_a}{\partial\mathbf{v}}\frac{I_{\mathbf{k}}^{\sigma T}}{2}
$$

$$
-\pi\,\sigma\omega_{\mathbf{k}}^{T}\sum_{\sigma',\sigma''}\int d\mathbf{k}'\,\frac{e^2}{32\,m_e^2\,\omega_{pe}^2}\frac{(\mathbf{k}\times\mathbf{k}')^2}{k^2\,k'^2\,|\mathbf{k}-\mathbf{k}'|^2}
$$

$$
\times\left(\frac{k'^2}{\sigma'\omega_{\mathbf{k}'}^{L}}-\frac{|\mathbf{k}-\mathbf{k}'|^2}{\sigma''\omega_{\mathbf{k}-\mathbf{k}'}^{L}}\right)^2\left(\sigma'\omega_{\mathbf{k}'}^{L}\,I_{\mathbf{k}-\mathbf{k}'}^{\sigma''L}\,\frac{I_{\mathbf{k}}^{\sigma T}}{2}+\sigma''\omega_{\mathbf{k}-\mathbf{k}'}^{L}\,I_{\mathbf{k}'}^{\sigma' L}\,\frac{I_{\mathbf{k}}^{\sigma T}}{2}\right.
$$

$$
\left.-\,\sigma\omega_{\mathbf{k}}^{T}\,I_{\mathbf{k}-\mathbf{k}'}^{\sigma''L}\,I_{\mathbf{k}'}^{\sigma' L}\right)\underbrace{\delta(\sigma\omega_{\mathbf{k}}^{T}-\sigma'\omega_{\mathbf{k}'}^{L}-\sigma''\omega_{\mathbf{k}-\mathbf{k}'}^{L})}_{\boxed{\text{decay } T\leftrightarrow L+L}}
$$

$$
-\pi\,\sigma\omega_{\mathbf{k}}^{T}\sum_{\sigma',\sigma''}\int d\mathbf{k}'\,\frac{e^2}{8T_e^2}\frac{\mu_{\mathbf{k}'}\,(\mathbf{k}\times\mathbf{k}')^2}{k^2\,k'^2\,|\mathbf{k}-\mathbf{k}'|^2}\left(\sigma'\omega_{\mathbf{k}'}^{L}\,I_{\mathbf{k}-\mathbf{k}'}^{\sigma''L}\,\frac{I_{\mathbf{k}}^{\sigma T}}{2}\right.
$$

$$
\left.+\,\sigma''\omega_{\mathbf{k}-\mathbf{k}'}^{L}\,\frac{I_{\mathbf{k}'}^{\sigma' S}}{\mu_{\mathbf{k}'}}\,\frac{I_{\mathbf{k}}^{\sigma T}}{2}-\sigma\omega_{\mathbf{k}}^{T}\,I_{\mathbf{k}-\mathbf{k}'}^{\sigma''L}\,\frac{I_{\mathbf{k}'}^{\sigma' S}}{\mu_{\mathbf{k}'}}\right)\underbrace{\delta(\sigma\omega_{\mathbf{k}}^{T}-\sigma'\omega_{\mathbf{k}'}^{S}-\sigma''\omega_{\mathbf{k}-\mathbf{k}'}^{L})}_{\boxed{\text{decay } T\leftrightarrow L+S}}
$$

$$
-\pi\,\sigma\omega_{\mathbf{k}}^{T}\sum_{\sigma',\sigma''}\int d\mathbf{k}'\,\frac{e^2}{8T_e^2}\frac{\mu_{\mathbf{k}-\mathbf{k}'}\,(\mathbf{k}\times\mathbf{k}')^2}{k^2\,k'^2\,|\mathbf{k}-\mathbf{k}'|^2}\left(\sigma'\omega_{\mathbf{k}'}^{L}\,\frac{I_{\mathbf{k}-\mathbf{k}'}^{\sigma''S}}{\mu_{\mathbf{k}-\mathbf{k}'}}\,\frac{I_{\mathbf{k}}^{\sigma T}}{2}\right.
$$

$$
\left.+\,\sigma''\omega_{\mathbf{k}-\mathbf{k}'}^{L}\,I_{\mathbf{k}'}^{\sigma' L}\,\frac{I_{\mathbf{k}}^{\sigma T}}{2}-\sigma\omega_{\mathbf{k}}^{T}\,\frac{I_{\mathbf{k}-\mathbf{k}'}^{\sigma''S}}{\mu_{\mathbf{k}-\mathbf{k}'}}\,I_{\mathbf{k}'}^{\sigma' L}\right)\underbrace{\delta(\sigma\omega_{\mathbf{k}}^{T}-\sigma'\omega_{\mathbf{k}'}^{L}-\sigma''\omega_{\mathbf{k}-\mathbf{k}'}^{S})}_{\boxed{\text{decay } T\leftrightarrow L+S}}
$$

$$
-\pi\,\sigma\omega_{\mathbf{k}}^{T}\sum_{\sigma',\sigma''}\int d\mathbf{k}'\,\frac{1}{4}\frac{e^2}{m_e^2}\frac{|\mathbf{k}-\mathbf{k}'|^2}{(\sigma\omega_{\mathbf{k}}^{T})^2(\sigma'\omega_{\mathbf{k}'}^{T})^2}\left(1+\frac{(\mathbf{k}\cdot\mathbf{k}')^2}{k^2k'^2}\right)
$$

$$\times \left(\sigma' \omega_{\mathbf{k}'}^{T} I_{\mathbf{k}-\mathbf{k}'}^{\sigma'' L} \frac{I_{\mathbf{k}}^{\sigma T}}{2} + \sigma'' \omega_{\mathbf{k}-\mathbf{k}'}^{L} \frac{I_{\mathbf{k}'}^{\sigma' T}}{2} \frac{I_{\mathbf{k}}^{\sigma T}}{2} \right.$$

$$\left. - \sigma \omega_{\mathbf{k}}^{T} I_{\mathbf{k}-\mathbf{k}'}^{\sigma'' L} \frac{I_{\mathbf{k}'}^{\sigma' T}}{2} \right) \underbrace{\delta(\sigma \omega_{\mathbf{k}}^{T} - \sigma' \omega_{\mathbf{k}'}^{T} - \sigma'' \omega_{\mathbf{k}-\mathbf{k}'}^{L})}_{\boxed{\text{decay } T \leftrightarrow L + T}}$$

$$+ \frac{1}{2} \frac{\pi}{\omega_{pe}^{2}} \frac{e^{2}}{m_e m_i} \sum_{\sigma'=\pm 1} \int d\mathbf{k}' \int d\mathbf{v} \, (\mathbf{k} - \mathbf{k}') \cdot \frac{\partial F_i}{\partial \mathbf{v}} \frac{(\mathbf{k} \times \mathbf{k}')^2}{k^2 k'^2}$$

$$\times \underbrace{\delta[\sigma \omega_{\mathbf{k}}^{T} - \sigma' \omega_{\mathbf{k}'}^{L} - (\mathbf{k} - \mathbf{k}') \cdot \mathbf{v}]}_{\boxed{\text{ind. scatt. } T \leftrightarrow L + i}} I_{\mathbf{k}'}^{\sigma' L} \frac{I_{\mathbf{k}}^{\sigma T}}{2}. \tag{8.66}$$

The particle and wave kinetic equations summarized in (8.63)–(8.66) constitute the self-consistent system of equations for electromagnetic Vlasov weak turbulence theory. These equations are not complete since not all terms appear in balanced form between the spontaneous and induced processes (for the waves), and the particle kinetic equations do not contain the velocity space friction (or drag) term. In order to supplement the set of equations in (8.63)–(8.66), we need to include the effects owing to particle discreteness, which we will turn to in Part IV next.

Part IV

Klimontovich Weak Turbulence Theory: Electromagnetic Formalism

9

Electromagnetic Klimontovich Weak Turbulence Theory

In Part IV, we finalize the discourse by formulating the electromagnetic weak turbulence theory that includes the discrete particle effects, thereby completing the formalism.

9.1 Nonlinear Electrodynamic Equations in Plasmas

The convenient starting point of the discussion is the set of equations already derived in Chapter 4, namely, (4.48), which was derived on the basis of Klimontovich equation combined with Maxwell's equation. In what follows, it is convenient to re-express the result by making use of the momentum space representation, as such a representation naturally lends itself to relativistic formalism:

$$
\begin{aligned}
\frac{\partial f_a}{\partial t} &= -e_a \frac{\partial}{\partial p_i} \int d\mathbf{k} \int \frac{d\omega}{\omega} \left[(\omega - \mathbf{k}\cdot\mathbf{v})\, \delta_{ij} + v_j\, k_i \right] \\
&\quad \times \left\langle \delta E^j_{-\mathbf{k},-\omega}\, \delta N^a_{\mathbf{k},\omega}(\mathbf{p}) \right\rangle, \\
\left(\delta_{ij} - \frac{c^2}{\omega^2} \left(k^2 \delta_{ij} - k_i k_j \right) \right) \delta E^j_{\mathbf{k},\omega} &= -\frac{4\pi i}{\omega} \sum_a e_a \int d\mathbf{p}\; v_i\, \delta N^a_{\mathbf{k},\omega}(\mathbf{p}), \\
(\omega - \mathbf{k}\cdot\mathbf{v}) \left[\delta N^a_{\mathbf{k},\omega}(\mathbf{p}) - \delta N^{a0}_{\mathbf{k},\omega}(\mathbf{p}) \right] &= -\frac{i e_a}{\omega}\, \delta E^i_{\mathbf{k},\omega} \left[(\omega - \mathbf{k}\cdot\mathbf{v})\, \delta_{ij} + v_i k_j \right] \frac{\partial f_a}{\partial p_j} \\
&\quad - i\, e_a \frac{\partial}{\partial p_j} \int d\mathbf{k}' \int \frac{d\omega'}{\omega'} \left[(\omega' - \mathbf{k}'\cdot\mathbf{v})\, \delta_{ij} + v_i k'_j \right] \\
&\quad \times \left[\delta E^i_{\mathbf{k}',\omega'}\, \delta N^a_{\mathbf{k}-\mathbf{k}',\omega-\omega'}(\mathbf{p}) \right. \\
&\quad \left. - \left\langle \delta E^i_{\mathbf{k}',\omega'}\, \delta N^a_{\mathbf{k}-\mathbf{k}',\omega-\omega'}(\mathbf{p}) \right\rangle \right],
\end{aligned}
\tag{9.1}
$$

where

$$
\left\langle \delta N^0_a(\mathbf{p})\, \delta N^0_b(\mathbf{p}') \right\rangle_{\mathbf{k},\omega} = (2\pi)^{-3} \delta_{ab}\, \delta(\mathbf{p} - \mathbf{p}')\, \delta(\omega - \mathbf{k}\cdot\mathbf{v})\, f_a(\mathbf{p}). \tag{9.2}
$$

Making use of definition (7.10), which is repeated here,

$$g_{\mathbf{k},\omega} = \frac{1}{\omega - \mathbf{k}\cdot\mathbf{v} + i0},$$
$$L^i_{\mathbf{k},\omega} = -ie_a\left[\left(1 - \frac{\mathbf{k}\cdot\mathbf{v}}{\omega}\right)\delta_{ij} + \frac{v_i k_j}{\omega}\right]\frac{\partial}{\partial p_j}, \tag{9.3}$$

and omitting δ in front of perturbed quantities, the equation for perturbed distribution δN_a in (9.1) is re-expressed as follows:

$$N^a_{\mathbf{k},\omega} = N^{a0}_{\mathbf{k},\omega} + g_{\mathbf{k},\omega}\, L^i_{\mathbf{k},\omega}\, f_a\, E^i_{\mathbf{k},\omega}$$
$$+ g_{\mathbf{k},\omega}\int d\mathbf{k}'\int d\omega'\, L^i_{\mathbf{k}',\omega'}\left(N^a_{\mathbf{k}-\mathbf{k}',\omega-\omega'}\, E^i_{\mathbf{k}',\omega'} - \langle N^a_{\mathbf{k}-\mathbf{k}',\omega-\omega'}\, E^i_{\mathbf{k}',\omega'}\rangle\right). \tag{9.4}$$

This equation directly generalizes (7.11). The iterative solution that generalizes (7.12) is

$$N^{a(1)}_{\mathbf{k},\omega} = N^{a0}_{\mathbf{k},\omega} + g_{\mathbf{k},\omega}\, L^i_{\mathbf{k},\omega}\, f_a\, E^i_{\mathbf{k},\omega},$$
$$N^{a(2)}_{\mathbf{k},\omega} = g_{\mathbf{k},\omega}\int d\mathbf{k}'\int d\omega'\, L^i_{\mathbf{k}',\omega'}\left(N^{a(1)}_{\mathbf{k}-\mathbf{k}',\omega-\omega'}\, E^i_{\mathbf{k}',\omega'} - \left\langle N^{a(1)}_{\mathbf{k}-\mathbf{k}',\omega-\omega'}\, E^i_{\mathbf{k}',\omega'}\right\rangle\right),$$
$$N^{a(3)}_{\mathbf{k},\omega} = g_{\mathbf{k},\omega}\int d\mathbf{k}'\int d\omega'\, L^i_{\mathbf{k}',\omega'}\left(N^{a(2)}_{\mathbf{k}-\mathbf{k}',\omega-\omega'}\, E^i_{\mathbf{k}',\omega'} - \left\langle N^{a(2)}_{\mathbf{k}-\mathbf{k}',\omega-\omega'}\, E^i_{\mathbf{k}',\omega'}\right\rangle\right), \tag{9.5}$$

etc. Note that (9.5) is similar to (7.12), except for the "source" term $N^{a0}_{\mathbf{k},\omega}$. Successively inserting the lower-order solution to the next-order solution, and expressing the results in terms of the shorthand notations introduced in (7.14), we have

$$N^{a(1)}_K = N^{a0}_K + g_K\, L^i_K\, f_a\, E^i_K,$$
$$N^{a(2)}_K = \int dK'\, g_K\, L^i_{K'}\left(N^{a0}_{K-K'}\, E^i_{K'} - \langle N^{a0}_{K-K'}\, E^i_{K'}\rangle\right)$$
$$+ \int dK'\, g_K\, L^i_{K'}\, g_{K-K'}\, L^j_{K-K'}\, f_a\left(E^i_{K'}\, E^j_{K-K'} - \left\langle E^i_{K'}\, E^j_{K-K'}\right\rangle\right),$$
$$N^{a(3)}_K = \int dK'\int dK''\, g_K\, L^i_{K'}\, g_{K-K'}\, L^j_{K''}\left(N^{a0}_{K-K'-K''}\, E^i_{K'}\, E^j_{K''}\right.$$
$$\left. - E^i_{K'}\left\langle N^{a0}_{K-K'-K''}\, E^j_{K''}\right\rangle - \left\langle E^i_{K'}\, N^{a0}_{K-K'-K''}\, E^j_{K''}\right\rangle\right)$$
$$+ \int dK'\int dK''\, g_K\, L^i_{K'}\, g_{K-K'}\, L^j_{K''}\, g_{K-K'-K''}\, L^l_{K-K'-K''}\, f_a$$
$$\times\left(E^i_{K'}\, E^j_{K''}\, E^l_{K-K'-K''} - E^i_{K'}\left\langle E^j_{K''}\, E^l_{K-K'-K''}\right\rangle - \left\langle E^i_{K'}\, E^j_{K''}\, E^l_{K-K'-K''}\right\rangle\right), \tag{9.6}$$

etc. The symmetrized iterative solution thus emerges,

$$
\begin{aligned}
N_K^a &= N_K^{a0} + g_K\, L_K^i\, f_a\, E_K^i \\
&+ \frac{1}{2} \sum_{1+2=K} g_{1+2} \left(L_1^i\, g_2\, L_2^j + L_2^j\, g_1\, L_1^i \right) f_a \left(E_1^i\, E_2^j - \left\langle E_1^i\, E_2^j \right\rangle \right) \\
&+ \frac{1}{2} \sum_{1+2+3=K} g_{1+2+3}\, L_1^i\, g_{2+3} \left(L_2^j\, g_3\, L_3^l + L_3^l\, g_2\, L_2^j \right) f_a \\
&\times \left(E_1^i\, E_2^j\, E_3^l - E_1^i \left\langle E_2^j\, E_3^l \right\rangle - \left\langle E_1^i\, E_2^j\, E_3^l \right\rangle \right). \qquad (9.7)
\end{aligned}
$$

This generalizes (7.16), and constitutes the perturbative solution of nonlinear Klimontovich equation for perturbed distribution function.

Substituting (9.7) to the right-hand side of the wave equation in (9.1), we arrive at

$$
\begin{aligned}
&\Lambda_{ij}(K)\, E_j(K) + \sum_{1+2=K} \chi_{ijk}^{(2)}(1|2) \left[E_j(1)\, E_k(2) - \left\langle E_j(1)\, E_k(2) \right\rangle \right] \\
&+ \sum_{1+2+3=K} \chi_{ijkl}^{(3)}(1|2|3) \left[E_j(1)\, E_k(2)\, E_l(3) - E_j(1)\, \langle E_k(2)\, E_l(3) \rangle \right. \\
&\left. - \left\langle E_j(1)\, E_k(2)\, E_l(3) \right\rangle \right] = -i \sum_a \frac{4\pi e_a}{\omega} \int d\mathbf{p}\, v_i\, N_K^{a0}, \qquad (9.8)
\end{aligned}
$$

where $\Lambda_{ij}(K)$, $\chi_{ijk}^{(2)}(1|2)$, and $\chi_{ijkl}^{(3)}(1|2|3)$ are linear dielectric response tensor and second- and third-order nonlinear susceptibility tensors already defined in (7.20). Upon taking the dot product of (9.8) with $E_i(K')$, we have

$$
\begin{aligned}
&\Lambda_{ij}(K) \left\langle E_i\, E_j \right\rangle_K \delta(K + K') + \sum_1 \chi_{ijk}^{(2)}(1|K-1) \left\langle E_i(K')\, E_j(1)\, E_k(K-1) \right\rangle \\
&\quad + 2\,\delta(K + K') \sum_1 \chi_{ijkl}^{(3)}(1|-1|K)\, \langle E_i\, E_l \rangle_K \left\langle E_j\, E_k \right\rangle_1 \\
&= -i \sum_a \frac{4\pi e_a}{\omega} \int d\mathbf{p}\, v_i \left\langle N_K^{a0}\, E_i(K') \right\rangle. \qquad (9.9)
\end{aligned}
$$

This generalizes the Vlasov weak turbulence formalism given in (7.22) by including the source term on the right-hand side. Equation (9.9) also generalizes the electrostatic Klimontovich formalism described by (4.56).

To determine the third-body cumulant, $\left\langle E_i(K')\, E_j(1)\, E_k(K-1) \right\rangle$, and thereby close the infinite hierarchy of correlations, we write $E_i(K) = E_i^{(0)}(K) + E_i^{(1)}(K)$ and express

$$\left\langle E_i(K')\,E_j(1)\,E_k(K-1)\right\rangle \approx \left\langle E_i^{(1)}(K')\,E_j^{(0)}(1)\,E_k^{(0)}(K-1)\right\rangle + \left\langle E_i^{(0)}(K')\,E_j^{(1)}(1)\,E_k^{(0)}(K-1)\right\rangle + \left\langle E_i^{(0)}(K')\,E_j^{(0)}(1)\,E_k^{(1)}(K-1)\right\rangle,$$

as before. We then drop the superscript (0) later. Quantities with superscript (1) can be obtained from (9.8) by ignoring the third-order nonlinearity,

$$E_i^{(1)}(K) = -i\,\Lambda_{ij}^{-1}(K)\sum_a \frac{4\pi e_a}{\omega}\int d\mathbf{p}\,v_j\,N_K^{a0} - \Lambda_{ij}^{-1}(K)\sum_{1+2=K}\chi_{jkl}^{(2)}(1|2)\left[E_k^{(0)}(1)\,E_l^{(0)}(2) - \left\langle E_k^{(0)}(1)\,E_l^{(0)}(2)\right\rangle\right], \tag{9.10}$$

which generalizes (7.23). Recall that (7.23) pertains to collisionless electromagnetic formalism. Equation (9.10) also generalizes (4.62), which relates to electrostatic formalism with discrete particle effects. From (9.10), it is rather straightforward to obtain other related quantities, $E_i^{(1)}(K')$, $E_j^{(1)}(1)$, and $E_k^{(1)}(K-1)$, by permutation of the arguments. The result is given in (9.11), where we have dropped the superscript (0):

$$\begin{aligned}
&\langle E_i(K')\,E_j(1)\,E_k(K-1)\rangle \\
&= -i\,\Lambda_{il}^{-1}(K')\sum_a \frac{4\pi e_a}{\omega'}\int d\mathbf{p}\,v_l\,\langle N_{K'}^{a0}\,E_j(1)\,E_k(K-1)\rangle \\
&\quad -i\,\Lambda_{jl}^{-1}(1)\sum_a \frac{4\pi e_a}{\omega_1}\int d\mathbf{p}\,v_l\,\langle E_i(K')\,N_{K_1}^{a0}\,E_k(K-1)\rangle \\
&\quad -i\,\Lambda_{kl}^{-1}(K-1)\sum_a \frac{4\pi e_a}{\omega-\omega_1}\int d\mathbf{p}\,v_l\,\langle E_i(K')\,E_j(1)\,N_{K-1}^{a0}\rangle \\
&\quad -2\,\delta(K+K')\,\Lambda_{il}^{-1}(K')\,\chi_{lmn}^{(2)}(1|K'-1)\,\langle E_m\,E_j\rangle_1\,\langle E_n\,E_k\rangle_{K'-1} \\
&\quad -2\,\delta(K+K')\,\Lambda_{jl}^{-1}(1)\,\chi_{lmn}^{(2)}(-K'|K'+1)\,\langle E_i\,E_m\rangle_{K'}\,\langle E_k\,E_n\rangle_{K-1} \\
&\quad -2\,\delta(K+K')\,\Lambda_{kl}^{-1}(K-1)\,\chi_{lmn}^{(2)}(-K'|K'+K-1)\,\langle E_i\,E_m\rangle_{K'}\,\langle E_j\,E_n\rangle_1.
\end{aligned} \tag{9.11}$$

Inserting (9.11) to (9.9), we obtain

$$\begin{aligned}
&\Lambda_{ij}(K)\,\langle E_i\,E_j\rangle_K \\
&\quad -2\sum_{K'}\chi_{ijk}^{(2)}(K'|K-K')\Big[\Big(\Lambda_{jl}^{-1}(K')\,\chi_{lmn}^{(2)}(-K+K'|K)\,\langle E_k\,E_m\rangle_{K-K'} \\
&\quad +\Lambda_{kl}^{-1}(K-K')\,\chi_{lmn}^{(2)}(-K'|K)\,\langle E_j\,E_m\rangle_{K'}\Big)\,\langle E_i\,E_n\rangle_K
\end{aligned}$$

$$+ \Lambda_{il}^{-1}(-K)\, \chi_{lmn}^{(2)}(-K'| - K + K')\, \langle E_j\, E_m \rangle_{K'}\, \langle E_k\, E_n \rangle_{K-K'} \Big]$$

$$+ 2 \sum_{K'} \chi_{ijkl}^{(3)}(K'| - K'|K)\, \langle E_i\, E_l \rangle_K\, \langle E_j\, E_k \rangle_{K'}$$

$$= - i \sum_a \frac{4\pi e_a}{\omega} \int d\mathbf{p}\; v_i\, \langle N_K^{a0}(\mathbf{p})\, E_i(-K) \rangle$$

$$+ i \sum_a 4\pi e_a \sum_{K'} \int d\mathbf{p}\; \chi_{ijk}^{(2)}(K'|K - K')$$

$$\times \Big(\Lambda_{jl}^{-1}(K')\, \frac{v_l}{\omega'}\, \langle E_i(-K)\, E_k(K - K')\, N_{K'}^{a0}(\mathbf{p}) \rangle$$

$$+ \Lambda_{kl}^{-1}(K - K')\, \frac{v_l}{\omega - \omega'}\, \langle E_i(-K)\, E_j(K')\, N_{K-K'}^{a0}(\mathbf{p}) \rangle$$

$$- \Lambda_{il}^{-1}(-K)\, \frac{v_l}{\omega}\, \langle E_j(K')\, E_k(K - K')\, N_{-K}^{a0}(\mathbf{p}) \rangle \Big)\,. \tag{9.12}$$

To compute the terms on the right-hand side of (9.12), we return to (9.8), multiply $N_{-K}^{a0}(\mathbf{p})$, and take the average,

$$\Lambda_{ij}(-K)\, \langle E_j(-K)\, N_K^{a0}(\mathbf{p}) \rangle$$

$$= i \sum_b \frac{4\pi e_b}{\omega} \int d\mathbf{p}'\; v_i'\, \langle N_{-K}^{b0}(\mathbf{p}')\, N_K^{a0}(\mathbf{p}) \rangle$$

$$- \sum_{K'} \chi_{ijk}^{(2)}(-K'| - K + K')\, \langle E_j(-K')\, E_k(-K + K')\, N_K^{a0}(\mathbf{p}) \rangle$$

$$- 2 \sum_{K'} \chi_{ijkl}^{(3)}(-K'|K'| - K)\, \langle E_j\, E_k \rangle_{K'}\, \langle N_K^{a0}(\mathbf{p})\, E_l(-K) \rangle, \tag{9.13}$$

where we have replaced K by $-K$.

Making use of (4.47), or to reiterate,

$$\langle N_{-K}^{b0}(\mathbf{p}')\, N_K^{a0}(\mathbf{p}) \rangle = \langle N_a^0(\mathbf{p})\, N_b^0(\mathbf{p}') \rangle_K$$

$$= (2\pi)^{-3}\, \delta_{ab}\, \delta(\mathbf{p} - \mathbf{p}')\, \delta(\omega - \mathbf{k} \cdot \mathbf{v})\, f_a(\mathbf{p}), \tag{9.14}$$

we have

$$\langle N_K^{a0}(\mathbf{p})\, E_i(-K) \rangle = i\, \frac{4\pi e_a}{(2\pi)^3\, \omega}\, \Lambda_{ij}^{-1}(-K)\, v_j\, \delta(\omega - \mathbf{k} \cdot \mathbf{v})\, f_a(\mathbf{p})$$

$$- \Lambda_{ij}^{-1}(-K) \sum_{K'} \chi_{jkl}^{(2)}(-K'| - K + K')$$

$$\times\, \langle E_k(-K')\, E_l(-K + K')\, N_K^{a0}(\mathbf{p}) \rangle$$

$$- 2\Lambda_{ij}^{-1}(-K) \sum_{K'} \chi_{jklm}^{(3)}(-K'|K'|-K) \times \langle E_k\, E_l \rangle_{K'} \langle N_K^{a0}(\mathbf{p})\, E_m(-K)\rangle. \tag{9.15}$$

In the last term of (9.15), we may use the iterative solution,

$$\langle N_K^{a0}(\mathbf{p})\, E_m(-K)\rangle = i\,\frac{4\pi e_a}{(2\pi)^3\,\omega}\,\Lambda_{mn}^{-1}(-K)\, v_n\, \delta(\omega - \mathbf{k}\cdot\mathbf{v})\, f_a(\mathbf{p}), \tag{9.16}$$

to obtain

$$\begin{aligned}\langle N_K^{a0}(\mathbf{p})\, E_i(-K)\rangle &= i\,\frac{4\pi e_a}{(2\pi)^3\,\omega}\,\Lambda_{ij}^{-1}(-K)\Big(v_j + 2\sum_{K'} \chi_{jklm}^{(3)}(-K'|K'|-K)\\ &\quad\times \langle E_k\, E_l\rangle_{K'}\,\Lambda_{mn}^{-1}(-K)\, v_n\Big)\,\delta(\omega-\mathbf{k}\cdot\mathbf{v})\, f_a(\mathbf{p})\\ &\quad - \Lambda_{ij}^{-1}(-K)\sum_{K'}\chi_{jkl}^{(2)}(-K'|-K+K')\\ &\quad\times \langle E_k(-K')\, E_l(-K+K')\, N_K^{a0}(\mathbf{p})\rangle.\end{aligned} \tag{9.17}$$

The remaining quantities to evaluate are

$$\begin{aligned}&\langle E_k(-K')\, E_l(-K+K')\, N_K^{a0}(\mathbf{p})\rangle,\\ &\langle E_i(-K)\, E_k(K-K')\, N_{K'}^{a0}(\mathbf{p})\rangle,\\ &\langle E_i(-K)\, E_j(K')\, N_{K-K'}^{a0}(\mathbf{p})\rangle,\\ &\langle E_j(K')\, E_k(K-K')\, N_{-K}^{a0}(\mathbf{p})\rangle,\end{aligned}$$

all of which are of a generic form,

$$\langle E_a(K_1)\, E_b(-K_1+K_2)\, N_{-K_2}^{a0}(\mathbf{p})\rangle.$$

To evaluate this generic quantity, we return to (9.8) without the third-order term. We again resort to the iterative method and make use of (9.10). We then write

$$\begin{aligned}\langle E_a(K_1)\, E_b(-K_1+K_2)\, N_{-K_2}^{a0}(\mathbf{p})\rangle &= \langle E_a^{(1)}(K_1)\, E_b^{(0)}(-K_1+K_2)\, N_{-K_2}^{a0}(\mathbf{p})\rangle\\ &\quad + \langle E_a^{(0)}(K_1)\, E_b^{(1)}(-K_1+K_2)\, N_{-K_2}^{a0}(\mathbf{p})\rangle.\end{aligned} \tag{9.18}$$

Making use of (9.10) and the associated quantities with alternative arguments, we may evaluate the right-hand side of (9.18). In doing so, we ignore terms such as $\langle E\, N^0\, N^0 \rangle$. After some manipulations, we obtain, upon dropping the superscript (0),

$$
\begin{aligned}
&\langle E_a(K_1)\, E_b(-K_1+K_2)\, N^{a0}_{-K_2}(\mathbf{p})\rangle \\
&\quad = -\Lambda^{-1}_{al}(K_1) \sum_{K'} \chi^{(2)}_{lmn}(K'|K_1-K') \\
&\qquad \times [\, \langle E_m(K')\, E_n(K_1-K')\, E_b(-K_1+K_2) N^{a0}_{-K_2}(\mathbf{p})\rangle \\
&\qquad - \langle E_m(K')\, E_n(K_1-K')\rangle\, \langle E_b(-K_1+K_2) N^{a0}_{-K_2}(\mathbf{p})\rangle \,] \\
&\qquad - \Lambda^{-1}_{bl}(-K_1+K_2) \sum_{K'} \chi^{(2)}_{lmn}(K'|-K_1+K_2-K') \\
&\qquad \times [\, \langle E_m(K')\, E_n(-K_1+K_2-K')\, E_a(K_1)\, N^{a0}_{-K_2}(\mathbf{p})\rangle \\
&\qquad - \langle E_m(K')\, E_n(-K_1+K_2-K')\rangle\, \langle E_a(K_1)\, N^{a0}_{-K_2}(\mathbf{p})\rangle \,] \qquad (9.19) \\
&\quad = -2\Lambda^{-1}_{al}(K_1)\, \chi^{(2)}_{lmn}(K_1-K_2|K_2)\, \langle E_m\, E_b\rangle_{K_1-K_2}\, \langle E_n(K_2) N^{a0}_{-K_2}(\mathbf{p})\rangle \\
&\qquad - 2\Lambda^{-1}_{bl}(-K_1+K_2)\, \chi^{(2)}_{lmn}(-K_1|K_2)\, \langle E_m\, E_a\rangle_{K_1}\, \langle E_n(K_2) N^{a0}_{-K_2}(\mathbf{p})\rangle.
\end{aligned}
$$

Making use of the lowest-order solution – see (9.16),

$$
\langle E_n(K_2)\, N^{a0}_{-K_2}(\mathbf{p})\rangle = -i\, \frac{4\pi\, e_a}{(2\pi)^3\, \omega_2}\, \Lambda^{-1}_{np}(K_2)\, v_p\, \delta(\omega_2 - \mathbf{k}_2\cdot\mathbf{v})\, f_a(\mathbf{p}), \qquad (9.20)
$$

the right-hand side of (9.19) can be evaluated:

$$
\begin{aligned}
&\langle E_a(K_1)\, E_b(-K_1+K_2)\, N^{a0}_{-K_2}(\mathbf{p})\rangle \\
&\quad = \frac{8\pi\, e_a i}{(2\pi)^3\, \omega_2} \Big(\Lambda^{-1}_{al}(K_1)\, \chi^{(2)}_{lmn}(K_1-K_2|K_2)\, \langle E_m\, E_b\rangle_{K_1-K_2} \\
&\qquad + \Lambda^{-1}_{bl}(-K_1+K_2)\, \chi^{(2)}_{lmn}(-K_1|K_2)\, \langle E_m\, E_a\rangle_{K_1} \Big) \\
&\qquad \times \Lambda^{-1}_{np}(K_2)\, v_p\, \delta(\omega_2 - \mathbf{k}_2\cdot\mathbf{v})\, f_a(\mathbf{p}). \qquad (9.21)
\end{aligned}
$$

Identifying $a = k$, $b = l$, $K_1 = -K'$, and $K_2 = -K$, we have

$$
\begin{aligned}
&\langle E_k(-K')\, E_l(-K+K')\, N^{a0}_{K}(\mathbf{p})\rangle \\
&\quad = -\frac{8\pi\, e_a i}{(2\pi)^3\, \omega} \Big(\Lambda^{-1}_{km}(-K')\, \chi^{(2)}_{mnp}(-K'+K|-K)\, \langle E_n\, E_l\rangle_{K-K'} \\
&\qquad + \Lambda^{-1}_{lm}(K'-K)\, \chi^{(2)}_{mnp}(K'|-K)\, \langle E_n\, E_k\rangle_{K'} \Big) \\
&\qquad \times \Lambda^{-1}_{pq}(-K)\, v_q\, \delta(\omega - \mathbf{k}\cdot\mathbf{v})\, f_a(\mathbf{p}). \qquad (9.22)
\end{aligned}
$$

With the choice of $a = i$, $b = k$, $K_1 = -K$, and $K_2 = -K'$, we have

$$\begin{aligned}
&\langle E_i(-K)\, E_k(K-K')\, N_{K'}^{a0}(\mathbf{p})\rangle \\
&= -\frac{8\pi e_a i}{(2\pi)^3\,\omega'}\Big(\Lambda_{im}^{-1}(-K)\,\chi_{mnp}^{(2)}(-K+K'|-K')\,\langle E_n\,E_k\rangle_{K-K'} \\
&\quad + \Lambda_{km}^{-1}(K-K')\,\chi_{mnp}^{(2)}(K|-K')\,\langle E_n\,E_i\rangle_K\Big) \\
&\quad \times \Lambda_{pq}^{-1}(-K')\,v_q\,\delta(\omega'-\mathbf{k}'\cdot\mathbf{v})\,f_a(\mathbf{p}).
\end{aligned} \tag{9.23}$$

Next, upon identifying $a = i$, $b = j$, $K_1 = -K$, and $K_2 = -K + K'$, we obtain

$$\begin{aligned}
&\langle E_i(-K)\, E_j(K')\, N_{K-K'}^{a0}(\mathbf{p})\rangle \\
&= -\frac{8\pi e_a i}{(2\pi)^3\,(\omega-\omega')}\Big(\Lambda_{im}^{-1}(-K)\,\chi_{mnp}^{(2)}(-K'|-K+K')\,\langle E_n\,E_j\rangle_{K'} \\
&\quad + \Lambda_{jm}^{-1}(K')\,\chi_{mnp}^{(2)}(K|-K+K')\,\langle E_n\,E_i\rangle_K\Big) \\
&\quad \times \Lambda_{pq}^{-1}(-K+K')\,v_q\,\delta[\omega-\omega'-(\mathbf{k}-\mathbf{k}')\cdot\mathbf{v}]\,f_a(\mathbf{p}).
\end{aligned} \tag{9.24}$$

Finally, for $a = j$, $b = k$, $K_1 = K'$, and $K_2 = K$, we arrive at

$$\begin{aligned}
&\langle E_j(K')\, E_k(K-K')\, N_{-K}^{a0}(\mathbf{p})\rangle \\
&= \frac{8\pi e_a i}{(2\pi)^3\,\omega}\Big(\Lambda_{jm}^{-1}(K')\,\chi_{mnp}^{(2)}(K'-K|K)\,\langle E_n\,E_k\rangle_{K-K'} \\
&\quad + \Lambda_{km}^{-1}(-K'+K)\,\chi_{mnp}^{(2)}(-K'|K)\,\langle E_n\,E_j\rangle_{K'}\Big) \\
&\quad \times \Lambda_{pq}^{-1}(K)\,v_q\,\delta(\omega-\mathbf{k}\cdot\mathbf{v})\,f_a(\mathbf{p}).
\end{aligned} \tag{9.25}$$

Inserting (9.22) to (9.17), we obtain the closed-form expression for the source term,

$$\begin{aligned}
\langle N_K^{a0}(\mathbf{p})\,E_i(-K)\rangle &= \frac{4\pi e_a i}{(2\pi)^3\,\omega}\,\Lambda_{ij}^{-1}(-K) \\
&\quad \times \Bigg[v_j + 2\sum_{K'}\chi_{jkl}^{(2)}(-K'|-K+K') \\
&\quad \times \Big(\Lambda_{km}^{-1}(-K')\,\chi_{mnp}^{(2)}(-K'+K|-K)\,\langle E_n\,E_l\rangle_{K-K'} \\
&\quad + \Lambda_{lm}^{-1}(K'-K)\,\chi_{mnp}^{(2)}(K'|-K)\,\langle E_n\,E_k\rangle_{K'}\Big)\,\Lambda_{pq}^{-1}(-K)\,v_q \\
&\quad + 2\sum_{K'}\chi_{jklm}^{(3)}(-K'|K'|-K)\,\langle E_k\,E_l\rangle_{K'}\,\Lambda_{mn}^{-1}(-K)\,v_n\Bigg] \\
&\quad \times \delta(\omega-\mathbf{k}\cdot\mathbf{v})\,f_a(\mathbf{p}).
\end{aligned} \tag{9.26}$$

Substituting (9.23)–(9.26) to the right-hand side of (9.12), we obtain the intermediate form of nonlinear spectral balance equation including the effects of particle discreteness:

$$
\begin{aligned}
&\Lambda_{ij}(K)\,\langle E_i\,E_j\rangle_K - 2\sum_{K'}\chi^{(2)}_{ijk}(K'|K-K') \\
&\quad\times\Bigg[\Bigg(\Lambda^{-1}_{jl}(K')\,\chi^{(2)}_{lmn}(-K+K'|K)\,\langle E_k\,E_m\rangle_{K-K'} \\
&\quad+\Lambda^{-1}_{kl}(K-K')\,\chi^{(2)}_{lmn}(-K'|K)\,\langle E_j\,E_m\rangle_{K'}\Bigg)\langle E_i\,E_n\rangle_K \\
&\quad+\Lambda^{-1}_{il}(-K)\,\chi^{(2)}_{lmn}(-K'|-K+K')\,\langle E_j\,E_m\rangle_{K'}\,\langle E_k\,E_n\rangle_{K-K'}\Bigg] \\
&\quad+2\sum_{K'}\chi^{(3)}_{ijkl}(K'|-K'|K)\,\langle E_i\,E_l\rangle_K\,\langle E_j\,E_k\rangle_{K'} \\
&=\sum_a\frac{(4\pi e_a)^2}{(2\pi)^3\,\omega^2}\int d\mathbf{p}\;v_i\,\Lambda^{-1}_{ij}(-K)\,v_j\;\delta(\omega-\mathbf{k}\cdot\mathbf{v})\;f_a(\mathbf{p}) \\
&\quad+2\sum_{K'}\sum_a\frac{(4\pi e_a)^2}{(2\pi)^3\,\omega'^2}\int d\mathbf{p}\;\chi^{(2)}_{ijk}(K'|K-K')\,\Lambda^{-1}_{jl}(K')\,v_l \\
&\quad\times\Bigg(\Lambda^{-1}_{im}(-K)\,\chi^{(2)}_{mnp}(-K+K'|-K')\,\langle E_n\,E_k\rangle_{K-K'} \\
&\quad+\Lambda^{-1}_{km}(K-K')\,\chi^{(2)}_{mnp}(K|-K')\,\langle E_n\,E_i\rangle_K\Bigg)\Lambda^{-1}_{pq}(-K')\,v_q\;\delta(\omega'-\mathbf{k}'\cdot\mathbf{v})\;f_a(\mathbf{p}) \\
&\quad+2\sum_{K'}\sum_a\frac{(4\pi e_a)^2}{(2\pi)^3\,(\omega-\omega')^2}\int d\mathbf{p}\;\chi^{(2)}_{ijk}(K'|K-K')\,\Lambda^{-1}_{kl}(K-K')\,v_l \\
&\quad\times\Bigg(\Lambda^{-1}_{im}(-K)\,\chi^{(2)}_{mnp}(-K'|-K+K')\,\langle E_n\,E_j\rangle_{K'} \\
&\quad+\Lambda^{-1}_{jm}(K')\,\chi^{(2)}_{mnp}(K|-K+K')\,\langle E_n\,E_i\rangle_K\Bigg) \\
&\quad\times\Lambda^{-1}_{pq}(-K+K')\,v_q\;\delta[\omega-\omega'-(\mathbf{k}-\mathbf{k}')\cdot\mathbf{v}]\;f_a(\mathbf{p}).
\end{aligned}
\tag{9.27}
$$

This can be simplified by making use of the symmetry properties – see Section 7.3,

$$
\begin{aligned}
\Lambda_{ij}(-K) &= \Lambda^*_{ij}(K), \\
\chi^{(2)}_{ijk}(-K_1|-K_2) &= \chi^{(2)*}_{ijk}(K_1|K_2), \\
\chi^{(3)}_{ijkl}(-K_1|-K_2|-K_3) &= \chi^{(3)*}_{ijkl}(K_1|K_2|K_3), \\
\chi^{(2)}_{ijk}(K_1|K_2) &= \chi^{(2)}_{ikj}(K_2|K_1),
\end{aligned}
$$

$$\chi^{(3)}_{ijkl}(K_1|K_2|K_3) = \chi^{(3)}_{ijlk}(K_1|K_3|K_2),$$
$$\chi^{(2)}_{ijk}(K_1+K_2|-K_2,-K_2) = -\chi^{a(2)}_{jik}(K_1|K_2). \tag{9.28}$$

Then the nonlinear spectral balance equation (9.27) reduces to

$$\begin{aligned}
&\Lambda_{ij}(\mathbf{k},\omega)\,\langle\delta E_i\,\delta E_j\rangle_{\mathbf{k},\omega} - 2\int d\mathbf{k}'\int d\omega'\,\chi^{(2)}_{ijk}(\mathbf{k}',\omega'|\mathbf{k}-\mathbf{k}',\omega-\omega')\\
&\quad\times\bigg[\bigg(-\Lambda^{-1}_{jl}(\mathbf{k}',\omega')\,\chi^{(2)}_{nlm}(\mathbf{k}',\omega'|\mathbf{k}-\mathbf{k}',\omega-\omega')\,\langle\delta E_k\,\delta E_m\rangle_{\mathbf{k}-\mathbf{k}',\omega-\omega'}\\
&\quad-\Lambda^{-1}_{kl}(\mathbf{k}-\mathbf{k}',\omega-\omega')\,\chi^{(2)}_{nml}(\mathbf{k}',\omega'|\mathbf{k}-\mathbf{k}',\omega-\omega')\,\langle\delta E_j\,\delta E_m\rangle_{\mathbf{k}',\omega'}\bigg)\\
&\quad\times\langle\delta E_i\,\delta E_n\rangle_{\mathbf{k},\omega} + \Lambda^{*\,-1}_{il}(\mathbf{k},\omega)\,\chi^{(2)*}_{lmn}(\mathbf{k}',\omega'|\mathbf{k}-\mathbf{k}',\omega-\omega')\\
&\quad\times\langle\delta E_j\,\delta E_m\rangle_{\mathbf{k}',\omega'}\,\langle\delta E_k\,\delta E_n\rangle_{\mathbf{k}-\mathbf{k}',\omega-\omega'}\bigg]\\
&\quad+2\int d\mathbf{k}'\int d\omega'\,\chi^{(3)}_{ijkl}(\mathbf{k}',\omega'|-\mathbf{k}',-\omega'|\mathbf{k},\omega)\\
&\quad\times\langle\delta E_i\,\delta E_l\rangle_{\mathbf{k},\omega}\,\langle\delta E_j\,\delta E_k\rangle_{\mathbf{k}',\omega'}\\
&=\sum_a\frac{(4\pi e_a)^2}{(2\pi)^3\,\omega^2}\int d\mathbf{p}\;v_i\,\Lambda^{*\,-1}_{ij}(\mathbf{k},\omega)\,v_j\,\delta(\omega-\mathbf{k}\cdot\mathbf{v})\,f_a(\mathbf{p})\\
&\quad+2\int d\mathbf{k}'\int d\omega'\sum_a\frac{(4\pi e_a)^2}{(2\pi)^3\,\omega'^2}\int d\mathbf{p}\;\chi^{(2)}_{ijk}(\mathbf{k}',\omega'|\mathbf{k}-\mathbf{k}',\omega-\omega')\\
&\quad\times\Lambda^{-1}_{jl}(\mathbf{k}',\omega')\,v_l\bigg(\Lambda^{*\,-1}_{im}(\mathbf{k},\omega)\,\chi^{(2)*}_{mpn}(\mathbf{k}',\omega'|\mathbf{k}-\mathbf{k}',\omega-\omega')\\
&\quad\times\langle\delta E_n\,\delta E_k\rangle_{\mathbf{k}-\mathbf{k}',\omega-\omega'} - \Lambda^{-1}_{km}(\mathbf{k}-\mathbf{k}',\omega-\omega')\\
&\quad\times\chi^{(2)}_{npm}(\mathbf{k}',\omega'|\mathbf{k}-\mathbf{k}',\omega-\omega')\,\langle\delta E_n\,\delta E_i\rangle_{\mathbf{k},\omega}\bigg)\\
&\quad\times\Lambda^{*\,-1}_{pq}(\mathbf{k}',\omega')\,v_q\,\delta(\omega'-\mathbf{k}'\cdot\mathbf{v})\,f_a(\mathbf{p})\\
&\quad+2\int d\mathbf{k}'\int d\omega'\sum_a\frac{(4\pi e_a)^2}{(2\pi)^3\,(\omega-\omega')^2}\\
&\quad\times\int d\mathbf{p}\;\chi^{(2)}_{ijk}(\mathbf{k}',\omega'|\mathbf{k}-\mathbf{k}',\omega-\omega')\,\Lambda^{-1}_{kl}(\mathbf{k}-\mathbf{k}',\omega-\omega')\,v_l\\
&\quad\times\bigg(\Lambda^{*\,-1}_{im}(\mathbf{k},\omega)\,\chi^{(2)*}_{mnp}(\mathbf{k}',\omega'|\mathbf{k}-\mathbf{k}',\omega-\omega')\,\langle\delta E_n\,\delta E_j\rangle_{\mathbf{k}',\omega'}\\
&\quad-\Lambda^{-1}_{jm}(\mathbf{k}',\omega')\,\chi^{(2)}_{nmp}(\mathbf{k}',\omega'|\mathbf{k}-\mathbf{k}',\omega-\omega')\,\langle\delta E_n\,\delta E_i\rangle_{\mathbf{k},\omega}\bigg)\\
&\quad\times\Lambda^{*\,-1}_{pq}(\mathbf{k}-\mathbf{k}',\omega-\omega')\,v_q\,\delta[\omega-\omega'-(\mathbf{k}-\mathbf{k}')\cdot\mathbf{v}]\,f_a(\mathbf{p}),
\end{aligned}\tag{9.29}$$

where we have written down the final result in terms of the longhand notation.

Let us further manipulate (9.29) by making use of (7.32), (7.33), and (7.36). By taking the series of steps similar to that which led to (7.34), it is possible to show, albeit the intermediate steps are quite lengthy and tedious, that (9.29) reduces to

$$
\begin{aligned}
&\epsilon_\parallel(\mathbf{k},\omega)\,\langle E_\parallel^2\rangle_{\mathbf{k},\omega} + \Lambda_\perp(\mathbf{k},\omega)\,\langle E_\perp^2\rangle_{\mathbf{k},\omega} \\
&\quad + 2\int d\mathbf{k}'\int d\omega'\left[\frac{k_j'\,k_l'}{k'^2}\,\frac{\chi_{ijk}^{(2)}(\mathbf{k}',\omega'|\mathbf{k}-\mathbf{k}',\omega-\omega')\,\chi_{nlm}^{(2)}(\mathbf{k}',\omega'|\mathbf{k}-\mathbf{k}',\omega-\omega')}{\epsilon_\parallel(\mathbf{k}',\omega')}\right. \\
&\quad + \left.\left(\delta_{jl}-\frac{k_j'\,k_l'}{k'^2}\right)\frac{\chi_{ijk}^{(2)}(\mathbf{k}',\omega'|\mathbf{k}-\mathbf{k}',\omega-\omega')\,\chi_{nlm}^{(2)}(\mathbf{k}',\omega'|\mathbf{k}-\mathbf{k}',\omega-\omega')}{\Lambda_\perp(\mathbf{k}',\omega')}\right] \\
&\quad \times\left[\frac{k_i\,k_n}{k^2}\,\frac{(\mathbf{k}-\mathbf{k}')_k\,(\mathbf{k}-\mathbf{k}')_m}{(\mathbf{k}-\mathbf{k}')^2}\,\langle E_\parallel^2\rangle_{\mathbf{k}-\mathbf{k}',\omega-\omega'}\,\langle E_\parallel^2\rangle_{\mathbf{k},\omega}\right. \\
&\quad + \frac{1}{2}\frac{k_i\,k_n}{k^2}\left(\delta_{km}-\frac{(\mathbf{k}-\mathbf{k}')_k\,(\mathbf{k}-\mathbf{k}')_m}{(\mathbf{k}-\mathbf{k}')^2}\right)\langle E_\perp^2\rangle_{\mathbf{k}-\mathbf{k}',\omega-\omega'}\,\langle E_\parallel^2\rangle_{\mathbf{k},\omega} \\
&\quad + \frac{1}{2}\left(\delta_{in}-\frac{k_i\,k_n}{k^2}\right)\frac{(\mathbf{k}-\mathbf{k}')_k\,(\mathbf{k}-\mathbf{k}')_m}{(\mathbf{k}-\mathbf{k}')^2}\,\langle E_\parallel^2\rangle_{\mathbf{k}-\mathbf{k}',\omega-\omega'}\,\langle E_\perp^2\rangle_{\mathbf{k},\omega} \\
&\quad + \left.\frac{1}{4}\left(\delta_{in}-\frac{k_i\,k_n}{k^2}\right)\left(\delta_{km}-\frac{(\mathbf{k}-\mathbf{k}')_k\,(\mathbf{k}-\mathbf{k}')_m}{(\mathbf{k}-\mathbf{k}')^2}\right)\langle E_\perp^2\rangle_{\mathbf{k}-\mathbf{k}',\omega-\omega'}\,\langle E_\perp^2\rangle_{\mathbf{k},\omega}\right] \\
&\quad + 2\int d\mathbf{k}'\int d\omega'\left[\frac{(\mathbf{k}-\mathbf{k}')_k\,(\mathbf{k}-\mathbf{k}')_l}{(\mathbf{k}-\mathbf{k}')^2}\right. \\
&\quad \times\frac{\chi_{ijk}^{(2)}(\mathbf{k}',\omega'|\mathbf{k}-\mathbf{k}',\omega-\omega')\,\chi_{nml}^{(2)}(\mathbf{k}',\omega'|\mathbf{k}-\mathbf{k}',\omega-\omega')}{\epsilon_\parallel(\mathbf{k}-\mathbf{k}',\omega-\omega')} \\
&\quad + \left(\delta_{kl}-\frac{(\mathbf{k}-\mathbf{k}')_k\,(\mathbf{k}-\mathbf{k}')_l}{(\mathbf{k}-\mathbf{k}')^2}\right) \\
&\quad \times\left.\frac{\chi_{ijk}^{(2)}(\mathbf{k}',\omega'|\mathbf{k}-\mathbf{k}',\omega-\omega')\,\chi_{nml}^{(2)}(\mathbf{k}',\omega'|\mathbf{k}-\mathbf{k}',\omega-\omega')}{\Lambda_\perp(\mathbf{k}-\mathbf{k}',\omega-\omega')}\right] \\
&\quad \times\left[\frac{k_i\,k_n}{k^2}\frac{k_j'\,k_m'}{k'^2}\,\langle E_\parallel^2\rangle_{\mathbf{k}',\omega'}\,\langle E_\parallel^2\rangle_{\mathbf{k},\omega} + \frac{1}{2}\frac{k_i\,k_n}{k^2}\left(\delta_{jm}-\frac{k_j'\,k_m'}{k'^2}\right)\langle E_\perp^2\rangle_{\mathbf{k}',\omega'}\,\langle E_\parallel^2\rangle_{\mathbf{k},\omega}\right. \\
&\quad + \frac{1}{2}\left(\delta_{in}-\frac{k_i\,k_n}{k^2}\right)\frac{k_j'\,k_m'}{k'^2}\,\langle E_\parallel^2\rangle_{\mathbf{k}',\omega'}\,\langle E_\perp^2\rangle_{\mathbf{k},\omega} \\
&\quad + \left.\frac{1}{4}\left(\delta_{in}-\frac{k_i\,k_n}{k^2}\right)\left(\delta_{jm}-\frac{k_j'\,k_m'}{k'^2}\right)\langle E_\perp^2\rangle_{\mathbf{k}',\omega'}\,\langle E_\perp^2\rangle_{\mathbf{k},\omega}\right] \\
&\quad - 2\int d\mathbf{k}'\int d\omega'\left[\frac{k_i\,k_l}{k^2}\,\frac{\chi_{ijk}^{(2)}(\mathbf{k}',\omega'|\mathbf{k}-\mathbf{k}',\omega-\omega')\,\chi_{lmn}^{(2)*}(\mathbf{k}',\omega'|\mathbf{k}-\mathbf{k}',\omega-\omega')}{\epsilon_\parallel^*(\mathbf{k},\omega)}\right. \\
&\quad + \left.\left(\delta_{il}-\frac{k_i\,k_l}{k^2}\right)\frac{\chi_{ijk}^{(2)}(\mathbf{k}',\omega'|\mathbf{k}-\mathbf{k}',\omega-\omega')\,\chi_{lmn}^{(2)*}(\mathbf{k}',\omega'|\mathbf{k}-\mathbf{k}',\omega-\omega')}{\Lambda_\perp^*(\mathbf{k},\omega)}\right]
\end{aligned}
$$

$$
\times \left[\frac{k'_j\,k'_m}{k'^2} \frac{(\mathbf{k}-\mathbf{k}')_k\,(\mathbf{k}-\mathbf{k}')_n}{(\mathbf{k}-\mathbf{k}')^2} \langle E_\parallel^2\rangle_{\mathbf{k}',\omega'} \langle E_\parallel^2\rangle_{\mathbf{k}-\mathbf{k}',\omega-\omega'} \right.
$$

$$
+ \frac{1}{2} \frac{k'_j\,k'_m}{k'^2} \left(\delta_{kn} - \frac{(\mathbf{k}-\mathbf{k}')_k\,(\mathbf{k}-\mathbf{k}')_n}{(\mathbf{k}-\mathbf{k}')^2} \right) \langle E_\parallel^2\rangle_{\mathbf{k}',\omega'} \langle E_\perp^2\rangle_{\mathbf{k}-\mathbf{k}',\omega-\omega'}
$$

$$
+ \frac{1}{2} \left(\delta_{jm} - \frac{k'_j\,k'_m}{k'^2} \right) \frac{(\mathbf{k}-\mathbf{k}')_k\,(\mathbf{k}-\mathbf{k}')_n}{(\mathbf{k}-\mathbf{k}')^2} \langle E_\perp^2\rangle_{\mathbf{k}',\omega'} \langle E_\parallel^2\rangle_{\mathbf{k}-\mathbf{k}',\omega-\omega'}
$$

$$
+ \frac{1}{4} \left(\delta_{jm} - \frac{k'_j\,k'_m}{k'^2} \right) \left(\delta_{kn} - \frac{(\mathbf{k}-\mathbf{k}')_k\,(\mathbf{k}-\mathbf{k}')_n}{(\mathbf{k}-\mathbf{k}')^2} \right)
$$

$$
\left. \times \langle E_\perp^2\rangle_{\mathbf{k}',\omega'} \langle E_\perp^2\rangle_{\mathbf{k}-\mathbf{k}',\omega-\omega'} \right] + 2 \int d\mathbf{k}' \int d\omega'\, \chi^{(3)}_{ijkl}(\mathbf{k}',\omega'| -\mathbf{k}', -\omega'|\mathbf{k},\omega)
$$

$$
\times \left[\frac{k_i\,k_l}{k^2} \frac{k'_j\,k'_k}{k'^2} \langle E_\parallel^2\rangle_{\mathbf{k}',\omega'} \langle E_\parallel^2\rangle_{\mathbf{k},\omega} + \frac{1}{2} \frac{k_i\,k_l}{k^2} \left(\delta_{jk} - \frac{k'_j\,k'_k}{k'^2} \right) \langle E_\perp^2\rangle_{\mathbf{k}',\omega'} \langle E_\parallel^2\rangle_{\mathbf{k},\omega} \right.
$$

$$
+ \frac{1}{2} \left(\delta_{il} - \frac{k_i\,k_l}{k^2} \right) \frac{k'_j\,k'_k}{k'^2} \langle E_\parallel^2\rangle_{\mathbf{k}',\omega'} \langle E_\perp^2\rangle_{\mathbf{k},\omega}
$$

$$
\left. + \frac{1}{4} \left(\delta_{il} - \frac{k_i\,k_l}{k^2} \right) \left(\delta_{jk} - \frac{k'_j\,k'_k}{k'^2} \right) \langle E_\perp^2\rangle_{\mathbf{k}',\omega'} \langle E_\perp^2\rangle_{\mathbf{k},\omega} \right]
$$

$$
= \sum_a \frac{(4\pi e_a)^2}{(2\pi)^3\,\omega^2} \int d\mathbf{p} \left[\frac{(\mathbf{k}\cdot\mathbf{v})^2}{k^2} \frac{1}{\epsilon_\parallel^*(\mathbf{k},\omega)} + \frac{(\mathbf{k}\times\mathbf{v})^2}{k^2} \frac{1}{\Lambda_\perp^*(\mathbf{k},\omega)} \right]
$$

$$
\times \delta(\omega - \mathbf{k}\cdot\mathbf{v})\, f_a(\mathbf{p}) + 2 \int d\mathbf{k}' \int d\omega' \sum_a \frac{(4\pi e_a)^2}{(2\pi)^3\,\omega'^2} \int d\mathbf{p}
$$

$$
\times \left[\frac{k'_j\,k'_l}{k'^2} \frac{1}{\epsilon_\parallel(\mathbf{k}',\omega')} + \left(\delta_{jl} - \frac{k'_j\,k'_l}{k'^2} \right) \frac{1}{\Lambda_\perp(\mathbf{k}',\omega')} \right]
$$

$$
\times \left[\frac{k_i\,k_m}{k^2} \frac{1}{\epsilon_\parallel^*(\mathbf{k},\omega)} + \left(\delta_{im} - \frac{k_i\,k_m}{k^2} \right) \frac{1}{\Lambda_\perp^*(\mathbf{k},\omega)} \right]
$$

$$
\times \chi^{(2)}_{ijk}(\mathbf{k}',\omega'|\mathbf{k}-\mathbf{k}',\omega-\omega')\, \chi^{(2)*}_{mpn}(\mathbf{k}',\omega'|\mathbf{k}-\mathbf{k}',\omega-\omega')
$$

$$
\times \left[\frac{k'_p\,k'_q}{k'^2} \frac{1}{\epsilon_\parallel^*(\mathbf{k}',\omega')} + \left(\delta_{pq} - \frac{k'_p\,k'_q}{k'^2} \right) \frac{1}{\Lambda_\perp^*(\mathbf{k}',\omega')} \right]
$$

$$
\times \left[\frac{(\mathbf{k}-\mathbf{k}')_n\,(\mathbf{k}-\mathbf{k}')_k}{(\mathbf{k}-\mathbf{k}')^2} \langle E_\parallel^2\rangle_{\mathbf{k}-\mathbf{k}',\omega-\omega'} \right.
$$

$$
\left. + \frac{1}{2} \left(\delta_{nk} - \frac{(\mathbf{k}-\mathbf{k}')_n\,(\mathbf{k}-\mathbf{k}')_k}{(\mathbf{k}-\mathbf{k}')^2} \right) \langle E_\perp^2\rangle_{\mathbf{k}-\mathbf{k}',\omega-\omega'} \right] v_l\, v_q\, \delta(\omega' - \mathbf{k}'\cdot\mathbf{v})\, f_a(\mathbf{p})
$$

$$
- 2 \int d\mathbf{k}' \int d\omega' \sum_a \frac{(4\pi e_a)^2}{(2\pi)^3\,\omega'^2} \int d\mathbf{p}
$$

$$
\times \left[\frac{k'_j\,k'_l}{k'^2} \frac{1}{\epsilon_\parallel(\mathbf{k}',\omega')} + \left(\delta_{jl} - \frac{k'_j\,k'_l}{k'^2} \right) \frac{1}{\Lambda_\perp(\mathbf{k}',\omega')} \right]
$$

$$\times \left[\frac{(\mathbf{k}-\mathbf{k}')_k\,(\mathbf{k}-\mathbf{k}')_m}{(\mathbf{k}-\mathbf{k}')^2} \frac{1}{\epsilon_\parallel(\mathbf{k}-\mathbf{k}',\omega-\omega')} \right.$$

$$\left. + \left(\delta_{km} - \frac{(\mathbf{k}-\mathbf{k}')_k\,(\mathbf{k}-\mathbf{k}')_m}{(\mathbf{k}-\mathbf{k}')^2} \right) \frac{1}{\Lambda_\perp(\mathbf{k}-\mathbf{k}',\omega-\omega')} \right]$$

$$\times\, \chi^{(2)}_{ijk}(\mathbf{k}',\omega'|\mathbf{k}-\mathbf{k}',\omega-\omega')\; \chi^{(2)}_{npm}(\mathbf{k}',\omega'|\mathbf{k}-\mathbf{k}',\omega-\omega')$$

$$\times \left[\frac{k'_p\,k'_q}{k'^2} \frac{1}{\epsilon^*_\parallel(\mathbf{k}',\omega')} + \left(\delta_{pq} - \frac{k'_p\,k'_q}{k'^2} \right) \frac{1}{\Lambda^*_\perp(\mathbf{k}',\omega')} \right]$$

$$\times \left[\frac{k_n\,k_i}{k^2}\, \langle E^2_\parallel \rangle_{\mathbf{k},\omega} + \frac{1}{2}\left(\delta_{ni} - \frac{k_n\,k_i}{k^2} \right) \langle E^2_\perp \rangle_{\mathbf{k},\omega} \right] v_l\, v_q\; \delta(\omega' - \mathbf{k}'\cdot\mathbf{v})\; f_a(\mathbf{p})$$

$$+ 2 \int d\mathbf{k}' \int d\omega' \sum_a \frac{(4\pi e_a)^2}{(2\pi)^3\,(\omega-\omega')^2} \int d\mathbf{p}$$

$$\times \left[\frac{(\mathbf{k}-\mathbf{k}')_k\,(\mathbf{k}-\mathbf{k}')_l}{(\mathbf{k}-\mathbf{k}')^2} \frac{1}{\epsilon_\parallel(\mathbf{k}-\mathbf{k}',\omega-\omega')} \right.$$

$$\left. + \left(\delta_{kl} - \frac{(\mathbf{k}-\mathbf{k}')_k\,(\mathbf{k}-\mathbf{k}')_l}{(\mathbf{k}-\mathbf{k}')^2} \right) \frac{1}{\Lambda_\perp(\mathbf{k}-\mathbf{k}',\omega-\omega')} \right]$$

$$\times \left[\frac{k_i\,k_m}{k^2} \frac{1}{\epsilon^*_\parallel(\mathbf{k},\omega)} + \left(\delta_{im} - \frac{k_i\,k_m}{k^2} \right) \frac{1}{\Lambda^*_\perp(\mathbf{k},\omega)} \right]$$

$$\times\, \chi^{(2)}_{ijk}(\mathbf{k}',\omega'|\mathbf{k}-\mathbf{k}',\omega-\omega')\; \chi^{(2)*}_{mnp}(\mathbf{k}',\omega'|\mathbf{k}-\mathbf{k}',\omega-\omega')$$

$$\times \left[\frac{(\mathbf{k}-\mathbf{k}')_p\,(\mathbf{k}-\mathbf{k}')_q}{(\mathbf{k}-\mathbf{k}')^2} \frac{1}{\epsilon^*_\parallel(\mathbf{k}-\mathbf{k}',\omega-\omega')} \right.$$

$$\left. + \left(\delta_{pq} - \frac{(\mathbf{k}-\mathbf{k}')_p\,(\mathbf{k}-\mathbf{k}')_q}{(\mathbf{k}-\mathbf{k}')^2} \right) \frac{1}{\Lambda^*_\perp(\mathbf{k}-\mathbf{k}',\omega-\omega')} \right]$$

$$\times \left[\frac{k'_n\,k'_j}{k'^2}\, \langle E^2_\parallel \rangle_{\mathbf{k}',\omega'} + \frac{1}{2}\left(\delta_{nj} - \frac{k'_n\,k'_j}{k'^2} \right) \langle E^2_\perp \rangle_{\mathbf{k}',\omega'} \right]$$

$$\times\, v_l\, v_q\; \delta[\omega - \omega' - (\mathbf{k}-\mathbf{k}')\cdot\mathbf{v}]\; f_a(\mathbf{p})$$

$$- 2 \int d\mathbf{k}' \int d\omega' \sum_a \frac{(4\pi e_a)^2}{(2\pi)^3\,(\omega-\omega')^2} \int d\mathbf{p}$$

$$\times \left[\frac{(\mathbf{k}-\mathbf{k}')_k\,(\mathbf{k}-\mathbf{k}')_l}{(\mathbf{k}-\mathbf{k}')^2} \frac{1}{\epsilon_\parallel(\mathbf{k}-\mathbf{k}',\omega-\omega')} \right.$$

$$\left. + \left(\delta_{kl} - \frac{(\mathbf{k}-\mathbf{k}')_k\,(\mathbf{k}-\mathbf{k}')_l}{(\mathbf{k}-\mathbf{k}')^2} \right) \frac{1}{\Lambda_\perp(\mathbf{k}-\mathbf{k}',\omega-\omega')} \right]$$

$$\times \left[\frac{k'_j\,k'_m}{k'^2} \frac{1}{\epsilon_\parallel(\mathbf{k}',\omega')} + \left(\delta_{jm} - \frac{k'_j\,k'_m}{k'^2} \right) \frac{1}{\Lambda_\perp(\mathbf{k}',\omega')} \right]$$

$$\times\, \chi^{(2)}_{ijk}(\mathbf{k}',\omega'|\mathbf{k}-\mathbf{k}',\omega-\omega')\; \chi^{(2)}_{nmp}(\mathbf{k}',\omega'|\mathbf{k}-\mathbf{k}',\omega-\omega')$$

$$\times\left[\frac{(\mathbf{k}-\mathbf{k}')_p\,(\mathbf{k}-\mathbf{k}')_q}{(\mathbf{k}-\mathbf{k}')^2}\,\frac{1}{\epsilon^*_\parallel(\mathbf{k}-\mathbf{k}',\omega-\omega')}\right.$$
$$\left.+\left(\delta_{pq}-\frac{(\mathbf{k}-\mathbf{k}')_p\,(\mathbf{k}-\mathbf{k}')_q}{(\mathbf{k}-\mathbf{k}')^2}\right)\frac{1}{\Lambda^*_\perp(\mathbf{k}-\mathbf{k}',\omega-\omega')}\right]$$
$$\times\left[\frac{k_n\,k_i}{k^2}\,\langle E^2_\parallel\rangle_{\mathbf{k},\omega}+\frac{1}{2}\left(\delta_{ni}-\frac{k_n\,k_i}{k^2}\right)\langle E^2_\perp\rangle_{\mathbf{k},\omega}\right]$$
$$\times\, v_l\, v_q\;\delta[\omega-\omega'-(\mathbf{k}-\mathbf{k}')\cdot\mathbf{v}]\;f_a(\mathbf{p}). \tag{9.30}$$

This generalizes (7.35), and if we ignore everything to the right of equality, then we recover (7.35). Equation (9.30) can further be separated into longitudinal and transverse parts once we write the electric field spectra into eigenmode forms, $\langle E^2_\parallel\rangle_{\mathbf{k},\omega}=\sum_{\alpha=L,S}\sum_{\sigma=\pm1} I^{\sigma\alpha}_{\mathbf{k}}\,\delta(\omega-\sigma\omega^\alpha_{\mathbf{k}})$ and $\langle E^2_\perp\rangle_{\mathbf{k},\omega}=\sum_{\sigma=\pm1} I^{\sigma T}_{\mathbf{k}}\,\delta(\omega-\sigma\omega^T_{\mathbf{k}})$ – see (7.48), and thus derive separate nonlinear spectral balance equations for longitudinal and transverse modes, as it was done in Sections 7.2.1 and 7.2.2.

For the moment, we discuss the particle kinetic equation. We again take the simple view that the quasilinear approximation is valid. As a consequence, we adopt the following formal particle kinetic equation:

$$\frac{\partial f_a}{\partial t}=-e_a\,\frac{\partial}{\partial p_i}\int d\mathbf{k}\int d\omega\left[\left(1-\frac{\mathbf{k}\cdot\mathbf{v}}{\omega}\right)\delta_{ij}+\frac{v_j\,k_i}{\omega}\right]\left\langle N^a_{\mathbf{k},\omega}(\mathbf{p})\,E^j_{-\mathbf{k},-\omega}\right\rangle,$$
$$N^a_{\mathbf{k},\omega}(\mathbf{p})=N^{a0}_{\mathbf{k},\omega}(\mathbf{p})-i\,e_a\,\frac{E^k_{\mathbf{k},\omega}}{\omega-\mathbf{k}\cdot\mathbf{v}}\left[\left(1-\frac{\mathbf{k}\cdot\mathbf{v}}{\omega}\right)\delta_{kl}+\frac{v_k\,k_l}{\omega}\right]\frac{\partial f_a}{\partial p_l}. \tag{9.31}$$

We may construct the relevant quantity by ignoring the nonlinear term,

$$\begin{aligned}\left\langle N^a_{\mathbf{k},\omega}(\mathbf{p})\,E^j_{-\mathbf{k},-\omega}\right\rangle&=\left\langle N^{a0}_{\mathbf{k},\omega}(\mathbf{p})\,E^j_{-\mathbf{k},-\omega}\right\rangle\\&\quad-i\,e_a\,\frac{\langle E_j\,E_k\rangle_{\mathbf{k},\omega}}{\omega-\mathbf{k}\cdot\mathbf{v}}\left[\left(1-\frac{\mathbf{k}\cdot\mathbf{v}}{\omega}\right)\delta_{kl}+\frac{v_k\,k_l}{\omega}\right]\frac{\partial f_a}{\partial p_l}\\&=i\,\frac{4\pi\,e_a}{(2\pi)^3\,\omega}\,\Lambda^{-1}_{jk}(-\mathbf{k},\,-\omega)\,v_k\,\delta(\omega-\mathbf{k}\cdot\mathbf{v})\,f_a\\&\quad-i\,e_a\,\frac{\langle E_j\,E_k\rangle_{\mathbf{k},\omega}}{\omega-\mathbf{k}\cdot\mathbf{v}}\left[\left(1-\frac{\mathbf{k}\cdot\mathbf{v}}{\omega}\right)\delta_{kl}+\frac{v_k\,k_l}{\omega}\right]\frac{\partial f_a}{\partial p_l}.\end{aligned} \tag{9.32}$$

Inserting (9.32) to (9.31), we obtain

$$\frac{\partial f_a}{\partial t}=\pi e_a^2\,\frac{\partial}{\partial p_i}\int d\mathbf{k}\int d\omega\left[\left(1-\frac{\mathbf{k}\cdot\mathbf{v}}{\omega}\right)\delta_{ij}+\frac{v_j\,k_i}{\omega}\right]$$
$$\times\left\{\frac{1}{2\pi^3\,\omega}\left[\frac{k_j\,(\mathbf{k}\cdot\mathbf{v})}{k^2}\,\frac{\mathrm{Im}\,\epsilon_\parallel(\mathbf{k},\omega)}{|\epsilon_\parallel(\mathbf{k},\omega)|^2}\right.\right.$$

$$+\left(v_j - \frac{k_j\,(\mathbf{k}\cdot\mathbf{v})}{k^2}\right)\frac{\operatorname{Im}\Lambda_\perp(\mathbf{k},\omega)}{|\Lambda_\perp(\mathbf{k},\omega)|^2}\Bigg]\,\delta(\omega-\mathbf{k}\cdot\mathbf{v})\,f_a$$
$$+\,\delta(\omega-\mathbf{k}\cdot\mathbf{v})\left[\frac{k_j\,k_k}{k^2}\,\langle E_\parallel^2\rangle_{\mathbf{k},\omega}+\frac{1}{2}\left(\delta_{jk}-\frac{k_j\,k_k}{k^2}\right)\langle E_\perp^2\rangle_{\mathbf{k},\omega}\right]$$
$$\times\left[\left(1-\frac{\mathbf{k}\cdot\mathbf{v}}{\omega}\right)\delta_{kl}+\frac{v_k\,k_l}{\omega}\right]\frac{\partial f_a}{\partial p_l}\Bigg\}, \tag{9.33}$$

where we have made use of (7.32) and (7.33).

If we make explicit use of the resonance condition, $\omega=\mathbf{k}\cdot\mathbf{v}$, to write

$$\frac{(\mathbf{k}\cdot\mathbf{v})^2}{k^2}=\frac{\omega^2}{k^2}, \qquad v^2-\frac{(\mathbf{k}\cdot\mathbf{v})^2}{k^2}=\frac{(\mathbf{k}\times\mathbf{v})^2}{k^2},$$

then we obtain the formal particle kinetic equation,

$$\frac{\partial f_a}{\partial t}=\pi e_a^2\int d\mathbf{k}\int d\omega\left(\frac{\mathbf{k}}{k}\cdot\frac{\partial}{\partial\mathbf{p}}\right)\Bigg\{\delta(\omega-\mathbf{k}\cdot\mathbf{v})\Bigg[\frac{1}{2\pi^3}\left(\frac{\operatorname{Im}\epsilon_\parallel(\mathbf{k},\omega)}{k\,|\epsilon_\parallel(\mathbf{k},\omega)|^2}\right.$$
$$\left.+\frac{(\mathbf{k}\times\mathbf{v})^2}{(\mathbf{k}\cdot\mathbf{v})^2}\frac{\operatorname{Im}\Lambda_\perp(\mathbf{k},\omega)}{k\,|\Lambda_\perp(\mathbf{k},\omega)|^2}\right)f_a$$
$$+\left(\langle E_\parallel^2\rangle_{\mathbf{k},\omega}+\frac{1}{2}\frac{(\mathbf{k}\times\mathbf{v})^2}{(\mathbf{k}\cdot\mathbf{v})^2}\langle E_\perp^2\rangle_{\mathbf{k},\omega}\right)\frac{\mathbf{k}}{k}\cdot\frac{\partial f_a}{\partial\mathbf{p}}\Bigg]\Bigg\}. \tag{9.34}$$

Nonlinear spectral balance equation (9.30) and formal particle kinetic equation (9.34) constitute the foundational equations subject to further analysis, which is done next.

9.2 Formal Equations of Electromagnetic Klimontovich Weak Turbulence Theory

9.2.1 Electron and Ion Particle Kinetic Equations

For further reduction of particle kinetic equation, we start from (9.34) and take into account the contributions from linear eigenmodes. Electrons can resonate only with high-frequency modes, while ions resonate with low-frequency fluctuations and waves. Transverse electromagnetic modes cannot resonate with the particles since the wave speed is higher than the speed of light. As a consequence, we only need to retain the effects of L mode for electrons, while for ions only S mode needs to be taken into consideration. Making use of various properties of linear dielectric susceptibilities discussed in Section 7.3.1, it is possible to derive the particle kinetic equations for electrons and ions,

$$
\begin{aligned}
\frac{\partial f_e(\mathbf{p})}{\partial t} &= \pi e^2 \sum_{\sigma=\pm 1} \int d\mathbf{k}\, \frac{\mathbf{k}}{k} \cdot \frac{\partial}{\partial \mathbf{p}}\, \delta(\sigma\omega_{\mathbf{k}}^L - \mathbf{k}\cdot\mathbf{v}) \\
&\quad \times \left(\frac{\sigma\omega_{\mathbf{k}}^L}{4\pi^2 k}\, f_e(\mathbf{p}) + I_{\mathbf{k}}^{\sigma L}\, \frac{\mathbf{k}}{k} \cdot \frac{\partial f_e(\mathbf{p})}{\partial \mathbf{p}} \right), \\
\frac{\partial f_i(\mathbf{p})}{\partial t} &= \pi e^2 \sum_{\sigma=\pm 1} \int d\mathbf{k}\, \frac{\mathbf{k}}{k} \cdot \frac{\partial}{\partial \mathbf{p}}\, \delta(\sigma\omega_{\mathbf{k}}^S - \mathbf{k}\cdot\mathbf{v}) \\
&\quad \times \left(\frac{\sigma\mu_{\mathbf{k}}\omega_{\mathbf{k}}^L}{4\pi^2 k}\, f_i(\mathbf{p}) + I_{\mathbf{k}}^{\sigma S}\, \frac{\mathbf{k}}{k} \cdot \frac{\partial f_i(\mathbf{p})}{\partial \mathbf{p}} \right).
\end{aligned} \tag{9.35}
$$

When compared with the corresponding particle kinetic equation (8.63) derived under the Vlasov kinetic approach, it becomes evident that the forms in (9.35) are more general in that the right-hand side is characterized by the Fokker–Planck form in which the velocity friction (or drag) term appears in balanced form together with the velocity space diffusion term,

$$
\begin{aligned}
\frac{\partial f_a}{\partial t} &= \frac{\partial}{\partial \mathbf{p}} \cdot \left(\mathbf{A}_a\, f_a + \mathsf{D}_a \cdot \frac{\partial f_a}{\partial \mathbf{p}} \right), \qquad (a = e, i), \\
\mathbf{A}_e &= \frac{e^2}{4\pi} \int d\mathbf{k}\, \frac{\mathbf{k}}{k^2} \sum_{\sigma=\pm 1} \sigma\omega_{\mathbf{k}}^L\, \delta(\sigma\omega_{\mathbf{k}}^L - \mathbf{k}\cdot\mathbf{v}), \\
\mathbf{A}_i &= \frac{e^2}{4\pi} \int d\mathbf{k}\, \frac{\mathbf{k}}{k^2} \sum_{\sigma=\pm 1} \sigma\mu_{\mathbf{k}}\omega_{\mathbf{k}}^L\, \delta(\sigma\omega_{\mathbf{k}}^S - \mathbf{k}\cdot\mathbf{v}), \\
\mathsf{D}_e &= \pi e^2 \int d\mathbf{k}\, \frac{\mathbf{k}\mathbf{k}}{k^2} \sum_{\sigma=\pm 1} \delta(\sigma\omega_{\mathbf{k}}^L - \mathbf{k}\cdot\mathbf{v}) I_{\mathbf{k}}^{\sigma L}, \\
\mathsf{D}_i &= \pi e^2 \int d\mathbf{k}\, \frac{\mathbf{k}\mathbf{k}}{k^2} \sum_{\sigma=\pm 1} \delta(\sigma\omega_{\mathbf{k}}^S - \mathbf{k}\cdot\mathbf{v}) I_{\mathbf{k}}^{\sigma S}.
\end{aligned} \tag{9.36}
$$

The basic equation here is the same as that discussed in Chapter 4 without the nonlinear term – see (4.86). This shows that the electromagnetic generalization does not modify the particle kinetic equations. In Chapter 6, we made use of the form of electron particle kinetic equation in (9.36) – see (6.3) – together with electrostatic wave kinetic equations, in order to analyze the weak Langmuir turbulence problem and the formation of asymptotic electron kappa velocity distribution function.

9.2.2 Formal Wave Kinetic Equations

We now separate the nonlinear spectral balance equation (9.30) into longitudinal and transverse parts and derive the wave kinetic equations of electromagnetic weak turbulence theory at the formal level. We first reintroduce the slow time derivative as is customary,

$$\begin{aligned}
&\epsilon_\parallel(\mathbf{k},\omega)\,\langle E_\parallel^2\rangle_{\mathbf{k},\omega}+\Lambda_\perp(\mathbf{k},\omega)\,\langle E_\perp^2\rangle_{\mathbf{k},\omega}\\
&\quad\to \mathrm{Re}\,\epsilon_\parallel(\mathbf{k},\omega)\,\langle E_\parallel^2\rangle_{\mathbf{k},\omega}+\mathrm{Re}\,\Lambda_\perp(\mathbf{k},\omega)\,\langle E_\perp^2\rangle_{\mathbf{k},\omega}\\
&\quad+i\mathrm{Im}\,\epsilon_\parallel(\mathbf{k},\omega)\,\langle E_\parallel^2\rangle_{\mathbf{k},\omega}+i\mathrm{Im}\,\Lambda_\perp(\mathbf{k},\omega)\,\langle E_\perp^2\rangle_{\mathbf{k},\omega}\\
&\quad+\frac{i}{2}\,\epsilon_\parallel'(\mathbf{k},\omega)\,\frac{\partial}{\partial t}\,\langle E_\parallel^2\rangle_{\mathbf{k},\omega}+\frac{i}{2}\Lambda_\perp'(\mathbf{k},\omega)\,\frac{\partial}{\partial t}\,\langle E_\perp^2\rangle_{\mathbf{k},\omega},
\end{aligned}\tag{9.37}$$

where $\epsilon_\parallel'(\mathbf{k},\omega)=\partial[\mathrm{Re}\,\epsilon_\parallel(\mathbf{k},\omega)]/\partial\omega$ and $\Lambda_\perp'(\mathbf{k},\omega)=\partial[\mathrm{Re}\,\Lambda_\perp(\mathbf{k},\omega)]/\partial\omega$. With these, upon taking the linear eigenmode representation, the imaginary part of (9.30) can further be manipulated. Here, we write down the full expressions for the sake of completeness. In what follows, we designate various terms with boxed numerical identification. This is done in order to aid readers in tracing the steps in subsequent derivations.

$$\begin{aligned}
\boxed{1}\;\; 0 &= \frac{i}{2}\sum_\sigma\sum_\alpha \epsilon_\parallel'(\mathbf{k},\sigma\omega_\mathbf{k}^\alpha)\,\frac{\partial I_\mathbf{k}^{\sigma\alpha}}{\partial t}\,\underbrace{\delta(\omega-\sigma\omega_\mathbf{k}^\alpha)}\\
&+\frac{i}{2}\sum_\sigma \Lambda_\perp'(\mathbf{k},\sigma\omega_\mathbf{k}^T)\,\frac{\partial I_\mathbf{k}^{\sigma T}}{\partial t}\,\underbrace{\delta(\omega-\sigma\omega_\mathbf{k}^T)}\\
\boxed{2}\;\; &+i\sum_\sigma\sum_\alpha \mathrm{Im}\,\epsilon_\parallel(\mathbf{k},\sigma\omega_\mathbf{k}^\alpha)\,I_\mathbf{k}^{\sigma\alpha}\,\underbrace{\delta(\omega-\sigma\omega_\mathbf{k}^\alpha)}\\
&+i\sum_\sigma \mathrm{Im}\,\Lambda_\perp(\mathbf{k},\sigma\omega_\mathbf{k}^T)\,I_\mathbf{k}^{\sigma T}\,\underbrace{\delta(\omega-\sigma\omega_\mathbf{k}^T)}\\
\boxed{3}\;\; &+2i\sum_{\sigma,\sigma''}\sum_{\alpha,\gamma}\mathrm{Im}\int d\mathbf{k}'\left[\frac{k_i\,k_j'\,(\mathbf{k}-\mathbf{k}')_k}{k\,k'\,|\mathbf{k}-\mathbf{k}'|}\right.\\
&\times\chi_{ijk}^{(2)}(\mathbf{k}',\sigma\omega_\mathbf{k}^\alpha-\sigma''\omega_{\mathbf{k}-\mathbf{k}'}^\gamma|\mathbf{k}-\mathbf{k}',\,-\sigma''\omega_{\mathbf{k}-\mathbf{k}'}^\gamma)\\
&\times\frac{k_n\,k_l'\,(\mathbf{k}-\mathbf{k}')_m}{k\,k'\,|\mathbf{k}-\mathbf{k}'|}\,\chi_{nlm}^{(2)}(\mathbf{k}',\sigma\omega_\mathbf{k}^\alpha-\sigma''\omega_{\mathbf{k}-\mathbf{k}'}^\gamma|\mathbf{k}-\mathbf{k}',\,-\sigma''\omega_{\mathbf{k}-\mathbf{k}'}^\gamma)\\
&\times\mathscr{P}\,\frac{1}{\epsilon_\parallel(\mathbf{k}',\sigma\omega_\mathbf{k}^\alpha-\sigma''\omega_{\mathbf{k}-\mathbf{k}'}^\gamma)}+\left(\delta_{jl}-\frac{k_j'\,k_l'}{k'^2}\right)\frac{k_i\,(\mathbf{k}-\mathbf{k}')_k}{k\,|\mathbf{k}-\mathbf{k}'|}\\
&\times\chi_{ijk}^{(2)}(\mathbf{k}',\sigma\omega_\mathbf{k}^\alpha-\sigma''\omega_{\mathbf{k}-\mathbf{k}'}^\gamma|\mathbf{k}-\mathbf{k}',\,-\sigma''\omega_{\mathbf{k}-\mathbf{k}'}^\gamma)\\
&\times\frac{k_n\,(\mathbf{k}-\mathbf{k}')_m}{k\,|\mathbf{k}-\mathbf{k}'|}\,\chi_{nlm}^{(2)}(\mathbf{k}',\sigma\omega_\mathbf{k}^\alpha-\sigma''\omega_{\mathbf{k}-\mathbf{k}'}^\gamma|\mathbf{k}-\mathbf{k}',\,-\sigma''\omega_{\mathbf{k}-\mathbf{k}'}^\gamma)\\
&\mathscr{P}\,\frac{1}{\Lambda_\perp(\mathbf{k}',\sigma\omega_\mathbf{k}^\alpha-\sigma''\omega_{\mathbf{k}-\mathbf{k}'}^\gamma)}-i\pi\sum_{\sigma'}\sum_\beta\frac{k_i\,k_j'\,(\mathbf{k}-\mathbf{k}')_k}{k\,k'\,|\mathbf{k}-\mathbf{k}'|}
\end{aligned}$$

$$
\begin{aligned}
&\times \chi^{(2)}_{ijk}(\mathbf{k}',\sigma'\omega^{\beta}_{\mathbf{k}'}|\mathbf{k}-\mathbf{k}',\sigma''\omega^{\gamma}_{\mathbf{k}-\mathbf{k}'})\,\frac{k_n\,k'_l\,(\mathbf{k}-\mathbf{k}')_m}{k\,k'\,|\mathbf{k}-\mathbf{k}'|}\\
&\times \chi^{(2)}_{nlm}(\mathbf{k}',\sigma'\omega^{\beta}_{\mathbf{k}'}|\mathbf{k}-\mathbf{k}',\sigma''\omega^{\gamma}_{\mathbf{k}-\mathbf{k}'})\,\frac{\delta(\sigma\omega^{\alpha}_{\mathbf{k}}-\sigma'\omega^{\beta}_{\mathbf{k}'}-\sigma''\omega^{\gamma}_{\mathbf{k}-\mathbf{k}'})}{\epsilon'_{\parallel}(\mathbf{k}',\sigma'\omega^{\beta}_{\mathbf{k}'})}\\
&-i\pi\sum_{\sigma'}\left(\delta_{jl}-\frac{k'_j\,k'_l}{k'^2}\right)\frac{k_i\,(\mathbf{k}-\mathbf{k}')_k}{k\,|\mathbf{k}-\mathbf{k}'|}\,\chi^{(2)}_{ijk}(\mathbf{k}',\sigma'\omega^{T}_{\mathbf{k}'}|\mathbf{k}-\mathbf{k}',\sigma''\omega^{\gamma}_{\mathbf{k}-\mathbf{k}'})\\
&\times \frac{k_n\,(\mathbf{k}-\mathbf{k}')_m}{k\,|\mathbf{k}-\mathbf{k}'|}\,\chi^{(2)}_{nlm}(\mathbf{k}',\sigma'\omega^{T}_{\mathbf{k}'}|\mathbf{k}-\mathbf{k}',\sigma''\omega^{\gamma}_{\mathbf{k}-\mathbf{k}'})\,\frac{1}{\Lambda'_{\perp}(\mathbf{k}',\sigma'\omega^{T}_{\mathbf{k}'})}\\
&\times \delta(\sigma\omega^{\alpha}_{\mathbf{k}}-\sigma'\omega^{T}_{\mathbf{k}'}-\sigma''\omega^{\gamma}_{\mathbf{k}-\mathbf{k}'})\Bigg]\,I^{\sigma''\gamma}_{\mathbf{k}-\mathbf{k}'}\,I^{\sigma\alpha}_{\mathbf{k}}\,\underbrace{\delta(\omega-\sigma\omega^{\alpha}_{\mathbf{k}})}
\end{aligned}
$$

$$
\begin{aligned}
\boxed{4}\quad &+2i\sum_{\sigma,\sigma''}\sum_{\alpha}\mathrm{Im}\int d\mathbf{k}'\left[\left(\delta_{km}-\frac{(\mathbf{k}-\mathbf{k}')_k\,(\mathbf{k}-\mathbf{k}')_m}{(\mathbf{k}-\mathbf{k}')^2}\right)\frac{k_i\,k'_j}{k\,k'}\right.\\
&\times \chi^{(2)}_{ijk}(\mathbf{k}',\sigma\omega^{\alpha}_{\mathbf{k}}-\sigma''\omega^{T}_{\mathbf{k}-\mathbf{k}'}|\mathbf{k}-\mathbf{k}',\sigma''\omega^{T}_{\mathbf{k}-\mathbf{k}'})\,\frac{k_n\,k'_l}{k\,k'}\\
&\times \chi^{(2)}_{nlm}(\mathbf{k}',\sigma\omega^{\alpha}_{\mathbf{k}}-\sigma''\omega^{T}_{\mathbf{k}-\mathbf{k}'}|\mathbf{k}-\mathbf{k}',\sigma''\omega^{T}_{\mathbf{k}-\mathbf{k}'})\,\mathscr{P}\,\frac{1}{\epsilon_{\parallel}(\mathbf{k}',\sigma\omega^{\alpha}_{\mathbf{k}}-\sigma''\omega^{T}_{\mathbf{k}-\mathbf{k}'})}\\
&+\left(\delta_{jl}-\frac{k'_j\,k'_l}{k'^2}\right)\left(\delta_{km}-\frac{(\mathbf{k}-\mathbf{k}')_k\,(\mathbf{k}-\mathbf{k}')_m}{(\mathbf{k}-\mathbf{k}')^2}\right)\\
&\times \frac{k_i}{k}\,\chi^{(2)}_{ijk}(\mathbf{k}',\sigma\omega^{\alpha}_{\mathbf{k}}-\sigma''\omega^{T}_{\mathbf{k}-\mathbf{k}'}|\mathbf{k}-\mathbf{k}',\sigma''\omega^{T}_{\mathbf{k}-\mathbf{k}'})\\
&\times \frac{k_n}{k}\,\chi^{(2)}_{nlm}(\mathbf{k}',\sigma\omega^{\alpha}_{\mathbf{k}}-\sigma''\omega^{T}_{\mathbf{k}-\mathbf{k}'}|\mathbf{k}-\mathbf{k}',\sigma''\omega^{T}_{\mathbf{k}-\mathbf{k}'})\,\mathscr{P}\,\frac{1}{\Lambda_{\perp}(\mathbf{k}',\sigma\omega^{\alpha}_{\mathbf{k}}-\sigma''\omega^{T}_{\mathbf{k}-\mathbf{k}'})}\\
&-i\pi\sum_{\sigma'}\sum_{\beta}\left(\delta_{km}-\frac{(\mathbf{k}-\mathbf{k}')_k\,(\mathbf{k}-\mathbf{k}')_m}{(\mathbf{k}-\mathbf{k}')^2}\right)\frac{k_i\,k'_j}{k\,k'}\\
&\times \chi^{(2)}_{ijk}(\mathbf{k}',\sigma'\omega^{\beta}_{\mathbf{k}'}|\mathbf{k}-\mathbf{k}',\sigma''\omega^{T}_{\mathbf{k}-\mathbf{k}'})\,\frac{k_n\,k'_l}{k\,k'}\,\chi^{(2)}_{nlm}(\mathbf{k}',\sigma'\omega^{\beta}_{\mathbf{k}'}|\mathbf{k}-\mathbf{k}',\sigma''\omega^{T}_{\mathbf{k}-\mathbf{k}'})\\
&\times \frac{\delta(\sigma\omega^{\alpha}_{\mathbf{k}}-\sigma'\omega^{\beta}_{\mathbf{k}'}-\sigma''\omega^{T}_{\mathbf{k}-\mathbf{k}'})}{\epsilon'_{\parallel}(\mathbf{k}',\sigma'\omega^{\beta}_{\mathbf{k}'})}-i\pi\sum_{\sigma'}\left(\delta_{jl}-\frac{k'_j\,k'_l}{k'^2}\right)\\
&\times\left(\delta_{km}-\frac{(\mathbf{k}-\mathbf{k}')_k\,(\mathbf{k}-\mathbf{k}')_m}{(\mathbf{k}-\mathbf{k}')^2}\right)\frac{k_i}{k}\,\chi^{(2)}_{ijk}(\mathbf{k}',\sigma'\omega^{T}_{\mathbf{k}'}|\mathbf{k}-\mathbf{k}',\sigma''\omega^{T}_{\mathbf{k}-\mathbf{k}'})\\
&\left.\times \frac{k_n}{k}\,\chi^{(2)}_{nlm}(\mathbf{k}',\sigma'\omega^{T}_{\mathbf{k}'}|\mathbf{k}-\mathbf{k}',\sigma''\omega^{T}_{\mathbf{k}-\mathbf{k}'})\,\frac{\delta(\sigma\omega^{\alpha}_{\mathbf{k}}-\sigma'\omega^{T}_{\mathbf{k}'}-\sigma''\omega^{T}_{\mathbf{k}-\mathbf{k}'})}{\Lambda'_{\perp}(\mathbf{k}',\sigma'\omega^{T}_{\mathbf{k}'})}\right]\\
&\times \frac{1}{2}\,I^{\sigma''T}_{\mathbf{k}-\mathbf{k}'}\,I^{\sigma\alpha}_{\mathbf{k}}\,\underbrace{\delta(\omega-\sigma\omega^{\alpha}_{\mathbf{k}})}
\end{aligned}
$$

$$\boxed{5}\ +2i\sum_{\sigma,\sigma''}\sum_{\gamma}\mathrm{Im}\int d\mathbf{k}'\left[\frac{k'_j\,(\mathbf{k}-\mathbf{k}')_k}{k'\,|\mathbf{k}-\mathbf{k}'|}\left(\delta_{in}-\frac{k_i\,k_n}{k^2}\right)\right.$$
$$\times\,\chi^{(2)}_{ijk}(\mathbf{k}',\sigma\omega^T_{\mathbf{k}}-\sigma''\omega^\gamma_{\mathbf{k}-\mathbf{k}'}|\mathbf{k}-\mathbf{k}',\sigma''\omega^\gamma_{\mathbf{k}-\mathbf{k}'})\,\frac{k'_l\,(\mathbf{k}-\mathbf{k}')_m}{k'\,|\mathbf{k}-\mathbf{k}'|}$$
$$\times\,\chi^{(2)}_{nlm}(\mathbf{k}',\sigma\omega^T_{\mathbf{k}}-\sigma''\omega^\gamma_{\mathbf{k}-\mathbf{k}'}|\mathbf{k}-\mathbf{k}',\sigma''\omega^\gamma_{\mathbf{k}-\mathbf{k}'})\,\mathscr{P}\,\frac{1}{\epsilon_\parallel(\mathbf{k}',\sigma\omega^T_{\mathbf{k}}-\sigma''\omega^\gamma_{\mathbf{k}-\mathbf{k}'})}$$
$$+\left(\delta_{jl}-\frac{k'_j\,k'_l}{k'^2}\right)\left(\delta_{in}-\frac{k_i\,k_n}{k^2}\right)\frac{(\mathbf{k}-\mathbf{k}')_k}{|\mathbf{k}-\mathbf{k}'|}$$
$$\times\,\chi^{(2)}_{ijk}(\mathbf{k}',\sigma\omega^T_{\mathbf{k}}-\sigma''\omega^\gamma_{\mathbf{k}-\mathbf{k}'}|\mathbf{k}-\mathbf{k}',\sigma''\omega^\gamma_{\mathbf{k}-\mathbf{k}'})\,\frac{(\mathbf{k}-\mathbf{k}')_m}{|\mathbf{k}-\mathbf{k}'|}$$
$$\times\,\chi^{(2)}_{nlm}(\mathbf{k}',\sigma\omega^T_{\mathbf{k}}-\sigma''\omega^\gamma_{\mathbf{k}-\mathbf{k}'}|\mathbf{k}-\mathbf{k}',\sigma''\omega^\gamma_{\mathbf{k}-\mathbf{k}'})\,\mathscr{P}\,\frac{1}{\Lambda_\perp(\mathbf{k}',\sigma\omega^T_{\mathbf{k}}-\sigma''\omega^\gamma_{\mathbf{k}-\mathbf{k}'})}$$
$$-\,i\pi\sum_{\sigma'}\sum_{\beta}\frac{k'_j\,(\mathbf{k}-\mathbf{k}')_k}{k'\,|\mathbf{k}-\mathbf{k}'|}\left(\delta_{in}-\frac{k_i\,k_n}{k^2}\right)\chi^{(2)}_{ijk}(\mathbf{k}',\sigma'\omega^\beta_{\mathbf{k}'}|\mathbf{k}-\mathbf{k}',\sigma''\omega^\gamma_{\mathbf{k}-\mathbf{k}'})$$
$$\times\,\frac{k'_l\,(\mathbf{k}-\mathbf{k}')_m}{k'\,|\mathbf{k}-\mathbf{k}'|}\,\chi^{(2)}_{nlm}(\mathbf{k}',\sigma'\omega^\beta_{\mathbf{k}'}|\mathbf{k}-\mathbf{k}',\sigma''\omega^\gamma_{\mathbf{k}-\mathbf{k}'})\,\frac{\delta(\sigma\omega^T_{\mathbf{k}}-\sigma'\omega^\beta_{\mathbf{k}'}-\sigma''\omega^\gamma_{\mathbf{k}-\mathbf{k}'})}{\epsilon'_\parallel(\mathbf{k}',\sigma'\omega^\beta_{\mathbf{k}'})}$$
$$-\,i\pi\sum_{\sigma'}\left(\delta_{jl}-\frac{k'_j\,k'_l}{k'^2}\right)\left(\delta_{in}-\frac{k_i\,k_n}{k^2}\right)\frac{(\mathbf{k}-\mathbf{k}')_k}{|\mathbf{k}-\mathbf{k}'|}\,\chi^{(2)}_{ijk}(\mathbf{k}',\sigma'\omega^T_{\mathbf{k}'}|\mathbf{k}-\mathbf{k}',\sigma''\omega^\gamma_{\mathbf{k}-\mathbf{k}'})$$
$$\left.\times\,\frac{(\mathbf{k}-\mathbf{k}')_m}{|\mathbf{k}-\mathbf{k}'|}\,\chi^{(2)}_{nlm}(\mathbf{k}',\sigma'\omega^T_{\mathbf{k}'}|\mathbf{k}-\mathbf{k}',\sigma''\omega^\gamma_{\mathbf{k}-\mathbf{k}'})\,\frac{\delta(\sigma\omega^T_{\mathbf{k}}-\sigma'\omega^T_{\mathbf{k}'}-\sigma''\omega^\gamma_{\mathbf{k}-\mathbf{k}'})}{\Lambda'_\perp(\mathbf{k}',\sigma'\omega^T_{\mathbf{k}'})}\right]$$
$$\times\,\frac{1}{2}\,I^{\sigma''\gamma}_{\mathbf{k}-\mathbf{k}'}\,I^{\sigma T}_{\mathbf{k}}\,\underbrace{\delta(\omega-\sigma\omega^T_{\mathbf{k}})}$$

$$\boxed{6}\ +2i\sum_{\sigma,\sigma''}\mathrm{Im}\int d\mathbf{k}'\left[\frac{k'_j}{k'}\left(\delta_{in}-\frac{k_i\,k_n}{k^2}\right)\chi^{(2)}_{ijk}(\mathbf{k}',\sigma\omega^T_{\mathbf{k}}-\sigma''\omega^T_{\mathbf{k}-\mathbf{k}'}|\mathbf{k}-\mathbf{k}',\sigma''\omega^T_{\mathbf{k}-\mathbf{k}'})\right.$$
$$\times\,\frac{k'_l}{k'}\left(\delta_{km}-\frac{(\mathbf{k}-\mathbf{k}')_k\,(\mathbf{k}-\mathbf{k}')_m}{(\mathbf{k}-\mathbf{k}')^2}\right)\chi^{(2)}_{nlm}(\mathbf{k}',\sigma\omega^T_{\mathbf{k}}-\sigma''\omega^T_{\mathbf{k}-\mathbf{k}'}|\mathbf{k}-\mathbf{k}',\sigma''\omega^T_{\mathbf{k}-\mathbf{k}'})$$
$$\times\,\mathscr{P}\,\frac{1}{\epsilon_\parallel(\mathbf{k}',\sigma\omega^T_{\mathbf{k}}-\sigma''\omega^T_{\mathbf{k}-\mathbf{k}'})}+\left(\delta_{in}-\frac{k_i\,k_n}{k^2}\right)\left(\delta_{jl}-\frac{k'_j\,k'_l}{k'^2}\right)$$
$$\times\,\chi^{(2)}_{ijk}(\mathbf{k}',\sigma\omega^T_{\mathbf{k}}-\sigma''\omega^T_{\mathbf{k}-\mathbf{k}'}|\mathbf{k}-\mathbf{k}',\sigma''\omega^T_{\mathbf{k}-\mathbf{k}'})\left(\delta_{km}-\frac{(\mathbf{k}-\mathbf{k}')_k\,(\mathbf{k}-\mathbf{k}')_m}{(\mathbf{k}-\mathbf{k}')^2}\right)$$
$$\times\,\chi^{(2)}_{nlm}(\mathbf{k}',\sigma\omega^T_{\mathbf{k}}-\sigma''\omega^T_{\mathbf{k}-\mathbf{k}'}|\mathbf{k}-\mathbf{k}',\sigma''\omega^T_{\mathbf{k}-\mathbf{k}'})\,\mathscr{P}\,\frac{1}{\Lambda_\perp(\mathbf{k}',\sigma\omega^T_{\mathbf{k}}-\sigma''\omega^T_{\mathbf{k}-\mathbf{k}'})}$$
$$-\,i\pi\sum_{\sigma'}\sum_{\beta}\frac{k'_j}{k'}\left(\delta_{in}-\frac{k_i\,k_n}{k^2}\right)\chi^{(2)}_{ijk}(\mathbf{k}',\sigma'\omega^\beta_{\mathbf{k}'}|\mathbf{k}-\mathbf{k}',\sigma''\omega^T_{\mathbf{k}-\mathbf{k}'})$$

$$\times \frac{k'_l}{k'}\left(\delta_{km} - \frac{(\mathbf{k}-\mathbf{k}')_k\,(\mathbf{k}-\mathbf{k}')_m}{(\mathbf{k}-\mathbf{k}')^2}\right)\chi^{(2)}_{nlm}(\mathbf{k}',\sigma'\omega^\beta_{\mathbf{k}'}|\mathbf{k}-\mathbf{k}',\sigma''\omega^T_{\mathbf{k}-\mathbf{k}'})$$

$$\times \frac{\delta(\sigma\omega^T_{\mathbf{k}} - \sigma'\omega^\beta_{\mathbf{k}'} - \sigma''\omega^T_{\mathbf{k}-\mathbf{k}'})}{\epsilon'_\parallel(\mathbf{k}',\sigma'\omega^\beta_{\mathbf{k}'})} - i\pi\sum_{\sigma'}\left(\delta_{in} - \frac{k_i\,k_n}{k^2}\right)\left(\delta_{jl} - \frac{k'_j\,k'_l}{k'^2}\right)$$

$$\times \chi^{(2)}_{ijk}(\mathbf{k}',\sigma'\omega^T_{\mathbf{k}'}|\mathbf{k}-\mathbf{k}',\sigma''\omega^T_{\mathbf{k}-\mathbf{k}'})\left(\delta_{km} - \frac{(\mathbf{k}-\mathbf{k}')_k\,(\mathbf{k}-\mathbf{k}')_m}{(\mathbf{k}-\mathbf{k}')^2}\right)$$

$$\times \chi^{(2)}_{nlm}(\mathbf{k}',\sigma'\omega^T_{\mathbf{k}'}|\mathbf{k}-\mathbf{k}',\sigma''\omega^T_{\mathbf{k}-\mathbf{k}'})\,\frac{\delta(\sigma\omega^T_{\mathbf{k}} - \sigma'\omega^T_{\mathbf{k}'} - \sigma''\omega^T_{\mathbf{k}-\mathbf{k}'})}{\Lambda'_\perp(\mathbf{k}',\sigma'\omega^T_{\mathbf{k}'})}\Bigg]$$

$$\times \frac{1}{4}\,I^{\sigma''T}_{\mathbf{k}-\mathbf{k}'}\,I^{\sigma T}_{\mathbf{k}}\;\underbrace{\delta(\omega - \sigma\omega^T_{\mathbf{k}})}$$

7 $$+2i\sum_{\sigma,\sigma'}\sum_{\alpha,\beta}\mathrm{Im}\int d\mathbf{k}'\left[\frac{k_i\,k'_j\,(\mathbf{k}-\mathbf{k}')_k}{k\,k'\,|\mathbf{k}-\mathbf{k}'|}\,\chi^{(2)}_{ijk}(\mathbf{k}',\sigma'\omega^\beta_{\mathbf{k}'}|\mathbf{k}-\mathbf{k}',\sigma\omega^\alpha_{\mathbf{k}} - \sigma'\omega^\beta_{\mathbf{k}'})\right.$$

$$\times \frac{k_n\,k'_m\,(\mathbf{k}-\mathbf{k}')_l}{k\,k'\,|\mathbf{k}-\mathbf{k}'|}\,\chi^{(2)}_{nml}(\mathbf{k}',\sigma'\omega^\beta_{\mathbf{k}'}|\mathbf{k}-\mathbf{k}',\sigma\omega^\alpha_{\mathbf{k}} - \sigma'\omega^\beta_{\mathbf{k}'})$$

$$\times \mathscr{P}\,\frac{1}{\epsilon_\parallel(\mathbf{k}-\mathbf{k}',\sigma\omega^\alpha_{\mathbf{k}} - \sigma'\omega^\beta_{\mathbf{k}'})} + \frac{k_i\,k'_j}{k\,k'}\left(\delta_{kl} - \frac{(\mathbf{k}-\mathbf{k}')_k\,(\mathbf{k}-\mathbf{k}')_l}{(\mathbf{k}-\mathbf{k}')^2}\right)$$

$$\times \chi^{(2)}_{ijk}(\mathbf{k}',\sigma'\omega^\beta_{\mathbf{k}'}|\mathbf{k}-\mathbf{k}',\sigma\omega^\alpha_{\mathbf{k}} - \sigma'\omega^\beta_{\mathbf{k}'})\,\frac{k_n\,k'_m}{k\,k'}$$

$$\times \chi^{(2)}_{nml}(\mathbf{k}',\sigma'\omega^\beta_{\mathbf{k}'}|\mathbf{k}-\mathbf{k}',\sigma\omega^\alpha_{\mathbf{k}} - \sigma'\omega^\beta_{\mathbf{k}'})\,\mathscr{P}\,\frac{1}{\Lambda_\perp(\mathbf{k}-\mathbf{k}',\sigma\omega^\alpha_{\mathbf{k}} - \sigma'\omega^\beta_{\mathbf{k}'})}$$

$$-i\pi\sum_{\sigma''}\sum_{\gamma}\frac{k_i\,k'_j\,(\mathbf{k}-\mathbf{k}')_k}{k\,k'\,|\mathbf{k}-\mathbf{k}'|}\,\chi^{(2)}_{ijk}(\mathbf{k}',\sigma'\omega^\beta_{\mathbf{k}'}|\mathbf{k}-\mathbf{k}',\sigma''\omega^\gamma_{\mathbf{k}-\mathbf{k}'})$$

$$\times \frac{k_n\,k'_m\,(\mathbf{k}-\mathbf{k}')_l}{k\,k'\,|\mathbf{k}-\mathbf{k}'|}\,\chi^{(2)}_{nml}(\mathbf{k}',\sigma'\omega^\beta_{\mathbf{k}'}|\mathbf{k}-\mathbf{k}',\sigma''\omega^\gamma_{\mathbf{k}-\mathbf{k}'})$$

$$\times \frac{\delta(\sigma\omega^\alpha_{\mathbf{k}} - \sigma'\omega^\beta_{\mathbf{k}'} - \sigma''\omega^\gamma_{\mathbf{k}-\mathbf{k}'})}{\epsilon'_\parallel(\mathbf{k}-\mathbf{k}',\sigma''\omega^\gamma_{\mathbf{k}-\mathbf{k}'})} - i\pi\sum_{\sigma''}\frac{k_i\,k'_j}{k\,k'}\left(\delta_{kl} - \frac{(\mathbf{k}-\mathbf{k}')_k\,(\mathbf{k}-\mathbf{k}')_l}{(\mathbf{k}-\mathbf{k}')^2}\right)$$

$$\times \chi^{(2)}_{ijk}(\mathbf{k}',\sigma'\omega^\beta_{\mathbf{k}'}|\mathbf{k}-\mathbf{k}',\sigma''\omega^T_{\mathbf{k}-\mathbf{k}'})\,\frac{k_n\,k'_m}{k\,k'}\,\chi^{(2)}_{nml}(\mathbf{k}',\sigma'\omega^\beta_{\mathbf{k}'}|\mathbf{k}-\mathbf{k}',\sigma''\omega^T_{\mathbf{k}-\mathbf{k}'})$$

$$\left.\times \frac{\delta(\sigma\omega^\alpha_{\mathbf{k}} - \sigma'\omega^\beta_{\mathbf{k}'} - \sigma''\omega^T_{\mathbf{k}-\mathbf{k}'})}{\Lambda'_\perp(\mathbf{k}-\mathbf{k}',\sigma''\omega^T_{\mathbf{k}-\mathbf{k}'})}\right]I^{\sigma'\beta}_{\mathbf{k}'}\,I^{\sigma\alpha}_{\mathbf{k}}\;\underbrace{\delta(\omega - \sigma\omega^\alpha_{\mathbf{k}})}$$

8 $$+2i\sum_{\sigma,\sigma'}\sum_{\alpha}\mathrm{Im}\int d\mathbf{k}'\left[\left(\delta_{jm} - \frac{k'_j\,k'_m}{k'^2}\right)\frac{k_i\,(\mathbf{k}-\mathbf{k}')_k}{k\,|\mathbf{k}-\mathbf{k}'|}\right.$$

$$\times \chi^{(2)}_{ijk}(\mathbf{k}',\sigma'\omega^T_{\mathbf{k}'}|\mathbf{k}-\mathbf{k}',\sigma\omega^\alpha_{\mathbf{k}} - \sigma'\omega^T_{\mathbf{k}'})\,\frac{k_n\,(\mathbf{k}-\mathbf{k}')_l}{k\,|\mathbf{k}-\mathbf{k}'|}$$

$$\times\, \chi^{(2)}_{nml}(\mathbf{k}',\sigma'\omega^T_{\mathbf{k}'}|\mathbf{k}-\mathbf{k}',\sigma\omega^\alpha_{\mathbf{k}}-\sigma'\omega^T_{\mathbf{k}'})\,\mathscr{P}\,\frac{1}{\epsilon_\parallel(\mathbf{k}-\mathbf{k}',\sigma\omega^\alpha_{\mathbf{k}}-\sigma'\omega^T_{\mathbf{k}'})}$$

$$+\frac{k_i}{k}\left(\delta_{jm}-\frac{k'_j\,k'_m}{k'^2}\right)\left(\delta_{kl}-\frac{(\mathbf{k}-\mathbf{k}')_k\,(\mathbf{k}-\mathbf{k}')_l}{(\mathbf{k}-\mathbf{k}')^2}\right)$$

$$\times\, \chi^{(2)}_{ijk}(\mathbf{k}',\sigma'\omega^T_{\mathbf{k}'}|\mathbf{k}-\mathbf{k}',\sigma\omega^\alpha_{\mathbf{k}}-\sigma'\omega^T_{\mathbf{k}'})\,\frac{k_n}{k}$$

$$\times\, \chi^{(2)}_{nml}(\mathbf{k}',\sigma'\omega^T_{\mathbf{k}'}|\mathbf{k}-\mathbf{k}',\sigma\omega^\alpha_{\mathbf{k}}-\sigma'\omega^T_{\mathbf{k}'})\,\mathscr{P}\,\frac{1}{\Lambda_\perp(\mathbf{k}-\mathbf{k}',\sigma\omega^\alpha_{\mathbf{k}}-\sigma'\omega^T_{\mathbf{k}'})}$$

$$-i\pi\sum_{\sigma''}\sum_{\gamma}\left(\delta_{jm}-\frac{k'_j\,k'_m}{k'^2}\right)\frac{k_i\,(\mathbf{k}-\mathbf{k}')_k}{k\,|\mathbf{k}-\mathbf{k}'|}\,\chi^{(2)}_{ijk}(\mathbf{k}',\sigma'\omega^T_{\mathbf{k}'}|\mathbf{k}-\mathbf{k}',\sigma''\omega^\gamma_{\mathbf{k}-\mathbf{k}'})$$

$$\times\,\frac{k_n\,(\mathbf{k}-\mathbf{k}')_l}{k\,|\mathbf{k}-\mathbf{k}'|}\,\chi^{(2)}_{nml}(\mathbf{k}',\sigma'\omega^T_{\mathbf{k}'}|\mathbf{k}-\mathbf{k}',\sigma''\omega^\gamma_{\mathbf{k}-\mathbf{k}'})\,\frac{\delta(\sigma\omega^\alpha_{\mathbf{k}}-\sigma'\omega^T_{\mathbf{k}'}-\sigma''\omega^\gamma_{\mathbf{k}-\mathbf{k}'})}{\epsilon'_\parallel(\mathbf{k}-\mathbf{k}',\sigma''\omega^\gamma_{\mathbf{k}-\mathbf{k}'})}$$

$$-i\pi\sum_{\sigma''}\frac{k_i}{k}\left(\delta_{jm}-\frac{k'_j\,k'_m}{k'^2}\right)\left(\delta_{kl}-\frac{(\mathbf{k}-\mathbf{k}')_k\,(\mathbf{k}-\mathbf{k}')_l}{(\mathbf{k}-\mathbf{k}')^2}\right)$$

$$\times\, \chi^{(2)}_{ijk}(\mathbf{k}',\sigma'\omega^T_{\mathbf{k}'}|\mathbf{k}-\mathbf{k}',\sigma''\omega^T_{\mathbf{k}-\mathbf{k}'})\,\frac{k_n}{k}\,\chi^{(2)}_{nml}(\mathbf{k}',\sigma'\omega^T_{\mathbf{k}'}|\mathbf{k}-\mathbf{k}',\sigma''\omega^T_{\mathbf{k}-\mathbf{k}'})$$

$$\times\,\frac{\delta(\sigma\omega^\alpha_{\mathbf{k}}-\sigma'\omega^T_{\mathbf{k}'}-\sigma''\omega^T_{\mathbf{k}-\mathbf{k}'})}{\Lambda'_\perp(\mathbf{k}-\mathbf{k}',\sigma''\omega^T_{\mathbf{k}-\mathbf{k}'})}\Bigg]\,\frac{1}{2}\,I^{\sigma'T}_{\mathbf{k}'}\,I^{\sigma\alpha}_{\mathbf{k}}\,\underbrace{\delta(\omega-\sigma\omega^\alpha_{\mathbf{k}})}$$

9 $$+2i\sum_{\sigma,\sigma'}\sum_{\beta}\mathrm{Im}\int d\mathbf{k}'\left[\left(\delta_{in}-\frac{k_i\,k_n}{k^2}\right)\frac{k'_j\,(\mathbf{k}-\mathbf{k}')_k}{k'\,|\mathbf{k}-\mathbf{k}'|}\right.$$

$$\times\, \chi^{(2)}_{ijk}(\mathbf{k}',\sigma'\omega^\beta_{\mathbf{k}'}|\mathbf{k}-\mathbf{k}',\sigma\omega^T_{\mathbf{k}}-\sigma'\omega^\beta_{\mathbf{k}'})\,\frac{k'_m\,(\mathbf{k}-\mathbf{k}')_l}{k'\,|\mathbf{k}-\mathbf{k}'|}$$

$$\times\, \chi^{(2)}_{nml}(\mathbf{k}',\sigma'\omega^\beta_{\mathbf{k}'}|\mathbf{k}-\mathbf{k}',\sigma\omega^T_{\mathbf{k}}-\sigma'\omega^\beta_{\mathbf{k}'})\,\mathscr{P}\,\frac{1}{\epsilon_\parallel(\mathbf{k}-\mathbf{k}',\sigma\omega^T_{\mathbf{k}}-\sigma'\omega^\beta_{\mathbf{k}'})}$$

$$+\left(\delta_{in}-\frac{k_i\,k_n}{k^2}\right)\frac{k'_j\,k'_m}{k'^2}\left(\delta_{kl}-\frac{(\mathbf{k}-\mathbf{k}')_k\,(\mathbf{k}-\mathbf{k}')_l}{(\mathbf{k}-\mathbf{k}')^2}\right)$$

$$\times\, \chi^{(2)}_{ijk}(\mathbf{k}',\sigma'\omega^\beta_{\mathbf{k}'}|\mathbf{k}-\mathbf{k}',\sigma\omega^T_{\mathbf{k}}-\sigma'\omega^\beta_{\mathbf{k}'})\,\chi^{(2)}_{nml}(\mathbf{k}',\sigma'\omega^\beta_{\mathbf{k}'}|\mathbf{k}-\mathbf{k}',\sigma\omega^T_{\mathbf{k}}-\sigma'\omega^\beta_{\mathbf{k}'})$$

$$\times\,\mathscr{P}\,\frac{1}{\Lambda_\perp(\mathbf{k}-\mathbf{k}',\sigma\omega^T_{\mathbf{k}}-\sigma'\omega^\beta_{\mathbf{k}'})}-i\pi\sum_{\sigma''}\sum_{\gamma}\left(\delta_{in}-\frac{k_i\,k_n}{k^2}\right)\frac{k'_j\,(\mathbf{k}-\mathbf{k}')_k}{k'\,|\mathbf{k}-\mathbf{k}'|}$$

$$\times\, \chi^{(2)}_{ijk}(\mathbf{k}',\sigma'\omega^\beta_{\mathbf{k}'}|\mathbf{k}-\mathbf{k}',\sigma''\omega^\gamma_{\mathbf{k}-\mathbf{k}'})\,\frac{k'_m\,(\mathbf{k}-\mathbf{k}')_l}{k'\,|\mathbf{k}-\mathbf{k}'|}$$

$$\times\, \chi^{(2)}_{nml}(\mathbf{k}',\sigma'\omega^\beta_{\mathbf{k}'}|\mathbf{k}-\mathbf{k}',\sigma''\omega^\gamma_{\mathbf{k}-\mathbf{k}'})\,\frac{\delta(\sigma\omega^T_{\mathbf{k}}-\sigma'\omega^\beta_{\mathbf{k}'}-\sigma''\omega^\gamma_{\mathbf{k}-\mathbf{k}'})}{\epsilon'_\parallel(\mathbf{k}-\mathbf{k}',\sigma''\omega^\gamma_{\mathbf{k}-\mathbf{k}'})}$$

$$
\begin{aligned}
&-i\pi\sum_{\sigma''}\left(\delta_{in}-\frac{k_i\,k_n}{k^2}\right)\frac{k'_j\,k'_m}{k'^2}\left(\delta_{kl}-\frac{(\mathbf{k}-\mathbf{k}')_k\,(\mathbf{k}-\mathbf{k}')_l}{(\mathbf{k}-\mathbf{k}')^2}\right)\\
&\times\chi^{(2)}_{ijk}(\mathbf{k}',\sigma'\omega^\beta_{\mathbf{k}'}|\mathbf{k}-\mathbf{k}',\sigma''\omega^T_{\mathbf{k}-\mathbf{k}'})\,\chi^{(2)}_{nml}(\mathbf{k}',\sigma'\omega^\beta_{\mathbf{k}'}|\mathbf{k}-\mathbf{k}',\sigma''\omega^T_{\mathbf{k}-\mathbf{k}'})\\
&\times\frac{\delta(\sigma\omega^T_{\mathbf{k}}-\sigma'\omega^\beta_{\mathbf{k}'}-\sigma''\omega^T_{\mathbf{k}-\mathbf{k}'})}{\Lambda'_\perp(\mathbf{k}-\mathbf{k}',\sigma''\omega^T_{\mathbf{k}-\mathbf{k}'})}\Bigg]\frac{1}{2}\,I^{\sigma'\beta}_{\mathbf{k}'}\,I^{\sigma T}_{\mathbf{k}}\,\underbrace{\delta(\omega-\sigma\omega^T_{\mathbf{k}})}
\end{aligned}
$$

10

$$
\begin{aligned}
&+2i\sum_{\sigma,\sigma'}\mathrm{Im}\int d\mathbf{k}'\Bigg[\left(\delta_{in}-\frac{k_i\,k_n}{k^2}\right)\left(\delta_{jm}-\frac{k'_j\,k'_m}{k'^2}\right)\frac{(\mathbf{k}-\mathbf{k}')_k}{|\mathbf{k}-\mathbf{k}'|}\\
&\times\chi^{(2)}_{ijk}(\mathbf{k}',\sigma'\omega^T_{\mathbf{k}'}|\mathbf{k}-\mathbf{k}',\sigma\omega^T_{\mathbf{k}}-\sigma'\omega^T_{\mathbf{k}'})\,\frac{(\mathbf{k}-\mathbf{k}')_l}{|\mathbf{k}-\mathbf{k}'|}\\
&\times\chi^{(2)}_{nml}(\mathbf{k}',\sigma'\omega^T_{\mathbf{k}'}|\mathbf{k}-\mathbf{k}',\sigma\omega^T_{\mathbf{k}}-\sigma'\omega^T_{\mathbf{k}'})\,\mathscr{P}\,\frac{1}{\epsilon_\parallel(\mathbf{k}-\mathbf{k}',\sigma\omega^T_{\mathbf{k}}-\sigma'\omega^T_{\mathbf{k}'})}\\
&+\left(\delta_{in}-\frac{k_i\,k_n}{k^2}\right)\left(\delta_{jm}-\frac{k'_j\,k'_m}{k'^2}\right)\left(\delta_{kl}-\frac{(\mathbf{k}-\mathbf{k}')_k\,(\mathbf{k}-\mathbf{k}')_l}{(\mathbf{k}-\mathbf{k}')^2}\right)\\
&\times\chi^{(2)}_{ijk}(\mathbf{k}',\sigma'\omega^T_{\mathbf{k}'}|\mathbf{k}-\mathbf{k}',\sigma\omega^T_{\mathbf{k}}-\sigma'\omega^T_{\mathbf{k}'})\\
&\times\chi^{(2)}_{nml}(\mathbf{k}',\sigma'\omega^T_{\mathbf{k}'}|\mathbf{k}-\mathbf{k}',\sigma\omega^T_{\mathbf{k}}-\sigma'\omega^T_{\mathbf{k}'})\,\mathscr{P}\,\frac{1}{\Lambda_\perp(\mathbf{k}-\mathbf{k}',\sigma\omega^T_{\mathbf{k}}-\sigma'\omega^T_{\mathbf{k}'})}\\
&-i\pi\sum_{\sigma''}\sum_{\gamma}\left(\delta_{in}-\frac{k_i\,k_n}{k^2}\right)\left(\delta_{jm}-\frac{k'_j\,k'_m}{k'^2}\right)\frac{(\mathbf{k}-\mathbf{k}')_k}{|\mathbf{k}-\mathbf{k}'|}\\
&\times\chi^{(2)}_{ijk}(\mathbf{k}',\sigma'\omega^T_{\mathbf{k}'}|\mathbf{k}-\mathbf{k}',\sigma''\omega^\gamma_{\mathbf{k}-\mathbf{k}'})\,\frac{(\mathbf{k}-\mathbf{k}')_l}{|\mathbf{k}-\mathbf{k}'|}\\
&\times\chi^{(2)}_{nml}(\mathbf{k}',\sigma'\omega^T_{\mathbf{k}'}|\mathbf{k}-\mathbf{k}',\sigma''\omega^\gamma_{\mathbf{k}-\mathbf{k}'})\,\frac{\delta(\sigma\omega^T_{\mathbf{k}}-\sigma'\omega^T_{\mathbf{k}'}-\sigma''\omega^\gamma_{\mathbf{k}-\mathbf{k}'})}{\epsilon'_\parallel(\mathbf{k}-\mathbf{k}',\sigma''\omega^\gamma_{\mathbf{k}-\mathbf{k}'})}\\
&-i\pi\sum_{\sigma''}\left(\delta_{in}-\frac{k_i\,k_n}{k^2}\right)\left(\delta_{jm}-\frac{k'_j\,k'_m}{k'^2}\right)\left(\delta_{kl}-\frac{(\mathbf{k}-\mathbf{k}')_k\,(\mathbf{k}-\mathbf{k}')_l}{(\mathbf{k}-\mathbf{k}')^2}\right)\\
&\times\chi^{(2)}_{ijk}(\mathbf{k}',\sigma'\omega^T_{\mathbf{k}'}|\mathbf{k}-\mathbf{k}',\sigma''\omega^T_{\mathbf{k}-\mathbf{k}'})\,\chi^{(2)}_{nml}(\mathbf{k}',\sigma'\omega^T_{\mathbf{k}'}|\mathbf{k}-\mathbf{k}',\sigma''\omega^T_{\mathbf{k}-\mathbf{k}'})\\
&\times\frac{\delta(\sigma\omega^T_{\mathbf{k}}-\sigma'\omega^T_{\mathbf{k}'}-\sigma''\omega^T_{\mathbf{k}-\mathbf{k}'})}{\Lambda'_\perp(\mathbf{k}-\mathbf{k}',\sigma''\omega^T_{\mathbf{k}-\mathbf{k}'})}\Bigg]\frac{1}{4}\,I^{\sigma' T}_{\mathbf{k}'}\,I^{\sigma T}_{\mathbf{k}}\,\underbrace{\delta(\omega-\sigma\omega^T_{\mathbf{k}})}
\end{aligned}
$$

11

$$
\begin{aligned}
&-2i\sum_{\sigma',\sigma''}\sum_{\beta,\gamma}\mathrm{Im}\int d\mathbf{k}'\Bigg[i\pi\sum_{\sigma}\sum_{\alpha}\frac{k_i\,k'_j\,(\mathbf{k}-\mathbf{k}')_k}{k\,k'\,|\mathbf{k}-\mathbf{k}'|}\\
&\times\chi^{(2)}_{ijk}(\mathbf{k}',\sigma'\omega^\beta_{\mathbf{k}'}|\mathbf{k}-\mathbf{k}',\sigma''\omega^\gamma_{\mathbf{k}-\mathbf{k}'})\,\frac{k_l\,k'_m\,(\mathbf{k}-\mathbf{k}')_n}{k\,k'\,|\mathbf{k}-\mathbf{k}'|}\\
&\times\chi^{(2)*}_{lmn}(\mathbf{k}',\sigma'\omega^\beta_{\mathbf{k}'}|\mathbf{k}-\mathbf{k}',\sigma''\omega^\gamma_{\mathbf{k}-\mathbf{k}'})
\end{aligned}
$$

$$\times \frac{\delta(\sigma\omega_{\mathbf{k}}^{\alpha} - \sigma'\omega_{\mathbf{k}'}^{\beta} - \sigma''\omega_{\mathbf{k}-\mathbf{k}'}^{\gamma})}{\epsilon_{\parallel}'(\mathbf{k},\sigma\omega_{\mathbf{k}}^{\alpha})} \underbrace{\delta(\omega - \sigma\omega_{\mathbf{k}}^{\alpha})}$$

$$+ i\pi \sum_{\sigma} \frac{k_j' (\mathbf{k}-\mathbf{k}')_k}{k' |\mathbf{k}-\mathbf{k}'|} \left(\delta_{il} - \frac{k_i k_l}{k^2}\right) \chi_{ijk}^{(2)}(\mathbf{k}',\sigma'\omega_{\mathbf{k}'}^{\beta}|\mathbf{k}-\mathbf{k}',\sigma''\omega_{\mathbf{k}-\mathbf{k}'}^{\gamma})$$

$$\times \frac{k_m' (\mathbf{k}-\mathbf{k}')_n}{k' |\mathbf{k}-\mathbf{k}'|} \chi_{lmn}^{(2)*}(\mathbf{k}',\sigma'\omega_{\mathbf{k}'}^{\beta}|\mathbf{k}-\mathbf{k}',\sigma''\omega_{\mathbf{k}-\mathbf{k}'}^{\gamma})$$

$$\times \frac{\delta(\sigma\omega_{\mathbf{k}}^{T} - \sigma'\omega_{\mathbf{k}'}^{\beta} - \sigma''\omega_{\mathbf{k}-\mathbf{k}'}^{\gamma})}{\Lambda_{\perp}'(\mathbf{k},\sigma\omega_{\mathbf{k}}^{T})} \underbrace{\delta(\omega - \sigma\omega_{\mathbf{k}}^{T})}\Bigg] I_{\mathbf{k}'}^{\sigma'\beta} I_{\mathbf{k}-\mathbf{k}'}^{\sigma''\gamma}$$

12 $$-2i \sum_{\sigma',\sigma''} \sum_{\beta} \mathrm{Im} \int d\mathbf{k}' \Bigg[i\pi \sum_{\sigma} \sum_{\alpha} \frac{k_i k_j'}{k k'}$$

$$\times \left(\delta_{kn} - \frac{(\mathbf{k}-\mathbf{k}')_k (\mathbf{k}-\mathbf{k}')_n}{(\mathbf{k}-\mathbf{k}')^2}\right) \chi_{ijk}^{(2)}(\mathbf{k}',\sigma'\omega_{\mathbf{k}'}^{\beta}|\mathbf{k}-\mathbf{k}',\sigma''\omega_{\mathbf{k}-\mathbf{k}'}^{T})$$

$$\times \frac{k_l k_m'}{k k'} \chi_{lmn}^{(2)*}(\mathbf{k}',\sigma'\omega_{\mathbf{k}'}^{\beta}|\mathbf{k}-\mathbf{k}',\sigma''\omega_{\mathbf{k}-\mathbf{k}'}^{T})$$

$$\times \frac{\delta(\sigma\omega_{\mathbf{k}}^{\alpha} - \sigma'\omega_{\mathbf{k}'}^{\beta} - \sigma''\omega_{\mathbf{k}-\mathbf{k}'}^{T})}{\epsilon_{\parallel}'(\mathbf{k},\sigma\omega_{\mathbf{k}}^{\alpha})} \underbrace{\delta(\omega - \sigma\omega_{\mathbf{k}}^{\alpha})}$$

$$+ i\pi \sum_{\sigma} \frac{k_j'}{k'} \left(\delta_{kn} - \frac{(\mathbf{k}-\mathbf{k}')_k (\mathbf{k}-\mathbf{k}')_n}{(\mathbf{k}-\mathbf{k}')^2}\right) \left(\delta_{il} - \frac{k_i k_l}{k^2}\right)$$

$$\times \chi_{ijk}^{(2)}(\mathbf{k}',\sigma'\omega_{\mathbf{k}'}^{\beta}|\mathbf{k}-\mathbf{k}',\sigma''\omega_{\mathbf{k}-\mathbf{k}'}^{T}) \frac{k_m'}{k'} \chi_{lmn}^{(2)*}(\mathbf{k}',\sigma'\omega_{\mathbf{k}'}^{\beta}|\mathbf{k}-\mathbf{k}',\sigma''\omega_{\mathbf{k}-\mathbf{k}'}^{T})$$

$$\times \frac{\delta(\sigma\omega_{\mathbf{k}}^{T} - \sigma'\omega_{\mathbf{k}'}^{\beta} - \sigma''\omega_{\mathbf{k}-\mathbf{k}'}^{T})}{\Lambda_{\perp}'(\mathbf{k},\sigma\omega_{\mathbf{k}}^{T})} \underbrace{\delta(\omega - \sigma\omega_{\mathbf{k}}^{T})}\Bigg] \frac{1}{2} I_{\mathbf{k}'}^{\sigma'\beta} I_{\mathbf{k}-\mathbf{k}'}^{\sigma''T}$$

13 $$-2i \sum_{\sigma',\sigma''} \sum_{\gamma} \mathrm{Im} \int d\mathbf{k}' \Bigg[i\pi \sum_{\sigma} \sum_{\alpha} \left(\delta_{jm} - \frac{k_j' k_m'}{k'^2}\right) \frac{k_i (\mathbf{k}-\mathbf{k}')_k}{k |\mathbf{k}-\mathbf{k}'|}$$

$$\times \chi_{ijk}^{(2)}(\mathbf{k}',\sigma'\omega_{\mathbf{k}'}^{T}|\mathbf{k}-\mathbf{k}',\sigma''\omega_{\mathbf{k}-\mathbf{k}'}^{\gamma}) \frac{k_l (\mathbf{k}-\mathbf{k}')_n}{k |\mathbf{k}-\mathbf{k}'|}$$

$$\times \chi_{lmn}^{(2)*}(\mathbf{k}',\sigma'\omega_{\mathbf{k}'}^{T}|\mathbf{k}-\mathbf{k}',\sigma''\omega_{\mathbf{k}-\mathbf{k}'}^{\gamma}) \frac{\delta(\sigma\omega_{\mathbf{k}}^{\alpha} - \sigma'\omega_{\mathbf{k}'}^{T} - \sigma''\omega_{\mathbf{k}-\mathbf{k}'}^{\gamma})}{\epsilon_{\parallel}'(\mathbf{k},\sigma\omega_{\mathbf{k}}^{\alpha})} \underbrace{\delta(\omega - \sigma\omega_{\mathbf{k}}^{\alpha})}$$

$$+ i\pi \sum_{\sigma} \left(\delta_{jm} - \frac{k_j' k_m'}{k'^2}\right) \frac{(\mathbf{k}-\mathbf{k}')_k}{|\mathbf{k}-\mathbf{k}'|} \left(\delta_{il} - \frac{k_i k_l}{k^2}\right)$$

$$\times \chi_{ijk}^{(2)}(\mathbf{k}',\sigma'\omega_{\mathbf{k}'}^{T}|\mathbf{k}-\mathbf{k}',\sigma''\omega_{\mathbf{k}-\mathbf{k}'}^{\gamma}) \frac{(\mathbf{k}-\mathbf{k}')_n}{|\mathbf{k}-\mathbf{k}'|} \chi_{lmn}^{(2)*}(\mathbf{k}',\sigma'\omega_{\mathbf{k}'}^{T}|\mathbf{k}-\mathbf{k}',\sigma''\omega_{\mathbf{k}-\mathbf{k}'}^{\gamma})$$

$$\times \frac{\delta(\sigma\omega_{\mathbf{k}}^{T} - \sigma'\omega_{\mathbf{k}'}^{T} - \sigma''\omega_{\mathbf{k}-\mathbf{k}'}^{\gamma})}{\Lambda_{\perp}'(\mathbf{k},\sigma\omega_{\mathbf{k}}^{T})} \underbrace{\delta(\omega - \sigma\omega_{\mathbf{k}}^{T})}\Bigg] \frac{1}{2} I_{\mathbf{k}'}^{\sigma'T} I_{\mathbf{k}-\mathbf{k}'}^{\sigma''\gamma}$$

$$
\boxed{14}\;\; -2i \sum_{\sigma',\sigma''} \mathrm{Im} \int d\mathbf{k}' \Bigg[i\pi \sum_{\sigma} \sum_{\alpha} \frac{k_i}{k} \left(\delta_{jm} - \frac{k'_j\, k'_m}{k'^2} \right)
$$

$$
\times \left(\delta_{kn} - \frac{(\mathbf{k}-\mathbf{k}')_k\, (\mathbf{k}-\mathbf{k}')_n}{(\mathbf{k}-\mathbf{k}')^2} \right) \chi^{(2)}_{ijk}(\mathbf{k}', \sigma'\omega^T_{\mathbf{k}'} | \mathbf{k}-\mathbf{k}', \sigma''\omega^T_{\mathbf{k}-\mathbf{k}'}) \frac{k_l}{k}
$$

$$
\times \chi^{(2)*}_{lmn}(\mathbf{k}', \sigma'\omega^T_{\mathbf{k}'} | \mathbf{k}-\mathbf{k}', \sigma''\omega^T_{\mathbf{k}-\mathbf{k}'}) \frac{\delta(\sigma\omega^\alpha_{\mathbf{k}} - \sigma'\omega^T_{\mathbf{k}'} - \sigma''\omega^T_{\mathbf{k}-\mathbf{k}'})}{\epsilon'_\parallel(\mathbf{k}, \sigma\omega^\alpha_{\mathbf{k}})} \underbrace{\delta(\omega - \sigma\omega^\alpha_{\mathbf{k}})}
$$

$$
+ i\pi \sum_{\sigma} \left(\delta_{jm} - \frac{k'_j\, k'_m}{k'^2} \right) \left(\delta_{kn} - \frac{(\mathbf{k}-\mathbf{k}')_k\, (\mathbf{k}-\mathbf{k}')_n}{(\mathbf{k}-\mathbf{k}')^2} \right) \left(\delta_{il} - \frac{k_i\, k_l}{k^2} \right)
$$

$$
\times \chi^{(2)}_{ijk}(\mathbf{k}', \sigma'\omega^T_{\mathbf{k}'} | \mathbf{k}-\mathbf{k}', \sigma''\omega^T_{\mathbf{k}-\mathbf{k}'})\, \chi^{(2)*}_{lmn}(\mathbf{k}', \sigma'\omega^T_{\mathbf{k}'} | \mathbf{k}-\mathbf{k}', \sigma''\omega^T_{\mathbf{k}-\mathbf{k}'})
$$

$$
\times \frac{\delta(\sigma\omega^T_{\mathbf{k}} - \sigma'\omega^T_{\mathbf{k}'} - \sigma''\omega^T_{\mathbf{k}-\mathbf{k}'})}{\Lambda'_\perp(\mathbf{k}, \sigma\omega^T_{\mathbf{k}})} \underbrace{\delta(\omega - \sigma\omega^T_{\mathbf{k}})} \Bigg] \frac{1}{4} I^{\sigma' T}_{\mathbf{k}'}\, I^{\sigma'' T}_{\mathbf{k}-\mathbf{k}'}
$$

$$
\boxed{15}\;\; +2i \sum_{\sigma,\sigma'} \sum_{\alpha,\beta} \mathrm{Im} \int d\mathbf{k}' \frac{k_i\, k_l}{k^2} \frac{k'_j\, k'_k}{k'^2} \chi^{(3)}_{ijkl}(\mathbf{k}', \sigma'\omega^\beta_{\mathbf{k}'} | -\mathbf{k}', -\sigma'\omega^\beta_{\mathbf{k}'} | \mathbf{k}, \sigma\omega^\alpha_{\mathbf{k}})
$$

$$
\times I^{\sigma'\beta}_{\mathbf{k}'}\, I^{\sigma\alpha}_{\mathbf{k}} \underbrace{\delta(\omega - \sigma\omega^\alpha_{\mathbf{k}})} + 2i \sum_{\sigma,\sigma'} \sum_{\alpha} \mathrm{Im} \int d\mathbf{k}' \frac{1}{2} \frac{k_i\, k_l}{k^2} \left(\delta_{jk} - \frac{k'_j\, k'_k}{k'^2} \right)
$$

$$
\times \chi^{(3)}_{ijkl}(\mathbf{k}', \sigma'\omega^T_{\mathbf{k}'} | -\mathbf{k}', -\sigma'\omega^T_{\mathbf{k}'} | \mathbf{k}, \sigma\omega^\alpha_{\mathbf{k}})\, I^{\sigma' T}_{\mathbf{k}'}\, I^{\sigma\alpha}_{\mathbf{k}} \underbrace{\delta(\omega - \sigma\omega^\alpha_{\mathbf{k}})}
$$

$$
+ 2i \sum_{\sigma,\sigma'} \sum_{\beta} \mathrm{Im} \int d\mathbf{k}' \frac{1}{2} \left(\delta_{il} - \frac{k_i\, k_l}{k^2} \right) \frac{k'_j\, k'_k}{k'^2}
$$

$$
\times \chi^{(3)}_{ijkl}(\mathbf{k}', \sigma'\omega^\beta_{\mathbf{k}'} | -\mathbf{k}', -\sigma'\omega^\beta_{\mathbf{k}'} | \mathbf{k}, \sigma\omega^T_{\mathbf{k}})\, I^{\sigma'\beta}_{\mathbf{k}'}\, I^{\sigma T}_{\mathbf{k}} \underbrace{\delta(\omega - \sigma\omega^T_{\mathbf{k}})}
$$

$$
+ 2i \sum_{\sigma,\sigma'} \mathrm{Im} \int d\mathbf{k}' \frac{1}{4} \left(\delta_{il} - \frac{k_i\, k_l}{k^2} \right) \left(\delta_{jk} - \frac{k'_j\, k'_k}{k'^2} \right)
$$

$$
\times \chi^{(3)}_{ijkl}(\mathbf{k}', \sigma'\omega^T_{\mathbf{k}'} | -\mathbf{k}', -\sigma'\omega^T_{\mathbf{k}'} | \mathbf{k}, \sigma\omega^T_{\mathbf{k}})\, I^{\sigma' T}_{\mathbf{k}'}\, I^{\sigma T}_{\mathbf{k}} \underbrace{\delta(\omega - \sigma\omega^T_{\mathbf{k}})}
$$

$$
\boxed{16}\;\; = i\, \mathrm{Im} \sum_a \frac{(4\pi e_a)^2}{(2\pi)^3} \int d\mathbf{p} \Bigg[i\pi \sum_{\sigma} \sum_{\alpha} \frac{1}{k^2\, \epsilon'_\parallel(\mathbf{k}, \sigma\omega^\alpha_{\mathbf{k}})}
$$

$$
\times \delta(\sigma\omega^\alpha_{\mathbf{k}} - \mathbf{k}\cdot\mathbf{v})\, f_a(\mathbf{p}) \underbrace{\delta(\omega - \sigma\omega^\alpha_{\mathbf{k}})} \Bigg]
$$

$$
\boxed{17}\;\; +2i \sum_{\sigma,\sigma''} \sum_{\alpha,\gamma} \mathrm{Im} \int d\mathbf{k}' \sum_a \frac{(4\pi e_a)^2}{(2\pi)^3} \int d\mathbf{p} \frac{1}{(\mathbf{k}'\cdot\mathbf{v})^2}
$$

$$
\times \Bigg[\frac{k'_j\, (\mathbf{k}'\cdot\mathbf{v})}{k'^2} \frac{1}{\epsilon_\parallel(\mathbf{k}', \sigma\omega^\alpha_{\mathbf{k}} - \sigma''\omega^\gamma_{\mathbf{k}-\mathbf{k}'})}
$$

$$+\left(v_j - \frac{k'_j\,(\mathbf{k}'\cdot\mathbf{v})}{k'^2}\right)\frac{1}{\Lambda_\perp(\mathbf{k}',\sigma\omega_\mathbf{k}^\alpha - \sigma''\omega_{\mathbf{k}-\mathbf{k}'}^\gamma)}\Bigg]$$

$$\times\left[\frac{k'_p\,(\mathbf{k}'\cdot\mathbf{v})}{k'^2}\frac{1}{\epsilon_\parallel^*(\mathbf{k}',\sigma\omega_\mathbf{k}^\alpha - \sigma''\omega_{\mathbf{k}-\mathbf{k}'}^\gamma)}\right.$$

$$+\left(v_p - \frac{k'_p\,(\mathbf{k}'\cdot\mathbf{v})}{k'^2}\right)\frac{1}{\Lambda_\perp^*(\mathbf{k}',\sigma\omega_\mathbf{k}^\alpha - \sigma''\omega_{\mathbf{k}-\mathbf{k}'}^\gamma)}\Bigg]$$

$$\times\frac{k_i\,(\mathbf{k}-\mathbf{k}')_k}{k\,|\mathbf{k}-\mathbf{k}'|}\,\chi_{ijk}^{(2)}(\mathbf{k}',\sigma\omega_\mathbf{k}^\alpha - \sigma''\omega_{\mathbf{k}-\mathbf{k}'}^\gamma|\mathbf{k}-\mathbf{k}',\sigma''\omega_{\mathbf{k}-\mathbf{k}'}^\gamma)$$

$$\times\frac{k_m\,(\mathbf{k}-\mathbf{k}')_n}{k\,|\mathbf{k}-\mathbf{k}'|}\,\chi_{mpn}^{(2)*}(\mathbf{k}',\sigma\omega_\mathbf{k}^\alpha - \sigma''\omega_{\mathbf{k}-\mathbf{k}'}^\gamma|\mathbf{k}-\mathbf{k}',\sigma''\omega_{\mathbf{k}-\mathbf{k}'}^\gamma)$$

$$\times\left[i\pi\,I_{\mathbf{k}-\mathbf{k}'}^{\sigma''\gamma}\,\frac{\delta(\sigma\omega_\mathbf{k}^\alpha - \sigma''\omega_{\mathbf{k}-\mathbf{k}'}^\gamma - \mathbf{k}'\cdot\mathbf{v})}{\epsilon'_\parallel(\mathbf{k},\sigma\omega_\mathbf{k}^\alpha)}\,f_a(\mathbf{p})\right]\underbrace{\delta(\omega-\sigma\omega_\mathbf{k}^\alpha)}$$

18 $$+2i\sum_{\sigma,\sigma''}\sum_{\gamma}\mathrm{Im}\int d\mathbf{k}'\,\sum_a\frac{(4\pi e_a)^2}{(2\pi)^3}\int d\mathbf{p}\,\frac{1}{(\mathbf{k}'\cdot\mathbf{v})^2}$$

$$\times\left[\frac{k'_j\,(\mathbf{k}'\cdot\mathbf{v})}{k'^2}\frac{1}{\epsilon_\parallel(\mathbf{k}',\sigma\omega_\mathbf{k}^T - \sigma''\omega_{\mathbf{k}-\mathbf{k}'}^\gamma)}\right.$$

$$+\left(v_j - \frac{k'_j\,(\mathbf{k}'\cdot\mathbf{v})}{k'^2}\right)\frac{1}{\Lambda_\perp(\mathbf{k}',\sigma\omega_\mathbf{k}^T - \sigma''\omega_{\mathbf{k}-\mathbf{k}'}^\gamma)}\Bigg]$$

$$\times\left[\frac{k'_p\,(\mathbf{k}'\cdot\mathbf{v})}{k'^2}\frac{1}{\epsilon_\parallel^*(\mathbf{k}',\sigma\omega_\mathbf{k}^T - \sigma''\omega_{\mathbf{k}-\mathbf{k}'}^\gamma)}\right.$$

$$+\left(v_p - \frac{k'_p\,(\mathbf{k}'\cdot\mathbf{v})}{k'^2}\right)\frac{1}{\Lambda_\perp^*(\mathbf{k}',\sigma\omega_\mathbf{k}^T - \sigma''\omega_{\mathbf{k}-\mathbf{k}'}^\gamma)}\Bigg]$$

$$\times\left(\delta_{im} - \frac{k_i\,k_m}{k^2}\right)\frac{(\mathbf{k}-\mathbf{k}')_k}{|\mathbf{k}-\mathbf{k}'|}\,\chi_{ijk}^{(2)}(\mathbf{k}',\sigma\omega_\mathbf{k}^T - \sigma''\omega_{\mathbf{k}-\mathbf{k}'}^\gamma|\mathbf{k}-\mathbf{k}',\sigma''\omega_{\mathbf{k}-\mathbf{k}'}^\gamma)$$

$$\times\frac{(\mathbf{k}-\mathbf{k}')_n}{|\mathbf{k}-\mathbf{k}'|}\,\chi_{mpn}^{(2)*}(\mathbf{k}',\sigma\omega_\mathbf{k}^T - \sigma''\omega_{\mathbf{k}-\mathbf{k}'}^\gamma|\mathbf{k}-\mathbf{k}',\sigma''\omega_{\mathbf{k}-\mathbf{k}'}^\gamma)$$

$$\times\left[i\pi\,I_{\mathbf{k}-\mathbf{k}'}^{\sigma''\gamma}\,\frac{\delta(\sigma\omega_\mathbf{k}^T - \sigma''\omega_{\mathbf{k}-\mathbf{k}'}^\gamma - \mathbf{k}'\cdot\mathbf{v})}{\Lambda'_\perp(\mathbf{k},\sigma\omega_\mathbf{k}^T)}\,f_a(\mathbf{p})\right]\underbrace{\delta(\omega-\sigma\omega_\mathbf{k}^T)}$$

19 $$+2i\sum_{\sigma,\sigma''}\sum_{\alpha}\mathrm{Im}\int d\mathbf{k}'\,\sum_a\frac{(4\pi e_a)^2}{(2\pi)^3}\int d\mathbf{p}\,\frac{1}{(\mathbf{k}'\cdot\mathbf{v})^2}$$

$$\times\left[\frac{k'_j\,(\mathbf{k}'\cdot\mathbf{v})}{k'^2}\frac{1}{\epsilon_\parallel(\mathbf{k}',\sigma\omega_\mathbf{k}^\alpha - \sigma''\omega_{\mathbf{k}-\mathbf{k}'}^T)}\right.$$

$$+\left(v_j - \frac{k'_j\,(\mathbf{k}'\cdot\mathbf{v})}{k'^2}\right)\frac{1}{\Lambda_\perp(\mathbf{k}',\sigma\omega_\mathbf{k}^\alpha - \sigma''\omega_{\mathbf{k}-\mathbf{k}'}^T)}\Bigg]$$

$$\times\left[\frac{k'_p\,(\mathbf{k}'\cdot\mathbf{v})}{k'^2}\frac{1}{\epsilon_\parallel^*(\mathbf{k}',\sigma\omega_\mathbf{k}^\alpha-\sigma''\omega_{\mathbf{k}-\mathbf{k}'}^T)}\right.$$
$$\left.+\left(v_p-\frac{k'_p\,(\mathbf{k}'\cdot\mathbf{v})}{k'^2}\right)\frac{1}{\Lambda_\perp^*(\mathbf{k}',\sigma\omega_\mathbf{k}^\alpha-\sigma''\omega_{\mathbf{k}-\mathbf{k}'}^T)}\right]\frac{k_i}{k}$$
$$\times\left(\delta_{nk}-\frac{(\mathbf{k}-\mathbf{k}')_n\,(\mathbf{k}-\mathbf{k}')_k}{(\mathbf{k}-\mathbf{k}')^2}\right)\chi_{ijk}^{(2)}(\mathbf{k}',\sigma\omega_\mathbf{k}^\alpha-\sigma''\omega_{\mathbf{k}-\mathbf{k}'}^T|\mathbf{k}-\mathbf{k}',\sigma''\omega_{\mathbf{k}-\mathbf{k}'}^T)$$
$$\times\frac{k_m}{k}\,\chi_{mpn}^{(2)*}(\mathbf{k}',\sigma\omega_\mathbf{k}^\alpha-\sigma''\omega_{\mathbf{k}-\mathbf{k}'}^T|\mathbf{k}-\mathbf{k}',\sigma''\omega_{\mathbf{k}-\mathbf{k}'}^T)$$
$$\times\left[i\pi\,\frac{1}{\epsilon'_\parallel(\mathbf{k},\sigma\omega_\mathbf{k}^\alpha)}\,\frac{1}{2}\,I_{\mathbf{k}-\mathbf{k}'}^{\sigma''T}\,\delta(\sigma\omega_\mathbf{k}^\alpha-\sigma''\omega_{\mathbf{k}-\mathbf{k}'}^T-\mathbf{k}'\cdot\mathbf{v})\;f_a(\mathbf{p})\right]\underbrace{\delta(\omega-\sigma\omega_\mathbf{k}^\alpha)}$$

20 $$+2i\sum_{\sigma,\sigma'}\mathrm{Im}\int d\mathbf{k}'\,\sum_a\frac{(4\pi e_a)^2}{(2\pi)^3}\int d\mathbf{p}\,\frac{1}{(\mathbf{k}'\cdot\mathbf{v})^2}$$
$$\times\left[\frac{k'_j\,(\mathbf{k}'\cdot\mathbf{v})}{k'^2}\frac{1}{\epsilon_\parallel(\mathbf{k}',\sigma\omega_\mathbf{k}^T-\sigma''\omega_{\mathbf{k}-\mathbf{k}'}^T)}\right.$$
$$\left.+\left(v_j-\frac{k'_j\,(\mathbf{k}'\cdot\mathbf{v})}{k'^2}\right)\frac{1}{\Lambda_\perp(\mathbf{k}',\sigma\omega_\mathbf{k}^T-\sigma''\omega_{\mathbf{k}-\mathbf{k}'}^T)}\right]$$
$$\times\left[\frac{k'_p\,(\mathbf{k}'\cdot\mathbf{v})}{k'^2}\frac{1}{\epsilon_\parallel^*(\mathbf{k}',\sigma\omega_\mathbf{k}^T-\sigma''\omega_{\mathbf{k}-\mathbf{k}'}^T)}\right.$$
$$\left.+\left(v_p-\frac{k'_p\,(\mathbf{k}'\cdot\mathbf{v})}{k'^2}\right)\frac{1}{\Lambda_\perp^*(\mathbf{k}',\sigma\omega_\mathbf{k}^T-\sigma''\omega_{\mathbf{k}-\mathbf{k}'}^T)}\right]$$
$$\times\left(\delta_{im}-\frac{k_i\,k_m}{k^2}\right)\chi_{ijk}^{(2)}(\mathbf{k}',\sigma\omega_\mathbf{k}^T-\sigma''\omega_{\mathbf{k}-\mathbf{k}'}^T|\mathbf{k}-\mathbf{k}',\sigma''\omega_{\mathbf{k}-\mathbf{k}'}^T)$$
$$\times\left(\delta_{nk}-\frac{(\mathbf{k}-\mathbf{k}')_n\,(\mathbf{k}-\mathbf{k}')_k}{(\mathbf{k}-\mathbf{k}')^2}\right)\chi_{mpn}^{(2)*}(\mathbf{k}',\sigma\omega_\mathbf{k}^T-\sigma''\omega_{\mathbf{k}-\mathbf{k}'}^T|\mathbf{k}-\mathbf{k}',\sigma''\omega_{\mathbf{k}-\mathbf{k}'}^T)$$
$$\times\left[i\pi\,\frac{1}{\Lambda'_\perp(\mathbf{k},\sigma\omega_\mathbf{k}^T)}\,\frac{1}{2}\,I_{\mathbf{k}-\mathbf{k}'}^{\sigma''T}\,\delta(\sigma\omega_\mathbf{k}^T-\sigma''\omega_{\mathbf{k}-\mathbf{k}'}^T-\mathbf{k}'\cdot\mathbf{v})\;f_a(\mathbf{p})\right]\underbrace{\delta(\omega-\sigma\omega_\mathbf{k}^T)}$$

21 $$-2i\sum_{\sigma,\sigma''}\sum_{\alpha,\gamma}\mathrm{Im}\int d\mathbf{k}'\,\sum_a\frac{(4\pi e_a)^2}{(2\pi)^3}\int d\mathbf{p}\,\frac{1}{(\mathbf{k}'\cdot\mathbf{v})^2}$$
$$\times\left[\frac{k'_j\,(\mathbf{k}'\cdot\mathbf{v})}{k'^2}\frac{1}{\epsilon_\parallel(\mathbf{k}',\sigma\omega_\mathbf{k}^\alpha-\sigma''\omega_{\mathbf{k}-\mathbf{k}'}^\gamma)}\right.$$
$$\left.+\left(v_j-\frac{k'_j\,(\mathbf{k}'\cdot\mathbf{v})}{k'^2}\right)\frac{1}{\Lambda_\perp(\mathbf{k}',\sigma\omega_\mathbf{k}^\alpha-\sigma''\omega_{\mathbf{k}-\mathbf{k}'}^\gamma)}\right]$$
$$\times\left[\frac{k'_p\,(\mathbf{k}'\cdot\mathbf{v})}{k'^2}\frac{1}{\epsilon_\parallel^*(\mathbf{k}',\sigma\omega_\mathbf{k}^\alpha-\sigma''\omega_{\mathbf{k}-\mathbf{k}'}^\gamma)}\right.$$

$$+\left(v_p - \frac{k'_p\,(\mathbf{k}'\cdot\mathbf{v})}{k'^2}\right)\frac{1}{\Lambda^*_\perp(\mathbf{k}',\sigma\omega^\alpha_\mathbf{k}-\sigma''\omega^\gamma_{\mathbf{k}-\mathbf{k}'})}\Bigg]$$

$$\times\,\frac{k_i\,(\mathbf{k}-\mathbf{k}')_k}{k\,|\mathbf{k}-\mathbf{k}'|}\,\chi^{(2)}_{ijk}(\mathbf{k}',\sigma\omega^\alpha_\mathbf{k}-\sigma''\omega^\gamma_{\mathbf{k}-\mathbf{k}'}|\mathbf{k}-\mathbf{k}',\sigma''\omega^\gamma_{\mathbf{k}-\mathbf{k}'})$$

$$\times\,\frac{k_n\,(\mathbf{k}-\mathbf{k}')_m}{k\,|\mathbf{k}-\mathbf{k}'|}\,\chi^{(2)}_{npm}(\mathbf{k}',\sigma\omega^\alpha_\mathbf{k}-\sigma''\omega^\gamma_{\mathbf{k}-\mathbf{k}'}|\mathbf{k}-\mathbf{k}',\sigma''\omega^\gamma_{\mathbf{k}-\mathbf{k}'})$$

$$\times\Bigg[(-i\pi)\,\frac{1}{\epsilon'_\parallel(\mathbf{k}-\mathbf{k}',\sigma''\omega^\gamma_{\mathbf{k}-\mathbf{k}'})}\,I^{\sigma\alpha}_\mathbf{k}$$

$$\times\,\delta(\sigma\omega^\alpha_\mathbf{k}-\sigma''\omega^\gamma_{\mathbf{k}-\mathbf{k}'}-\mathbf{k}'\cdot\mathbf{v})\,f_a(\mathbf{p})\Bigg]\,\underbrace{\delta(\omega-\sigma\omega^\alpha_\mathbf{k})}$$

$$\boxed{22}\quad -2i\sum_{\sigma,\sigma''}\sum_\gamma \mathrm{Im}\int d\mathbf{k}'\,\sum_a\frac{(4\pi e_a)^2}{(2\pi)^3}\int d\mathbf{p}\,\frac{1}{(\mathbf{k}'\cdot\mathbf{v})^2}$$

$$\times\Bigg[\frac{k'_j\,(\mathbf{k}'\cdot\mathbf{v})}{k'^2}\,\frac{1}{\epsilon_\parallel(\mathbf{k}',\sigma\omega^T_\mathbf{k}-\sigma''\omega^\gamma_{\mathbf{k}-\mathbf{k}'})}$$

$$+\left(v_j - \frac{k'_j\,(\mathbf{k}'\cdot\mathbf{v})}{k'^2}\right)\frac{1}{\Lambda_\perp(\mathbf{k}',\sigma\omega^T_\mathbf{k}-\sigma''\omega^\gamma_{\mathbf{k}-\mathbf{k}'})}\Bigg]$$

$$\times\Bigg[\frac{k'_p\,(\mathbf{k}'\cdot\mathbf{v})}{k'^2}\,\frac{1}{\epsilon^*_\parallel(\mathbf{k}',\sigma\omega^T_\mathbf{k}-\sigma''\omega^\gamma_{\mathbf{k}-\mathbf{k}'})}$$

$$+\left(v_p - \frac{k'_p\,(\mathbf{k}'\cdot\mathbf{v})}{k'^2}\right)\frac{1}{\Lambda^*_\perp(\mathbf{k}',\sigma\omega^T_\mathbf{k}-\sigma''\omega^\gamma_{\mathbf{k}-\mathbf{k}'})}\Bigg]$$

$$\times\,\frac{(\mathbf{k}-\mathbf{k}')_k}{|\mathbf{k}-\mathbf{k}'|}\left(\delta_{ni}-\frac{k_n\,k_i}{k^2}\right)\chi^{(2)}_{ijk}(\mathbf{k}',\sigma\omega^T_\mathbf{k}-\sigma''\omega^\gamma_{\mathbf{k}-\mathbf{k}'}|\mathbf{k}-\mathbf{k}',\sigma''\omega^\gamma_{\mathbf{k}-\mathbf{k}'})$$

$$\times\,\frac{(\mathbf{k}-\mathbf{k}')_m}{|\mathbf{k}-\mathbf{k}'|}\,\chi^{(2)}_{npm}(\mathbf{k}',\sigma\omega^T_\mathbf{k}-\sigma''\omega^\gamma_{\mathbf{k}-\mathbf{k}'}|\mathbf{k}-\mathbf{k}',\sigma''\omega^\gamma_{\mathbf{k}-\mathbf{k}'})$$

$$\times\Bigg[(-i\pi)\,\frac{1}{\epsilon'_\parallel(\mathbf{k}-\mathbf{k}',\sigma''\omega^\gamma_{\mathbf{k}-\mathbf{k}'})}\,\frac{1}{2}\,I^{\sigma T}_\mathbf{k}$$

$$\times\,\delta(\sigma\omega^T_\mathbf{k}-\sigma''\omega^\gamma_{\mathbf{k}-\mathbf{k}'}-\mathbf{k}'\cdot\mathbf{v})\,f_a(\mathbf{p})\Bigg]\,\underbrace{\delta(\omega-\sigma\omega^T_\mathbf{k})}$$

$$\boxed{23}\quad -2i\sum_{\sigma,\sigma''}\sum_\alpha \mathrm{Im}\int d\mathbf{k}'\,\sum_a\frac{(4\pi e_a)^2}{(2\pi)^3}\int d\mathbf{p}\,\frac{1}{(\mathbf{k}'\cdot\mathbf{v})^2}$$

$$\times\Bigg[\frac{k'_j\,(\mathbf{k}'\cdot\mathbf{v})}{k'^2}\,\frac{1}{\epsilon_\parallel(\mathbf{k}',\sigma\omega^\alpha_\mathbf{k}-\sigma''\omega_{\mathbf{k}-\mathbf{k}'})}$$

$$+\left(v_j - \frac{k'_j\,(\mathbf{k}'\cdot\mathbf{v})}{k'^2}\right)\frac{1}{\Lambda_\perp(\mathbf{k}',\sigma\omega^\alpha_\mathbf{k}-\sigma''\omega_{\mathbf{k}-\mathbf{k}'})}\Bigg]$$

$$\times \left[\frac{k'_p\,(\mathbf{k}'\cdot\mathbf{v})}{k'^2} \frac{1}{\epsilon^*_\parallel(\mathbf{k}',\sigma\omega^\alpha_\mathbf{k} - \sigma''\omega_{\mathbf{k}-\mathbf{k}'})} + \left(v_p - \frac{k'_p\,(\mathbf{k}'\cdot\mathbf{v})}{k'^2} \right) \frac{1}{\Lambda^*_\perp(\mathbf{k}',\sigma\omega^\alpha_\mathbf{k} - \sigma''\omega_{\mathbf{k}-\mathbf{k}'})} \right]$$

$$\times \frac{k_i}{k} \left(\delta_{km} - \frac{(\mathbf{k}-\mathbf{k}')_k\,(\mathbf{k}-\mathbf{k}')_m}{(\mathbf{k}-\mathbf{k}')^2} \right) \chi^{(2)}_{ijk}(\mathbf{k}',\sigma\omega^\alpha_\mathbf{k} - \sigma''\omega_{\mathbf{k}-\mathbf{k}'}|\mathbf{k}-\mathbf{k}',\sigma''\omega^T_{\mathbf{k}-\mathbf{k}'})$$

$$\times \frac{k_n}{k}\, \chi^{(2)}_{npm}(\mathbf{k}',\sigma\omega^\alpha_\mathbf{k} - \sigma''\omega_{\mathbf{k}-\mathbf{k}'}|\mathbf{k}-\mathbf{k}',\sigma''\omega^T_{\mathbf{k}-\mathbf{k}'})$$

$$\times \left[(-i\pi)\, \frac{1}{\Lambda'_\perp(\mathbf{k}-\mathbf{k}',\sigma''\omega^T_{\mathbf{k}-\mathbf{k}'})}\, I^{\sigma\alpha}_\mathbf{k} \times \delta(\sigma\omega^\alpha_\mathbf{k} - \sigma''\omega^T_{\mathbf{k}-\mathbf{k}'} - \mathbf{k}'\cdot\mathbf{v})\; f_a(\mathbf{p}) \right] \underbrace{\delta(\omega - \sigma\omega^\alpha_\mathbf{k})}$$

24 $$-2i \sum_{\sigma,\sigma''} \mathrm{Im} \int d\mathbf{k}' \sum_a \frac{(4\pi e_a)^2}{(2\pi)^3} \int d\mathbf{p}\, \frac{1}{(\mathbf{k}'\cdot\mathbf{v})^2} \left[\frac{k'_j\,(\mathbf{k}'\cdot\mathbf{v})}{k'^2} \times \frac{1}{\epsilon_\parallel(\mathbf{k}',\sigma\omega^T_\mathbf{k} - \sigma''\omega^T_{\mathbf{k}-\mathbf{k}'})} + \left(v_j - \frac{k'_j\,(\mathbf{k}'\cdot\mathbf{v})}{k'^2} \right) \frac{1}{\Lambda_\perp(\mathbf{k}',\sigma\omega^T_\mathbf{k} - \sigma''\omega^T_{\mathbf{k}-\mathbf{k}'})} \right]$$

$$\times \left[\frac{k'_p\,(\mathbf{k}'\cdot\mathbf{v})}{k'^2} \frac{1}{\epsilon^*_\parallel(\mathbf{k}',\sigma\omega^T_\mathbf{k} - \sigma''\omega^T_{\mathbf{k}-\mathbf{k}'})} + \left(v_p - \frac{k'_p\,(\mathbf{k}'\cdot\mathbf{v})}{k'^2} \right) \frac{1}{\Lambda^*_\perp(\mathbf{k}',\sigma\omega^T_\mathbf{k} - \sigma''\omega^T_{\mathbf{k}-\mathbf{k}'})} \right]$$

$$\times \left(\delta_{ni} - \frac{k_n\,k_i}{k^2} \right) \chi^{(2)}_{ijk}(\mathbf{k}',\sigma\omega^T_\mathbf{k} - \sigma''\omega^T_{\mathbf{k}-\mathbf{k}'}|\mathbf{k}-\mathbf{k}',\sigma''\omega^T_{\mathbf{k}-\mathbf{k}'})$$

$$\times \left(\delta_{km} - \frac{(\mathbf{k}-\mathbf{k}')_k\,(\mathbf{k}-\mathbf{k}')_m}{(\mathbf{k}-\mathbf{k}')^2} \right)$$

$$\times \chi^{(2)}_{npm}(\mathbf{k}',\sigma\omega^T_\mathbf{k} - \sigma''\omega^T_{\mathbf{k}-\mathbf{k}'}|\mathbf{k}-\mathbf{k}',\sigma''\omega^T_{\mathbf{k}-\mathbf{k}'})$$

$$\times \left[(-i\pi)\, \frac{1}{2} \frac{1}{\Lambda'_\perp(\mathbf{k}-\mathbf{k}',\sigma''\omega^T_{\mathbf{k}-\mathbf{k}'})}\, I^{\sigma T}_\mathbf{k} \times \delta(\sigma\omega^T_\mathbf{k} - \sigma''\omega^T_{\mathbf{k}-\mathbf{k}'} - \mathbf{k}'\cdot\mathbf{v})\; f_a(\mathbf{p}) \right] \underbrace{\delta(\omega - \sigma\omega^T_\mathbf{k})}$$

25 $$+2i \sum_{\sigma,\sigma'} \sum_{\alpha,\beta} \mathrm{Im} \int d\mathbf{k}' \sum_a \frac{(4\pi e_a)^2}{(2\pi)^3} \int d\mathbf{p}\, \frac{1}{[(\mathbf{k}-\mathbf{k}')\cdot\mathbf{v}]^2}$$

$$\times \left[\frac{(\mathbf{k}-\mathbf{k}')_k\,[(\mathbf{k}-\mathbf{k}')\cdot\mathbf{v}]}{(\mathbf{k}-\mathbf{k}')^2} \frac{1}{\epsilon_\parallel(\mathbf{k}-\mathbf{k}',\sigma\omega^\alpha_\mathbf{k} - \sigma'\omega^\beta_{\mathbf{k}'})} \right.$$

$$
\begin{aligned}
&+\left(v_k-\frac{(\mathbf{k}-\mathbf{k}')_k\,[(\mathbf{k}-\mathbf{k}')\cdot\mathbf{v}]}{(\mathbf{k}-\mathbf{k}')^2}\right)\frac{1}{\Lambda_\perp(\mathbf{k}-\mathbf{k}',\sigma\omega_\mathbf{k}^\alpha-\sigma'\omega_{\mathbf{k}'}^\beta)}\Bigg]\\
&\times\left[\frac{(\mathbf{k}-\mathbf{k}')_p\,[(\mathbf{k}-\mathbf{k}')\cdot\mathbf{v}]}{(\mathbf{k}-\mathbf{k}')^2}\frac{1}{\epsilon_\parallel^*(\mathbf{k}-\mathbf{k}',\sigma\omega_\mathbf{k}^\alpha-\sigma'\omega_{\mathbf{k}'}^\beta)}\right.\\
&\left.+\left(v_p-\frac{(\mathbf{k}-\mathbf{k}')_p\,[(\mathbf{k}-\mathbf{k}')\cdot\mathbf{v}]}{(\mathbf{k}-\mathbf{k}')^2}\right)\frac{1}{\Lambda_\perp^*(\mathbf{k}-\mathbf{k}',\sigma\omega_\mathbf{k}^\alpha-\sigma'\omega_{\mathbf{k}'}^\beta)}\right]\\
&\times\frac{k_i\,k_j'}{k\,k'}\,\chi_{ijk}^{(2)}(\mathbf{k}',\sigma'\omega_{\mathbf{k}'}^\beta|\mathbf{k}-\mathbf{k}',\sigma\omega_\mathbf{k}^\alpha-\sigma'\omega_{\mathbf{k}'}^\beta)\\
&\times\frac{k_m\,k_n'}{k\,k'}\,\chi_{mnp}^{(2)*}(\mathbf{k}',\sigma'\omega_{\mathbf{k}'}^\beta|\mathbf{k}-\mathbf{k}',\sigma\omega_\mathbf{k}^\alpha-\sigma'\omega_{\mathbf{k}'}^\beta)\\
&\times\left[(i\pi)I_{\mathbf{k}'}^{\sigma'\beta}\,\frac{1}{\epsilon_\parallel'(\mathbf{k},\sigma\omega_\mathbf{k}^\alpha)}\,\delta[\sigma\omega_\mathbf{k}^\alpha-\sigma'\omega_{\mathbf{k}'}^\beta-(\mathbf{k}-\mathbf{k}')\cdot\mathbf{v}]\;f_a(\mathbf{p})\right]\underbrace{\delta(\omega-\sigma\omega_\mathbf{k}^\alpha)}
\end{aligned}
$$

26

$$
\begin{aligned}
&+2i\sum_{\sigma,\sigma'}\sum_\beta\mathrm{Im}\int d\mathbf{k}'\,\sum_a\frac{(4\pi e_a)^2}{(2\pi)^3}\int d\mathbf{p}\,\frac{1}{[(\mathbf{k}-\mathbf{k}')\cdot\mathbf{v}]^2}\\
&\times\left[\frac{(\mathbf{k}-\mathbf{k}')_k\,[(\mathbf{k}-\mathbf{k}')\cdot\mathbf{v}]}{(\mathbf{k}-\mathbf{k}')^2}\frac{1}{\epsilon_\parallel(\mathbf{k}-\mathbf{k}',\sigma\omega_\mathbf{k}^T-\sigma'\omega_{\mathbf{k}'}^\beta)}\right.\\
&\left.+\left(v_k-\frac{(\mathbf{k}-\mathbf{k}')_k\,[(\mathbf{k}-\mathbf{k}')\cdot\mathbf{v}]}{(\mathbf{k}-\mathbf{k}')^2}\right)\frac{1}{\Lambda_\perp(\mathbf{k}-\mathbf{k}',\sigma\omega_\mathbf{k}^T-\sigma'\omega_{\mathbf{k}'}^\beta)}\right]\\
&\times\left[\frac{(\mathbf{k}-\mathbf{k}')_p\,[(\mathbf{k}-\mathbf{k}')\cdot\mathbf{v}]}{(\mathbf{k}-\mathbf{k}')^2}\frac{1}{\epsilon_\parallel^*(\mathbf{k}-\mathbf{k}',\sigma\omega_\mathbf{k}^T-\sigma'\omega_{\mathbf{k}'}^\beta)}\right.\\
&\left.+\left(v_p-\frac{(\mathbf{k}-\mathbf{k}')_p\,[(\mathbf{k}-\mathbf{k}')\cdot\mathbf{v}]}{(\mathbf{k}-\mathbf{k}')^2}\right)\frac{1}{\Lambda_\perp^*(\mathbf{k}-\mathbf{k}',\sigma\omega_\mathbf{k}^T-\sigma'\omega_{\mathbf{k}'}^\beta)}\right]\\
&\times\frac{k_j'}{k'}\left(\delta_{im}-\frac{k_i\,k_m}{k^2}\right)\chi_{ijk}^{(2)}(\mathbf{k}',\sigma'\omega_{\mathbf{k}'}^\beta|\mathbf{k}-\mathbf{k}',\sigma\omega_\mathbf{k}^T-\sigma'\omega_{\mathbf{k}'}^\beta)\\
&\times\frac{k_n'}{k'}\,\chi_{mnp}^{(2)*}(\mathbf{k}',\sigma'\omega_{\mathbf{k}'}^\beta|\mathbf{k}-\mathbf{k}',\sigma\omega_\mathbf{k}^T-\sigma'\omega_{\mathbf{k}'}^\beta)\\
&\times\left[(i\pi)I_{\mathbf{k}'}^{\sigma'\beta}\,\frac{1}{\Lambda_\perp'(\mathbf{k},\sigma\omega_\mathbf{k}^T)}\delta[\sigma\omega_\mathbf{k}^T-\sigma'\omega_{\mathbf{k}'}^\beta-(\mathbf{k}-\mathbf{k}')\cdot\mathbf{v}]\;f_a(\mathbf{p})\right]\underbrace{\delta(\omega-\sigma\omega_\mathbf{k}^T)}
\end{aligned}
$$

27

$$
\begin{aligned}
&+2i\sum_{\sigma,\sigma'}\sum_\alpha\mathrm{Im}\int d\mathbf{k}'\,\sum_a\frac{(4\pi e_a)^2}{(2\pi)^3}\int d\mathbf{p}\,\frac{1}{[(\mathbf{k}-\mathbf{k}')\cdot\mathbf{v}]^2}\\
&\times\left[\frac{(\mathbf{k}-\mathbf{k}')_k\,[(\mathbf{k}-\mathbf{k}')\cdot\mathbf{v}]}{(\mathbf{k}-\mathbf{k}')^2}\frac{1}{\epsilon_\parallel(\mathbf{k}-\mathbf{k}',\sigma\omega_\mathbf{k}^\alpha-\sigma'\omega_{\mathbf{k}'}^T)}\right.\\
&\left.+\left(v_k-\frac{(\mathbf{k}-\mathbf{k}')_k\,[(\mathbf{k}-\mathbf{k}')\cdot\mathbf{v}]}{(\mathbf{k}-\mathbf{k}')^2}\right)\frac{1}{\Lambda_\perp(\mathbf{k}-\mathbf{k}',\sigma\omega_\mathbf{k}^\alpha-\sigma'\omega_{\mathbf{k}'}^T)}\right]
\end{aligned}
$$

$$\times\left[\frac{(\mathbf{k}-\mathbf{k}')_p\,[(\mathbf{k}-\mathbf{k}')\cdot\mathbf{v}]}{(\mathbf{k}-\mathbf{k}')^2}\frac{1}{\epsilon_\parallel^*(\mathbf{k}-\mathbf{k}',\sigma\omega_\mathbf{k}^\alpha-\sigma'\omega_{\mathbf{k}'}^T)}\right.$$
$$\left.+\left(v_p-\frac{(\mathbf{k}-\mathbf{k}')_p\,[(\mathbf{k}-\mathbf{k}')\cdot\mathbf{v}]}{(\mathbf{k}-\mathbf{k}')^2}\right)\frac{1}{\Lambda_\perp^*(\mathbf{k}-\mathbf{k}',\sigma\omega_\mathbf{k}^\alpha-\sigma'\omega_{\mathbf{k}'}^T)}\right]$$
$$\times\frac{k_i}{k}\left(\delta_{nj}-\frac{k_n'\,k_j'}{k'^2}\right)\chi_{ijk}^{(2)}(\mathbf{k}',\sigma'\omega_{\mathbf{k}'}^T|\mathbf{k}-\mathbf{k}',\sigma\omega_\mathbf{k}^\alpha-\sigma'\omega_{\mathbf{k}'}^T)$$
$$\times\frac{k_m}{k}\,\chi_{mnp}^{(2)*}(\mathbf{k}',\sigma'\omega_{\mathbf{k}'}^T|\mathbf{k}-\mathbf{k}',\sigma\omega_\mathbf{k}^\alpha-\sigma'\omega_{\mathbf{k}'}^T)$$
$$\times\left[(i\pi)\frac{1}{2}\,I_{\mathbf{k}'}^{\sigma'T}\,\frac{1}{\epsilon_\parallel'(\mathbf{k},\sigma\omega_\mathbf{k}^\alpha)}\delta[\sigma\omega_\mathbf{k}^\alpha-\sigma'\omega_{\mathbf{k}'}^T-(\mathbf{k}-\mathbf{k}')\cdot\mathbf{v}]\;f_a(\mathbf{p})\right]\underbrace{\delta(\omega-\sigma\omega_\mathbf{k}^\alpha)}$$

28
$$+2i\sum_{\sigma,\sigma'}\mathrm{Im}\int d\mathbf{k}'\sum_a\frac{(4\pi e_a)^2}{(2\pi)^3}\int d\mathbf{p}\,\frac{1}{[(\mathbf{k}-\mathbf{k}')\cdot\mathbf{v}]^2}$$
$$\times\left[\frac{(\mathbf{k}-\mathbf{k}')_k\,[(\mathbf{k}-\mathbf{k}')\cdot\mathbf{v}]}{(\mathbf{k}-\mathbf{k}')^2}\frac{1}{\epsilon_\parallel(\mathbf{k}-\mathbf{k}',\sigma\omega_\mathbf{k}^T-\sigma'\omega_{\mathbf{k}'}^T)}\right.$$
$$\left.+\left(v_k-\frac{(\mathbf{k}-\mathbf{k}')_k\,[(\mathbf{k}-\mathbf{k}')\cdot\mathbf{v}]}{(\mathbf{k}-\mathbf{k}')^2}\right)\frac{1}{\Lambda_\perp(\mathbf{k}-\mathbf{k}',\sigma\omega_\mathbf{k}^T-\sigma'\omega_{\mathbf{k}'}^T)}\right]$$
$$\times\left[\frac{(\mathbf{k}-\mathbf{k}')_p\,[(\mathbf{k}-\mathbf{k}')\cdot\mathbf{v}]}{(\mathbf{k}-\mathbf{k}')^2}\frac{1}{\epsilon_\parallel^*(\mathbf{k}-\mathbf{k}',\sigma\omega_\mathbf{k}^T-\sigma'\omega_{\mathbf{k}'}^T)}\right.$$
$$\left.+\left(v_p-\frac{(\mathbf{k}-\mathbf{k}')_p\,[(\mathbf{k}-\mathbf{k}')\cdot\mathbf{v}]}{(\mathbf{k}-\mathbf{k}')^2}\right)\frac{1}{\Lambda_\perp^*(\mathbf{k}-\mathbf{k}',\sigma\omega_\mathbf{k}^T-\sigma'\omega_{\mathbf{k}'}^T)}\right]$$
$$\times\left(\delta_{im}-\frac{k_i\,k_m}{k^2}\right)\chi_{ijk}^{(2)}(\mathbf{k}',\sigma'\omega_{\mathbf{k}'}^T|\mathbf{k}-\mathbf{k}',\sigma\omega_\mathbf{k}^T-\sigma'\omega_{\mathbf{k}'}^T)$$
$$\times\left(\delta_{nj}-\frac{k_n'\,k_j'}{k'^2}\right)\chi_{mnp}^{(2)*}(\mathbf{k}',\sigma'\omega_{\mathbf{k}'}^T|\mathbf{k}-\mathbf{k}',\sigma\omega_\mathbf{k}^T-\sigma'\omega_{\mathbf{k}'}^T)$$
$$\times\left[(i\pi)I_{\mathbf{k}'}^{\sigma'T}\,\frac{1}{\Lambda_\perp'(\mathbf{k},\sigma\omega_\mathbf{k}^T)}\delta[\sigma\omega_\mathbf{k}^T-\sigma'\omega_{\mathbf{k}'}^T-(\mathbf{k}-\mathbf{k}')\cdot\mathbf{v}]\;f_a(\mathbf{p})\right]\underbrace{\delta(\omega-\sigma\omega_\mathbf{k}^T)}$$

29
$$-2i\sum_{\sigma,\sigma''}\sum_{\alpha,\beta}\mathrm{Im}\int d\mathbf{k}'\sum_a\frac{(4\pi e_a)^2}{(2\pi)^3}\int d\mathbf{p}\,\frac{1}{[(\mathbf{k}-\mathbf{k}')\cdot\mathbf{v}]^2}$$
$$\times\left[\frac{(\mathbf{k}-\mathbf{k}')_k\,[(\mathbf{k}-\mathbf{k}')\cdot\mathbf{v}]}{(\mathbf{k}-\mathbf{k}')^2}\frac{1}{\epsilon_\parallel(\mathbf{k}-\mathbf{k}',\sigma\omega_\mathbf{k}^\alpha-\sigma'\omega_{\mathbf{k}'}^\beta)}\right.$$
$$\left.+\left(v_k-\frac{(\mathbf{k}-\mathbf{k}')_k\,[(\mathbf{k}-\mathbf{k}')\cdot\mathbf{v}]}{(\mathbf{k}-\mathbf{k}')^2}\right)\frac{1}{\Lambda_\perp(\mathbf{k}-\mathbf{k}',\sigma\omega_\mathbf{k}^\alpha-\sigma'\omega_{\mathbf{k}'}^\beta)}\right]$$
$$\times\left[\frac{(\mathbf{k}-\mathbf{k}')_p\,[(\mathbf{k}-\mathbf{k}')\cdot\mathbf{v}]}{(\mathbf{k}-\mathbf{k}')^2}\frac{1}{\epsilon_\parallel^*(\mathbf{k}-\mathbf{k}',\sigma\omega_\mathbf{k}^\alpha-\sigma'\omega_{\mathbf{k}'}^\beta)}\right.$$

$$+\left(v_p - \frac{(\mathbf{k}-\mathbf{k}')_p\,[(\mathbf{k}-\mathbf{k}')\cdot\mathbf{v}]}{(\mathbf{k}-\mathbf{k}')^2}\right)\frac{1}{\Lambda_\perp^*(\mathbf{k}-\mathbf{k}',\sigma\omega_\mathbf{k}^\alpha-\sigma'\omega_{\mathbf{k}'}^\beta)}\Bigg]$$

$$\times \frac{k_i\,k_j'}{k\,k'}\,\chi_{ijk}^{(2)}(\mathbf{k}',\sigma'\omega_{\mathbf{k}'}^\beta|\mathbf{k}-\mathbf{k}',\sigma\omega_\mathbf{k}^\alpha-\sigma'\omega_{\mathbf{k}'}^\beta)$$

$$\times \frac{k_n\,k_m'}{k\,k'}\,\chi_{nmp}^{(2)}(\mathbf{k}',\sigma'\omega_{\mathbf{k}'}^\beta|\mathbf{k}-\mathbf{k}',\sigma\omega_\mathbf{k}^\alpha-\sigma'\omega_{\mathbf{k}'}^\beta)$$

$$\times\left[(-i\pi)\frac{1}{\epsilon_\parallel'(\mathbf{k}',\sigma'\omega_{\mathbf{k}'}^\beta)}\,I_\mathbf{k}^{\sigma\alpha}\delta[\sigma\omega_\mathbf{k}^\alpha-\sigma'\omega_{\mathbf{k}'}^\beta-(\mathbf{k}-\mathbf{k}')\cdot\mathbf{v}]\;f_a(\mathbf{p})\right]\underbrace{\delta(\omega-\sigma\omega_\mathbf{k}^\alpha)}$$

[30] $$-2i\sum_{\sigma,\sigma''}\sum_\alpha \mathrm{Im}\int d\mathbf{k}'\sum_a\frac{(4\pi e_a)^2}{(2\pi)^3}\int d\mathbf{p}\,\frac{1}{[(\mathbf{k}-\mathbf{k}')\cdot\mathbf{v}]^2}$$

$$\times\Bigg[\frac{(\mathbf{k}-\mathbf{k}')_k\,[(\mathbf{k}-\mathbf{k}')\cdot\mathbf{v}]}{(\mathbf{k}-\mathbf{k}')^2}\frac{1}{\epsilon_\parallel(\mathbf{k}-\mathbf{k}',\sigma\omega_\mathbf{k}^\alpha-\sigma'\omega_{\mathbf{k}'}^T)}$$

$$+\left(v_k - \frac{(\mathbf{k}-\mathbf{k}')_k\,[(\mathbf{k}-\mathbf{k}')\cdot\mathbf{v}]}{(\mathbf{k}-\mathbf{k}')^2}\right)\frac{1}{\Lambda_\perp(\mathbf{k}-\mathbf{k}',\sigma\omega_\mathbf{k}^\alpha-\sigma'\omega_{\mathbf{k}'}^T)}\Bigg]$$

$$\times\Bigg[\frac{(\mathbf{k}-\mathbf{k}')_p\,[(\mathbf{k}-\mathbf{k}')\cdot\mathbf{v}]}{(\mathbf{k}-\mathbf{k}')^2}\frac{1}{\epsilon_\parallel^*(\mathbf{k}-\mathbf{k}',\sigma\omega_\mathbf{k}^\alpha-\sigma'\omega_{\mathbf{k}'}^T)}$$

$$+\left(v_p - \frac{(\mathbf{k}-\mathbf{k}')_p\,[(\mathbf{k}-\mathbf{k}')\cdot\mathbf{v}]}{(\mathbf{k}-\mathbf{k}')^2}\right)\frac{1}{\Lambda_\perp^*(\mathbf{k}-\mathbf{k}',\sigma\omega_\mathbf{k}^\alpha-\sigma'\omega_{\mathbf{k}'}^T)}\Bigg]$$

$$\times\frac{k_i}{k}\left(\delta_{jm}-\frac{k_j'\,k_m'}{k'^2}\right)\chi_{ijk}^{(2)}(\mathbf{k}',\sigma'\omega_{\mathbf{k}'}^T|\mathbf{k}-\mathbf{k}',\sigma\omega_\mathbf{k}^\alpha-\sigma'\omega_{\mathbf{k}'}^T)$$

$$\times\frac{k_n}{k}\,\chi_{nmp}^{(2)}(\mathbf{k}',\sigma'\omega_{\mathbf{k}'}^T|\mathbf{k}-\mathbf{k}',\sigma\omega_\mathbf{k}^\alpha-\sigma'\omega_{\mathbf{k}'}^T)\Bigg[(-i\pi)$$

$$\times\frac{1}{\Lambda_\perp'(\mathbf{k}',\sigma'\omega_{\mathbf{k}'}^T)}\,I_\mathbf{k}^{\sigma\alpha}\;\delta[\sigma\omega_\mathbf{k}^\alpha-\sigma'\omega_{\mathbf{k}'}^T-(\mathbf{k}-\mathbf{k}')\cdot\mathbf{v}]\;f_a(\mathbf{p})\Bigg]\underbrace{\delta(\omega-\sigma\omega_\mathbf{k}^\alpha)}$$

[31] $$-2i\sum_{\sigma,\sigma''}\sum_\beta \mathrm{Im}\int d\mathbf{k}'\sum_a\frac{(4\pi e_a)^2}{(2\pi)^3}\int d\mathbf{p}\,\frac{1}{[(\mathbf{k}-\mathbf{k}')\cdot\mathbf{v}]^2}$$

$$\times\Bigg[\frac{(\mathbf{k}-\mathbf{k}')_k\,[(\mathbf{k}-\mathbf{k}')\cdot\mathbf{v}]}{(\mathbf{k}-\mathbf{k}')^2}\frac{1}{\epsilon_\parallel(\mathbf{k}-\mathbf{k}',\sigma\omega_\mathbf{k}^T-\sigma'\omega_{\mathbf{k}'}^\beta)}$$

$$+\left(v_k - \frac{(\mathbf{k}-\mathbf{k}')_k\,[(\mathbf{k}-\mathbf{k}')\cdot\mathbf{v}]}{(\mathbf{k}-\mathbf{k}')^2}\right)\frac{1}{\Lambda_\perp(\mathbf{k}-\mathbf{k}',\sigma\omega_\mathbf{k}^T-\sigma'\omega_{\mathbf{k}'}^\beta)}\Bigg]$$

$$\times\Bigg[\frac{(\mathbf{k}-\mathbf{k}')_p\,[(\mathbf{k}-\mathbf{k}')\cdot\mathbf{v}]}{(\mathbf{k}-\mathbf{k}')^2}\frac{1}{\epsilon_\parallel^*(\mathbf{k}-\mathbf{k}',\sigma\omega_\mathbf{k}^T-\sigma'\omega_{\mathbf{k}'}^\beta)}$$

$$+\left(v_p - \frac{(\mathbf{k}-\mathbf{k}')_p\,[(\mathbf{k}-\mathbf{k}')\cdot\mathbf{v}]}{(\mathbf{k}-\mathbf{k}')^2}\right)\frac{1}{\Lambda_\perp^*(\mathbf{k}-\mathbf{k}',\sigma\omega_\mathbf{k}^T-\sigma'\omega_{\mathbf{k}'}^\beta)}\Bigg]$$

$$\times \frac{k'_j}{k'}\left(\delta_{ni} - \frac{k_n k_i}{k^2}\right) \chi^{(2)}_{ijk}(\mathbf{k}', \sigma'\omega^\beta_{\mathbf{k}'}|\mathbf{k}-\mathbf{k}', \sigma\omega^T_\mathbf{k} - \sigma'\omega^\beta_{\mathbf{k}'})$$

$$\times \frac{k'_m}{k'} \chi^{(2)}_{nmp}(\mathbf{k}', \sigma'\omega^\beta_{\mathbf{k}'}|\mathbf{k}-\mathbf{k}', \sigma\omega^T_\mathbf{k} - \sigma'\omega^\beta_{\mathbf{k}'}) \frac{1}{2}\Bigg[(-i\pi)$$

$$\times \frac{1}{\epsilon'_\parallel(\mathbf{k}', \sigma'\omega^\beta_{\mathbf{k}'})} I^{\sigma T}_\mathbf{k}\, \delta[\sigma\omega^T_\mathbf{k} - \sigma'\omega^\beta_{\mathbf{k}'} - (\mathbf{k}-\mathbf{k}')\cdot\mathbf{v}]\, f_a(\mathbf{p})\Bigg] \underbrace{\delta(\omega - \sigma\omega^T_\mathbf{k})}$$

$$\boxed{32}\ -2i \sum_{\sigma,\sigma''} \mathrm{Im} \int d\mathbf{k}' \sum_a \frac{(4\pi e_a)^2}{(2\pi)^3} \int d\mathbf{p}\, \frac{1}{[(\mathbf{k}-\mathbf{k}')\cdot\mathbf{v}]^2}$$

$$\times \Bigg[\frac{(\mathbf{k}-\mathbf{k}')_k\,[(\mathbf{k}-\mathbf{k}')\cdot\mathbf{v}]}{(\mathbf{k}-\mathbf{k}')^2} \frac{1}{\epsilon_\parallel(\mathbf{k}-\mathbf{k}', \sigma\omega^T_\mathbf{k} - \sigma'\omega^T_{\mathbf{k}'})}$$

$$+\left(v_k - \frac{(\mathbf{k}-\mathbf{k}')_k\,[(\mathbf{k}-\mathbf{k}')\cdot\mathbf{v}]}{(\mathbf{k}-\mathbf{k}')^2}\right) \frac{1}{\Lambda_\perp(\mathbf{k}-\mathbf{k}', \sigma\omega^T_\mathbf{k} - \sigma'\omega^T_{\mathbf{k}'})}\Bigg]$$

$$\times \Bigg[\frac{(\mathbf{k}-\mathbf{k}')_p\,[(\mathbf{k}-\mathbf{k}')\cdot\mathbf{v}]}{(\mathbf{k}-\mathbf{k}')^2} \frac{1}{\epsilon^*_\parallel(\mathbf{k}-\mathbf{k}', \sigma\omega^T_\mathbf{k} - \sigma'\omega^T_{\mathbf{k}'})}$$

$$+\left(v_p - \frac{(\mathbf{k}-\mathbf{k}')_p\,[(\mathbf{k}-\mathbf{k}')\cdot\mathbf{v}]}{(\mathbf{k}-\mathbf{k}')^2}\right) \frac{1}{\Lambda^*_\perp(\mathbf{k}-\mathbf{k}', \sigma\omega^T_\mathbf{k} - \sigma'\omega^T_{\mathbf{k}'})}\Bigg]$$

$$\times \left(\delta_{ni} - \frac{k_n k_i}{k^2}\right) \chi^{(2)}_{ijk}(\mathbf{k}', \sigma'\omega^T_{\mathbf{k}'}|\mathbf{k}-\mathbf{k}', \sigma\omega^T_\mathbf{k} - \sigma'\omega^T_{\mathbf{k}'})$$

$$\times \left(\delta_{jm} - \frac{k'_j k'_m}{k'^2}\right) \chi^{(2)}_{nmp}(\mathbf{k}', \sigma'\omega^T_{\mathbf{k}'}|\mathbf{k}-\mathbf{k}', \sigma\omega^T_\mathbf{k} - \sigma'\omega^T_{\mathbf{k}'}) \frac{1}{2}\Bigg[(-i\pi) \tag{9.38}$$

$$\times \frac{1}{\Lambda'_\perp(\mathbf{k}', \sigma'\omega^T_{\mathbf{k}'})} I^{\sigma T}_\mathbf{k}\, \delta[\sigma\omega^T_\mathbf{k} - \sigma'\omega^T_{\mathbf{k}'} - (\mathbf{k}-\mathbf{k}')\cdot\mathbf{v}]\, f_a(\mathbf{p})\Bigg] \underbrace{\delta(\omega - \sigma\omega^T_\mathbf{k})}.$$

In (9.38), the various terms identified by boxed numerical scheme contain individual terms identified by either $\delta(\omega - \sigma\omega^\alpha_\mathbf{k})$ or $\delta(\omega - \sigma\omega^T_\mathbf{k})$. We have indicated these by making use of under-braces, in order to highlight these delta functions. This means that one may collect all the terms containing either of the two delta functions and separate the formal equations into formal longitudinal versus transverse wave kinetic equations. We discuss these two equations separately next.

Collecting various terms containing the factor $\delta(\omega - \sigma\omega^\alpha_\mathbf{k})$ and removing the common factor $(i/2)\sum_\sigma \sum_\alpha$, we obtain the following general form of wave kinetic equation, arranged according to the types of interaction. In deriving the final form, we have ignored nonlinear scattering terms associated with the factor of the type $1/\Lambda_\perp$. The result is

$$
\begin{aligned}
\frac{\partial I_{\mathbf{k}}^{\sigma\alpha}}{\partial t} &= -\frac{2\,\mathrm{Im}\,\epsilon_\parallel(\mathbf{k},\sigma\omega_{\mathbf{k}}^\alpha)}{\epsilon_\parallel'(\mathbf{k},\sigma\omega_{\mathbf{k}}^\alpha)}\, I_{\mathbf{k}}^{\sigma\alpha} \qquad \boxed{\text{induced emission}} \\
&+ \sum_a \frac{4e_a^2}{k^2|\epsilon_\parallel'(\mathbf{k},\sigma\omega_{\mathbf{k}}^\alpha)|^2}\int d\mathbf{p}\;\delta(\sigma\omega_{\mathbf{k}}^\alpha - \mathbf{k}\cdot\mathbf{v})\, f_a(\mathbf{p}) \qquad \boxed{\text{spontaneous emission}} \\
&- \frac{4\pi}{\epsilon_\parallel'(\mathbf{k},\sigma\omega_{\mathbf{k}}^\alpha)}\sum_{\sigma',\sigma''}\sum_{\beta,\gamma}\int d\mathbf{k}'\;\frac{k_i\,k_j'\,(\mathbf{k}-\mathbf{k}')_k}{k\,k'\,|\mathbf{k}-\mathbf{k}'|}\,\chi_{ijk}^{(2)}(\mathbf{k}',\sigma'\omega_{\mathbf{k}'}^\beta|\mathbf{k}-\mathbf{k}',\sigma''\omega_{\mathbf{k}-\mathbf{k}'}^\gamma) \\
&\times \frac{k_n\,k_l'\,(\mathbf{k}-\mathbf{k}')_m}{k\,k'\,|\mathbf{k}-\mathbf{k}'|}\,\chi_{nlm}^{(2)*}(\mathbf{k}',\sigma'\omega_{\mathbf{k}'}^\beta|\mathbf{k}-\mathbf{k}',\sigma''\omega_{\mathbf{k}-\mathbf{k}'}^\gamma) \\
&\times \left[\frac{I_{\mathbf{k}-\mathbf{k}'}^{\sigma''\gamma}\,I_{\mathbf{k}}^{\sigma\alpha}}{\epsilon_\parallel'(\mathbf{k}',\sigma'\omega_{\mathbf{k}'}^\beta)} + \frac{I_{\mathbf{k}'}^{\sigma'\beta}\,I_{\mathbf{k}}^{\sigma\alpha}}{\epsilon_\parallel'(\mathbf{k}-\mathbf{k}',\sigma''\omega_{\mathbf{k}-\mathbf{k}'}^\gamma)}\right. \\
&\left. - \frac{I_{\mathbf{k}'}^{\sigma'\beta}\,I_{\mathbf{k}-\mathbf{k}'}^{\sigma''\gamma}}{\epsilon_\parallel'(\mathbf{k},\sigma\omega_{\mathbf{k}}^\alpha)}\right]\delta(\sigma\omega_{\mathbf{k}}^\alpha - \sigma'\omega_{\mathbf{k}'}^\beta - \sigma''\omega_{\mathbf{k}-\mathbf{k}'}^\gamma) \qquad \boxed{\text{decay }(\alpha,\beta,\gamma)} \\
&- \frac{8\pi}{\epsilon_\parallel'(\mathbf{k},\sigma\omega_{\mathbf{k}}^\alpha)}\sum_{\sigma',\sigma''}\sum_{\beta}\int d\mathbf{k}'\;\frac{k_i\,k_j'}{k\,k'}\,\chi_{ijk}^{(2)}(\mathbf{k}',\sigma'\omega_{\mathbf{k}'}^\beta|\mathbf{k}-\mathbf{k}',\sigma''\omega_{\mathbf{k}-\mathbf{k}'}^T) \\
&\times \frac{k_n\,k_l'}{k\,k'}\left(\delta_{km} - \frac{(\mathbf{k}-\mathbf{k}')_k\,(\mathbf{k}-\mathbf{k}')_m}{(\mathbf{k}-\mathbf{k}')^2}\right)\chi_{nlm}^{(2)*}(\mathbf{k}',\sigma'\omega_{\mathbf{k}'}^\beta|\mathbf{k}-\mathbf{k}',\sigma''\omega_{\mathbf{k}-\mathbf{k}'}^T) \\
&\times \left[\frac{1}{2}\frac{I_{\mathbf{k}-\mathbf{k}'}^{\sigma''T}\,I_{\mathbf{k}}^{\sigma\alpha}}{\epsilon_\parallel'(\mathbf{k}',\sigma'\omega_{\mathbf{k}'}^\beta)} + \frac{I_{\mathbf{k}'}^{\sigma'\beta}\,I_{\mathbf{k}}^{\sigma\alpha}}{\Lambda_\perp'(\mathbf{k}-\mathbf{k}',\sigma''\omega_{\mathbf{k}-\mathbf{k}'}^T)}\right. \\
&\left. - \frac{1}{2}\frac{I_{\mathbf{k}-\mathbf{k}'}^{\sigma''T}\,I_{\mathbf{k}'}^{\sigma'\beta}}{\epsilon_\parallel'(\mathbf{k},\sigma\omega_{\mathbf{k}}^\alpha)}\right]\delta(\sigma\omega_{\mathbf{k}}^\alpha - \sigma'\omega_{\mathbf{k}'}^\beta - \sigma''\omega_{\mathbf{k}-\mathbf{k}'}^T) \qquad \boxed{\text{decay }(\alpha,\beta,T)} \\
&- \frac{4\pi}{\epsilon_\parallel'(\mathbf{k},\sigma\omega_{\mathbf{k}}^\alpha)}\sum_{\sigma',\sigma''}\int d\mathbf{k}'\;\frac{k_i}{k}\left(\delta_{jl} - \frac{k_j'\,k_l'}{k'^2}\right)\chi_{ijk}^{(2)}(\mathbf{k}',\sigma'\omega_{\mathbf{k}'}^T|\mathbf{k}-\mathbf{k}',\sigma''\omega_{\mathbf{k}-\mathbf{k}'}^T) \\
&\times \frac{k_n}{k}\left(\delta_{km} - \frac{(\mathbf{k}-\mathbf{k}')_k\,(\mathbf{k}-\mathbf{k}')_m}{(\mathbf{k}-\mathbf{k}')^2}\right)\chi_{nlm}^{(2)*}(\mathbf{k}',\sigma'\omega_{\mathbf{k}'}^T|\mathbf{k}-\mathbf{k}',\sigma''\omega_{\mathbf{k}-\mathbf{k}'}^T) \\
&\times \left[\frac{1}{2}\frac{I_{\mathbf{k}-\mathbf{k}'}^{\sigma''T}\,I_{\mathbf{k}}^{\sigma\alpha}}{\Lambda_\perp'(\mathbf{k}',\sigma'\omega_{\mathbf{k}'}^T)} + \frac{1}{2}\frac{I_{\mathbf{k}'}^{\sigma'T}\,I_{\mathbf{k}}^{\sigma\alpha}}{\Lambda_\perp'(\mathbf{k}-\mathbf{k}',\sigma''\omega_{\mathbf{k}-\mathbf{k}'}^T)}\right. \\
&\left. - \frac{1}{4}\frac{I_{\mathbf{k}'}^{\sigma'T}\,I_{\mathbf{k}-\mathbf{k}'}^{\sigma''T}}{\epsilon_\parallel'(\mathbf{k},\sigma\omega_{\mathbf{k}}^\alpha)}\right]\delta(\sigma\omega_{\mathbf{k}}^\alpha - \sigma'\omega_{\mathbf{k}'}^T - \sigma''\omega_{\mathbf{k}-\mathbf{k}'}^T) \qquad \boxed{\text{decay }(\alpha,T,T)} \\
&- \frac{4}{\epsilon_\parallel'(\mathbf{k},\sigma\omega_{\mathbf{k}}^\alpha)}\sum_{\sigma'}\sum_{\beta}\mathrm{Im}\int d\mathbf{k}' \\
&\times \left[\frac{k_i\,k_j'\,(\mathbf{k}-\mathbf{k}')_k}{k\,k'\,|\mathbf{k}-\mathbf{k}'|}\,\chi_{ijk}^{(2)}(\mathbf{k}',\sigma'\omega_{\mathbf{k}'}^\beta|\mathbf{k}-\mathbf{k}',\sigma\omega_{\mathbf{k}}^\alpha - \sigma'\omega_{\mathbf{k}'}^\beta)\right.
\end{aligned}
$$

$$
\begin{aligned}
&\times \frac{k_n\, k'_m\, (\mathbf{k}-\mathbf{k}')_l}{k\, k'\, |\mathbf{k}-\mathbf{k}'|}\, \chi^{(2)}_{nml}(\mathbf{k}',\sigma'\omega^\beta_{\mathbf{k}'}|\mathbf{k}-\mathbf{k}',\sigma\omega^\alpha_{\mathbf{k}}-\sigma'\omega^\beta_{\mathbf{k}'}) \\
&\times \mathscr{P}\,\frac{2}{\epsilon_\parallel(\mathbf{k}-\mathbf{k}',\sigma\omega^\alpha_{\mathbf{k}}-\sigma'\omega^\beta_{\mathbf{k}'})} + \frac{k_i\,k_l}{k^2}\frac{k'_j\,k'_k}{k'^2} \\
&\times \chi^{(3)}_{ijkl}(\mathbf{k}',\sigma'\omega^\beta_{\mathbf{k}'}|-\mathbf{k}',-\sigma'\omega^\beta_{\mathbf{k}'}|\mathbf{k},\sigma\omega^\alpha_{\mathbf{k}})\Bigg]\, I^{\sigma'\beta}_{\mathbf{k}'}\, I^{\sigma\alpha}_{\mathbf{k}} \qquad \boxed{\text{ind. scatt. }(\alpha,\beta)} \\
&+ \frac{4}{\epsilon'_\parallel(\mathbf{k},\sigma\omega^\alpha_{\mathbf{k}})}\sum_{\sigma'}\sum_{\beta}\int d\mathbf{k}'\sum_a\int d\mathbf{p}\,\frac{4e_a^2}{|\mathbf{k}-\mathbf{k}'|^2\,|\epsilon_\parallel(\mathbf{k}-\mathbf{k}',\sigma\omega^\alpha_{\mathbf{k}}-\sigma'\omega^\beta_{\mathbf{k}'})|^2} \\
&\times \frac{k_i\,k'_j\,(\mathbf{k}-\mathbf{k}')_k}{k\,k'\,|\mathbf{k}-\mathbf{k}'|}\,\chi^{(2)}_{ijk}(\mathbf{k}',\sigma'\omega^\beta_{\mathbf{k}'}|\mathbf{k}-\mathbf{k}',\sigma\omega^\alpha_{\mathbf{k}}-\sigma'\omega^\beta_{\mathbf{k}'}) \\
&\times \frac{k_m\,k'_n\,(\mathbf{k}-\mathbf{k}')_p}{k\,k'\,|\mathbf{k}-\mathbf{k}'|}\,\chi^{(2)*}_{mnp}(\mathbf{k}',\sigma'\omega^\beta_{\mathbf{k}'}|\mathbf{k}-\mathbf{k}',\sigma\omega^\alpha_{\mathbf{k}}-\sigma'\omega^\beta_{\mathbf{k}'}) \\
&\times \left(\frac{I^{\sigma'\beta}_{\mathbf{k}'}}{\epsilon'_\parallel(\mathbf{k},\sigma\omega^\alpha_{\mathbf{k}})} - \frac{I^{\sigma\alpha}_{\mathbf{k}}}{\epsilon'_\parallel(\mathbf{k}',\sigma'\omega^\beta_{\mathbf{k}'})}\right)\delta[\sigma\omega^\alpha_{\mathbf{k}}-\sigma'\omega^\beta_{\mathbf{k}'}-(\mathbf{k}-\mathbf{k}')\cdot\mathbf{v}]\, f_a(\mathbf{p}) \\
&\qquad \boxed{\text{spont. scatt. }(\alpha,\beta)} \\
&- \frac{4}{\epsilon'_\parallel(\mathbf{k},\sigma\omega^\alpha_{\mathbf{k}})}\sum_{\sigma'}\mathrm{Im}\int d\mathbf{k}'\Bigg[\left(\delta_{jm}-\frac{k'_j\,k'_m}{k'^2}\right)\frac{k_i\,(\mathbf{k}-\mathbf{k}')_k}{k\,|\mathbf{k}-\mathbf{k}'|} \\
&\times \chi^{(2)}_{ijk}(\mathbf{k}',\sigma'\omega^T_{\mathbf{k}'}|\mathbf{k}-\mathbf{k}',\sigma\omega^\alpha_{\mathbf{k}}-\sigma'\omega^T_{\mathbf{k}'})\,\frac{k_n\,(\mathbf{k}-\mathbf{k}')_l}{k\,|\mathbf{k}-\mathbf{k}'|} \\
&\times \chi^{(2)}_{nml}(\mathbf{k}',\sigma'\omega^T_{\mathbf{k}'}|\mathbf{k}-\mathbf{k}',\sigma\omega^\alpha_{\mathbf{k}}-\sigma'\omega^T_{\mathbf{k}'})\,\mathscr{P}\,\frac{2}{\epsilon_\parallel(\mathbf{k}-\mathbf{k}',\sigma\omega^\alpha_{\mathbf{k}}-\sigma'\omega^T_{\mathbf{k}'})} \\
&+ \frac{k_i\,k_l}{k^2}\left(\delta_{jk}-\frac{k'_j\,k'_k}{k'^2}\right)\chi^{(3)}_{ijkl}(\mathbf{k}',\sigma'\omega^T_{\mathbf{k}'}|-\mathbf{k}',-\sigma'\omega^T_{\mathbf{k}'}|\mathbf{k},\sigma\omega^\alpha_{\mathbf{k}})\Bigg]\,\frac{I^{\sigma'T}_{\mathbf{k}'}}{2}\, I^{\sigma\alpha}_{\mathbf{k}} \\
&\qquad \boxed{\text{ind. scatt. }(\alpha,T)} \\
&+ \frac{4}{\epsilon'_\parallel(\mathbf{k},\sigma\omega^\alpha_{\mathbf{k}})}\sum_{\sigma'}\int d\mathbf{k}'\sum_a\int d\mathbf{p}\,\frac{4e_a^2}{(\mathbf{k}-\mathbf{k}')^2\,|\epsilon_\parallel(\mathbf{k}-\mathbf{k}',\sigma\omega^\alpha_{\mathbf{k}}-\sigma'\omega^T_{\mathbf{k}'})|} \\
&\times \frac{k_i\,(\mathbf{k}-\mathbf{k}')_k}{k\,|\mathbf{k}-\mathbf{k}'|}\left(\delta_{nj}-\frac{k'_n\,k'_j}{k'^2}\right)\chi^{(2)}_{ijk}(\mathbf{k}',\sigma'\omega^T_{\mathbf{k}'}|\mathbf{k}-\mathbf{k}',\sigma\omega^\alpha_{\mathbf{k}}-\sigma'\omega^T_{\mathbf{k}'}) \\
&\times \frac{k_m\,(\mathbf{k}-\mathbf{k}')_p}{k\,|\mathbf{k}-\mathbf{k}'|}\,\chi^{(2)*}_{mnp}(\mathbf{k}',\sigma'\omega^T_{\mathbf{k}'}|\mathbf{k}-\mathbf{k}',\sigma\omega^\alpha_{\mathbf{k}}-\sigma'\omega^T_{\mathbf{k}'}) \qquad (9.39) \\
&\times \left[\frac{1}{2}\frac{I^{\sigma'T}_{\mathbf{k}'}}{\epsilon'_\parallel(\mathbf{k},\sigma\omega^\alpha_{\mathbf{k}})} - \frac{I^{\sigma\alpha}_{\mathbf{k}}}{\Lambda'_\perp(\mathbf{k}',\sigma'\omega^T_{\mathbf{k}'})}\right]\delta[\sigma\omega^\alpha_{\mathbf{k}}-\sigma'\omega^T_{\mathbf{k}'}-(\mathbf{k}-\mathbf{k}')\cdot\mathbf{v}]\, f_a(\mathbf{p}). \\
&\qquad \boxed{\text{spont. scatt. }(\alpha,T)}
\end{aligned}
$$

Taking $\alpha = L$ or S leads to the wave kinetic equation for L or S mode, which will be discussed later. At this point, the formalism is still given in terms of various susceptibilities.

Next we consider the transverse mode associated with the overall factor $\delta(\omega - \sigma\omega_{\mathbf{k}}^T)$:

$$
\begin{aligned}
\frac{\partial I_{\mathbf{k}}^{\sigma T}}{\partial t} &= -\frac{2\,\mathrm{Im}\,\Lambda_\perp(\mathbf{k},\sigma\omega_{\mathbf{k}}^T)}{\Lambda'_\perp(\mathbf{k},\sigma\omega_{\mathbf{k}}^T)}\, I_{\mathbf{k}}^{\sigma T} \qquad \boxed{\text{induced emission}} \\
&\quad + \sum_a \frac{4e_a^2}{|\omega_{\mathbf{k}}^T \Lambda'_\perp(\mathbf{k},\sigma\omega_{\mathbf{k}}^T)|^2} \int d\mathbf{p}\, \frac{(\mathbf{k}\times\mathbf{v})^2}{k^2}\, \delta(\sigma\omega_{\mathbf{k}}^T - \mathbf{k}\cdot\mathbf{v})\, f_a(\mathbf{p}) \\
&\qquad \boxed{\text{spontaneous emission}} \\
&\quad - \frac{4\pi}{\Lambda'_\perp(\mathbf{k},\sigma\omega_{\mathbf{k}}^T)} \sum_{\sigma',\sigma''}\sum_{\beta,\gamma} \int d\mathbf{k}'\, \frac{k'_j\,(\mathbf{k}-\mathbf{k}')_k}{k'\,|\mathbf{k}-\mathbf{k}'|} \left(\delta_{in} - \frac{k_i\,k_n}{k^2}\right) \\
&\quad \times \chi_{ijk}^{(2)}(\mathbf{k}',\sigma'\omega_{\mathbf{k}'}^\beta|\mathbf{k}-\mathbf{k}',\sigma''\omega_{\mathbf{k}-\mathbf{k}'}^\gamma)\, \frac{k'_l\,(\mathbf{k}-\mathbf{k}')_m}{k'\,|\mathbf{k}-\mathbf{k}'|} \\
&\quad \times \chi_{nlm}^{(2)*}(\mathbf{k}',\sigma'\omega_{\mathbf{k}'}^\beta|\mathbf{k}-\mathbf{k}',\sigma''\omega_{\mathbf{k}-\mathbf{k}'}^\gamma) \left(\frac{1}{2}\frac{I_{\mathbf{k}-\mathbf{k}'}^{\sigma''\gamma}\, I_{\mathbf{k}}^{\sigma T}}{\epsilon'_\parallel(\mathbf{k}',\sigma'\omega_{\mathbf{k}'}^\beta)} + \frac{1}{2}\frac{I_{\mathbf{k}'}^{\sigma'\beta}\, I_{\mathbf{k}}^{\sigma T}}{\epsilon'_\parallel(\mathbf{k}-\mathbf{k}',\sigma''\omega_{\mathbf{k}-\mathbf{k}'}^\gamma)} \right. \\
&\quad \left. - \frac{I_{\mathbf{k}'}^{\sigma'\beta}\, I_{\mathbf{k}-\mathbf{k}'}^{\sigma''\gamma}}{\Lambda'_\perp(\mathbf{k},\sigma\omega_{\mathbf{k}}^T)}\right) \delta(\sigma\omega_{\mathbf{k}}^T - \sigma'\omega_{\mathbf{k}'}^\beta - \sigma''\omega_{\mathbf{k}-\mathbf{k}'}^\gamma) \qquad \boxed{\text{decay } (T,\beta,\gamma)} \\
&\quad - \frac{8\pi}{\Lambda'_\perp(\mathbf{k},\sigma\omega_{\mathbf{k}}^T)} \sum_{\sigma',\sigma''}\sum_{\beta} \int d\mathbf{k}' \left(\delta_{jl} - \frac{k'_j k'_l}{k'^2}\right)\left(\delta_{in} - \frac{k_i k_n}{k^2}\right) \\
&\quad \times \frac{(\mathbf{k}-\mathbf{k}')_k(\mathbf{k}-\mathbf{k}')_m}{|\mathbf{k}-\mathbf{k}'|^2}\, \chi_{ijk}^{(2)}(\mathbf{k}',\sigma'\omega_{\mathbf{k}'}^T|\mathbf{k}-\mathbf{k}',\sigma''\omega_{\mathbf{k}-\mathbf{k}'}^\beta) \\
&\quad \times \chi_{nlm}^{(2)*}(\mathbf{k}',\sigma'\omega_{\mathbf{k}'}^T|\mathbf{k}-\mathbf{k}',\sigma''\omega_{\mathbf{k}-\mathbf{k}'}^\beta) \left(\frac{I_{\mathbf{k}'}^{\sigma' T}\, I_{\mathbf{k}-\mathbf{k}'}^{\sigma''\beta}}{\Lambda'_\perp(\mathbf{k},\sigma\omega_{\mathbf{k}}^T)} - \frac{I_{\mathbf{k}-\mathbf{k}'}^{\sigma''\beta}\, I_{\mathbf{k}}^{\sigma T}}{\Lambda'_\perp(\mathbf{k}',\sigma'\omega_{\mathbf{k}'}^T)} \right. \\
&\quad \left. - \frac{1}{2}\frac{I_{\mathbf{k}'}^{\sigma' T}\, I_{\mathbf{k}}^{\sigma T}}{\epsilon'_\parallel(\mathbf{k}-\mathbf{k}',\sigma''\omega_{\mathbf{k}-\mathbf{k}'}^\beta)}\right) \delta(\sigma\omega_{\mathbf{k}}^T - \sigma'\omega_{\mathbf{k}'}^T - \sigma''\omega_{\mathbf{k}-\mathbf{k}'}^\beta) \qquad \boxed{\text{decay } (T,T,\beta)} \\
&\quad - \frac{4}{\Lambda'_\perp(\mathbf{k},\sigma\omega_{\mathbf{k}}^T)} \sum_{\sigma'}\sum_{\beta} \mathrm{Im} \int d\mathbf{k}' \left[\left(\delta_{in} - \frac{k_i\,k_n}{k^2}\right) \frac{k'_j\,(\mathbf{k}-\mathbf{k}')_k}{k'\,|\mathbf{k}-\mathbf{k}'|}\right. \\
&\quad \times \chi_{ijk}^{(2)}(\mathbf{k}',\sigma'\omega_{\mathbf{k}'}^\beta|\mathbf{k}-\mathbf{k}',\sigma\omega_{\mathbf{k}}^T - \sigma'\omega_{\mathbf{k}'}^\beta)\, \frac{k'_m\,(\mathbf{k}-\mathbf{k}')_l}{k'\,|\mathbf{k}-\mathbf{k}'|} \\
&\quad \times \chi_{nml}^{(2)}(\mathbf{k}',\sigma'\omega_{\mathbf{k}'}^\beta|\mathbf{k}-\mathbf{k}',\sigma\omega_{\mathbf{k}}^T - \sigma'\omega_{\mathbf{k}'}^\beta)\, \mathscr{P}\, \frac{2}{\epsilon_\parallel(\mathbf{k}-\mathbf{k}',\sigma\omega_{\mathbf{k}}^T - \sigma'\omega_{\mathbf{k}'}^\beta)}
\end{aligned}
$$

$$+\left(\delta_{il}-\frac{k_i\,k_l}{k^2}\right)\frac{k'_j\,k'_k}{k'^2}\,\chi^{(3)}_{ijkl}(\mathbf{k}',\sigma'\omega^\beta_{\mathbf{k}'}|-\mathbf{k}',\,-\sigma'\omega^\beta_{\mathbf{k}'}|\mathbf{k},\sigma\omega^T_{\mathbf{k}})\Bigg]\,I^{\sigma'\beta}_{\mathbf{k}'}\,\frac{I^{\sigma T}_{\mathbf{k}}}{2}$$

$$\boxed{\text{ind. scatt. }(T,\beta)}$$

$$+\frac{4}{\Lambda'_\perp(\mathbf{k},\sigma\omega^T_{\mathbf{k}})}\sum_{\sigma'}\sum_{\beta}\int d\mathbf{k}'\sum_a\int d\mathbf{p}\,\frac{4e_a^2}{(\mathbf{k}-\mathbf{k}')^2|\epsilon_\parallel(\mathbf{k}-\mathbf{k}',\sigma\omega^T_{\mathbf{k}}-\sigma'\omega^\beta_{\mathbf{k}'})|^2}$$

$$\times\frac{k'_j\,(\mathbf{k}-\mathbf{k}')_k}{k'\,|\mathbf{k}-\mathbf{k}'|}\left(\delta_{im}-\frac{k_i\,k_m}{k^2}\right)\chi^{(2)}_{ijk}(\mathbf{k}',\sigma'\omega^\beta_{\mathbf{k}'}|\mathbf{k}-\mathbf{k}',\sigma\omega^T_{\mathbf{k}}-\sigma'\omega^\beta_{\mathbf{k}'})$$

$$\times\frac{k'_n\,(\mathbf{k}-\mathbf{k}')_p}{k'\,|\mathbf{k}-\mathbf{k}'|}\,\chi^{(2)*}_{mnp}(\mathbf{k}',\sigma'\omega^\beta_{\mathbf{k}'}|\mathbf{k}-\mathbf{k}',\sigma\omega^T_{\mathbf{k}}-\sigma'\omega^\beta_{\mathbf{k}'}) \tag{9.40}$$

$$\times\left(\frac{I^{\sigma'\beta}_{\mathbf{k}'}}{\Lambda'_\perp(\mathbf{k},\sigma\omega^T_{\mathbf{k}})}-\frac{1}{2}\frac{I^{\sigma T}_{\mathbf{k}}}{\epsilon'_\parallel(\mathbf{k}',\sigma'\omega^\beta_{\mathbf{k}'})}\right)\delta[\sigma\omega^T_{\mathbf{k}}-\sigma'\omega^\beta_{\mathbf{k}'}-(\mathbf{k}-\mathbf{k}')\cdot\mathbf{v}]\,f_a(\mathbf{p}).$$

$$\boxed{\text{spont. scatt. }(T,\beta)}$$

(9.40) describes nonlinear interaction and time evolution of transverse wave. Together with the equation for longitudinal modes (9.39), the wave kinetic equations of electromagnetic Klimontovich weak turbulence theory are formally expressed in terms of linear and nonlinear susceptibilities. Next we evaluate various response functions by approximate means, as in Section 8.2. The resulting set of equations constitutes the fully general electromagnetic weak turbulence formalism that can be readily solved by numerical means or that readily lend them to analytical treatment. Such a formalism that incorporates discrete particle effects under fully electromagnetic framework is the culmination of all the discussions presented in this book.

9.3 Electromagnetic Klimontovich Weak Turbulence Theory: Summary

To derive the desired final and most general equations for electromagnetic Klimontovich weak turbulence theory involving plasma eigenmodes, that is, longitudinal eigenmodes, L and S, and transverse mode, T, all we need to do is evaluate extra terms that arise from the particle discreteness effects that were missing in the equations derived in Chapter 8.

For Langmuir wave, we already discussed the induced emission terms in Section 8.2.1. The first extra term relates to the spontaneous emission,

$$\left.\frac{\partial I^{\sigma L}_{\mathbf{k}}}{\partial t}\right|_{\text{spont.em.}}=\frac{(\omega^L_{\mathbf{k}})^2}{k^2}\sum_a e_a^2\int d\mathbf{p}\;\delta(\sigma\omega^L_{\mathbf{k}}-\mathbf{k}\cdot\mathbf{v})\;f_a(\mathbf{p}). \tag{9.41}$$

This term is the counterpart to (8.35). The decay terms for L mode are already derived, and the result is (8.40). The discrete particle effects do not influence the decay terms. Induced scattering term for L mode is given by (8.58). The spontaneous scattering term is new, and is given formally by

$$
\begin{aligned}
\left.\frac{\partial I_{\mathbf{k}}^{\sigma L}}{\partial t}\right|_{\text{spont.scatt.}} &= \frac{4}{\epsilon_\parallel'(\mathbf{k},\sigma\omega_{\mathbf{k}}^L)}\sum_{\sigma'}\int d\mathbf{k}'\sum_a\int d\mathbf{p} \\
&\times \frac{4e_a^2}{|\mathbf{k}-\mathbf{k}'|^2\,|\epsilon_\parallel(\mathbf{k}-\mathbf{k}',\sigma\omega_{\mathbf{k}}^\alpha-\sigma'\omega_{\mathbf{k}'}^L)|^2} \\
&\times \frac{k_i\,k_j'\,(\mathbf{k}-\mathbf{k}')_k}{k\,k'\,|\mathbf{k}-\mathbf{k}'|}\,\chi_{ijk}^{(2)}(\mathbf{k}',\sigma'\omega_{\mathbf{k}'}^L|\mathbf{k}-\mathbf{k}',\sigma\omega_{\mathbf{k}}^L-\sigma'\omega_{\mathbf{k}'}^L) \\
&\times \frac{k_n\,k_m'\,(\mathbf{k}-\mathbf{k}')_l}{k\,k'\,|\mathbf{k}-\mathbf{k}'|}\,\chi_{nml}^{(2)*}(\mathbf{k}',\sigma'\omega_{\mathbf{k}'}^L|\mathbf{k}-\mathbf{k}',\sigma\omega_{\mathbf{k}}^L-\sigma'\omega_{\mathbf{k}'}^L) \\
&\times \left(\frac{I_{\mathbf{k}'}^{\sigma' L}}{\epsilon_\parallel'(\mathbf{k},\sigma\omega_{\mathbf{k}}^L)}-\frac{I_{\mathbf{k}}^{\sigma L}}{\epsilon_\parallel'(\mathbf{k}',\sigma'\omega_{\mathbf{k}'}^L)}\right) \\
&\times \delta[\sigma\omega_{\mathbf{k}}^L-\sigma'\omega_{\mathbf{k}'}^L-(\mathbf{k}-\mathbf{k}')\cdot\mathbf{v}]\;f_a(\mathbf{p}) \\
&+ \frac{4}{\epsilon_\parallel'(\mathbf{k},\sigma\omega_{\mathbf{k}}^L)}\sum_{\sigma'}\int d\mathbf{k}' \\
&\times \sum_a\int d\mathbf{p}\,\frac{4e_a^2}{(\mathbf{k}-\mathbf{k}')^2\,|\epsilon_\parallel(\mathbf{k}-\mathbf{k}',\sigma\omega_{\mathbf{k}}^L-\sigma'\omega_{\mathbf{k}'}^T)|} \\
&\times \frac{k_i\,(\mathbf{k}-\mathbf{k}')_k}{k\,|\mathbf{k}-\mathbf{k}'|}\left(\delta_{jn}-\frac{k_j'\,k_n'}{k'^2}\right)\chi_{ijk}^{(2)}(\mathbf{k}',\sigma'\omega_{\mathbf{k}'}^T|\mathbf{k}-\mathbf{k}',\sigma\omega_{\mathbf{k}}^L-\sigma'\omega_{\mathbf{k}'}^T) \\
&\times \frac{k_m\,(\mathbf{k}-\mathbf{k}')_l}{k\,|\mathbf{k}-\mathbf{k}'|}\,\chi_{mnl}^{(2)*}(\mathbf{k}',\sigma'\omega_{\mathbf{k}'}^T|\mathbf{k}-\mathbf{k}',\sigma\omega_{\mathbf{k}}^L-\sigma'\omega_{\mathbf{k}'}^T) \qquad (9.42) \\
&\times \left(\frac{1}{2}\frac{I_{\mathbf{k}'}^{\sigma' T}}{\epsilon_\parallel'(\mathbf{k},\sigma\omega_{\mathbf{k}}^\alpha)}-\frac{I_{\mathbf{k}}^{\sigma\alpha}}{\Lambda_\perp'(\mathbf{k}',\sigma'\omega_{\mathbf{k}'}^T)}\right) \\
&\times \delta[\sigma\omega_{\mathbf{k}}^\alpha-\sigma'\omega_{\mathbf{k}'}^T-(\mathbf{k}-\mathbf{k}')\cdot\mathbf{v}]\;f_a(\mathbf{p}).
\end{aligned}
$$

Making use of

- $$
\begin{aligned}
&\frac{k_i\,k_j'\,(\mathbf{k}-\mathbf{k}')_k}{k\,k'\,|\mathbf{k}-\mathbf{k}'|}\,\chi_{ijk}^{(2)}(\mathbf{k}',\sigma'\omega_{\mathbf{k}'}^L|\mathbf{k}-\mathbf{k}',\sigma\omega_{\mathbf{k}}^L-\sigma'\omega_{\mathbf{k}'}^L) \\
&\times \frac{k_n\,k_m'\,(\mathbf{k}-\mathbf{k}')_l}{k\,k'\,|\mathbf{k}-\mathbf{k}'|}\,\chi_{nml}^{(2)}(\mathbf{k}',\sigma'\omega_{\mathbf{k}'}^L|\mathbf{k}-\mathbf{k}',\sigma\omega_{\mathbf{k}}^L-\sigma'\omega_{\mathbf{k}'}^L) \\
&= -\frac{e^2}{4m_e^2}\frac{(\mathbf{k}\cdot\mathbf{k}')^2}{k^2\,k'^2}\frac{(\mathbf{k}-\mathbf{k}')^2}{\omega_{pe}^4}\,[\chi_\parallel^e(\mathbf{k}-\mathbf{k}',\sigma\omega_{\mathbf{k}}^L-\sigma'\omega_{\mathbf{k}'}^L)]^2,
\end{aligned}
$$
- $$
\left(\delta_{jm}-\frac{k_j'\,k_m'}{k'^2}\right)\frac{k_i\,(\mathbf{k}-\mathbf{k}')_k}{k\,|\mathbf{k}-\mathbf{k}'|}\,\chi_{ijk}^{(2)}(\mathbf{k}',\sigma'\omega_{\mathbf{k}'}^T|\mathbf{k}-\mathbf{k}',\sigma\omega_{\mathbf{k}}^L-\sigma'\omega_{\mathbf{k}'}^T)
$$

$$\times \frac{k_n\,(\mathbf{k}-\mathbf{k}')_l}{k\,|\mathbf{k}-\mathbf{k}'|}\,\chi^{(2)*}_{nml}(\mathbf{k}',\sigma'\omega^T_{\mathbf{k}'}|\mathbf{k}-\mathbf{k}',\sigma\omega^L_{\mathbf{k}}-\sigma'\omega^T_{\mathbf{k}'})$$
$$\times \frac{1}{|\epsilon_\parallel(\mathbf{k}-\mathbf{k}',\sigma\omega^L_{\mathbf{k}}-\sigma'\omega^T_{\mathbf{k}'})|^2} = \frac{e^2}{4m_e^2}\frac{(\mathbf{k}\times\mathbf{k}')^2}{k^2\,k'^2}\frac{(\mathbf{k}-\mathbf{k}')^2}{\omega^4_{pe}}, \tag{9.43}$$

we arrive at

$$\begin{aligned}\left.\frac{\partial I^{\sigma L}_{\mathbf{k}}}{\partial t}\right|_{\text{spont.scatt.}} &= -\frac{\sigma\omega^L_{\mathbf{k}}}{16\pi^2 n^2}\sum_{\sigma'}\int d\mathbf{k}'\int d\mathbf{p}\,\frac{(\mathbf{k}\cdot\mathbf{k}')^2}{k^2\,k'^2}\\ &\quad\times\delta[\sigma\omega^L_{\mathbf{k}}-\sigma'\omega^L_{\mathbf{k}'}-(\mathbf{k}-\mathbf{k}')\cdot\mathbf{v}]\\ &\quad\times\left(\sigma'\omega^L_{\mathbf{k}'}\,I^{\sigma L}_{\mathbf{k}}-\sigma\omega^L_{\mathbf{k}}\,I^{\sigma' L}_{\mathbf{k}'}\right)[f_e(\mathbf{p})+f_i(\mathbf{p})]\\ &\quad-\frac{\sigma\omega^L_{\mathbf{k}}}{16\pi^2 n^2}\sum_{\sigma'}\int d\mathbf{k}'\int d\mathbf{p}\,\frac{(\mathbf{k}\times\mathbf{k}')^2}{k^2\,k'^2}\delta[\sigma\omega^L_{\mathbf{k}}-\sigma'\omega^T_{\mathbf{k}'}-(\mathbf{k}-\mathbf{k}')\cdot\mathbf{v}]\\ &\quad\times\left(\sigma'\omega^T_{\mathbf{k}'}\,I^{\sigma L}_{\mathbf{k}}-\sigma\omega^L_{\mathbf{k}}\,\frac{I^{\sigma' T}_{\mathbf{k}'}}{2}\right)[f_e(\mathbf{p})+f_i(\mathbf{p})].\end{aligned} \tag{9.44}$$

Next, we take $\alpha = S$. Among the new terms is the spontaneous emission term for S mode,

$$\left.\frac{\partial}{\partial t}\right|_{\text{spont.emiss.}}\frac{I^{\sigma S}_{\mathbf{k}}}{\mu_{\mathbf{k}}} = \sum_a e_a^2\,\frac{\mu_{\mathbf{k}}(\omega^L_{\mathbf{k}})^2}{k^2}\int d\mathbf{p}\,\delta(\sigma\omega^S_{\mathbf{k}}-\mathbf{k}\cdot\mathbf{v})\,f_a(\mathbf{p}). \tag{9.45}$$

This is the counterpart to the induced emission term, given in (8.65). For ion-sound mode, nonlinear scattering processes are generally slow, so we may ignore such processes.

The spontaneous emission term in the transverse wave kinetic equation is given by

$$\left.\frac{\partial}{\partial t}\right|_{\text{spont.emiss.}}\frac{I^{\sigma T}_{\mathbf{k}}}{2} = \sum_a \frac{e_a^2}{2}\int d\mathbf{p}\,\frac{(\mathbf{k}\times\mathbf{v})^2}{k^2}\,\delta(\sigma\omega^T_{\mathbf{k}}-\mathbf{k}\cdot\mathbf{v})\,f_a(\mathbf{p}). \tag{9.46}$$

The spontaneous scattering term for T mode is specified by

$$\begin{aligned}\left.\frac{\partial}{\partial t}\right|_{\text{spont.scatt.}}\frac{I^{\sigma T}_{\mathbf{k}}}{2} &= \sigma\omega^T_{\mathbf{k}}\sum_{\sigma'}\int d\mathbf{k}'\int d\mathbf{p}\,\frac{2e^2}{(\mathbf{k}-\mathbf{k}')^2\,|\epsilon_\parallel(\mathbf{k}-\mathbf{k}',\sigma\omega^T_{\mathbf{k}}-\sigma'\omega^L_{\mathbf{k}'})|^2}\\ &\quad\times\frac{k'_j\,(\mathbf{k}-\mathbf{k}')_k}{k'\,|\mathbf{k}-\mathbf{k}'|}\left(\delta_{im}-\frac{k_i k_m}{k^2}\right)\chi^{(2)}_{ijk}(\mathbf{k}',\sigma'\omega^L_{\mathbf{k}'}|\mathbf{k}-\mathbf{k}',\sigma\omega^T_{\mathbf{k}}-\sigma'\omega^L_{\mathbf{k}'})\\ &\quad\times\frac{k'_n\,(\mathbf{k}-\mathbf{k}')_p}{k'\,|\mathbf{k}-\mathbf{k}'|}\,\chi^{(2)*}_{mnp}(\mathbf{k}',\sigma'\omega^L_{\mathbf{k}'}|\mathbf{k}-\mathbf{k}',\sigma\omega^T_{\mathbf{k}}-\sigma'\omega^L_{\mathbf{k}'})\,\mu^T_{\mathbf{k}}\,\mu^L_{\mathbf{k}'}\end{aligned}$$

$$\times \left(\sigma\omega_{\mathbf{k}}^T \frac{I_{\mathbf{k}'}^{\sigma' L}}{\mu_{\mathbf{k}'}^L} - \sigma'\omega_{\mathbf{k}'}^L \frac{I_{\mathbf{k}}^{\sigma T}}{2\mu_{\mathbf{k}}^T}\right) \delta[\sigma\omega_{\mathbf{k}}^T - \sigma'\omega_{\mathbf{k}'}^L - (\mathbf{k}-\mathbf{k}')\cdot\mathbf{v}]$$
$$\times [f_e(\mathbf{p}) + f_i(\mathbf{p})]. \tag{9.47}$$

Making use of

$$\frac{k_j'\,(\mathbf{k}-\mathbf{k}')_k}{k'\,|\mathbf{k}-\mathbf{k}'|}\left(\delta_{im} - \frac{k_i\,k_m}{k^2}\right) \chi_{ijk}^{(2)}(\mathbf{k}',\sigma'\omega_{\mathbf{k}'}^L|\mathbf{k}-\mathbf{k}',\sigma\omega_{\mathbf{k}}^T - \sigma'\omega_{\mathbf{k}'}^L)$$
$$\times \frac{k_n'\,(\mathbf{k}-\mathbf{k}')_p}{k'\,|\mathbf{k}-\mathbf{k}'|} \chi_{mnp}^{(2)*}(\mathbf{k}',\sigma'\omega_{\mathbf{k}'}^L|\mathbf{k}-\mathbf{k}',\sigma\omega_{\mathbf{k}}^T - \sigma'\omega_{\mathbf{k}'}^L)$$
$$= \frac{e^2}{4m_e^2}\frac{(\mathbf{k}-\mathbf{k}')^2}{\omega_{pe}^4}\frac{(\mathbf{k}\times\mathbf{k}')^2}{k^2\,k'^2}\,|\chi_\parallel^e(\mathbf{k}-\mathbf{k}',\sigma\omega_{\mathbf{k}}^T - \sigma'\omega_{\mathbf{k}'}^L)|^2, \tag{9.48}$$

we obtain the desired expression for spontaneous scattering term within the T wave kinetic equation,

$$\left.\frac{\partial}{\partial t}\right|_{\text{spont.scatt.}} \frac{I_{\mathbf{k}}^{\sigma T}}{2} = -\frac{\sigma\omega_{\mathbf{k}}^T}{32\pi^2 n^2}\sum_{\sigma'}\int d\mathbf{k}'\int d\mathbf{p}\,\frac{(\mathbf{k}\times\mathbf{k}')^2}{k^2\,k'^2}$$
$$\times\,\delta[\sigma\omega_{\mathbf{k}}^T - \sigma'\omega_{\mathbf{k}'}^L - (\mathbf{k}-\mathbf{k}')\cdot\mathbf{v}]$$
$$\times\left(\sigma'\omega_{\mathbf{k}'}^L\frac{I_{\mathbf{k}}^{\sigma T}}{2} - \sigma\omega_{\mathbf{k}}^T\,I_{\mathbf{k}'}^{\sigma' L}\right)[f_e(\mathbf{p}) + f_i(\mathbf{p})]. \tag{9.49}$$

Gathering all the results we have obtained thus far, we summarize the equations of Klimontovich weak turbulence theory that incorporate the full electromagnetic effects and particle discreteness effects. The following displayed equations list the wave kinetic equations for L, S, and T modes with each wave-particle and wave-wave process indicated, as well as the particle equation:

- $$\frac{\partial I_{\mathbf{k}}^{\sigma L}}{\partial t} = \sigma\omega_{\mathbf{k}}^L\,\frac{4\pi^2 e^2}{k^2}\int d\mathbf{p}\,\delta(\sigma\omega_{\mathbf{k}}^L - \mathbf{k}\cdot\mathbf{v})\underbrace{\left(\frac{\sigma\omega_{\mathbf{k}}^L}{4\pi^2}\,f_e(\mathbf{p}) + \mathbf{k}\cdot\frac{\partial f_e(\mathbf{p})}{\partial\mathbf{p}}\,I_{\mathbf{k}}^{\sigma L}\right)}_{\boxed{\text{spont. \& ind. emiss. } L\leftrightarrow e}}$$
$$+\pi\sigma\,\omega_{\mathbf{k}}^L\,\frac{e^2}{2T_e^2}\sum_{\sigma',\sigma''}\int d\mathbf{k}'\,\frac{\mu_{\mathbf{k}-\mathbf{k}'}\,(\mathbf{k}\cdot\mathbf{k}')^2}{k^2\,k'^2\,|\mathbf{k}-\mathbf{k}'|^2}\left(\sigma\omega_{\mathbf{k}}^L\,I_{\mathbf{k}'}^{\sigma' L}\,\frac{I_{\mathbf{k}-\mathbf{k}'}^{\sigma'' S}}{\mu_{\mathbf{k}-\mathbf{k}'}}\right.$$
$$\left.-\,\sigma'\omega_{\mathbf{k}'}^L\,\frac{I_{\mathbf{k}-\mathbf{k}'}^{\sigma'' S}}{\mu_{\mathbf{k}-\mathbf{k}'}}\,I_{\mathbf{k}}^{\sigma L} - \sigma''\omega_{\mathbf{k}-\mathbf{k}'}^L\,I_{\mathbf{k}'}^{\sigma' L}\,I_{\mathbf{k}}^{\sigma L}\right)\underbrace{\delta(\sigma\omega_{\mathbf{k}}^L - \sigma'\omega_{\mathbf{k}'}^L - \sigma''\omega_{\mathbf{k}-\mathbf{k}'}^S)}_{\boxed{\text{decay } L\leftrightarrow L+S}}$$
$$+\pi\sigma\mu_{\mathbf{k}}^L\,\omega_{\mathbf{k}}^L\,\frac{e^2}{8\,m_e^2\,\omega_{pe}^2}\sum_{\sigma',\sigma''}\int d\mathbf{k}'\,\frac{(\mathbf{k}\times\mathbf{k}')^2}{k^2\,k'^2\,|\mathbf{k}-\mathbf{k}'|^2}\left(\frac{k'^2}{\sigma'\omega_{\mathbf{k}'}^L} + \frac{k^2}{\sigma\omega_{\mathbf{k}}^L}\right)^2$$

$$
\begin{aligned}
&\times\left(\sigma\omega_{\mathbf{k}}^{L}\,\frac{I_{\mathbf{k}-\mathbf{k}'}^{\sigma''T}}{2}\,I_{\mathbf{k}'}^{\sigma'L}-\sigma'\omega_{\mathbf{k}'}^{L}\,\frac{I_{\mathbf{k}-\mathbf{k}'}^{\sigma''T}}{2}\,I_{\mathbf{k}}^{\sigma L}-\sigma''\omega_{\mathbf{k}-\mathbf{k}'}^{T}\,I_{\mathbf{k}'}^{\sigma'L}\,I_{\mathbf{k}}^{\sigma L}\right)\\
&\times\underbrace{\delta(\sigma\omega_{\mathbf{k}}^{L}-\sigma'\omega_{\mathbf{k}'}^{L}-\sigma''\omega_{\mathbf{k}-\mathbf{k}'}^{T})}_{\boxed{\text{decay } L\leftrightarrow L+T}}\\
&+\pi\sigma\,\omega_{\mathbf{k}}^{L}\,\frac{e^2}{2T_e^2}\sum_{\sigma',\sigma''}\int d\mathbf{k}'\,\frac{\mu_{\mathbf{k}'}\,(\mathbf{k}\times\mathbf{k}')^2}{k^2\,k'^2\,|\mathbf{k}-\mathbf{k}'|^2}\left(\sigma\omega_{\mathbf{k}}^{L}\,\frac{I_{\mathbf{k}-\mathbf{k}'}^{\sigma''T}}{2}\frac{I_{\mathbf{k}'}^{\sigma'S}}{\mu_{\mathbf{k}'}}\right.\\
&\left.-\sigma'\omega_{\mathbf{k}'}^{L}\,\frac{I_{\mathbf{k}-\mathbf{k}'}^{\sigma''T}}{2}\,I_{\mathbf{k}}^{\sigma L}-\sigma''\omega_{\mathbf{k}-\mathbf{k}'}^{T}\,\frac{I_{\mathbf{k}'}^{\sigma'S}}{\mu_{\mathbf{k}'}}\,I_{\mathbf{k}}^{\sigma L}\right)\underbrace{\delta(\sigma\omega_{\mathbf{k}}^{L}-\sigma'\omega_{\mathbf{k}'}^{S}-\sigma''\omega_{\mathbf{k}-\mathbf{k}'}^{T})}_{\boxed{\text{decay } L\leftrightarrow S+T}}\\
&+\pi\sigma\omega_{\mathbf{k}}^{L}\,\frac{e^2}{4m_e^2}\sum_{\sigma',\sigma''}\int d\mathbf{k}'\,\frac{k^2}{(\omega_{\mathbf{k}'}^{T})^2(\omega_{\mathbf{k}-\mathbf{k}'}^{T})^2}\left(1+\frac{[\mathbf{k}'\cdot(\mathbf{k}-\mathbf{k}')]^2}{k^2|\mathbf{k}-\mathbf{k}'|^2}\right)\\
&\times\left(\sigma\omega_{\mathbf{k}}^{L}\,\frac{I_{\mathbf{k}'}^{\sigma'T}}{2}\frac{I_{\mathbf{k}-\mathbf{k}'}^{\sigma''T}}{2}-\sigma'\omega_{\mathbf{k}'}^{T}\,\frac{I_{\mathbf{k}-\mathbf{k}'}^{\sigma''T}}{2}\,I_{\mathbf{k}}^{\sigma L}-\sigma''\omega_{\mathbf{k}-\mathbf{k}'}^{T}\,\frac{I_{\mathbf{k}'}^{\sigma'T}}{2}\,I_{\mathbf{k}}^{\sigma L}\right)\\
&\times\underbrace{\delta(\sigma\omega_{\mathbf{k}}^{L}-\sigma'\omega_{\mathbf{k}'}^{T}-\sigma''\omega_{\mathbf{k}-\mathbf{k}'}^{T})}_{\boxed{\text{decay } L\leftrightarrow T+T}}\\
&-\frac{\sigma\omega_{\mathbf{k}}^{L}}{4n^2}\sum_{\sigma'}\int d\mathbf{k}'\int d\mathbf{p}\,\frac{(\mathbf{k}\cdot\mathbf{k}')^2}{k^2\,k'^2}\,\delta[\sigma\omega_{\mathbf{k}}^{L}-\sigma'\omega_{\mathbf{k}'}^{L}-(\mathbf{k}-\mathbf{k}')\cdot\mathbf{v}]\\
&\times\Bigg[\underbrace{\frac{1}{4\pi^2}\left(\sigma'\omega_{\mathbf{k}'}^{L}\,I_{\mathbf{k}}^{\sigma L}-\sigma\omega_{\mathbf{k}}^{L}\,I_{\mathbf{k}'}^{\sigma'L}\right)[f_e(\mathbf{p})+f_i(\mathbf{p})]}_{\boxed{\text{spont. scatt. } L\leftrightarrow L+p}}\\
&\underbrace{-\,I_{\mathbf{k}'}^{\sigma'L}\,I_{\mathbf{k}}^{\sigma L}\,(\mathbf{k}-\mathbf{k}')\cdot\frac{\partial f_i(\mathbf{p})}{\partial\mathbf{p}}}_{\boxed{\text{ind. scatt. } L\leftrightarrow L+i}}\Bigg]\\
&-\frac{\sigma\omega_{\mathbf{k}}^{L}}{4n^2}\sum_{\sigma'}\int d\mathbf{k}'\int d\mathbf{p}\,\frac{(\mathbf{k}\times\mathbf{k}')^2}{k^2\,k'^2}\,\delta[\sigma\omega_{\mathbf{k}}^{L}-\sigma'\omega_{\mathbf{k}'}^{T}-(\mathbf{k}-\mathbf{k}')\cdot\mathbf{v}]\\
&\times\Bigg[\underbrace{\frac{1}{4\pi^2}\left(\sigma'\omega_{\mathbf{k}'}^{T}\,I_{\mathbf{k}}^{\sigma L}-\sigma\omega_{\mathbf{k}}^{L}\,\frac{I_{\mathbf{k}'}^{\sigma'T}}{2}\right)[f_e(\mathbf{p})+f_i(\mathbf{p})]}_{\boxed{\text{spont. scatt. } L\leftrightarrow T+p}}\\
&\underbrace{-\,\frac{I_{\mathbf{k}'}^{\sigma'T}}{2}\,I_{\mathbf{k}}^{\sigma L}\,(\mathbf{k}-\mathbf{k}')\cdot\frac{\partial f_i(\mathbf{p})}{\partial\mathbf{p}}}_{\boxed{\text{ind. scatt. } L\leftrightarrow T+i}}\Bigg]; \qquad (9.50)
\end{aligned}
$$

$$\bullet\ \frac{\partial}{\partial t}\frac{I_{\mathbf{k}}^{\sigma S}}{\mu_{\mathbf{k}}^{S}} = \mu_{\mathbf{k}}\sigma\omega_{\mathbf{k}}^{L}\,\frac{4\pi^2 e^2}{k^2}\int d\mathbf{p}\,\delta(\sigma\omega_{\mathbf{k}}^{S} - \mathbf{k}\cdot\mathbf{v})$$

$$\times\Bigg[\underbrace{\frac{\sigma\omega_{\mathbf{k}}^{L}}{4\pi^2}\,[f_e(\mathbf{p}) + f_i(\mathbf{p})]}+\underbrace{\left(\mathbf{k}\cdot\frac{\partial}{\partial\mathbf{p}}\right)[f_e(\mathbf{p}) + f_i(\mathbf{p})]\,\frac{I_{\mathbf{k}}^{\sigma S}}{\mu_{\mathbf{k}}}}\Bigg]$$

$$\boxed{\text{spont. \& ind. emiss. } S \leftrightarrow p}$$

$$+\pi\sigma\omega_{\mathbf{k}}^{L}\,\frac{e^2}{4T_e^2}\sum_{\sigma',\sigma''}\int d\mathbf{k}'\,\frac{\mu_{\mathbf{k}}\,[\mathbf{k}'\cdot(\mathbf{k}-\mathbf{k}')]^2}{k^2\,k'^2\,|\mathbf{k}-\mathbf{k}'|^2}\Bigg(\sigma\omega_{\mathbf{k}}^{L}\,I_{\mathbf{k}'}^{\sigma' L}\,I_{\mathbf{k}-\mathbf{k}'}^{\sigma'' L}$$

$$-\sigma'\omega_{\mathbf{k}'}^{L}\,I_{\mathbf{k}-\mathbf{k}'}^{\sigma'' L}\,\frac{I_{\mathbf{k}}^{\sigma S}}{\mu_{\mathbf{k}}} - \sigma''\omega_{\mathbf{k}-\mathbf{k}'}^{L}\,I_{\mathbf{k}'}^{\sigma' L}\,\frac{I_{\mathbf{k}}^{\sigma S}}{\mu_{\mathbf{k}}}\Bigg)\,\underbrace{\delta(\sigma\omega_{\mathbf{k}}^{S} - \sigma'\omega_{\mathbf{k}'}^{L} - \sigma''\omega_{\mathbf{k}-\mathbf{k}'}^{L})}$$

$$\boxed{\text{decay } S \leftrightarrow L + L}$$

$$+\pi\sigma\omega_{\mathbf{k}}^{L}\,\frac{e^2}{2T_e^2}\sum_{\sigma',\sigma''}\int d\mathbf{k}'\,\frac{\mu_{\mathbf{k}}\,(\mathbf{k}\times\mathbf{k}')^2}{k^2\,k'^2\,|\mathbf{k}-\mathbf{k}'|^2}\Bigg(\sigma\omega_{\mathbf{k}}^{L}\,\frac{I_{\mathbf{k}-\mathbf{k}'}^{\sigma'' T}}{2}\,I_{\mathbf{k}'}^{\sigma' L} \tag{9.51}$$

$$-\sigma'\omega_{\mathbf{k}'}^{L}\,\frac{I_{\mathbf{k}-\mathbf{k}'}^{\sigma'' T}}{2}\,\frac{I_{\mathbf{k}}^{\sigma S}}{\mu_{\mathbf{k}}} - \sigma''\omega_{\mathbf{k}-\mathbf{k}'}^{T}\,I_{\mathbf{k}'}^{\sigma' L}\,\frac{I_{\mathbf{k}}^{\sigma S}}{\mu_{\mathbf{k}}}\Bigg)\,\underbrace{\delta(\sigma\omega_{\mathbf{k}}^{S} - \sigma'\omega_{\mathbf{k}'}^{L} - \sigma''\omega_{\mathbf{k}-\mathbf{k}'}^{T})};$$

$$\boxed{\text{decay } S \leftrightarrow L + T}$$

$$\bullet\ \frac{\partial}{\partial t}\frac{I_{\mathbf{k}}^{\sigma T}}{2} = \sum_a \frac{4\pi^2 e_a^2}{\sigma\omega_{\mathbf{k}}^{T}}\int d\mathbf{p}\,\frac{(\mathbf{k}\times\mathbf{v})^2}{k^2}\,\delta(\sigma\omega_{\mathbf{k}}^{T} - \mathbf{k}\cdot\mathbf{v})$$

$$\times\Bigg(\underbrace{\frac{\sigma\omega_{\mathbf{k}}^{T}}{8\pi^2}\,f_a(\mathbf{p})}+\underbrace{\mathbf{k}\cdot\frac{\partial f_a(\mathbf{p})}{\partial\mathbf{p}}\,\frac{I_{\mathbf{k}}^{\sigma T}}{2}}\Bigg)$$

$$\boxed{\text{spont. \& ind. emiss. } T \leftrightarrow p}$$

$$+\pi\,\sigma\omega_{\mathbf{k}}^{T}\,\frac{e^2}{32\,m_e^2\,\omega_{pe}^2}\sum_{\sigma',\sigma''}\int d\mathbf{k}'\,\frac{(\mathbf{k}\times\mathbf{k}')^2}{k^2\,k'^2\,|\mathbf{k}-\mathbf{k}'|^2}$$

$$\times\left(\frac{k'^2}{\sigma'\omega_{\mathbf{k}'}^{L}} - \frac{|\mathbf{k}-\mathbf{k}'|^2}{\sigma''\omega_{\mathbf{k}-\mathbf{k}'}^{L}}\right)^2\Bigg(\sigma\omega_{\mathbf{k}}^{T}\,I_{\mathbf{k}'}^{\sigma' L}\,I_{\mathbf{k}-\mathbf{k}'}^{\sigma'' L} - \sigma'\omega_{\mathbf{k}'}^{L}\,I_{\mathbf{k}-\mathbf{k}'}^{\sigma'' L}\,\frac{I_{\mathbf{k}}^{\sigma T}}{2}$$

$$-\sigma''\omega_{\mathbf{k}-\mathbf{k}'}^{L}\,I_{\mathbf{k}'}^{\sigma' L}\,\frac{I_{\mathbf{k}}^{\sigma T}}{2}\Bigg)\,\underbrace{\delta(\sigma\omega_{\mathbf{k}}^{T} - \sigma'\omega_{\mathbf{k}'}^{L} - \sigma''\omega_{\mathbf{k}-\mathbf{k}'}^{L})}$$

$$\boxed{\text{decay } T \leftrightarrow L + L}$$

$$+\pi\,\sigma\omega_{\mathbf{k}}^{T}\,\frac{e^2}{4T_e^2}\sum_{\sigma',\sigma''}\int d\mathbf{k}'\,\frac{\mu_{\mathbf{k}-\mathbf{k}'}\,(\mathbf{k}\times\mathbf{k}')^2}{k^2\,k'^2\,|\mathbf{k}-\mathbf{k}'|^2}\Bigg(\sigma\omega_{\mathbf{k}}^{T}\,I_{\mathbf{k}'}^{\sigma' L}\,\frac{I_{\mathbf{k}-\mathbf{k}'}^{\sigma'' S}}{\mu_{\mathbf{k}-\mathbf{k}'}}$$

$$
\begin{aligned}
&- \sigma' \omega_{\mathbf{k}'}^{L} \frac{I_{\mathbf{k}-\mathbf{k}'}^{\sigma'' S}}{\mu_{\mathbf{k}-\mathbf{k}'}} \frac{I_{\mathbf{k}}^{\sigma T}}{2} - \sigma'' \omega_{\mathbf{k}-\mathbf{k}'}^{L} I_{\mathbf{k}'}^{\sigma' L} \frac{I_{\mathbf{k}}^{\sigma T}}{2} \Bigg) \underbrace{\delta(\sigma \omega_{\mathbf{k}}^{T} - \sigma' \omega_{\mathbf{k}'}^{L} - \sigma'' \omega_{\mathbf{k}-\mathbf{k}'}^{S})}_{\boxed{\text{decay } T \leftrightarrow L + S}} \\
&+ \pi\, \sigma \omega_{\mathbf{k}}^{T} \frac{e^2}{4m_e^2} \sum_{\sigma',\sigma''} \int d\mathbf{k}' \, \frac{|\mathbf{k}-\mathbf{k}'|^2}{(\omega_{\mathbf{k}}^{T})^2 (\omega_{\mathbf{k}'}^{T})^2} \left(1 + \frac{(\mathbf{k}\cdot\mathbf{k}')^2}{k^2 k'^2} \right) \\
&\times \Bigg(\sigma \omega_{\mathbf{k}}^{T} I_{\mathbf{k}-\mathbf{k}'}^{\sigma'' L} \frac{I_{\mathbf{k}'}^{\sigma' T}}{2} - \sigma' \omega_{\mathbf{k}'}^{T} I_{\mathbf{k}-\mathbf{k}'}^{\sigma'' L} \frac{I_{\mathbf{k}}^{\sigma T}}{2} \\
&- \sigma'' \omega_{\mathbf{k}-\mathbf{k}'}^{L} \frac{I_{\mathbf{k}'}^{\sigma' T}}{2} \frac{I_{\mathbf{k}}^{\sigma T}}{2} \Bigg) \underbrace{\delta(\sigma \omega_{\mathbf{k}}^{T} - \sigma' \omega_{\mathbf{k}'}^{T} - \sigma'' \omega_{\mathbf{k}-\mathbf{k}'}^{L})}_{\boxed{\text{decay } T \leftrightarrow T + L}} \\
&- \frac{\sigma \omega_{\mathbf{k}}^{T}}{8n^2} \sum_{\sigma'} \int d\mathbf{k}' \int d\mathbf{p} \, \frac{(\mathbf{k}\times\mathbf{k}')^2}{k^2 k'^2} \, \delta[\sigma \omega_{\mathbf{k}}^{T} - \sigma' \omega_{\mathbf{k}'}^{L} - (\mathbf{k}-\mathbf{k}')\cdot\mathbf{v}] \\
&\times \Bigg[\underbrace{\frac{1}{4\pi^2} \left(\sigma' \omega_{\mathbf{k}'}^{L} \frac{I_{\mathbf{k}}^{\sigma T}}{2} - \sigma \omega_{\mathbf{k}}^{T} I_{\mathbf{k}'}^{\sigma' L} \right) [f_e(\mathbf{p}) + f_i(\mathbf{p})]}_{\boxed{\text{spont. scatt. } T \leftrightarrow L + p}} \\
&- \underbrace{I_{\mathbf{k}'}^{\sigma' L} \frac{I_{\mathbf{k}}^{\sigma T}}{2} (\mathbf{k}-\mathbf{k}') \cdot \frac{\partial f_i(\mathbf{p})}{\partial \mathbf{p}}}_{\boxed{\text{ind. scatt. } T \leftrightarrow L + i}} \Bigg]; \qquad (9.52)
\end{aligned}
$$

- $$
\begin{aligned}
\frac{\partial f_e(\mathbf{p})}{\partial t} &= \pi e^2 \sum_{\sigma=\pm1} \int d\mathbf{k} \, \frac{\mathbf{k}}{k} \cdot \frac{\partial}{\partial \mathbf{p}} \, \delta(\sigma \omega_{\mathbf{k}}^{L} - \mathbf{k}\cdot\mathbf{v}) \\
&\times \left(\frac{\sigma \omega_{\mathbf{k}}^{L}}{4\pi^2 k} f_e(\mathbf{p}) + I_{\mathbf{k}}^{\sigma L} \frac{\mathbf{k}}{k} \cdot \frac{\partial f_e(\mathbf{p})}{\partial \mathbf{p}} \right),
\end{aligned}
$$
- $$
\begin{aligned}
\frac{\partial f_i(\mathbf{p})}{\partial t} &= \pi e^2 \sum_{\sigma=\pm1} \int d\mathbf{k} \, \frac{\mathbf{k}}{k} \cdot \frac{\partial}{\partial \mathbf{p}} \mu_{\mathbf{k}} \, \delta(\sigma \omega_{\mathbf{k}}^{S} - \mathbf{k}\cdot\mathbf{v}) \\
&\times \left(\frac{\sigma \omega_{\mathbf{k}}^{L}}{4\pi^2 k} f_i(\mathbf{p}) + \frac{I_{\mathbf{k}}^{\sigma S}}{\mu_{\mathbf{k}}} \frac{\mathbf{k}}{k} \cdot \frac{\partial f_i(\mathbf{p})}{\partial \mathbf{p}} \right). \qquad (9.53)
\end{aligned}
$$

- Auxiliary Definitions

$$
\omega_{\mathbf{k}}^{L} = \omega_{pe} \left(1 + \frac{3}{2} k^2 \lambda_{De}^2 \right), \qquad \omega_{-\mathbf{k}}^{L} = -\omega_{\mathbf{k}}^{L},
$$

$$
\omega_{\mathbf{k}}^{S} = k c_S \sqrt{\frac{1 + 3T_i/T_e}{1 + k^2 \lambda_{De}^2}}, \qquad \omega_{-\mathbf{k}}^{S} = -\omega_{\mathbf{k}}^{S}, \qquad c_S = \sqrt{\frac{T_i}{m_e}},
$$

$$\omega_{\mathbf{k}}^{T} = \omega_{pe}\sqrt{1+\frac{c^2k^2}{\omega_{pe}^2}}, \qquad \omega_{-\mathbf{k}}^{T} = -\omega_{\mathbf{k}}^{T},$$

$$\mu_{\mathbf{k}} = |k|^3\lambda_{De}^3\sqrt{\frac{m_e}{m_i}\left(1+\frac{3T_i}{T_e}\right)}, \tag{9.54}$$

The summary of equations presented in (9.54) constitutes the general weak turbulence theory where all the terms appear in balanced form between spontaneous and induced processes. In Chapter 10, the final chapter, we discuss the applications of the electromagnetic weak turbulence theory.

10

Applications of Electromagnetic Klimontovich Weak Turbulence Theory

In Chapters 5 and 6 we applied the electrostatic Klimontovich weak turbulence theory to two specific problems. First was on the spontaneous emission of electrostatic fluctuations in thermal equilibrium plasma, which includes the emission of plasma eigenmodes as well as non-eignemodes. The discussion also pertained to plasmas that are slightly out of equilibrium. Collisional kinetic theory becomes relevant for such plasma states. It was shown that collisional processes bring the system back to thermal equilibrium. We then made another application to the plasma that can be considered as being far from equilibrium, and one in which collective excitation of plasma eigenmodes via instability dominates the system. Thus, in Chapter 6, we discussed the weak turbulence theory of gentle electron beam-plasma interaction and the excitation of Langmuir instability. The early saturation of Langmuir instability was shown to be well described by quasilinear theory, but for time scales longer than the quasilinear relaxation period, the mode coupling interaction involving decay and scattering enters the picture. It was shown by numerical analysis that the long time scale evolution of Langmuir and ion-sound turbulence involves the generation of energetic tail population in the electron velocity distribution function. We subsequently discussed the analytical theory of asymptotically steady-state electron distribution and Langmuir turbulence intensity, and showed that the solution corresponds to the electron kappa distribution. It was also shown that the electron kappa distribution is intimately related to the solar wind electrons measured by artificial spacecraft.

In this chapter, we discuss the consequence of retaining the electromagnetic effects. The first example is the generation of electromagnetic fluctuations in thermal plasmas. For plasmas in thermal equilibrium, electrostatic spontaneous emission theory explains the natural emission and absorption of Langmuir and ion-sound waves. Electromagnetic generalization, however, cannot explain the emission of transverse electromagnetic radiation in the high-frequency regime, but instead, such a theory does predict the emission of magnetic field fluctuations in the low-frequency regime. The non-propagating, or aperiodic, magnetic field fluctuations spontaneously generated in thermal unmagnetized plasmas have potentially significant cosmological implications, as such a mechanism may explain the origin of large-scale magnetic field in the early universe, a topic that was explored by Schlickeiser (2012). The emission of high-frequency electromagnetic waves, that is, the radiation, however, requires nonlinear correction, and it will be shown that such a correction indeed constitutes a new radiation emission mechanism.

For systems slightly out of equilibrium, we already discussed that the collisional kinetic theory is applicable. Under electromagnetic formalism, the standard Balescu–Lenard–Landau equation (Landau, 1937; Balescu, 1960; Guernsey, 1960; Lenard, 1960) is generalized to Beliaev–Budker equation (Beliaev and Budker, 1956) relevant to fully electromagnetic collisional process. Beliaev–Budker equation becomes important for relativistic plasmas.

The final application of fully electromagnetic Klimontovich weak turbulence theory is again on the gentle or weak electron beam-plasma interaction. The inclusion of electromagnetic effects leads to a new physical process, namely, the emission of transverse electromagnetic radiation at the plasma frequency and/or its harmonics. Such a non-thermal radiation emission process, which can be viewed as a process of partially converting the energy associated with the beam-generated Langmuir turbulence into the radiation energy by nonlinear processes, is the basic radiation mechanism responsible for type II and type III solar radio bursts. Such a radio emission process is generally called the "plasma emission," and it has been extensively discussed in the solar radio physics context. Nevertheless, complete numerical solution starting from the electron beam-plasma induced Langmuir instability, mode coupling, and partial conversion to electromagnetic radiation has not been demonstrated until quite recently (Ziebell et al., 2014a, 2015). We will overview such a solution in the present chapter.

10.1 Spontaneous Emission of Magnetic Field Fluctuations

Consider the general spectral balance equation (9.30). For thermal plasmas, we may ignore nonlinear terms such that we may simply balance the linear response terms with the spontaneous emission terms, which, in the non-relativistic limit, is expressed by

$$
\begin{aligned}
&\epsilon_\parallel(\mathbf{k},\omega)\left\langle\delta E_\parallel^2\right\rangle_{\mathbf{k},\omega}+\Lambda_\perp(\mathbf{k},\omega)\left\langle\delta E_\perp^2\right\rangle_{\mathbf{k},\omega}\\
&\quad=\sum_a\frac{(4\pi e_a)^2}{(2\pi)^3\omega^2}\int d\mathbf{v}\left[\frac{(\mathbf{k}\cdot\mathbf{v})^2}{k^2}\frac{1}{\epsilon_\parallel^*(\mathbf{k},\omega)}+\frac{(\mathbf{k}\times\mathbf{v})^2}{k^2}\frac{1}{\Lambda_\perp^*(\mathbf{k},\omega)}\right]\\
&\quad\times\delta(\omega-\mathbf{k}\cdot\mathbf{v})f_a(\mathbf{v}),
\end{aligned}
\tag{10.1}
$$

where $\Lambda_\perp(\mathbf{k},\omega)$, defined in (7.32), is given by

$$\Lambda_\perp(\mathbf{k},\omega) = \epsilon_\perp(\mathbf{k},\omega) - \frac{c^2k^2}{\omega^2}, \tag{10.2}$$

and the longitudinal and transverse components of the linear dielectric tensor are defined by

$$\begin{aligned}
\epsilon_\parallel(\mathbf{k},\omega) &= 1 + \sum_a \frac{4\pi e_a^2}{m_a k^2} \int d\mathbf{v}\, \frac{\mathbf{k}\cdot\partial f_a(\mathbf{v})/\partial\mathbf{v}}{\omega - \mathbf{k}\cdot\mathbf{v} + i0}, \\
\epsilon_\perp(\mathbf{k},\omega) &= 1 + \sum_a \frac{2\pi e_a^2}{m_a\omega} \int \frac{d\mathbf{v}}{\omega - \mathbf{k}\cdot\mathbf{v} + i0} \frac{(\mathbf{k}\times\mathbf{v})\times\mathbf{k}}{k^2} \cdot \frac{\partial f_a(\mathbf{v})}{\partial\mathbf{v}}.
\end{aligned} \tag{10.3}$$

From (10.1) we readily obtain

$$\begin{aligned}
\left\langle \delta E_\parallel^2 \right\rangle_{\mathbf{k},\omega} &= \sum_a \frac{(4\pi e_a)^2}{(2\pi)^3\omega^2|\epsilon_\parallel(\mathbf{k},\omega)|^2} \int d\mathbf{v}\, \frac{(\mathbf{k}\cdot\mathbf{v})^2}{k^2}\, \delta(\omega - \mathbf{k}\cdot\mathbf{v})\, f_a(\mathbf{v}), \\
\left\langle \delta E_\perp^2 \right\rangle_{\mathbf{k},\omega} &= \sum_a \frac{(4\pi e_a)^2}{(2\pi)^3\omega^2|\Lambda_\perp(\mathbf{k},\omega)|^2} \int d\mathbf{v}\, \frac{(\mathbf{k}\times\mathbf{v})^2}{k^2}\, \delta(\omega - \mathbf{k}\cdot\mathbf{v})\, f_a(\mathbf{v}).
\end{aligned} \tag{10.4}$$

For thermal equilibrium we have

$$\begin{aligned}
\epsilon_\parallel(\mathbf{k},\omega) &= 1 - \sum_a \frac{4\pi e_a^2}{T\omega} \int d\mathbf{v}\, \frac{1}{\omega - \mathbf{k}\cdot\mathbf{v} + i0} \frac{\omega(\mathbf{k}\cdot\mathbf{v})}{k^2}\, f_a(\mathbf{v}) \\
&= 1 + \sum_a \frac{2\omega_{pa}^2}{k^2 v_{Ta}^2}\left[1 + \frac{\omega}{kv_{Ta}}\, Z\left(\frac{\omega}{kv_{Ta}}\right)\right], \\
\epsilon_\perp(\mathbf{k},\omega) &= 1 - \sum_a \frac{2\pi e_a^2}{T\omega} \int d\mathbf{v}\, \frac{1}{\omega - \mathbf{k}\cdot\mathbf{v} + i0} \frac{(\mathbf{k}\times\mathbf{v})^2}{k^2}\, f_a(\mathbf{v}) \\
&= 1 + \sum_a \frac{\omega_{pa}^2}{\omega k v_{Ta}}\, Z\left(\frac{\omega}{kv_{Ta}}\right),
\end{aligned} \tag{10.5}$$

where $v_{Ta} = (2T/m_a)^{1/2}$ and $Z(\zeta)$ is the plasma dispersion function – see (5.5). As a consequence, we have

$$\begin{aligned}
\operatorname{Im}\epsilon_\parallel(\mathbf{k},\omega) &= \sum_a \frac{4\pi^2 e_a^2}{T\omega} \int d\mathbf{v}\, \delta(\omega - \mathbf{k}\cdot\mathbf{v})\, \frac{(\mathbf{k}\cdot\mathbf{v})^2}{k^2}\, f_a(\mathbf{v}), \\
\operatorname{Im}\epsilon_\perp(\mathbf{k},\omega) &= \sum_a \frac{2\pi^2 e_a^2}{T\omega} \int d\mathbf{v}\, \delta(\omega - \mathbf{k}\cdot\mathbf{v})\, \frac{(\mathbf{k}\times\mathbf{v})^2}{k^2}\, f_a(\mathbf{v}),
\end{aligned} \tag{10.6}$$

from which we obtain

$$\left\langle \delta E_{\parallel}^{2} \right\rangle_{\mathbf{k},\omega} = \frac{T}{2\pi^{3}\omega} \frac{\operatorname{Im}\epsilon_{\parallel}(\mathbf{k},\omega)}{|\epsilon_{\parallel}(\mathbf{k},\omega)|^{2}},$$

$$\left\langle \delta E_{\perp}^{2} \right\rangle_{\mathbf{k},\omega} = \frac{T}{\pi^{3}\omega} \frac{\operatorname{Im}\epsilon_{\perp}(\mathbf{k},\omega)}{|\epsilon_{\perp}(\mathbf{k},\omega) - c^{2}k^{2}/\omega^{2}|^{2}}. \tag{10.7}$$

Note that the magnetic field fluctuations also follow from (10.7),

$$\left\langle \delta B^{2} \right\rangle_{\mathbf{k},\omega} = \frac{c^{2}k^{2}}{\omega^{2}} \frac{T}{\pi^{3}\,\omega} \frac{\operatorname{Im}\epsilon_{\perp}(\mathbf{k},\omega)}{|\epsilon_{\perp}(\mathbf{k},\omega) - c^{2}k^{2}/\omega^{2}|^{2}}. \tag{10.8}$$

The first expression in (10.7) is the same as (5.5), and it describes the spectrum of spontaneously emitted electrostatic fluctuations. According to (10.7), however, transverse electromagnetic fluctuations satisfying the condition for superluminosity, or fast wave condition,

$$ck/\omega < 1, \tag{10.9}$$

cannot be spontaneously emitted even for non-relativistic treatment. For relativistic case, the emission is strictly forbidden, since no particle speed can exceed the speed of light *in vacuo*, c. Even for non-relativistic case, fast mode emission will be virtually non-existent. This is because for thermal equilibrium distribution function,

$$f_{a}(\mathbf{v}) = \frac{n_{a}}{(2\pi T/m_{a})^{3/2}} \exp\left(-\frac{m_{a}v^{2}}{2T}\right), \tag{10.10}$$

the superluminal region will be characterized by a negligible fluctuation intensity since the overall emission intensity is proportional to

$$\exp\left(-\frac{m_{a}\omega^{2}}{2k^{2}T}\right) = \exp\left(-\frac{c^{2}}{v_{Ta}^{2}} \frac{\omega^{2}}{c^{2}k^{2}}\right) \ll 1. \tag{10.11}$$

For sub-luminal regime,

$$ck/\omega > 1, \tag{10.12}$$

on the other hand, thermal plasma may emit electromagnetic fluctuations. Sub-luminous electromagnetic fluctuations in unmagnetized plasmas have been discussed by Tajima et al. (1992); Yoon (2007); Tautz and Schlickeiser (2007); Schlickeiser and Yoon (2012); Felten et al. (2013); and Lazar et al. (2012). Shown in Figure 10.1 is the contour plot of the sub-luminous magnetic field fluctuations described by (10.8). The spontaneous emission of fluctuating B field is closely associated with the Weibel instability. Recall that the unstable Weibel mode in the presence of temperature anisotropy is a purely growing mode with $\operatorname{Re}\omega = 0$. The spontaneously emitted magnetic field fluctuation spectrum has peak intensity for $\operatorname{Re}\omega = 0$, which implies that the magnetic field fluctuations can be described as being quasistationary. This property is identical to that of Weibel instability. Thus, the spontaneously emitted B field fluctuations for thermal plasmas provide the seed perturbation, which will be amplified as the Weibel instability when excessive perpendicular free energy source becomes available.

Here, we should note that the Weibel mode is characterized by purely aperiodic transverse fluctuations with $\mathrm{Re}\,\omega = 0$ and $\mathrm{Im}\,\omega =$ finite. Since the basic conjecture of the main body of this book is based upon the assumption of weak wave growth or damping, $\mathrm{Re}\,\omega \gg |\mathrm{Im}\,\omega|$, the Weibel mode, whether it is stable or unstable, does not naturally lend itself to the formalism discussed in the main body of this book. The Weibel mode is a classic example of the "reactive" mode – see the discussion in Appendix E, where the weak turbulence theory of reactively unstable mode is considered under electrostatic approximation. The natural question is whether it is possible to extend the electromagnetic weak turbulence formalism for reactively unstable modes, which includes aperiodic Weibel mode. Schlickeiser and Yoon (2012) and Yoon et al. (2014) discussed the re-formulation of quasilinear kinetic theory that includes the effects of spontaneously emitted and reabsorbed Weibel mode. Schlickeiser (2012) applied the theory of aperiodic magnetic field fluctuations in order to put forth a novel theory for origin of cosmic magnetic fields in the early universe. In standard astrophysics, the origin of cosmic magnetic fields is closely associated with the so-called Biermann battery mechanism (Hao et al., 2008), combined with some magnetic dynamo action (Kulsrud and Zweibel, 2008). Schlickeiser (2012), on the other hand, suggested a novel idea based upon the plasma kinetic theory.

While the linear formulae for transverse electric and magnetic field fluctuations (10.7) and (10.8) are appropriate for describing the sub-luminous "Weibel" mode branch of the electromagnetic wave spectrum, these linear theoretical formulae cannot describe the emission of fast electromagnetic waves in unmagnetized plasmas. That is, according to the linear theory, unmagnetized plasmas cannot radiate electromagnetic waves spontaneously. However, the situation changes when we include nonlinear terms in the general weak turbulence theoretical formalism.

10.2 Electromagnetic Radiation in Thermal Plasmas

To begin the discussion, let us focus our attention on the transverse eigenmode, satisfying the dispersion relation, $\omega = \omega_{\mathbf{k}}^{T}$. In general, spontaneous emission can take place for all ω including the eigenmode as well as those ω's that do not satisfy the linear dispersion relation. We focus our attention on the eigenmode solution near the transverse mode only, however, since we are interested in the emission of radiation in thermal plasmas. This means that the equations of electromagnetic Klimontovich weak turbulence theory, (9.50)–(9.53), are applicable. For thermal

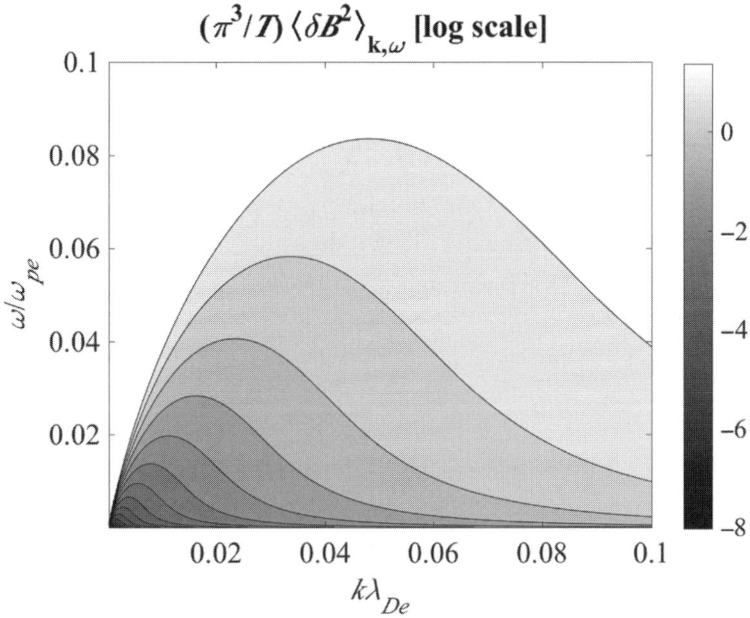

Figure 10.1 Spontaneously emitted magnetic field fluctuation spectrum, (10.8), plotted against normalized frequency ω/ω_p and dimensionless wave number $k\lambda_{De}$.

plasmas, however, all physical quantities are in stationary state, $\partial/\partial t = 0$. Consequently, we may assume that the particle distribution function and wave intensities are stationary and isotropic in that we may take $f_a(\mathbf{v}) = f_a(v)$, $a = e, i$, $I_{\mathbf{k}}^{\sigma\alpha} = I_k^{\alpha}$, and $I_{\mathbf{k}}^{\sigma T} = I_k^T$. We may further assume that the particle distributions are given by thermal Maxwell–Boltzmann distribution. In the wave kinetic equations for longitudinal electrostatic waves, we may ignore nonlinear mode coupling terms, since the wave intensities for thermal equilibrium plasmas are low so that generally nonlinear terms are inconsequential. However, in the transverse wave equation, since the linear response term is absent owing to the fact that linear wave-particle resonance cannot be satisfied between the particles and the fast radiation mode, nonlinear mode coupling terms become essential. Among the nonlinear mode coupling terms, we ignore decay processes since these terms only involve wave energy and momentum exchanges but do not involve particles. In the steady-state solution that we seek, we pay attention to asymptotic quasi-equilibrium between the electrons and electrostatic and electromagnetic fluctuations. Hence, we may ignore low-frequency ion-sound mode at the outset.

The starting point is the transverse wave kinetic equation (9.52) in which spontaneous and induced emission terms as well as various three-wave interaction terms depicting various decay processes are ignored. Resorting to the velocity (rather than momentum) representation, the relevant equation is

$$
\begin{aligned}
\frac{\partial}{\partial t}\frac{I_{\mathbf{k}}^{\sigma T}}{2} &= \sigma\omega_{\mathbf{k}}^{T}\frac{e^2}{2m_e^2\omega_{pe}^2}\sum_{\sigma'}\int d\mathbf{k}'\int d\mathbf{v}\,\frac{(\mathbf{k}\times\mathbf{k}')^2}{k^2k'^2}\\
&\quad\times\delta[\sigma\omega_{\mathbf{k}}^{T}-\sigma'\omega_{\mathbf{k}'}^{L}-(\mathbf{k}-\mathbf{k}')\cdot\mathbf{v}]\\
&\quad\times\left[\frac{ne^2}{\omega_{pe}^2}\left(\sigma\omega_{\mathbf{k}}^{T}\,I_{\mathbf{k}'}^{\sigma' L}-\sigma'\omega_{\mathbf{k}'}^{L}\,\frac{I_{\mathbf{k}}^{\sigma T}}{2}\right)[F_e(\mathbf{v})+F_i(\mathbf{v})]\right.\\
&\quad\left.+\pi\,\frac{m_e}{m_i}\,I_{\mathbf{k}'}^{\sigma' L}\,\frac{I_{\mathbf{k}}^{\sigma T}}{2}\,(\mathbf{k}-\mathbf{k}')\cdot\frac{\partial F_i(\mathbf{v})}{\partial\mathbf{v}}\right].
\end{aligned}
\tag{10.13}
$$

Making use of $(\mathbf{k}\times\mathbf{k}')^2 = k^2k'^2-(\mathbf{k}\cdot\mathbf{k}')^2$, and assuming Maxwellian particle distributions, we have

$$
\begin{aligned}
0 &= \int d\mathbf{k}'\,\frac{k^2k'^2-(\mathbf{k}\cdot\mathbf{k}')^2}{k^2k'^2}\int d\mathbf{v}\,\delta[\omega_k^T-\omega_{k'}^L+|\mathbf{k}-\mathbf{k}'|\,v_z]\\
&\quad\times\left\{\frac{T}{4\pi^2}\left(\omega_k^T\,I_{k'}^L-\omega_{k'}^L\,\frac{I_k^T}{2}\right)\left[e^{-m_e(v_x^2+v_y^2+v_z^2)/2T}+\left(\frac{m_i}{m_e}\right)^{3/2}e^{-m_i(v_x^2+v_y^2+v_z^2)/2T}\right]\right.\\
&\quad\left.-I_{k'}^L\,\frac{I_k^T}{2}\,(\omega_k^T-\omega_{k'}^L)\left(\frac{m_i}{m_e}\right)^{3/2}e^{-m_i(v_x^2+v_y^2+v_z^2)/2T}\right\},
\end{aligned}
\tag{10.14}
$$

where we have assumed steady state, $\partial/\partial t = 0$, and have assumed that the vector $\mathbf{k}-\mathbf{k}'$ lies along z axis without loss of generality. We have also ignored $\sigma' = -\sigma$ at the outset and ignored the σ dependence associated with the wave intensities, since they are assumed to be isotropic, hence independent of the propagation direction. We have then chosen $\sigma = \sigma' = 1$.

Integrations along v_x and v_y are trivial. Upon carrying out the v_z integration by virtue of the delta function resonance condition, we have

$$
\begin{aligned}
0 &= \int d\mathbf{k}'\,\frac{k^2k'^2-(\mathbf{k}\cdot\mathbf{k}')^2}{k^2k'^2\,|\mathbf{k}-\mathbf{k}'|}\\
&\quad\times\left\{\frac{T}{4\pi^2}\left(\omega_k^T\,I_{k'}^L-\omega_{k'}^L\,\frac{I_k^T}{2}\right)\exp\!\left(-\frac{m_e}{2T}\frac{(\omega_k^T-\omega_{k'}^L)^2}{|\mathbf{k}-\mathbf{k}'|^2}\right)\right.\\
&\quad+\left[\omega_k^T\left(\frac{T}{4\pi^2}-\frac{I_k^T}{2}\right)I_{k'}^L-\omega_{k'}^L\left(\frac{T}{4\pi^2}-I_{k'}^L\right)\frac{I_k^T}{2}\right]\\
&\quad\left.\times\left(\frac{m_i}{m_e}\right)^{1/2}\exp\!\left(-\frac{m_i}{2T}\frac{(\omega_k^T-\omega_{k'}^L)^2}{|\mathbf{k}-\mathbf{k}'|^2}\right)\right\}.
\end{aligned}
\tag{10.15}
$$

At this point, we note that the most important contribution to the $\mathbf{k}'$ integral should come from k' that satisfies $\omega_k^T \sim \omega_{k'}^L$, or $k \sim k_*$, where

$$k_*^2 = \frac{2}{3}\frac{c^2}{v_{Te}^2}\,k^2. \tag{10.16}$$

As a consequence, we may simply consider the quantity within the square bracket associated with the ion thermal distribution, and also the quantities within the square bracket associated with the electron thermal distribution, and set them equal to zero separately. Upon replacing k' by k_*, we thus have

$$\begin{aligned}
&\omega_k^T\, I_{k_*}^L - \omega_{k_*}^L\,\frac{I_k^T}{2} = 0,\\
&\omega_k^T\left(\frac{T}{4\pi^2} - \frac{I_k^T}{2}\right) I_{k_*}^L - \omega_{k_*}^L\left(\frac{T}{4\pi^2} - I_{k_*}^L\right)\frac{I_k^T}{2} = 0.
\end{aligned} \tag{10.17}$$

Upon making use of

$$3k_*^2\lambda_{De}^2 = \frac{c^2k^2}{\omega_{pe}^2}, \tag{10.18}$$

and making use of (see [6.19])

$$I_k^L = \frac{T}{4\pi^2}\frac{1}{1+3k^2\lambda_{De}^2}, \tag{10.19}$$

we immediately obtain

$$I_k^T = \frac{T}{2\pi^2}\frac{1}{1+c^2k^2/\omega_{pe}^2} = \frac{T}{2\pi^2}\frac{\omega_{pe}^2}{(\omega_k^T)^2}, \tag{10.20}$$

for both equations in (10.17). This is the electric field spectral intensity for spontaneously emitted radiation in thermal plasmas. In short, thermal plasmas spontaneously emit (and reabsorb) all the eigenmodes naturally excited in plasmas, including Langmuir, ion-sound, and transverse electromagnetic modes, but in order to discuss the fast electromagnetic branch, one must include nonlinear effects. For longitudinal modes (L and S), linear treatment is sufficient, as we have already shown in Section 5.1. The spontaneous emission of electromagnetic radiation discussed in this section was first discovered by Ziebell et al. (2014b), who also confirmed the theoretical findings by numerical particle-in-cell (PIC) simulation.

Next, we turn to another problem associated with quasi-equilibrium state. If the plasma is close to but not exactly at thermal equilibrium state, then collisional processes bring such a plasma back to thermal equilibrium state. This was discussed in Chapter 5 under electrostatic formalism. However, when the plasma is relativistic, the simple electrostatic approximation is no longer applicable and one must resort to fully electromagnetic formalism for collisional relaxation process. We next discuss the derivation of collisional kinetic equation for relativistic plasma.

10.3 Electromagnetic Collisional Kinetic Equation

To discuss the collisional processes for relativistic plasmas slightly out of equilibrium, we return to (9.34), which is repeated here for the sake of convenience:

$$\frac{\partial f_a}{\partial t} = \pi e_a^2 \int d\mathbf{k} \int d\omega \left(\frac{\mathbf{k}}{k}\cdot\frac{\partial}{\partial \mathbf{p}}\right)\Bigg\{\delta(\omega-\mathbf{k}\cdot\mathbf{v})\Bigg[\frac{1}{2\pi^3}\Bigg(\frac{\mathrm{Im}\,\epsilon_\parallel(\mathbf{k},\omega)}{k\,|\epsilon_\parallel(\mathbf{k},\omega)|^2} + \frac{(\mathbf{k}\times\mathbf{v})^2\omega^2\,\mathrm{Im}\,\epsilon_\perp(\mathbf{k},\omega)}{k\,|\omega^2\epsilon_\perp(\mathbf{k},\omega)-c^2k^2|^2}\Bigg) f_a + \left(\langle\delta E_\parallel^2\rangle_{\mathbf{k},\omega} + \frac{1}{2}\frac{(\mathbf{k}\times\mathbf{v})^2}{\omega^2}\langle\delta E_\perp^2\rangle_{\mathbf{k},\omega}\right)\frac{\mathbf{k}}{k}\cdot\frac{\partial f_a}{\partial \mathbf{p}}\Bigg]\Bigg\}. \tag{10.21}$$

If the plasma is slightly out of equilibrium, yet there is no free energy source to excite any collective modes, that is, in the absence of instabilities, then one does not need to solve the wave kinetic equation, but instead, we may employ the electric field spectra taken from the spontaneous emission theory, namely (10.4), which is now represented in momentum formalism,

$$\langle\delta E_\parallel^2\rangle_{\mathbf{k},\omega} = \sum_b \frac{2e_b^2}{\pi\omega^2|\epsilon_\parallel(\mathbf{k},\omega)|^2}\int d\mathbf{p}'\,\frac{(\mathbf{k}\cdot\mathbf{v}')^2}{k^2}\,\delta(\omega-\mathbf{k}\cdot\mathbf{v}')\,f_b(\mathbf{p}'),$$

$$\langle\delta E_\perp^2\rangle_{\mathbf{k},\omega} = \sum_b \frac{2e_b^2\omega^2}{\pi\,|\omega^2\epsilon_\perp(\mathbf{k},\omega)-c^2k^2|^2} \times \int d\mathbf{p}'\,\frac{(\mathbf{k}\times\mathbf{v}')^2}{k^2}\,\delta(\omega-\mathbf{k}\cdot\mathbf{v}')\,f_b(\mathbf{p}'), \tag{10.22}$$

where the relativistic momentum-velocity relationship is assumed, $\mathbf{p} = m_a\gamma\mathbf{v}$ or $\mathbf{v} = \mathbf{p}/(m_a\gamma)$ with the relativistic mass factor defined by $\gamma = (1 - v^2/c^2)^{-1/2} = [1 + (p/m_a c)^2]^{1/2}$. The dielectric constants and their imaginary parts are given in momentum representation by

$$\epsilon_\parallel(\mathbf{k},\omega) = 1 + \sum_b \frac{4\pi e_b^2}{k^2}\int d\mathbf{p}'\,\frac{1}{\omega-\mathbf{k}\cdot\mathbf{v}'+i0}\,\mathbf{k}\cdot\frac{\partial f_b(\mathbf{p}')}{\partial \mathbf{p}'},$$

$$\epsilon_\perp(\mathbf{k},\omega) = 1 + \sum_b \frac{2\pi e_b^2}{\omega}\int d\mathbf{p}'\,\frac{1}{\omega-\mathbf{k}\cdot\mathbf{v}'+i0}\,\frac{(\mathbf{k}\times\mathbf{v}')\times\mathbf{k}}{k^2} \cdot\left[\frac{\partial f_b(\mathbf{p}')}{\partial \mathbf{p}'} + \frac{\mathbf{k}}{\omega}\times\left(\mathbf{v}'\times\frac{\partial f_b(\mathbf{p}')}{\partial \mathbf{p}'}\right)\right], \tag{10.23}$$

and

$$\operatorname{Im}\epsilon_{\parallel}(\mathbf{k},\omega) = -\sum_b \frac{4\pi^2 e_b^2}{k^2}\int d\mathbf{p}'\,\delta(\omega-\mathbf{k}\cdot\mathbf{v}')\,\mathbf{k}\cdot\frac{\partial f_b(\mathbf{p}')}{\partial\mathbf{p}'},\tag{10.24}$$

$$\operatorname{Im}\epsilon_{\perp}(\mathbf{k},\omega) = -\sum_b \frac{2\pi^2 e_b^2}{\omega^2}\int d\mathbf{p}'\,\delta(\omega-\mathbf{k}\cdot\mathbf{v}')\,\frac{[(\mathbf{k}\times\mathbf{v}')\times\mathbf{k}]^2}{k^4}\,\mathbf{k}\cdot\frac{\partial f_b(\mathbf{p}')}{\partial\mathbf{p}'}.$$

Substituting (10.22) and (10.24) to (10.21) and carrying out the ω integration, one has

$$\begin{aligned}\frac{\partial f_a(\mathbf{p})}{\partial t} &= \sum_b e_a^2 e_b^2\int d\mathbf{p}'\int d\mathbf{k}\left(\frac{\mathbf{k}}{k}\cdot\frac{\partial}{\partial\mathbf{p}}\right)\delta(\mathbf{k}\cdot\mathbf{v}-\mathbf{k}\cdot\mathbf{v}')\\ &\times\left(\frac{2}{k^2|\epsilon_{\parallel}(\mathbf{k},\mathbf{k}\cdot\mathbf{v})|^2}+\frac{(\mathbf{k}\times\mathbf{v})^2(\mathbf{k}\times\mathbf{v}')^2}{k^2|(\mathbf{k}\cdot\mathbf{v})^2\epsilon_{\perp}(\mathbf{k},\mathbf{k}\cdot\mathbf{v})-c^2k^2|^2}\right)\\ &\times\left(\frac{\mathbf{k}}{k}\cdot\frac{\partial f_a(\mathbf{p})}{\partial\mathbf{p}}\,f_b(\mathbf{p}')-\frac{\mathbf{k}}{k}\cdot\frac{\partial f_b(\mathbf{p}')}{\partial\mathbf{p}'}\,f_a(\mathbf{p})\right),\end{aligned}\tag{10.25}$$

where we have invoked $(\mathbf{k}\times\mathbf{v})^2 = [(\mathbf{k}\times\mathbf{v})\times\mathbf{k}]^2/k^2$. Under the assumption that $\int d\mathbf{p}\,(\mathbf{k}\times\mathbf{v})\,f_a(\mathbf{p}) = 0$, we have a useful identity

$$(\mathbf{k}\times\mathbf{v})^2(\mathbf{k}\times\mathbf{v}')^2 = 2[(\mathbf{k}\times\mathbf{v})\cdot(\mathbf{k}\times\mathbf{v}')]^2.\tag{10.26}$$

This can easily be checked by direct substitution for $\mathbf{k} = k\hat{\mathbf{z}}$. Then one can easily show that

$$\begin{aligned}(\mathbf{k}\times\mathbf{v})^2(\mathbf{k}\times\mathbf{v}')^2 &= k^4(v_x^2 v'^2_x + v_y^2 v'^2_y + v_x^2 v'^2_y + v_y^2 v'^2_x),\\ [(\mathbf{k}\times\mathbf{v})\cdot(\mathbf{k}\times\mathbf{v}')]^2 &= k^4(v_x^2 v'^2_x + v_y^2 v'^2_y + 2v_x v_y v'_x v'_y).\end{aligned}$$

Upon making use of

$$\begin{aligned}\int dp'_x\int dp'_y\, v'^2_x f_b(\mathbf{p}') &= \int dp'_x\int dp'_y\, v'^2_y f_b(\mathbf{p}'),\\ \int dp'_x\int dp'_y\, v'_x f_b(\mathbf{p}') = 0 &= \int dp'_x\int dp'_y\, v'_y f_b(\mathbf{p}'),\end{aligned}$$

one can prove the relationship (10.26). Thus, making use of (10.26), we obtain

$$\begin{aligned}\frac{\partial f_a(\mathbf{p})}{\partial t} &= \frac{\partial}{\partial p_i}\int d\mathbf{p}'\,Q_{ij}\left(\frac{\partial f_a(\mathbf{p})}{\partial p_j}\,f_b(\mathbf{p}')-\frac{\partial f_b(\mathbf{p}')}{\partial p'_j}\,f_a(\mathbf{p})\right),\\ Q_{ij} &= 2\sum_b e_a^2 e_b^2\int d\mathbf{k}\,\frac{k_i k_j}{k^4}\,\delta(\mathbf{k}\cdot\mathbf{v}-\mathbf{k}\cdot\mathbf{v}')\\ &\times\left(\frac{1}{|\epsilon_{\parallel}(\mathbf{k},\mathbf{k}\cdot\mathbf{v})|^2}+\frac{[(\mathbf{k}\times\mathbf{v})\cdot(\mathbf{k}\times\mathbf{v}')]^2}{|(\mathbf{k}\cdot\mathbf{v})^2\epsilon_{\perp}(\mathbf{k},\mathbf{k}\cdot\mathbf{v})-c^2k^2|^2}\right).\end{aligned}\tag{10.27}$$

This is the electromagnetic, or, equivalently, relativistic, generalization of the Balescu–Lenard equation (5.22).

One can also derive the generalization to the Landau equation (5.31), which was first derived by Beliaev and Budker (1956). In this approximation, we simply take

$$\epsilon_{\parallel} \to 1, \qquad \epsilon_{\perp} \to 1. \tag{10.28}$$

Then we get

$$Q_{ij} = 2\sum_b e_a^2 e_b^2 \int d\mathbf{k}\, \frac{k_i k_j}{k^4}\, \delta(\mathbf{k}\cdot\mathbf{v} - \mathbf{k}\cdot\mathbf{v}')\left(1 + \frac{[(\mathbf{k}\times\mathbf{v})\cdot(\mathbf{k}\times\mathbf{v}')]^2}{|(\mathbf{k}\cdot\mathbf{v})^2 - c^2k^2|^2}\right). \tag{10.29}$$

As with the Landau equation, we must replace the limit of k integral from $k_{\min} = \lambda_{De}^{-1}$ to $k_{\max} = T/e^2$. Making use of

$$(\mathbf{k}\times\mathbf{v})\cdot(\mathbf{k}\times\mathbf{v}') = k^2(\mathbf{v}\cdot\mathbf{v}') - (\mathbf{k}\cdot\mathbf{v})(\mathbf{k}\cdot\mathbf{v}'), \tag{10.30}$$

we obtain

$$\begin{aligned}\frac{\partial f_a(\mathbf{p})}{\partial t} &= 2\sum_b e_a^2 e_b^2 \frac{\partial}{\partial p_i}\int d\mathbf{p}' \int d\mathbf{k}\,[c^2 - (\mathbf{v}\cdot\mathbf{v}')]^2\,\delta(\mathbf{k}\cdot\mathbf{v} - \mathbf{k}\cdot\mathbf{v}')\\ &\quad\times \frac{k_i k_j}{[c^2k^2 - (\mathbf{k}\cdot\mathbf{v})^2]^2}\left(\frac{\partial}{\partial p_j} - \frac{\partial}{\partial p_j'}\right) f_a(\mathbf{p}) f_b(\mathbf{p}').\end{aligned} \tag{10.31}$$

This is the relativistic version of the Landau collisional kinetic equation, with appropriate cutoffs in the k integral implicitly assumed.

10.4 Plasma Emission by EM Weak Turbulence Process

Solar radio burst phenomena in the meter wavelengths were identified and subsequently classified into several types since the decade of 1950s (Wild, 1950, 1951; Wild and McCready, 1950a,b; Wild et al., 1954, 1959). Among the five different types of solar radio bursts, type III radio bursts are the most studied and well-understood category in a comparative sense. Most models agree that fast electrons escaping from active regions in the Sun, that is, solar flares, first excite Langmuir instability by electron beam-plasma interaction. Subsequently, nonlinear processes partially convert the electron beam energy into radiation. The details of the conversion process can be described within the framework of electromagnetic weak turbulence theory. A similar process may also be operative for type II emissions (Roberts, 1959; Wild et al., 1959). Type II emissions are associated with the electrons accelerated at the interplanetary shock wave, known as the coronal mass ejection, generated during solar active events.

For general solar and coronal physics, the readers may refer to an excellent monograph by Aschwanden (2005). For the physics of solar wind, an excellent reference may be the monograph by Meyer-Vernet (2007). For general space plasma physics, among excellent books are those by Baumjohann and Treumann (1997); Treumann and Baumjohann (1997); Parks (2004); Cravens (1997); and Kivelson and Russell (1995), just to mention a few.

Both type II and III radio bursts are often associated with the fundamental/harmonic two-band structure, although it is more common for type II bursts than for type III emissions. It is commonly believed that the fundamental/harmonic refers to the plasma frequency or its harmonic, hence the process involved in these solar radio bursts is called the "plasma emission." As the type II or type III radio generating electrons travel outward from the solar coronal source region into the interplanetary space, they encounter decreasing background number density, or, equivalently, decreasing local plasma frequency. As a result, these emissions exhibit characteristic descending tone in frequency-time plot.

There are excellent review articles and monographs on the subject of type III radio bursts, as well as some representative research papers. There are too numerous references so that it is practically impossible to cite them all – see, e.g., Wild et al. (1963); Kundu (1965); Zheleznyakov (1970); Zaitsev et al. (1972); Wild and Smerd (1972); Stewart (1974); Fainberg and Stone (1974); Lin et al. (1973, 1981, 1986); Smerd (1976); Rosenberg (1976); Melrose (1980a,b, 1986); Goldman (1983); Suzuki and Dulk (1985); Dulk (1985); Reiner et al. (1992); Ergun et al. (1998); Robinson and Cairns (1998a,b,c); Mann et al. (1999); Cane et al. (2002); Reiner et al. (1992, 2009); Saint-Hilaire et al. (2013); Reid and Ratcliffe (2014), etc. – but perhaps one of the most comprehensive reviews of the progress, which was made on the topic of solar radio emission up to the decade of 1980s, may be the monograph edited by McLean and Labrum (1985). If one is interested in detailed solar radio physics from the perspective of plasma astrophysics, one is referred to the excellent two-volume monograph by Melrose (1980a).

The first theoretical model of plasma emission was proposed by Ginzburg and Zheleznyakov (1958), which has undergone considerable improvements ever since, and some alternative models have also been proposed (Tsytovich, 1967; Kaplan and Tsytovich, 1968; Zheleznyakov and Zaitsev, 1970a,b; Papadopoulos et al., 1974; Bardwell and Goldman, 1976; Magelssen and Smith, 1977; Nicholson et al., 1978; Goldstein et al., 1979; Smith et al., 1979; Goldman and Dubois, 1982; Melrose, 1982, 1987; Goldman, 1983; Cairns, 1987a,b,c; Robinson and Cairns, 1998a,b,c; Wu et al., 2002; Li et al., 2008a,b, 2009; Tsiklauri, 2011; Graham et al., 2012; Thejappa et al., 2012; Krafft et al., 2013). In this section the plasma emission will be discussed on the basis of EM weak turbulence theory (Yoon, 2006; Yoon et al., 2012b; Ziebell et al., 2014a). Over the past several decades, the plasma emission was discussed within the framework of reduced or partial EM weak turbulence theory (Li et al., 2008a,b, 2009; Schmidt and Cairns, 2012a,b, 2014), but the complete numerical solution of the entire set of EM weak turbulence equations has not been done until quite recently (Ziebell et al., 2014a, 2015). The beam-plasma interaction and plasma emission can also be studied by means of direct EM PIC simulation (Kasaba et al., 2001; Karlický and Vandas, 2007; Rhee et al., 2009b,a; Umeda, 2010; Ganse et al., 2012a,b). The analytical EM weak turbulence theory can be employed to interpret the results of PIC simulations.

To investigate the problem, we start from the basic equations of EM weak turbulence theory, namely, (9.50)–(9.54). In employing the equations, however, we make note of the fact that the radiation emission is a byproduct of the beam-plasma instability process. The generalized L and S mode wave kinetic equations (9.50) and (9.51) contain various extra terms that describe the back-reaction of L mode in response to the radiation generation, but these effects are generally weak. Consequently, in describing the wave dynamics of L and S modes, let us resort to the simpler electrostatic description as in (6.4) and (6.5). Here, we repeat the wave and particle kinetic equations for the sake of convenience and completeness:

$$
\begin{aligned}
\frac{\partial I_{\mathbf{k}}^{\sigma L}}{\partial t} &= \frac{4\pi e^2}{m_e k^2} \int d\mathbf{v}\,\delta(\sigma\omega_{\mathbf{k}}^L - \mathbf{k}\cdot\mathbf{v}) \left(ne^2 F_e(\mathbf{v}) + \pi\sigma\omega_{\mathbf{k}}^L \mathbf{k}\cdot\frac{\partial F_e(\mathbf{v})}{\partial \mathbf{v}} I_{\mathbf{k}}^{\sigma L} \right) \\
&+ \sum_{\sigma',\sigma''=\pm 1} \int d\mathbf{k}'\, V_{\mathbf{k},\mathbf{k}'}^L \left(\frac{\sigma\omega_{\mathbf{k}}^L I_{\mathbf{k}'}^{\sigma' L} I_{\mathbf{k}-\mathbf{k}'}^{\sigma'' S}}{\mu_{\mathbf{k}-\mathbf{k}'}} - \frac{\sigma'\omega_{\mathbf{k}'}^L I_{\mathbf{k}-\mathbf{k}'}^{\sigma'' S} I_{\mathbf{k}}^{\sigma L}}{\mu_{\mathbf{k}-\mathbf{k}'}} \right. \\
&\left. - \sigma''\omega_{\mathbf{k}-\mathbf{k}'}^L I_{\mathbf{k}'}^{\sigma' L} I_{\mathbf{k}}^{\sigma L} \right) \delta(\sigma\omega_{\mathbf{k}}^L - \sigma'\omega_{\mathbf{k}'}^L - \sigma''\omega_{\mathbf{k}-\mathbf{k}'}^S) \\
&+ \sum_{\sigma'} \int d\mathbf{k}' \int d\mathbf{v}\, U_{\mathbf{k},\mathbf{k}'}^L \left[\frac{ne^2}{\omega_{pe}^2} \left(\sigma\omega_{\mathbf{k}}^L I_{\mathbf{k}'}^{\sigma' L} - \sigma'\omega_{\mathbf{k}'}^L I_{\mathbf{k}}^{\sigma L} \right) [F_e(\mathbf{v}) + F_i(\mathbf{v})] \right. \\
&\left. + \frac{\pi m_e}{m_i} I_{\mathbf{k}'}^{\sigma' L} I_{\mathbf{k}}^{\sigma L} (\mathbf{k}-\mathbf{k}')\cdot\frac{\partial F_i(\mathbf{v})}{\partial \mathbf{v}} \right] \delta[\sigma\omega_{\mathbf{k}}^L - \sigma'\omega_{\mathbf{k}'}^L - (\mathbf{k}-\mathbf{k}')\cdot\mathbf{v}],
\end{aligned} \tag{10.32}
$$

$$
\begin{aligned}
\frac{\partial}{\partial t} \frac{I_{\mathbf{k}}^{\sigma S}}{\mu_{\mathbf{k}}} &= \frac{4\pi \mu_{\mathbf{k}} e^2}{m_e k^2} \int d\mathbf{v}\,\delta(\sigma\omega_{\mathbf{k}}^S - \mathbf{k}\cdot\mathbf{v}) \left[ne^2 [f_e(\mathbf{v}) + f_i(\mathbf{v})] \right. \\
&\left. + \pi\sigma\omega_{\mathbf{k}}^L \left(\mathbf{k}\cdot\frac{\partial f_e(\mathbf{v})}{\partial \mathbf{v}} + \frac{m_e}{m_i}\mathbf{k}\cdot\frac{\partial f_i(\mathbf{v})}{\partial \mathbf{v}} \right) \frac{I_{\mathbf{k}}^{\sigma S}}{\mu_{\mathbf{k}}} \right] \\
&+ \sum_{\sigma',\sigma''} \int d\mathbf{k}'\, V_{\mathbf{k},\mathbf{k}'}^S \left(\sigma\omega_{\mathbf{k}}^L I_{\mathbf{k}'}^{\sigma' L} I_{\mathbf{k}-\mathbf{k}'}^{\sigma'' L} - \frac{\sigma'\omega_{\mathbf{k}'}^L I_{\mathbf{k}-\mathbf{k}'}^{\sigma'' L} I_{\mathbf{k}}^{\sigma S}}{\mu_{\mathbf{k}}} - \frac{\sigma''\omega_{\mathbf{k}-\mathbf{k}'}^L I_{\mathbf{k}'}^{\sigma' L} I_{\mathbf{k}}^{\sigma S}}{\mu_{\mathbf{k}}} \right) \\
&\times \delta(\sigma\omega_{\mathbf{k}}^S - \sigma'\omega_{\mathbf{k}'}^L - \sigma''\omega_{\mathbf{k}-\mathbf{k}'}^L),
\end{aligned} \tag{10.33}
$$

$$\begin{aligned}
\frac{\partial}{\partial t}\frac{I_{\mathbf{k}}^{\sigma T}}{2} &= \sum_{\sigma',\sigma''}\int d\mathbf{k}' V_{\mathbf{k},\mathbf{k}'}^{TLL}\left(\sigma\omega_{\mathbf{k}}^{T} I_{\mathbf{k}'}^{\sigma' L} I_{\mathbf{k}-\mathbf{k}'}^{\sigma'' L} - \frac{\sigma'\omega_{\mathbf{k}'}^{L} I_{\mathbf{k}-\mathbf{k}'}^{\sigma'' L} I_{\mathbf{k}}^{\sigma T}}{2}\right.\\
&\quad \left. - \frac{\sigma''\omega_{\mathbf{k}-\mathbf{k}'}^{L} I_{\mathbf{k}'}^{\sigma' L} I_{\mathbf{k}}^{\sigma T}}{2}\right)\delta(\sigma\omega_{\mathbf{k}}^{T} - \sigma'\omega_{\mathbf{k}'}^{L} - \sigma''\omega_{\mathbf{k}-\mathbf{k}'}^{L})\\
&\quad + \sum_{\sigma',\sigma''}\int d\mathbf{k}' V_{\mathbf{k},\mathbf{k}'}^{TLS}\left(\frac{\sigma\omega_{\mathbf{k}}^{T} I_{\mathbf{k}'}^{\sigma' L} I_{\mathbf{k}-\mathbf{k}'}^{\sigma'' S}}{\mu_{\mathbf{k}-\mathbf{k}'}} - \frac{\sigma'\omega_{\mathbf{k}'}^{L} I_{\mathbf{k}-\mathbf{k}'}^{\sigma'' S} I_{\mathbf{k}}^{\sigma T}}{2\mu_{\mathbf{k}-\mathbf{k}'}}\right.\\
&\quad \left. - \frac{\sigma''\omega_{\mathbf{k}-\mathbf{k}'}^{L} I_{\mathbf{k}'}^{\sigma' L} I_{\mathbf{k}}^{\sigma T}}{2}\right)\delta(\sigma\omega_{\mathbf{k}}^{T} - \sigma'\omega_{\mathbf{k}'}^{L} - \sigma''\omega_{\mathbf{k}-\mathbf{k}'}^{S}) \qquad (10.34)\\
&\quad + \sum_{\sigma',\sigma''}\int d\mathbf{k}' V_{\mathbf{k},\mathbf{k}'}^{TTL}\left(\frac{\sigma\omega_{\mathbf{k}}^{T} I_{\mathbf{k}'}^{\sigma' T} I_{\mathbf{k}-\mathbf{k}'}^{\sigma'' L}}{2} - \frac{\sigma'\omega_{\mathbf{k}'}^{T} I_{\mathbf{k}-\mathbf{k}'}^{\sigma'' L} I_{\mathbf{k}}^{\sigma T}}{2}\right.\\
&\quad \left. - \frac{\sigma''\omega_{\mathbf{k}-\mathbf{k}'}^{L} I_{\mathbf{k}'}^{\sigma' T} I_{\mathbf{k}}^{\sigma T}}{4}\right)\delta(\sigma\omega_{\mathbf{k}}^{T} - \sigma'\omega_{\mathbf{k}'}^{T} - \sigma''\omega_{\mathbf{k}-\mathbf{k}'}^{L})\\
&\quad + \sum_{\sigma'}\int d\mathbf{k}'\int d\mathbf{v}\, U_{\mathbf{k},\mathbf{k}'}^{T}\left[\frac{\hat{n}e^2}{\omega_{pe}^2}\left(\sigma\omega_{\mathbf{k}}^{T} I_{\mathbf{k}'}^{\sigma' L} - \sigma'\omega_{\mathbf{k}'}^{L}\frac{I_{\mathbf{k}}^{\sigma T}}{2}\right)[F_e(\mathbf{v}) + F_i(\mathbf{v})]\right.\\
&\quad \left. + \pi\frac{m_e}{m_i} I_{\mathbf{k}'}^{\sigma' L}\frac{I_{\mathbf{k}}^{\sigma T}}{2}(\mathbf{k}-\mathbf{k}')\cdot\frac{\partial F_i(\mathbf{v})}{\partial \mathbf{v}}\right]\delta[\sigma\omega_{\mathbf{k}}^{T} - \sigma'\omega_{\mathbf{k}'}^{L} - (\mathbf{k}-\mathbf{k}')\cdot\mathbf{v}],\\
\frac{\partial F_e(\mathbf{v})}{\partial t} &= \frac{\pi e^2}{m_e^2}\sum_{\sigma=\pm1}\int d\mathbf{k}\,\frac{\mathbf{k}}{k}\cdot\frac{\partial}{\partial \mathbf{v}}\,\delta(\sigma\omega_{\mathbf{k}}^{L} - \mathbf{k}\cdot\mathbf{v})\\
&\quad \times\left(\frac{m_e\sigma\omega_{\mathbf{k}}^{L}}{4\pi^2 k}F_e(\mathbf{v}) + I_{\mathbf{k}}^{\sigma L}\frac{\mathbf{k}}{k}\cdot\frac{\partial F_e(\mathbf{v})}{\partial \mathbf{v}}\right), \qquad (10.35)
\end{aligned}$$

where

$$\begin{aligned}
V_{\mathbf{k},\mathbf{k}'}^{L} &= \frac{\pi e^2\sigma\omega_{\mathbf{k}}^{L}}{2T_e^2}\frac{\mu_{\mathbf{k}-\mathbf{k}'}(\mathbf{k}\cdot\mathbf{k}')^2}{k^2k'^2|\mathbf{k}-\mathbf{k}'|^2},\\
V_{\mathbf{k},\mathbf{k}'}^{S} &= \frac{\pi e^2\sigma\omega_{\mathbf{k}}^{L}}{4T_e^2}\frac{\mu_{\mathbf{k}}[\mathbf{k}'\cdot(\mathbf{k}-\mathbf{k}')]^2}{k^2k'^2|\mathbf{k}-\mathbf{k}'|^2}.\\
U_{\mathbf{k},\mathbf{k}'}^{L} &= \frac{\sigma\omega_{\mathbf{k}}^{L}e^2}{n_e m_e^2\omega_{pe}^2}\frac{(\mathbf{k}\cdot\mathbf{k}')^2}{k^2k'^2},\\
V_{\mathbf{k},\mathbf{k}'}^{TLL} &= \frac{\pi e^2\sigma\omega_{\mathbf{k}}^{T}}{32m_e^2\omega_{pe}^2}\frac{(\mathbf{k}\times\mathbf{k}')^2}{k^2k'^2|\mathbf{k}-\mathbf{k}'|^2}\left(\frac{k'^2}{\sigma'\omega_{\mathbf{k}'}^{L}} - \frac{|\mathbf{k}-\mathbf{k}'|^2}{\sigma''\omega_{\mathbf{k}-\mathbf{k}'}^{L}}\right)^2,
\end{aligned}$$

$$
\begin{aligned}
V^{TLS}_{\mathbf{k},\mathbf{k}'} &= \frac{\pi e^2 \sigma \omega^T_{\mathbf{k}}}{4T_e^2} \frac{\mu_{\mathbf{k}-\mathbf{k}'}(\mathbf{k}\times\mathbf{k}')^2}{k^2 k'^2 |\mathbf{k}-\mathbf{k}'|^2}, \\
V^{TTL}_{\mathbf{k},\mathbf{k}'} &= \frac{\pi e^2 \sigma \omega^T_{\mathbf{k}}}{4m_e^2} \frac{|\mathbf{k}-\mathbf{k}'|^2}{(\omega^T_{\mathbf{k}})^2(\omega^T_{\mathbf{k}'})^2} \left(1 + \frac{(\mathbf{k}\cdot\mathbf{k}')^2}{k^2 k'^2}\right), \\
U^T_{\mathbf{k},\mathbf{k}'} &= \frac{\sigma \omega^T_{\mathbf{k}} e^2}{2\hat{n} m_e^2 \omega_{pe}^2} \frac{(\mathbf{k}\times\mathbf{k}')^2}{k^2 k'^2}.
\end{aligned}
\tag{10.36}
$$

Ziebell et al. (2014a, 2015) solved the set of equations (10.32)–(10.36) by numerical means for initial configuration in which the ions are considered stationary and distributed according to thermal equilibrium model with temperature T_i. The electrons are assumed to be initially composed of an isotropic thermal background plus a Gaussian distribution of streaming component, as in Chapter 6. As discussed in Section 10.2, T mode waves are spontaneously emitted via nonlinear spontaneous scattering process. As a consequence, even in the absence of electron beam there is a background level of radiation given by formula (10.19). Ziebell et al. (2014a, 2015) thus initiated their numerical analysis with a finite level of initial T mode intensity,

$$
I^{\sigma T}_{\mathbf{k}}(0) = \frac{T_e}{2\pi^2} \frac{1}{1 + c^2 k^2/\omega_{pe}^2}. \tag{10.37}
$$

The input parameter g is related to the particle discreteness, and in their study, Ziebell et al. (2014a, 2015) simply chose g in accordance with the plasma parameter $1/(n\lambda_D^3) = 5 \times 10^{-3}$. Other input parameters are the background to total electron density ratio $n_b/n_0 = 10^{-3}$, equal Gaussian beam and background temperatures $T_e = T_{\rm beam}$, the ion to electron temperature ratio $T_i/T_e = 1/7$, and the normalized beam speed V_b/v_{Te}, which they varied in their study. Here, we show an example of $V_b/v_{Te} = 6$.

In Figures 10.2–10.5, we show the numerical solution obtained by Ziebell et al. (2015). Figure 10.2 shows snapshots of electron velocity distribution function at two different time intervals corresponding to $\omega_{pe}t = 100$ and 1000. For relatively early time, namely $\omega_{pe}t = 100$, one can still see the drifting electron beam distribution feature, but as time progresses, the positive slope in the parallel velocity is reduced as a result of the well-known quasilinear velocity diffusion process.

As a side note, it should be mentioned that quasilinear relaxation of the electron beam associated with the type III radio source has been the focus of intense research in the literature. The challenge was how to explain the persistence of observed beam feature associated with type III events (Lin, 1970; Frank and Gurnett, 1972; Kane, 1972; Lin et al., 1973, 1981, 1986; Ergun et al., 1998; Gosling et al., 2003; Krucker et al., 2007; Wang et al., 2012b) in spite of rapid relaxation by quasilinear plateau formation. This was known in the early literature as *Sturrock's dilemma*

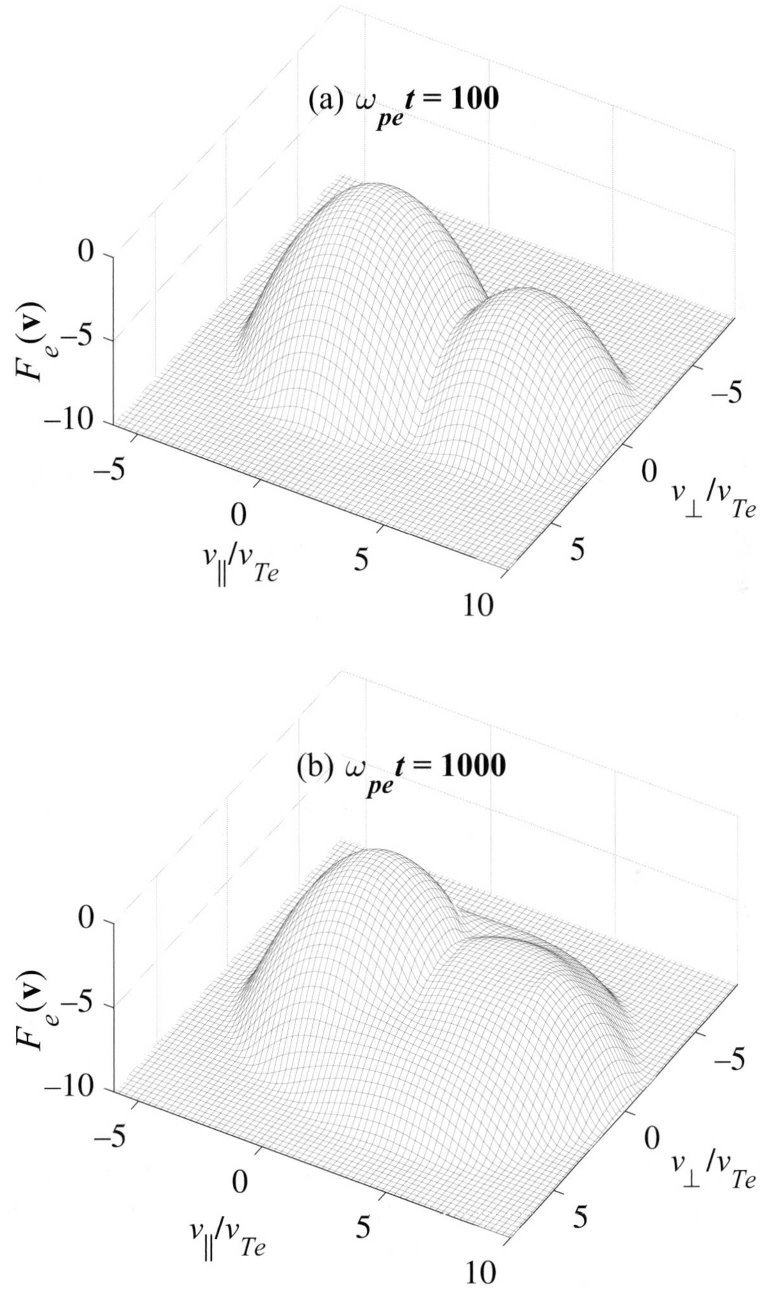

Figure 10.2 Time evolution of the electron velocity distribution function F_e versus $v_\perp/v_{Te}$ and $v_\parallel/v_{Te}$ format. Panel (a) shows F_e at $\omega_{pe}t = 100$ and panel (b) is for $\omega_{pe}t = 1000$.

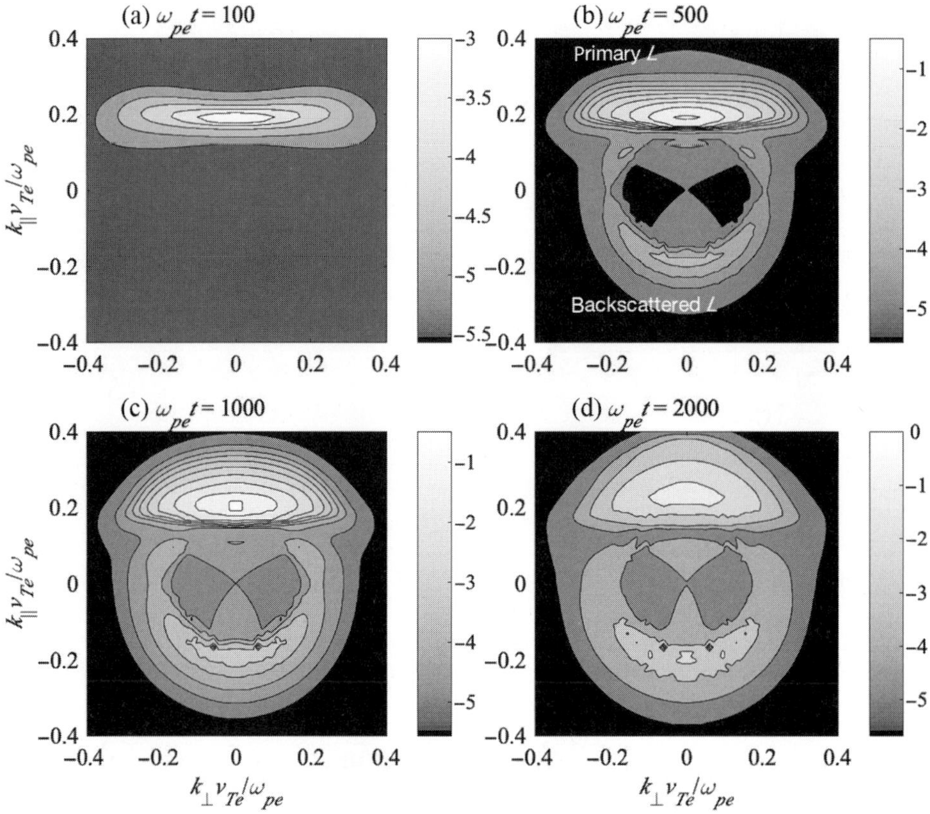

Figure 10.3 Time evolution of Langmuir wave spectral intensity $I_L(\mathbf{k})$ versus $k_\perp v_{Te}/\omega_{pe}$ and $k_\parallel v_{Te}/\omega_{pe}$. Panel (a) shows $I_L(\mathbf{k})$ at $\omega_{pe}t = 100$; panel (b) is for $\omega_{pe}t = 500$; panel (c) is for $\omega_{pe}t = 1000$; and the final panel (d) shows $I_L(\mathbf{k})$ at $\omega_{pe}t = 2000$ time step.

(Sturrock, 1964). Later, however, the issue was resolved in the context of finite physical dimension of the source size, which leads to the beam reformation via time-of-flight effects. In short, the problem related to the propagation of type III electrons has received extensive discussions in the literature. This includes the study of propagation of finite-sized type III electron source in homogeneous and inhomogeneous media (Sturrock, 1964; Zheleznyakov and Zaitsev, 1970b; Zaitsev et al., 1972; Papadopoulos et al., 1974; Takakura and Shibahashi, 1976; Magelssen and Smith, 1977; Goldstein et al., 1979; Smith et al., 1979; Grognard, 1982; Muschietti, 1990; Mel'nik et al., 1999; Kontar and Pécseli, 2002; Foroutan et al., 2007; Ratcliffe et al., 2012; Voshchepynets and Krasnoselskikh, 2015). This problem is, however, beyond the scope of the present monograph. Instead, we focus on the detailed radiation generation mechanism.

The Langmuir turbulence spectrum is shown in Figure 10.3. The portion of the 2D $\mathbf{k}$ space corresponding to $k_\parallel > 0$ is for forward-propagating L mode, $I_{\mathbf{k}}^{+L}$, while the other half space with $k_\parallel < 0$ belongs to the backward L mode, $I_{\mathbf{k}}^{-L}$. For

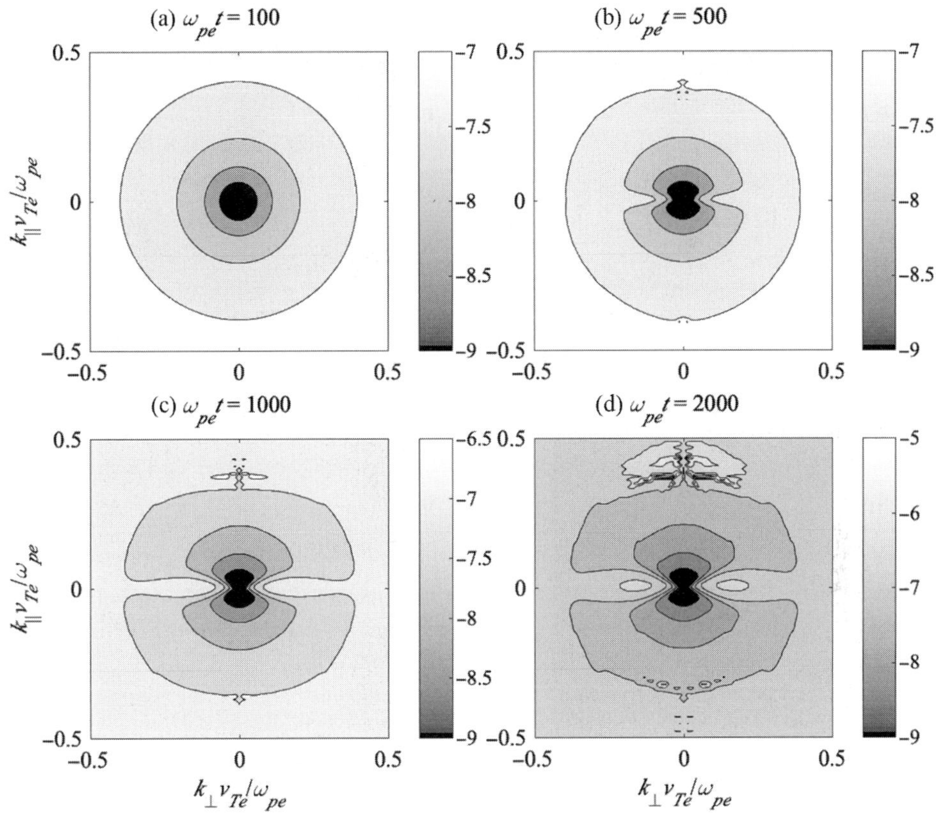

Figure 10.4 Time evolution of ion-sound wave spectral intensity $I_S(\mathbf{k})$ versus $k_\perp v_{Te}/\omega_{pe}$ and $k_\| v_{Te}/\omega_{pe}$ in the same format as Figure 10.3.

relatively early time, $\omega_{pe}t = 100$, the enhanced forward-propagating component (the primary L) can be seen to be excited, which is the result of initial bump-on-tail instability (panel a). For $\omega_{pe}t = 500$, the backscattering of the primary L mode into the oppositely traveling L mode, via a combined three-wave decay process and nonlinear scattering off ions, becomes visible. For $\omega_{pe}t = 1000$, the primary L mode begins to evolve into a semi-ark shape spectrum. The backscattered L mode was already generated in the form of ring spectrum even at $\omega_{pe}t = 500$. For the final time step $\omega_{pe}t = 2000$, the total L mode spectrum attains more and more ring-like feature, which was first noticed in earlier electrostatic weak turbulence calculations (Ziebell et al., 2008a,b). The numerical solution compares favorably with the simulated spectrum based upon the two-dimensional EM PIC methodology – see, e.g., figure 2 of the paper by Rhee et al. (2009a).

Nonlinear decay processes that generate the backscattered L mode is accompanied by the production of ion-sound mode by decay instability, as indicated in Figure 10.4, which shows the time evolution of the S mode spectral intensity in the same format as Figure 10.3.

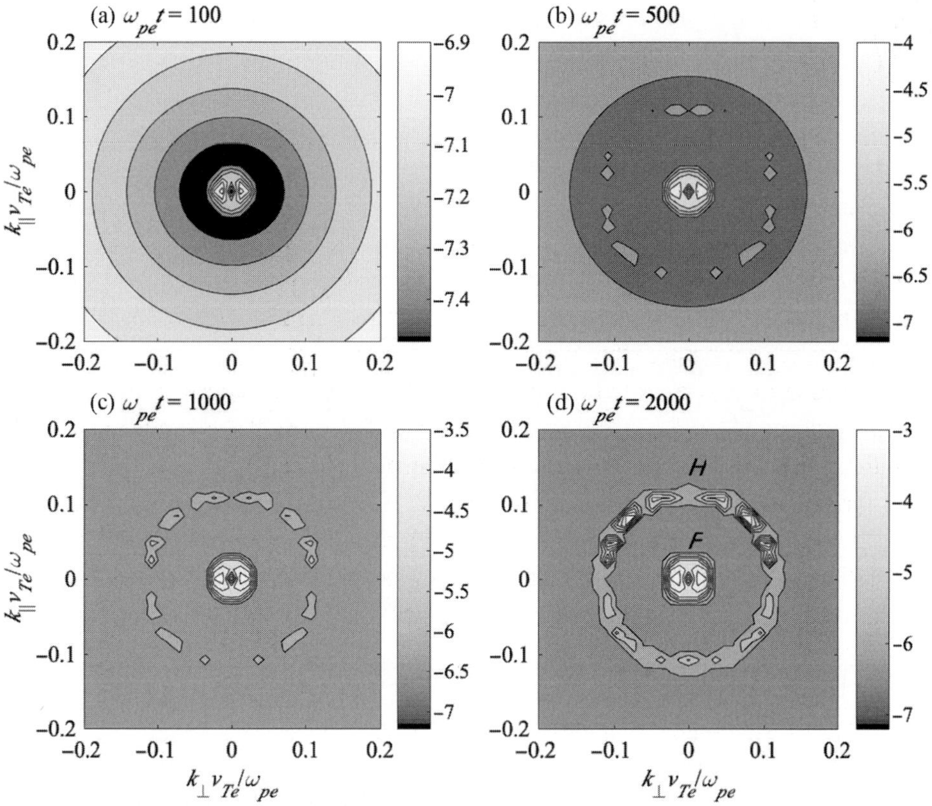

Figure 10.5 Time evolution of the radiation spectral intensity $I_T(\mathbf{k})$ in the same format as Figure 10.3.

Figure 10.5 plots the transverse EM radiation (T mode) spectrum. In Figure 10.5d, the generation of fundamental/second-harmonic pair of emissions with frequency in the vicinity of ω_{pe} and $2\omega_{pe}$ are clearly visible.

The physics of electron-beam, Langmuir, and ion-sound turbulence processes, and also the radiation generation processes based upon reduced methods, have been investigated in detail (Ziebell et al., 2001, 2008a,b; Kontar and Pécseli, 2002; Foroutan et al., 2007; Li et al., 2008a,b, 2009; Ratcliffe et al., 2012; Schmidt and Cairns, 2012a,b, 2014), but these works employ various simplifying assumptions. In contrast, the complete weak turbulence formalism contains all the terms, and hence is suitable for a detailed analysis (Ziebell et al., 2014a, 2015).

According to (10.34), the fundamental emission should be dictated by the three-wave process involving L mode decaying into T and S mode and the scattering process involving the beating of L and T mode mediated by the particles. Of the two processes, the decay mechanism ($L + S \to T$) is governed by the overall nonlinear coupling coefficient,

$$\frac{(\mathbf{k}\times\mathbf{k}')^2}{k^2k'^2} \propto \sin^2\vartheta. \tag{10.38}$$

If we associate $\mathbf{k}'$ and $\omega^L_{\mathbf{k}'}$ with the Langmuir wave, and $\mathbf{k}$ and $\omega^T_{\mathbf{k}}$ with the transverse radiation (which stems from the resonance condition $\sigma\omega^T_{\mathbf{k}} - \sigma'\omega^L_{\mathbf{k}'} - \sigma''\omega^S_{\mathbf{k}-\mathbf{k}'} = 0$), then if we allow $\mathbf{k}'$ vector to lie along the direction of beam propagation, it can be seen that the fundamental emission along the direction specified by $\vartheta = 0$ should be prohibited. Of course, the L mode spectrum has a broad range of angles in the present case of two-dimensional physics so that the fundamental emission by the decay process ($L + S \to T$) is allowed even for $\vartheta = 0$. Nevertheless, (10.38) predicts that the overall F emission by the decay mechanism should possess a dipolar radiation angular distribution. Figure 10.5a bears such a feature out. Note that the scattering process is also dictated by the same nonlinear coupling coefficient (10.38), so that the fundamental plasma emission having the dipolar structure with maximum radiation intensity in directions orthogonal to the direction of beam propagation direction is consistent.

For the second-harmonic emission, the basic emission mechanism is dictated by the coupling coefficient associated with the decay process $L + L \to T$,

$$\frac{(\mathbf{k}\times\mathbf{k}')^2}{k^2k'^2}\left(k'^2 - |\mathbf{k}-\mathbf{k}'|^2\right)^2 \sim \sin^2\vartheta\left(k^2 - 2kk'\cos\vartheta\right)^2 \propto \sin^2\vartheta\cos^2\vartheta, \tag{10.39}$$

where we identify k with the transverse mode and k' with the Langmuir mode, so that one may approximate $k \ll k'$. According to this picture, the second-harmonic ($\omega \sim 2\omega_{pe}$) emission should possess a quadrupole angular distribution. Indeed, Figure 10.5 shows that the second-harmonic emission peaks at 45° angle. The quadrupole emission is not symmetric with respect to $k_\parallel$ axis, featuring stronger forward emission, stronger than backward emission.

The third-harmonic emission is not apparent for the beam speed $V_b/v_{Te} = 6$, but for higher beam speed it is possible. The third- or higher-harmonic emission takes place via decay process $L + T \to T$. The coupling coefficient is given by

$$\left(1 + \frac{(\mathbf{k}\cdot\mathbf{k}')^2}{k^2k'^2}\right) \sim 1 + \cos^2\vartheta. \tag{10.40}$$

As this indicates, the third-harmonic plasma emission should be largely isotropic. In the solar type III bursts third harmonic is only rarely detected (Takakura and Yousef, 1974; Brazhenko et al., 2012). There are reports of occasional third harmonic of type II bursts in the literature (Zlotnik et al., 1998). The theory of $3H$ emission was first developed by Zheleznyakov and Zlotnik (1974), and also discussed in the paper trilogy by Cairns (1987a,b,c) – see also the paper by Kliem et al. (1992). The present theory allows for F, $2H$, $3H$ as well as even higher-harmonic emissions, and as such it generalizes previous theories.

In the paper by Ziebell et al. (2015), further analysis is carried out by turning certain terms in the T mode wave kinetic equation on or off artificially, thereby identifying the role of each term. By doing so, they were able to pinpoint the exact radiation generation mechanism for each harmonic mode. This is the advantage of the theoretical approach to study a complicated nonlinear problem in plasmas, as opposed to direct numerical simulation. Simulations are supposed to be more rigorous, but it is not so straightforward to interpret the result since all physical processes take place simultaneously, and it is not always possible to single out certain processes and turn them on or off.

Epilogue

In this book the author attempted to present a detailed discussion regarding the weak turbulence methodology for unmagnetized plasmas, starting from a simple Vlasov–Poisson system, moving on to the Klimontovich–Poisson system of equations. Then, the discussion returned to the Vlasov equation, but this time full Maxwell's equation is employed. Finally, the fully electromagnetic problem is extended to the Klimontovich framework, thus concluding the mathematical formalism. In all these discussions, however, the effects of ambient magnetic field were not considered, which limited the applicability in an appreciable way.

At present, the weak turbulence theory for magnetized plasmas is not at a completely satisfactory state. In the literature, some early works attempted to extend the weak turbulence theory to plasmas immersed in a constant applied magnetic field. Among them are the works by Tsytovich and Shvartsburg (1966), Melrose and Sy (1972), and Pustovalov and Silin (1975), for instance. Among these, the work by Pustovalov and Silin (1975), for which a monumental effort must have been devoted, deserves a close look. However, their result is very formal, and the ramifications thereof are not entirely evident. In short, these early works are preliminary, and further developments are necessary. If one makes simplifying assumptions, such as electrostatic approximation (Ellis and Porkolab, 1968; Porkolab and Chang, 1978; Mikhajlenko and Stepanov, 1981), or drift kinetic approximation, which amounts to ignoring cyclotron interactions (Ganguli et al., 2010; Rudakov et al., 2011; Crabtree et al., 2012; Mithaiwala et al., 2012), then one may find in the literature more concrete works that employ weak turbulence systemology in order to address complex plasma phenomena. The author of this book also made some efforts to extend the existing method of studying the weak plasma turbulence to include the effects of ambient magnetic field, but thus far, the discussions are limited to turbulence propagating along the magnetic field (Yoon, 2015a,b) or exactly perpendicular to it (Yoon, 2015c) – for a discussion of nonlinear wave interactions taking place in perpendicular directions, see also the work by Brodin and Stenflo (2015).

For turbulent processes taking place in macroscopic dimensions, plasmas immersed in a background magnetic field may be regarded as a magnetized electro-fluid so that mathematical and conceptual tools of the magnetohydrodynamic (MHD) turbulence theory may be brought forth (Biskamp, 2003; Bruno and Carbone, 2016). However, for kinetic scale turbulent phenomena, the tools and methodology of weak turbulence theory are called for, but at present, the standard framework of plasma weak turbulence theory has not been fully implemented for magnetized plasmas. It is the author's wish that not only the readers may find the material presented in this book useful and educational but also the content of this book may form a foundational structure based upon which future fully developed weak turbulence theory for magnetized plasmas may be constructed.

Appendix A

Time Irreversible Small Amplitude Perturbations

Linearized Vlasov equation for electrostatic perturbation in unmagnetized plasmas is given by

$$\left(\frac{\partial}{\partial t} + \mathbf{v} \cdot \frac{\partial}{\partial \mathbf{r}}\right)\delta f_a(\mathbf{r}, \mathbf{v}, t) + \frac{e_a}{m_a}\,\delta\mathbf{E}(\mathbf{r}, t) \cdot \frac{f_a(\mathbf{v})}{\partial \mathbf{v}} = 0,$$
$$\frac{\partial}{\partial \mathbf{r}} \cdot \delta\mathbf{E}(\mathbf{r}, t) = 4\pi \sum_a e_a \int d\mathbf{v}\, \delta f_a(\mathbf{r}, \mathbf{v}, t). \tag{A.1}$$

Applying the spatial Fourier transformation to the perturbed Vlasov equation, we have

$$\left(\frac{\partial}{\partial t} + i\mathbf{k} \cdot \mathbf{v}\right)\delta f_{\mathbf{k}}^a(\mathbf{v}, t) + \frac{e_a}{m_a}\,\delta\mathbf{E}_{\mathbf{k}}(t) \cdot \frac{\partial f_a(\mathbf{v})}{\partial \mathbf{v}} = 0,$$
$$i\mathbf{k} \cdot \delta\mathbf{E}_{\mathbf{k}}(t) = 4\pi \sum_a e_a \int d\mathbf{v}\, \delta f_{\mathbf{k}}^a(\mathbf{v}, t). \tag{A.2}$$

The next logical step is to take a similar integral transformation in time. However, physical phenomena do not operate in infinite domain, $-\infty < t < \infty$, but rather the problem always has an initial time $t = 0$, so the domain of time is $0 < t < \infty$. This means that we are only interested in forward time, or, equivalently, causal processes. Consequently, the type of integral transformation of a temporal function $g(t)$ must be of the form

$$g_\omega = (2\pi)^{-1} \int_0^\infty dt\; g(t)\, e^{i\omega t}. \tag{A.3}$$

The question is whether this integral transformation is well defined in the mathematical sense. As a concrete example, take, for instance, an exponential function

$$g(t) = e^{at}. \tag{A.4}$$

The transformed function is

$$g_\omega = \frac{1}{2\pi}\left(\frac{1}{a + i\omega} - \lim_{t \to \infty} \frac{e^{(a+i\omega)\,t}}{a + i\omega}\right). \tag{A.5}$$

If $a < 0$, then the limit vanishes, but if $a > 0$, then the limit diverges; hence, the integral transformation is not defined. Nevertheless, one may salvage the situation if we assume that ω is complex:

$$\omega = \mathrm{Re}\,\omega + i\,\mathrm{Im}\,\omega, \tag{A.6}$$

which is, of course, the Laplace transformation, and this is the approach that Landau took (Landau, 1946) in contrast to Vlasov (1938), who restricted the discussion to only real ω. Returning to the toy problem (A.4), as long as $\mathrm{Im}\,\omega > a$, the integral transformation is well defined, and the result is

$$g_\omega = \frac{1}{2\pi}\frac{1}{a+i\omega}. \tag{A.7}$$

From this, a general condition for the validity of integral transformation is that, for any function $g(t)$, integral transformation (A.3) is defined as long as ω is defined above all singular points associated with g_ω in complex ω space. This leads to the proper definition of inverse transformation. The naive definition

$$g(t) = \int_{-\infty}^{\infty} d\omega\, g_\omega\, e^{-i\omega t}$$

is inappropriate since ω must be higher than all singularities associated with g_ω in complex plane. Consequently, we take the infinite integration along a path L that is taken above the highest singular point associated with g_ω in complex plane,

$$g(t) = \int_L d\omega\;\, g_\omega\, e^{-i\omega t}. \tag{A.8}$$

The spectral transformation (A.3) and its inverse (A.8) constitute the Laplace transformation. The path L does not have to be a straight horizontal line – see Figure A.1 – as long as it is above the singular points. For instance, we may take a deformed integration contour as shown on the right of Figure A.1.

Applying the Laplace transformation to (A.2) under the assumption that $\lim_{t\to\infty}\delta f_{\mathbf{k}}^a(\mathbf{v},t) = 0$, we have

$$\frac{1}{2\pi}\,\delta f_{\mathbf{k}}^a(\mathbf{v},0) - i\,(\omega - \mathbf{k}\cdot\mathbf{v})\,\delta f_{\mathbf{k},\omega}^a(\mathbf{v}) + \frac{e_a}{m_a}\,\delta\mathbf{E}_{\mathbf{k},\omega}\cdot\frac{\partial f_a(\mathbf{v})}{\partial\mathbf{v}} = 0. \tag{A.9}$$

One may also take the Laplace transformation of Poisson's equation, and take $\delta\mathbf{E}_{\mathbf{k},\omega} = (\mathbf{k}/k)\,\delta E_{\mathbf{k},\omega}$, in order to arrive at

$$\epsilon(\mathbf{k},\omega)\,\delta E_{\mathbf{k},\omega} = -\sum_a \frac{2e_a}{k}\int d\mathbf{v}\,\frac{\delta f_{\mathbf{k}}^a(\mathbf{v},0)}{\omega - \mathbf{k}\cdot\mathbf{v}}, \tag{A.10}$$

Figure A.1 Integration paths for inverse Laplace transformation (A.8). The left panel shows a straight horizontal path, but it can be deformed as shown on the right, as long as the path is above the singular points.

where $\epsilon(\mathbf{k},\omega)$ is the longitudinal dielectric response function (1.55). For the sake of convenience, let us assume that $\mathbf{k} = \hat{\mathbf{z}}k$ and let us work with the reduced distribution function $\hat{F}_a(v_z) = 2\pi\int_0^\infty dv_\perp\, v_\perp\, F_a(\mathbf{v})$, where $f_a = nF_a$. Then (A.10) becomes

$$\delta E_{k\omega} = \frac{S(k,\omega)}{\epsilon(k,\omega)}, \tag{A.11}$$

where

$$S(k,\omega) = -\sum_a \frac{2e_a}{k}\int d\mathbf{v}\,\frac{\delta f_k(\mathbf{v},0)}{\omega - kv_z},$$
$$\epsilon(k,\omega) = 1 + \sum_a \frac{\omega_{pa}^2}{k}\int_{-\infty}^{\infty} dv_z\,\frac{d\hat{F}_a(v_z)/dv_z}{\omega - kv_z}. \tag{A.12}$$

We are interested in the temporal behavior of perturbed electric field, which can be discussed by considering the inverse Laplace transformation of perturbed electric field:

$$\delta E_k(t) = \int_L d\omega\, e^{-i\omega t}\,\frac{S(k,\omega)}{\epsilon(k,\omega)}, \tag{A.13}$$

where contour L is defined above the highest singularity associated with the integrand $S(k,\omega)/\epsilon(k,\omega)$. The singularity may come from any possible divergences associated with $S(k,\omega)$ or from the zeros of denominator $\epsilon(k,\omega)$. Let us focus on the zeros of $\epsilon(k,\omega)$, which are none other than the dispersion relation satisfying $\epsilon(k,\omega) = 0$, or $\omega = \omega_k + i\gamma_k$. We also assume that the zeros of $\epsilon(k,\omega)$ are made of poles rather than branch cuts. The task is to perform the integral along the path L. The question is how to perform such an integral. In general, Cauchy's residue theorem cannot be applied in analytically closed form since the general forms of $S(k,\omega)$ and $\epsilon(k,\omega)$ do not lend themselves to such a mathematical manipulation, and generally there is no guarantee that integrals converge asymptotically. However, we may carry out the contour integration in an approximate (that is, in an asymptotic) sense, by employing the analytic continuation (Landau, 1946). The idea is to deform the contour L down along the imaginary axis, as shown on the right-hand side of Figure A.2, past the real axis, but avoiding the singularities by encircling around each pole.

Suppose that we deform the contour down the imaginary ω axis sufficiently far and close the contour in an asymptotic sense. Let us denote the outer contour Γ and the closed complete contour C,

$$\int_C d\omega\, e^{-i\omega t}\,\frac{S(k,\omega)}{\epsilon(k,\omega)} = \int_L d\omega\, e^{-i\omega t}\,\frac{S(k,\omega)}{\epsilon(k,\omega)} + \int_\Gamma d\omega\, e^{-i\omega t}\,\frac{S(k,\omega)}{\epsilon(k,\omega)}, \tag{A.14}$$

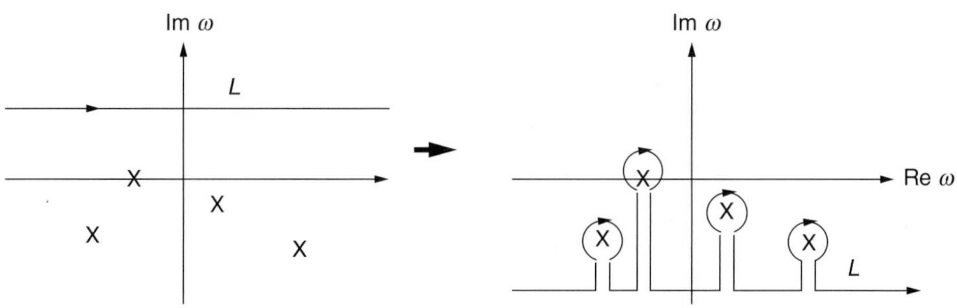

Figure A.2 Analytic continuation of integration path avoiding the singularities.

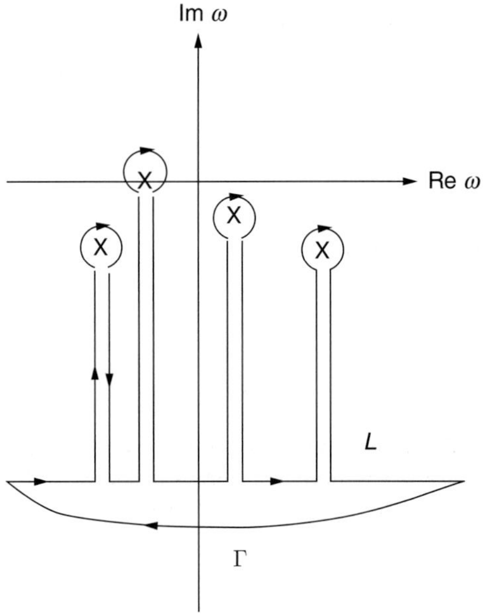

Figure A.3 Deformed contour L, including the encircled poles, cancelled out vertical paths connecting the poles, and outer contour Γ. Complete contour C thus lends itself to Cauchy's residue theorem.

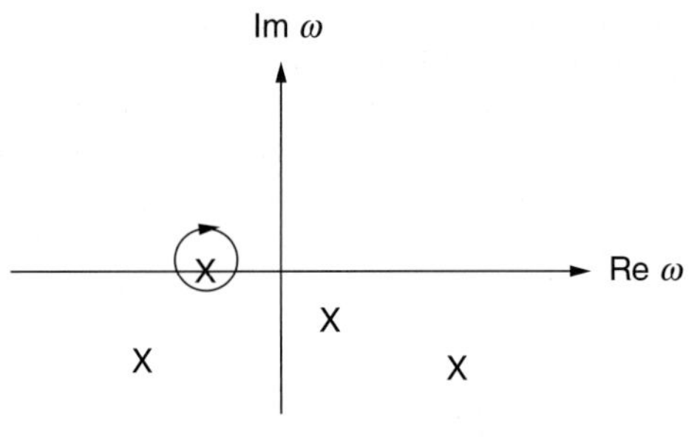

Figure A.4 In the time asymptotic sense, only the least damped mode will be important so that the residue contribution from that mode needs to be included in the inverse Laplace transformation.

as shown in Figure A.3. Each circle around the pole makes the residue contribution in the sense of Cauchy's theorem, but vertical paths cancel out. In the time asymptotic sense, $t \gg 1$, the contribution from the least damped mode with lowest value of $|\gamma_k|$ in magnitude will dominate (here we assume damping with $\gamma_k \leq 0$), as shown in Figure A.4.

Suppose that the most important mode, that is, the least damped mode, is given by

$$\omega = \omega_k + i\gamma_k = \omega_k - i\,|\gamma_k| = \omega_s. \tag{A.15}$$

Then, we may employ the Taylor series expansion, and, by invoking the residue theorem, obtain

$$\int_C d\omega\, e^{-i\omega t}\,\frac{S(k,\omega)}{\epsilon(k,\omega)} = \int_C d\omega\, e^{-i\omega t}\,\frac{S(k,\omega)}{(\omega-\omega_s)\,\partial\epsilon(k,\omega_s)/\partial\omega_s}$$
$$= -2\pi i\,\frac{S(k,\omega_k+i\gamma_k)\,e^{-i\omega_k t+\gamma_k t}}{\partial\epsilon(k,\omega_k+i\gamma_k)/\partial(\omega_k+i\gamma_k)}, \tag{A.16}$$

where we have used the fact that $\epsilon(k,\omega_k+i\gamma_k)=\epsilon(k,\omega_s)=0$ by definition. Here, the minus sign is because the contour loops around in the clockwise sense. Making use of (A.16), we have

$$\int_L d\omega\, e^{-i\omega t}\,\frac{S(k,\omega)}{\epsilon(k,\omega)} = -2\pi i\,\frac{S(k,\omega_k+i\gamma_k)\,e^{-i\omega_k t+\gamma_k t}}{\partial\epsilon(k,\omega_k+i\gamma_k)/\partial(\omega_k+i\gamma_k)}$$
$$-\int_\Gamma d\omega\, e^{-i\omega t}\,\frac{S(k,\omega)}{\epsilon(k,\omega)}. \tag{A.17}$$

As noted already, we generally do not know how to evaluate the integration around the path Γ. However, in the asymptotic sense, if we push the contour down infinitely low along the imaginary axis, then, in the end, we are not concerned with the contribution from Γ segment, so the result is

$$\int_L d\omega\, e^{-i\omega t}\,\frac{S(k,\omega)}{\epsilon(k,\omega)} \approx -2\pi i\,\frac{S(k,\omega_k+i\gamma_k)\,e^{-i\omega_k t-|\gamma_k|t}}{\partial\epsilon(k,\omega_k+i\gamma_k)/\partial(\omega_k+i\gamma_k)}. \tag{A.18}$$

From this it is seen that the initial fluctuation $\delta f_k(\mathbf{v},0)$ exponentially damps away,

$$\delta f_k(\mathbf{v},0)\,e^{-|\gamma_k|t}, \tag{A.19}$$

over time in the asymptotic sense. This is the well known *Landau damping*, which represents dissipation in collisionless plasmas. In general, dissipation is associated with collisions and irreversibility, and since collisionless plasmas are supposed to be dissipation less, the basic system of equations are time reversible. That is, Vlasov and Maxwell's equations remain unchanged if we change t to $-t$. From this, it might appear that the Landau damping is counter-intuitive and paradoxical. However, one should keep in mind that by employing the Laplace transformation, one is only concerned with physically meaningful causal solutions. Had we chosen the anti-causal Laplace transformation (with integral $\int_{-\infty}^{0} dt$), one would have had an exponentially increasing solution for positive time. In short, Landau damping *can be* time reversible. It is just that Landau restricted his discussion to physically meaningful causal solution.

Appendix B

Resonant Velocity Integral

We pick up where we left off regarding the inverse Laplace transformation,

$$\delta E_k(t) = \int_L d\omega\, e^{-i\omega t} \frac{S(k,\omega)}{\epsilon(k,\omega)}. \tag{B.1}$$

In (B.1), the denominator is made of $\epsilon(k,\omega)$, which is defined with the velocity integral

$$\epsilon(k,\omega) = 0 = 1 - \sum_a \frac{\omega_{pa}^2}{k^2} \int_{-\infty}^{\infty} dv_z \frac{d\hat{F}_a/dv_z}{v_z - \omega/k}. \tag{B.2}$$

According to the causal Laplace transformation, the quantity ω that appears in the integrand in (B.1) is defined along the contour L, so that it lies above the highest root. Suppose that all roots have $\gamma_k \leq 0$. Suppose also that the highest root is the weakly damped mode with γ_k close to 0. This means that the quantity ω that appears within the definition of $\epsilon(k,\omega)$ must be above that weakly damped root. The implication is that while the velocity integration is along real v_z, the pole ω/k lies above imaginary v_z axis – Figure B.1.

As the contour L is deformed (analytic continuation), however, the pole $v_z = \omega/k$, which originally lay above the real v_z axis, will now start to move down along imaginary v_z axis, until it crosses the real axis. Except for the loop around each pole, ω generally lies below the real axis, and, in fact, in the asymptotic sense, it has a very large imaginary part, implying that $v_z = \omega/k$ is also associated with large imaginary part. Accordingly, the velocity integral path must be deformed in such a way as to ensure that the pole $v_z = \omega/k$ always stays above the velocity integration path, as shown in Figure B.2.

If γ_k is infinitesimally small and negative ($\gamma_k = -0$), the velocity integration can be broken into the principal part and the half residue,

$$\frac{1}{v_z - \omega_k/k} = \mathscr{P}\frac{1}{v_z - \omega_k/k} + i\pi\delta\left(v_z - \frac{\omega_k}{k}\right), \tag{B.3}$$

as shown in Figure B.3. The relationship in (B.3) is known as the *Plemelj formula*, which is more generally expressed as

$$\frac{1}{x \pm i0} = \mathscr{P}\frac{1}{x} \mp i\pi\,\delta(x). \tag{B.4}$$

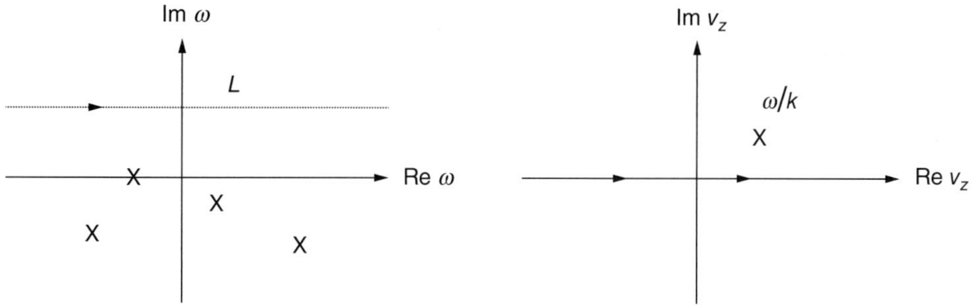

Figure B.1 The Laplace contour L, taken above all singularities associated with the integrands, and the corresponding velocity integral path, which lies below the pole ω/k with an infinitesimally positive imaginary part.

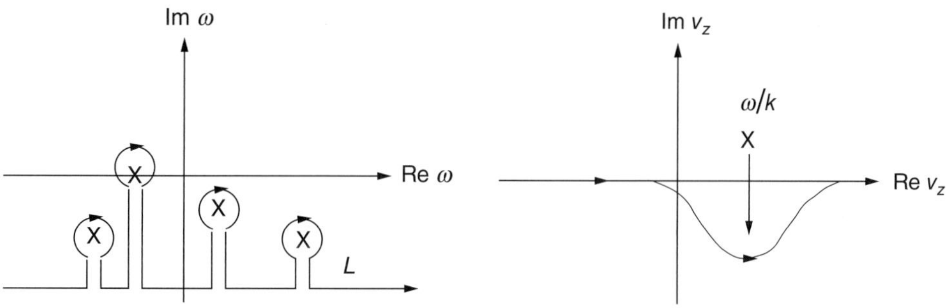

Figure B.2 Deformed Laplace contour L and the correspondingly deformed velocity integral path.

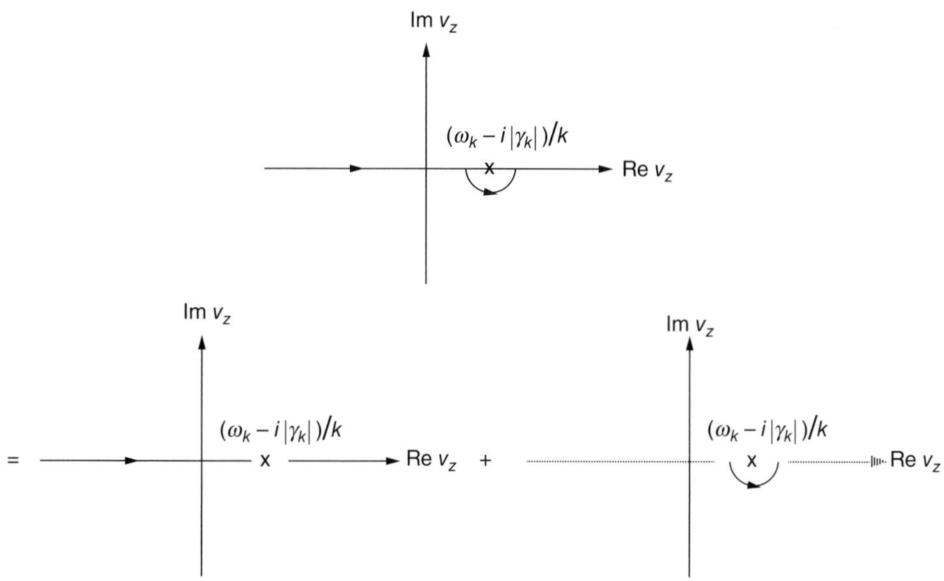

Figure B.3 Velocity integration for weakly damped mode (upper panel), which can be broken into the principal part plus one-half residue contribution.

In adopting the prescription (B.3) for treating the resonant velocity integration, we should keep in mind that the sign of γ_k can be either positive or negative depending on whether one is interested in instability or damping, but before one actually computes γ_k, one must always apply the Plemelj formula for positive sign associated with ω. This is the origin of the factor $+i0$ in the resonant denominator (1.42),

$$\frac{1}{\omega - \mathbf{k} \cdot \mathbf{v} + i0}. \tag{B.5}$$

Appendix C

Nonlinear Dispersion Relations

In this appendix, we discuss the consequence of retaining the nonlinear correction terms in the real part of nonlinear spectral balance equation. Two main results come out as consequences of retaining the nonlinear correction terms. One is the existence of harmonics of plasma frequency, which are nonlinear eigenmodes of the system, and the other is the nonlinear frequency shift of linear eigenmodes.

C.1 Nonlinear Eigenmodes

An interesting phenomenon unveiled through computer simulations (Klimas, 1983, 1990; Umeda et al., 2003) is that the Langmuir wave instability driven by a weak electron beam is accompanied by the generation of harmonic modes of fundamental plasma frequency, but the harmonics of ω_{pe} are not the normal modes of a plasma according to the textbook theory of small amplitude plasma waves; hence their existence cannot be discussed on the basis of linear Vlasov theory, nor can they be explained in terms of the plasma emission scenario of Chapter 10, as the excited modes are longitudinal. On the other hand, such a phenomenon *can* be explained by a theory of nonlinear dispersion relation. Let us focus on the first harmonic Langmuir wave with frequency close to twice the plasma oscillation frequency. Let us denote the new solution by

$$\omega = \pm\omega_{\mathbf{k}}^{N}. \tag{C.1}$$

Since the nonlinear eigenmode is presupposed to possess a wave frequency near $2\omega_{pe}$, while $\omega_{\mathbf{k}'}^{L} \sim \omega_{pe}$, and $\omega_{\mathbf{k}}^{N} - \omega_{\mathbf{k}'}^{L} \sim \omega_{pe}$, all three waves possess the fast-wave conditions,

$$|\omega_{\mathbf{k}}^{N}| \gg k v_{Te}, \quad |\omega_{\mathbf{k}'}^{L}| \gg k' v_{Te}, \quad |\omega_{\mathbf{k}}^{N} - \omega_{\mathbf{k}'}^{L}| \gg |\mathbf{k} - \mathbf{k}'|\, v_{Te}. \tag{C.2}$$

This means that the second-order susceptibility is dominated by the electron response, which can be approximated by (see 1.128)

$$\begin{aligned}
&\chi^{(2)}(\mathbf{k}', \pm\,\omega_{\mathbf{k}'}^{L}|\mathbf{k} - \mathbf{k}', \pm\,\omega_{\mathbf{k}}^{N} \mp \omega_{\mathbf{k}'}^{L}) \\
&= \frac{i}{2}\frac{e}{m_e}\frac{\omega_{pe}^2}{\omega_{\mathbf{k}}^{N}\omega_{\mathbf{k}'}^{L}(\omega_{\mathbf{k}}^{N} - \omega_{\mathbf{k}'}^{L})}\frac{1}{k\,k'\,|\mathbf{k} - \mathbf{k}'|} \\
&\times \left(\frac{k'^2}{\omega_{\mathbf{k}'}^{L}}\mathbf{k}\cdot(\mathbf{k} - \mathbf{k}') + \frac{|\mathbf{k} - \mathbf{k}'|^2}{\omega_{\mathbf{k}}^{N} - \omega_{\mathbf{k}'}^{L}}\mathbf{k}\cdot\mathbf{k}' + \frac{k^2}{\omega_{\mathbf{k}}^{N}}\mathbf{k}'\cdot(\mathbf{k} - \mathbf{k}')\right) \\
&\approx \frac{ie}{2m_e\omega_{pe}^2}\frac{2k'^2\,\mathbf{k}\cdot(\mathbf{k} - \mathbf{k}') + 2|\mathbf{k} - \mathbf{k}'|^2\,\mathbf{k}\cdot\mathbf{k}' + k^2\,\mathbf{k}'\cdot(\mathbf{k} - \mathbf{k}')}{4\,k\,k'\,|\mathbf{k} - \mathbf{k}'|}.
\end{aligned} \tag{C.3}$$

Let us introduce the following expression for notational simplicity:

$$a_{\mathbf{k},\mathbf{k}'} = \frac{\{k^2\,\mathbf{k}'\cdot(\mathbf{k}-\mathbf{k}') + 2k'^2\,\mathbf{k}\cdot(\mathbf{k}-\mathbf{k}') + 2|\mathbf{k}-\mathbf{k}'|^2\,(\mathbf{k}\cdot\mathbf{k}')\}^2}{16k^2\,k'^2\,|\mathbf{k}-\mathbf{k}'|^2}. \tag{C.4}$$

The nonlinear dispersion equation (1.90), upon ignoring the off diagonal terms involving $\langle\delta E^2\rangle_{\mathbf{k}',\omega'}\langle\delta E^2\rangle_{\mathbf{k}-\mathbf{k}',\omega-\omega'}$ and also ignoring the third-order nonlinear response, is given by

$$0 = \mathrm{Re}\left\{\epsilon(\mathbf{k},\omega) + 2\int d\mathbf{k}'\int d\omega'\left[\{\chi^{(2)}(\mathbf{k}',\omega'|\mathbf{k}-\mathbf{k}',\omega-\omega')\}^2\right.\right.$$
$$\left.\left.\times\left(\frac{\langle\delta E^2\rangle_{\mathbf{k}-\mathbf{k}',\omega-\omega'}}{\epsilon(\mathbf{k}',\omega')} + \frac{\langle\delta E^2\rangle_{\mathbf{k}',\omega'}}{\epsilon(\mathbf{k}-\mathbf{k}',\omega-\omega')}\right)\right]\right\}\langle\delta E^2\rangle_{\mathbf{k},\omega}. \tag{C.5}$$

We treat ω' and $\omega-\omega'$ as belonging to the fundamental Langmuir wave branch, while treating ω as the harmonic Langmuir wave. Expressing both L and N mode wave spectrum in the form of eigenmode intensities,

$$\begin{aligned}
\langle\delta E^2\rangle_{\mathbf{k}',\omega'} &= I^L_{\mathbf{k}'}\,\delta(\omega-\omega^L_{\mathbf{k}'}),\\
\langle\delta E^2\rangle_{\mathbf{k}-\mathbf{k}',\omega-\omega'} &= I^L_{\mathbf{k}-\mathbf{k}'}\,\delta(\omega-\omega'-\omega^L_{\mathbf{k}-\mathbf{k}'}),\\
\langle\delta E^2\rangle_{\mathbf{k},\omega} &= I^N_{\mathbf{k}}\,\delta(\omega-\omega^N_{\mathbf{k}}),
\end{aligned} \tag{C.6}$$

where only forward propagation is considered without loss of generality, and, making a trivial change of integral variables, we have

$$0 = \mathrm{Re}\left(\epsilon(\mathbf{k},\omega^N_{\mathbf{k}}) - \frac{e^2}{m_e^2\omega_{pe}^4}\int d\mathbf{k}'\,\frac{a_{\mathbf{k},\mathbf{k}'}\,I^L_{\mathbf{k}'}}{\epsilon(\mathbf{k}-\mathbf{k}',\omega^N_{\mathbf{k}}-\omega^L_{\mathbf{k}'})}\right)I^N_{\mathbf{k}}. \tag{C.7}$$

Making use of the following – see (1.115),

$$\begin{aligned}
\mathrm{Re}\,\epsilon(\mathbf{k},\omega^N_{\mathbf{k}}) &= 1 - \frac{\omega_{pe}^2}{\omega^{N2}_{\mathbf{k}}}\left(1 + 3k^2\lambda_{De}^2\,\frac{\omega_{pe}^2}{\omega^{N2}_{\mathbf{k}}}\right),\\
\mathrm{Re}\,\epsilon(\mathbf{k}-\mathbf{k}',\omega^N_{\mathbf{k}}-\omega^L_{\mathbf{k}'}) &= 1 - \frac{\omega^{L2}_{\mathbf{k}-\mathbf{k}'}}{(\omega^N_{\mathbf{k}}-\omega^L_{\mathbf{k}'})^2},
\end{aligned} \tag{C.8}$$

where $\omega^{L2}_{\mathbf{k}} = \omega_{pe}^2(1+3k^2\lambda_{De}^2)$, $\omega^{L2}_{\mathbf{k}'} = \omega_{pe}^2(1+3k'^2\lambda_{De}^2)$, and $\omega^{L2}_{\mathbf{k}-\mathbf{k}'} = \omega_{pe}^2(1+3|\mathbf{k}-\mathbf{k}'|^2\lambda_{De}^2)$, we obtain

$$0 = 1 - \frac{\omega_{pe}^2}{\omega^{N2}_{\mathbf{k}}}\left(1 + 3k^2\lambda_{De}^2\,\frac{\omega_{pe}^2}{\omega^{N2}_{\mathbf{k}}}\right)$$
$$-\frac{e^2}{m_e^2\omega_{pe}^4}\int d\mathbf{k}'\,\frac{a_{\mathbf{k},\mathbf{k}'}\,(\omega^N_{\mathbf{k}}-\omega^L_{\mathbf{k}'})^2\,I^L_{\mathbf{k}'}}{(\omega^N_{\mathbf{k}}-\omega^L_{\mathbf{k}'})^2-\omega^{L2}_{\mathbf{k}-\mathbf{k}'}}. \tag{C.9}$$

If we ignore thermal effects then we have

$$0 = \omega^N_{\mathbf{k}} + \omega_{pe} - \frac{\omega^N_{\mathbf{k}}\,(\omega^N_{\mathbf{k}}-\omega_{pe})}{(\omega^N_{\mathbf{k}}-2\omega_{pe})}\,\frac{e^2}{m_e^2\omega_{pe}^4}\int d\mathbf{k}'\,a_{\mathbf{k},\mathbf{k}'}\,I^L_{\mathbf{k}'}. \tag{C.10}$$

Upon approximating $\omega_{\mathbf{k}}^N \sim 2\omega_{pe}$ everywhere except in the denominator, we have

$$\omega_{\mathbf{k}}^N = 2\omega_{pe} + \frac{2e^2}{3m_e^2\omega_{pe}^3} \int d\mathbf{k}' \, a_{\mathbf{k},\mathbf{k}'} \, I_{\mathbf{k}'}^L. \tag{C.11}$$

This is the desired (first) harmonic Langmuir wave dispersion relation, which has a peculiar property of persisting even in the limit of $I_{\mathbf{k}'}^L \to 0$.

One can extend the discussion to higher harmonic Langmuir waves. Let us designate the harmonic eigenmodes including the fundamental and first harmonic mode by

$$\omega = \omega_{\mathbf{k}}^{Ln}, \qquad n = 1, 2, 3, \ldots. \tag{C.12}$$

Then, the wave intensity is given by

$$\langle \delta E^2 \rangle_{\mathbf{k}\omega} = \sum_{n=1,2,3,\ldots} I_{\mathbf{k}}^{Ln} \, \delta(\omega - \omega_{\mathbf{k}}^{Ln}). \tag{C.13}$$

One may insert (C.13) to (C.5). In direct generalization of (C.3), we may evaluate the second-order susceptibility,

$$\begin{aligned}
\left\{\chi^{(2)}(\mathbf{k}', \omega_{\mathbf{k}'}^{L(n-1)} | \mathbf{k}-\mathbf{k}', \omega - \omega_{\mathbf{k}'}^{L(n-1)})\right\}^2 &\approx -\frac{e^2}{4m_e^2\,\omega_{pe}^4}\, a_{\mathbf{k},\mathbf{k}'}^n, \\
a_{\mathbf{k},\mathbf{k}'}^n &= \frac{1}{n^4(n-1)^4\, k^2\, k'^2\, |\mathbf{k}-\mathbf{k}'|^2} \{ (n-1)\, k^2\, [\mathbf{k}' \cdot (\mathbf{k}-\mathbf{k}')] \\
&\quad + n\, k'^2\, [\mathbf{k} \cdot (\mathbf{k}-\mathbf{k}')] + n(n-1)\, |\mathbf{k}-\mathbf{k}'|^2\, (\mathbf{k}\cdot\mathbf{k}')\, \}^2.
\end{aligned} \tag{C.14}$$

This leads to

$$0 = \mathrm{Re}\left(\epsilon(\mathbf{k}, \omega) - \frac{e^2}{m_e^2\,\omega_{pe}^4} \int d\mathbf{k}' \, \frac{a_{\mathbf{k},\mathbf{k}'}^n \, I_{\mathbf{k}'}^{L(n-1)}}{\epsilon(\mathbf{k}-\mathbf{k}', \omega - \omega_{\mathbf{k}'}^{L(n-1)})} \right). \tag{C.15}$$

Approximately writing the quantities $\mathrm{Re}\,\epsilon(\mathbf{k}, \omega_{\mathbf{k}}^{Ln})$ and $\mathrm{Re}\,\epsilon(\mathbf{k}-\mathbf{k}', \omega_{\mathbf{k}}^{Ln} - \omega_{\mathbf{k}'}^{L(n-1)})$ as

$$\begin{aligned}
\mathrm{Re}\,\epsilon(\mathbf{k}, \omega_{\mathbf{k}}^{Ln}) &\approx 1 - \frac{\omega_{pe}^2}{(\omega_{\mathbf{k}}^{Ln})^2}\left(1 + 3k^2\lambda_{De}^2\, \frac{\omega_{pe}^2}{(\omega_{\mathbf{k}}^{Ln})^2}\right), \\
\mathrm{Re}\,\epsilon(\mathbf{k}-\mathbf{k}', \omega_{\mathbf{k}}^{Ln} - \omega_{\mathbf{k}'}^{L(n-1)}) &\approx \frac{(\omega_{\mathbf{k}}^{Ln} - \omega_{\mathbf{k}'}^{L(n-1)})^2 - (\omega_{\mathbf{k}-\mathbf{k}'}^{L1})^2}{(\omega_{\mathbf{k}}^{Ln} - \omega_{\mathbf{k}'}^{L(n-1)})^2},
\end{aligned} \tag{C.16}$$

we have

$$\begin{aligned}
0 = 1 &- \frac{\omega_{pe}^2}{(\omega_{\mathbf{k}}^{Ln})^2}\left(1 + 3k^2\lambda_{De}^2\, \frac{\omega_{pe}^2}{(\omega_{\mathbf{k}}^{Ln})^2}\right) \\
&- \frac{e^2}{m_e^2\omega_{pe}^4} \int d\mathbf{k}' \, \frac{a_{\mathbf{k},\mathbf{k}'}^n (\omega_{\mathbf{k}}^{Ln} - \omega_{\mathbf{k}'}^{L(n-1)})^2 I_{\mathbf{k}'}^{L(n-1)}}{(\omega_{\mathbf{k}}^{Ln} - \omega_{\mathbf{k}'}^{L(n-1)})^2 - (\omega_{\mathbf{k}-\mathbf{k}'}^{L1})^2}.
\end{aligned} \tag{C.17}$$

Ignoring thermal effects, we have

$$0 = \frac{(\omega_{\mathbf{k}}^{Ln})^2 - \omega_{pe}^2}{(\omega_{\mathbf{k}}^{Ln})^2} - \frac{[\omega_{\mathbf{k}}^{Ln} - (n-1)\omega_{pe}]^2}{[\omega_{\mathbf{k}}^{Ln} - (n-1)\omega_{pe}]^2 - \omega_{pe}^2} \frac{e^2}{m_e^2 \omega_{pe}^4} \int d\mathbf{k}'\, a_{\mathbf{k},\mathbf{k}'}^{n} I_{\mathbf{k}'}^{L(n-1)}. \tag{C.18}$$

Replacing $\omega_{\mathbf{k}}^{Ln}$ by $n\omega_{pe}$ everywhere except in the denominator associated with the second term on the right-hand side, we readily obtain (Yoon et al., 2003a)

$$\omega_{\mathbf{k}}^{Ln} = n\omega_{pe} + \frac{e^2}{2m_e^2\omega_{pe}^4} \frac{n^2}{n^2-1} \int d\mathbf{k}'\, a_{\mathbf{k},\mathbf{k}'}^{n} I_{\mathbf{k}'}^{L(n-1)}, \tag{C.19}$$

for $n \geq 2$. This is the desired multiple harmonic Langmuir wave dispersion relation when thermal corrections are ignored. For $n = 2$, we recover (C.11). In (C.19), the nonlinear dispersion relation depends on the harmonic mode intensities, $I_{\mathbf{k}}^{Ln}$. This implies that the discussion of harmonic Langmuir waves is incomplete without computing the intensities of these modes. Gaelzer et al. (2003) formulated the quasilinear theory of harmonic Langmuir waves with which the wave intensities can be calculated. Interested readers may refer to that paper.

C.2 Nonlinear Frequency Shift

We now discuss the classical problem of nonlinear frequency shift. For this situation, we only consider the linear eigenmodes and the frequency shifts thereof when the underlying plasma is weakly turbulent. We decompose the spectral intensity by considering only linear eigenmodes in the manner of (1.95), $\langle \delta E^2 \rangle_{\mathbf{k}\omega} = \sum_{\sigma=\pm1} \sum_{\alpha=L,S} I_{\mathbf{k}}^{\sigma\alpha}\, \delta(\omega - \sigma\omega_{\mathbf{k}}^{\alpha})$. Inserting this to (1.90), we have

$$\begin{aligned}
0 = {} & \sum_{\sigma=\pm1} \sum_{\alpha} \mathrm{Re}\, \epsilon(\mathbf{k}, \sigma\omega_{\mathbf{k}}^{\alpha})\, I_{\mathbf{k}}^{\alpha}\, \delta(\omega - \sigma\omega_{\mathbf{k}}^{\alpha}) \\
& + 2\mathrm{Re} \int d\mathbf{k}'\, d\omega' \Bigg[\sum_{\sigma,\sigma''=\pm1} \sum_{\alpha,\gamma} \frac{\{\chi^{(2)}(\mathbf{k}', \sigma\omega_{\mathbf{k}}^{\alpha} - \sigma''\omega_{\mathbf{k}-\mathbf{k}'}^{\gamma} | \mathbf{k}-\mathbf{k}', \sigma''\omega_{\mathbf{k}-\mathbf{k}'}^{\gamma})\}^2}{\epsilon(\mathbf{k}', \sigma\omega_{\mathbf{k}}^{\alpha} - \sigma''\omega_{\mathbf{k}-\mathbf{k}'}^{\gamma})} \\
& \times I_{\mathbf{k}}^{\sigma\alpha} I_{\mathbf{k}-\mathbf{k}'}^{\sigma''\gamma}\, \delta(\omega - \sigma\omega_{\mathbf{k}}^{\alpha})\, \delta(\omega' - \sigma\omega_{\mathbf{k}}^{\alpha} + \sigma''\omega_{\mathbf{k}-\mathbf{k}'}^{\gamma}) \\
& + \sum_{\sigma,\sigma'=\pm1} \sum_{\alpha,\beta} \frac{\{\chi^{(2)}(\mathbf{k}', \sigma'\omega_{\mathbf{k}'}^{\beta} | \mathbf{k}-\mathbf{k}', \sigma\omega_{\mathbf{k}}^{\alpha} - \sigma'\omega_{\mathbf{k}'}^{\beta})\}^2}{\epsilon(\mathbf{k}-\mathbf{k}', \sigma\omega_{\mathbf{k}}^{\alpha} - \sigma'\omega_{\mathbf{k}'}^{\beta})} \\
& \times I_{\mathbf{k}}^{\sigma\alpha} I_{\mathbf{k}'}^{\sigma'\beta}\, \delta(\omega - \sigma\omega_{\mathbf{k}}^{\alpha})\, \delta(\omega' - \sigma'\omega_{\mathbf{k}'}^{\beta}) \\
& - \sum_{\sigma',\sigma''=\pm1} \sum_{\beta,\gamma} \frac{|\chi^{(2)}(\mathbf{k}', \sigma'\omega_{\mathbf{k}'}^{\beta} | \mathbf{k}-\mathbf{k}', \sigma''\omega_{\mathbf{k}-\mathbf{k}'}^{\gamma})|^2}{\epsilon^*(\mathbf{k}, \sigma'\omega_{\mathbf{k}'}^{\beta} + \sigma''\omega_{\mathbf{k}-\mathbf{k}'}^{\gamma})} \\
& \times I_{\mathbf{k}'}^{\sigma'\beta} I_{\mathbf{k}-\mathbf{k}'}^{\sigma''\gamma}\, \delta(\omega' - \sigma'\omega_{\mathbf{k}'}^{\beta})\, \delta(\omega - \sigma'\omega_{\mathbf{k}'}^{\beta} - \sigma''\omega_{\mathbf{k}-\mathbf{k}'}^{\gamma}) \\
& - \sum_{\sigma,\sigma'=\pm1} \sum_{\alpha,\beta} \bar{\chi}^{(3)}(\mathbf{k}', \sigma'\omega_{\mathbf{k}'}^{\beta} | -\mathbf{k}', -\sigma'\omega_{\mathbf{k}'}^{\beta} | \mathbf{k}, \sigma\omega_{\mathbf{k}}^{\alpha}) \\
& \times I_{\mathbf{k}'}^{\sigma'\beta} I_{\mathbf{k}}^{\sigma\alpha}\, \delta(\omega - \sigma\omega_{\mathbf{k}}^{\alpha})\, \delta(\omega' - \sigma'\omega_{\mathbf{k}'}^{\beta}) \Bigg].
\end{aligned} \tag{C.20}$$

To the leading order, $\{\chi^{(2)}(\mathbf{k}',\omega'|\mathbf{k}-\mathbf{k}',\omega-\omega')\}^2 \approx -|\chi^{(2)}(\mathbf{k}',\omega'|\mathbf{k}-\mathbf{k}',\omega-\omega')|^2$. Consequently, the real part of second-order nonlinear terms comes from the principal parts of inverse dielectric functions. The principal part of the third term on the right-hand side within the second-order nonlinear term becomes unimportant if the condition $\omega = \sigma\omega_{\mathbf{k}}^{\alpha}$ is satisfied, since the principal part excludes the point $\sigma'\omega_{\mathbf{k}'}^{\beta} + \sigma''\omega_{\mathbf{k}-\mathbf{k}'}^{\gamma} = \omega$, by definition, when ω is equal to $\sigma\omega_{\mathbf{k}}^{\alpha}$. The first and second terms within the second-order nonlinear correction terms are equivalent, which can be seen by a simple interchange of dummy integral and summation variables. As a consequence, (C.20) reduces to

$$
\begin{aligned}
0 &= \mathrm{Re}\left(\epsilon(\mathbf{k},\sigma\omega_{\mathbf{k}}^{\alpha}) + \sum_{\sigma'=\pm1}\sum_{\beta}\int d\mathbf{k}'\, A_{\alpha,\beta}^{\sigma,\sigma'}(\mathbf{k},\mathbf{k}')\, I_{\mathbf{k}'}^{\sigma'\beta}\right) I_{\mathbf{k}}^{\sigma\alpha},\\
A_{\alpha,\beta}^{\sigma,\sigma'}(\mathbf{k},\mathbf{k}') &= 2\left(\mathscr{P}\,\frac{2\{\chi^{(2)}(\mathbf{k}',\sigma'\omega_{\mathbf{k}'}^{\beta}|\mathbf{k}-\mathbf{k}',\sigma\omega_{\mathbf{k}}^{\alpha}-\sigma'\omega_{\mathbf{k}'}^{\beta})\}^2}{\epsilon(\mathbf{k}-\mathbf{k}',\sigma\omega_{\mathbf{k}}^{\alpha}-\sigma'\omega_{\mathbf{k}'}^{\beta})}\right.\\
&\qquad \left. -\bar{\chi}^{(3)}(\mathbf{k}',\sigma'\omega_{\mathbf{k}'}^{\beta}|-\mathbf{k}',-\sigma'\omega_{\mathbf{k}'}^{\beta}|\mathbf{k},\sigma\omega_{\mathbf{k}}^{\alpha})\right).
\end{aligned}
\tag{C.21}
$$

C.2.1 Nonlinear Frequency Shift of Langmuir Waves

For Langmuir mode, $\alpha = L$, we have

$$
\begin{aligned}
0 &= \mathrm{Re}\,\epsilon(\mathbf{k},\sigma\omega_{\mathbf{k}}^{L})\\
&\quad + \mathrm{Re}\sum_{\sigma'=\pm1}\int d\mathbf{k}'\left(A_{L,L}^{\sigma,\sigma'}(\mathbf{k},\mathbf{k}')\, I_{\mathbf{k}'}^{\sigma'L} + A_{L,S}^{\sigma,\sigma'}(\mathbf{k},\mathbf{k}')\, I_{\mathbf{k}'}^{\sigma'S}\right).
\end{aligned}
\tag{C.22}
$$

Making note of the inequalities, $\omega_{\mathbf{k}}^{L} \gg kv_{Te}$, $\omega_{\mathbf{k}'}^{L} \gg k'v_{Te}$, $\omega_{\mathbf{k}}^{L} \gg kv_{Ti}$, and $\omega_{\mathbf{k}'}^{L} \gg k'v_{Ti}$, we may write

$$
\begin{aligned}
&\bar{\chi}^{(3)}(\mathbf{k}',\sigma'\omega_{\mathbf{k}'}^{L}|-\mathbf{k}',-\sigma'\omega_{\mathbf{k}'}^{L}|\mathbf{k},\sigma\omega_{\mathbf{k}}^{L})\\
&\quad\approx -\frac{1}{2}\frac{e^2}{m_e^2}\frac{(\mathbf{k}\cdot\mathbf{k}')^2}{k^2k'^2}\frac{|\mathbf{k}-\mathbf{k}'|^2}{(\sigma\omega_{\mathbf{k}}^{L})^3(\sigma'\omega_{\mathbf{k}'}^{L})}\,\chi_e(\mathbf{k}-\mathbf{k}',\sigma\omega_{\mathbf{k}}^{L}-\sigma'\omega_{\mathbf{k}'}^{L}),\\
&\{\chi^{(2)}(\mathbf{k}',\sigma'\omega_{\mathbf{k}'}^{L}|\mathbf{k}-\mathbf{k}',\sigma\omega_{\mathbf{k}}^{L}-\sigma'\omega_{\mathbf{k}'}^{L})\}^2\\
&\quad\approx -\frac{1}{4}\frac{e^2}{m_e^2}\frac{(\mathbf{k}\cdot\mathbf{k}')^2}{k^2k'^2}\frac{|\mathbf{k}-\mathbf{k}'|^2}{\omega_{\mathbf{k}}^{L2}\,\omega_{\mathbf{k}'}^{L2}}\{\chi_e(\mathbf{k}-\mathbf{k}',\sigma\omega_{\mathbf{k}}^{L}-\sigma'\omega_{\mathbf{k}'}^{L})\}^2.
\end{aligned}
\tag{C.23}
$$

Note that the electrons dominate nonlinear responses since the lighter and mobile electrons respond to nonlinear perturbations much more rapidly than the sluggish and heavier ions.

Denoting

$$
\begin{aligned}
\epsilon(\mathbf{k}-\mathbf{k}',\sigma\omega_{\mathbf{k}}^{L}-\sigma'\omega_{\mathbf{k}'}^{L}) &= 1+\chi_e(\mathbf{k}-\mathbf{k}',\sigma\omega_{\mathbf{k}}^{L}-\sigma'\omega_{\mathbf{k}'}^{L})\\
&\quad +\chi_i(\mathbf{k}-\mathbf{k}',\sigma\omega_{\mathbf{k}}^{L}-\sigma'\omega_{\mathbf{k}'}^{L}),
\end{aligned}
\tag{C.24}
$$

we have

$$
\begin{aligned}
\sum_{\sigma'=\pm1}\mathrm{Re}\,A_{L,L}^{\sigma,\sigma'}(\mathbf{k},\mathbf{k}')\, I_{\mathbf{k}'}^{\sigma'L} &= -\frac{e^2}{m_e^2}\frac{(\mathbf{k}\cdot\mathbf{k}')^2}{k^2k'^2}\sum_{\sigma'=\pm1}\frac{|\mathbf{k}-\mathbf{k}'|^2}{(\sigma\omega_{\mathbf{k}}^{L})^3(\omega_{\mathbf{k}'}^{L})^2}\\
&\quad\times\mathrm{Re}\left(\sigma\omega_{\mathbf{k}}^{L}\frac{\chi_e^2}{1+\chi_e+\chi_i}-\sigma'\omega_{\mathbf{k}'}^{L}\chi_e\right) I_{\mathbf{k}'}^{\sigma'L}.
\end{aligned}
\tag{C.25}
$$

In (C.25), χ_e and χ_i are shorthand notations for the full expression given by (C.24). From this, it can be seen that when the enhanced turbulence is primarily of Langmuir waves, the appropriate nonlinear dispersion equation is given by

$$
\begin{aligned}
0 = \mathrm{Re}\,\epsilon(\mathbf{k},\sigma\omega_{\mathbf{k}}^L) - \frac{e^2}{m_e^2\,\omega_{pe}^4}\int d\mathbf{k}'\,\frac{(\mathbf{k}\cdot\mathbf{k}')^2\,|\mathbf{k}-\mathbf{k}'|^2}{k^2\,k'^2} \\
\times \sum_{\sigma'=\pm 1}\mathrm{Re}\left(\frac{\chi_e^2}{1+\chi_e+\chi_i} - \sigma\sigma'\,\frac{\omega_{\mathbf{k}'}^L}{\omega_{\mathbf{k}}^L}\,\chi_e\right) I_{\mathbf{k}'}^{\sigma' L}.
\end{aligned}
\tag{C.26}
$$

If σ and σ' are of the same sign, then we may approximate

$$
\begin{aligned}
\mathrm{Re}\,\chi_e &\approx \frac{1}{|\mathbf{k}-\mathbf{k}'|^2\,\lambda_{De}^2}, \\
\mathrm{Re}\,\chi_i &\approx -\frac{\omega_{pi}^2}{(\omega_{\mathbf{k}}^L - \omega_{\mathbf{k}'}^L)^2} \approx -\frac{m_e}{m_i}\,\frac{1}{[(3/2)(k^2-k'^2)\,\lambda_{De}^2]^2}.
\end{aligned}
\tag{C.27}
$$

We further approximate

$$
\frac{\omega_{\mathbf{k}'}^L}{\omega_{\mathbf{k}}^L} \approx 1 - \frac{3}{2}\,(k^2 - k'^2)\,\lambda_{De}^2. \tag{C.28}
$$

Upon ignoring the ion response, we have

$$
\mathrm{Re}\left(\frac{\chi_e^2}{1+\chi_e+\chi_i} - \frac{\omega_{\mathbf{k}'}^L}{\omega_{\mathbf{k}}^L}\,\chi_e\right) \approx \frac{3}{2}\,\frac{k^2-k'^2}{|\mathbf{k}-\mathbf{k}'|^2} - 1. \tag{C.29}
$$

On the other hand, if σ and σ' are of the opposite sign, then we have

$$
\mathrm{Re}\,\chi_e \approx \frac{3}{4}, \qquad \mathrm{Re}\,\chi_i \approx -\frac{m_e}{2m_i}, \tag{C.30}
$$

from which we obtain

$$
\mathrm{Re}\left(\frac{\chi_e^2}{1+\chi_e+\chi_i} + \frac{\omega_{\mathbf{k}'}^L}{\omega_{\mathbf{k}}^L}\,\chi_e\right) \approx \frac{15}{14}. \tag{C.31}
$$

As a result, (C.26) reduces to

$$
\begin{aligned}
\mathrm{Re}\,\epsilon(\mathbf{k},\sigma\omega_{\mathbf{k}}^L) = \frac{\lambda_{De}^2}{4\pi\hat{n}T_e}\int d\mathbf{k}'\,\frac{(\mathbf{k}\cdot\mathbf{k}')^2}{k^2\,k'^2}\left[\left(\frac{3}{2}\,(k^2-k'^2) - |\mathbf{k}-\mathbf{k}'|^2\right) I_{\mathbf{k}'}^{\sigma L}\right. \\
\left. + \frac{15}{14}\,|\mathbf{k}-\mathbf{k}'|^2\,I_{\mathbf{k}'}^{-\sigma L}\right].
\end{aligned}
\tag{C.32}
$$

This leads to the modified Langmuir wave dispersion relation

$$
\begin{aligned}
\omega_{\mathbf{k}}^L &= \omega_{pe}\left(1 + \frac{3}{2}\,k^2\lambda_{De}^2\right) + \delta\omega_{\mathbf{k}}^L, \\
\delta\omega_{\mathbf{k}}^L &= \frac{3\omega_{pe}\lambda_{De}^2}{16\pi\hat{n}_e T_e}\int d\mathbf{k}'\,\frac{(\mathbf{k}\cdot\mathbf{k}')^2}{k^2\,k'^2} \\
&\quad\times\left[\left(k^2 - k'^2 - \frac{2}{3}\,|\mathbf{k}-\mathbf{k}'|^2\right) I_{\mathbf{k}'}^{+L} + \frac{5}{7}\,|\mathbf{k}-\mathbf{k}'|^2\,I_{\mathbf{k}'}^{-L}\right].
\end{aligned}
\tag{C.33}
$$

The quantity $\delta\omega_{\mathbf{k}}^{L}$ is the nonlinear frequency shift of Langmuir wave caused by Langmuir self turbulence.

Next, we consider the contribution to the nonlinear frequency shift from ion-sound waves. For this case, the appropriate inequalities are $\omega_{\mathbf{k}}^{L} \gg k\, v_{Te}$, $\omega_{\mathbf{k}'}^{S} \ll k'\, v_{Te}$, $|\omega_{\mathbf{k}}^{L} - \omega_{\mathbf{k}'}^{S}| \gg |\mathbf{k}-\mathbf{k}'|\, v_{Te}$, $\omega_{\mathbf{k}}^{L} \gg k\, v_{Ti}$, $\omega_{\mathbf{k}'}^{S} > k'\, v_{Ti}$, and $|\omega_{\mathbf{k}}^{L} - \omega_{\mathbf{k}'}^{S}| \gg |\mathbf{k}-\mathbf{k}'|\, v_{Ti}$. This leads to the following evaluation of nonlinear responses:

$$\begin{aligned}
&\chi^{(2)}(\mathbf{k}',\sigma'\omega_{\mathbf{k}'}^{S}|\mathbf{k}-\mathbf{k}',\sigma\omega_{\mathbf{k}}^{L}-\sigma'\omega_{\mathbf{k}'}^{S}) \\
&\quad\approx -\frac{i}{2}\frac{e}{m_e}\frac{k'}{\sigma\omega_{\mathbf{k}}^{L}(\sigma\omega_{\mathbf{k}}^{L}-\sigma'\omega_{\mathbf{k}'}^{S})}\frac{\mathbf{k}\cdot(\mathbf{k}-\mathbf{k}')}{k\,|\mathbf{k}-\mathbf{k}'|}\,\chi_e(\mathbf{k}',\sigma'\omega_{\mathbf{k}'}^{S}), \\
&\bar{\chi}^{(3)}(\mathbf{k}',\sigma'\omega_{\mathbf{k}'}^{S}|-\mathbf{k}',-\sigma'\omega_{\mathbf{k}'}^{S}|\mathbf{k},\sigma\omega_{\mathbf{k}}^{L}) \\
&\quad\approx -\frac{1}{2(\sigma\omega_{\mathbf{k}}^{L})^3}\frac{e^2}{m_e^2}\frac{\mathbf{k}\cdot\mathbf{k}'}{k^2}\left(\frac{2k^2}{\sigma\omega_{\mathbf{k}}^{L}}+\frac{\mathbf{k}\cdot(\mathbf{k}-\mathbf{k}')}{\sigma\omega_{\mathbf{k}}^{L}-\sigma'\omega_{\mathbf{k}'}^{S}}\right)\chi_e(\mathbf{k}',\sigma'\omega_{\mathbf{k}'}^{S}).
\end{aligned} \tag{C.34}$$

If we further approximate

$$\mathrm{Re}\,\chi_e(\mathbf{k}',\sigma'\omega_{\mathbf{k}'}^{S}) \approx \frac{1}{k'^2\lambda_{De}^2}, \tag{C.35}$$

then the nonlinear dispersion equation for Langmuir waves subject to ion-sound turbulence is

$$\begin{aligned}
\mathrm{Re}\,\epsilon(\mathbf{k},\sigma\omega_{\mathbf{k}}^{L}) &= \frac{e^2}{m_e^2\omega_{pe}^4}\int d\mathbf{k}'\,\mathrm{Re}\sum_{\sigma'=\pm1}\frac{1}{k'^2\lambda_{De}^2} \\
&\quad\times\left(\frac{k'^2\,[\mathbf{k}\cdot(\mathbf{k}-\mathbf{k}')]^2}{k^2\,|\mathbf{k}-\mathbf{k}'|^2}\frac{1}{k'^2\lambda_{De}^2}\,\mathscr{P}\frac{1}{\epsilon(\mathbf{k}-\mathbf{k}',\sigma\omega_{\mathbf{k}}^{L}-\sigma'\omega_{\mathbf{k}'}^{S})}\right. \\
&\quad\left.-\frac{(\mathbf{k}\cdot\mathbf{k}')(3k^2-\mathbf{k}\cdot\mathbf{k}')}{k^2}\right)I_{\mathbf{k}'}^{\sigma' S}.
\end{aligned} \tag{C.36}$$

In (C.36), the term associated with the denominator, $|\epsilon(\mathbf{k}-\mathbf{k}',\sigma\omega_{\mathbf{k}}^{L}-\sigma'\omega_{\mathbf{k}'}^{S})|^2$, is the dominant term, which simplifies (C.36). Upon replacing $\sigma\omega_{\mathbf{k}}^{L}$ by a generic ω, we obtain

$$\begin{aligned}
\mathrm{Re}\,\epsilon(\mathbf{k},\omega) &= \frac{1}{4\pi\hat{n}T_e}\int d\mathbf{k}'\,\mathrm{Re}\sum_{\sigma'=\pm1}\frac{[\mathbf{k}\cdot(\mathbf{k}-\mathbf{k}')]^2}{k^2\,|\mathbf{k}-\mathbf{k}'|^2} \\
&\quad\times\mathscr{P}\frac{1}{\epsilon(\mathbf{k}-\mathbf{k}',\omega-\sigma'\omega_{\mathbf{k}'}^{S})}\frac{I_{\mathbf{k}'}^{\sigma' S}}{k'^2\lambda_{De}^2}.
\end{aligned} \tag{C.37}$$

Further approximating the denominator,

$$\mathrm{Re}\,\epsilon(\mathbf{k}-\mathbf{k}',\omega-\sigma'\omega_{\mathbf{k}'}^{S}) \approx \frac{\omega^2-\omega_{\mathbf{k}-\mathbf{k}'}^{L2}}{\omega^2}, \tag{C.38}$$

we have

$$\begin{aligned}
0 &= 1-\frac{\omega_{\mathbf{k}}^{L}}{\omega}-\int d\mathbf{k}'\,\beta_{\mathbf{k},\mathbf{k}'}\,\frac{\omega_{pe}}{\omega-\omega_{\mathbf{k}-\mathbf{k}'}^{L}}, \\
\beta_{\mathbf{k},\mathbf{k}'} &= \frac{1}{16\pi\hat{n}T_e}\sum_{\sigma'=\pm1}\frac{[\mathbf{k}\cdot(\mathbf{k}-\mathbf{k}')]^2}{k^2\,|\mathbf{k}-\mathbf{k}'|^2}\frac{I_{\mathbf{k}'}^{\sigma' S}}{k'^2\lambda_{De}^2}.
\end{aligned} \tag{C.39}$$

One may iteratively solve this equation to obtain the Langmuir wave dispersion relation subject to the ion sound turbulence:

$$\omega_{\mathbf{k}}^{L} = \omega_{pe}\left(1 + \frac{3}{2}\,k^{2}\lambda_{De}^{2}\right) + \delta\omega_{\mathbf{k}}^{L},$$
$$\delta\omega_{\mathbf{k}}^{L} = \omega_{pe}\left(\int d\mathbf{k}'\,\beta_{\mathbf{k},\mathbf{k}'}\right)^{1/2}. \tag{C.40}$$

The nonlinear frequency shift of Langmuir waves, $\delta\omega_{\mathbf{k}}^{L}$, given above, is owing to the ion sound turbulence, which may be compared with (C.33), where it arises due to Langmuir self turbulence.

In the expression for $\beta_{\mathbf{k},\mathbf{k}'}$, the quantity $I_{\mathbf{k}'}^{\sigma' S}/(k'^{2}\,\lambda_{De}^{2})$ may appear to diverge at first sight. However, if we define "plasmon" number density by – see (2.20)

$$N_{\mathbf{k}}^{\sigma\alpha} = \frac{I_{\mathbf{k}}^{\sigma\alpha}}{\hbar(\sigma\omega_{\mathbf{k}}^{\alpha})\,[\partial \mathrm{Re}\epsilon(\mathbf{k},\sigma\omega_{\mathbf{k}}^{\alpha})/\partial(\sigma\omega_{\mathbf{k}}^{\alpha})]}, \tag{C.41}$$

where $\hbar$ is the Planck constant, and make use of (C.8), specifically,

$$\frac{1}{\partial \mathrm{Re}\epsilon(\mathbf{k},\sigma\omega_{\mathbf{k}}^{S})/\partial(\sigma\omega_{\mathbf{k}}^{S})} \approx \mu_{\mathbf{k}}\,\frac{(\sigma\omega_{\mathbf{k}}^{L})}{2}, \tag{C.42}$$

where $\mu_{\mathbf{k}}$ is defined in (C.8), then it is more convenient to absorb $\mu_{\mathbf{k}}$ into the definition of the ion-sound wave spectrum,

$$I_{\mathbf{k}}^{\sigma S} \rightarrow \frac{I_{\mathbf{k}}^{\sigma S}}{\mu_{\mathbf{k}}}. \tag{C.43}$$

Then in the new scheme, the quantity $\beta_{\mathbf{k},\mathbf{k}'}$ is well behaved mathematically.

C.2.2 Nonlinear Frequency Shift of Ion-Sound Waves

Let us consider the ion-sound mode, $\alpha = S$, in (C.21):

$$0 = \mathrm{Re}\,\epsilon(\mathbf{k},\sigma\omega_{\mathbf{k}}^{S})$$
$$+\mathrm{Re}\sum_{\sigma'=\pm1}\int d\mathbf{k}'\left(A_{S,L}^{\sigma,\sigma'}(\mathbf{k},\mathbf{k}')\,I_{\mathbf{k}'}^{\sigma' L} + A_{S,S}^{\sigma,\sigma'}(\mathbf{k},\mathbf{k}')\,I_{\mathbf{k}'}^{\sigma' S}\right). \tag{C.44}$$

For this case, we have the following inequalities:

$$\omega_{\mathbf{k}}^{S} \ll k\,v_{Te}, \quad \omega_{\mathbf{k}'}^{L} \gg k'\,v_{Te}, \quad |\omega_{\mathbf{k}}^{S} - \omega_{\mathbf{k}'}^{L}| \gg |\mathbf{k}-\mathbf{k}'|\,v_{Te},$$
$$\omega_{\mathbf{k}}^{S} > k\,v_{Ti}, \quad \omega_{\mathbf{k}'}^{L} \gg k'\,v_{Ti}, \quad |\omega_{\mathbf{k}}^{S} - \omega_{\mathbf{k}'}^{L}| \gg |\mathbf{k}-\mathbf{k}'|\,v_{Ti},$$
$$\omega_{\mathbf{k}}^{S} \ll k\,v_{Te}, \quad \omega_{\mathbf{k}'}^{S} \ll k'\,v_{Te}, \quad |\omega_{\mathbf{k}}^{S} - \omega_{\mathbf{k}'}^{S}| \ll |\mathbf{k}-\mathbf{k}'|\,v_{Te},$$
$$\omega_{\mathbf{k}}^{S} > k\,v_{Ti}, \quad \omega_{\mathbf{k}'}^{S} > k'\,v_{Ti}, \quad |\omega_{\mathbf{k}}^{S} - \omega_{\mathbf{k}'}^{S}| > |\mathbf{k}-\mathbf{k}'|\,v_{Ti}. \tag{C.45}$$

Note that ion-sound waves can exist only in a plasma where $T_e \gg T_i$. Therefore, $\omega_{\mathbf{k}}^{S} \sim kc_S$, $\omega_{\mathbf{k}'}^{S} \sim k'c_S$, $\omega_{\mathbf{k}}^{S} - \omega_{\mathbf{k}'}^{S} \sim (k-k')\,c_S$, and $c_S/v_{Ti} \sim (T_e/T_i)^{1/2} \gg 1$.

Of the various nonlinear response functions within the coefficient $A_{S,L}^{\sigma,\sigma'}(\mathbf{k},\mathbf{k}')$, the second-order electron response is the most important one. Others can be ignored.

$$\chi_e^{(2)}(\mathbf{k}',\sigma'\omega_{\mathbf{k}'}^L|\mathbf{k}-\mathbf{k}',\sigma\omega_{\mathbf{k}}^S-\sigma'\omega_{\mathbf{k}'}^L) \approx -\frac{i}{2}\frac{e}{m_e}\frac{\omega_{pe}^2}{(\sigma'\omega_{\mathbf{k}'}^L)(\sigma\omega_{\mathbf{k}}^S-\sigma'\omega_{\mathbf{k}'}^L)}\frac{\mathbf{k}'\cdot(\mathbf{k}-\mathbf{k}')}{k\,k'\,|\mathbf{k}-\mathbf{k}'|}\frac{2}{v_{Te}^2}. \tag{C.46}$$

Consequently, we have

$$A_{S,L}^{\sigma,\sigma'}(\mathbf{k},\mathbf{k}') \approx -\frac{e^2}{T_e^2}\frac{[\mathbf{k}'\cdot(\mathbf{k}-\mathbf{k}')]^2}{k^2\,k'^2\,|\mathbf{k}-\mathbf{k}'|^2}\,\mathscr{P}\frac{1}{\epsilon(\sigma\omega_{\mathbf{k}}^S-\sigma'\omega_{\mathbf{k}'}^L)}. \tag{C.47}$$

Among the nonlinear response functions within the coefficient $A_{S,S}^{\sigma,\sigma'}(\mathbf{k},\mathbf{k}')$, while it is true that in general the third-order susceptibility can be ignored, the ion second-order response can be of the same order as the electron term,

$$\chi_e^{(2)}(\mathbf{k}',\sigma'\omega_{\mathbf{k}'}^S|\mathbf{k}-\mathbf{k}',\sigma\omega_{\mathbf{k}}^S-\sigma'\omega_{\mathbf{k}'}^S) \approx \frac{ie}{T_e}\frac{\omega_{pe}^2}{k\,k'\,|\mathbf{k}-\mathbf{k}'|}\frac{1}{v_{Te}^2},$$

$$\begin{aligned}\chi_i^{(2)}&(\mathbf{k}',\sigma'\omega_{\mathbf{k}'}^S|\mathbf{k}-\mathbf{k}',\sigma\omega_{\mathbf{k}}^S-\sigma'\omega_{\mathbf{k}'}^S)\\ &\approx -\frac{i}{2}\frac{e}{m_i}\frac{\omega_{pi}^2}{(\sigma\omega_{\mathbf{k}}^S)(\sigma'\omega_{\mathbf{k}'}^S)(\sigma\omega_{\mathbf{k}}^S-\sigma'\omega_{\mathbf{k}'}^S)}\frac{1}{k\,k'\,|\mathbf{k}-\mathbf{k}'|}\\ &\times\left(\frac{k^2}{\sigma\omega_{\mathbf{k}}^S}\mathbf{k}'\cdot(\mathbf{k}-\mathbf{k}')+\frac{k'^2}{\sigma'\omega_{\mathbf{k}'}^S}\mathbf{k}\cdot(\mathbf{k}-\mathbf{k}')+\frac{|\mathbf{k}-\mathbf{k}'|^2}{\sigma\omega_{\mathbf{k}}^S-\sigma'\omega_{\mathbf{k}'}^S}\mathbf{k}\cdot\mathbf{k}'\right).\end{aligned} \tag{C.48}$$

Combining both electron and ion responses, we have

$$\begin{aligned}\chi^{(2)}&(\mathbf{k}',\sigma'\omega_{\mathbf{k}'}^S|\mathbf{k}-\mathbf{k}',\sigma\omega_{\mathbf{k}}^S-\sigma'\omega_{\mathbf{k}'}^S)\\ &= -\sigma\sigma'\frac{ie}{T_e}\frac{m_e}{2T_e}\frac{\omega_{pe}^2\,(\mathbf{k}\cdot\mathbf{k}')}{k^2\,k'^2\,|\mathbf{k}-\mathbf{k}'|}\left(1+\frac{|\mathbf{k}-\mathbf{k}'|^2}{(\sigma k-\sigma'k')^2}\right).\end{aligned} \tag{C.49}$$

This leads to

$$\begin{aligned}A_{S,S}^{\sigma,\sigma'}(\mathbf{k},\mathbf{k}') &= -\frac{e^2}{T_e^2}\frac{m_e^2}{T_e^2}\frac{\omega_{pe}^4\,(\mathbf{k}\cdot\mathbf{k}')^2}{k^4\,k'^4\,|\mathbf{k}-\mathbf{k}'|^2}\left(1+\frac{\omega_{\mathbf{k}-\mathbf{k}'}^{S2}}{(\sigma\omega_{\mathbf{k}}^S-\sigma'\omega_{\mathbf{k}'}^S)^2}\right)^2\\ &\times\mathscr{P}\frac{1}{\epsilon(\mathbf{k}-\mathbf{k}',\sigma\omega_{\mathbf{k}}^S-\sigma'\omega_{\mathbf{k}'}^S)}.\end{aligned} \tag{C.50}$$

Collecting the results we have obtained thus far, we arrive at

$$\begin{aligned}\mathrm{Re}\,\epsilon(\mathbf{k},\omega) &= \frac{1}{4\pi\hat{n}T_e}\sum_{\sigma'=\pm1}\int d\mathbf{k}'\left[\frac{[\mathbf{k}'\cdot(\mathbf{k}-\mathbf{k}')]^2}{k'^2\,|\mathbf{k}-\mathbf{k}'|^2}\frac{I_{\mathbf{k}'}^{\sigma'L}}{k^2\lambda_{De}^2}\,\mathscr{P}\frac{1}{\epsilon(\mathbf{k}-\mathbf{k}',\omega-\sigma'\omega_{\mathbf{k}'}^L)}\right.\\ &+\frac{(\mathbf{k}\cdot\mathbf{k}')^2}{k^2\,k'^2}\left(1+\frac{\omega_{\mathbf{k}-\mathbf{k}'}^{S2}}{(\omega-\sigma'\omega_{\mathbf{k}'}^S)^2}\right)^2\\ &\left.\times\frac{I_{\mathbf{k}'}^{\sigma'S}}{k^2\,k'^2\,|\mathbf{k}-\mathbf{k}'|^2\lambda_{De}^6}\,\mathscr{P}\frac{1}{\epsilon(\mathbf{k}-\mathbf{k}',\omega-\sigma'\omega_{\mathbf{k}'}^S)}\right],\end{aligned} \tag{C.51}$$

where we have replaced the argument $\sigma\omega_{\mathbf{k}}^S$ by a generic variable, ω. Since we are looking for solutions in the ion-sound mode range, $\omega \sim \omega_{\mathbf{k}}^S$, we approximate various linear responses by

$$\begin{aligned}
\mathrm{Re}\,\epsilon(\mathbf{k},\omega) &\approx \frac{2(1+k^2\lambda_{De}^2)}{k^2\lambda_{De}^2}\left(1-\frac{\omega_{\mathbf{k}}^S}{\omega}\right),\\
\mathrm{Re}\,\epsilon(\mathbf{k}-\mathbf{k}',\omega-\sigma'\omega_{\mathbf{k}'}^L) &\approx -\frac{2\sigma'}{\omega_{pe}}(\omega-\sigma'\omega_{\mathbf{k}'}^L+\sigma'\omega_{\mathbf{k}-\mathbf{k}'}^L),\\
\mathrm{Re}\,\epsilon(\mathbf{k}-\mathbf{k}',\omega-\sigma'\omega_{\mathbf{k}'}^S) &\approx \frac{1+|\mathbf{k}-\mathbf{k}'|^2\lambda_{De}^2}{|\mathbf{k}-\mathbf{k}'|^2\lambda_{De}^2}\left(1-\frac{\omega_{\mathbf{k}-\mathbf{k}'}^{S2}}{(\omega-\sigma'\omega_{\mathbf{k}'}^S)^2}\right),
\end{aligned} \tag{C.52}$$

where $\omega_{\mathbf{k}}^{S2}$ is defined in (C.6). This simplifies (C.52):

$$\begin{aligned}
0 = \omega - \omega_{\mathbf{k}}^S &+ \frac{\omega_{\mathbf{k}}^S}{8\pi\hat{n}T_e}\sum_{\sigma'=\pm1}\int d\mathbf{k}'\left[\frac{\omega_{pe}}{2}\frac{[\mathbf{k}'\cdot(\mathbf{k}-\mathbf{k}')]^2}{k'^2\,|\mathbf{k}-\mathbf{k}'|^2}\frac{\sigma' I_{\mathbf{k}'}^{\sigma' L}}{\omega-\sigma'\omega_{\mathbf{k}'}^L+\sigma'\omega_{\mathbf{k}-\mathbf{k}'}^L}\right.\\
&- k'\lambda_{De}\left(\frac{m_e}{m_i}\right)^{1/2}\left(1+\frac{3T_i}{T_e}\right)^{1/2}\frac{(\mathbf{k}\cdot\mathbf{k}')^2}{k^2\,k'^2}\left(1+\frac{|\mathbf{k}-\mathbf{k}'|^2}{(k-\sigma'k')^2}\right)^2 I_{\mathbf{k}'}^{\sigma' S}\\
&\left.\times\frac{(k-\sigma'k')^2c_S^2}{(\omega-\sigma'k'c_S)^2-(\mathbf{k}-\mathbf{k}')^2c_S^2}\right],
\end{aligned} \tag{C.53}$$

where we have redefined the ion-sound spectrum by absorbing the factor $\mu_{\mathbf{k}'}$ in the denominator.

In general, (C.53) requires numerical solution. If we ignore contribution from the ion-sound spectrum, and assume isotropic Langmuir turbulence, $I_{\mathbf{k}'}^{+L} = I_{\mathbf{k}'}^{-L} = I_{\mathbf{k}'}$, then we obtain

$$\begin{aligned}
\omega &= \omega_{\mathbf{k}}^S + \delta\omega_{\mathbf{k}}^S,\\
\delta\omega_{\mathbf{k}}^S &= \frac{\omega_{\mathbf{k}}^S}{12\pi\hat{n}T_e}\int d\mathbf{k}'\,\frac{[\mathbf{k}'\cdot(\mathbf{k}-\mathbf{k}')]^2}{k'^2\,|\mathbf{k}-\mathbf{k}'|^2}\frac{I_{\mathbf{k}'}}{(2\mathbf{k}\cdot\mathbf{k}'-k^2)\,\lambda_{De}^2},
\end{aligned} \tag{C.54}$$

where we have assumed $\omega_{\mathbf{k}}^S \ll \omega_{pe}$ in order to ignore $\omega_{\mathbf{k}}^S$ the denominators. Here, $\delta\omega_{\mathbf{k}}^S$ represents the nonlinear frequency shift of ion-sound waves by background high-frequency Langmuir turbulence.

Appendix D

Plasma Dispersion Function

The plasma dispersion or Fried–Conte function (Fried and Conte, 1961), or simply Z function, is defined by

$$Z(\zeta) = \begin{cases} \dfrac{1}{\pi^{1/2}} \displaystyle\int_{-\infty}^{\infty} dx \, \frac{e^{-x^2}}{x-\zeta}, & \mathrm{Im}\,\zeta > 0 \\ \dfrac{1}{\pi^{1/2}} \mathscr{P} \displaystyle\int_{-\infty}^{\infty} dx \, \frac{e^{-x^2}}{x-\zeta} + i\,\pi^{1/2} e^{-\zeta^2}, & \mathrm{Im}\,\zeta = 0 \\ \dfrac{1}{\pi^{1/2}} \displaystyle\int_{-\infty}^{\infty} dx \, \frac{e^{-x^2}}{x-\zeta} + 2i\,\pi^{1/2} e^{-\zeta^2}, & \mathrm{Im}\,\zeta < 0 \end{cases} . \tag{D.1}$$

• **Alternative Expression**

For $\mathrm{Im}\,\zeta > 0$, we may express

$$\frac{1}{x-\zeta} = i \int_0^{\infty} d\tau \, e^{-i(x-\zeta)\tau}, \tag{D.2}$$

while for $\mathrm{Im}\,\zeta < 0$, we may use

$$\frac{1}{x-\zeta} = -i \int_{-\infty}^{0} d\tau \, e^{-i(x-\zeta)\tau}. \tag{D.3}$$

Consequently, we have

$$Z(\zeta) = \begin{cases} \dfrac{i}{\pi^{1/2}} \displaystyle\int_0^{\infty} d\tau \int_{-\infty}^{\infty} dx \, e^{-x^2 - i(x-\zeta)\tau}, & \mathrm{Im}\,\zeta > 0 \\ -\dfrac{i}{\pi^{1/2}} \displaystyle\int_{-\infty}^{0} d\tau \int_{-\infty}^{\infty} dx \, e^{-x^2 - i(x-\zeta)\tau} + 2i\,\pi^{1/2} e^{-\zeta^2}, & \mathrm{Im}\,\zeta < 0 \end{cases} . \tag{D.4}$$

Carrying out the x integration, we obtain

$$Z(\zeta) = \begin{cases} 2i \displaystyle\int_0^{\infty} d\tau \, e^{-\tau^2 + 2i\zeta\tau}, & \mathrm{Im}\,\zeta > 0 \\ -2i \displaystyle\int_{-\infty}^{0} d\tau \, e^{-\tau^2 + 2i\zeta\tau} + 2i\,\pi^{1/2} e^{-\zeta^2}, & \mathrm{Im}\,\zeta < 0 \end{cases} , \tag{D.5}$$

which can be further manipulated as

$$Z(\zeta) = 2i\, e^{-\zeta^2} \int_{-\infty}^{i\zeta} d\tau\, e^{-\tau^2}. \tag{D.6}$$

Note that we have only discussed the case of $\mathrm{Im}\,\zeta > 0$ and $\mathrm{Im}\,\zeta < 0$, but not the marginal case $\mathrm{Im}\,\zeta = 0$. This is because for $\mathrm{Im}\,\zeta = 0$, the identity (D.3) is not valid. However, since the alternative expression in (D.6) for the Z function is valid for both signs of $\mathrm{Im}\,\zeta$, and since $\mathrm{Im}\,\zeta = 0$ is a singular case, the expression in (D.6) is extended to the singular case as well.

• Series Expansion

Since the alternative expression for Z function (D.6) is valid for both signs of $\mathrm{Im}\,\zeta$, it is permissible to consider only the case of $\mathrm{Im}\,\zeta > 0$ without loss of generality. From (D.5), for $\mathrm{Im}\,\zeta > 0$, namely,

$$Z(\zeta) = 2i \int_0^\infty d\tau\, e^{-\tau^2 + 2i\,\zeta\,\tau}, \tag{D.7}$$

we employ the Taylor series expansion for $e^{2i\,\zeta\,\tau}$, to obtain

$$Z(\zeta) = 2i \sum_{n=0}^\infty \frac{2^n}{n!} i^n \int_0^\infty d\tau\, e^{-\tau^2}\, \tau^n\, \zeta^n. \tag{D.8}$$

Making use of the integral identity

$$\int_0^\infty d\tau\, e^{-\tau^2}\, \tau^n = \left(\begin{array}{ll} (2a-1)!!\, \sqrt{\pi}/2^{a+1} & n = 2a \\ b!/2 & n = 2b+1 \end{array} \right., \tag{D.9}$$

we have

$$Z(\zeta) = i\sqrt{\pi} \sum_{a=0}^\infty \frac{(-\zeta^2)^a}{a!} - 2\zeta \sum_{b=0}^\infty \frac{2^b\, b!}{(2b+1)!} (-2\,\zeta^2)^b. \tag{D.10}$$

The imaginary term is but the Taylor series expansion of the exponential function. Consequently, we have

$$\begin{aligned} Z(\zeta) &= i\sqrt{\pi}\, e^{-\zeta^2} - 2\zeta \sum_{n=0}^\infty \frac{2^n\, n!}{(2n+1)!} (-2\,\zeta^2)^n \\ &= i\sqrt{\pi}\, e^{-\zeta^2} - 2\zeta \left(1 - \frac{2}{3}\zeta^2 + \frac{4}{15}\zeta^4 - \frac{8}{105}\zeta^6 + \cdots + a_n\,\zeta^{2n} + \cdots \right). \end{aligned} \tag{D.11}$$

The coefficients a_n can be automatically generated from the recursion relation:

$$a_n = -\frac{2}{2n+1} a_{n-1}, \qquad n = 1, 2, 3, \ldots, \qquad a_0 = 1. \tag{D.12}$$

• Asymptotic Expansion

For asymptotic expansion, we start from

$$\begin{aligned} Z(\zeta) &= \pi^{-1/2}\, \mathscr{P} \int_{-\infty}^\infty dx\, \frac{e^{-x^2}}{x - \zeta} + i\sigma\pi^{1/2}\, e^{-\zeta^2}, \\ \sigma &= \left\{ \begin{array}{ll} 0 & \mathrm{Im}\,\zeta > 0 \\ 1 & \mathrm{Im}\,\zeta = 0 \\ 2 & \mathrm{Im}\,\zeta < 0 \end{array} \right. . \end{aligned} \tag{D.13}$$

For large ζ, we expand the factor $1/(x-\zeta)$ to obtain

$$\int_{-\infty}^{\infty} dx\, \frac{e^{-x^2}}{x-\zeta} = -\frac{1}{\zeta}\int_{-\infty}^{\infty} dx\, e^{-x^2}\left(1+\frac{x^2}{\zeta^2}+\frac{x^4}{\zeta^4}+\cdots+\frac{x^{2n}}{\zeta^{2n}}+\cdots\right). \tag{D.14}$$

Making use of

$$\int_{-\infty}^{\infty} dx\, e^{-x^2}\, x^{2n} = \frac{(2n)!\,\sqrt{\pi}}{n!\,2^{2n}}, \tag{D.15}$$

we have

$$\begin{aligned} Z(\zeta) &= -\frac{1}{\zeta}\left(1+\frac{1}{2\zeta^2}+\frac{3}{4\zeta^4}+\cdots+\frac{a_n}{\zeta^{2n}}+\cdots\right)+i\sigma\pi^{\frac{1}{2}}\,e^{-\zeta^2}, \\ a_n &= \frac{\Gamma\left(n+\frac{1}{2}\right)}{\pi^{\frac{1}{2}}} = \frac{(2n)!}{n!\,2^{2n}}, \qquad \sigma = \begin{cases} 0 & \mathrm{Im}\,\zeta > 0 \\ 1 & \mathrm{Im}\,\zeta = 0 \\ 2 & \mathrm{Im}\,\zeta < 0 \end{cases}. \end{aligned} \tag{D.16}$$

In general, one may automatically generate the coefficients using the recursion relation:

$$a_n = \frac{2n-1}{2}\,a_{n-1}, \qquad n = 1,2,3,\ldots, \qquad a_0 = 1. \tag{D.17}$$

- **Differential Equation**

If we take the derivative of the dispersion function, we have

$$\frac{d}{d\zeta}Z(\zeta) = \pi^{-1/2}\frac{d}{d\zeta}\int_{-\infty}^{\infty} dx\, \frac{e^{-x^2}}{x-\zeta} = -2-2\zeta\, Z(\zeta). \tag{D.18}$$

From this, we obtain the differential equation for $Z(\zeta)$,

$$Z'(\zeta) = -2\,[1+\zeta\, Z(\zeta)]\,. \tag{D.19}$$

- **Continued Fraction**

The continued fraction expression for plasma dispersion function can be obtained based upon the asymptotic expansion. The conversion of an infinite series to a continued fraction is standard. For a given series of the form

$$s = \frac{1}{a_0}+\frac{1}{a_1}+\frac{1}{a_2}+\frac{1}{a_3}+\frac{1}{a_4}+\cdots, \tag{D.20}$$

the continued fraction can be generated by

$$s = \cfrac{1}{a_0+\cfrac{a_0^2}{-(a_1+a_0)+\cfrac{a_1^2}{(a_2+a_1)+\cfrac{a_2^2}{-(a_3+a_2)+\cfrac{a_3^2}{(a_4+a_3)+\cdots}}}}}. \tag{D.21}$$

From this, it follows that the continued fraction representation of Z function is given by

$$Z(\zeta) = -\cfrac{\zeta}{\zeta^2 + \cfrac{-\frac{1}{2}\zeta^2}{\zeta^2 + \frac{1}{2} + \cfrac{-\frac{3}{2}\zeta^2}{\zeta^2 + \frac{3}{2} + \cfrac{-\frac{5}{2}\zeta^2}{\zeta^2 + \frac{5}{2} + \cdots}}}}$$

$$= -\frac{\zeta}{\zeta^2+}\,\frac{-\frac{1}{2}\zeta^2}{\zeta^2 + \frac{1}{2}+}\,\frac{-\frac{3}{2}\zeta^2}{\zeta^2 + \frac{3}{2}+}\,\frac{-\frac{5}{2}\zeta^2}{\zeta^2 + \frac{5}{2} + \cdots} + i\sigma\pi^{1/2}\,e^{-\zeta^2}. \tag{D.22}$$

Some computer programs can generate the continued fraction representation by the following algorithm:

$$\begin{aligned}
Z(\zeta) &= -\frac{\zeta}{\zeta^2 + P} + i\sigma\pi^{1/2}\,e^{-\zeta^2}, \qquad P = \lim_{n\to\infty}\frac{A_n}{B_n},\\
A_{n+1} &= b_{n+1}\,A_n + a_{n+1}\,A_{n-1},\\
B_{n+1} &= b_{n+1}\,B_n + a_{n+1}\,B_{n-1}, \qquad (n = 0, 1, 2, \ldots),\\
A_0 &= 0, \qquad A_{-1} = 1,\\
B_0 &= 1, \qquad B_{-1} = 0,\\
a_n &= -\frac{(2n-1)\,\zeta^2}{2}, \qquad b_n = \zeta^2 + \frac{2n-1}{2}.
\end{aligned} \tag{D.23}$$

Appendix E

Weak Turbulence Theory for Reactive Instabilities

In the present Appendix, we discuss a possible method to extend the standard weak turbulence theory to a more general case. For the sake of simplicity, we restrict our discussion to one-dimensional situation. In the standard weak turbulence theory, which is the focus of the main part of this monograph, the plasma eigenmode is determined solely by setting the real part of the dielectric response function equal to zero:

$$\mathrm{Re}\,\epsilon(k,\omega_k) = 0. \tag{E.1}$$

However, for reactive instability, it is no longer appropriate to consider just the real part of the dielectric response function in order to evaluate the dispersion relation, but rather, both the real and imaginary parts of $\epsilon(k,\omega)$ must be set equal to zero. Moreover, it is no longer valid to consider that ω is real, since the root of the equation $\epsilon(k,\omega) = 0$, which is generally given by a transcendental function, is complex:

$$\epsilon(k,\omega_k + i\gamma_k) = \mathrm{Re}\,\epsilon(k,\omega_k + i\gamma_k) + i\,\mathrm{Im}\,\epsilon(k,\omega_k + i\gamma_k) = 0. \tag{E.2}$$

Linear and quasilinear theories of waves and instabilities for generally complex ω are well known. However, perturbative nonlinear theory that goes beyond the quasilinear method is not fully developed. In what follows, we discuss one possible way to formulate the weak turbulence theory for reactively unstable plasmas.

Let us begin the discourse starting from one-dimensional Vlasov–Poisson system of equations,

$$\left(\frac{\partial}{\partial t} + v\,\frac{\partial}{\partial x} + \frac{e_a}{m_a}\,E(x,t)\,\frac{\partial}{\partial v}\right) f_a(x,v,t) = 0,$$

$$\frac{\partial E(x,t)}{\partial x} = 4\pi \sum_a e_a \int dv\; f_a(x,v,t). \tag{E.3}$$

Separating the total particle distribution and electric field as is customary, $f_a(x,v,t) = n_a F_a(v,t) + \delta f_a(x,v,t)$ and $E(x,t) = \delta E(x,t)$, we have

$$\frac{\partial n_a F_a(v,t)}{\partial t} = -\frac{e_a}{m_a}\frac{\partial}{\partial v}\int dk\;\langle \delta E_k^*(t)\,\delta f_k^a(v,t)\rangle,$$

$$ik\,\delta E_k(t) = 4\pi \sum_a e_a \int dv\;\delta f_k^a(v,t),$$

$$\left(\frac{\partial}{\partial t} + ikv\right)\delta f_k^a(v,t) = -\frac{e_a}{m_a}\,\delta E_k(t)\,\frac{\partial n_a F_a(v,t)}{\partial v}$$
$$-\frac{e_a}{m_a}\frac{\partial}{\partial v}\int dk'\;[\,\delta E_{k'}(t)\,\delta f_{k-k'}^a(v,t) - \langle \delta E_{k'}(t)\,\delta f_{k-k'}^a(v,t)\rangle\,], \tag{E.4}$$

where we have made use of the spatial Fourier spectral representation,

$$\delta f_k^a(v,t) = \frac{1}{2\pi}\int dx\,\delta f_a(x,v,t)e^{-ikx}, \qquad \delta f_a(x,v,t) = \int dk\,\delta f_k^a(v,t)\,e^{ikx},$$
$$\delta E_k(t) = \frac{1}{2\pi}\int dx\,\delta E(x,t)e^{\ ikx}, \qquad \delta E(x,t) = \int dk\,\delta E_k(t)\,e^{ikx}, \tag{E.5}$$

and have made use of the fact that $\delta E_{-k} = \delta E_k^*$.

E.1 Generalized Quasi-Linear Theory

Let us first review the textbook quasi-linear theory (Sagdeev and Galeev, 1969; Davidson, 1972), which is the lowest-order nonlinear theory. In the purely quasi-linear picture, the spectral amplitudes do not depend on time, and we ignore nonlinear mode coupling, so that we have

$$\frac{\partial n_a F_a(v,t)}{\partial t} = -\frac{e_a}{m_a}\frac{\partial}{\partial v}\int dk\int d\omega\int d\omega'\,\langle \delta E_{k,\omega'}^*\,\delta f_{k,\omega}^a(v)\rangle\,e^{i\,(\omega'-\omega)\,t},$$
$$ik\,\delta E_{k\omega} = 4\pi\sum_a e_a\int dv\,\delta f_{k\omega}^a(v),$$
$$-i\,(\omega-kv)\,\delta f_{k,\omega}^a(v) = -\frac{e_a}{m_a}\,\delta E_{k,\omega}\,\frac{\partial n_a F_a(v,t)}{\partial v}. \tag{E.6}$$

In writing (E.6), we have represented the fluctuating quantities in terms of temporal spectral representation (Laplace transformation),

$$\delta f_{k,\omega}^a(v,t) = \frac{1}{(2\pi)^2}\int dx\,\delta f_a(x,v,t)e^{-ikx+i\omega t},$$
$$\delta f_a(x,v,t) = \int d\omega\int dk\,\delta f_{k,\omega}^a(v,t)\,e^{ikx-i\omega t},$$
$$\delta E_{k,\omega}(t) = \frac{1}{(2\pi)^2}\int dx\,\delta E(x,t)e^{-ikx+i\omega t},$$
$$\delta E(x,t) = \int d\omega\int dk\,\delta E_{k,\omega}(t)\,e^{ikx-i\omega t}, \tag{E.7}$$

where the adiabatic time dependence of the spectral amplitude is assumed, but, unlike in the main body of this book, ω is assumed to be fully complex, and the integrals $\int d\omega$ and $\int d\omega'$ are along the usual paths L, defined above all the singularities associated with the integrands.

Solving for $\delta f_{k,\omega}^a(v)$, we have

$$\delta f_{k,\omega}^a(v) = -\frac{ie_a}{m_a}\frac{\delta E_{k\omega}}{\omega-kv}\frac{\partial n_a F_a(v,t)}{\partial v}. \tag{E.8}$$

In (E.8), the resonant denominator $\omega - kv$ is no longer approximately treated by $\omega_r - kv + i0$, but we assume that ω is fully complex, $\omega = \omega_r + i\gamma$, except that we assume $\gamma > 0$ in the sense of causality requirement. Inserting (E.8) to Poisson equation, we obtain

$$\epsilon(k,\omega)\,\delta E_{k,\omega} = 0, \tag{E.9}$$

where the linear dielectric response function is given by

$$\epsilon(k,\omega) = 1 + \sum_a \frac{\omega_{pa}^2}{k}\int dv\,\frac{\partial F_a(v,t)/\partial v}{\omega - kv}. \tag{E.10}$$

We may designate the solution to the dispersion relation $\epsilon(k,\omega) = 0$ by

$$\omega = \omega_k + i\gamma_k, \tag{E.11}$$

and write the spectral electric field amplitude as

$$\delta E_{k,\omega} = \delta E_k \, \delta(\omega - \omega_k - i\gamma_k). \tag{E.12}$$

From this, upon inserting (E.12) to the definition of spatial Fourier representation of electric field in (E.5), we may construct the spectral wave energy density:

$$\langle \delta E_k^*(t) \, \delta E_k(t) \rangle = I_k \, e^{2\gamma_k t}. \tag{E.13}$$

where we have defined $\left\langle \delta E^2 \right\rangle_k = I_k$. Here, we reiterate that in the lowest-order nonlinear theory, that is, purely quasi-linear picture, the amplitude I_k does not depend on time. Instead, the time dependence of the wave energy density comes solely from the exponential factor $e^{2\gamma_k t}$. Upon taking the time derivative of (E.13), we obtain the quasilinear wave kinetic equation,

$$\frac{\partial}{\partial t}\left(I_k e^{2\gamma_k t}\right) = 2\gamma_k I_k e^{2\gamma_k t}. \tag{E.14}$$

Inserting the perturbed distribution $\delta f_{k,\omega}^a(v)$ (E.8) into the particle kinetic equation, that is, the first equation in (E.6), we obtain the customary quasilinear particle kinetic equation,

$$\frac{\partial F_a(v,t)}{\partial t} = \frac{e_a^2}{m_a^2}\frac{\partial}{\partial v}\int dk \frac{\gamma_k I_k e^{2\gamma_k t}}{(\omega_k - kv)^2 + \gamma_k^2}\frac{\partial F_a(v,t)}{\partial v}. \tag{E.15}$$

If we take the limit of $\gamma_k \to 0$, and make use of

$$\lim_{\gamma_k \to 0} \frac{\gamma_k}{(\omega_k - kv)^2 + \gamma_k^2} = \pi \, \delta(\omega_k - kv), \tag{E.16}$$

then we recover the customary form of quasilinear particle kinetic equation for weakly damped or growing mode,

$$\frac{\partial F_a(v,t)}{\partial t} = \frac{\pi e_a^2}{m_a^2}\frac{\partial}{\partial v}\int dk \, I_k \, \delta(\omega_k - kv)\,\frac{\partial F_a(v,t)}{\partial v}. \tag{E.17}$$

However, for reactively unstable plasmas, the quasilinear particle kinetic equation is (E.15), rather than (E.17).

E.2 Generalized Weak Turbulence Theory

Moving on to the next order theory, the key aspect concerns the time dependence of wave amplitude. In quasilinear theory, the spectral amplitudes $\delta f_{k,\omega}$ and $\delta E_{k,\omega}$ do not depend on adiabatic time variable. The time dependence of the wave energy density arises purely by virtue of the exponential factor $e^{2i\gamma_k t}$. In contrast, in the next-order weak turbulence scheme, we now allow the slow adiabatic time dependence on these amplitudes, which will be accounted for on the basis of nonlinear mode coupling effects. Thus, if we allow the slow-time dependence on the spectral wave energy density, then the quasi-linear wave kinetic equation (E.14) must be generalized to

$$\frac{\partial}{\partial t}\left(I_k e^{2\gamma_k t}\right) = 2\gamma_k I_k e^{2\gamma_k t} + \frac{\partial I_k}{\partial t}\, e^{2\gamma_k t}. \tag{E.18}$$

The extra factor, $\partial I_k/\partial t$, is to be determined from the nonlinear terms. In order to accomplish that, we solve the nonlinear perturbed Vlasov equation in the usual manner, by absorbing the slow-time derivative into the complex wave frequency, $\omega \to \omega + i\,\partial/\partial t$, and reintroducing it later. The spectral form of nonlinear equation for the perturbed distribution function is thus given by

$$\delta f^a_{k,\omega} = -\frac{ie_a}{m_a}\frac{\partial n_a F_a/\partial v}{\omega - kv}\,\delta E_{k,\omega} - \frac{ie_a}{m_a}\frac{1}{\omega - kv}\frac{\partial}{\partial v}\int dk'\int d\omega' \times \left(\delta E_{k'\omega'}\,\delta f^a_{k-k',\omega-\omega'} - \left\langle \delta E_{k'\omega'}\,\delta f^a_{k-k',\omega-\omega'}\right\rangle\right). \quad \text{(E.19)}$$

The iterative solution of (E.19) may be proceeded in the same manner as expounded in the main part of this book, where solutions up to third order were discussed. If we simplify the matter by restricting to just second order, then it is a straightforward exercise to obtain

$$0 = \epsilon(k,\omega)\left\langle \delta E_{k\omega}\,\delta E^*_{k\omega}\right\rangle + \int dk'\int d\omega'\,\chi(k',\omega'|k-k',\omega-\omega')\left\langle \delta E^*_{k\omega}\,\delta E_{k'\omega'}\,\delta E_{k-k',\omega-\omega'}\right\rangle, \quad \text{(E.20)}$$

where the linear dielectric constant is already defined, and the second-order nonlinear response function is defined in the customary manner, namely,

$$\chi(k',\omega'|k-k',\omega-\omega') = -\frac{i}{2}\sum_a \frac{e_a}{m_a}\frac{\omega_{pa}^2}{k} \times \int dv\,\frac{1}{\omega - kv}\frac{\partial}{\partial v}\left(\frac{1}{\omega' - k'v} + \frac{1}{\omega - \omega' - (k-k')\,v}\right)\frac{\partial F_a}{\partial v}, \quad \text{(E.21)}$$

except that ω and ω' are complex. The third-order cumulant may also be obtained in the same manner:

$$\begin{aligned}&\left\langle \delta E^*_{k\omega}\,\delta E_{k'\omega'}\,\delta E_{k-k',\omega-\omega'}\right\rangle \\ &= -2\left(\frac{\chi^*(k',\omega'|k-k',\omega-\omega')}{\epsilon^*(k,\omega)}\left\langle \delta E_{k',\omega}\,\delta E^*_{k',\omega'}\right\rangle\left\langle \delta E_{k-k',\omega-\omega'}\,\delta E^*_{k-k',\omega-\omega'}\right\rangle\right. \\ &\quad + \frac{\chi(k,\omega|-k+k',-\omega+\omega')}{\epsilon(k',\omega')}\left\langle \delta E_{k\omega}\,\delta E^*_{k\omega}\right\rangle\left\langle \delta E_{k-k',\omega-\omega'}\,\delta E_{-k+k',-\omega+\omega'}\right\rangle \\ &\quad \left. + \frac{\chi(k,\omega|-k',-\omega')}{\epsilon(k-k',\omega-\omega')}\left\langle \delta E_{k\omega}\,\delta E^*_{k\omega}\right\rangle\left\langle \delta E_{k',\omega'}\,\delta E_{-k',-\omega'}\right\rangle\right). \end{aligned} \quad \text{(E.22)}$$

Upon reintroducing the adiabatic time derivative, and denoting the complex root of the dispersion relation by

$$\varpi_k \equiv \omega_k + i\gamma_k, \quad \text{(E.23)}$$

one arrives at

$$\begin{aligned}\frac{\partial I_k}{\partial t} &= \mathrm{Im}\,\frac{4}{\partial\epsilon(k,\varpi_k)/\partial\varpi_k} \\ &\quad \times \int dk'\left(\frac{|\chi(k',\varpi_{k'}|k-k',\varpi_{k-k'})|^2}{\epsilon^*(k,\varpi_{k'}+\varpi_{k-k'})}\,\delta(\varpi_k - \varpi_{k'} - \varpi_{k-k'})\,I_{k'}\,I_{k-k'}\right. \\ &\quad + \frac{\chi(k',\varpi_k - \varpi_{k-k'}|k-k',\varpi_{k-k'})\,\chi(k,\varpi_k|-k+k',-\varpi_{k-k'})}{\epsilon(k',\varpi_k - \varpi_{k-k'})}\,I_k\,I_{k-k'} \\ &\quad \left. + \frac{\chi(k',\varpi_{k'}|k-k',\varpi_k - \varpi_{k'})\,\chi(k,\varpi_k|-k',-\varpi_{k'})}{\epsilon(k-k',\varpi_k - \varpi_{k'})}\,I_k\,I_{k'}\right). \end{aligned} \quad \text{(E.24)}$$

This equation can be substituted to (E.18), which, in principle, completes the derivation of wave kinetic equation. The resulting generalized weak turbulence theory is not so straightforward to analyze, either by analytical means or by direct numerical computation, since the nonlinear susceptibility involves complex velocity integrals with linear and nonlinear resonant denominators. For this reason, the weak turbulence theory of the type discussed in this Appendix has not been fully developed in the existing literature, although Yoon (2010); Yoon and Umeda (2010) applied the method outlined in this Appendix to the problem of Buneman instability. We close the present Appendix by considering the dielectric response functions for thermal distribution.

E.3 Generalized Second-Order Nonlinear Susceptibility

Let us consider the Maxwellian velocity distribution function. After a straightforward but somewhat tedious manipulations, it can be shown that

$$\begin{aligned}\chi(k',\omega'|k-k',\omega-\omega') = -i\sum_a \frac{e_a}{m_a}\frac{\omega_{pa}^2}{k^2 v_{Ta}^4}\int\frac{dx}{\pi^{1/2}}\Bigg[\frac{1}{k'}\Bigg(2+\frac{2\,\zeta_{k,\omega}^a}{x-\zeta_{k,\omega}^a}+\frac{2\zeta_{k',\omega'}^a}{x-\zeta_{k,\omega}^a}\\ +\frac{2\,(\zeta_{k',\omega'}^a)^2}{x-\zeta_{k',\omega'}^a}\frac{1}{x-\zeta_{k,\omega}^a}+\frac{\zeta_{k',\omega'}^a}{(x-\zeta_{k',\omega'}^a)^2}\frac{1}{x-\zeta_{k,\omega}^a}\Bigg)\\ +\frac{1}{k-k'}\Bigg(2+\frac{2\,\zeta_{k,\omega}^a}{x-\zeta_{k,\omega}^a}+\frac{2\zeta_{k-k',\omega-\omega'}^a}{x-\zeta_{k,\omega}^a}\\ +\frac{2\,(\zeta_{k-k',\omega-\omega'}^a)^2}{x-\zeta_{k-k',\omega-\omega'}^a}\frac{1}{x-\zeta_{k,\omega}^a}+\frac{\zeta_{k-k',\omega-\omega'}^a}{(x-\zeta_{k-k',\omega-\omega'}^a)^2}\frac{1}{x-\zeta_{k,\omega}^a}\Bigg)\Bigg]e^{-x^2},\end{aligned} \tag{E.25}$$

where

$$\zeta_{k',\omega'}^a=\frac{\omega'}{k'v_{Ta}},\qquad \zeta_{k-k',\omega-\omega'}^a=\frac{\omega-\omega'}{(k-k')\,v_{Ta}}. \tag{E.26}$$

Carrying out the x integration by making use of the definition for Z function, one can show that

$$\begin{aligned}\chi(k',\omega'|k-k',\omega-\omega') = i\sum_a \frac{e_a}{4T_a}\frac{\omega_{pa}^2}{(k\omega'-k'\omega)^2}\Big[k\,Z'(\zeta_{k\omega}^a)\\ -k'\,Z'(\zeta_{k'\omega'}^a)-(k-k')\,Z'(\zeta_{k-k',\omega-\omega'}^a)\Big].\end{aligned} \tag{E.27}$$

Appendix F

On Higher-Order Perturbative Expansion

If we employ perturbative solution for $\delta f^a_{\mathbf{k}\omega}$ beyond the third order truncation,

$$\delta f^a_{\mathbf{k}\omega} = \delta f^{a(1)}_{\mathbf{k}\omega} + \delta f^{a(2)}_{\mathbf{k}\omega} + \delta f^{a(3)}_{\mathbf{k}\omega} + \delta f^{a(4)}_{\mathbf{k}\omega} + \delta f^{a(5)}_{\mathbf{k}\omega} + \cdots, \tag{F.1}$$

then after some tedious but otherwise straightforward algebraic manipulations, it is possible to obtain the iterative solution, which is explicitly written down up to fifth order,

$$\begin{aligned}
f^a_{\mathbf{k}\omega} &= \alpha(K)\, n_a F_a\, \phi_K \\
&+ \sum_{1+2=K} \alpha^{(2)}(1|2)\, n_a F_a\, (\phi_1\,\phi_2 - \langle\phi_1\,\phi_2\rangle) \\
&+ \sum_{1+2+3=K} \alpha^{(3)}(1|2|3)\, n_a F_a\, (\phi_1\,\phi_2\,\phi_3 - \phi_1\,\langle\phi_2\,\phi_3\rangle - \langle\phi_1\,\phi_2\,\phi_3\rangle) \\
&+ \sum_{1+2+3+4=K} \alpha^{(4)}(1|2|3|4)\, n_a F_a\, (\phi_1\,\phi_2\,\phi_3\,\phi_4 - \phi_1\,\phi_2\,\langle\phi_3\,\phi_4\rangle \\
&- \phi_1\,\langle\phi_2\,\phi_3\,\phi_4\rangle - \langle\phi_1\,\phi_2\,\phi_3\,\phi_4\rangle + \langle\phi_1\,\phi_2\rangle\,\langle\phi_3\,\phi_4\rangle) \\
&+ \sum_{1+2+34+5=K} \alpha^{(5)}(1|2|3|4|5)\, n_a F_a\, (\phi_1\,\phi_2\,\phi_3\,\phi_4\,\phi_5 - \phi_1\,\phi_2\,\phi_3\,\langle\phi_4\,\phi_5\rangle \\
&- \phi_1\,\phi_2\,\langle\phi_3\,\phi_4\,\phi_5\rangle - \phi_1\,\langle\phi_2\,\phi_3\,\phi_4\,\phi_5\rangle + \phi_1\,\langle\phi_2\,\phi_3\rangle\,\langle\phi_4\,\phi_5\rangle \\
&- \langle\phi_1\,\phi_2\,\phi_3\,\phi_4\,\phi_5\rangle + \langle\phi_1\,\phi_2\,\phi_3\rangle\,\langle\phi_4\,\phi_5\rangle + \langle\phi_1\,\phi_2\rangle\,\langle\phi_3\,\phi_4\,\phi_5\rangle), \\
&\cdots,
\end{aligned} \tag{F.2}$$

where short-hand notations, $K \equiv (\mathbf{k},\omega)$, $1 \equiv (\mathbf{k}_1,\omega_1)$, $2 \equiv (\mathbf{k}_2,\omega_2)$, etc., are employed, δ is omitted for $\delta f_{\mathbf{k},\omega}$ and $\delta\phi_{\mathbf{k},\omega}$, and

$$\begin{aligned}
\alpha(K) &= \mathbf{k}\cdot\mathbf{g}_{\mathbf{k}\omega}, \\
\alpha^{(2)}(1|2) &= \frac{1}{2}\,[(\mathbf{k}_1\cdot\mathbf{g}_{\mathbf{k}_1+\mathbf{k}_2,\omega_1+\omega_2})\,(\mathbf{k}_2\cdot\mathbf{g}_{\mathbf{k}_2\omega_2}) \\
&\quad + (\mathbf{k}_2\cdot\mathbf{g}_{\mathbf{k}_1+\mathbf{k}_2,\omega_1+\omega_2})\,(\mathbf{k}_1\cdot\mathbf{g}_{\mathbf{k}_1\omega_1})], \\
\alpha^{(3)}(1|2|3) &= \frac{1}{2}\,(\mathbf{k}_1\cdot\mathbf{g}_{\mathbf{k}_1+\mathbf{k}_2+\mathbf{k}_3,\omega_1+\omega_2+\omega_3}) \\
&\quad \times [\,(\mathbf{k}_2\cdot\mathbf{g}_{\mathbf{k}_2+\mathbf{k}_3,\omega_2+\omega_3})\,(\mathbf{k}_3\cdot\mathbf{g}_{\mathbf{k}_3\omega_3}) + (\mathbf{k}_3\cdot\mathbf{g}_{\mathbf{k}_2+\mathbf{k}_3,\omega_2+\omega_3})\,(\mathbf{k}_2\cdot\mathbf{g}_{\mathbf{k}_2\omega_2})\,],
\end{aligned}$$

$$
\begin{aligned}
\alpha^{(4)}(1|2|3|4) &= \frac{1}{2}\,(\mathbf{k}_1\cdot\mathbf{g}_{\mathbf{k}_1+\mathbf{k}_2+\mathbf{k}_3+\mathbf{k}_4,\omega_1+\omega_2+\omega_3+\omega_4})\\
&\quad\times(\mathbf{k}_2\cdot\mathbf{g}_{\mathbf{k}_2+\mathbf{k}_3+\mathbf{k}_4,\omega_2+\omega_3+\omega_4})\,[\,(\mathbf{k}_3\cdot\mathbf{g}_{\mathbf{k}_3+\mathbf{k}_4,\omega_3+\omega_4})\,(\mathbf{k}_4\cdot\mathbf{g}_{\mathbf{k}_4,\omega_4})\\
&\quad+(\mathbf{k}_4\cdot\mathbf{g}_{\mathbf{k}_3+\mathbf{k}_4,\omega_3+\omega_4})\,(\mathbf{k}_3\cdot\mathbf{g}_{\mathbf{k}_3,\omega_3})\,],\\
&\alpha^{(5)}(\mathbf{k}_1,\omega_1|\mathbf{k}_2,\omega_2|\mathbf{k}_3,\omega_3|\mathbf{k}_4,\omega_4|\mathbf{k}_5,\omega_5)\\
&= \frac{1}{2}\,(\mathbf{k}_1\cdot\mathbf{g}_{\mathbf{k}_1+\mathbf{k}_2+\mathbf{k}_3+\mathbf{k}_4+\mathbf{k}_5,\omega_1+\omega_2+\omega_3+\omega_4+\omega_5})\\
&\quad\times(\mathbf{k}_2\cdot\mathbf{g}_{\mathbf{k}_2+\mathbf{k}_3+\mathbf{k}_4+\mathbf{k}_5,\omega_2+\omega_3+\omega_4+\omega_5})(\mathbf{k}_3\cdot\mathbf{g}_{\mathbf{k}_3+\mathbf{k}_4+\mathbf{k}_5,\omega_3+\omega_4+\omega_5})\\
&\quad\times[\,(\mathbf{k}_4\cdot\mathbf{g}_{\mathbf{k}_4+\mathbf{k}_5,\omega_4+\omega_5})(\mathbf{k}_5\cdot\mathbf{g}_{\mathbf{k}_5,\omega_5})+(\mathbf{k}_5\cdot\mathbf{g}_{\mathbf{k}_4+\mathbf{k}_5,\omega_4+\omega_5})(\mathbf{k}_4\cdot\mathbf{g}_{\mathbf{k}_4,\omega_4})\,],\\
&\quad\cdots.
\end{aligned}
\tag{F.3}
$$

The quantity of relevance is

$$
\begin{aligned}
\langle\delta f^a_{\mathbf{k}\omega}\,\delta\phi_{\mathbf{k}'\omega'}\rangle &= \alpha(K)\,\langle K K'\rangle\, n_a F_a\\
&\quad+\sum_{1+2=K}\alpha^{(2)}(1|2)\,\langle 12K'\rangle\, n_a F_a\\
&\quad+\sum_{1+2+3=K}\alpha^{(3)}(1|2|3)\Big[\langle 123K'\rangle-\langle 1K'\rangle\,\langle 23\rangle\Big]n_a F_a\\
&\quad+\sum_{1+2+3+4=K}\alpha^{(4)}(1|2|3|4)\Big[\langle 1234K'\rangle-\langle 12K'\rangle\,\langle 34\rangle-\langle 1K'\rangle\,\langle 234\rangle\Big]n_a F_a\\
&\quad+\sum_{1+2+3+4+5=K}\alpha^{(5)}(1|2|3|4|5)\Big[\langle 12345K'\rangle-\langle 123K'\rangle\,\langle 45\rangle\\
&\quad-\langle 12K'\rangle\,\langle 345\rangle-\langle 1K'\rangle\langle 2345\rangle+\langle 1K'\rangle\langle 23\rangle\langle 45\rangle\Big]n_a F_a\\
&\quad\cdots,
\end{aligned}
\tag{F.4}
$$

where we have made use of even more compact notations, $\langle 12\rangle=\langle\phi_1\phi_2\rangle$, etc. Some careful considerations show that many-body cumulants of electrostatic potential can be expressed as various sums of two- and three-body cumulants, as follows:

$$
\begin{aligned}
&\langle 123K'\rangle-\langle 1K'\rangle\langle 23\rangle = 2\langle 12\rangle\langle 3K'\rangle,\\
&\langle 1234K'\rangle-\langle 12K'\rangle\langle 34\rangle-\langle 1K'\rangle\langle 234\rangle\\
&\quad= 2\big(\langle 123\rangle\langle 4K'\rangle+\langle 13K'\rangle\langle 24\rangle+\langle 23K'\rangle\langle 14\rangle\big)\\
&\qquad+\langle 134\rangle\langle 2K'\rangle+\langle 34K'\rangle\langle 12\rangle,\\
&\langle 12345K'\rangle-\langle 123K'\rangle\langle 45\rangle-\langle 12K'\rangle\langle 345\rangle-\langle 1K'\rangle\langle 2345\rangle+\langle 1K'\rangle\langle 23\rangle\langle 45\rangle\\
&\quad= 6\big(\langle 12\rangle\langle 34\rangle\langle 5K'\rangle+\langle 13\rangle\langle 24\rangle\langle 5K'\rangle+\langle 14\rangle\langle 23\rangle\langle 5K'\rangle\\
&\qquad+\langle 14\rangle\langle 25\rangle\langle 3K'\rangle+\langle 14\rangle\langle 35\rangle\langle 2K'\rangle\big)\\
&\qquad+2\big(\langle 12\rangle\langle 45\rangle\langle 3K'\rangle+\langle 13\rangle\langle 45\rangle\langle 2K'\rangle+2\langle 1K'\rangle\langle 24\rangle\langle 35\rangle+\langle 1K'\rangle\langle 23\rangle\langle 45\rangle\big)\\
&\qquad+2\big(\langle 124\rangle\langle 35K'\rangle+\langle 134\rangle\langle 25K'\rangle+\langle 14K'\rangle\langle 235\rangle\big)\\
&\qquad+\langle 123\rangle\langle 45K'\rangle+\langle 145\rangle\langle 23K'\rangle+\langle 13K'\rangle\langle 245\rangle.
\end{aligned}
\tag{F.5}
$$

Putting all the results together, after somewhat lengthy but otherwise straightforward manipulations, one arrives at

$$
\begin{aligned}
\langle f_K^a \phi_{K'} \rangle &= \delta(K+K') \Bigg(\alpha(K) + 2\sum_1 \alpha^{(3)}(1|-1|K)\,\langle\phi^2\rangle_1 \\
&+ 2\sum_1\sum_2 \Big[3\alpha^{(5)}(1|-1|2|-2|K) + 3\alpha^{(5)}(1|2|-1|-2|K) \\
&+ 3\alpha^{(5)}(1|2|-2|-1|K) + 3\alpha^{(5)}(1|2|K|-1|-2) \\
&+ 3\alpha^{(5)}(1|K|2|-1|-2) + \alpha^{(5)}(1|-1|K|2|-2) \\
&+ \alpha^{(5)}(1|K|-1|2|-2) + \alpha^{(5)}(K|1|-1|2|-2) \\
&+ 2\alpha^{(5)}(K|1|2|-1|-2) \Big] \langle\phi^2\rangle_1 \langle\phi^2\rangle_2 + \cdots \Bigg) \langle\phi^2\rangle_K\, n_a F_a \\
&+ \sum_1 \alpha^{(2)}(1|K-1)\,\langle\phi_1\phi_{K-1}\phi_{K'}\rangle\, n_a F_a \\
&+ \sum_1\sum_2 \Big[2\alpha^{(4)}(1|2|K+K'-1-2|-K')\langle\phi_1\phi_2\phi_{K+K'-1-2}\rangle\langle\phi^2\rangle_{K'} \\
&+ 2\alpha^{(4)}(1|2|K-1|-2)\langle\phi_1\phi_{K-1}\phi_{K'}\rangle\langle\phi^2\rangle_2 \\
&+ 2\alpha^{(4)}(1|2|K-2|-1)\langle\phi_2\phi_{K-2}\phi_{K'}\rangle\langle\phi^2\rangle_1 \\
&+ \alpha^{(4)}(1|-K'|2|K+K'-1-2)\langle\phi_1\phi_2\phi_{K+K'-1-2}\rangle\langle\phi^2\rangle_{K'} \\
&+ \alpha^{(4)}(1|-1|2|K-2)\langle\phi_2\phi_{K-2}\phi_{K'}\rangle\langle\phi^2\rangle_1 + \cdots \Big] n_a F_a \\
&+ \sum_1\sum_2\sum_3\sum_4 \alpha^{(5)}(1|2|3|4|K-1-2-3-4) \\
&\times \Big[2\langle\phi_1\phi_2\phi_4\rangle\langle\phi_3\phi_{K-1-2-3-4}\phi_{K'}\rangle \\
&+ 2\langle\phi_1\phi_3\phi_4\rangle\langle\phi_2\phi_{K-1-2-3-4}\phi_{K'}\rangle + 2\langle\phi_1\phi_4\phi_{K'}\rangle\langle\phi_2\phi_3\phi_{K-1-2-3-4}\rangle \\
&+ \langle\phi_1\phi_2\phi_3\rangle\langle\phi_4\phi_{K-1-2-3-4}\phi_{K'}\rangle + \langle\phi_1\phi_4\phi_{K-1-2-3-4}\rangle\langle\phi_2\phi_3\phi_{K'}\rangle \\
&+ \langle\phi_1\phi_3\phi_{K'}\rangle\langle\phi_2\phi_4\phi_{K-1-2-3-4}\rangle \Big] n_a F_a \\
&+ \cdots .
\end{aligned}
\tag{F.6}
$$

In (F.6) all the terms on the right-hand side are expressed in terms of two-body correlations of the type $\langle\phi^2\rangle$, three-body correlations $\langle\phi^3\rangle$, and their products. If we approximately express the three-body cumulants, following the closure scheme (1.78),

$$
\begin{aligned}
\langle\phi_1\phi_2\phi_3\rangle &= \delta(1+2+3)\Big[\beta(1|2|3)\,\langle\phi^2\rangle_2\langle\phi^2\rangle_3 \\
&\quad + \beta(2|1|3)\,\langle\phi^2\rangle_1\langle\phi^2\rangle_3 + \beta(3|1|2)\,\langle\phi^2\rangle_1\langle\phi^2\rangle_2\Big], \\
\beta(1|2|3) &= 2i\,\frac{k_2k_3\,\chi^{(2)}(-2|-3)}{k_1\epsilon(1)},
\end{aligned}
\tag{F.7}
$$

then after another quite lengthy manipulations, it can be shown that the result simplifies as follows:

$$
\begin{aligned}
\langle f_K^a\phi_{-K}\rangle &= \alpha(K)\langle\phi^2\rangle_K\, n_aF_a \\
&+ \sum_{K'} \Big\{ \alpha^{(2)}(K'|K-K')\Big[\beta(K'|K-K'|-K)\,\langle\phi^2\rangle_{K-K'}\langle\phi^2\rangle_K
\end{aligned}
$$

$$
\begin{aligned}
&+\beta(K-K'|K'|-K)\,\langle\phi^2\rangle_{K'}\langle\phi^2\rangle_K + \beta(-K|K'|K-K')\,\langle\phi^2\rangle_{K'}\langle\phi^2\rangle_{K-K'}\Big]\\
&+2\alpha^{(3)}(K'|-K'|K)\,\langle\phi^2\rangle_{K'}\langle\phi^2\rangle_K\Big\}\,n_a F_a\\
&+\sum_{K'}\sum_{K''}\Big\{\Big[2\alpha^{(4)}(K'|K''|-K'-K''|K)+\alpha^{(4)}(K'|K|K''|-K'-K'')\Big]\\
&\times\Big[\beta(K'|K''|-K'-K'')\langle\phi^2\rangle_{K''}\langle\phi^2\rangle_{K'+K''}\\
&+\beta(K''|K'|-K'-K'')\langle\phi^2\rangle_{K'}\langle\phi^2\rangle_{K'+K''}\\
&+\beta(-K'-K''|K'|K'')\langle\phi^2\rangle_{K'}\langle\phi^2\rangle_{K''}\Big]\langle\phi^2\rangle_K\\
&+\Big[2\alpha^{(4)}(K'|K''|K-K'|-K'')+2\alpha^{(4)}(K''|K'|K-K'|-K'')\\
&+\alpha^{(4)}(K''|-K''|K'|K-K')\Big]\Big[\beta(K'|K-K'|-K)\langle\phi^2\rangle_{K-K'}\langle\phi^2\rangle_K\\
&+\beta(K-K'|K'|-K)\langle\phi^2\rangle_{K'}\langle\phi^2\rangle_K\\
&+\beta(-K|K'|K-K')\langle\phi^2\rangle_{K'}\langle\phi^2\rangle_{K-K'}\Big]\langle\phi^2\rangle_{K''}\Big\}\,n_a F_a\\
&+2\sum_{K'}\sum_{K''}\Big[3\alpha^{(5)}(K'|-K'|K''|-K''|K)\\
&+3\alpha^{(5)}(K'|K''|-K'|-K''|K)+3\alpha^{(5)}(K'|K''|-K''|-K'|K)\\
&+3\alpha^{(5)}(K'|K''|K|-K'|-K'')+3\alpha^{(5)}(K'|K|K''|-K'|-K'')\\
&+\alpha^{(5)}(K'|-K'|K|K''|-K'')+\alpha^{(5)}(K'|K|-K'|K''|-K'')\\
&+\alpha^{(5)}(K|K'|-K'|K''|-K'')+2\alpha^{(5)}(K|K'|K''|-K'|-K'')\Big]\\
&\times\langle\phi^2\rangle_{K'}\langle\phi^2\rangle_{K''}\langle\phi^2\rangle_K\Big\}\,n_a F_a\\
&+\sum_{K'}\sum_{K''}\sum_{K'''}\Big[2\alpha^{(5)}(K''|K'''|K'|-K''-K'''|K-K')\\
&+2\alpha^{(5)}(K''|K'|K'''|-K''-K'''|K-K')\\
&+2\alpha^{(5)}(K'|K''|K'''|K-K'|-K''-K''')\\
&+\alpha^{(5)}(K'|K''|K-K'|K'''|-K''-K''')\\
&+\alpha^{(5)}(K''|K'|K-K'|K'''|-K''-K''')\\
&+\alpha^{(5)}(K'''|K''|-K''-K'''|K'|K-K')\Big]\\
&\times\Big[\beta(K''|K'''|-K''-K''')\langle\phi^2\rangle_{K'''}\langle\phi^2\rangle_{K''+K'''}\\
&+\beta(K'''|K''|-K''-K''')\langle\phi^2\rangle_{K''}\langle\phi^2\rangle_{K''+K'''}\\
&+\beta(-K''-K'''|K''|K''')\langle\phi^2\rangle_{K''}\langle\phi^2\rangle_{K'''}\Big]\\
&\times\Big[\beta(K'|K-K'|-K)\langle\phi^2\rangle_{K-K'}\langle\phi^2\rangle_K\\
&+\beta(K-K'|K'|-K)\langle\phi^2\rangle_{K'}\langle\phi^2\rangle_K\\
&+\beta(-K|K'|K-K')\langle\phi^2\rangle_{K'}\langle\phi^2\rangle_{K-K'}\Big]\,n_a F_a\\
&+\cdots. \qquad\qquad (\text{F.8})
\end{aligned}
$$

It is instructive to rewrite (F.8) in terms of the electric field amplitudes via $\langle E^2\rangle_K = \langle k^2\phi^2\rangle_K$, and make use of explicit definitions for various α's:

$$
\begin{aligned}
\langle f_K^a \phi_{-K}\rangle &= \frac{\mathbf{k}\cdot\mathbf{g}_K}{k^2}\,\langle E^2\rangle_K\, n_a F_a \\
&+\sum_{K'}\Bigg\{\frac{1}{2}\,[\mathbf{k}'\cdot\mathbf{g}_K\,(\mathbf{k}-\mathbf{k}')\cdot\mathbf{g}_{K-K'} + (\mathbf{k}-\mathbf{k}')\cdot\mathbf{g}_K\,\mathbf{k}'\cdot\mathbf{g}_{K'}] \\
&\times\Bigg[\frac{\beta(K'|K-K'|-K)}{k^2|\mathbf{k}-\mathbf{k}'|^2}\,\langle E^2\rangle_{K-K'}\langle E^2\rangle_K + \frac{\beta(K-K'|K'|-K)}{k^2k'^2}\,\langle E^2\rangle_{K'}\langle E^2\rangle_K \\
&+\frac{\beta(-K|K'|K-K')}{k'^2|\mathbf{k}-\mathbf{k}'|^2}\,\langle E^2\rangle_{K'}\langle E^2\rangle_{K-K'}\Bigg] \\
&-\mathbf{k}'\cdot\mathbf{g}_K\,(\mathbf{k}\cdot\mathbf{g}_K\,\mathbf{k}'\cdot\mathbf{g}_{K-K'} - \mathbf{k}\cdot\mathbf{g}_{K-K'}\,\mathbf{k}'\cdot\mathbf{g}_{K'})\,\frac{\langle E^2\rangle_{K'}\langle E^2\rangle_K}{k^2k'^2}\Bigg\}\, n_a F_a \\
&+\sum_{K'}\sum_{K''}\frac{1}{k^2}\Bigg[-\mathbf{k}\cdot\mathbf{g}_K\,\mathbf{k}'\cdot\mathbf{g}_K\,\mathbf{k}''\cdot\mathbf{g}_{K-K'}\,(\mathbf{k}'+\mathbf{k}'')\cdot\mathbf{g}_{K-K'-K''} \\
&+\mathbf{k}'\cdot\mathbf{g}_K\,\mathbf{k}''\cdot\mathbf{g}_{K-K'}\,(\mathbf{k}'+\mathbf{k}'')\cdot\mathbf{g}_{K'+K''}\,\mathbf{k}\cdot\mathbf{g}_{K-K'-K''} \\
&-\frac{1}{2}\,\mathbf{k}'\cdot\mathbf{g}_K\,\mathbf{k}''\cdot\mathbf{g}_{K'}\,\mathbf{k}\cdot\mathbf{g}_{K-K'}\,(\mathbf{k}'+\mathbf{k}'')\cdot\mathbf{g}_{K'+K''} \\
&+\frac{1}{2}\,\mathbf{k}'\cdot\mathbf{g}_K\,(\mathbf{k}'+\mathbf{k}'')\cdot\mathbf{g}_{K'}\,\mathbf{k}''\cdot\mathbf{g}_{K''}\,\mathbf{k}\cdot\mathbf{g}_{K-K'}\Bigg] \\
&\times\Bigg[\frac{\beta(K'|K''|-K'-K'')}{k''^2|\mathbf{k}'+\mathbf{k}''|^2}\,\langle E^2\rangle_{K''}\langle E^2\rangle_{K'+K''} \\
&+\frac{\beta(K''|K'|-K'-K'')}{k'^2|\mathbf{k}'+\mathbf{k}''|^2}\langle E^2\rangle_{K'}\langle E^2\rangle_{K'+K''} \\
&+\frac{\beta(-K'-K''|K'|K'')}{k'^2k''^2}\,\langle E^2\rangle_{K'}\langle E^2\rangle_{K''}\Bigg]\langle E^2\rangle_K\, n_a F_a \\
&+\sum_{K'}\sum_{K''}\frac{1}{k''^2}\Bigg[\mathbf{k}'\cdot\mathbf{g}_K\,\mathbf{k}''\cdot\mathbf{g}_{K''}\,\mathbf{k}''\cdot\mathbf{g}_{K-K'}\,(\mathbf{k}-\mathbf{k}')\cdot\mathbf{g}_{K-K'-K''} \\
&+\mathbf{k}'\cdot\mathbf{g}_K\,\mathbf{k}''\cdot\mathbf{g}_{K-K'}\,(\mathbf{k}-\mathbf{k}')\cdot\mathbf{g}_{K-K'}\,\mathbf{k}''\cdot\mathbf{g}_{K-K'-K''} \\
&+\mathbf{k}''\cdot\mathbf{g}_K\,\mathbf{k}''\cdot\mathbf{g}_{K''}\,\mathbf{k}'\cdot\mathbf{g}_{K-K''}\,(\mathbf{k}-\mathbf{k}')\cdot\mathbf{g}_{K-K'-K''} \\
&-\mathbf{k}''\cdot\mathbf{g}_K\,(\mathbf{k}-\mathbf{k}')\cdot\mathbf{g}_{K-K'}\,\mathbf{k}'\cdot\mathbf{g}_{K-K''}\,\mathbf{k}''\cdot\mathbf{g}_{K-K'-K''} \\
&-\frac{1}{2}\,\mathbf{k}'\cdot\mathbf{g}_K\,\mathbf{k}''\cdot\mathbf{g}_K\,(\mathbf{k}-\mathbf{k}')\cdot\mathbf{g}_{K-K'}\,\mathbf{k}''\cdot\mathbf{g}_{K-K''} \\
&-\frac{1}{2}\,\mathbf{k}''\cdot\mathbf{g}_K\,(\mathbf{k}-\mathbf{k}')\cdot\mathbf{g}_K\,\mathbf{k}'\cdot\mathbf{g}_{K'}\,\mathbf{k}''\cdot\mathbf{g}_{K-K''}\Bigg] \\
&\times\Bigg[\frac{\beta(K'|K-K'|-K)}{k^2|\mathbf{k}-\mathbf{k}'|^2}\,\langle E^2\rangle_{K-K'}\langle E^2\rangle_K + \frac{\beta(K-K'|K'|-K)}{k^2k'^2}\,\langle E^2\rangle_{K'}\langle E^2\rangle_K \\
&+\frac{\beta(-K|K'|K-K')}{k'^2|\mathbf{k}-\mathbf{k}'|^2}\,\langle E^2\rangle_{K'}\langle E^2\rangle_{K-K'}\Bigg]\langle E^2\rangle_{K''}\, n_a F_a \\
&+2\sum_{K'}\sum_{K''}\frac{1}{k^2k'^2k''^2}\Bigg[\frac{3}{2}\,\mathbf{k}\cdot\mathbf{g}_K\,\mathbf{k}'\cdot\mathbf{g}_K\,\mathbf{k}''\cdot\mathbf{g}_K\,\mathbf{k}'\cdot\mathbf{g}_{K-K'}\,\mathbf{k}''\cdot\mathbf{g}_{K-K''}
\end{aligned}
$$

$$
\begin{aligned}
&-\frac{3}{2}\mathbf{k}'\cdot\mathbf{g}_K\,\mathbf{k}''\cdot\mathbf{g}_K\,\mathbf{k}''\cdot\mathbf{g}_{K''}\,\mathbf{k}'\cdot\mathbf{g}_{K-K'}\,\mathbf{k}\cdot\mathbf{g}_{K-K''}\\
&+\frac{3}{2}\mathbf{k}\cdot\mathbf{g}_K\,\mathbf{k}'\cdot\mathbf{g}_K\,\mathbf{k}''\cdot\mathbf{g}_{K-K'}\,\mathbf{k}''\cdot\mathbf{g}_{K-K''}\,\mathbf{k}'\cdot\mathbf{g}_{K-K'-K''}\\
&-\frac{3}{2}\mathbf{k}'\cdot\mathbf{g}_K\,\mathbf{k}''\cdot\mathbf{g}_{K''}\,\mathbf{k}''\cdot\mathbf{g}_{K-K'}\,\mathbf{k}\cdot\mathbf{g}_{K-K''}\,\mathbf{k}'\cdot\mathbf{g}_{K-K'-K''}\\
&+\frac{3}{2}\mathbf{k}\cdot\mathbf{g}_K\,\mathbf{k}'\cdot\mathbf{g}_K\,\mathbf{k}'\cdot\mathbf{g}_{K-K'}\,\mathbf{k}''\cdot\mathbf{g}_{K-K'}\,\mathbf{k}''\cdot\mathbf{g}_{K-K'-K''}\\
&-\frac{3}{2}\mathbf{k}'\cdot\mathbf{g}_K\,\mathbf{k}'\cdot\mathbf{g}_{K'}\,\mathbf{k}\cdot\mathbf{g}_{K-K'}\,\mathbf{k}''\cdot\mathbf{g}_{K-K'}\,\mathbf{k}''\cdot\mathbf{g}_{K-K'-K''}\\
&+\frac{3}{2}\mathbf{k}'\cdot\mathbf{g}_K\,\mathbf{k}''\cdot\mathbf{g}_{K''}\,\mathbf{k}''\cdot\mathbf{g}_{K-K'}\,\mathbf{k}'\cdot\mathbf{g}_{K'+K''}\,\mathbf{k}\cdot\mathbf{g}_{K-K'-K''}\\
&+\frac{3}{2}\mathbf{k}'\cdot\mathbf{g}_K\,\mathbf{k}'\cdot\mathbf{g}_{K'}\,\mathbf{k}''\cdot\mathbf{g}_{K-K'}\,\mathbf{k}''\cdot\mathbf{g}_{K'+K''}\,\mathbf{k}\cdot\mathbf{g}_{K-K'-K''}\\
&-\frac{3}{2}\mathbf{k}'\cdot\mathbf{g}_K\,\mathbf{k}''\cdot\mathbf{g}_{K'}\,\mathbf{k}''\cdot\mathbf{g}_{K''}\,\mathbf{k}\cdot\mathbf{g}_{K-K'}\,\mathbf{k}'\cdot\mathbf{g}_{K'+K''}\\
&-\frac{3}{2}\mathbf{k}'\cdot\mathbf{g}_K\,\mathbf{k}'\cdot\mathbf{g}_{K'}\,\mathbf{k}''\cdot\mathbf{g}_{K'}\,\mathbf{k}\cdot\mathbf{g}_{K-K'}\,\mathbf{k}''\cdot\mathbf{g}_{K'+K''}\\
&-\mathbf{k}'\cdot\mathbf{g}_0\,\mathbf{k}\cdot\mathbf{g}_K\,\mathbf{k}''\cdot\mathbf{g}_{K'}\,\mathbf{k}''\cdot\mathbf{g}_{K''}\,\mathbf{k}'\cdot\mathbf{g}_{K'+K''}\\
&-\mathbf{k}'\cdot\mathbf{g}_0\,\mathbf{k}\cdot\mathbf{g}_K\,\mathbf{k}'\cdot\mathbf{g}_{K'}\,\mathbf{k}''\cdot\mathbf{g}_{K'}\,\mathbf{k}''\cdot\mathbf{g}_{K'+K''}\Big]\\
&\times\langle E^2\rangle_{K'}\langle E^2\rangle_{K''}\langle E^2\rangle_K\,n_aF_a\\
&+\cdots. \qquad\qquad (\text{F.9})
\end{aligned}
$$

The perturbation series in (F.9) diverges when various resonance conditions are satisfied. For instance, if we are interested in the velocity space that satisfies the linear wave-particle resonance condition, $\omega-\mathbf{k}\cdot\mathbf{v}=0$, then, by noting that

$$
\mathbf{g}_K\propto\frac{1}{\omega-\mathbf{k}\cdot\mathbf{v}}, \qquad (\text{F.10})
$$

we may rearrange the various terms in (F.9) as an ascending series in $\mathbf{g}_K$. If we keep only those terms that contain this factor, and retain the most important contributions from each term within the perturbation expansion in powers of wave field energy, then the partial series becomes

$$
\begin{aligned}
\langle f_K^a\,\phi_{-K}\rangle=\mathbf{k}\cdot\mathbf{g}_K\,\frac{\langle E^2\rangle_K}{k^2}\Bigg[&1-\sum_{K'}\mathbf{k}'\cdot\mathbf{g}_K\,\mathbf{k}'\cdot\mathbf{g}_{K-K'}\,\frac{\langle E^2\rangle_{K'}}{k'^2}\\
&+3\left(\sum_{K'}\mathbf{k}'\cdot\mathbf{g}_K\,\mathbf{k}'\cdot\mathbf{g}_{K-K'}\,\frac{\langle E^2\rangle_{K'}}{k'^2}\right)^2+\mathcal{O}(g_K^3)+\cdots\Bigg]n_aF_a. \qquad (\text{F.11})
\end{aligned}
$$

This shows that the straightforward perturbation series diverges beyond the first couple of terms, and that the entire treatment requires some sort of renormalization. The situation is reminiscent of the mathematical asymptotic series, where retaining more terms does not necessarily improve the approximation, but may actually make the situation worse. Note that (F.11) has a qualitative form of a series given by

$$
\frac{\mathcal{O}\left(\left\langle E^2\right\rangle\right)}{\omega-\mathbf{k}\cdot\mathbf{v}}+\frac{\mathcal{O}\left[\left(\left\langle E^2\right\rangle\right)^2\right]}{(\omega-\mathbf{k}\cdot\mathbf{v})^2}+\frac{\mathcal{O}\left[\left(\left\langle E^2\right\rangle\right)^3\right]}{(\omega-\mathbf{k}\cdot\mathbf{v})^3}+\cdots. \qquad (\text{F.12})
$$

This implies that for velocity space that satisfies the linear wave-particle resonance condition, $\omega - \mathbf{k} \cdot \mathbf{v} \approx 0$, the straightforward perturbative expansion appears to fail, but if the series in (F.12) can be summed, then it must be given by the form

$$\frac{\mathcal{O}\left(\left\langle E^2 \right\rangle\right)}{\omega - \mathbf{k} \cdot \mathbf{v} - \mathcal{O}\left(\left\langle E^2 \right\rangle\right)}. \tag{F.13}$$

This points to the need for a renormalization scheme, where the influence of wave intensity must be incorporated into the resonance condition,

$$\omega - \mathbf{k} \cdot \mathbf{v} \rightarrow \omega - \mathbf{k} \cdot \mathbf{v} + i\eta_{\mathbf{k},\omega}, \tag{F.14}$$

where

$$i\eta_{\mathbf{k},\omega} \propto \mathcal{O}\left(\left\langle \delta E^2 \right\rangle_{\mathbf{k},\omega}\right). \tag{F.15}$$

One particular method to achieve such a renormalization procedure is discussed in Appendix G.

Appendix G

On Renormalized Kinetic Turbulence Theory

Appendix F pointed to the fact that partial summation of leading terms in the infinite series implies the necessity for a correction of the resonant denominator where the wave intensity modifies the wave-particle resonance condition. This provides a clue as to how to systematically modify the resonance factor at the outset. The following discussion is based on the work carried out by Rudakov and Tsytovich (1971). We return to the Vlasov–Poisson system of equations

$$\left(\frac{\partial}{\partial t} + \mathbf{v} \cdot \frac{\partial}{\partial \mathbf{r}} - \frac{e_a}{m_a} \frac{\partial \delta\phi}{\partial \mathbf{r}} \cdot \frac{\partial}{\partial \mathbf{v}}\right) f_a = 0,$$

$$\frac{\partial^2 \delta\phi}{\partial \mathbf{r}^2} = -4\pi \sum_a e_a \int d\mathbf{v}\, \delta f_a. \tag{G.1}$$

The relevant equations for averaged and perturbed quantities are given by

$$\delta\phi_{\mathbf{k},\omega} = \sum_a \frac{4\pi e_a}{k^2} \int d\mathbf{v}\, \delta f^a_{\mathbf{k},\omega},$$

$$\frac{\partial n_a F_a}{\partial t} = \frac{i e_a}{m_a} \int d\mathbf{k} \int d\omega\, \mathbf{k} \cdot \frac{\partial}{\partial \mathbf{v}} \langle \delta\phi_{-\mathbf{k},-\omega}\, \delta f^a_{\mathbf{k},\omega} \rangle,$$

$$\begin{aligned}\left(\omega - \mathbf{k} \cdot \mathbf{v} + i\, \frac{\partial}{\partial t}\right) \delta f^a_{\mathbf{k},\omega} = & -\frac{e_a}{m_a}\, \delta\phi_{\mathbf{k},\omega}\, \mathbf{k} \cdot \frac{\partial n_a F_a}{\partial \mathbf{v}} \\ & - \frac{e_a}{m_a} \int d\mathbf{k}' \int d\omega'\, \mathbf{k}' \\ & \cdot \frac{\partial}{\partial \mathbf{v}} \left[\delta\phi_{\mathbf{k}',\omega'}\, \delta f^a_{\mathbf{k}-\mathbf{k}',\omega-\omega'} - \langle \delta\phi_{\mathbf{k}',\omega'}\, \delta f^a_{\mathbf{k}-\mathbf{k}',\omega-\omega'} \rangle\right].\end{aligned} \tag{G.2}$$

In the straightforward weak turbulence scheme, the perturbed distribution $\delta f^a_{\mathbf{k},\omega}$ is iteratively solved in terms of the field strength $\delta\phi_{\mathbf{k},\omega}$ as the series expansion parameter.

G.1 Renormalization of Kinetic Equation

In the renormalized kinetic theory by Rudakov and Tsytovich (1971), first, a quantity $i\eta^a_{\mathbf{k},\omega}\, \delta f^a_{\mathbf{k},\omega}$ is added to both sides of the equation for $\delta f^a_{\mathbf{k},\omega}$, which is to be determined later, resulting in the expression

$$(\omega - \mathbf{k}\cdot\mathbf{v} + i\eta^a_{\mathbf{k},\omega})\,\delta f^a_{\mathbf{k},\omega} = -\frac{e_a}{m_a}\,\delta\phi_{\mathbf{k},\omega}\,\mathbf{k}\cdot\frac{\partial n_a F_a}{\partial \mathbf{v}} + i\eta^a_{\mathbf{k},\omega}\,\delta f^a_{\mathbf{k},\omega}$$
$$- \frac{e_a}{m_a}\int d\mathbf{k}'\int d\omega'\,\mathbf{k}'$$
$$\cdot\frac{\partial}{\partial \mathbf{v}}\left[\delta\phi_{\mathbf{k}',\omega'}\,\delta f^u_{\mathbf{k}-\mathbf{k}',\omega-\omega'} - \langle\delta\phi_{\mathbf{k}',\omega'}\,\delta f^u_{\mathbf{k}-\mathbf{k}',\omega-\omega'}\rangle\right]. \tag{G.3}$$

Note that (G.3) is formally identical to the original equation in (G.2). Recall that the form of resonance factor prescribed in (G.3), namely, $\omega - \mathbf{k}\cdot\mathbf{v} + i\eta^a_{\mathbf{k},\omega}$, is given in the form of (F.14). Then an operator $g^a_{\mathbf{k},\omega}(\mathbf{v})$ is introduced, which satisfies the equation

$$g^a_{\mathbf{k},\omega}(\mathbf{v})\,(\omega - \mathbf{k}\cdot\mathbf{v} + i\eta^a_{\mathbf{k},\omega})\,F^a_{\mathbf{k},\omega} = F^a_{\mathbf{k},\omega}, \tag{G.4}$$

where $F^a_{\mathbf{k},\omega}$ is an arbitrary function, or, equivalently, we may formally write

$$g^a_{\mathbf{k},\omega}(\mathbf{v}) = (\omega - \mathbf{k}\cdot\mathbf{v} + i\eta^a_{\mathbf{k},\omega})^{-1}. \tag{G.5}$$

Evidently, $g^a_{\mathbf{k},\omega}(\mathbf{v})$ represents the renormalized resonant velocity denominator.

The formal solution to (G.3) is then given by

$$\delta f^a_{\mathbf{k},\omega} = -\frac{e_a}{m_a}\,\delta\phi_{\mathbf{k},\omega}\,g^a_{\mathbf{k},\omega}(\mathbf{v})\,\mathbf{k}\cdot\frac{\partial n_a F_a}{\partial \mathbf{v}} + i\,g^a_{\mathbf{k},\omega}(\mathbf{v})\,\eta^a_{\mathbf{k},\omega}\,\delta f^a_{\mathbf{k},\omega}$$
$$- \frac{e_a}{m_a}\,g^a_{\mathbf{k},\omega}(\mathbf{v})\int d\mathbf{k}'\int d\omega'\,\mathbf{k}'\cdot\frac{\partial}{\partial \mathbf{v}}\left[\delta\phi_{\mathbf{k}',\omega'}\,\delta f^a_{\mathbf{k}-\mathbf{k}',\omega-\omega'}\right.$$
$$\left. - \langle\delta\phi_{\mathbf{k}',\omega'}\,\delta f^a_{\mathbf{k}-\mathbf{k}',\omega-\omega'}\rangle\right]. \tag{G.6}$$

Changing the spectral argument from $(\mathbf{k},\omega)$ to $(\mathbf{k}-\mathbf{k}',\omega-\omega')$, we also have

$$\delta f^a_{\mathbf{k}-\mathbf{k}',\omega-\omega'} = -\frac{e_a}{m_a}\,\delta\phi_{\mathbf{k}-\mathbf{k}',\omega-\omega'}\,g^a_{\mathbf{k}-\mathbf{k}',\omega-\omega'}(\mathbf{v})\,(\mathbf{k}-\mathbf{k}')\cdot\frac{\partial n_a F_a}{\partial \mathbf{v}}$$
$$+ i\,g^a_{\mathbf{k}-\mathbf{k}',\omega-\omega'}(\mathbf{v})\,\eta^a_{\mathbf{k}-\mathbf{k}',\omega-\omega'}\,\delta f^a_{\mathbf{k}-\mathbf{k}',\omega-\omega'}$$
$$- \frac{e_a}{m_a}\,g^a_{\mathbf{k}-\mathbf{k}',\omega-\omega'}(\mathbf{v})\int d\mathbf{k}''\int d\omega''\,\mathbf{k}''\cdot\frac{\partial}{\partial \mathbf{v}}$$
$$\times\left[\delta\phi_{\mathbf{k}'',\omega''}\,\delta f^a_{\mathbf{k}-\mathbf{k}'-\mathbf{k}'',\omega-\omega'-\omega''} - \langle\delta\phi_{\mathbf{k}'',\omega''}\,\delta f^a_{\mathbf{k}-\mathbf{k}'-\mathbf{k}'',\omega-\omega'-\omega''}\rangle\right]. \tag{G.7}$$

Let us insert (G.7) to the right-hand side of (G.6):

$$(\omega - \mathbf{k}\cdot\mathbf{v} + i\eta^a_{\mathbf{k},\omega})\,\delta f^a_{\mathbf{k},\omega} = -\frac{e_a}{m_a}\,\delta\phi_{\mathbf{k},\omega}\,\mathbf{k}\cdot\frac{\partial n_a F_a}{\partial \mathbf{v}} + i\eta^a_{\mathbf{k},\omega}\,\delta f^a_{\mathbf{k},\omega}$$
$$+ \frac{e_a^2}{m_a^2}\int d\mathbf{k}'\int d\omega'\,\mathbf{k}'\cdot\frac{\partial}{\partial \mathbf{v}}\left(\delta\phi_{\mathbf{k}',\omega'}\,\delta\phi_{\mathbf{k}-\mathbf{k}',\omega-\omega'} - \left\langle\delta\phi_{\mathbf{k}',\omega'}\,\delta\phi_{\mathbf{k}-\mathbf{k}',\omega-\omega'}\right\rangle\right)$$
$$\times\, g^a_{\mathbf{k}-\mathbf{k}',\omega-\omega'}(\mathbf{v})\,(\mathbf{k}-\mathbf{k}')\cdot\frac{\partial n_a F_a}{\partial \mathbf{v}}$$
$$- i\,\frac{e_a}{m_a}\int d\mathbf{k}'\int d\omega'\,\mathbf{k}'\cdot\frac{\partial}{\partial \mathbf{v}}\left(g^a_{\mathbf{k}-\mathbf{k}',\omega-\omega'}(\mathbf{v})\,\eta^a_{\mathbf{k}-\mathbf{k}',\omega-\omega'}\,\delta\phi_{\mathbf{k}',\omega'}\,\delta f^a_{\mathbf{k}-\mathbf{k}',\omega-\omega'}\right.$$

$$
\begin{aligned}
&-\left\langle g^a_{\mathbf{k}-\mathbf{k}',\omega-\omega'}(\mathbf{v})\,\eta^a_{\mathbf{k}-\mathbf{k}',\omega-\omega'}\,\delta\phi_{\mathbf{k}',\omega'}\,\delta f^a_{\mathbf{k}-\mathbf{k}',\omega-\omega'}\right\rangle\Big)\\
&+\frac{e_a^2}{m_a^2}\int d\mathbf{k}'\int d\omega'\int d\mathbf{k}''\int d\omega''\,\mathbf{k}'\cdot\frac{\partial}{\partial\mathbf{v}}\,g^a_{\mathbf{k}-\mathbf{k}',\omega-\omega'}(\mathbf{v})\,\mathbf{k}''\cdot\frac{\partial}{\partial\mathbf{v}}\\
&\times\Big(\delta\phi_{\mathbf{k}',\omega'}\,\delta\phi_{\mathbf{k}'',\omega''}\,\delta f^a_{\mathbf{k}-\mathbf{k}'-\mathbf{k}'',\omega-\omega'-\omega''}-\delta\phi_{\mathbf{k}',\omega'}\left\langle\delta\phi_{\mathbf{k}'',\omega''}\,\delta f^a_{\mathbf{k}-\mathbf{k}'-\mathbf{k}'',\omega-\omega'-\omega''}\right\rangle\\
&-\left\langle\delta\phi_{\mathbf{k}',\omega'}\,\delta\phi_{\mathbf{k}'',\omega''}\,\delta f^a_{\mathbf{k}-\mathbf{k}'-\mathbf{k}'',\omega-\omega'-\omega''}\right\rangle\Big).
\end{aligned}
\tag{G.8}
$$

This equation is exact and no approximation is made.

Rudakov and Tsytovich (1971) make the observation that in the last term, which involves double integral, $\int d\mathbf{k}'\int d\omega'\int d\mathbf{k}''\int d\omega''$, if we choose the singular term where $\mathbf{k}''=-\mathbf{k}'$ and $\omega''=-\omega'$, namely,

$$
-\frac{e_a^2}{m_a^2}\int d\mathbf{k}'\int d\omega'\,\mathbf{k}'\cdot\frac{\partial}{\partial\mathbf{v}}\,g^a_{\mathbf{k}-\mathbf{k}',\omega-\omega'}(\mathbf{v})\,\mathbf{k}'\cdot\frac{\partial}{\partial\mathbf{v}}\,\delta\phi_{\mathbf{k}',\omega}\,\delta\phi_{-\mathbf{k}',-\omega'}\,\delta f^a_{\mathbf{k},\omega}, \tag{G.9}
$$

then this term can be equated with $i\eta^a_{\mathbf{k},\omega}\,\delta f^a_{\mathbf{k},\omega}$. Thus the quantity $\eta^a_{\mathbf{k},\omega}$ is identified formally as

$$
i\eta^a_{\mathbf{k},\omega}\,\delta f^a_{\mathbf{k},\omega}=-\frac{e_a^2}{m_a^2}\frac{\partial}{\partial v_i}\int d\mathbf{k}'\int d\omega'\,k_i'\,k_j'\,|\delta\phi_{\mathbf{k}',\omega'}|^2\,g^a_{\mathbf{k}-\mathbf{k}',\omega-\omega'}\,\frac{\partial\delta f^a_{\mathbf{k},\omega}}{\partial v_j}. \tag{G.10}
$$

Note that the quantity $|\delta\phi_{\mathbf{k}',\omega}|^2$ is equivalent to the ensemble average $\langle\delta\phi^2\rangle_{\mathbf{k}',\omega}$ for stationary and homogeneous turbulence with random phases. When (G.10) is inserted back to (G.8), then the term $i\eta^a_{\mathbf{k},\omega}\,\delta f^a_{\mathbf{k},\omega}$ no longer appears on the right-hand side:

$$
\begin{aligned}
(\omega-\mathbf{k}\cdot\mathbf{v}+i\eta^a_{\mathbf{k},\omega})\,\delta f^a_{\mathbf{k},\omega}=&-\frac{e_a}{m_a}\,\delta\phi_{\mathbf{k},\omega}\,\mathbf{k}\cdot\frac{\partial n_aF_a}{\partial\mathbf{v}}\\
&+\frac{e_a^2}{m_a^2}\sum_{\substack{\mathbf{k}_1+\mathbf{k}_2=\mathbf{k}\\ \omega_1+\omega_2=\omega}}\mathbf{k}_1\cdot\frac{\partial}{\partial\mathbf{v}}\Big(\delta\phi_{\mathbf{k}_1,\omega_1}\,\delta\phi_{\mathbf{k}_2,\omega_2}\\
&-\left\langle\delta\phi_{\mathbf{k}_1,\omega_1}\,\delta\phi_{\mathbf{k}_2,\omega_2}\right\rangle\Big)\,g^a_{\mathbf{k}_2,\omega_2}\,\mathbf{k}_2\cdot\frac{\partial n_aF_a}{\partial\mathbf{v}}\\
&-i\,\frac{e_a}{m_a}\sum_{\substack{\mathbf{k}_1+\mathbf{k}_2=\mathbf{k}\\ \omega_1+\omega_2=\omega}}\mathbf{k}_1\cdot\frac{\partial}{\partial\mathbf{v}}\Big(g^a_{\mathbf{k}_2,\omega_2}\,\eta^a_{\mathbf{k}_2,\omega_2}\,\delta\phi_{\mathbf{k}_1,\omega_1}\,\delta f^a_{\mathbf{k}_2,\omega_2}\\
&-\left\langle g^a_{\mathbf{k}_2,\omega_2}\,\delta\eta^a_{\mathbf{k}_2,\omega_2}\,\delta\phi_{\mathbf{k}_1,\omega_1}\,\delta f^a_{\mathbf{k}_2,\omega_2}\right\rangle\Big)\\
&+\frac{e_a^2}{m_a^2}\sum_{\substack{\mathbf{k}_1+\mathbf{k}_2+\mathbf{k}_3=\mathbf{k}\\ \omega_1+\omega_2+\omega_3=\omega}}\mathbf{k}_1\cdot\frac{\partial}{\partial\mathbf{v}}\,g^a_{\mathbf{k}_2+\mathbf{k}_3,\omega_2+\omega_3}\,\mathbf{k}_2\cdot\frac{\partial}{\partial\mathbf{v}}\\
&\times\Big(\delta\phi_{\mathbf{k}_1,\omega_1}\,\delta\phi_{\mathbf{k}_2,\omega_2}\,\delta f^a_{\mathbf{k}_3,\omega_3}-\delta\phi_{\mathbf{k}_1,\omega_1}\left\langle\delta\phi_{\mathbf{k}_2,\omega_2}\,\delta f^a_{\mathbf{k}_3,\omega_3}\right\rangle\\
&-\left\langle\delta\phi_{\mathbf{k}_1,\omega_1}\,\delta\phi_{\mathbf{k}_2,\omega_2}\,\delta f^a_{\mathbf{k}_3,\omega_3}\right\rangle\Big).
\end{aligned}
\tag{G.11}
$$

In (G.11), of course, the singular term

$$
-\frac{e_a^2}{m_a^2}\sum_{\mathbf{k}_1,\omega_1}\mathbf{k}_1\cdot\frac{\partial}{\partial\mathbf{v}}\,g^a_{\mathbf{k}-\mathbf{k}_1,\omega-\omega_1}\,\mathbf{k}_1\cdot\frac{\partial}{\partial\mathbf{v}}\,\delta\phi_{\mathbf{k}_1,\omega_1}\,\delta\phi_{-\mathbf{k}_1,-\omega_1}\,\delta f^a_{\mathbf{k},\omega}
$$

should be excluded since this term was already singled out in order to eliminate the term $i\eta^a_{\mathbf{k},\omega}\, f^a_{\mathbf{k},\omega}$. However, since this term is but a term of measure zero within the integral, $\sum_{\mathbf{k}_1+\mathbf{k}_2+\mathbf{k}_3=\mathbf{k}} \sum_{\omega_1+\omega_2+\omega_3=\omega} = \int d\mathbf{k}_1 \int d\mathbf{k}_2 \int d\mathbf{k}_3\, \delta(\mathbf{k}_1+\mathbf{k}_2+\mathbf{k}_3-\mathbf{k}) \int d\omega_1 \int d\omega_2 \int d\omega_3\, \delta(\omega_1+\omega_2+\omega_3-\omega)$, when averaged over the ensemble, the presence or absence of this term does not matter. Besides, in what follows, we will truncate (G.11) by ignoring explicit nonlinear terms. Consequently, in the renormalization scheme by Rudakov and Tsytovich (1971), the nonlinear coupling term serves the purpose of providing the means to identify the term $i\eta^a_{\mathbf{k},\omega}$, but in the subsequent application, explicit nonlinear mode coupling terms are not taken into account.

G.2 Renormalized Quasilinear Theory

Let us thus write the truncated solution for perturbed distribution,

$$\delta f^a_{\mathbf{k},\omega} = -\frac{e_a}{m_a}\, g^a_{\mathbf{k},\omega}\, \mathbf{k}\cdot\frac{\partial n_a F_a}{\partial \mathbf{v}}\, \delta\phi_{\mathbf{k},\omega}, \tag{G.12}$$

where we have only kept the first term on the right-hand side of (G.11). Inserting this solution to Poisson's equation, we have

$$\epsilon(\mathbf{k},\omega)\, \delta\phi_{\mathbf{k},\omega} = 0, \tag{G.13}$$

where $\epsilon(\mathbf{k},\omega)$ is the renormalized dielectric response function,

$$\epsilon(\mathbf{k},\omega) = 1 + \sum_a \frac{\omega_{pa}^2}{k^2} \int d\mathbf{v}\, g^a_{\mathbf{k},\omega}\, \mathbf{k}\cdot\frac{\partial F_a}{\partial \mathbf{v}}. \tag{G.14}$$

Multiplying $\phi_{-\mathbf{k},-\omega}$ and taking the ensemble average, we obtain

$$\epsilon(\mathbf{k},\omega)\, \langle\delta\phi^2\rangle_{\mathbf{k},\omega} = 0. \tag{G.15}$$

The operator $g^a_{\mathbf{k},\omega}$ is determined from (G.4), and, in turn, the quantity $i\eta^a_{\mathbf{k},\omega}$ is defined by (G.10). Combining the two equations, we have

$$g^a_{\mathbf{k},\omega}(\mathbf{v}) \left(\omega - \mathbf{k}\cdot\mathbf{v} - \frac{e_a^2}{m_a^2}\frac{\partial}{\partial \mathbf{v}}\cdot \int d\mathbf{k}' \int d\omega'\, \mathbf{k}'\mathbf{k}' \right.$$
$$\left. \times \left\langle\delta\phi^2\right\rangle_{\mathbf{k}',\omega'} g^a_{\mathbf{k}-\mathbf{k}',\omega-\omega'}(\mathbf{v})\cdot\frac{\partial}{\partial \mathbf{v}}\right) F^a_{\mathbf{k},\omega}(\mathbf{v}) = F^a_{\mathbf{k},\omega}(\mathbf{v}). \tag{G.16}$$

We reiterate that $F^a_{\mathbf{k},\omega}(\mathbf{v})$ is an arbitrary function.

Inserting (G.12) to the formal particle kinetic equation in (G.2), we have

$$\frac{\partial F_a}{\partial t} = -\mathrm{Im}\,\frac{\partial}{\partial \mathbf{v}}\cdot \mathsf{D}(\mathbf{v})\cdot\frac{\partial F_a}{\partial \mathbf{v}},$$
$$\mathsf{D}(\mathbf{v}) = \frac{e_a^2}{m_a^2}\int d\mathbf{k} \int d\omega\, \mathbf{k}\mathbf{k} \left\langle\delta\phi^2\right\rangle_{\mathbf{k},\omega} g^a_{\mathbf{k},\omega}(\mathbf{v}). \tag{G.17}$$

Upon reintroducing the slow-time derivative associated with ω, we also obtain the wave kinetic equation from (G.15), upon taking the imaginary part,

$$\frac{\partial}{\partial t}\left\langle\delta\phi^2\right\rangle_{\mathbf{k},\omega} = 2\mathrm{Im}\,\frac{\epsilon(\mathbf{k},\omega)}{\partial\epsilon(\mathbf{k},\omega)/\partial\omega}\left\langle\delta\phi^2\right\rangle_{\mathbf{k},\omega}. \tag{G.18}$$

To sum up, the renormalized quasilinear kinetic theory comprises of the particle kinetic equation (G.17), wave kinetic equation (G.18), dispersion relation (G.15) with the dielectric response function defined by (G.14), and the equation for renormalized resonant velocity denominator, $g^a_{\mathbf{k},\omega}$, given by (G.16). The equations of renormalized quasilinear theory are formal, but their solution is not so straightforward to obtain. In what follows, we introduce an approximate analysis. Before we do that, however, let us introduce Green's function for the resonant velocity denominator.

G.3 Green's Function for Resonant Velocity Denominator

At this point Rudakov and Tsytovich (1971) introduce another quantity, $G^a_{\mathbf{k},\omega}(\mathbf{v},\mathbf{v}')$, defined by

$$g^a_{\mathbf{k},\omega}(\mathbf{v})\,H(\mathbf{v}) = \int d\mathbf{v}'\;G^a_{\mathbf{k},\omega}(\mathbf{v},\mathbf{v}')\,H(\mathbf{v}'). \tag{G.19}$$

If we take $H(\mathbf{v}) = \delta f^a_{\mathbf{k},\omega}(\mathbf{v})$, then we have

$$g^a_{\mathbf{k},\omega}(\mathbf{v})\,\delta f^a_{\mathbf{k},\omega}(\mathbf{v}) = \int d\mathbf{v}'\;G^a_{\mathbf{k},\omega}(\mathbf{v},\mathbf{v}')\,\delta f^a_{\mathbf{k},\omega}(\mathbf{v}'). \tag{G.20}$$

Upon taking $F^a_{\mathbf{k},\omega}(\mathbf{v}) = G^a_{\mathbf{k},\omega}(\mathbf{v},\mathbf{v}')$ in (G.16), we obtain

$$\begin{aligned} g^a_{\mathbf{k},\omega}(\mathbf{v}) \Bigg((\omega - \mathbf{k}\cdot\mathbf{v}) - \frac{e_a^2}{m_a^2}\frac{\partial}{\partial \mathbf{v}} \cdot \int d\mathbf{k}' \int d\omega'\,\mathbf{k}'\mathbf{k}' \left\langle \delta\phi^2 \right\rangle_{\mathbf{k}',\omega'} \\ \times\; g^a_{\mathbf{k}-\mathbf{k}',\omega-\omega'}(\mathbf{v}) \cdot \frac{\partial}{\partial \mathbf{v}} \Bigg) G^a_{\mathbf{k},\omega}(\mathbf{v},\mathbf{v}') = G^a_{\mathbf{k},\omega}(\mathbf{v},\mathbf{v}'). \end{aligned} \tag{G.21}$$

Upon choosing $H(\mathbf{v}) = (\omega - \mathbf{k}\cdot\mathbf{v})\,G^a_{\mathbf{k},\omega}(\mathbf{v},\mathbf{v}')$ in (G.19), it follows that

$$g^a_{\mathbf{k},\omega}(\mathbf{v})\,(\omega - \mathbf{k}\cdot\mathbf{v})\,G^a_{\mathbf{k},\omega}(\mathbf{v},\mathbf{v}') = \int d\mathbf{v}''\;G^a_{\mathbf{k},\omega}(\mathbf{v},\mathbf{v}'')\,(\omega - \mathbf{k}\cdot\mathbf{v}'')\,G^a_{\mathbf{k},\omega}(\mathbf{v}'',\mathbf{v}'). \tag{G.22}$$

By simply making use of different spectral arguments in the definition (G.19), namely, $g^a_{\mathbf{k}-\mathbf{k}',\omega-\omega'}(\mathbf{v})$ $H(\mathbf{v}) = \int d\mathbf{v}'\;G^a_{\mathbf{k}-\mathbf{k}',\omega-\omega'}(\mathbf{v},\mathbf{v}')\,H(\mathbf{v}')$, and making the choice of

$$H(\mathbf{v}) = \mathbf{k}'\cdot\frac{\partial}{\partial \mathbf{v}}\,G^a_{\mathbf{k},\omega}(\mathbf{v},\mathbf{v}'), \tag{G.23}$$

we obtain

$$g^a_{\mathbf{k}-\mathbf{k}',\omega-\omega'}(\mathbf{v})\,\mathbf{k}'\cdot\frac{\partial}{\partial \mathbf{v}}\,G^a_{\mathbf{k},\omega}(\mathbf{v},\mathbf{v}') = \int d\mathbf{v}''\;G^a_{\mathbf{k}-\mathbf{k}',\omega-\omega'}(\mathbf{v},\mathbf{v}'')\,\mathbf{k}'\cdot\frac{\partial}{\partial \mathbf{v}''}\,G^a_{\mathbf{k},\omega}(\mathbf{v}'',\mathbf{v}'). \tag{G.24}$$

We make use of (G.19) again, but this time we choose

$$H(\mathbf{v}) = \mathbf{k}'\cdot\frac{\partial}{\partial \mathbf{v}}\,g^a_{\mathbf{k}-\mathbf{k}',\omega-\omega'}(\mathbf{v})\,\mathbf{k}'\cdot\frac{\partial}{\partial \mathbf{v}}\,G^a_{\mathbf{k},\omega}(\mathbf{v},\mathbf{v}'), \tag{G.25}$$

in order to arrive at

$$
\begin{aligned}
& g^a_{\mathbf{k},\omega}(\mathbf{v})\,\mathbf{k}'\cdot\frac{\partial}{\partial\mathbf{v}}\,g^a_{\mathbf{k}-\mathbf{k}',\omega-\omega'}(\mathbf{v})\,\mathbf{k}'\cdot\frac{\partial}{\partial\mathbf{v}}\,G^a_{\mathbf{k},\omega}(\mathbf{v},\mathbf{v}') \\
&\quad = \int d\mathbf{v}''\,G^a_{\mathbf{k},\omega}(\mathbf{v},\mathbf{v}'')\,\mathbf{k}'\cdot\frac{\partial}{\partial\mathbf{v}''}\,g^a_{\mathbf{k}-\mathbf{k}',\omega-\omega'}(\mathbf{v}'')\,\mathbf{k}'\cdot\frac{\partial}{\partial\mathbf{v}''}\,G^a_{\mathbf{k},\omega}(\mathbf{v}'',\mathbf{v}') \\
&\quad = \int d\mathbf{v}''\,G^a_{\mathbf{k},\omega}(\mathbf{v},\mathbf{v}'')\,\mathbf{k}'\cdot\frac{\partial}{\partial\mathbf{v}''} \\
&\qquad \times \int d\mathbf{v}'''\,G^a_{\mathbf{k}-\mathbf{k}',\omega-\omega'}(\mathbf{v}'',\mathbf{v}''')\,\mathbf{k}'\cdot\frac{\partial}{\partial\mathbf{v}'''}\,G^a_{\mathbf{k},\omega}(\mathbf{v}''',\mathbf{v}'),
\end{aligned} \tag{G.26}
$$

where we have made use of (G.24) in the last equality.

Making use of (G.22) and (G.26), (G.21) can be rewritten as follows:

$$
\begin{aligned}
& \int d\mathbf{v}'\left((\omega-\mathbf{k}\cdot\mathbf{v})\,\delta(\mathbf{v}-\mathbf{v}') - \frac{e_a^2}{m_a^2}\frac{\partial}{\partial\mathbf{v}}\cdot\int d\mathbf{k}'\int d\omega'\,\mathbf{k}'\mathbf{k}'\left\langle\delta\phi^2\right\rangle_{\mathbf{k}',\omega'}\right. \\
&\qquad \left.\times\, G^a_{\mathbf{k}-\mathbf{k}',\omega-\omega'}(\mathbf{v},\mathbf{v}')\cdot\frac{\partial}{\partial\mathbf{v}'}\right) G^a_{\mathbf{k},\omega}(\mathbf{v}',\mathbf{v}'') = \delta(\mathbf{v}-\mathbf{v}''),
\end{aligned} \tag{G.27}
$$

where we have made some trivial replacements of dummy integral variables. This equation constitutes the differential equation for Green's function $G^a_{\mathbf{k},\omega}(\mathbf{v},\mathbf{v}')$, which will be coupled to the perturbed distribution function $\delta f^a_{\mathbf{k},\omega}(\mathbf{v})$ via (G.20).

G.4 Approximate Analysis

To solve (G.27), let us consider the solution near linear wave-particle resonance regime,

$$
\eta \equiv \omega - \mathbf{k}\cdot\mathbf{v} \approx 0. \tag{G.28}
$$

It turns out that the solution $G^a_{\mathbf{k},\omega}(\mathbf{v},\mathbf{v}')$ is proportional to $\delta(\mathbf{v}-\mathbf{v}')$ for this case. With this *a posteriori* knowledge, we may expand $G^a_{\mathbf{k},\omega}(\mathbf{v}',\mathbf{v}'')$ by

$$
G^a_{\mathbf{k},\omega}(\mathbf{v},\mathbf{v}') = G^{a(0)}_{\mathbf{k},\omega}(\mathbf{v})\,\delta(\mathbf{v}-\mathbf{v}') + \eta\,G^{a(1)}_{\mathbf{k},\omega}(\mathbf{v}) + \cdots. \tag{G.29}
$$

Inserting this to (G.27), we have

$$
\begin{aligned}
& \int d\mathbf{v}'\left[\eta\,\delta(\mathbf{v}-\mathbf{v}') - \frac{e_a^2}{m_a^2}\frac{\partial}{\partial\mathbf{v}}\cdot\int d\mathbf{k}'\int d\omega'\,\mathbf{k}'\mathbf{k}'\left\langle\delta\phi^2\right\rangle_{\mathbf{k}',\omega'}\right. \\
&\qquad \left.\times\, G^{a(0)}_{\mathbf{k}-\mathbf{k}',\omega-\omega'}(\mathbf{v})\,\delta(\mathbf{v}-\mathbf{v}')\cdot\frac{\partial}{\partial\mathbf{v}}\right] G^{a(0)}_{\mathbf{k},\omega}(\mathbf{v}')\,\delta(\mathbf{v}'-\mathbf{v}'') \\
&\quad = \delta(\mathbf{v}-\mathbf{v}'').
\end{aligned} \tag{G.30}
$$

Upon carrying out the trivial velocity integration, we have

$$
\begin{aligned}
& \left(\eta - \frac{e_a^2}{m_a^2}\frac{\partial}{\partial\mathbf{v}}\cdot\int d\mathbf{k}'\int d\omega'\,\mathbf{k}'\mathbf{k}'\left\langle\delta\phi^2\right\rangle_{\mathbf{k}',\omega'}\right. \\
&\qquad \left.\times\, G^{a(0)}_{\mathbf{k}-\mathbf{k}',\omega-\omega'}(\mathbf{v})\cdot\frac{\partial}{\partial\mathbf{v}}\right) G^{a(0)}_{\mathbf{k},\omega}(\mathbf{v})\,\delta(\mathbf{v}-\mathbf{v}'') = \delta(\mathbf{v}-\mathbf{v}'').
\end{aligned} \tag{G.31}
$$

Making note of the fact that

$$\frac{\partial}{\partial \mathbf{v}} = -\mathbf{k}\,\frac{\partial}{\partial \eta}, \qquad \mathbf{k}'\cdot\frac{\partial}{\partial \mathbf{v}} = -(\mathbf{k}\cdot\mathbf{k}')\,\frac{\partial}{\partial \eta}. \tag{G.32}$$

we obtain

$$\left(\eta + \frac{\partial}{\partial \eta}\,D(\eta)\,\frac{\partial}{\partial \eta}\right) G^{a(0)}_{\mathbf{k},\omega}(\eta) = 1,$$

$$D(\eta) = -\frac{e_a^2}{m_a^2}\int d\mathbf{k}'\int d\omega'\left\langle\delta\phi^2\right\rangle_{\mathbf{k}',\omega'}\,(\mathbf{k}\cdot\mathbf{k}')^2\,G^{a(0)}_{\mathbf{k}-\mathbf{k}',\omega-\omega'}(\eta). \tag{G.33}$$

This is a nonlinear equation, but if we keep in mind that the present scheme is valid for $\eta \approx 0$, then we may simply replace $G^{a(0)}_{\mathbf{k}-\mathbf{k}',\omega-\omega'}(\eta)$ that appears in $D(\eta)$ by $G^{a(0)}_{\mathbf{k}-\mathbf{k}',\omega-\omega'}(0)$. Let us call the resultant diffusion coefficient associated with this approximation, $-i\mathscr{D}_0$,

$$D(\eta) = D(0) + \eta\,\frac{\partial D(\eta)}{\partial \eta}\bigg|_{\eta=0} + \cdots,$$

$$D(0) \equiv -i\mathscr{D}_0,$$

$$\mathscr{D}_0 = i\,\frac{e_a^2}{m_a^2}\int d\mathbf{k}'\int d\omega'\left\langle\delta\phi^2\right\rangle_{\mathbf{k}',\omega'}\,(\mathbf{k}\cdot\mathbf{k}')^2\,G^{a(0)}_{\mathbf{k}-\mathbf{k}',\omega-\omega'}(0). \tag{G.34}$$

Upon inserting (G.34) to (G.33), we have

$$\left(\eta - i\mathscr{D}_0\,\frac{\partial^2}{\partial \eta^2}\right) G^{a(0)}_{\mathbf{k},\omega}(\eta) = 1. \tag{G.35}$$

To solve this equation, let us express the solution as

$$G^{a(0)}_{\mathbf{k},\omega}(\eta) = \frac{1}{2\pi}\int_{-\infty}^{\infty} d\tau\,\hat{G}^{a(0)}_{\mathbf{k},\omega}(\tau)\,\exp(i\eta\tau),$$

$$\hat{G}^{a(0)}_{\mathbf{k},\omega}(\tau) = 0, \quad \tau < 0. \tag{G.36}$$

Then we have

$$\frac{\partial^2}{\partial \eta^2}\,G^{a(0)}_{\mathbf{k},\omega}(\eta) = -\frac{1}{2\pi}\int_0^{\infty} d\tau\,\tau^2\,\hat{G}^{a(0)}_{\mathbf{k},\omega}(\tau)\,\exp(i\eta\tau),$$

$$\eta\,G^{a(0)}_{\mathbf{k},\omega}(\eta) = \frac{i}{2\pi}\int_0^{\infty} d\tau\,\exp(i\eta\tau)\,\frac{\partial}{\partial \tau}\,\hat{G}^{a(0)}_{\mathbf{k},\omega}(\tau),$$

$$1 = \frac{1}{2\pi}\int_0^{\infty} d\tau\,2\pi\,\delta(\tau). \tag{G.37}$$

From this we have

$$\frac{\partial \hat{G}^{a(0)}_{\mathbf{k},\omega}(\tau)}{\partial \tau} + \tau^2\,\mathscr{D}_0\,\hat{G}^{a(0)}_{\mathbf{k},\omega}(\tau) = -2\pi i\,\delta(\tau). \tag{G.38}$$

For $\tau \neq 0$, the solution is obtained as follows:

$$\hat{G}^{a(0)}_{\mathbf{k},\omega} = g_0\,\exp\left(-\mathscr{D}_0\,\frac{\tau^3}{3}\right), \qquad g_0 = \text{const}. \tag{G.39}$$

To obtain the coefficient g_0, let us integrate (G.38) from $\tau = -\epsilon$ to $\tau = \epsilon$, where $\epsilon \ll 1$:

$$\int_{-\epsilon}^{\epsilon} d\tau \, \frac{\partial \hat{G}^{a(0)}_{\mathbf{k},\omega}(\tau)}{\partial \tau} + \int_{-\epsilon}^{\epsilon} d\tau \, \tau^2 \, \mathscr{D}_0 \, \hat{G}^{a(0)}_{\mathbf{k},\omega}(\tau) = -2\pi i \int_{-\epsilon}^{\epsilon} d\tau \, \delta(\tau), \tag{G.40}$$

or

$$\hat{G}^{a(0)}_{\mathbf{k},\omega}(\tau = \epsilon) - \hat{G}^{a(0)}_{\mathbf{k},\omega}(\tau = -\epsilon) = -2\pi i. \tag{G.41}$$

Since $\hat{G}^{a(0)}_{\mathbf{k},\omega}(\tau = -\epsilon) = 0$, or $\hat{G}^{a(0)}_{\mathbf{k},\omega}(\tau) = 0$, for $\tau < 0$, we have

$$\lim_{\epsilon \to 0} \hat{G}^{a(0)}_{\mathbf{k},\omega}(\tau = \epsilon) = g_0 = -2\pi i. \tag{G.42}$$

Inserting the solution (G.39) to (G.36), we arrive at

$$G^{a(0)}_{\mathbf{k},\omega}(\mathbf{v}) = -i \int_0^\infty d\tau \, \exp\left(i\,(\omega - \mathbf{k}\cdot\mathbf{v})\,\tau - \mathscr{D}_0 \, \frac{\tau^3}{3} \right). \tag{G.43}$$

Inserting this result to (G.29), and inserting the resultant expression to (G.20), one arrives at

$$g^a_{\mathbf{k},\omega}(\mathbf{v}) = -i \int_0^\infty d\tau \, \exp\left(i\,(\omega - \mathbf{k}\cdot\mathbf{v})\,\tau - \mathscr{D}_0(\mathbf{k},\omega) \, \frac{\tau^3}{3} \right). \tag{G.44}$$

If we set $\mathscr{D}_0 = 0$, then we have the customary expression for the resonant velocity denominator that characterizes the weak turbulence theory. The quantity $\mathscr{D}_0$ is yet to be determined, but it can be formally solved from (G.33) and (G.43):

$$\begin{aligned} \mathscr{D}_0(\mathbf{k},\omega) = \frac{e_a^2}{m_a^2} \int d\mathbf{k}' \int d\omega' \left\langle \delta\phi^2 \right\rangle_{\mathbf{k}',\omega'} (\mathbf{k}\cdot\mathbf{k}')^2 \\ \times \int_0^\infty d\tau \, \exp\left(i\,[\omega - \omega' - (\mathbf{k}-\mathbf{k}')\cdot\mathbf{v}]\,\tau - \mathscr{D}_0(\mathbf{k}-\mathbf{k}',\omega-\omega') \, \frac{\tau^3}{3} \right). \end{aligned} \tag{G.45}$$

This formal equation does not enjoy any closed-form solution. Consequently, we need further approximation.

G.5 Resonance-Broadening Approximation

As a further approximation, let us ignore $\mathscr{D}_0(\mathbf{k}-\mathbf{k}',\omega-\omega')$ in (G.45) to write

$$\begin{aligned} \mathscr{D}_0(\mathbf{k},\omega) = \frac{e_a^2}{m_a^2} \int d\mathbf{k}' \int d\omega' \left\langle \delta\phi^2 \right\rangle_{\mathbf{k}',\omega'} (\mathbf{k}\cdot\mathbf{k}')^2 \\ \times \int_0^\infty d\tau \, \exp\left\{ i\,[\omega - \omega' - (\mathbf{k}-\mathbf{k}')\cdot\mathbf{v}]\,\tau \right\}. \end{aligned} \tag{G.46}$$

Carrying out the τ integration and taking the real part, we have

$$\mathscr{D}_0(\mathbf{k},\omega) = \frac{\pi e_a^2}{m_a^2} \int d\mathbf{k}' \int d\omega' \left\langle \delta\phi^2 \right\rangle_{\mathbf{k}',\omega'} (\mathbf{k}\cdot\mathbf{k}')^2 \, \delta[\omega - \omega' - (\mathbf{k}-\mathbf{k}')\cdot\mathbf{v}]. \tag{G.47}$$

Upon carrying out the ω' integration by virtue of the delta function, we obtain

$$\mathscr{D}_0(\mathbf{k},\omega) = \frac{\pi e_a^2}{m_a^2} \int d\mathbf{k}' \left\langle \delta\phi^2 \right\rangle_{\mathbf{k}',\omega-\mathbf{k}\cdot\mathbf{v}+\mathbf{k}'\cdot\mathbf{v}} (\mathbf{k}\cdot\mathbf{k}')^2. \tag{G.48}$$

Inserting this result to (G.44), we arrive at the renormalized resonant velocity denominator obtained under the resonance-broadening approximation,

$$g^a_{\mathbf{k},\omega}(\mathbf{v}) = -i \int_0^\infty d\tau \, \exp\Bigg(i\,(\omega - \mathbf{k}\cdot\mathbf{v})\,\tau - \frac{\pi e_a^2}{m_a^2} \int d\mathbf{k}' \left\langle \delta\phi^2 \right\rangle_{\mathbf{k}',\omega-\mathbf{k}\cdot\mathbf{v}+\mathbf{k}'\cdot\mathbf{v}} (\mathbf{k}\cdot\mathbf{k}')^2 \frac{\tau^3}{3} \Bigg). \tag{G.49}$$

With this approximate result, we now have a set of formal equations for renormalized quasilinear theory,

$$\begin{aligned}
\frac{\partial F_a(\mathbf{v})}{\partial t} &= -\mathrm{Im}\, \frac{e_a^2}{m_a^2} \int d\mathbf{k} \int d\omega \, \mathbf{k}\cdot\frac{\partial}{\partial \mathbf{v}} \left(g^a_{\mathbf{k},\omega}(\mathbf{v}) \left\langle \delta\phi^2 \right\rangle_{\mathbf{k},\omega} \mathbf{k}\cdot\frac{\partial F_a(\mathbf{v})}{\partial \mathbf{v}} \right), \\
\frac{\partial \left\langle \delta\phi^2 \right\rangle_{\mathbf{k},\omega}}{\partial t} &= -\mathrm{Im}\, \frac{2\epsilon(\mathbf{k},\omega)}{\partial\epsilon(\mathbf{k},\omega)/\partial\omega} \left\langle \delta\phi^2 \right\rangle_{\mathbf{k},\omega}, \\
\epsilon(\mathbf{k},\omega) &= 1 + \sum_a \frac{\omega_{pa}^2}{k^2} \int d\mathbf{v}\, g^a_{\mathbf{k},\omega}(\mathbf{v})\, \mathbf{k}\cdot\frac{\partial F_a}{\partial \mathbf{v}} = 0.
\end{aligned} \tag{G.50}$$

In order to render the formal theory in (G.50) to a more concrete form, we must consider the eigenmode solution, or, equivalently, the dispersion relation. Suppose that the eigenmode $\omega = \omega_{\mathbf{k}}^\alpha$ can be approximately obtained from the real part of the renormalized dielectric constant, $\mathrm{Re}\,\epsilon(\mathbf{k},\omega_{\mathbf{k}}^\alpha) = 0$, then we may write the eigenmode amplitude as

$$k^2 \left\langle \delta\phi^2 \right\rangle_{\mathbf{k},\omega} = \sum_\alpha I_{\mathbf{k}}^\alpha \, \delta(\omega - \omega_{\mathbf{k}}^\alpha). \tag{G.51}$$

Then we have

$$\begin{aligned}
\frac{\partial F_a(\mathbf{v})}{\partial t} &= -\frac{\partial}{\partial v_i} \left(\mathrm{Im}\, \frac{e_a^2}{m_a^2} \sum_\alpha \int d\mathbf{k}\, \frac{k_i k_j}{k^2}\, g_{\mathbf{k}}^{\alpha a}(\mathbf{v})\, I_{\mathbf{k}}^\alpha \right) \frac{\partial F_a(\mathbf{v})}{\partial v_j}, \\
\frac{\partial I_{\mathbf{k}}^\alpha}{\partial t} &= -\frac{2\,\mathrm{Im}\epsilon(\mathbf{k},\omega_{\mathbf{k}}^\alpha)}{\partial\, \mathrm{Re}\,\epsilon(\mathbf{k},\omega_{\mathbf{k}}^\alpha)/\partial\omega_{\mathbf{k}}^\alpha} I_{\mathbf{k}}^\alpha, \\
\epsilon(\mathbf{k},\omega_{\mathbf{k}}^\alpha) &= 1 + \sum_a \frac{\omega_{pa}^2}{k^2} \int d\mathbf{v}\, g_{\mathbf{k}}^{\alpha a}(\mathbf{v})\, \mathbf{k}\cdot\frac{\partial F_a}{\partial \mathbf{v}} = 0. \\
g_{\mathbf{k}}^{\alpha a}(\mathbf{v}) &= -i \int_0^\infty d\tau\, \exp\Bigg(i\,(\omega_{\mathbf{k}}^\alpha - \mathbf{k}\cdot\mathbf{v})\,\tau \\
&\quad - \frac{\pi e_a^2}{m_a^2} \sum_\beta \int d\mathbf{k}'\, I_{\mathbf{k}'}^\beta\, \delta[\omega_{\mathbf{k}}^\alpha - \omega_{\mathbf{k}'}^\beta - (\mathbf{k}-\mathbf{k}')\cdot\mathbf{v}] \frac{(\mathbf{k}\cdot\mathbf{k}')^2}{k'^2} \frac{\tau^3}{3} \Bigg).
\end{aligned} \tag{G.52}$$

This result represents the renormalized quasilinear theory. While the result is self-consistent, the final set of equations is not so straightforward to analyze.

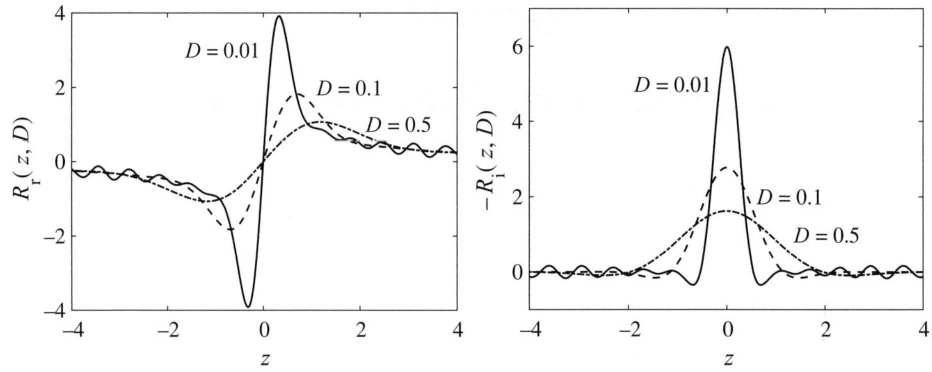

Figure G.1 Real (left) and imaginary (right) parts of the broadened resonance function $R(z, D)$ for several values of D.

We close the present Appendix by considering the resonance function,

$$R(z, D) \equiv -i \int_0^\infty d\tau \, \exp\left(iz\tau - D\frac{\tau^3}{3} \right), \qquad \mathrm{Im}\, z > 0, \tag{G.53}$$

which is related to the function $g_{\mathbf{k}}^{\alpha a}(\mathbf{v})$ defined in (G.52). If $D \to 0$, then we simply obtain

$$R(z, D \to 0) = \frac{1}{z} \approx \mathscr{P}\frac{1}{z} - i\pi\, \delta(z), \tag{G.54}$$

for small $\mathrm{Im}\, z$. For finite D, the function $R(z, D)$ depicts the broadened resonance function. Numerical plots of real and imaginary parts of $R(z, D)$ are shown in Figure G.1 for $D = 0.01$, 0.1, and 0.5.

Appendix H

One-Dimensional Normalized Equations

In this Appendix, we further consider the normalized equations (6.13)–(6.15) and take the limit of one-dimensional situation. We assume that the ions are stationary. The velocity space diffusion coefficient D is of interest to us,

$$
\begin{aligned}
D &= \int_{-\infty}^{\infty} dq \sum_{\sigma=\pm 1} \delta(\sigma x_q^L - qu)\, \mathscr{E}_{\sigma L}(q) \\
&= \frac{2}{|u|} \int_0^{\infty} dq \left[\delta\left(q - \frac{x_q^L}{u}\right) \mathscr{E}_{+L}(q) + \delta\left(q + \frac{x_q^L}{u}\right) \mathscr{E}_{-L}(q) \right].
\end{aligned}
\tag{H.1}
$$

Consequently, we have

$$
\begin{aligned}
\left.\frac{\partial \Phi_e}{\partial T}\right|_{\text{diff.}} &= \frac{\partial}{\partial u}\left(D\, \frac{\partial \Phi_e}{\partial u}\right), \\
D &= \frac{2}{|u|} \left[\Theta(u)\, \mathscr{E}_{+L}(q) + \Theta(-u)\, \mathscr{E}_{-L}(q)\right]_{q=x_q^L/|u|},
\end{aligned}
\tag{H.2}
$$

where

$$
x_q^L = \begin{cases} 1 + 3q^2/4 & (q > 0) \\ -\left(1 + 3q^2/4\right) & (q < 0) \end{cases}.
\tag{H.3}
$$

The velocity drag term is likewise manipulated as

$$
\left.\frac{\partial \Phi_e}{\partial T}\right|_{\text{drag}} = 2g\, \frac{\partial}{\partial u} \left[\Theta(u) - \Theta(-u)\right] \Phi_e(u).
\tag{H.4}
$$

Spontaneous and induced emission terms in normalized one-dimensional form are given by

$$
\begin{aligned}
\left.\frac{\partial \mathscr{E}_{\pm L}(q)}{\partial T}\right|_{\text{emiss.}} &= \frac{\pi}{q^2} \left(\frac{g}{q}\, \Phi_e(u) \pm x_q^L\, \mathscr{E}_{\pm L}(q)\, \frac{\partial \Phi_e(u)}{\partial u}\right)_{u=\pm x_q^L/q}, \\
\left.\frac{\partial \mathscr{E}_{\pm S}(q)}{\partial T}\right|_{\text{emiss.}} &= \frac{\pi \mu_q}{q^2} \left[\frac{g}{q}\, [\Phi_e(u) + \Phi_i(u)] \right. \\
&\quad \left. \pm x_q^L\, \mathscr{E}_{\pm S}(q) \left(\frac{\partial \Phi_e(u)}{\partial u} - \frac{2u}{\tau}\, \Phi_i(u)\right)\right]_{u=\pm x_q^S/q},
\end{aligned}
\tag{H.5}
$$

where

$$x_q^S = \begin{cases} Aq/(1+q^2/2)^{1/2} & (q>0) \\ -Aq/(1+q^2/2)^{1/2} & (q<0) \end{cases},$$

$$A = \frac{(1+3\tau)^{1/2}}{(2M)^{1/2}}. \tag{H.6}$$

The decay terms are considered next. Consider the decay terms for L mode,

$$\begin{aligned}\left.\frac{\partial \mathscr{E}_{\sigma L}(q)}{\partial T}\right|_{\text{decay}} &= \left(\frac{1+3\tau}{2M}\right)^{1/2} \sum_{\sigma',\sigma''=\pm 1} \sigma x_q^L \int_{-\infty}^{\infty} dq'\, |q-q'| \\ &\times \delta\left(\sigma x_q^L - \sigma' x_{q'}^L - \sigma'' x_{q-q'}^S\right) \left[\sigma x_q^L\, \mathscr{E}_{\sigma' L}(q')\, \mathscr{E}_{\sigma'' S}(q-q')\right. \\ &\left. -\, \sigma' x_{q'}^L\, \mathscr{E}_{\sigma'' S}(q-q')\, \mathscr{E}_{\sigma L}(q) - \sigma'' x_{q-q'}^L\, \mathscr{E}_{\sigma' L}(q')\, \mathscr{E}_{\sigma L}(q)\right].\end{aligned} \tag{H.7}$$

In order to explicitly carry out the q' integration via three wave resonance condition, we first rewrite the q' integral over positive range by invoking the symmetry property, $x_{-q} = -x_q$, then we write out the double summations over indices $\sigma' = \pm 1$ and $\sigma'' = \pm 1$, and evaluate the following three wave resonance delta functions of all possible combinations:

$$\begin{aligned}
\delta\left(x_q^L - x_{q'}^L - x_{q-q'}^S\right) &= \delta\left(\frac{3}{4}(q^2 - q'^2) - A(q-q')\right) \\
&= \frac{4/3}{|2q - 4A/3|}\left[\Theta\left(\frac{4A}{3} - q\right)\delta\left(q' + q - \frac{4A}{3}\right) + \delta(q'-q)\right], \\
\delta\left(x_q^L - x_{q'}^L + x_{q-q'}^S\right) &= \frac{4/3}{|2q + 4A/3|}\,\delta(q'-q), \\
\delta\left(x_q^L + x_{q'}^L - x_{q-q'}^S\right) &= \delta\left(\frac{3}{4}(q^2 + q'^2) - A(q-q')\right) : \text{ no root}, \\
\delta\left(x_q^L + x_{q'}^L + x_{q-q'}^S\right) &= \delta\left(\frac{3}{4}(q^2 + q'^2) + A(q-q')\right) : \text{ no root}, \\
\delta\left(x_q^L + x_{q'}^L - x_{q+q'}^S\right) &= \delta\left(\frac{3}{4}(q^2 + q'^2) - A(q+q')\right) : \text{ no root}, \\
\delta\left(x_q^L + x_{q'}^L + x_{q+q'}^S\right) &= \delta\left(\frac{3}{4}(q^2 + q'^2) + A(q+q')\right) : \text{ no root}, \\
\delta\left(x_q^L - x_{q'}^L - x_{q+q'}^S\right) &= \frac{4/3}{|2q - 4A/3|}\,\Theta\left(q - \frac{4A}{3}\right)\delta\left(q' - q + \frac{4A}{3}\right), \\
\delta\left(x_q^L - x_{q'}^L + x_{q+q'}^S\right) &= \frac{4/3}{|2q + 4A/3|}\,\delta\left(q' - q - \frac{4A}{3}\right).
\end{aligned} \tag{H.8}$$

Making use of (H.8), (H.7) is now written explicitly as

$$\begin{aligned}\left.\frac{\partial \mathscr{E}_{\pm L}(q)}{\partial T}\right|_{\text{decay}} &= q_* x_q^L \left\{x_q^L\, \mathscr{E}_{\mp L}(q+q_*)\, \mathscr{E}_{\mp S}(2q+q_*)\right. \\ &\left. - \left[x_{q+q_*}^L\, \mathscr{E}_{\mp S}(2q+q_*) - x_{2q+q_*}^L\, \mathscr{E}_{\mp L}(q+q_*)\right] \mathscr{E}_{\pm L}(q)\right\} \\ &+ q_*\, \Theta(q-q_*)\, x_q^L \left\{x_q^L\, \mathscr{E}_{\mp L}(q-q_*)\, \mathscr{E}_{\pm S}(2q-q_*)\right.\end{aligned}$$

$$
\begin{aligned}
&- \left[x^L_{q-q_*}\,\mathscr{E}_{\pm S}(2q-q_*) + x^L_{2q-q_*}\,\mathscr{E}_{\mp L}(q-q_*)\right]\mathscr{E}_{\pm L}(q)\Big\} \\
&+ q_*\,\Theta(q_*-q)\,\Theta(2q-q_*)\,x^L_q\,\Big\{x^L_q\,\mathscr{E}_{\pm L}(q_*-q)\,\mathscr{E}_{\pm S}(2q-q_*) \\
&- \left[x^L_{q_*-q}\,\mathscr{E}_{\pm S}(2q-q_*) + x^L_{2q-q_*}\,\mathscr{E}_{\pm L}(q_*-q)\right]\mathscr{E}_{\pm L}(q)\Big\} \\
&+ q_*\,\Theta(q_*-q)\,\Theta(q_*-2q)\,x^L_q\,\Big\{x^L_q\,\mathscr{E}_{\pm L}(q_*-q)\,\mathscr{E}_{\pm S}(q_*-2q) \\
&- \left[x^L_{q_*-q}\,\mathscr{E}_{\pm S}(q_*-2q) - x^L_{q_*-2q}\,\mathscr{E}_{\pm L}(q_*-q)\right]\mathscr{E}_{\pm L}(q)\Big\},
\end{aligned}
$$

$$
q_* = \frac{4}{3}\left(\frac{1+3\tau}{2M}\right)^{1/2}. \tag{H.9}
$$

We may likewise manipulate for the S-mode, where the various possible three wave resonance conditions are given below:

$$
\begin{aligned}
\delta\left(x^S_q - x^L_{q'} - x^L_{q-q'}\right) &= \frac{2/3}{|q|}\,\Theta(q'-q)\,\delta\left(q' - \frac{q+4A/3}{2}\right) \\
&= \frac{2/3}{|q|}\,\Theta\left(\frac{4A}{3}-q\right)\delta\left(q' - \frac{q+4A/3}{2}\right), \\
\delta\left(x^S_q - x^L_{q'} + x^L_{q-q'}\right) &= \frac{2/3}{|q|}\,\Theta\left(q-\frac{4A}{3}\right)\delta\left(q' - \frac{q+4A/3}{2}\right), \\
\delta\left(x^S_q + x^L_{q'} - x^L_{q-q'}\right) &= \frac{2/3}{|q|}\,\Theta\left(k-\frac{4A}{3}\right)\delta\left(q' - \frac{q-4A/3}{2}\right), \\
\delta\left(x^S_q + x^L_{q'} + x^L_{q-q'}\right) &= \frac{2/3}{|q|}\,\Theta(q'-q)\,\Theta\left(q-\frac{4A}{3}\right)\delta\left(q' - \frac{q-4A/3}{2}\right) = 0, \\
\delta\left(x^S_q + x^L_{q'} - x^L_{q+q'}\right) &= \frac{2/3}{|q|}\,\Theta\left(\frac{4A}{3}-q\right)\delta\left(q' - \frac{4A/3-q}{2}\right), \\
\delta\left(x^S_q + x^L_{q'} + x^L_{q+q'}\right) &= 0, \\
\delta\left(x^S_q - x^L_{q'} - x^L_{q+q'}\right) &= 0, \\
\delta\left(x^S_q - x^L_{q'} + x^L_{q+q'}\right) &= \delta\left(Aq + \frac{3}{4}q^2 + \frac{3}{2}qq'\right) = 0.
\end{aligned} \tag{H.10}
$$

In (H.10), those delta functions designated as zeros either have no contributions or they cannot be satisfied. Making use of (H.10), we have

$$
\begin{aligned}
\left.\frac{\partial \mathscr{E}_{\pm S}(q)}{\partial T}\right|_{\text{decay}} &= \frac{q_*}{2}\,\Theta(q-q_*)\,x^L_q\,\Big\{x^L_q\,\mathscr{E}_{\pm L}\left(\frac{q+q_*}{2}\right)\mathscr{E}_{\mp L}\left(\frac{q-q_*}{2}\right) \\
&- \left[x^L_{(q+q_*)/2}\,\mathscr{E}_{\mp L}\left(\frac{q-q_*}{2}\right) - x^L_{(q-q_*)/2}\,\mathscr{E}_{\pm L}\left(\frac{q+q_*}{2}\right)\right]\mathscr{E}_{\pm S}(q)\Big\} \\
&+ \frac{q_*}{2}\,\Theta(q_*-q)\,x^L_q\,\Big\{x^L_q\,\mathscr{E}_{\pm L}\left(\frac{q+q_*}{2}\right)\mathscr{E}_{\pm L}\left(\frac{q_*-q}{2}\right) \\
&- \left[x^L_{(q+q_*)/2}\,\mathscr{E}_{\pm L}\left(\frac{q_*-q}{2}\right) - x^L_{(q_*-q)/2}\,\mathscr{E}_{\pm L}\left(\frac{q+q_*}{2}\right)\right]\mathscr{E}_{\pm S}(q)\Big\}.
\end{aligned} \tag{H.11}
$$

For induced scattering involving L mode, we have

$$\left.\frac{\partial \mathscr{E}_{\sigma L}(q)}{\partial T}\right|_{\text{ind.sc.}} = -\sum_{\sigma'=\pm 1}\int_{-\infty}^{\infty} dq' \int_{-\infty}^{\infty} du\, \delta\Big[\sigma x_q^L - \sigma' x_{q'}^L - (q-q')u\Big]$$
$$\times\, \mathscr{E}_{\sigma' L}(q')\, \mathscr{E}_{\sigma L}(q)\, (q-q')\, \frac{\partial}{\partial u}\left((\sigma x_q^L - \sigma' x_{q'}^L)\, \Phi_e(u) - \frac{\sigma x_q^L}{M}\, \Phi_i(u)\right). \tag{H.12}$$

Again, re-expressing the q' integral over positive range only by making use of the symmetry property, $x_{-q} = -x_q$, and explicitly writing out the summation over the index $\sigma' = \pm 1$, one may obtain

$$\left.\frac{\partial \mathscr{E}_{\pm L}(q)}{\partial T}\right|_{\text{ind.sc.}} = \mp \int_0^{\infty} dq'\, u\left(|q-q'|\,\frac{\partial \Phi_e(u)}{\partial u}\right.$$
$$\left. + \frac{2}{\tau}\, x_q^L\, \frac{q-q'}{|q-q'|}\, \Phi_i(u)\right)_{u=\pm 3(q+q')/4} \mathscr{E}_{\pm L}(q')\, \mathscr{E}_{\pm L}(q)$$
$$\mp \int_0^{\infty} dq'\, u\left(|q+q'|\,\frac{\partial \Phi_e(u)}{\partial u}\right.$$
$$\left. + \frac{2}{\tau}\, x_q^L\, \frac{q+q'}{|q+q'|}\, \Phi_i(u)\right)_{u=\pm 3(q-q')/4} \mathscr{E}_{\mp L}(q')\, \mathscr{E}_{\pm L}(q). \tag{H.13}$$

The spontaneous scattering term may also be evaluated likewise:

$$\left.\frac{\partial \mathscr{E}_{\sigma L}(q)}{\partial T}\right|_{\text{spt.scatt.}} = -g\sum_{\sigma'=\pm 1}\int_{-\infty}^{\infty} dq' \int_{-\infty}^{\infty} du\, \delta\Big[\sigma x_q^L - \sigma' x_{q'}^L - (q-q')u\Big]$$
$$\times\, \sigma x_q^L \Big[\sigma' x_{q'}^L\, \mathscr{E}_{\sigma L}(q) - \sigma x_q^L\, \mathscr{E}_{\sigma' L}(q')\Big]\Big[\Phi_e(u) + \Phi_i(u)\Big]$$
$$= -g\, x_q^L \int_0^{\infty} dq' \left\{ \frac{x_{q'}^L\, \mathscr{E}_{\pm L}(q) - x_q^L\, \mathscr{E}_{\pm L}(q')}{|q-q'|}\, \big[\Phi_e(u) + \Phi_i(u)\big]_{u=\pm 3(q+q')/4}\right.$$
$$\left. + \frac{x_{q'}^L\, \mathscr{E}_{\pm L}(q) - x_q^L\, \mathscr{E}_{\pm L}(q')}{|q+q'|}\, \big[\Phi_e(u) + \Phi_i(u)\big]_{u=\pm 3(q-q')/4}\right\}. \tag{H.14}$$

Putting all the terms together, we obtain a set of equations that may be solved readily by numerical finite difference scheme, subject to the initial condition specified in Chapter 6:

$$\frac{\partial \Phi_e(u)}{\partial T} = A\Phi_e(u) + \frac{\partial}{\partial u}\left(D\,\frac{\partial \Phi_e(u)}{\partial u}\right),$$
$$A = 2g\,\frac{\partial}{\partial u}\left[\Theta(u) - \Theta(-u)\right],$$
$$D = \frac{2}{|u|}\left[\Theta(u)\, \mathscr{E}_{+L}(q) + \Theta(-u)\, \mathscr{E}_{-L}(q)\right]_{q=x_q^L/|u|},$$
$$\frac{\partial \mathscr{E}_{\pm L}(q)}{\partial T} = \frac{\pi}{q^2}\left(\frac{g}{q}\, \Phi_e(u) \pm x_q^L\, \mathscr{E}_{\pm L}(q)\, \frac{\partial \Phi_e(u)}{\partial u}\right)_{u=\pm x_q^L/q}$$
$$+ q_*\, x_q^L \Big\{ x_q^L\, \mathscr{E}_{\mp L}(q+q_*)\, \mathscr{E}_{\mp S}(2q+q_*)$$

$$
\begin{aligned}
&- \left[x^L_{q+q_*} \mathscr{E}_{\mp S}(2q+q_*) - x^L_{2q+q_*} \mathscr{E}_{\mp L}(q+q_*) \right] \mathscr{E}_{\pm L}(q) \Big\} \\
&+ q_* \, \Theta(q-q_*) \, x^L_q \Big\{ x^L_q \, \mathscr{E}_{\mp L}(q-q_*) \, \mathscr{E}_{\pm S}(2q-q_*) \\
&- \left[x^L_{q-q_*} \mathscr{E}_{\pm S}(2q-q_*) + x^L_{2q-q_*} \mathscr{E}_{\mp L}(q-q_*) \right] \mathscr{E}_{\pm L}(q) \Big\} \\
&+ q_* \, \Theta(q_*-q) \, \Theta(2q-q_*) \, x^L_q \Big\{ x^L_q \, \mathscr{E}_{\pm L}(q_*-q) \, \mathscr{E}_{\pm S}(2q-q_*) \\
&- \left[x^L_{q_*-q} \mathscr{E}_{\pm S}(2q-q_*) + x^L_{2q-q_*} \mathscr{E}_{\pm L}(q_*-q) \right] \mathscr{E}_{\pm L}(q) \Big\} \\
&+ q_* \, \Theta(q_*-q) \, \Theta(q_*-2q) \, x^L_q \Big\{ x^L_q \, \mathscr{E}_{\pm L}(q_*-q) \, \mathscr{E}_{\pm S}(q_*-2q) \\
&- \left[x^L_{q_*-q} \mathscr{E}_{\pm S}(q_*-2q) - x^L_{q_*-2q} \mathscr{E}_{\pm L}(q_*-q) \right] \mathscr{E}_{\pm L}(q) \Big\} \\
&\mp \int_0^\infty dq' \, u \left(|q-q'| \, \frac{\partial \Phi_e(u)}{\partial u} \right. \\
&\left. + \frac{2}{\tau} \, x^L_q \, \frac{q-q'}{|q-q'|} \, \Phi_i(u) \right)_{u=\pm 3(q+q')/4} \mathscr{E}_{\pm L}(q') \, \mathscr{E}_{\pm L}(q) \\
&\mp \int_0^\infty dq' \, u \left(|q+q'| \, \frac{\partial \Phi_e(u)}{\partial u} \right. \\
&\left. + \frac{2}{\tau} \, x^L_q \, \frac{q+q'}{|q+q'|} \, \Phi_i(u) \right)_{u=\pm 3(q-q')/4} \mathscr{E}_{\mp L}(q') \, \mathscr{E}_{\pm L}(q) \\
&- g \, x^L_q \int_0^\infty dq' \left\{ \frac{x^L_{q'} \, \mathscr{E}_{\pm L}(q) - x^L_q \, \mathscr{E}_{\pm L}(q')}{|q-q'|} \left[\Phi_e(u) + \Phi_i(u) \right]_{u=\pm 3(q+q')/4} \right. \\
&\left. + \frac{x^L_{q'} \, \mathscr{E}_{\pm L}(q) - x^L_q \, \mathscr{E}_{\pm L}(q')}{|q+q'|} \left[\Phi_e(u) + \Phi_i(u) \right]_{u=\pm 3(q-q')/4} \right\},
\end{aligned}
$$

$$
\begin{aligned}
\frac{\partial \mathscr{E}_{\pm S}(q)}{\partial T} &= \frac{\pi \mu_q}{q^2} \left[\frac{g}{q} \left[\Phi_e(u) + \Phi_i(u) \right] \right. \\
&\left. \pm \, x^L_q \, \mathscr{E}_{\pm S}(q) \left(\frac{\partial \Phi_e(u)}{\partial u} - \frac{2u}{\tau} \, \Phi_i(u) \right) \right]_{u=\pm x^S_q/q} \\
&+ \frac{q_*}{2} \, \Theta(q-q_*) \, x^L_q \left\{ x^L_q \, \mathscr{E}_{\pm L}\left(\frac{q+q_*}{2} \right) \mathscr{E}_{\mp L}\left(\frac{q-q_*}{2} \right) \right. \\
&\left. - \left[x^L_{(q+q_*)/2} \, \mathscr{E}_{\mp L}\left(\frac{q-q_*}{2} \right) - x^L_{(q-q_*)/2} \, \mathscr{E}_{\pm L}\left(\frac{q+q_*}{2} \right) \right] \mathscr{E}_{\pm S}(q) \right\} \\
&+ \frac{q_*}{2} \, \Theta(q_*-q) \, x^L_q \left\{ x^L_q \, \mathscr{E}_{\pm L}\left(\frac{q+q_*}{2} \right) \mathscr{E}_{\pm L}\left(\frac{q_*-q}{2} \right) \right. \\
&\left. - \left[x^L_{(q+q_*)/2} \, \mathscr{E}_{\pm L}\left(\frac{q_*-q}{2} \right) - x^L_{(q_*-q)/2} \, \mathscr{E}_{\pm L}\left(\frac{q+q_*}{2} \right) \right] \mathscr{E}_{\pm S}(q) \right\}, \\
q_* &= \frac{4}{3} \left(\frac{1+3\tau}{2M} \right)^{1/2}. \qquad \text{(H.15)}
\end{aligned}
$$

References

Akhiezer, A. I., Akhiezer, I. A., Polovin, R. V., Sitenko, A. G., and Stepanov, K. N. 1975. *Plasma Electrodynamics. Volume 1 – Linear Theory. Volume 2 – Non-Linear Theory and Fluctuations*. Pergamon, New York.

Alexandrov, A. F., Bogdankevich, L. S., and Rukhadze, A. A. 1984. *Principles of Plasma Electrodynamics*. Springer-Verlag, Berlin.

Appert, K., Tran, T. M., and Vaclavik. 1976. Two-dimensional quasi-linear evolution of the electron-beam-plasma instability. *Phys. Rev. Lett.*, **37**, 502.

Armstrong, T. P., Pasonessa, M. T., Bell II, E. V., and Krimigis, S. M. 1983. Voyager observations of Saturnian ion and electron phase space densities. *J. Geophys. Res.*, **88**, 8893.

Aschwanden, M. J. 2005. *Physics of the Solar Corona*. Springer-Verlag, Berlin.

Balescu, R. 1960. Irreversible processes in ionized gases. *Phys. Fluids*, **3**, 52.

Bardwell, S., and Goldman, M. V. 1976. Three-dimensional Langmuir wave instabilities in type III solar radio bursts. *Astrophys. J.*, **209**, 912.

Baumjohann, W., and Treumann, R. A. 1997. *Basic Space Plasma Physics*. Imperial College Press, London.

Beliaev, S. T., and Budker, G. I. 1956. The relativistic kinetic equation. *Sov. Phys. Dok.*, **1**, 218.

Bernstein, I. B., and Engelmann, F. 1966. Quasilinear theory of plasma waves. *Phys. Fluids*, **9**, 937.

Biskamp, D. 2003. *Magnetohydrodynamic Turbulence*. Cambridge University Press, Cambridge.

Brazhenko, A. I., Melnik, V. N., Konovalenko, A. A., Pylaev, O. S., Frantsuzenko, A. V., Dorovskyy, V. V., Vashchishin, R. V., and Rucker, H. O. 2012. Search for the third harmonic of type III bursts radio emission at decameter wavelengths. *Odessa Astron. Publ.*, **25**, 181.

Brodin, G., and Stenflo, L. 2015. Three-wave coupling coefficients for perpendicular wave propagation in a magnetized plasma. *Phys. Plasmas*, **22**, 104503.

Bruno, R., and Carbone, V. 2016. *Turbulence in the Solar Wind*. Springer International Publishing, Switzerland.

Cairns, I. H. 1987a. Fundamental plasma emission involving ion sound waves. *J. Plasma Phys.*, **28**, 169.

Cairns, I. H. 1987b. Second harmonic plasma emission involving ion sound waves. *J. Plasma Phys.*, **28**, 179.

Cairns, I. H. 1987c. Third and higher harmonic plasma emission due to Raman scattering. *J. Plasma Phys.*, **28**, 199.

Cane, H. V., Erickson, W. C., and Prestage, N. P. 2002. Solar flares, type III radio bursts, coronal mass ejections, and energetic particles. *J. Geophys. Res.*, **107**, 1315.

Cap, F. F. 1976. *Handbook on Plasma Instabilities*. Academic Press, New York.

Chen, L. 1987. *Waves and Instabilities in Plasmas*. World Scientific, Singapore.

Crabtree, C., Rudakov, L., Ganguli, G., Mithaiwala, M., Galinsky, V., and Shevchenko, V. 2012. Weak turbulence in the magnetosphere: Formation of whistler wave cavity by nonlinear scattering. *Phys. Plasmas*, **19**, 032903.

Cravens, T. E. 1997. *Physics of Solar System Plasmas*. Cambridge University Press, Cambridge.

Davidson, R. C. 1972. *Methods in Nonlinear Plasma Theory*. Academic Press, New York.

Dawson, J. M., and Shanny, R. 1968. Some investigations of nonlinear behavior in one-dimensional plasmas. *Phys. Fluids*, **11**, 1506.

Diamond, P. H., Itoh, S.-I., and Itoh, K. 2010. *Modern Plasma Phsyics, Volume 1: Physical Kinetics of Turbulent Plasmas*. Cambridge University Press, Cambridge.

Drummond, W. E., and Pines, D. 1962. Nonlinear stability of plasma oscillations. *Nucl. Fusion Suppl.*, **2**, 1049.

Dulk, G. A. 1985. Radio emission from the sun and stars. *Ann. Rev. Astron. Astrophys.*, **23**, 169.

Dum, C. T. 1990. Simulation studies of plasma waves in the electron foreshock: The generation of Langmuir waves by a gentle bump-on-tail electron distribution. *J. Geophys. Res.*, **95**, 8095.

Dum, C. T., and Nishikawa, K.-I. 1994. Two-dimensional simulation studies of the electron beam-plasma instability. *Phys. Plasmas*, **1**, 1821.

Dupree, T. H. 1966. A perturbation theory for strong plasma turbulence. *Phys. Fluids*, **9**, 1773.

Ellis, R. A., and Porkolab, M. 1968. Nonlinear interations of cyclotron harmonic plasma waves. *Phys. Rev. Lett.*, **21**, 529.

Ergun, R. E., Larson, D., Lin, R. P., McFadden, J. P., Carlson, C. W., Anderson, K. A., Muschietti, L., McCarthy, M., Parks, G. K., Reme, H., Bosqued, J. M., D'Uston, C., Sanderson, T. R., Wenzel, K. P., Kaiser, M., Lepping, R. P., Bale, S. D., Kellogg, P., and Bougeret, J.-L. 1998. Wind spacecraft observations of solar impulsive electron events associated with solar type III radio bursts. *Astrophys. J.*, **503**, 435.

Fainberg, J., and Stone, R. G. 1974. Satellite observations of type III solar radio bursts at low frequencies. *Space Sci. Rev.*, **16**, 145.

Feldman, W. C., Asbridge, J. R., Bame, S. J., Montgomery, M. D., and Gary, S. P. 1975. Solar wind electrons. *J. Geophys. Res.*, **80**, 4181.

Felten, T., Schlickeiser, R., Yoon, P. H., and Lazar, M. 2013. Spontaneous electromagnetic fluctuations in unmagnetized plasmas. II. Relativistic form factors of aperiodic thermal modes. *Phys. Plasmas*, **20**, 052113.

Foroutan, G., Robinson, P. A., Zahed, H., Li, B., and Cairns, I. H. 2007. Quasilinear dynamics of a cloud of hot electrons propagating through a plasma in the presence of an externally applied uniform electric field. *Phys. Plasmas*, **14**, 122902.

Frank, L. A., and Gurnett, D. A. 1972. Direct observations of low-energy solar electrons associated with a type III solar radio burst. *Solar Phys.*, **27**, 446.

Fried, D. F., and Conte, S. D. 1961. *Plasma Dispersion Function: The Hilbert Transform of the Gaussian*. Academic Press, New York.

Frieman, E. A., and Rutherford, P. 1964. Kinetic theory of weakly unstable plasma. *Ann. Phys.*, **28**, 134.

Gaelzer, R., Yoon, P. H., Umeda, T., Omura, Y., and Matsumoto, H. 2003. Harmonic Langmuir waves: II. Turbulence spectrum. *Phys. Plasmas*, **10**, 373.

Ganguli, G., Rudakov, L., Scales, W., Wang, J., and Mithaiwala, M. 2010. Three dimensional character of whistler turbulence. *Phys. Plasmas*, **17**, 052310.

Ganse, U., Kilian, P., Vainio, R., and Spanier, F. 2012a. Emission of type II radio bursts – Single-beam *versus* two-beam scenario. *Solar Phys.*, **280**(2), 551–560.

Ganse, U., Kilian, P., Spanier, F., and Vainio, R. 2012b. Nonlinear wave interactions as emission process of type II radio bursts. *Astrophys. J.*, **751**(2), 145.

Gary, S. P. 1993. *Theory of Space Plasma Microinstabilities*. Cambridge University Press, Cambridge.

Ginzburg, C. L., and Zheleznyakov, V. V. 1958. On the possible mechanisms of sporadic solar radio emission (radiation in an isotropic plasma). *Sov. Astron. – AJ*, **2**, 653.

Goldman, M. V. 1983. Progress and problems in the theory of type III solar radio emission. *Solar Phys.*, **89**, 403.

Goldman, M. V., and Dubois, D. F. 1982. Beam-plasma instability in the presence of low-frequency turbulence. *Phys. Fluids*, **25**, 1062.

Goldstein, M. L., Smith, R. A., and Papadopoulos, K. 1979. Nonlinear stability of solar type III radio bursts, II. Application to observations near 1 AU. *Astrophys. J.*, **234**, 683.

Gorbunov, L. M., and Silin, V. P. 1965. Nonlinear interaction between plasma waves. *Sov. Phys. JETP*, **20**, 135.

Gorbunov, L. M., Pustovalov, V. V., and Silin, V. P. 1965. Nonlinear interaction of electromagnetic waves in a plasma. *Sov. Phys. JETP*, **20**, 967.

Gosling, J. T., Asbridge, J. R., Bame, S. J., Feldman, W. C., Zwicki, R. D., Paschmann, G., Sckopke, N., and Hynds, R. J. 1981. Interplanetary ions during an energetic storm particle event – The distribution function from solar wind thermal energies to 1.6 MeV. *J. Geophys. Res.*, **86**, 547.

Gosling, J. T., Skoug, R. M., and McComas, D. J. 2003. Solar electron bursts at very low energies: Evidence for acceleration in the high corona? *Geophys. Res. Lett.*, **30**, 1697.

Graham, D. B., Cairns, I. H., Malaspina, D. M., and Ergun, R. E. 2012. Evidence against the oscillating two-streaming instability and spatial collapse of Langmuir waves in solar type III radio bursts. *Astrophys. J. Lett.*, **753**, L18.

Grognard, R. J.-M. 1982. Numerical simulation of the weak turbulence excited by a beam of electrons in the interplanetary plasma. *Solar Phys.*, **81**, 173.

Guernsey, R. L. 1960. The kinetic theory of fully ionized gases. *DOE Tech. Rep.*, NP-9067.

Gurnett, D. A., Marsch, E., Pilipp, W., Schwenn, R., and Rosenbauer, H. 1979. Ion acoustic waves and related plasma observations in the solar wind. *J. Geophys. Res.*, **84**, 2029.

Hao, X., O'Shea, B. W., Collins, D. C., Norman, M. L., Li, H., and Li, S. 2008. The Biermann battery in cosmological MHD simulations of population III star formation. *Astrophys. J. Lett.*, **688**, L57.

Hasegawa, A. 1975. *Instabilities and Nonlinear Effects*. Springer, New York.

Hasegawa, A., Mima, K., and Duong-van, M. 1985. Plasma distribution function in a superthermal radiation field. *Phys. Rev. Lett.*, **54**, 2608.

Helander, P., and Sigmar, D. J. 2002. *Collisional Transport in Magnetized Plasmas*. Cambridge University Press, Cambridge.

Horton, W., and Choi, D.-I. 1979. Renormalized turbulence theory for the ion acoustic problem. *Phys. Rep.*, **49**, 273.

Huba, J. D. 2009. *NRL Plasma Formulary*. Naval Research Laboratory, Washington, DC.

Ichimaru, S. 1973. *Basic Principles of Plasma Physics*. Benjamin, New York.

Ishihara, O., and Hirose, A. 1981. Quasilinear mechanism of high-energy ion-tail formation in the ion-acoustic instability. *Phys. Rev. Lett.*, **46**, 771.

Ishihara, O., Hirose, A., and Langdon, A. B. 1980. Nonlinear saturation of the Buneman instability. *Phys. Rev. Lett.*, **44**, 1404.

Ishihara, O., Hirose, A., and Langdon, A. B. 1981. Nonlinear evolution of Buneman instability. *Phys. Fluids*, **24**, 452.

Itoh, S.-I., and Itoh, K. 2009. Kinetic description of nonlinear plasma turbulence. *J. Phys. Soc. Jap.*, **78**, 124502.

Ivanov, A. A., Soboleva, T. K., and Yushmanov, P. N. 1976. Three-dimensional quasi-linear relaxation. *Sov. Phys. JETP*, **42**, 1027.

Joyce, G., Knorr, G., and Burns, T. 1971. Nonlinear behavior of the one-dimensional weak beam plasma system. *Phys. Fluids*, **14**, 797.

Kadomtsev, B. B. 1965. *Plasma Turbulence*. Academic Press, New York.

Kane, S. R. 1972. Evidence for a common origin of the electrons responsible for the impulsive X-ray and type III radio bursts. *Solar Phys.*, **27**, 174.

Kaplan, S. A., and Tsytovich, V. N. 1968. Radio emission from beams of fast particles under cosmic conditions. *Sov. Astron. – AJ*, **11**, 956.

Kaplan, S. A., and Tsytovich, V. N. 1973. *Plasma Astrophysics*. Pergamon Press, New York.

Karlický, M., and Vandas, M. 2007. Shock drift electron acceleration and generation of waves. *Plant. Space Sci.*, **55**(15), 2336–2339.

Kasaba, Y., Matsumoto, H., and Omura, Y. 2001. One- and two-dimensional simulations of electron beam instability: Generation of electrostatic and electromagnetic 2f/sub p/ waves. *J. Geophys. Res.*, **106**(A9), 18693–18711.

Kim, S., Schlickeiser, R., Yoon, P. H., López, R. A., and Lazar, M. 2017. Spontaneous emission of electromagnetic fluctuations in Kappa magnetized plasmas. *Plasma Phys. Control. Fusion*, **59**, 125003.

Kivelson, M. G., and Russell, C. T. 1995. *Introduction to Space Physics*. Cambridge University Press, Cambridge.

Kliem, B., Krüger, A., and Treumann, R. A. 1992. Third plasma harmonic radiation in type II bursts. *Solar Phys.*, **140**, 149.

Klimas, A. J. 1983. A mechanism for plasma waves at the harmonics of the plasma frequency in the electron foreshock boundary. *J. Geophys. Res.*, **88**, 9081.

Klimas, A. J. 1990. Trapping saturation of the bump-on-tail instability and electrostatic harmonic excitation in Earth's foreshock. *J. Geophys. Res.*, **95**, 14905.

Klimontovich, Yu. L. 1967. *The Statistical Theory of Non-Equilibrium Processes in a Plasma*. Pergamon Press, New York.

Klimontovich, Yu. L. 1982. *Kinetic Theory of Nonideal Gases and Nonideal Plasmas*. Pergamon Press, New York.

Kono, M., and Ichikawa, H. 1973. Renormalization of the wave-particle interaction in weakly turbulent plasmas. *Prog. Theor. Phys.*, **49**, 754.

Kono, M., and Škorić, M. 2010. *Nonlinear Physics of Plasmas*. Springer, Heidelberg.

Kontar, E. P., and Pécseli, H. L. 2002. Nonlinear development of electron-beam-driven weak turbulence in an inhomogeneous plasma. *Phys. Rev. E*, **65**, 066408.

Kovrizhnykh, L. M. 1965. On the theory of a turbulent plasma. *Sov. Phys. JETP*, **21**, 744.

Kovrizhnykh, L. M., and Tsytovich, V. N. 1964. Interaction of longitudinal and transverse waves in a plasma. *Sov. Phys. JETP*, **19**, 1494.

Kovrizhnykh, L. M., and Tsytovich, V. N. 1965. Effects of transverse electromagnetic wave decay in a plasma. *Sov. Phys. JETP*, **20**, 978.

Krafft, C., Volokitin, A. S., and Krasnoselskikh, V. V. 2013. Interaction of energetic particles with waves in strongly inhomogeneous solar wind plasmas. *Astrophys. J.*, **778**, 111.

Krall, N. A., and Trivelpiece, A. W. 1973. *Principles of Plasma Physics*. McGraw-Hill, New York.

Krommes, J. A. 2002. Fundamental statistical descriptions of plasma turbulence in magnetic fields. *Phys. Rep.*, **360**, 1.

Krucker, S., Kontar, E. P., Christe, S., and Lin, R. P. 2007. Solar flare electron spectra at the Sun and near the Earth. *Astrophys. J.*, **663**, L109.

Kulsrud, R. M., and Zweibel, E. G. 2008. On the origin of cosmic magnetic fields. *Rep. Prog. Phys.*, **71**, 046901.

Kundu, M. R. 1965. *Solar Radio Astronomy*. Interscience, New York.

Landau, L. D. 1937. The kinetic equation in the case of Coulomb interaction. *Zh. Eksp. Teor. Fiz.*, **7**, 2.

Landau, L. D. 1946. On the vibrations of the electronic plasma. *Zh. Eksp. Teor. Fiz.*, **16**, 574.

Lazar, M., Yoon, P. H., and Schlickeiser, R. 2012. Spontaneous electromagnetic fluctuations in unmagnetized plasmas. III. Generalized Kappa distributions. *Phys. Plasmas*, **19**, 122108.

Le Chat, G., Issautier, K., Meyer-Vernet, N., Zouganelis, I., and Maksimovic, M. 2009. Quasi-thermal noise in space plasma: "kappa" distributions. *Phys. Plasmas*, **16**, 102903.

Lenard, A. 1960. On Bogoliubov's kinetic equation for a spatially homogeneous plasma. *Ann. Phys.*, **10**, 390.

Leubner, M. P. 2002. A nonextensive entropy approach to kappa-distributions. *Astrophys. Space Sci.*, **282**, 573.

Li, B., Cairns, I. H., and Robinson, P. A. 2008a. Simulations of coronal type III solar radio bursts: 1. Simulation model. *J. Geophys. Res.*, **113**, A06104.

Li, B., Cairns, I. H., and Robinson, P. A. 2008b. Simulations of coronal type III solar radio bursts: 2. Dynamic spectrum for typical parameters. *J. Geophys. Res.*, **113**, A06105.

Li, B., Cairns, I. H., and Robinson, P. A. 2009. Simulations of coronal type III solar radio bursts: 3. Effects of beam and coronal parameters. *J. Geophys. Res.*, **114**, A02104.

Lin, R. P. 1970. The emission and propagation of $\sim$40keV solar flare electrons. *Solar Phys.*, **12**, 266.

Lin, R. P. 1998. WIND observations of suprathermal electrons in the interplanetary medium. *Space Sci. Rev.*, **86**, 61.

Lin, R. P., Evans, L. G., and Fainberg, J. 1973. Simultaneous observations of fast solar electrons and type III radio burst emission near 1 AU. *Astrophys. Lett.*, **14**, 191.

Lin, R. P., Potter, D. W., Gurnett, D. A., and Scarf, F. L. 1981. Energetic electrons and plasma waves associated with a solar type III radio burst. *Astrophys. J.*, **251**, 364.

Lin, R. P., Levedahl, W. K., Lotko, W., Gurnett, D. A., and Scarf, F. L. 1986. Evidence for nonlinear wave-wave interactions in solar type III radio bursts. *Astrophys. J.*, **308**, 954.

Livadiotis, G. 2017. *Kappa Distributions*. Elsevier Amsterdam.

Livadiotis, G., and McComas, D. J. 2009. Beyond kappa distributions: Exploiting Tsallis statistical mechanics in space plasmas. *J. Geophys. Res.*, **114**, A11105.

López, R. A., and Yoon, P. H. 2017. Simulation of electromagnetic fluctuations in thermal magnetized plasma. *Plasma Phys. Control. Fusion*, **59**, 115003.

Mace, R. L., Hellberg, M. A., and Treumann, R. A. 1998. Electrostatic fluctuations in plasmas containing suprathermal particles. *J. Plasma Phys.*, **59**, 393.

Magelssen, G. R., and Smith, D. F. 1977. Nonrelativistic electron stream propagation in the solar atmosphere and type III radio bursts. *Solar Phys.*, **55**, 211.

Maksimovic, M., Hoang, S., Meyer-Vernet, N., Moncuquet, M., and Bougeret, J.-L. 1995. Solar wind electron parameters from quasi-thermal noise spectroscopy and comparison with other measurements on Ulysses. *J. Geophys. Res.*, **199**, 19881.

Maksimovic, M., Pierrard, V., and Riley, P. 1997. Ulysses electron distributions fitted with Kappa functions. *Geophys. Res. Lett.*, **24**, 1151.

Maksimovic, M., Zouganelis, I., Chaufray, J.-Y., Issautier, K., Scime, E. E., Littleton, J. E., Marsch, E., McComas, D. J., Salem, C., Lin, R. P., and Elliott, H. 2005. Radial evolution of the electron distribution functions in the fast solar wind between 0.3 and 1.5 AU. *J. Geophys. Res.*, **110**, A09104.

Mann, G., Jansen, F., MacDowall, R. J., Kaiser, M. L., and Stone, R. G. 1999. A heliospheric density model and type III radio bursts. *Astron. Astrophys.*, **348**, 614.

McLean, D. J., and Labrum, N. R. 1985. *Solar Radiophysics: Studies of Emission from the Sun at Metre Wavelengths*. Cambridge University Press, Cambridge.

Mel'nik, V., Lapshin, V., and Kontar, E. 1999. Propagation of a monoenergetic electron beam in the solar corona. *Solar Phys.*, **184**, 353.

Melrose, D. B. 1980a. *Plasma Astrophysics*, Vol. 1 & 2. Gordon and Breach, New York.

Melrose, D. B. 1980b. The emission mechanisms for solar radio bursts. *Space Sci. Rev.*, **26**, 3.

Melrose, D. B. 1982. 'Plasma emission' without Langmuir waves. *Austr. J. Phys.*, **35**, 67.

Melrose, D. B. 1986. *Instabilities in Space and Laboratory Plasmas*. Cambridge University Press, Cambridge.

Melrose, D. B. 1987. Plasma emission – A review. *Solar Phys.*, **111**, 89.

Melrose, D. B., and Sy, W. 1972. Scattering of waves in a magnetoactive plasma. *Astrophys. Space Sci.*, **17**, 343.

Meyer-Vernet, N. 1979. On natural noises detected by antennas in plasmas. *J. Geophys. Res.*, **84**, 5373.

Meyer-Vernet, N. 2007. *Basics of the Solar Wind*. Cambridge University Press, Cambridge.

Meyer-Vernet, N., Couturier, P., Hoang, S., Perche, C., Steinberg, J. L., Fainberg, J., and Meetre, C. 1986. Plasma diagnosis from thermal noise and limits on dust flux or mass in comet Giacobini-Zinner. *Science*, **232**, 370.

Meyer-Vernet, N., Hoang, S., and Moncuquet, M. 1993. Bernstein waves in the Io plasma torus: A novel kind of electron temperature sensor. *J. Geophys. Res.*, **98**, 21163.

Meyer-Vernet, N., Issautier, K., and Moncuquet, M. 2017. Quasi-thermal noise spectroscopy: The art and the practice. *J. Geophys. Res.*, **122**, 7925.

Mikhailovskii, A. B. 1974. *Theory of Plasma Instabilities*, Vol. 1 & 2. Consultants Bureau, New York.

Mikhajlenko, V. S., and Stepanov, K. N. 1981. Weak ion-cyclotron turbulence theory of a plasma with a transverse current. *Plasma Phys.*, **23**, 1165.

Mithaiwala, M., Rudakov, L., Crabtree, C., and Ganguli, G. 2012. Co-existence of whistler waves with kinetic Alfven wave turbulence for the high-beta solar wind plasma. *Phys. Plasmas*, **19**, 102902.

Montgomery, D. C., and Tidman, D. A. 1964. *Plasma Kinetic Theory*. McGraw-Hill, New York.

Morse, R. L., and Nielson, C. W. 1969. Numerical simulation of warm two-beam plasma. *Phys. Fluids*, **12**, 2418.

Muschietti, L. 1990. Electron beam formation and stability. *Solar Phys.*, **130**, 201.

Muschietti, L., and Dum, D. T. 1991. Nonlinear wave scattering and electron beam relaxation. *Phys. Fluids B*, **3**, 1968.

Musher, S. L., Rubenchik, A. M., and Zakharov, V. E. 1995. Weak Langmuir turbulence. *Phys. Report*, **252**, 177.

Newbury, J. A., Russell, C. T., Phillips, J. L., and Gary, S. P. 1998. Electron temperature in the ambient solar wind: Typical properties and a lower bound at 1 AU. *J. Geophys. Res.*, **103**, 9553.

Nicholson, D. R. 1983. *Introduction to Plasma Theory*. John Wiley & Sons, New York.

Nicholson, D. R., Goldman, M. V., Hoyng, P., and Weatherall, J. C. 1978. Nonlinear Langmuir waves during type III solar radio bursts. *Astrophys. J.*, **223**, 605.

Nishikawa, K.-I., and Cairns, I. H. 1991. Simulation of the nonlinear evolution of electron plasma waves. *J. Geophys. Res.*, **96**, 19343.

Olbert, S. 1968. *Physics of the Magnetosphere*, eds. R. D. L. Carovillano & J. F. McClay. Astrophys. Space Sci. Libr., Vol. 10 Springer, Dordrecht p. 641.

Papadopoulos, K., Geldstein, M. L., and Smith, R. A. 1974. Stabilization of electron streams in type III solar radio bursts. *Astrophys. J.*, **190**, 175.

Parks, G. K. 2004. *Physics of Space Plasmas*. Westview Press, Boulder.

Pezzi, O., Valentini, F., and Veltri, P. 2016. Collisional relaxation of fine velocity structures in plasmas. *Phys. Rev. Lett.*, **116**, 145001.

Pitaevskii, L. P., and Lifshitz, E. M. 1981. *Physical Kinetics: Volume 10 (Course of Theoretical Physics)*. Pergamon, New York.

Porkolab, M., and Chang, R. P. H. 1978. Nonlinear wave effects in laboratory plasmas: A comparison between theory and experiment. *Rev. Mod. Phys.*, **50**, 745.

Pustovalov, V. V., and Silin, V. P. 1975. Nonlinear theory of the interaction of waves in a plasma. Page 37 of: *Theory of Plasma* (ed. D. V. Skobel'tsyn). Proc. (Trudy) of the P. N. Lebedev Physics Institute, Vol. 61. Consultants Bureau, New York, p. 37.

Ratcliffe, H., Bian, N. H., and Kontar, E. P. 2012. Density fluctuations and the acceleration of electrons by beam-generated Langmuir waves in the solar corona. *Astrophys. J.*, **761**, 176.

Reid, H. A. S., and Ratcliffe, H. 2014. A review of solar type III radio bursts. *Res. Astron. Astrophys.*, **14**, 773.

Reiner, M. J., Stone, R. G., and Fainberg, J. 1992. Detection of fundamental and harmonic type III radio emission and the associated Langmuir waves at the source region. *Astrophys. J.*, **394**, 340.

Reiner, M. J., Goetz, K., Fainberg, J., Kaiser, M. L., Maksimovic, M., Cecconi, B., Hoang, S., Bale, S. D., and Bougeret, J.-L. 2009. Multipoint observations of solar type III radio bursts from STEREO and Wind. *Solar Phys.*, **259**, 255.

Rha, K., Ryu, C.-M., and Yoon, P. H. 2013. Asymmetric electron distributions in the solar wind. *Astrophys. J. Lett.*, **775**, L21.

Rhee, T., Ryu, C.-M., Woo, M. Kaang, H. H., Yi, S., and Yoon, P. H. 2009a. Multiple harmonic plasma emission. *Astrophys. J.*, **694**(1), 618–625.

Rhee, T., Woo, M., and Ryu, C.-M. 2009b. Simulation study of plasma emission in beam-plasma interactions. *J. Kor. Phys. Soc.*, **54**(1), 313–316.

Roberson, C., Gentle, K. W., and Nielson, P. 1971. Experimental test of quasilinear theory. *Phys. Rev. Lett.*, **26**, 226.

Roberts, J. A. 1959. Solar radio bursts of spectral type II. *Austr. J. Phys.*, **12**, 327.

Robinson, P. A., and Cairns, I. H. 1998a. Fundamental and harmonic emission in type III solar radio bursts – I. Emission at a single location or frequency. *Solar Phys.*, **181**, 363.

Robinson, P. A., and Cairns, I. H. 1998b. Fundamental and harmonic emission in type III solar radio bursts – II. Dominant modes and dynamic spectra. *Solar Phys.*, **181**, 395.

Robinson, P. A., and Cairns, I. H. 1998c. Fundamental and harmonic emission in type III solar radio bursts – III. Heliocentric variation of interplanetary beam and source parameters. *Solar Phys.*, **181**, 429.

Rogister, A., and Oberman, C. 1968. On the kinetic theory of stable and weakly unstable plasma. Part 1. *J. Plasma Phys.*, **2**, 33.

Rogister, A., and Oberman, C. 1969. On the kinetic theory of stable and weakly unstable plasma. Part 2. *J. Plasma Phys.*, **3**, 119.

Rosenberg, H. 1976. Solar radio observations and interpretations. *Phil. Trans. R. Soc. London, Ser. A.*, **281**, 461.

Rosenbluth, M. N., MacDonald, W. M., and Judd, D. L. 1957. Fokker-Planck equation for an inverse-square force. *Phys. Rev.*, **107**, 1.

Rudakov, L. I., and Tsytovich, V. N. 1971. The theory of plasma turbulence for strong wave-particle interaction. *Plasma Phys.*, **13**, 213.

Rudakov, L. I., Mithaiwala, M., Ganguli, G., and Crabtree, C. 2011. Linear and nonlinear Landau resonance of kinetic Alfvn waves: Consequences for electron distribution and wave spectrum in the solar wind. *Phys. Plasmas*, **18**, 012307.

Ryu, C.-M., Rhee, T., Umeda, T., Yoon, P. H., and Omura, Y. 2007. Turbulent acceleration of superthermal electrons. *Phys. Plasmas*, **14**, 100701.

Sagdeev, R. Z., and Galeev, A. A. 1969. *Nonlinear Plasma Theory*. Benjamin, New York.

Saint-Hilaire, P., Vilmer, N., and Kerdraon, A. 2013. A decade of solar type III radio bursts observed by the Nançay radioheliograph 1998–2008. *Astrophys. J.*, **762**, 60.

Schlickeiser, R. 2012. Cosmic magnetization: From spontaneously emitted aperiodic turbulent to ordered equipartition fields. *Phys. Rev. Lett.*, **109**, 261101.

Schlickeiser, R., and Yoon, P. H. 2012. Spontaneous electromagnetic fluctuations in unmagnetized plasmas I: General theory and nonrelativistic limit. *Phys. Plasmas*, **19**, 022105.

Schmidt, J. M., and Cairns, I. H. 2012a. Type II radio bursts: 1. New entirely analytic formalism for the electron beams, Langmuir waves, and radio emission. *J. Geophys. Res.*, **117**, A04106.

Schmidt, J. M., and Cairns, I. H. 2012b. Type II radio bursts: 2. Application of the new analytic formalism. *J. Geophys. Res.*, **117**, A11104.

Schmidt, J. M., and Cairns, I. H. 2014. Type II solar radio bursts predicted by 3D MHD CME and kinetic radio emission simulations. *J. Geophys. Res.*, **119**, 69.

Sentman, D. D. 1982. Thermal fluctuations and the diffuse electrostatic emissions. *J. Geophys. Res.*, **87**, 1455.

Sitenko, A. G. 1967. *Electromagnetic Fluctuations in Plasmas*. Academic Press, New York.

Sitenko, A. G. 1982. *Fluctuations and Nonlinear Wave Interactions in Plasmas*. Pergamon, New York.

Sitenko, A. G., and Malnev, V. 1995. *Plasma Physics Theory*. Chapman and Hall, New York.

Smerd, S. F. 1976. Fundamental and harmonic radiation in solar type III bursts. *Solar Phys.*, **46**, 493.

Smith, R. A., Goldstein, M. K., and Papadopoulos, K. 1979. Nonlinear stability of solar type III radio bursts. I. Theory. *Astrophys. J.*, **234**, 348.

Stewart, R. T. 1974. Ground-based observations of type III bursts. Page 161 of: *Coronal Disturbances* (ed. G. Newkirk). Proc. IAU Symp., Vol. 57. Reidel.

Stix, T. H. 1992. *Waves in Plasmas*. Springer-Verlag, New York.

Sturrock, P. A. 1964. Type III solar radio bursts. Page 357 of: *Proc. AAS-NASA Symposium on the Physics of Solar Flares*, Vol. NASA SP-50. NASA, Washington, DC.

Štverák, V., Maksimovic, M., Trávníček, P. M., Marsch, E., Fazakerley, A. N., and Scime, E. E. 2009. Radial evolution of nonthermal electron populations in the low-latitude solar wind: Helios, Cluster, and Ulysses Observations. *J. Geophys. Res.*, **114**, A05104.

Suzuki, S., and Dulk, G. A. 1985. Bursts of type III and type V. Page 289 of: *in Solar Radiophysics* (ed. D. J. McLean and N. R. Labrum). Cambridge University Press, Cambridge.

Swanson, D. G. 1989. *Plasma Waves*. Academic Press, New York.

Tajima, T., Cable, S., and Kulsrud, R. M. 1992. On zero-frequency magnetic fluctuations in plasmas. *Phys. Fluids B*, **4**, 2338.

Takakura, T., and Shibahashi, H. 1976. Dynamics of a cloud of fast electrons travelling through the plasma. *Solar Phys.*, **46**, 323.

Takakura, T., and Yousef, S. 1974. The third harmonic of type III solar radio bursts. *Solar Phys.*, **36**, 451.

Tautz, R. C., and Schlickeiser, R. 2007. Spontaneous emission of Weibel fluctuations by anisotropic distributions. *Phys. Plasmas*, **14**, 102102.

Thejappa, G., MacDowall, R. J., Bergamo, M., and Papadopoulos, K. 2012. Evidence for the oscillating two stream instability and spatial collapse of Langmuir waves in a solar type III radio bursts. *Astrophys. J. Lett.*, **747**, L1.

Tigik, S. F., Ziebell, L. F., Yoon, P. H., and Kontar, E. P. 2016. Two-dimensional time evolution of beam-plasma instability in the presence of binary collisions. *Astron. Astrophys.*, **586**, A19.

Treumann, R. A. 1999a. Kinetic theoretical foundation of Lorentzian statistical mechanics. *Phys. Scr.*, **59**, 19.

Treumann, R. A. 1999b. Generalized-Lorentzian thermodynamics. *Phys. Scr.*, **59**, 204.

Treumann, R. A., and Baumjohann, W. 1997. *Advanced Space Plasma Physics*. Imperial College Press, London.

Tsallis, C. 1988. Possible generalization of Boltzmann-Gibbs statistics. *J. Stat. Phys.*, **52**, 479.

Tsallis, C. 2009. *Introduction to Nonextensive Statistical Mechanics*. Springer, New York.

Tsiklauri, D. 2011. An alternative to the plasma emission model: Particle-in-cell, self-consistent electromagnetic wave emission simulations of solar type III radio bursts. *Phys. Plasmas*, **18**, 052903.

Tsunoda, S. I., Doveil, F., and Malmberg, J. H. 1987. Experimental test of the quasilinear theory of the interaction between a weak warm electron beam and a spectrum of waves. *Phys. Rev. Lett.*, **58**, 1112.

Tsytovich, V. N. 1967. Reviews of topical problems: Nonlinear effects in a plasma. *Sov. Phys. Usp.*, **9**, 805.

Tsytovich, V. N. 1970. *Nonlinear Effects in a Plasma*. Plenum Press, New York.

Tsytovich, V. N. 1977a. *An Introduction to the Theory of Plasma Turbulence*. Pergamon, New York.

Tsytovich, V. N. 1977b. *Theory of Turbulent Plasma*. Consultants Bureau, New York.

Tsytovich, V. N. 1995. *Lectures on Nonlinear Plasma Kinetics*. Springer, Berlin.

Tsytovich, V. N., and Shvartsburg, A. B. 1966. Contribution to the theory of nonlinear interaction of waves in a magnetoactive anisotropic plasma. *Sov. Phys. JETP*, **22**, 554.

Umeda, T. 2010. Electromagnetic plasma emission during beam-plasma interaction: Parametric decay versus induced scattering. *J. Geophys. Res.*, **115**, A01204.

Umeda, T., Omura, Y., Yoon, P. H., Gaelzer, R., and Matsumoto, H. 2003. Harmonic Langmuir waves: III. Vlasov simulation. *Phys. Plasmas*, **10**, 382.

Vahala, G., and Montgomery, D. 1970. Discrete spectra and damped waves in quasilinear theory. *J. Plasma Phys.*, **4**, 677.

Vasyliunas, V. M. 1968. A survey of low-energy electrons in the evening sector of the magnetosphere with OGO 1 and OGO 3. *J. Geophys. Res.*, **73**, 2839.

Vedenov, A. A., and Velikhov, E. P. 1962. Quasilinear approximation in the kinetic theory of a low-density plasma. *Sov. Phys. JETP*, **16**, 682.

Vlasov, A. A. 1938. On vibration properties of electron gas. *Zh. Eksp. Teor. Fiz.*, **8**, 25.

Voshchepynets, A., and Krasnoselskikh, V. 2015. Probabilistic model of beam-plasma interaction in randomly inhomogeneous solar wind. *J. Geophys. Res.*, **120**, 10139.

Wang, L., Lin, R. P., Salem, C., Pulupa, M., Larson, D. E., Yoon, P. H., and Luhmann, J. G. 2012a. Quiet-time interplanetary $|sim$ 2-20 keV superhalo electrons at solar minimum. *Astrophys. J. Lett.*, **753**, L23.

Wang, L., Lin, R. P., Krucker, S., and Mason, G. M. 2012b. A Statistical study of solar electron events over one solar cycle. *Astrophys. J. Lett.*, **759**, 69.

Weibel, E. S. 1959. Spontaneously growing transverse waves in a plasma due to an anisotropic velocity distribution. *Phys. Rev. Lett.*, **2**, 83.

Weinstock, J. 1969. Formulation of a statistical theory of strong plasma turbulence. *Phys. Fluids*, **12**, 1045.

Wild, J. P. 1950. Observations of the spectrum of high-intensity solar radiation at metre wavelengths. III. Isolated bursts. *Austr. J. Sci. Res. A*, **3**, 541.

Wild, J. P. 1951. Observations of the spectrum of high-intensity solar radiation at metre wavelengths. IV. Enhanced radiation. *Austr. J. Sci. Res. A*, **4**, 36.

Wild, J. P., and McCready, L. L. 1950a. Observations of the spectrum of high-intensity solar radiation at metre wavelengths. I. The apparatus and spectral types of solar burst observed. *Austr. J. Sci. Res. A*, **3**, 387.

Wild, J. P., and McCready, L. L. 1950b. Observations of the spectrum of high-intensity solar radiation at metre wavelengths. II. Outbursts. *Austr. J. Sci. Res. A*, **3**, 399.

Wild, J. P., and Smerd, S. F. 1972. Radio bursts from the solar corona. *Ann. Rev. Astron. Astrophys.*, **10**, 159.

Wild, J. P., Murray, J. D., and Rowe, W. C. 1954. Harmonics in the spectra of solar radio disturbances. *Austr. J. Phys.*, **7**, 439.

Wild, J. P., Sheridan, K. V., and Neylan, A. A. 1959. An investigation of the speed of the solar disturbances responsible for type III radio bursts. *Austr. J. Phys.*, **12**, 369.

Wild, J. P., Smerd, S. F., and Weiss, A. A. 1963. Solar bursts. *Ann. Rev. Astron. Astrophys.*, **1**, 291.

Wu, C. S., Wang, C. B., Yoon, P. H., Zheng, H. N., and Wang, S. 2002. Generation of type III solar radio bursts in the low corona by direct amplification. *Astrophys. J.*, **575**, 1094.

Yoon, P. H. 2006. Statistical theory of electromagnetic weak turbulence. *Phys. Plasmas*, **13**, 022302.

Yoon, P. H. 2007. Relativistic Weibel instability. *Phys. Plasmas*, **14**, 064504.

Yoon, P. H. 2010. Weak turbulence theory for reactive instability. *Phys. Plasmas*, **17**, 112316.

Yoon, P. H. 2014. Electron kappa distribution and quasi-thermal noise. *J. Geophys. Res.*, **119**, 70774.

Yoon, P. H. 2015a. Kinetic theory of turbulence for parallel propagation revisited: Formal results. *Phys. Plasmas*, **22**, 082309.

Yoon, P. H. 2015b. Kinetic theory of turbulence for parallel propagation revisited: Low-to-intermediate frequency regime. *Phys. Plasmas*, **22**, 092307.

Yoon, P. H. 2015c. Kinetic theory of weak turbulence in magnetized plasmas: Perpendicular propagation. *Phys. Plasmas*, **22**, 082310.

Yoon, P. H., and Umeda, T. 2010. Nonlinear turbulence theory and simulation of Buneman instability. *Phys. Plasmas*, **17**, 112317.

Yoon, P. H., Gaelzer, R., Umeda, T., Omura, Y., and Matsumoto, H. 2003a. Harmonic Langmuir waves: I. Nonlinear dispersion relation. *Phys. Plasmas*, **10**, 364.

Yoon, P. H., Ryu, C.-M., and Rhee, T. 2003b. Particle kinetic equation including weakly turbulent mode coupling. *Phys. Plasmas*, **10**, 3881.

Yoon, P. H., Rhee, T., and Ryu, C.-M. 2005. Self-consistent generation of superthermal electrons by beam-plasma interaction. *Phys. Rev. Lett.*, **95**, 215003.

Yoon, P. H., Hong, J., Kim, S., Lee, J., Lee, J., Park, J., Park, K. S., and Seough, J. J. 2012a. Asymmetric solar wind electron distributions. *Astrophys. J.*, **755**, 112.

Yoon, P. H., Ziebell, L. F., Gaelzer, R., and Pavan, J. 2012b. Electromagnetic weak turbulence theory revisited. *Phys. Plasmas*, **19**, 102303.

Yoon, P. H., Schlickesier, R., and Kolberg, U. 2014. Thermal fluctuation levels of magnetic and electric fields in unmagnetized plasma: The rigorous relativistic kinetic theory. *Phys. Plasmas*, **21**, 032109.

Zaitsev, V. V., Mityakov, N. A., and Rappoport, V. O. 1972. A dynamic theory of type III solar radio bursts. *Solar Phys.*, **24**, 444.

Zank, G. P. 2014. *Transport Processes in Space Physics and Astrophysics*. Springer, Heidelberg.

Zheleznyakov, V. V. 1970. *Radio Emission of the Sun and Planets*. Pergamon, New York.

Zheleznyakov, V. V., and Zaitsev, V. V. 1970a. Contribution to the theory of type III solar radio bursts. I. *Sov. Astron.AJ*, **14**, 47.

Zheleznyakov, V. V., and Zaitsev, V. V. 1970b. The theory of type III solar radio bursts. II. *Sov. Astron. AJ*, **14**, 250.

Zheleznyakov, V. V., and Zlotnik, E. Ya. 1974. On the third harmonic in solar radio bursts. *Solar Phys.*, **36**, 451.

Ziebell, L. F., Gaelzer, R., and Yoon, P. H. 2001. Nonlinear development of weak beam-plasma instability. *Phys. Plasmas*, **8**, 3982.

Ziebell, L. F., Yoon, P. H., Gaelzer, R., and Pavan, J. 2008a. Langmuir condensation by spontaneous scattering off electrons in two dimensions. *Plasmas Phys. Control. Fusion*, **54**, 055012.

Ziebell, L. F., Gaelzer, R., Pavan, J., and Yoon, P. H. 2008b. Two-dimensional nonlinear dynamics of beam-plasma instability. *Plasmas Phys. Control. Fusion*, **50**, 085011.

Ziebell, L. F., Yoon, P. H., Gaelzer, R., and Pavan, J. 2014a. Plasma emission by weak turbulence processes. *Astrophys. J. Lett.*, **795**, L32.

Ziebell, L. F., Yoon, P. H., Simões Jr., F. J. R., Gaelzer, R., and Pavan, J. 2014b. Spontaneous emission of electromagnetic radiation in turbulent plasmas. *Phys. Plasmas*, **21**, 010701.

Ziebell, L. F., Yoon, P. H., Petruzzellis, L. T., Gaelzer, R., and Pavan, J. 2015. Plasma emission by nonlinear electromagnetic processes. *Astrophy. J.*, **806**, 237.

Zlotnik, E. Ya., Klassen, A., Klein, K.-L., Aurass, H., and Mann, G. 1998. Third harmonic plasma emission in solar type II radio bursts. *Astron. Astrophy.*, **331**, 1087.

Zouganelis, I. 2008. Measuring suprathermal electron parameters in space plasmas: Implementation of the quasi-thermal noise spectroscopy with kappa distributions using in situ Ulysses/URAP radio measurements in the solar wind. *J. Geophys. Res.*, **113**, A08111.

Index